Lecture Notes in Computer Science 14679

The series Lecture Notes in Computer Science (LNCS), including its subseries Lecture Notes in Artificial Intelligence (LNAI) and Lecture Notes in Bioinformatics (LNBI), has established itself as a medium for the publication of new developments in computer science and information technology research, teaching, and education.

LNCS enjoys close cooperation with the computer science R & D community, the series counts many renowned academics among its volume editors and paper authors, and collaborates with prestigious societies. Its mission is to serve this international community by providing an invaluable service, mainly focused on the publication of conference and workshop proceedings and postproceedings. LNCS commenced publication in 1973.

Jens Vygen · Jarosław Byrka
Editors

Integer Programming and Combinatorial Optimization

25th International Conference, IPCO 2024
Wrocław, Poland, July 3–5, 2024
Proceedings

Springer

Editors
Jens Vygen
University of Bonn
Bonn, Germany

Jarosław Byrka
University of Wrocław
Wrocław, Poland

ISSN 0302-9743 ISSN 1611-3349 (electronic)
Lecture Notes in Computer Science
ISBN 978-3-031-59834-0 ISBN 978-3-031-59835-7 (eBook)
https://doi.org/10.1007/978-3-031-59835-7

This Springer imprint is published by the registered company Springer Nature Switzerland AG
The registered company address is: Gewerbestrasse 11, 6330 Cham, Switzerland

Paper in this product is recyclable.

Preface

This volume contains the 33 extended abstracts presented at IPCO 2024, the 25th Conference on Integer Programming and Combinatorial Optimization, held July 3–5, 2024, in Wrocław, Poland.

The IPCO conference is under the auspices of the Mathematical Optimization Society. It is held every year. IPCO brings together researchers and practitioners working on various aspects of integer programming and combinatorial optimization. It has become an important forum for presenting recent developments in theory, computation, and applications in these areas. Traditionally, IPCO consists of three days of non-parallel sessions, with no invited talks. More information on IPCO and its history can be found at www.mathopt.org/?nav=ipco.

This year, we received 101 submissions, none of which was withdrawn. In a single-blind review process, each submission was reviewed by at least three Program Committee members, often with the help of external reviewers. All 15 PC members met in Aussois in January 2024 and, after thorough discussions, selected 33 papers to be presented at IPCO 2024 and included in this volume. Again, many excellent submissions could not be accepted.

As usual, this volume is edited jointly by the PC chair (Jens Vygen) and the main local organizer (Jarosław Byrka, also a PC member). We expect the full versions of the accepted extended abstracts that appear in this volume to be published in refereed journals. A special issue of Mathematical Programming B containing such full papers is planned by us.

The Program Committee decided that the Best Paper Award for IPCO 2024 is given to Gérard Cornuéjols, Siyue Liu, and R. Ravi for their paper "Approximately Packing Dijoins via Nowhere-Zero Flows". Congratulations!

We would like to thank:

- All authors who submitted extended abstracts to IPCO; we are happy how active all areas of integer programming and combinatorial optimization are;
- The members of the Program Committee, who graciously gave their time and energy;
- The external reviewers, whose expertise was instrumental in guiding our decisions;
- The organizers of the Aussois workshop: Karen Aardal, Britta Peis, and Rico Zenklusen, for inviting us and allowing us to hold the PC meeting in Aussois;
- The EasyChair developers for their excellent platform making many things so much easier;
- Springer for their efficient cooperation in producing this volume and for sponsoring the Best Paper Award;
- The sponsors: Cardinal Operations, FICO, MOSEK, and The Optimization Firm;
- The members of the Organizing Committee, and all people in Wroclaw who helped to make this conference possible;
- The speakers of the summer school preceding IPCO: Sophie Huiberts, Neil Olver, and Vera Traub;

– The Mathematical Optimization Society and in particular the members of its IPCO Steering Committee: Karen Aardal, Jochen Könemann, and Giacomo Zambelli, for their help and advice.

March 2024

Jens Vygen
Jarosław Byrka

Organization

Program Committee

Ahmad Abdi	London School of Economics and Political Science, UK
Kristóf Bérczi	Eötvös Loránd University, Hungary
Niv Buchbinder	Tel Aviv University, Israel
Jarosław Byrka	University of Wrocław, Poland
Karthekeyan Chandrasekaran	University of Illinois at Urbana-Champaign, USA
Sanjeeb Dash	IBM, USA
Sami Davies	University of California, Berkeley, USA
Samuel Fiorini	Université Libre de Bruxelles, Belgium
Zachary Friggstad	University of Alberta, Canada
Swati Gupta	Massachusetts Institute of Technology, USA
Sophie Huiberts	CNRS/Clermont Auvergne University, France
Jon Lee	University of Michigan, USA
Gonzalo Muñoz	Universidad de O'Higgins, Chile
Ola Svensson	École Polytechnique Fédérale de Lausanne, Switzerland
Jens Vygen (Chair)	University of Bonn, Germany

Local Organizing Committee

Marek Adamczyk	University of Wrocław, Poland
Marcin Bieńkowski	University of Wrocław, Poland
Martin Böhm	University of Wrocław, Poland
Jarosław Byrka (Chair)	University of Wrocław, Poland
Łukasz Jeż	University of Wrocław, Poland

Additional Reviewers

Achterberg, Tobias
Adamczyk, Marek
Aliev, Iskander
Aouad, Ali
Aprile, Manuel
Atamtürk, Alper
Au, Gary
Augustino, Brandon
Averkov, Gennadiy
Balkanski, Eric
Bamas, Etienne
Banerjee, Sandip
Bansal, Ishan
Barahona, Francisco
Basu, Amitabh
Battista, Federico
Beck, Matthias
Belotti, Pietro
Bhaskar, Umang
Black, Alexander
Blauth, Jannis
Bobbio, Federico
Borst, Sander
Bourgais, Mathieu
Bucarey, Víctor
Böhm, Martin
Cassis, Alejandro
Castro, Margarita
Chekuri, Chandra
Chen, Rui
Chen, Yu
Cheung, Wang Chi
Chicoisne, Renaud
Chmiela, Antonia
Cook, William
Cory-Wright, Ryan
Dadush, Daniel
Dahan, Mathieu
Dalirrooyfard, Mahsa
De Loera, Jesus
Del Pia, Alberto
Dey, Santanu
Di Summa, Marco
Dong, Sally
Drygala, Marina
Dubey, Yatharth
Eisenbrand, Friedrich
Faenza, Yuri
Feldman, Jacob
Feldman, Moran
Fujishige, Satoru
Fukasawa, Ricardo
Garzón, Gabriel Ruiz
Ghadiri, Mehrdad
Ghuge, Rohan
Grandoni, Fabrizio
Gribling, Sander
Gupta, Anupam
Gupta, Varun
Haslebacher, Sebastian
Hathcock, Daniel
Held, Stephan
Hertrich, Christoph
Hildebrand, Robert
Hojny, Christopher
Hougardy, Stefan
Ibrahimpur, Sharat
Jabal Ameli, Afrouz
Jabbarzade, Peyman
Jain, Rhea
Jiang, Hongyi
Kafer, Sean
Kaibel, Volker
Kakimura, Naonori
Kao, Mong-Jen
Kapralov, Michael
Karimi, Mehdi
Kazachkov, Aleksandr M.
Khajavirad, Aida
Király, Csaba
Király, Tamás
Kobayashi, Yusuke
Kober, Stefan
Koutecky, Martin
Kulkarni, Shubhang
Kumar, Amit

Laekhanukit, Bundit
Laurent, Monique
Lee, Dabeen
Letchford, Adam
Levin, Roie
Li, Shi
Liao, Jianmin
Liers, Frauke
Lintzmayer, Carla Negri
Livanos, Vasilis
Ljubic, Ivana
Lodi, Andrea
Lu, Xinhang
Lucarelli, Giorgio
Martinez Mori, Juan Carlos
Megow, Nicole
Molinaro, Marco
Monaci, Michele
Moondra, Jai
Moran, Diego
Mousavi, Ramin
Mucha, Marcin
Nagarajan, Viswanath
Nannicini, Giacomo
Naor, Seffi
Natura, Bento
Nägele, Martin
Nöbel, Christian
Olver, Neil
Ostrowski, James
Paat, Joseph
Padmanabhan, Swati
Paluch, Katarzyna
Pan, Yuchong
Pashkovich, Kanstantsin
Peis, Britta
Perez-Salazar, Sebastian
Pfetsch, Marc
Polak, Adam
Puhlmann, Luise
Pálvölgyi, Dömötör
Rothvoss, Thomas
Ryan, Chris
Salavatipour, Mohammad
Sambharya, Rajiv
Sanità, Laura
Schiffer, Benjamin
Schlomberg, Niklas
Schlotter, Ildikó
Schröder, Marc
Schwarcz, Tamás
Schymura, Matthias
Sebő, András
Serrano, Felipe
Shan, Liren
Shirley, Morgan
Sinnl, Markus
Skutella, Martin
Song, Dogyoon
Souza Brito, Samuel
Speakman, Emily
Spoerhase, Joachim
Stouras, Miltiadis
Sun, Shengding
Swamy, Chaitanya
Tawarmalani, Mohit
Thiery, Théophile
Tiwary, Hans Raj
Traub, Vera
Tröbst, Thorben
Tunçel, Levent
Uchoa, Eduardo
Vargas Koch, Laura
Vargas, Luis
Vintan, Radu
Végh, László
Walter, Matthias
Wang, Guanyi
Wang, Yuyan
Weltge, Stefan
Wigal, Michael
Włodarczyk, Michał
Xu, Chao
Xu, Luze
Yamaguchi, Yutaro
Yan, Julia
Yuan, Weiqiang
Yuditsky, Yelena
Zavalnij, Bogdan
Zhang, Haixiang
Zhong, Mingxian
Zhou, Rudy

Contents

Sparsity and Integrality Gap Transference Bounds for Integer Programs

Iskander Aliev[1(✉)], Marcel Celaya[1], and Martin Henk[2]

[1] Cardiff University, Cardiff, UK
{alievi,celayam}@cardiff.ac.uk
[2] Technische Universität Berlin, Berlin, Germany
henk@math.tu-berlin.de

Abstract. We obtain new transference bounds that connect two active areas of research: proximity and sparsity of solutions to integer programs. Specifically, we study the additive integrality gap of the integer linear programs $\min\{\boldsymbol{c}\cdot\boldsymbol{x} : \boldsymbol{x} \in P \cap \mathbb{Z}^n\}$, where $P = \{\boldsymbol{x} \in \mathbb{R}^n : \boldsymbol{A}\boldsymbol{x} = \boldsymbol{b}, \boldsymbol{x} \geq \boldsymbol{0}\}$ is a polyhedron in the standard form determined by an integer $m \times n$ matrix $\boldsymbol{A}$ and an integer vector $\boldsymbol{b}$. The main result of the paper gives an upper bound for the integrality gap that drops exponentially in the size of support of the optimal solutions corresponding to the vertices of the integer hull of P. Additionally, we obtain a new proximity bound that estimates the ℓ_2-distance from a vertex of P to its nearest integer point in the polyhedron P. The proofs make use of the results from the geometry of numbers and convex geometry.

Keywords: Integrality gap · Proximity · Sparsity

1 Introduction and Main Results

The proximity and sparsity of solutions to integer programs are two active areas of research in the theory of mathematical programming.

Proximity-type results study the quality of approximation of the solutions to integer programs by the solutions of linear programming relaxations. This is a traditional research topic with the first contributions dated back at least to Gomory [17,18]. The most influential results in this area also include the proximity bounds by Cook et al. [13] and by Eisenbrand and Weismantel [16]. For more recent contributions, we refer the reader to Celaya et al. [11], Lee et al. [19,20], and Paat, Weismantel and Weltge [22].

Sparsity-type results study the size of support of solutions to integer programs. This area of research takes its origin from the integer Carathéodory theorems of Cook, Fonlupt and Schrijver [12] and Sebő [23] and later major contributions by Eisenbrand and Shmonin [15]. In recent years, this topic has been studied in numerous papers, including Abdi et al. [1], Aliev et al. [2,4,5], Berndt, Jansen and Klein [7], Dubey and Liu [14] and Oertel, Paat and Weismantel [21].

Recent works of Lee et al. [19] and Aliev et al. [3] show that the proximity and sparsity areas are highly interconnected. The paper [3] establishes *transference bounds* that link both areas in two *special cases*: for corner polyhedra and

J. Vygen and J. Byrka (Eds.): IPCO 2024, LNCS 14679, pp. 1–13, 2024.
https://doi.org/10.1007/978-3-031-59835-7_1

knapsacks with positive entries. Remarkably, this gives a drastic improvement on the previously known proximity bounds for knapsacks obtained in [6].

In this paper, we establish the first transference bounds that involve the integrality gap of integer linear programs and hold in the *general case*, addressing a future research question posed in [3]. The proofs explore a new geometric approach that combines Minkowski's geometry of numbers and box slicing inequalities.

Specifically, for a matrix $\boldsymbol{A} \in \mathbb{Z}^{m\times n}$ with $m < n$ and $\boldsymbol{b} \in \mathbb{Z}^m$, we define the polyhedron

$$P(\boldsymbol{A},\boldsymbol{b}) = \{\boldsymbol{x} \in \mathbb{R}^n_{\geq 0} : \boldsymbol{A}\boldsymbol{x} = \boldsymbol{b}\}$$

and assume that $P(\boldsymbol{A},\boldsymbol{b})$ contains integer points. Given a cost vector $\boldsymbol{c} \in \mathbb{R}^n$, we consider the integer linear programming problem

$$\min\{\boldsymbol{c}\cdot\boldsymbol{x} : \boldsymbol{x} \in P(\boldsymbol{A},\boldsymbol{b}) \cap \mathbb{Z}^n\}, \qquad (1)$$

where $\boldsymbol{c}\cdot\boldsymbol{x}$ stands for the standard inner product. A very successful and traditional approach for solving optimisation problems of the form (1) is based on solving its linear programming relaxation

$$\min\{\boldsymbol{c}\cdot\boldsymbol{x} : \boldsymbol{x} \in P(\boldsymbol{A},\boldsymbol{b})\}, \qquad (2)$$

obtained by dropping the integrality constraint. Subsequently, various methods are used to construct a feasible integer solution to (1) from a fractional solution of (2).

Suppose that (1) is feasible and bounded. Let $\mathrm{IP}(\boldsymbol{A},\boldsymbol{b},\boldsymbol{c})$ and $\mathrm{LP}(\boldsymbol{A},\boldsymbol{b},\boldsymbol{c})$ denote the optimal values of (1) and (2), respectively. We will focus on estimating the *(additive) integrality gap* $\mathrm{IG}(\boldsymbol{A},\boldsymbol{b},\boldsymbol{c})$ defined as

$$\mathrm{IG}(\boldsymbol{A},\boldsymbol{b},\boldsymbol{c}) = \mathrm{IP}(\boldsymbol{A},\boldsymbol{b},\boldsymbol{c}) - \mathrm{LP}(\boldsymbol{A},\boldsymbol{b},\boldsymbol{c}).$$

The integrality gap is a fundamental proximity characteristic of the problem (1) extensively studied in the literature. The upper bounds for $\mathrm{IG}(\boldsymbol{A},\boldsymbol{b},\boldsymbol{c})$ appear already in the work of Blair and Jeroslow [8,9]. To state the best currently known estimates, we will need to introduce the following notation. By $\|\cdot\|_1$ and $\|\cdot\|_\infty$ we denote the ℓ_1 and ℓ_∞ norms, respectively. Without loss of generality, we assume that $\boldsymbol{A}$ has rank m and, for $1 \leq r \leq m$, denote by $\Delta_r(\boldsymbol{A})$ the maximum absolute $r \times r$ subdeterminant of A, that is

$$\Delta_r(\boldsymbol{A}) = \max\{|\det \boldsymbol{B}| : \boldsymbol{B} \text{ is a } r \times r \text{ submatrix of } \boldsymbol{A}\}.$$

The sensitivity theorems of Cook et al. [13] (see also Celaya et al. [11] for further improvements) imply the bound

$$\mathrm{IG}(\boldsymbol{A},\boldsymbol{b},\boldsymbol{c}) \leq \|\boldsymbol{c}\|_1 (n-m)\Delta_m(\boldsymbol{A}). \qquad (3)$$

More recently, Eisenbrand and Weismantel [16] obtained the estimate

$$\mathrm{IG}(\boldsymbol{A},\boldsymbol{b},\boldsymbol{c}) \leq \|\boldsymbol{c}\|_\infty m(2m\Delta_1(\boldsymbol{A})+1)^m, \qquad (4)$$

which is, remarkably, independent of the dimension n.

Given a set $K \subset \mathbb{R}^n$ we will denote by $\text{conv}(K)$ the convex hull of K. The polyhedron

$$P_I(\boldsymbol{A}, \boldsymbol{b}) = \text{conv}(P(\boldsymbol{A}, \boldsymbol{b}) \cap \mathbb{Z}^n)$$

is traditionally referred to as the *integer hull* of $P(\boldsymbol{A}, \boldsymbol{b})$. Clearly, the problem (1) has at least one optimal solution which is a vertex of the integer hull $P_I(\boldsymbol{A}, \boldsymbol{b})$. Given such a solution $\boldsymbol{z}^*$, we obtain transference bounds that link the integrality gap $\text{IG}(\boldsymbol{A}, \boldsymbol{b}, \boldsymbol{c})$ with the size of the support of $\boldsymbol{z}^*$.

Observe that the integrality gap is positive homogeneous of degree one in $\boldsymbol{c}$, that is for $t > 0$

$$\text{IG}(\boldsymbol{A}, \boldsymbol{b}, t\boldsymbol{c}) = t\,\text{IG}(\boldsymbol{A}, \boldsymbol{b}, \boldsymbol{c}). \tag{5}$$

In what follows, we also use the notation $\|\cdot\|_2$ for the ℓ_2 norm. In view of (5), we may assume without loss of generality that $\boldsymbol{c}$ is a unit vector, that is $\|\boldsymbol{c}\|_2 = 1$.

Given a vector $\boldsymbol{x} = (x_1, \ldots, x_n)^\top \in \mathbb{R}^n$, we will denote by $\text{supp}(\boldsymbol{x})$ the *support* of $\boldsymbol{x}$, that is $\text{supp}(\boldsymbol{x}) = \{i : x_i \neq 0\}$. To measure the size of the support, we use the 0-*norm* $\|\boldsymbol{x}\|_0 = |\text{supp}(\boldsymbol{x})|$. The first result makes use of the quantity

$$\Delta(\boldsymbol{A}) = \sqrt{\det \boldsymbol{A}\boldsymbol{A}^\top}.$$

Geometrically, $\Delta(\boldsymbol{A})$ is the m-dimensional volume of the parallelepiped determined by the rows of $\boldsymbol{A}$. We will also denote by $\gcd(\boldsymbol{A})$ the greatest common divisor of all $m \times m$ subdeterminants of $\boldsymbol{A}$.

Theorem 1. *Let $\boldsymbol{A} \in \mathbb{Z}^{m\times n}$ be a matrix of rank m, $\boldsymbol{b} \in \mathbb{Z}^m$ and $\boldsymbol{c} \in \mathbb{R}^n$ be a unit cost vector. Suppose that (1) is feasible and bounded. Let $\boldsymbol{z}^*$ be an optimal solution to (1) which is a vertex of $P_I(\boldsymbol{A}, \boldsymbol{b})$. Then*

$$\text{IG}(\boldsymbol{A}, \boldsymbol{b}, \boldsymbol{c}) \leq \frac{s}{2^{s-m-1}} \cdot \frac{\Delta(\boldsymbol{A})}{\gcd(\boldsymbol{A})}, \tag{6}$$

where $s = \|\boldsymbol{z}^*\|_0$.

Hence, we obtain an upper bound for the integrality gap that drops *exponentially* in the size of support of any optimal solution to (1) which is a vertex of the integer hull $P_I(\boldsymbol{A}, \boldsymbol{b})$. We remark that [3] gives optimal transference bounds for positive knapsacks and corner polyhedra that connect the ℓ_∞-distance proximity and the size of the support of integer feasible solutions. Theorem 1 applies in the general case and a different setting; it connects the integrality gap and the size of the support of optimal solutions to (1).

Next, we obtain from Theorem 1 transference bounds in terms of the maximum absolute $m \times m$ subdeterminant $\Delta_m(\boldsymbol{A})$ and the maximum absolute entry $\Delta_1(\boldsymbol{A})$.

Corollary 2. *Assume the conditions of Theorem 1. Then the bounds*

$$\mathrm{IG}(\boldsymbol{A},\boldsymbol{b},\boldsymbol{c}) \leq \frac{s\binom{s+m}{m}^{1/2}}{2^{s-m-1}} \cdot \frac{\Delta_m(\boldsymbol{A})}{\gcd(\boldsymbol{A})} \tag{7}$$

and

$$\mathrm{IG}(\boldsymbol{A},\boldsymbol{b},\boldsymbol{c}) \leq \frac{s(s+m)^{m/2}}{2^{s-m-1}} \cdot \frac{(\Delta_1(\boldsymbol{A}))^m}{\gcd(\boldsymbol{A})} \tag{8}$$

hold.

For a unit cost vector $\boldsymbol{c}$, the bound (7) improves on (3) when $s \geq 4m$, and the bound (8) improves on (4) when $s \geq 6m$.

The proof of Theorem 1 makes use of results from convex geometry and Minkowski's geometry of numbers. Following a similar approach, we obtain a new upper bound for the ℓ_2-distance from a vertex of the polyhedron $P(\boldsymbol{A},\boldsymbol{b})$ to its nearest integer point in $P(\boldsymbol{A},\boldsymbol{b})$.

Theorem 3. *Let $\boldsymbol{A} \in \mathbb{Z}^{m\times n}$, $m < n$, be a matrix of rank m, $\boldsymbol{b} \in \mathbb{Z}^m$, and suppose that $P(\boldsymbol{A},\boldsymbol{b})$ contains integer points. Let $\boldsymbol{x}^*$ be a vertex of $P(\boldsymbol{A},\boldsymbol{b})$. There exists an integer point $\boldsymbol{z}^* \in P(\boldsymbol{A},\boldsymbol{b})$ such that*

$$\|\boldsymbol{x}^* - \boldsymbol{z}^*\|_2 \leq \frac{\Delta(\boldsymbol{A})}{\gcd(\boldsymbol{A})} - 1. \tag{9}$$

We remark that for certain matrices $\boldsymbol{A}$ the bound (9) is smaller than the ℓ_2-norm proximity bounds $\sqrt{n}(n-m)\Delta_m(\boldsymbol{A})$ and $m(2m\Delta_1(\boldsymbol{A})+1)^m$ that can be derived from [11] and [16], respectively. It is sufficient to observe that the ratio $\Delta(\boldsymbol{A})/\Delta_m(\boldsymbol{A})$ can be arbitrarily close to one and the ratio $\Delta(\boldsymbol{A})/\Delta_1(\boldsymbol{A})$ can be arbitrarily small.

Applying Theorem 3 to a vertex optimal solution $\boldsymbol{x}^*$ of (2) we obtain the following estimate.

Corollary 4. *Let $\boldsymbol{A} \in \mathbb{Z}^{m\times n}$, $m < n$, be a matrix of rank m, $\boldsymbol{b} \in \mathbb{Z}^m$ and $\boldsymbol{c} \in \mathbb{R}^n$ be a unit cost vector. Suppose that (1) is feasible and bounded. Then the bound*

$$\mathrm{IG}(\boldsymbol{A},\boldsymbol{b},\boldsymbol{c}) \leq \frac{\Delta(\boldsymbol{A})}{\gcd(\boldsymbol{A})} - 1 \tag{10}$$

holds.

For a unit cost vector $\boldsymbol{c}$, the bound (10) improves (3) when $\Delta(\boldsymbol{A})/\gcd(\boldsymbol{A}) < (n-m)\Delta_m(\boldsymbol{A})+1$ and improves (4) when $\Delta(\boldsymbol{A})/\gcd(\boldsymbol{A}) < mn^{-1/2}(2m\Delta_1(\boldsymbol{A})+1)^m+1$.

2 Volumes and Linear Transforms

In this section, we develop geometric tools needed for the proof of the transference bounds.

For a matrix $\boldsymbol{A} \in \mathbb{R}^{l\times r}$ we denote by $\text{lin}(\boldsymbol{A})$ the linear subspace of $\mathbb{R}^l$ spanned by the columns of A. Given a set $M \subset \mathbb{R}^r$ we let $\boldsymbol{A}M = \{\boldsymbol{A}\boldsymbol{x} \in \mathbb{R}^l : \boldsymbol{x} \in M\}$ and use the notation $[\boldsymbol{A}] = \boldsymbol{A}[0,1]^r$. For a set $X \subset \mathbb{R}^l$ and a linear subspace L of $\mathbb{R}^l$, we denote by $X|L$ the orthogonal projection of X onto L. Further, $\text{vol}_i(\cdot)$ denotes the i-dimensional volume.

Let S be an $(l-k)$-dimensional subspace of $\mathbb{R}^l$. Consider an orthonormal basis $\boldsymbol{s}_1, \ldots, \boldsymbol{s}_{l-k}, \boldsymbol{s}_{l-k+1}, \ldots, \boldsymbol{s}_l$ of $\mathbb{R}^l$ such that the first $l-k$ vectors form a basis of S. Let further $\boldsymbol{S}_{l-k} = (\boldsymbol{s}_1, \ldots, \boldsymbol{s}_{l-k}) \in \mathbb{R}^{l\times(l-k)}$ and $\boldsymbol{S}_k = (\boldsymbol{s}_{l-k+1}, \ldots, \boldsymbol{s}_l) \in \mathbb{R}^{l\times k}$.

Given a measurable set M in the subspace S, we are interested in the $(l-k)$-dimensional volume of its image $\boldsymbol{D}M$, where we assume that $\boldsymbol{D} \in \mathbb{R}^{l\times l}$ is an invertible matrix. The first result gives a general expression for $\text{vol}_{l-k}(\boldsymbol{D}M)$ in terms of $\text{vol}_{l-k}(M)$.

Lemma 1. *Let S be an $(l-k)$-dimensional subspace of $\mathbb{R}^l$. Let $M \subset S$ be measurable, $\boldsymbol{D} \in \mathbb{R}^{l\times l}$ be nonsingular and let the rows of $\boldsymbol{B} \in \mathbb{R}^{k\times l}$ form a basis of the subspace $(\boldsymbol{D}S)^\perp$, the orthogonal complement of the subspace $\boldsymbol{D}S = \text{lin}(\boldsymbol{D}\boldsymbol{S}_{l-k})$. Then*

$$\text{vol}_{l-k}(\boldsymbol{D}M) = |\det(\boldsymbol{D})|\sqrt{\frac{\det(\boldsymbol{B}\boldsymbol{B}^\top)}{\det(\boldsymbol{B}\boldsymbol{D}\boldsymbol{D}^\top\boldsymbol{B}^\top)}}\text{vol}_{l-k}(M).$$

Proof. By the elementary properties of volume, we have

$$\text{vol}_{l-k}(\boldsymbol{D}M) = \text{vol}_{l-k}([\boldsymbol{D}\boldsymbol{S}_{l-k}])\text{vol}_{l-k}(M). \tag{11}$$

On the other hand, we have

$$\begin{aligned} |\det(\boldsymbol{D})| &= \text{vol}_l([\boldsymbol{D}(\boldsymbol{S}_{l-k}, \boldsymbol{S}_k)]) \\ &= \text{vol}_{l-k}([\boldsymbol{D}\boldsymbol{S}_{l-k}])\text{vol}_k([\boldsymbol{D}\boldsymbol{S}_k]|\text{lin}(\boldsymbol{D}\boldsymbol{S}_{l-k})^\perp) \\ &= \text{vol}_{l-k}([\boldsymbol{D}\boldsymbol{S}_{l-k}])\text{vol}_k([\boldsymbol{D}\boldsymbol{S}_k]|\text{lin}(\boldsymbol{B}^\top)). \end{aligned} \tag{12}$$

Now, according to the definition of projections, we have

$$[\boldsymbol{D}\boldsymbol{S}_k]|\text{lin}(\boldsymbol{B}^\top) = \boldsymbol{B}^\top(\boldsymbol{B}\boldsymbol{B}^\top)^{-1}\boldsymbol{B}[\boldsymbol{D}\boldsymbol{S}_k]$$

and hence

$$\begin{aligned} \text{vol}_k([\boldsymbol{D}\boldsymbol{S}_k]|\text{lin}(\boldsymbol{B}^\top)) &= \sqrt{\det(\boldsymbol{S}_k^\top\boldsymbol{D}^\top\boldsymbol{B}^\top(\boldsymbol{B}\boldsymbol{B}^\top)^{-1}\boldsymbol{B}\boldsymbol{B}^\top(\boldsymbol{B}\boldsymbol{B}^\top)^{-1}\boldsymbol{B}\boldsymbol{D}\boldsymbol{S}_k)} \\ &= \sqrt{\det((\boldsymbol{B}\boldsymbol{B}^\top)^{-1})}\sqrt{\det(\boldsymbol{B}\boldsymbol{D}\boldsymbol{S}_k\boldsymbol{S}_k^\top\boldsymbol{D}^\top\boldsymbol{B}^\top)} \\ &= \sqrt{\frac{\det(\boldsymbol{B}\boldsymbol{D}\boldsymbol{D}^\top\boldsymbol{B}^\top)}{\det(\boldsymbol{B}\boldsymbol{B}^\top)}}. \end{aligned}$$

Substituting this expression in (12) gives along (11) the identity. □

The second result provides a lower bound for $\mathrm{vol}_{l-k}(\boldsymbol{D}M)$ that involves the eigenvalues of the matrix $\boldsymbol{D}^\top\boldsymbol{D}$ and $\mathrm{vol}_{l-k}(M)$.

Lemma 2. *Let $M \subset S$ be measurable, $\boldsymbol{D} \in \mathbb{R}^{l\times l}$ nonsingular and let $\lambda_1 \le \lambda_2 \le \cdots \le \lambda_l$ be the positive eigenvalues of $\boldsymbol{D}^\top\boldsymbol{D}$. Then*

$$\mathrm{vol}_{l-k}(\boldsymbol{D}M) \ge \left(\prod_{i=1}^{l-k}\sqrt{\lambda_i}\right)\mathrm{vol}_{l-k}(M).$$

Proof. According to (11), we need to estimate

$$\mathrm{vol}_{l-k}([\boldsymbol{D}\boldsymbol{S}_{l-k}]) = \sqrt{\det(\boldsymbol{S}_{l-k}^\top\boldsymbol{D}^\top\boldsymbol{D}\boldsymbol{S}_{l-k})}.$$

Let $\boldsymbol{O} \in \mathbb{R}^{l\times l}$ be a matrix such that its rows form an orthonormal basis consisting of eigenvectors of $\boldsymbol{D}^\top\boldsymbol{D}$. For convenience, we will denote by $\mathrm{diag}(\lambda_i)$ the diagonal matrix with the eigenvalues $\lambda_1, \ldots, \lambda_l$ on the main diagonal. Then we have $\boldsymbol{D}^\top\boldsymbol{D} = \boldsymbol{O}^\top\mathrm{diag}(\lambda_i)\boldsymbol{O}$.

Let $[l] = \{1, \ldots, l\}$ and let $\binom{[l]}{r}$ be the set of all r-element subsets of $[l]$. With $\widetilde{\boldsymbol{S}_{l-k}} = \boldsymbol{O}\boldsymbol{S}_{l-k}$ we get by the Cauchy-Binet formula

$$\begin{aligned}
\det(\boldsymbol{S}_{l-k}^\top\boldsymbol{D}^\top\boldsymbol{D}\boldsymbol{S}_{l-k}) &= \det(\widetilde{\boldsymbol{S}_{l-k}}^\top\mathrm{diag}(\lambda_i^{1/2})\mathrm{diag}(\lambda_i^{1/2})\widetilde{\boldsymbol{S}_{l-k}}) \\
&= \sum_{I\in\binom{[l]}{l-k}} \det(\mathrm{diag}(\lambda_i^{1/2})^I)^2(\det(\widetilde{\boldsymbol{S}_{l-k}}^I))^2 \\
&\ge \left(\prod_{i=1}^{l-k}\lambda_i\right)\sum_{I\in\binom{[l]}{l-k}}(\det(\widetilde{\boldsymbol{S}_{l-k}}^I))^2 \\
&= \left(\prod_{i=1}^{l-k}\lambda_i\right)\det(\widetilde{\boldsymbol{S}_{l-k}}^\top\widetilde{\boldsymbol{S}_{l-k}}) = \prod_{i=1}^{l-k}\lambda_i.
\end{aligned}$$

Here $\mathrm{diag}(\lambda_i^{1/2})^I$ is the $(l-k)\times(l-k)$ diagonal matrix with $\lambda_i^{1/2}$ indexed by I and $\widetilde{\boldsymbol{S}_{l-k}}^I$ is the $(l-k)\times(l-k)$ submatrix of $\widetilde{\boldsymbol{S}_{l-k}}$ with rows indexed by I. In the second to last identity we used again the Cauchy-Binet formula. □

From Lemma 2 we obtain the following corollary.

Corollary 5. *Let S be an $(l-k)$-dimensional subspace of $\mathbb{R}^l$ and let $\boldsymbol{D} = \mathrm{diag}(d_1, \ldots, d_l)$ with $0 < d_1 \le d_2 \le \cdots \le d_l$. Then*

$$\mathrm{vol}_{l-k}(\boldsymbol{D}(S\cap(-1,1)^l)) \ge 2^{l-k}\prod_{i=1}^{l-k}d_i\,. \tag{13}$$

Proof. By Vaaler's cube slicing inequality [25], we have $\mathrm{vol}_{l-k}(S\cap(-1,1)^l) \ge 2^{l-k}$. Now the bound (13) immediately follows from Lemma 2. □

3 Proofs of the Transference Bounds

We will derive Theorem 1 from Corollary 4 and from the following result.

Theorem 6. *Let $\boldsymbol{A} \in \mathbb{Z}^{m\times n}$, with $n > m+1$, be a matrix of rank m, $\boldsymbol{b} \in \mathbb{Z}^m$ and $\boldsymbol{c} \in \mathbb{R}^n$ be a unit cost vector. Suppose that (1) is feasible and bounded. Let $\boldsymbol{z}^* = (z_1^*, \dots, z_n^*)^\top$ be an optimal solution of (1) which is a vertex of $P_I(\boldsymbol{A}, \boldsymbol{b})$. Assuming without loss of generality $z_1^* \leq \cdots \leq z_n^*$, the bound*

$$\mathrm{IG}(\boldsymbol{A}, \boldsymbol{b}, \boldsymbol{c}) \leq \frac{(n-m)}{\prod_{i=1}^{n-m-1}(z_i^*+1)} \cdot \frac{\Delta(\boldsymbol{A})}{\gcd(\boldsymbol{A})} \tag{14}$$

holds.

Proof. Let $\boldsymbol{c}|\ker(\boldsymbol{A})$ denote the orthogonal projection of the vector $\boldsymbol{c}$ on the kernel subspace $\ker(\boldsymbol{A}) = \{\boldsymbol{x} \in \mathbb{R}^n : \boldsymbol{A}\boldsymbol{x} = \boldsymbol{0}\}$ of the matrix $\boldsymbol{A}$. Observe first that if $\boldsymbol{c}$ is orthogonal to $\ker(\boldsymbol{A})$, then $\mathrm{IG}(\boldsymbol{A}, \boldsymbol{b}, \boldsymbol{c}) = 0$ and the bound (14) holds. Hence, we may assume without loss of generality that $\boldsymbol{c}|\ker(\boldsymbol{A})$ is a nonzero vector.

Suppose, to derive a contradiction, that the bound (14) does not hold. Then there exists a vertex $\boldsymbol{x}^*$ of $P(\boldsymbol{A}, \boldsymbol{b})$ and a vertex $\boldsymbol{z}^*$ of $P_I(\boldsymbol{A}, \boldsymbol{b})$ optimising (1), such that, assuming $z_1^* \leq \cdots \leq z_n^*$, we have

$$\boldsymbol{c} \cdot (\boldsymbol{z}^* - \boldsymbol{x}^*) > \frac{(n-m)}{\prod_{i=1}^{n-m-1}(z_i^*+1)} \cdot \frac{\Delta(\boldsymbol{A})}{\gcd(\boldsymbol{A})}. \tag{15}$$

Let $d_i = z_i^* + 1$, $i \in [n]$, $\boldsymbol{D} = \mathrm{diag}(d_1, \dots, d_n)$ and let $\boldsymbol{B} \in \mathbb{Z}^{(m+1)\times n}$ be the matrix obtained by adding the $(m+1)$-st row $\boldsymbol{c}^\top$ to the matrix $\boldsymbol{A}$. Let further $V = \ker(\boldsymbol{B})$. Consider the box section

$$K = V \cap (-d_1, d_1) \times \cdots \times (-d_n, d_n).$$

We can write $K = \boldsymbol{D}M$, where M is a $(n-m-1)$-dimensional section of the cube $(-1,1)^n$. Hence, by Corollary 5,

$$\mathrm{vol}_{n-m-1}(K) \geq 2^{n-m-1} \prod_{i=1}^{n-m-1} d_i. \tag{16}$$

Consider the origin-symmetric convex set

$$L = \mathrm{conv}(\boldsymbol{x}^* - \boldsymbol{z}^*, K, \boldsymbol{z}^* - \boldsymbol{x}^*) \subset \ker(\boldsymbol{A}).$$

The set L is a bi-pyramid with apexes $\pm(\boldsymbol{x}^* - \boldsymbol{z}^*)$ and $(n-m-1)$-dimensional basis K. As $K \subset \ker(\boldsymbol{A}) \cap \ker(\boldsymbol{c})$ the height of $\boldsymbol{x}^* - \boldsymbol{z}^*$ over K is given by

$$\frac{\boldsymbol{c}|\ker(\boldsymbol{A}) \cdot (\boldsymbol{z}^* - \boldsymbol{x}^*)}{\|\boldsymbol{c}|\ker(\boldsymbol{A})\|_2} = \frac{\boldsymbol{c} \cdot (\boldsymbol{z}^* - \boldsymbol{x}^*)}{\|\boldsymbol{c}|\ker(\boldsymbol{A})\|_2}.$$

Hence, we have

$$\mathrm{vol}_{n-m}(L) = \frac{2\,\boldsymbol{c}\cdot(\boldsymbol{z}^* - \boldsymbol{x}^*)\,\mathrm{vol}_{n-m-1}(K)}{(n-m)\|\boldsymbol{c}|\ker(\boldsymbol{A})\|_2}.$$

Then, using (16) and (15) and noting that the assumption $\|\boldsymbol{c}\|_2 = 1$ implies $\|\boldsymbol{c}|\ker(\boldsymbol{A})\|_2 \leq 1$, we obtain the lower bound

$$\mathrm{vol}_{n-m}(L) > 2^{n-m}\frac{\Delta(\boldsymbol{A})}{\gcd(\boldsymbol{A})}. \tag{17}$$

Observe that the lattice $\Lambda(\boldsymbol{A}) = \ker(\boldsymbol{A}) \cap \mathbb{Z}^n$ has determinant

$$\det(\Lambda(\boldsymbol{A})) = \frac{\Delta(\boldsymbol{A})}{\gcd(\boldsymbol{A})} \tag{18}$$

(see e.g. [24, Chapter 1, §1]). The $(n-m)$-dimensional subspace $\ker(\boldsymbol{A})$ can be considered as a usual Euclidean $(n-m)$-dimensional space. Therefore, by (17), (18) and Minkowski's first fundamental theorem (in the form of Theorem II in Chapter III of [10]), applied to the set L and the lattice $\Lambda(\boldsymbol{A})$, there is a nonzero point $\boldsymbol{y} \in L \cap \Lambda(\boldsymbol{A})$.

Suppose first that $\boldsymbol{c}\cdot\boldsymbol{y} = 0$. Consider the points $\boldsymbol{y}^+ = \boldsymbol{z}^* + \boldsymbol{y}$ and $\boldsymbol{y}^- = \boldsymbol{z}^* - \boldsymbol{y}$. We have $\boldsymbol{y}^+, \boldsymbol{y}^- \in (\boldsymbol{z}^* + K)$ and, consequently, $\boldsymbol{y}^+, \boldsymbol{y}^- \in P(\boldsymbol{A}, \boldsymbol{b})$. Further, $\boldsymbol{z}^*$ is the midpoint of the segment with endpoints $\boldsymbol{y}^+$ and $\boldsymbol{y}^-$, contradicting the choice of $\boldsymbol{z}^*$ as a vertex of the integer hull $P_I(\boldsymbol{A}, \boldsymbol{b})$.

It remains to consider the case $\boldsymbol{c}\cdot\boldsymbol{y} \neq 0$. Since L is origin-symmetric, we may assume without loss of generality that $\boldsymbol{c}\cdot\boldsymbol{y} < 0$. Observe that the point $\boldsymbol{y}^+ = \boldsymbol{z}^* + \boldsymbol{y}$ is in the set $\mathrm{conv}(\boldsymbol{z}^* + K, \boldsymbol{x}^*)$ and hence $\boldsymbol{y}^+ \in P(\boldsymbol{A}, \boldsymbol{b})$. Now, it is sufficient to notice that $\boldsymbol{c}\cdot\boldsymbol{y}^+ < \boldsymbol{c}\cdot\boldsymbol{z}^*$, contradicting the optimality of $\boldsymbol{z}^*$. □

3.1 Proof of Theorem 1

Let $I = \{i_1, \ldots, i_k\} \subset [n]$ with $i_1 < i_2 < \cdots < i_k$. We will use the notation $\boldsymbol{A}_I$ for the $m \times k$ submatrix of $\boldsymbol{A}$ with columns indexed by I. In the same manner, given $\boldsymbol{x} \in \mathbb{R}^n$, we will denote by $\boldsymbol{x}_I$ the vector $(x_{i_1}, \ldots, x_{i_k})^\top$. By $\mathbb{R}^I$ we denote the k-dimensional real space with coordinates indexed by I.

Let $\boldsymbol{x}^*$ be a vertex optimal solution to (2). Clearly, we may assume that $\boldsymbol{x}^* \neq \boldsymbol{z}^*$. Let $I \subset [n]$ denote the set of indices i for which at least one of z_i^*, x_i^* is non-zero. We consider a new linear program

$$\min\{\boldsymbol{c}_I \cdot \boldsymbol{x}_I : \hat{\boldsymbol{A}}_I\boldsymbol{x}_I = \hat{\boldsymbol{b}}, \boldsymbol{x}_I \geq 0\}, \tag{19}$$

where $\hat{\boldsymbol{A}}_I \in \mathbb{Z}^{\hat{m}\times\hat{n}}$ is a full-row-rank matrix with $\gcd(\hat{\boldsymbol{A}}_I) = 1$ and $\hat{\boldsymbol{b}} \in \mathbb{Z}^{\hat{m}}$ such that $\hat{\boldsymbol{A}}_I\boldsymbol{x}_I = \hat{\boldsymbol{b}}$ and $\boldsymbol{A}_I\boldsymbol{x}_I = \boldsymbol{b}$ decribe the same affine subspace in $\mathbb{R}^I$. Note that $\boldsymbol{x}_I^*$ and $\boldsymbol{z}_I^*$ are optimal fractional and integral solutions, respectively, and that $\boldsymbol{z}_I^*$ is a vertex of the integer hull of $P(\hat{\boldsymbol{A}}_I, \hat{\boldsymbol{b}})$. Note also that

$$\Delta(\hat{\boldsymbol{A}}_I) \leq \frac{\Delta(\boldsymbol{A})}{\gcd(\boldsymbol{A})}, \tag{20}$$

one can see this by observing that the quantity on the left is a divisor of the volume of the orthogonal projection of a parallelepiped whose volume is given by the quantity on the right.

If $\hat{n} = \hat{m} + 1$, then the bound (6) immediately follows from the bound (10) in Corollary 4. Otherwise, suppose that $\hat{n} > \hat{m} + 1$. We have then that $\hat{\boldsymbol{A}}_I$, $\hat{\boldsymbol{b}}$, $\boldsymbol{c}_I / \|\boldsymbol{c}_I\|_2$, $\boldsymbol{x}_I^*$, $\boldsymbol{z}_I^*$ satisfy the hypotheses of Theorem 6. We therefore get

$$\mathrm{IG}(\boldsymbol{A}, \boldsymbol{b}, \boldsymbol{c}) = \mathrm{IG}(\hat{\boldsymbol{A}}_I, \hat{\boldsymbol{b}}, \boldsymbol{c}_I) \leq \|\boldsymbol{c}_I\|_2 \cdot \frac{\hat{n} - \hat{m}}{\prod_i (z_i^* + 1)} \cdot \Delta(\hat{\boldsymbol{A}}_I), \tag{21}$$

where the product in the denominator is over the $\hat{n} - \hat{m} - 1$ smallest coordinates of $\boldsymbol{z}_I^*$.

Now, $\hat{n}$ is equal to $\|\boldsymbol{z}^* + \boldsymbol{x}^*\|_0$. Also, $\boldsymbol{x}^*$ has support of size at most $\hat{m}$, since $\|\boldsymbol{x}^*\|_0 = \|\boldsymbol{x}_I^*\|_0$ and $\boldsymbol{x}_I^*$ is a vertex of the new linear program (19). Thus we get

$$\hat{n} - \hat{m} \leq \|\boldsymbol{z}^* + \boldsymbol{x}^*\|_0 - \|\boldsymbol{x}^*\|_0 \leq \|\boldsymbol{z}^*\|_0 = s. \tag{22}$$

On the other hand, we have the following lower bound for the product in the denominator of (21):

$$\prod_i (z_i^* + 1) \geq 2^{\hat{n} - \hat{m} - 1 - |I \setminus \mathrm{Supp}(\boldsymbol{z}_I^*)|} = 2^{s - \hat{m} - 1} \geq 2^{s - m - 1}. \tag{23}$$

Combining together (21), (22), (23), and the fact that $\|\boldsymbol{c}_I\|_2 \leq 1$, we get

$$\mathrm{IG}(\boldsymbol{A}, \boldsymbol{b}, \boldsymbol{c}) \leq \frac{s}{2^{s-m-1}} \cdot \Delta(\hat{\boldsymbol{A}}_I). \tag{24}$$

The desired conclusion (6) then follows from (20) and (24).

3.2 Proof of Corollary 2

Let I and $\hat{\boldsymbol{A}}_I$ be as in the proof of Theorem 1. We will derive the bounds (7) and (8) from the bound (24). Choose any $J \subset [n] \backslash I$ that is minimal with respect to the property that $\boldsymbol{A}_{I \cup J}$ has rank m. Thus, $|J| = m - \hat{m}$, and

$$|I \cup J| = \hat{n} + (m - \hat{m}) \leq s + \hat{m} + (m - \hat{m}) = s + m. \tag{25}$$

As with (20), we have

$$\Delta(\hat{\boldsymbol{A}}_I) \leq \frac{\Delta(\boldsymbol{A}_{I \cup J})}{\gcd(\boldsymbol{A}_{I \cup J})} \leq \frac{\Delta(\boldsymbol{A}_{I \cup J})}{\gcd(\boldsymbol{A})}. \tag{26}$$

By (25) and the Cauchy-Binet formula, we have

$$\Delta(\boldsymbol{A}_{I \cup J}) \leq \binom{|I \cup J|}{m}^{1/2} \Delta_m(\boldsymbol{A}_{I \cup J}) \leq \binom{s+m}{m}^{1/2} \Delta_m(\boldsymbol{A}). \tag{27}$$

Combining (24), (26), and (27), we obtain the bound (7). On the other hand, if we use (25) and Hadamard's inequality, we get

$$\Delta(\boldsymbol{A}_{I \cup J}) \leq (\sqrt{|I \cup J|} \cdot \Delta_1(\boldsymbol{A}_{I \cup J}))^m \leq (\sqrt{s+m} \cdot \Delta_1(\boldsymbol{A}))^m. \tag{28}$$

Combining (24), (26), and (28), we obtain the bound (8).

4 Proof of the ℓ_2-Distance Proximity Bound

First, we will prove two lemmas needed for the proof of Theorem 3.

Let $\boldsymbol{y} \in \mathbb{R}^n$ and let

$$C^n(\boldsymbol{y}) = \{\boldsymbol{x} \in \mathbb{R}^n : \|\boldsymbol{x} - \boldsymbol{y}\|_\infty < 1\}$$

be an open cube in $\mathbb{R}^n$ with edge length 2 centered at the point $\boldsymbol{y}$. Given two points $\boldsymbol{u}, \boldsymbol{v} \in \mathbb{R}^n$ we will consider an open set $D(\boldsymbol{u}, \boldsymbol{v})$ defined as

$$D(\boldsymbol{u}, \boldsymbol{v}) = \operatorname{conv}(C^n(\boldsymbol{u}), C^n(\boldsymbol{v})).$$

Lemma 3. *Let $\boldsymbol{u}, \boldsymbol{v} \in \mathbb{R}^n_{\geq 0}$. Then $D(\boldsymbol{u}, \boldsymbol{v}) \cap \mathbb{Z}^n = D(\boldsymbol{u}, \boldsymbol{v}) \cap \mathbb{Z}^n_{\geq 0}$.*

Proof. Suppose, to derive a contradiction, that there exists an integer point $\boldsymbol{z} = (z_1, \ldots, z_n)^\top \in D(\boldsymbol{u}, \boldsymbol{v})$ such that $z_j \leq -1$ for some $j \in [n]$. Then there exist points $\boldsymbol{x} = (x_1, \ldots, x_n)^\top \in C^n(\boldsymbol{u})$ and $\boldsymbol{y} = (y_1, \ldots, y_n)^\top \in C^n(\boldsymbol{v})$ such that for some $\lambda \in [0, 1]$

$$z_j = \lambda x_j + (1 - \lambda) y_j.$$

Therefore, since $x_j > -1$ and $y_j > -1$, we must have $z_j > -1$. □

Next, we consider an origin-symmetric open convex set $E = E(\boldsymbol{u}, \boldsymbol{v})$ defined as

$$E = \operatorname{conv}(C^n(\boldsymbol{u} - \boldsymbol{v}), C^n(\boldsymbol{v} - \boldsymbol{u})).$$

Notice that

$$E = (D(\boldsymbol{u}, \boldsymbol{v}) - \boldsymbol{v}) \cup (-D(\boldsymbol{u}, \boldsymbol{v}) + \boldsymbol{v}). \tag{29}$$

Lemma 4. *Suppose that $\boldsymbol{u}, \boldsymbol{v} \in P(\boldsymbol{A}, \boldsymbol{b})$. Then the bound*

$$\operatorname{vol}_{n-m}(E \cap \ker(\boldsymbol{A})) \geq 2^{n-m}(1 + \|\boldsymbol{u} - \boldsymbol{v}\|_2) \tag{30}$$

holds.

Proof. First, we will separately consider the case $n = m + 1$. Then $\ker(\boldsymbol{A})$ has dimension one and, noticing that $\boldsymbol{u} - \boldsymbol{v} \in \ker(\boldsymbol{A})$ we can write

$$\operatorname{vol}_1(E \cap \ker(\boldsymbol{A})) = \operatorname{vol}_1(C^n(\boldsymbol{0}) \cap \ker(\boldsymbol{A})) + 2\|\boldsymbol{u} - \boldsymbol{v}\|_2.$$

Since $\operatorname{vol}_1(C^n(\boldsymbol{0}) \cap \ker(\boldsymbol{A})) \geq 2$, we obtain the bound (30).

For the rest of the proof we assume that $n > m+1$. If $\boldsymbol{u} = \boldsymbol{v}$, then $E = C^n(\boldsymbol{0})$ and the bound (30) immediately follows from Vaaler's cube slicing inequality [25]. Hence, we may also assume without loss of generality that $\boldsymbol{u} - \boldsymbol{v} \neq \boldsymbol{0}$.

Let

$$S = C^n(\boldsymbol{0}) \cap \ker(\boldsymbol{A}) \cap \ker((\boldsymbol{u} - \boldsymbol{v})^\top).$$

The set S is a section of the open cube $C^n(\mathbf{0})$. Since $\boldsymbol{u} - \boldsymbol{v} \in \ker(\boldsymbol{A}) \setminus \{\mathbf{0}\}$, the section S has dimension $n - m - 1$. Let further

$$S^+ = \{\boldsymbol{x} \in C^n(\mathbf{0}) \cap \ker(\boldsymbol{A}) : (\boldsymbol{u} - \boldsymbol{v}) \cdot \boldsymbol{x} > 0\}$$

and

$$S^- = \{\boldsymbol{x} \in C^n(\mathbf{0}) \cap \ker(\boldsymbol{A}) : (\boldsymbol{u} - \boldsymbol{v}) \cdot \boldsymbol{x} < 0\}.$$

By construction, S^+, S^- and S do not overlap and $C^n(\mathbf{0}) \cap \ker(\boldsymbol{A}) = S^+ \cup S^- \cup S$. Further, by Vaaler's cube slicing inequality [25], we have

$$\mathrm{vol}_{n-m}(S^+) + \mathrm{vol}_{n-m}(S^-) = \mathrm{vol}_{n-m}(C^n(\mathbf{0}) \cap \ker(\boldsymbol{A})) \geq 2^{n-m} \tag{31}$$

and

$$\mathrm{vol}_{n-m-1}(S) \geq 2^{n-m-1}. \tag{32}$$

Observe that $E \cap \ker(\boldsymbol{A})$ contains the sets $S^+ + \boldsymbol{u} - \boldsymbol{v}$, $S^- + \boldsymbol{v} - \boldsymbol{u}$ and the cylinder $\mathrm{conv}(\boldsymbol{u} - \boldsymbol{v} + S, \boldsymbol{v} - \boldsymbol{u} + S)$. These three sets do not overlap and, using (31) and (32), we have

$$\begin{aligned}
\mathrm{vol}_{n-m}(E \cap \ker(\boldsymbol{A})) &\geq \mathrm{vol}_{n-m}(S^+) + \mathrm{vol}_{n-m}(S^-) \\
&\quad + \mathrm{vol}_{n-m}(\mathrm{conv}(\boldsymbol{u} - \boldsymbol{v} + S, \boldsymbol{v} - \boldsymbol{u} + S)) \\
&= \mathrm{vol}_{n-m}(C^n(\mathbf{0}) \cap \ker(\boldsymbol{A})) + 2\,\mathrm{vol}_{n-m-1}(S)\,\|\boldsymbol{u} - \boldsymbol{v}\|_2 \\
&\geq 2^{n-m}(1 + \|\boldsymbol{u} - \boldsymbol{v}\|_2).
\end{aligned}$$

Hence, we obtain the bound (30). □

4.1 Proof of Theorem 3

We will say that $B \subset [n]$ is a *basis* of $\boldsymbol{A}$ if $|B| = m$ and the submatrix $\boldsymbol{A}_B$ is nonsingular. Take any vertex $\boldsymbol{x}^* \in P(\boldsymbol{A}, \boldsymbol{b})$. There is a basis B of $\boldsymbol{A}$ such that, denoting by N the complement of B in $[n]$, we have

$$\boldsymbol{x}^*_B = \boldsymbol{A}_B^{-1}\boldsymbol{b} \text{ and } \boldsymbol{x}^*_N = \mathbf{0}_N.$$

Choose an integer point $\boldsymbol{z}^* \in P(\boldsymbol{A}, \boldsymbol{b})$ with the minimum possible distance between the points $\boldsymbol{x}^*_N = \mathbf{0}_N$ and $\boldsymbol{z}^*_N$. Then

$$\|\boldsymbol{z}^*_N\|_2 = \min\{\|\boldsymbol{y}_N\|_2 : \boldsymbol{y} \in P(\boldsymbol{A}, \boldsymbol{b}) \cap \mathbb{Z}^n\}. \tag{33}$$

Suppose, to derive a contradiction, that the bound (9) does not hold for the point $\boldsymbol{z}^*$. Then, using (18),

$$\|\boldsymbol{x}^* - \boldsymbol{z}^*\|_2 > \frac{\Delta(\boldsymbol{A})}{\gcd(\boldsymbol{A})} - 1 = \det(\Lambda(\boldsymbol{A})) - 1. \tag{34}$$

Recall that we denote by $\Lambda(\boldsymbol{A})$ the lattice formed by all integer points in the kernel subspace of the matrix $\boldsymbol{A}$.

The lower bound (34) and Lemma 4 imply that for $E = E(\boldsymbol{x}^*, \boldsymbol{z}^*)$ we have

$$\mathrm{vol}(E \cap \ker(\boldsymbol{A})) > 2^{n-m} \det(\Lambda(\boldsymbol{A})). \tag{35}$$

The $(n-m)$-dimensional subspace $\ker(\boldsymbol{A})$ can be considered as a usual Euclidean $(n-m)$-dimensional space. Noting the bound (35), Minkowski's first fundamental theorem (in the form of Theorem II in Chapter III of [10]) implies that the set $E \cap \ker(\boldsymbol{A})$ contains nonzero points $\pm\boldsymbol{z}$ of the lattice $\Lambda(\boldsymbol{A})$. Using (29), we may assume without loss of generality that we have $\boldsymbol{z} \in D(\boldsymbol{x}^*, \boldsymbol{z}^*) - \boldsymbol{z}^*$. Therefore, the point $\boldsymbol{w} = \boldsymbol{z} + \boldsymbol{z}^*$ is in the set $D(\boldsymbol{x}^*, \boldsymbol{z}^*) \cap (\ker(\boldsymbol{A}) + \boldsymbol{z}^*)$. By Lemma 3, $\boldsymbol{w} \in \mathbb{Z}^n_{\geq 0}$ and, hence, $\boldsymbol{w} \in P(\boldsymbol{A}, \boldsymbol{b})$.

Next, we will show that $\|\boldsymbol{w}_N\|_2 < \|\boldsymbol{z}^*_N\|_2$, contradicting (33). Notice first that for any $\boldsymbol{x} \in P(\boldsymbol{A}, \boldsymbol{b})$ we have $\boldsymbol{x}_B = \boldsymbol{A}_B^{-1}(\boldsymbol{b} - \boldsymbol{A}_N \boldsymbol{x}_N)$. Hence $\boldsymbol{w}_N = \boldsymbol{z}^*_N$ implies that $\boldsymbol{w} = \boldsymbol{z}^*$. Therefore, we may assume that $\boldsymbol{w}_N \neq \boldsymbol{z}^*_N$.

Take any index $j \in N$. Since $\boldsymbol{w} \in D(\boldsymbol{x}^*, \boldsymbol{z}^*)$, we have $w_j \leq z^*_j$. Hence there is at least one index $j_0 \in N$ with $w_{j_0} < z^*_{j_0}$. Therefore $\|\boldsymbol{w}_N\|_2 < \|\boldsymbol{z}^*_N\|_2$ and we obtain a contradiction with (33).

Acknowledgement. The authors thank the anonymous referees for their valuable comments and suggestions.

References

1. Abdi, A., Cornuéjols, G., Guenin, B., Tunçel, L.: Dyadic linear programming and extensions. arXiv:2309.04601 (2023)
2. Aliev, I., Averkov, G., De Loera, J.A., Oertel, T.: Sparse representation of vectors in lattices and semigroups. Math. Program. **192**, 519–546 (2022)
3. Aliev, I., Celaya, M., Henk, M., Williams, A.: Distance-sparsity transference for vertices of corner polyhedra. SIAM J. Optim. **31**(1), 200–216 (2021)
4. Aliev, I., De Loera, J.A., Eisenbrand, F., Oertel, T., Weismantel, R.: The support of integer optimal solutions. SIAM J. Optim. **28**(3), 2152–2157 (2018)
5. Aliev, I., De Loera, J.A., Oertel, T., O'Neill, C.: Sparse solutions of linear Diophantine equations. SIAM J. Appl. Algebra Geom. **1**(1), 239–253 (2017)
6. Aliev, I., Henk, M., Oertel, T.: Distances to lattice points in knapsack polyhedra. Math. Program. **182**(1–2), 175–198 (2020)
7. Berndt, S., Jansen, K., Klein, K.-M.: New bounds for the vertices of the integer hull. In: Symposium on Simplicity in Algorithms (SOSA), pp. 25–36. Society for Industrial and Applied Mathematics (SIAM), Philadelphia (2021)
8. Blair, C.E., Jeroslow, R.G.: The value function of a mixed integer program. I. Discrete Math. **19**(2), 121–138 (1977)
9. Blair, C.E., Jeroslow, R.G.: The value function of an integer program. Math. Program. **23**(3), 237–273 (1982)
10. Cassels, J.W.S.: An Introduction to the Geometry of Numbers. Classics in Mathematics. Springer, Heidelberg (1996). https://doi.org/10.1007/978-3-642-62035-5

11. Celaya, M., Kuhlmann, S., Paat, J., Weismantel, R.: Improving the Cook et al. proximity bound given integral valued constraints. In: Aardal, K., Sanitá, L. (eds.) IPCO 2022. LNCS, vol. 13265, pp. 84–97. Springer, Cham (2022). https://doi.org/10.1007/978-3-031-06901-7_7
12. Cook, W., Fonlupt, J., Schrijver, A.: An integer analogue of Carathéodory's theorem. J. Combin. Theory Ser. B **40**(1), 63–70 (1986)
13. Cook, W., Gerards, A.M.H., Schrijver, A., Tardos, É.: Sensitivity theorems in integer linear programming. Math. Program. **34**(3), 251–264 (1986)
14. Dubey, Y., Liu, S.: A short proof of tight bounds on the smallest support size of integer solutions to linear equations. arXiv:2307.08826 (2023)
15. Eisenbrand, F., Shmonin, G.: Carathéodory bounds for integer cones. Oper. Res. Lett. **34**(5), 564–568 (2006)
16. Eisenbrand, F., Weismantel, R.: Proximity results and faster algorithms for integer programming using the Steinitz lemma. In: Proceedings of the Twenty-Ninth Annual ACM-SIAM Symposium on Discrete Algorithms, pp. 808–816. Society for Industrial and Applied Mathematics (SIAM), Philadelphia (2018)
17. Gomory, R.E.: On the relation between integer and noninteger solutions to linear programs. Proc. Nat. Acad. Sci. U.S.A. **53**, 260–265 (1965)
18. Gomory, R.E.: Some polyhedra related to combinatorial problems. Linear Algebra Appl. **2**, 451–558 (1969)
19. Lee, J., Paat, J., Stallknecht, I., Xu, L.: Improving proximity bounds using sparsity. In: Baïou, M., Gendron, B., Günlük, O., Mahjoub, A.R. (eds.) ISCO 2020. LNCS, vol. 12176, pp. 115–127. Springer, Cham (2020). https://doi.org/10.1007/978-3-030-53262-8_10
20. Lee, J., Paat, J., Stallknecht, I., Xu, L.: Polynomial upper bounds on the number of differing columns of Δ-modular integer programs. Math. Oper. Res. **48**, 2267–2286 (2023)
21. Oertel, T., Paat, J., Weismantel, R.: The distributions of functions related to parametric integer optimization. SIAM J. Appl. Algebra Geom. **4**(3), 422–440 (2020)
22. Paat, J., Weismantel, R., Weltge, S.: Distances between optimal solutions of mixed-integer programs. Math. Program. **179**(1–2), 455–468 (2020)
23. Sebő, A.: Hilbert bases, Carathéodory's theorem and combinatorial optimization. In: Proceedings of the 1st Integer Programming and Combinatorial Optimization Conference, pp. 431–455. University of Waterloo Press (1990)
24. Skolem, T.: Diophantische Gleichungen. Ergebnisse der Mathematik, vol 5. Springer, Berlin (1938)
25. Vaaler, J.D.: A geometric inequality with applications to linear forms. Pacific J. Math. **83**(2), 543–553 (1979)

Separating k-MEDIAN from the Supplier Version

Aditya Anand(✉) and Euiwoong Lee

University of Michigan, Ann Arbor, USA
adanand@umich.edu

Abstract. Given a metric space (V, d) along with an integer k, the k-MEDIAN problem asks to open k centers $C \subseteq V$ to minimize $\sum_{v \in V} d(v, C)$, where $d(v, C) := \min_{c \in C} d(v, c)$. While the best-known approximation ratio 2.613 holds for the more general *supplier version* where an additional set $F \subseteq V$ is given with the restriction $C \subseteq F$, the best known hardness for these two versions are $1+1/e \approx 1.36$ and $1+2/e \approx 1.73$ respectively, using the same reduction from MAXIMUM k-COVERAGE. We prove the following two results separating them.

1. We give a 1.546-parameterized approximation algorithm that runs in time $f(k)n^{O(1)}$. Since $1+2/e$ is proved to be the optimal approximation ratio for the supplier version in the parameterized setting, this result separates the original k-MEDIAN from the supplier version.
2. We prove a 1.416-hardness for polynomial-time algorithms assuming the Unique Games Conjecture. This is achieved via a new fine-grained hardness of MAXIMUM k-COVERAGE for small set sizes.

Our upper bound and lower bound are derived from almost the same expression, with the only difference coming from the well-known separation between the powers of LP and SDP on (hypergraph) vertex cover.

Keywords: Approximation algorithms · Clustering · Parameterized approximation · Optimization · Hardness of approximation

1 Introduction

k-MEDIAN is perhaps the most well-studied clustering objective in finite general metrics. Given a (semi-)metric space (V, d) along with an integer k, the goal is to open k centers $C \subseteq V$ to minimize the objective function $\sum_{v \in V} d(v, C)$, where $d(v, C) := \min_{c \in C} d(v, c)$. It has been the subject of numerous papers introducing diverse algorithmic techniques, including filtering [32,33], metric embedding [7,9], local search [3], primal LP rounding [11,12], primal-dual [27], greedy with dual fitting analysis [10,26], and bipoint rounding [8,31]. The current best approximation ratio is slightly lower than 2.613 [16,20], achieved by the combination of greedy search analyzed dual fitting method and bipoint rounding.

E. Lee—Supported in part by NSF grant CCF-2236669 and Google.

J. Vygen and J. Byrka (Eds.): IPCO 2024, LNCS 14679, pp. 14–27, 2024.
https://doi.org/10.1007/978-3-031-59835-7_2

One can define a slightly more general version of k-MEDIAN by putting restrictions on which points can become centers; the input specifies a partition of V into *candidate centers* $F \subseteq V$ and *clients* $L \subseteq V$, and the goal is to choose $C \subseteq F$ with $|C| = k$ to minimize $\sum_{v \in L} d(v, C)$. Let us call this version the *supplier version* in this paper. This distinction between the original version and the supplier version is apparent in a similar clustering objective k-CENTER (where the objective function is $\max_{v \in C} d(v, C)$), whose supplier version has been called k-SUPPLIER and sometimes studied separately. k-CENTER and k-SUPPLIER admit 2 and 3 approximation algorithms respectively, which are optimal assuming $\mathbf{P} \neq \mathbf{NP}$, so we know there is a strict separation between these two.

For k-MEDIAN, this distinction has not been emphasized as much. Earlier papers define k-MEDIAN as the original version [7,9,11,32,33], but most papers in this century define k-MEDIAN as the supplier version. One reason for this transformation might be the influence of techniques coming from the FACILITY LOCATION problem. As the name suggests, FACILITY LOCATION originates from the planning perspective (just like k-SUPPLIER), so it is natural to assume that V is partitioned into the set of clients L and set of potential facility sites F with opening costs $f : F \to \mathbb{R}^+$ and the goal is to open $C \subseteq F$ (without any restriction on $|C|$) to minimize $f(C) + \sum_{v \in L} d(v, C)$. But once we interpret k-MEDIAN in the context of geometric clustering, such a restriction seems unnecessary, and it is more natural to study the original version directly.

Then the question is: will the original version of k-MEDIAN exhibit better approximability than the supplier version, just like k-CENTER and k-SUPPLIER? Algorithmically, no such separation is known, as many of the current algorithmic techniques, including the ones giving the current best approximation ratio, rely on the connection between k-MEDIAN and FACILITY LOCATION; the best approximation ratio for k-MEDIAN still holds for the supplier version. However, these two problems show different behaviors in terms of hardness. The reduction from MAXIMUM k-COVERAGE to FACILITY LOCATION by Guha and Khuller [23], adapted to k-MEDIAN by Jain et al. [26], shows that the supplier version of k-MEDIAN is NP-hard to approximate within a factor $(1 + 2/e) \approx 1.73$. The same reduction yields only $(1 + 1/e) \approx 1.36$ for the original k-MEDIAN. This separation in the current hardness, together with the separation between k-CENTER and k-SUPPLIER, suggest that the best approximation ratio achieved by a polynomial-time algorithm for the original k-MEDIAN might be strictly smaller than that of the supplier version. Of course, formally showing such a separation requires a polynomial-time algorithm for the original version with the approximation ratio strictly less than $1 + 2/e$, which seems hard to achieve now given the big gap between the upper and lower bounds for both versions.

A relatively new direction of *parameterized approximation*, which studies algorithms running in time $f(k) \cdot n^{O(1)}$ for any computable function $f(k)$, might be a good proxy for the polynomial-time approximation. In fact, Cohen-Addad et al. [14] showed that there exists a parameterized approximation algorithm for the supplier version that guarantees a $(1 + 2/e + \epsilon)$-approximation for any $\epsilon > 0$. Moreover, this additional running time does not seem to cross the NP-hardness boundary; assuming the Gap Exponential Time Hypothesis (Gap-ETH), they show that no FPT algorithm can achieve an approximation ratio $(1 + 2/e - \epsilon)$ for any $\epsilon > 0$. Many subsequent (and concurrent) papers studied parameterized approximability of many variants of k-MEDIAN and obtained improved approximation ratios over polynomial-time algorithms [1,2,5,6,15,19,21,22,25,29,35], but for quite a few of these problems, including k-MEDIAN and CAPACITATED k-MEDIAN, no parameterized approximation algorithm achieved a result impossible for their polynomial-time counterparts. Therefore, apart from its practical benefit when k is small, the study of parameterized approximation might yield a meanigful prediction for polynomial-time approximability.

Our main algorithmic result is the following parameterized approximation algorithm that strictly separates the original and supplier versions of k-MEDIAN. To the best of our knowledge, it is the first separation between the original and supplier versions of any variant of k-MEDIAN.

Theorem 1. *For any $\epsilon > 0$, there is an algorithm for k-MEDIAN that runs in time $f(k, \epsilon) \cdot n^{O(1)}$ and guarantees an approximation ratio of*

$$\min_{p\in[0,1]} \max_{d\in\mathbb{Z}_{\geq 1}} \left[1 + \left(1 - \frac{1-p}{d}\right)^d + \left(1 - \frac{pd + (1-p)}{d}\right)^d + \epsilon\right] \approx 1.546 + \epsilon,$$

where the optimal value is achieved with $p^* := (10-6\sqrt{2})/7 \approx 0.22$ and $d^* := 3$.

Our main hardness result is the following theorem against *polynomial-time* algorithms assuming the Unique Games Conjecture (UGC) [30], improving the best known lower bound of $(1 + 1/e) \approx 1.36$ [23,26].

Theorem 2. *Assuming the Unique Games Conjecture, for any $\epsilon > 0$, k-MEDIAN is NP-hard to approximate within a factor*

$$\max_{d\in\mathbb{Z}_{\geq 3}} \min_{p\in[0,1]} \left[1 + \left(1 - \frac{1-p}{d-1}\right)^d + \left(1 - \min(1, \frac{pd + (1-p)}{d-1})\right)^d - \epsilon\right] \approx 1.416 - \epsilon.$$

Somewhat surprisingly, this expression almost exactly matches the algorithmic guarantee except (1) denominators being $d - 1$ instead of d (which causes the $\min(1, .)$) and (2) $\max_d \min_p$ vs $\min_p \max_d$.

The difference between $d - 1$ and d comes from the well-known separation between the powers of LP and SDP on (hypergraph) vertex cover, elaborated more in the overview below. Furthermore, we observe that the max-min and the min-max values of the function in Theorem 1 coincide, so these two results might point towards the optimal approximability of k-MEDIAN, both for polynomial-time and parameterized algorithms.

Our hardness result proceeds via, and hence shows, more fine grained hardness for MAXIMUM k-COVERAGE. Given an instance of MAXIMUM k-COVERAGE with universe size n and a constant $d \geq 3$ so that each element appears in exactly d sets, we show that assuming the Unique Games Conjecture, it is NP-Hard to distinguish the YES-case where k sets cover all elements, and the NO-case where for any $\alpha \in [0,1]$, αk sets cover at most $1-(1-\frac{\alpha}{d-1})^d$ fraction of the elements where $k = \frac{n}{d-1}$. This is a new fine-grained version of the $(1-\frac{1}{e})$ hardness of [17] when $\alpha = 1$ for small d.

1.1 Overview

In this overview, we explain our intuition behind these two results and how they are related. For the sake of simplicity, let us ignore the arbitrarily small constant $\epsilon > 0$ in the overview.

Hardness. Our improved hardness, which eventually guided our algorithm, comes from observing the classical reduction of Guha and Khuller [23] and Jain et al. [26]. It is a reduction from the famous Maximum k-Coverage problem; given a hypergraph $H = (V_H, E_H)$, and $k \in \mathbb{N}$, choose $S \subseteq V_H$ of k vertices that intersects the most number of hyperedges.[1] The classical result of Feige [17] shows that it is NP-hard to distinguish between the (YES) case where k vertices intersect all hyperedges and the (NO) case where any choice of k vertices intersect at most $(1 - 1/e) \approx 0.63$ fraction of the hyperedges.

Given a hardness instance (H, k) for Maximum k-Coverage, the reduction to k-Median is simply to construct the vertex-hyperedge incidence graph $G = (V_G, E_G)$ where $V_G = V_H \cup E_H$ and a pair $(v, e) \in V_H \times E_H$ is in E_G if $v \in e$ in H. In the supplier version, one can simply let $F = V_H$ and $L = E_H$. Given $C \subseteq F$, for each $e \in L$, the distance $d(e, C)$ is 1 if it contains a vertex in C and at least 3 otherwise, thanks to the bipartiteness of G. Therefore, the average distance for clients in L becomes 1 in the YES case and at least $1 \cdot (1 - 1/e) + 3 \cdot (1/e) = 1 + 2/e$ in the NO case.

The same construction works for the original version without F and L with two differences. The first difference is that the objective function sums over points in V_H as well as E_H, but this can be handled by duplicating many points for each $e \in E_H$. The bigger issue is the fact that C might contain points from E_H, which implies that $d(e, C)$ can be possibly 2 even when C does not contain a vertex from e. Therefore, the hardness factor weakens to $1 \cdot (1 - 1/e) + 2 \cdot (1/e) = 1 + 1/e$.

The natural question to ask to improve the hardness is then: how much does placing centers at E_H help cover other points in E_H? Suppose that a solution C contains $(1-p)k$ centers from V_H and pk centers from E_H for some $p \in [0,1]$. Compared to the solution that puts all centers in V_H, putting some centers at E_H will hurt the ability to cover points in E_H at distance 1 (i.e., for $e \in E_H$, $d(e, C) = 1$ only if there exists $v \in (C \cap e)$), but it will help covering points in E_H at distance 2 (i.e., for $e \in E_H$, $d(e, C) \leq 2$ if there exists $f \in (C \cap E_H)$ with $e \cap f \neq \emptyset$). To bound the effect of the latter, one natural and conservative observation is, when the hypergraph H has uniformity d (i.e., each $e \in E_H$

[1] Typically, it is stated in terms of the dual set system where the input is a set system, and the goal is to choose k sets whose union size is maximized.

has $|e| = d$), letting $C' = (V_H \cap C) \cup (\cup_{e \in (E_H \cap C)} e)$, a hyperedge $e \in E_H$ has $d(e, C) \geq 3$ if it does not intersect C'. Since each $e \in C \cap E_H$ generates at most d new points in C', we can easily see $|C'| \leq ((1-p) + dp)k$. Therefore, it seems we need a good hardness of approximation for Maximum k-Coverage with small d in order to go beyond $1 + 1/e$. Quantitatively, for a d-uniform hypergraph H, the ideal hardness one can imagine is the following.

- In the YES case, $k := |V_H|/d$ vertices intersect (almost) all the hyperedges. For k-Median, there exist k centers that cover almost all points at distance 1.
- In the NO case, H is like a *random hypergraph*; for any $\gamma \in [0, 1]$, any choice of $\gamma|V_H|$ vertices intersect at most $1 - (1-\gamma)^d$ hyperedges. In other words, for fixed γ, the best solution is essentially to pick each vertex with probability γ. For k-Median, for any solution C with $|C \cap V_H| = (1-p)k$ and $|C \cap E_H| = pk$, we use the above guarantee twice to $(C \cap V_H)$ and C' to conclude that (1) at most a $1 - (1 - (1-p)/d)^d$ fraction of hyperedges can be covered by distance at most 1 and (2) at most $1 - (1 - (1-p+dp)/d)^d$ fraction of hyperedges can be covered by distance at most 2.

Such an ideal hardness turns out to be impossible to prove, but one can get close. For any integer $d \geq 3$, one can prove the above ideal hardness for d-uniform hypergraphs with the only difference being $k := |V_H|/d-1$ instead of $|V_H|/d$. This means that the denominator d is replaced by $d-1$ in the expressions $1-(1-(1-p)/d)^d$ and $1-(1-(1-p+dp)/d)^d$ above, which yields the hardness result of Theorem 2. If we had proved the ideal hardness above, then the hardness ratio would have exactly matched the algorithm of Theorem 1. We elaborate more about this gap in the discussion below.

Algorithms. Our algorithm follows the framework of Cohen-Addad et al. [14] for the supplier version. One can compute a *coreset* $P \subseteq V$, $|P| = O(k \log n)$ with weight function $w : P \to \mathbb{R}^+$ so that for any choice of k centers, the objective function for points in V is almost the same as that for weighted points in P. Therefore, one can focus on P as the set of points. Let $P_1^*, \ldots, P_k^*$ be the partition of P with respect to the optimal clustering. For each $i \in [k]$, let $c_i^* \in V$ be the optimal center corresponding to P_i^*, and $\ell_i = \text{argmin}_{\ell \in P_i^*} d(\ell, c_i^*)$ be the *leader* of P_i^*. Guessing $\ell_1, \ldots, \ell_k$ and (approximately) guessing $d(\ell_1, c_1^*), \ldots, d(\ell_k, c_k^*)$ takes a parameterized time $O(k \log n)^k$. Then, for each $i \in [k]$, let C_i be the points that have distance (almost) equal to $d(\ell_i, c_i^*)$. By definition $c_i^* \in C_i$.

We solve the standard LP relaxation for k-Median, with only the additional constraint that we fractionally open exactly one center from each C_i. (A standard trick can make sure that C_i's are disjoint.) The rounding algorithm opens exactly one center from C_i (different clusters are independent) according to LP values. Fix a point $v \in P$. A standard analysis for k-Median shows that with probability at least $1 - 1/e$, a center fractionally connected to v will open, and the expected distance from v to the closest open center conditioned on this event is at most its contribution to the LP objective. In the other event, if $v \in P_i^*$ for some $i \in [k]$, since one center $c \in C_i$ is open and

$$d(v,c) \le d(v,c_i^*) + d(\ell_i,c_i^*) + d(\ell_i,c),$$

the fact that $d(v,c_i^*) \ge d(\ell_i,c_i^*)$ (by definition of ℓ_i) and $d(\ell_i,c_i^*) \approx d(\ell_i,c)$ (by definition of C_i) ensures that this distance is almost at most $3d(v,c_i^*)$. Combining these two events shows that the total expected distance is at most $(1-{}^1\!/_e)LP + ({}^3\!/_e)OPT \le (1+{}^2\!/_e)OPT$.

In order to obtain a possibly better result over the supplier version, using the power that we can open centers anywhere, for some fixed $p \in [0,1]$, we do the following. For each $i \in [k]$, we simply open the leader ℓ_i as a center with probability p, and with the remaining probability $(1-p)$, open exactly one center from C_i as usual. For the analysis, let us fix $v \in P_i$ and consider its distance to the closest center in this rounding strategy. For simplicity, assume v is fractionally connected to $c_1,\dots,c_d$ with fraction ${}^1\!/_d$ each, where $c_i \in C_i$ for $i \in [d]$. Let us also assume that $d(v,c_i) = 1$ for each $i \in [d]$, and $d(\ell_i,c_i) = 1$ for each $i \in [d]$. Observe that c_i opens with probability ${}^{(1-p)}\!/_d$, ℓ_i opens with probability p, and some other center in C_i opens with probability ${}^{(1-p)(d-1)}\!/_d$. If at least one c_i, $i \in [d]$ is opened, then v's connection cost is at most 1. If at least one $\ell_i, i \in [d]$ is open, then v's connection cost is at most 2. It follows that the probability that no center at distance 1 from v opens is exactly $(1-{}^{(1-p)}\!/_d)^d$, and the probability that no center at distance ≤ 2 from v opens is exactly $({}^{(1-p)(d-1)}\!/_d)^d = (1-{}^{(1-p+pd)}\!/_d)^d$. Note they exactly match the analysis for the (ideal) hardness; the first probability $(1-{}^{(1-p)}\!/_d)^d$ corresponds to the case where we choose random $(1-p)k$ vertices in the hypergraph, and the second probability $(1-{}^{(1-p+pd)}\!/_d)^d$ corresponds to the case where we choose random $(1-p+pd)k$ vertices due to picking pk hyperedges!

Of course, the above intuition already assumes various regularities involving the LP values and distances. We show that such regular cases are indeed the worst case in terms of approximation ratio. The analysis involves a series of factor-revealing programs that try to find the worst-case configuration for a fixed point, which reveal simplifying structural properties of the optimal configuration via rank and convexity arguments.

Discussion and Future Work. The first question from this work might be: where is the difference between d and $d-1$ coming from, and how can we close this gap? As we saw, if we had been able to prove the "ideal" hardness, that is, the inability to distinguish a d-uniform hypergraph with vertex cover size ${}^{|V_H|}\!/_d$ and a random d-uniform hypergraph, this gap would not exist.

If we just want to *fool* the LP, since every d-uniform hypergraph admits a fractional covering of all hyperedges with a ${}^1\!/_d$ fraction of vertices, we believe that a random d-uniform hypergraph will exhibit such a property, giving an LP (even Sherali-Adams) gap with the exact same factor as the upper bound. Indeed, when $d = 2$, Charikar, Makarychev, and Makarychev [13] showed that even $O(n^\delta)$-levels of Sherali-Adams cannot distinguish a random graph and a nearly-bipartite graph.

However, for d-uniform hypergraphs with small d, SDPs are expected to strictly outperform the LP, and this is evident when $d = 2$ where the LP (even Sherali-Adams) guarantees at most $3/4$-approximation for MAXIMUM k-COVERAGE while the SDP guarantees a 0.929-approximation [34]. Our Unique Games Hardness is based on a standard way to construct an SDP gap based on pairwise independence, which is essentially from Håstad's seminal result for MAX 3LIN and MAX 3SAT when $d = 3$ [24]. It is reasonable to expect that SDPs give a strictly better guarantee for k-MEDIAN instances constructed from (the vertex-hyperedge incidence graph of) d-uniform hypergraphs, but reducing the general k-MEDIAN instance to such a case seems a significant technical challenge as our analysis relies on simple properties of the LP.

A more fundamental limitation of our work is that the upper bound is given by parameterized algorithms and the lower bound is only against polynomial-time algorithms. The biggest open question is to prove an optimal approximability for either polynomial-time or parameterized algorithms. Currently, it might be the case that the optimal thresholds for these two classes of algorithms coincide just like the supplier version.

Improving the best polynomial-time approximation ratio for any version of k-MEDIAN has been studied intensively. Matching the $(1 + 2/e)$-hardness for the supplier version or going below for the original version is a long-standing open problem, but we believe that giving an algorithm for the original version with the approximation ratio strictly better than the current known 2.613 [16,20] for the supplier version is a meaningful step. Such an algorithm is likely to yield insight into how to take advantage of opening centers anywhere in the context of primal-dual or dual-fitting analysis, whereas our improved parameterized algorithm uses a primal LP rounding algorithm.

From the parameterized hardness perspective, we first remark that MAXIMUM k-COVERAGE in d-uniform hypergraphs with constant d (or even $d = f(k)$) admits a parameterized approximation scheme [4,28]. However, there might be more sophisticated ways to understand the effect of opening centers anywhere. A better understanding of the hardness of LABEL COVER in the parameterized setting, combined with Feige's reduction to MAXIMUM k-COVERAGE, might be a way to achieve a strictly improvement over $(1 + 1/e)$.

2 Approximation Algorithm

In this section, we prove our main algorithmic result Theorem 1, restated below.

Theorem 1. *For any $\epsilon > 0$, there is an algorithm for k-MEDIAN that runs in time $f(k, \epsilon) \cdot n^{O(1)}$ and guarantees an approximation ratio of*

$$\min_{p \in [0,1]} \max_{d \in \mathbb{Z}_{\geq 1}} \left[1 + \left(1 - \frac{1-p}{d}\right)^d + \left(1 - \frac{pd + (1-p)}{d}\right)^d + \epsilon \right] \approx 1.546 + \epsilon,$$

where the optimal value is achieved with $p^* := (10-6\sqrt{2})/7 \approx 0.22$ and $d^* := 3$.

Section 2.1 presents our algorithm. Section 2.2 introduces the setup for our analysis, which tries to understand the worst-case configuration for a fixed point via a series of factor-revealing programs.

2.1 Algorithm

We begin by computing a *coreset*. For any choice of centers $C \subseteq V$ and a set of weighted points $V' \subseteq V$ where point $v \in V'$ has weight w_v, let $cost(V', C)$ denote the cost of assigning the points V' to the centers C so as to minimize the total weighted connection cost. Concretely, for every assignment function $f : V' \to C$, define the cost of f as $\sum_{v \in V'} w_v d(v, f(v))$. Then $cost(V', C)$ is the minimum cost among all such assignments f.

Definition 1. *Given a k-MEDIAN instance (V, d, k) and $\epsilon > 0$, an ϵ-coreset V' is a set of points with a weight w_v for each $v \in V'$, so that for any choice of centers $C \subseteq V$ with $|C| = k$, $cost(V', C) \in (1 - \epsilon, 1 + \epsilon)cost(V, C)$.*

Theorem 3 ([18]). *There is a polynomial time algorithm that when given a k-MEDIAN instance and $\epsilon > 0$, computes an ϵ-coreset $V' \subseteq V$ with $|V'| \leq \mathcal{O}(k \log n / \epsilon^2)$.*

We start by considering a natural LP relaxation and proceed in a similar manner to the algorithm of [14]. We first compute a coreset V' for the point set V of size $\mathcal{O}(\frac{k \log n}{\epsilon^2})$, using Theorem 3. It follows that for any choice of k centers $C \subseteq V$, we have $cost(V', C) \in (1-\epsilon, 1+\epsilon)cost(V, C)$. Henceforth we focus on minimizing $cost(V', C)$.

For the optimal partition $\{V_1^*, V_2^*, \ldots, V_k^*\}$ of V' and corresponding choice of centers $\{c_1^*, c_2^*, \ldots, c_k^*\}$, we guess for each $i \in [k]$ the leaders ℓ_i, which are the points in V_i^* closest to c_i^* for each $i \in [k]$. Since the coreset guarantees $|V'| \leq \mathcal{O}(\frac{k \log n}{\epsilon^2})$, this can be accomplished in time $\mathcal{O}((\frac{k \log n}{\epsilon^2})^k)$. We also guess the distances R_i^* of the leaders ℓ_i to the corresponding center c_i^* for each $i \in [k]$, rounded down to the nearest power of $(1+\epsilon)$. By the argument used in [14], we can assume without loss of generality that the distance between any two points of C is at least 1 and at most $n^{\mathcal{O}(1)}$: this means there are only $\mathcal{O}(\log(\frac{n}{\epsilon}))$ choices for the distances R_i^*, which in turn means that this step can be accomplished in time $\mathcal{O}((\log \frac{n}{\epsilon})^k)$. It can be easily shown that $\mathcal{O}((\log \frac{n}{\epsilon})^k)$ and $\mathcal{O}((\frac{k \log n}{\epsilon^2})^k)$ can be bounded above by $f(k, \frac{1}{\epsilon})n^{\mathcal{O}(1)}$ as follows. First, note that it is enough to upper bound $(\log \frac{n}{\epsilon})^{\mathcal{O}(k)}$. If $k > \frac{\log \frac{n}{\epsilon}}{\log \log \frac{n}{\epsilon}}$, then this expression is upper bounded by $2^{\mathcal{O}(k \log k)}$. Otherwise, $k < \frac{\log \frac{n}{\epsilon}}{\log \log \frac{n}{\epsilon}}$, and we get an upper bound of $(\log \frac{n}{\epsilon})^{\mathcal{O}\left(\frac{\log \frac{n}{\epsilon}}{\log \log \frac{n}{\epsilon}}\right)} \leq (\frac{n}{\epsilon})^{\mathcal{O}(1)}$.

Once we guess these quantities, let C_i be the subset of V which are at distance at most $R_i^*(1+\epsilon)$ from ℓ_i. Notice that we must have $c_i^* \in C_i$. For purely technical reasons, by adding a copy of ℓ_i in C_i, we will assume that $\ell_i \notin C_i$. Also by making copies of points, we will assume that $C_1, \ldots, C_\ell$ are disjoint.

Consider the following natural LP relaxation for the problem.

$$\begin{aligned}
\min \quad & \sum_{v \in V'} w_v \sum_{c \in V} d(v,c) x_{v,c} & \\
\text{s.t.} \quad & x_{v,c} \leq y_c & \forall v \in V', c \in V \\
& \sum_{c \in C_i} y_c = 1 & \forall i \in [k] \\
& \sum_{c \in V} x_{v,c} = 1 & \forall v \in V' \\
& x_{v,c} = 0 & \forall v \in V', i \in [k], c \in C_i \text{ such that } d(v,c) < d(\ell_i, c)
\end{aligned} \tag{1}$$

Clearly, this LP is a relaxation as witnessed by the canonical integral solution where one can assign each point in V_i^* to (a copy of) $c_i^* \in C_i$.

We solve the LP to obtain a (fractional) optimal solution $(\mathbf{x}, \mathbf{y})$ for the LP - we will also assume that $x_{v,c} = y_c$ for each $c \in V$ and each point $v \in V'$ with $x_{v,c} \neq 0$. This can be accomplished using a fairly straightforward construction by "splitting centers" which we describe below. Similar constructions have been used in previous work for other related problems.

Pick a point $c \in V$ for which there is a point v such that $x_{v,c} \neq 0$ and $x_{v,c} \neq y_c$. Let $v_1, v_2 \ldots v_t$ be the points v with $x_{v,c} \neq 0$, taken in non-decreasing order of $x_{v,c}$. Create $t+1$ many copies $c_1, c_2 \ldots c_{t+1}$ of the point c (with all distances from other points the same as the original point c). Set $y_{c_1} = x_{v_1,c}$. Next, set $y_{c_2} = x_{v_2,c} - x_{v_1,c}$, and so on, so that $y_{c_j} = x_{v_j,c} - x_{v_{j-1},c}$ for all $2 \leq j \leq t$. Set $y_{c_{t+1}} = 1 - \sum_{j=1}^{t} y_{c_j}$. Finally, for each point v_i, $i \in [t]$, we set $x_{v_i,c_j} = y_{c_j}$ for $j \in \{1, 2 \ldots i\}$. It is clear that the construction preserves feasibility and optimality, and $x_{v,c} = y_c$ whenever $x_{v,c} \neq 0$.

Our rounding algorithm is very simple: we pick a threshold $p \in [0,1]$ to be determined later by analysis, and for each $i \in [k]$, we pick ℓ_i into C with probability p, and with remaining probability $1-p$, we choose exactly one center c from C_i from the distribution given by the LP: more concretely, each center $t \in C_i$ is picked with probability y_t.

Algorithm 1. Rounding algorithm

Input: k-MEDIAN instance (V, d, k) with coreset V', the sets $C_1, C_2 \ldots C_k$ and leaders $\ell_1, \ell_2 \ldots \ell_k$. A parameter $p \in [0,1]$, and the optimal LP solution $(\mathbf{x}, \mathbf{y})$.
Output: An approximate solution $C \subseteq V$ with $|C| = k$.
for $i \in [k]$ **do**
  With probability p pick ℓ_i.
  With remaining probability $1-p$, pick exactly one center $t \in C_i$ with probability y_t.
end for

Observe that for every $i \in [k]$, the algorithm opens either a center in C_i, or the leader ℓ_i.

2.2 Analysis: Casting as Factor-Revealing Problems

Once we decide on the centers, each point $v \in V'$ can be assigned to the closest open center. For the sake of the analysis, we consider the following specific assignment of the points to the open centers; the cost only increases by doing so. Henceforth, we fix $v \in V'$. Let $I_v \subseteq [k]$ be the set of indices i for which there is a center $c \in C_i$ with $x_{v,c} \neq 0$, and suppose $|I_v| = \ell$. For the sake of simplicity and clarity of exposition, by re-numbering indices, let us assume that $I_v = [\ell]$.

For each $i \in [\ell]$, define the *flow* to cluster i as $\mu_i = \sum_{c \in C_i} x_{v,c}$. Let s_i be the weighted average distance to a center $c \in C_i$, where center c is given weight $x_{v,c}$, so that $s_i = (\sum_{c \in C_i} x_{v,c} d(v,c))/\mu_i$. Assume without loss of generality that $s_i \leq s_j$ whenever $i \leq j$ for $i, j \in [\ell]$. Note that s_i is the *expected* cost of assigning v to a random center $c \in C_i$ according to the probabilities $x_{v,c}$. Algorithm 2 describes our assignment algorithm.

Algorithm 2. Assignment algorithm

Input: k-MEDIAN instance with coreset V' and centers C opened by Algorithm 1, a point $v \in V'$, and the set $I_v = [\ell]$.
Output: An assignment of v to a center of C.
Let A be the set of indices $i \in [\ell]$ with $s_i \leq 1.5 s_1$.
Let D be the set of centers c satisfying $x_{v,c} \neq 0$, with $c \in C_i$ so that $s_i \leq 3s_1(1+\epsilon)$ for some $i \in [\ell]$.
if $D \cap C \neq \emptyset$ **then**
  Assign v to the center $c \in C_{i^*}$, where i^* is the first index with $C_{i^*} \cap D \cap C \neq \emptyset$.
else if there is a $j \in A$ such that ℓ_j is open **then**
  Assign v to ℓ_{j^*} where j^* is the first index in A such that ℓ_{j^*} is open.
else
  Assign v to any open center in C_1.
end if

First, we have the following lemma which shows that v will be always assigned to a relatively close center.

Lemma 1. *For any $j \in [\ell]$, we must have $d(v, \ell_j) \leq 2s_j$. Consequently, v will be always assigned an open center at distance at most* $3s_1(1+\epsilon)$.

Proof. Fix any index $j \in [\ell]$. Let $c'_j \in C_j$ be a point in C_j which minimizes $d(v,c)$ among all points $c \in C_j$ with $x_{v,c} \neq 0$. Clearly, $d(v, c'_j) \leq s_j$. By the triangle inequality, we have $d(v, \ell_j) \leq d(v, c'_j) + d(c'_j, \ell_j)$. Since $x_{v,c'_j} \neq 0$, the LP constraint (1) ensures $d(v, c'_j) \geq d(c'_j, \ell_j)$. Thus $d(v, \ell_j) \leq 2d(v, c'_j) \leq 2s_j$.

Observe that Algorithm 1 ensures that either ℓ_j is open, or some center $c_j \in C_j$ is open. Also $d(\ell_j, c_j) \leq d(\ell_j, c'_j)(1+\epsilon) \leq d(v, c'_j)(1+\epsilon) \leq s_j(1+\epsilon)$, where the first inequality follows by the definition of C_j, and the second inequality again follows from the LP constraint (1). Together, we obtain $d(v, c_j) \le d(v, \ell_j) + d(\ell_j, c_j) \le 2s_j + s_j(1+\epsilon) = 3s_j(1+\epsilon)$. Since this holds for any j, and in particular when $j = 1$, it is clear from the description of Algorithm 2 that v will be always assigned an open center at distance at most $3s_1(1+\epsilon)$ and the result follows.

Let $LP(v) := \sum_{c\in V} x_{v,c} d(v,c)$ be the connection cost of v in the LP. For each case of the assignment given by Algorithm 2, let us define $cost(v)$ to be the following upper bound on $d(v, C)$:

1. Some center $c \in D$ is open. In this case, the point goes to the open center $c_{i*} \in D \cap C_{i*}$, where i^* is the minimum index for which there is a center open in $D \cap C_{i^*}$. Given that there is a center open in $D \cap C_{i^*}$, observe that Algorithm 1 opens each center $c \in D \cap C_{i^*}$ with probability proportional to y_c. Thus the expected connection cost is exactly s_{i^*}. Let $cost(v) = s_{i^*}$ in this case.
2. No center in D is open, but some leader ℓ_j is open, for some $j \in A$. The point v is then assigned to the leader ℓ_{j^*}, where $j^* \in A$ is the minimum index j^* so that ℓ_{j^*} is open. Using Lemma 1, the connection cost for v is upper bounded by $2s_{j^*}$. Let $cost(v) = 2s_{j^*}$ in this case.
3. Any other scenario. Again, using Lemma 1, we upper bound the cost by $3s_1(1+\epsilon)$. Let $cost(v) = 3(1+\epsilon)s_1$ in this case.

The rest of the section will be devoted to upper bounding the ratio $\mathbb{E}[cost(v)]/LP(v)$, which will yield Theorem 1.

Theorem 4. *For every v, the ratio $(\mathbb{E}[cost(v)]/LP(v)) \le \alpha_{alg}(1+O(\epsilon)) \approx 1.546(1+O(\epsilon))$.*

As in Algorithm 2, let A be the set of indices $i \in [\ell]$ with $s_i \le 1.5s_1$, and let B be the set of indices $i \in [\ell]$ with $s_i > 1.5s_1$. Fix the "degrees" $d_A = |A|$, $d_B = |B|$, the "flows" $\mu_A = \sum_{i\in A} \mu_i$, $\mu_B = \sum_{i\in B} \mu_i$. By our previous assumption, since $s_1 \le s_2 \le s_3 \ldots \le s_\ell$, it follows that $A = [|d_A|]$ and $B = [\ell \setminus |d_A|]$.

Both $\mathbb{E}[cost(v)]$ and $LP(v)$ will be entirely determined by $p, d_A, d_B, (\mu_i)_{i\in[\ell]}$, and $(s_i)_{i\in[\ell]}$. So our main question is: for a given p (that our algorithm can control), which values for $d_A, d_B, (\mu_i)_{i\in[\ell]}$, and $(s_i)_{i\in[\ell]}$ will maximize the ratio $\mathbb{E}[cost(v)]/LP(v)$?

It is simple to see that $LP(v) = \sum_{i\in[\ell]} \mu_i s_i$, while $\mathbb{E}[cost(v)]$ is a complex function of $d_A, d_B, (\mu_i)_{i\in[\ell]}$, and $(s_i)_{i\in[\ell]}$. We will show that the values for these variables maximizing the ratio $\mathbb{E}[cost(v)]/LP(v)$ must exhibit a certain structure, which finally leads to the desired upper bound. This can be achieved via a series of factor-revealing programs that maximize the ratio over a subset of variables while the others are considered fixed. We will first consider such a program where $p, \mu_1, \ldots, \mu_\ell, d_A, d_B$ are fixed and vary $(s_i)_{i\in[\ell]}$.

Henceforth, we assume without loss of generality that each $s_i \le 3s_1(1+\epsilon)$. If this is not satisfied, then for every $i \in [\ell]$ with $s_i > 3s_1(1+\epsilon)$, we set $s_i =$

$3s_1(1+\epsilon)$. This operation can only increase $\mathbb{E}[cost(v)]$, and decrease $LP(v)$, and hence can only increase the ratio $\mathbb{E}[cost(v)]/LP(v)$.

First we show that $\mathbb{E}[cost(v)]$ is a linear function of $(s_i)_{i\in[\ell]}$.

Lemma 2. *$\mathbb{E}[cost(v)] = \sum_{i=1}^{\ell} \gamma_i s_i$ where each γ_i is a function of p, d_A, $d_B, (\mu_i)_{i\in[\ell]}$ only.*

It follows that to obtain the worst case ratio for a client, we need to maximize the ratio of two linear functions in the variables $(s_i)_{i\in[\ell]}$. This optimization problem itself can now be cast as a linear program, and we show that any extreme point optimal solution of this LP must satisfy $s_i \in \{s_1, 1.5s_1, 3s_1\}$ for each $i \in [\ell]$. Once we have this nice structural property of the worst case configuration, we optimize over the other variables using simple inequalities and convexity arguments to obtain our result.

References

1. Adamczyk, M., Byrka, J., Marcinkowski, J., Meesum, S.M., Wlodarczyk, M.: Constant-factor FPT approximation for capacitated k-median. In: 27th Annual European Symposium on Algorithms (ESA 2019). Schloss Dagstuhl-Leibniz-Zentrum fuer Informatik (2019)
2. Agrawal, A., Inamdar, T., Saurabh, S., Xue, J.: Clustering what matters: optimal approximation for clustering with outliers. J. Artif. Intell. Res. **78**, 143–166 (2023)
3. Arya, V., Garg, N., Khandekar, R., Meyerson, A., Munagala, K., Pandit, V.: Local search heuristic for k-median and facility location problems. In: Proceedings of the Thirty-Third Annual ACM Symposium on Theory of Computing, pp. 21–29 (2001)
4. Badanidiyuru, A., Kleinberg, R., Lee, H.: Approximating low-dimensional coverage problems. In: Proceedings of the Twenty-Eighth Annual Symposium on Computational Geometry, pp. 161–170 (2012)
5. Bandyapadhyay, S., Fomin, F.V., Golovach, P.A., Purohit, N., Simonov, K.: FPT approximation for fair minimum-load clustering. In: 17th International Symposium on Parameterized and Exact Computation (2022)
6. Bandyapadhyay, S., Lochet, W., Saurabh, S.: FPT constant-approximations for capacitated clustering to minimize the sum of cluster radii. In: 39th International Symposium on Computational Geometry (SoCG 2023). Schloss Dagstuhl-Leibniz-Zentrum für Informatik (2023)
7. Bartal, Y.: On approximating arbitrary metrices by tree metrics. In: Proceedings of the Thirtieth Annual ACM Symposium on Theory of Computing, pp. 161–168 (1998)
8. Byrka, J., Pensyl, T., Rybicki, B., Srinivasan, A., Trinh, K.: An improved approximation for k-median and positive correlation in budgeted optimization. ACM Trans. Algorithms (TALG) **13**(2), 1–31 (2017)
9. Charikar, M., Chekuri, C., Goel, A., Guha, S.: Rounding via trees: deterministic approximation algorithms for group Steiner trees and k-median. In: Proceedings of the Thirtieth Annual ACM Symposium on Theory of Computing, pp. 114–123 (1998)
10. Charikar, M., Guha, S.: Improved combinatorial algorithms for the facility location and k-median problems. In: 40th Annual Symposium on Foundations of Computer Science (Cat. No. 99CB37039), pp. 378–388. IEEE (1999)

11. Charikar, M., Guha, S., Tardos, É., Shmoys, D.B.: A constant-factor approximation algorithm for the k-median problem. In: Proceedings of the Thirty-First Annual ACM Symposium on Theory of Computing, pp. 1–10 (1999)
12. Charikar, M., Li, S.: A dependent LP-rounding approach for the k-median problem. In: Czumaj, A., Mehlhorn, K., Pitts, A., Wattenhofer, R. (eds.) ICALP 2012. LNCS, vol. 7391, pp. 194–205. Springer, Heidelberg (2012). https://doi.org/10.1007/978-3-642-31594-7_17
13. Charikar, M., Makarychev, K., Makarychev, Y.: Integrality gaps for Sherali-Adams relaxations. In: Proceedings of the Forty-First Annual ACM symposium on Theory of Computing, pp. 283–292 (2009)
14. Cohen-Addad, V., Gupta, A., Kumar, A., Lee, E., Li, J.: Tight FPT approximations for k-median and k-means. In: 46th International Colloquium on Automata, Languages, and Programming (ICALP 2019), vol. 132, pp. 42–1. Schloss Dagstuhl–Leibniz-Zentrum fuer Informatik (2019)
15. Cohen-Addad, V., Li, J.: On the fixed-parameter tractability of capacitated clustering. In: 46th International Colloquium on Automata, Languages, and Programming (ICALP 2019), vol. 132, p. 41-1. Schloss Dagstuhl–Leibniz-Zentrum fuer Informatik (2019)
16. Cohen-Addad Viallat, V., Grandoni, F., Lee, E., Schwiegelshohn, C.: Breaching the 2 LMP approximation barrier for facility location with applications to k-median. In: Proceedings of the 2023 Annual ACM-SIAM Symposium on Discrete Algorithms (SODA), pp. 940–986. SIAM (2023)
17. Feige, U.: A threshold of ln n for approximating set cover. J. ACM (JACM) **45**(4), 634–652 (1998)
18. Feldman, D., Langberg, M.: A unified framework for approximating and clustering data. In: Proceedings of the Forty-Third Annual ACM Symposium on Theory of Computing, pp. 569–578 (2011)
19. Feng, Q., Zhang, Z., Huang, Z., Xu, J., Wang, J.: A unified framework of FPT approximation algorithms for clustering problems. In: 31st International Symposium on Algorithms and Computation (ISAAC 2020). Schloss Dagstuhl-Leibniz-Zentrum für Informatik (2020)
20. Gowda, K.N., Pensyl, T., Srinivasan, A., Trinh, K.: Improved bi-point rounding algorithms and a golden barrier for k-median. In: Proceedings of the 2023 Annual ACM-SIAM Symposium on Discrete Algorithms (SODA), pp. 987–1011. SIAM (2023)
21. Goyal, D., Jaiswal, R.: Tight FPT approximation for socially fair clustering. Inf. Process. Lett. **182**, 106383 (2023)
22. Goyal, D., Jaiswal, R., Kumar, A.: FPT approximation for constrained metric k-median/means. In: 15th International Symposium on Parameterized and Exact Computation. p. 1 (2020)
23. Guha, S., Khuller, S.: Greedy strikes back: improved facility location algorithms. J. Algorithms **31**(1), 228–248 (1999)
24. Håstad, J.: Some optimal inapproximability results. J. ACM (JACM) **48**(4), 798–859 (2001)
25. Inamdar, T., Varadarajan, K.: Capacitated sum-of-radii clustering: an FPT approximation. In: 28th Annual European Symposium on Algorithms (ESA 2020). Schloss Dagstuhl-Leibniz-Zentrum für Informatik (2020)
26. Jain, K., Mahdian, M., Markakis, E., Saberi, A., Vazirani, V.V.: Greedy facility location algorithms analyzed using dual fitting with factor-revealing LP. J. ACM (JACM) **50**(6), 795–824 (2003)

27. Jain, K., Vazirani, V.V.: Approximation algorithms for metric facility location and k-median problems using the primal-dual schema and Lagrangian relaxation. J. ACM (JACM) **48**(2), 274–296 (2001)
28. Jain, P., et al.: Parameterized approximation scheme for biclique-free max k-weight SAT and max coverage. In: Proceedings of the 2023 Annual ACM-SIAM Symposium on Discrete Algorithms (SODA), pp. 3713–3733. SIAM (2023)
29. Jaiswal, R., Kumar, A.: Clustering what matters in constrained settings. In: 34th International Symposium on Algorithms and Computation (2023)
30. Khot, S.: On the power of unique 2-prover 1-round games. In: Proceedings of the Thirty-Fourth Annual ACM Symposium on Theory of Computing, pp. 767–775 (2002)
31. Li, S., Svensson, O.: Approximating k-median via pseudo-approximation. In: Proceedings of the Forty-Fifth Annual ACM Symposium on Theory of Computing, pp. 901–910 (2013)
32. Lin, J.H., Vitter, J.S.: Approximation algorithms for geometric median problems. Inf. Process. Lett. **44**(5), 245–249 (1992)
33. Lin, J.H., Vitter, J.S.: e-approximations with minimum packing constraint violation. In: Proceedings of the Twenty-Fourth Annual ACM Symposium on Theory of Computing, pp. 771–782 (1992)
34. Manurangsi, P.: A note on max k-vertex cover: faster FPT-AS, smaller approximate kernel and improved approximation. In: 2nd Symposium on Simplicity in Algorithms (2019)
35. Xu, Y., Möhring, R.H., Xu, D., Zhang, Y., Zou, Y.: A constant FPT approximation algorithm for hard-capacitated k-means. Optim. Eng. **21**, 709–722 (2020)

A Better-Than-1.6-Approximation for Prize-Collecting TSP

Jannis Blauth[1], Nathan Klein[2], and Martin Nägele[1](✉)

[1] Research Institute for Discrete Mathematics and Hausdorff Center for Mathematics, University of Bonn, Bonn, Germany
blauth@or.uni-bonn.de, mnaegele@uni-bonn.de

[2] Institute for Advanced Study, Princeton, NJ, USA
nklein@ias.edu

Abstract. Prize-Collecting TSP is a variant of the traveling salesperson problem where one may drop vertices from the tour at the cost of vertex-dependent penalties. The quality of a solution is then measured by adding the length of the tour and the sum of all penalties of vertices that are not visited. We present a polynomial-time approximation algorithm with an approximation guarantee slightly below 1.6, where the guarantee is with respect to the natural linear programming relaxation of the problem. This improves upon the previous best-known approximation ratio of 1.774. Our approach is based on a known decomposition for solutions of this linear relaxation into rooted trees. Our algorithm takes a tree from this decomposition and then performs a pruning step before doing parity correction on the remainder. Using a simple analysis, we bound the approximation guarantee of the proposed algorithm by $(1+\sqrt{5})/2 \approx 1.618$, the golden ratio. With some additional technical care we further improve it to 1.599.

Keywords: Combinatorial Optimization · Approxim. Algorithms · TSP

1 Introduction

The metric traveling salesperson problem (TSP) is one of the most fundamental problems in combinatorial optimization. In an instance of this problem, we are given a set V of n vertices along with their pairwise symmetric distances,

N. Klein was supported in part by Air Force Office of Scientific Research grant FA9550-20-1-0212 and NSF grants DGE-1762114 and CCF-1813135. M. Nägele was supported by the Swiss National Science Foundation (grant no. P500PT_206742) and the Deutsche Forschungsgemeinschaft (DFG, German Research Foundation) under Germany's Excellence Strategy – EXZ-2047/1 – 390685813.

J. Vygen and J. Byrka (Eds.): IPCO 2024, LNCS 14679, pp. 28–42, 2024.
https://doi.org/10.1007/978-3-031-59835-7_3

$c\colon V \times V \to \mathbb{R}_{\geq 0}$, which form a metric. The goal is to find a shortest possible Hamiltonian cycle. In the classical interpretation, there is a salesperson who needs to visit a set of cities V and wants to minimize the length of their tour. In this work, we study a variant known as *prize-collecting* TSP, in which the salesperson can decide whether or not to include each city besides the starting one[1] in their tour at the cost of a city-dependent penalty. Formally, the problem can be stated as follows.

Prize-Collecting TSP (PCTSP): Given a complete undirected graph $G = (V, E)$ with metric edge lengths $c_e \geq 0$ for all $e \in E$, a root $r \in V$, and penalties $\pi_v \geq 0$ for all $v \in V \setminus \{r\}$, the task is to find a cycle $C = (V_C, E_C)$ in G that contains the root r and minimizes

$$\sum_{e \in E_C} c_e + \sum_{v \in V \setminus V_C} \pi_v \ .$$

This is a very natural generalization of TSP (one can recover TSP by setting $\pi_v = \infty$ for all $v \in V \setminus \{r\}$), as from the salesperson's perspective some cities may not be worth visiting if they significantly increase the length of the tour. Indeed, in many real-world settings instances of TSP are actually prize-collecting.

As mentioned, PCTSP is at least as hard as TSP. Thus, it is NP-hard to approximate within a factor of $^{123}/_{122}$ [25]. On the positive side, the first constant-factor approximation algorithm for PCTSP was shown in the early'90 s [8], giving a ratio of 2.5. After a series of improvements [2,20,21], the best approximation factor is now slightly below 1.774 [10].

In TSP and many of its variants, such approximation guarantees typically rely on lower bounds obtained through *linear programming relaxations.* For PCTSP, the natural such formulation is the following:[2]

$$\begin{aligned} \min \sum_{e \in E} c_e x_e + \sum_{v \in V} \pi_v (1 - y_v) & \\ x(\delta(v)) &= 2 y_v \ \forall v \in V \setminus \{r\} \\ x(\delta(r)) &\leq 2 \\ x(\delta(S)) &\geq 2 y_v \ \forall S \subseteq V \setminus \{r\}, v \in S \\ y_r &= 1 \\ x_e &\geq 0 \quad \forall e \in E \\ y_v &\geq 0 \quad \forall v \in V. \end{aligned} \qquad \text{(PCTSP LP relaxation)}$$

One can see that $y_v \leq 1$ is implied by the above formulation, hence the variables y_v can be interpreted as the extent to which the vertex v is visited by the fractional solution. In this paper, we prove the following.

[1] "Rooted" and unrooted versions are reducible to one another while preserving approximability, as noted in, e.g., [2]. Here, we always require a root vertex r.

[2] We use $\pi_r := 0$ for convenience. For $S \subseteq V$, we denote $\delta(S) := \{e \in E \colon |e \cap S| = 1\}$; for $v \in V$, we use $\delta(v) := \delta(\{v\})$.

Theorem 1. *There is a polynomial-time LP-relative* 1.599*-approximation algorithm for PCTSP.*

To obtain this result, we exploit a known decomposition of solutions (x, y) to the above relaxation into trees (see Lemma 1 for the formal statement), which can be derived from an existential result on packing branchings in a directed multigraph by Bang-Jensen, Frank, and Jackson [6, Theorem 2.6] and was—in a generalized form—first used in the context of PCTSP by Blauth and Nägele [10]. The decomposition can be interpreted as a distribution μ over a polynomial number of trees $\mathcal{T}$ rooted at r such that for each tree $T \in \mathcal{T}$ (i) $\mathbb{E}_{T\sim\mu}[c(E[T])] \leq c^{\top}x$ and (ii) $\mathbb{P}_{T\sim\mu}[v \in V[T]] = y_v$ for all $v \in V$.[3]

Our algorithm proceeds as follows. We apply the decomposition to a slightly modified LP solution with $y_v = 0$ or $y_v \geq \delta$ for each $v \in V$ for some parameter δ. Then, for a tree T in the support of μ and a threshold γ, we prune the tree. Concretely, we find the inclusion-wise minimal subtree of T which spans all vertices $v \in V[T]$ with $y_v \geq \gamma$. Finally, we add the minimum cost matching on the odd degree vertices of this subtree. While our algorithm simply tries all possible trees T in the support of μ and all possible thresholds $\gamma = y_v$ for $v \in V$, our analysis is randomized: We sample the tree from μ and the threshold γ from a specific distribution, and prove the main result in expectation. Clearly, the same guarantee then holds for the best choice of T and γ.

After describing the algorithm in more detail in Sect. 2, we bound its approximation ratio by the golden ratio $(1+\sqrt{5})/2 \approx 1.618$ through a simple analysis in Sect. 3. In Sect. 4, we show that a minor adaption of the algorithm and a slightly more involved analysis allows us to push the approximation guarantee to 1.599 as in Theorem 1.

As mentioned, this improves upon the 1.774-approximation by Blauth and Nägele [10]. It remains open whether there is an efficient algorithm for PCTSP that matches—in terms of the approximation factor—the $3/2$-approximation for TSP by Christofides [15] and Serdyukov [34] (also see [14,36]), or the current best known approximation guarantee for TSP, which is just slightly below $3/2$ [23,24]. The ideal result for PCTSP would be an algorithm that given an α-approximation for TSP produces an α-approximation for PCTSP (or possibly an $(\alpha + \varepsilon)$-approximation for every $\varepsilon > 0$). Such a result was recently shown for Path TSP [35], and as approximation algorithms for PCTSP begin to approach the threshold $3/2$, this possibility feels less out of reach.

1.1 Prior Work on PCTSP

While Balas [5] was the first to study prize-collecting variations of TSP, the first constant-factor approximation algorithm for PCTSP was given by Bienstock, Goemans, Simchi-Levi, and Williamson [8] through a simple threshold rounding approach: Starting from a solution (x, y) of the PCTSP LP relaxation, the

---

[3] For a (sub-)graph $H = (V_H, E_H)$, we write $V[H] := V_H$ and $E[H] := E_H$. Moreover, for a set F of edges we abbreviate $c(F) := \sum_{e\in F} c_e$.

Christofides-Serdyukov algorithm is used to construct a tour on all vertices $v \in V$ with $y_v \geq 3/5$, giving an LP-relative $5/2$-approximation. Goemans and Williamson [21] later obtained a 2-approximation through a classical primal-dual approach. More precisely, they showed how to compute a tree T with $c(E[T]) \leq c^\top x$ and $\pi(V \setminus V[T]) \leq \pi^\top(1-y)$, so that doubling the tree yields the 2-approximation.[4]

The factor of 2 was first beaten by Archer, Bateni, Hajiaghayi, and Karloff [2]. As a black-box subroutine, they use an approximation algorithm for TSP which we assume has ratio ρ. They achieved a $2 - \left(\frac{2-\rho}{2+\rho}\right)^2$ approximation, which—for $\rho = 3/2$—is approximately 1.979. Their algorithm runs the primal-dual algorithm of Goemans and Williamson and the ρ-approximation algorithm for TSP on a carefully selected node set, and outputs the better of the two tours. Goemans [20] then observed that running both threshold rounding for different thresholds and the primal-dual algorithm, and choosing the best among the computed solutions yields an approximation guarantee of $1/(1 - \frac{1}{\beta}e^{1-2/\beta})$, where β denotes the approximation guarantee of an LP-relative approximation algorithm for TSP that is used in a black-box way. For $\beta = 3/2$, the guarantee of Goemans equals approximately 1.914. Goemans was the first to exploit a randomized analysis of threshold rounding, in which the threshold γ is chosen from a specific distribution.

Blauth and Nägele [10] refined the threshold rounding approach by sampling a connected subgraph such that each vertex $v \in V$ with $y_v \geq \gamma$ (again, γ denotes the threshold) is always contained in the vertex set of this subgraph, whereas each vertex $v \in V$ with $y_v < \gamma$ is contained with probability at least $\exp\left(-3y_v/4\gamma\right)$. Since each vertex below the threshold is guaranteed to have even degree in this subgraph, parities can be corrected at no extra cost, yielding an approximation guarantee of slightly below 1.774 through a randomized analysis. This guarantee beats those of the previously mentioned algorithms even for $\rho = 1$ and $\beta = 4/3$ (the integrality gap of the linear programming relaxation for TSP used in [20], the Held-Karp relaxation, is at least $4/3$). Although the high-level idea of [10] is not too complicated, it requires a good deal of technical care to sample this subgraph and analyze the expected penalty cost. The tree construction in our algorithm is much simpler, which is also reflected in the analysis.

1.2 Related Results

Alongside the general version targeted here, PCTSP was studied in special metric spaces. A PTAS is known for graph metrics in planar graphs [7] and in metrics with bounded doubling dimension [12]. For asymmetric edge costs satisfying the triangle inequality, a $\lceil \log(|V|) \rceil$-approximation is known [31].

Besides PCTSP, there is a wide class of other prize-collecting TSP variants, most of which originate from the work of Balas [5]. Although PCTSP can be seen

[4] As observed in [10], such a tree—and thus an algorithm matching the guarantee of the primal-dual approach—can immediately be obtained from the decomposition of solutions of the PCTSP LP relaxation that was mentioned earlier and that is also used in this paper. Therefore, there are two elementary ways to get a 2-approximation.

as the main variant in this problem class, there are other variants that include a lower bound on some minimum prize money that needs to be collected [3–5,19], or an upper bound on the distance that can be traveled [11,13,16,32,33].

Prize-collecting versions have also been studied for other classical combinatorial optimization problems. The prize-collecting Steiner tree problem admits a 1.968-approximation [2], thereby going beyond the integrality gap of 2 of the natural linear programming relaxation. Also the more general prize-collecting Steiner forest problem (see, e.g., [22]) admits an approximation guarantee going beyond the integrality gap of the natural LP relaxation [1]. Interestingly, for the prize-collecting Steiner forest problem, it is known that the integrality gap is strictly larger than the one of the Steiner forest problem [26], indicating that prize-collecting aspects may in some cases make the problem strictly harder to approximate. To date, no such separation is known for TSP and PCTSP.

2 Our Algorithm

Our algorithm follows the basic idea of the Christofides-Serdyukov algorithm for TSP, which is to combine a spanning tree T with a shortest odd(T)-join[5], and shortcut an Eulerian tour in the resulting even-degree graph to obtain a cycle. The operation of adding a shortest odd(T)-join to a tree T is also known as *parity correction*, as it results in a graph in which every vertex has even degree. Typically, for an even cardinality set Q (in particular, for $Q = \text{odd}(T)$), the cost of a shortest Q-join is bounded by the cost $c^\top z$ of a point z that is feasible for the dominant of the Q-join polytope, which is given by (see [17])

$$P^{\uparrow}_{Q\text{-join}} := \{x \in \mathbb{R}^E : x(\delta(S)) \geq 1 \ \forall S \subseteq V \text{ with } |S \cap Q| \text{ odd}\}. \tag{1}$$

We use this approach with two additional variations: First, because our setting allows to not include some vertices in the returned tour, we use trees T that may not span all of V. Second, we follow what is known as a Best-of-Many approach, i.e., we construct a polynomial-size set of trees, construct a tour from each of them, and return the best. For the analysis of such an approach, one typically provides a distribution over the involved trees and analyzes the expected cost of the returned tour—which implies the same bound on the best tour.

We base our tree construction on the following decomposition lemma, which we restate here in the form given in [10]. As mentioned earlier, this result is closely related to an existential statement on packing branchings in a directed multigraph [6, Theorem 2.6].

Lemma 1 (**[10, Lemma 12]**). *Let (x, y) be a feasible solution of the PCTSP LP relaxation. We can in polynomial time compute a set $\mathcal{T}$ of trees that all contain the root r, and weights $\mu \in [0,1]^{\mathcal{T}}$ such that $\sum_{T \in \mathcal{T}} \mu_T = 1$*[6]*,*

---

5 For a graph $G = (V, E)$, we denote by $\text{odd}(G) := \{v \in V : \deg_G(v) \text{ odd}\}$ the set of vertices with odd degree in G. Furthermore, for $Q \subseteq V$ of even cardinality, a Q-join is a set of edges that has odd degree precisely at vertices in Q. (For metric cost functions there is always a minimum cost Q-join that is a matching).

6 For a set $F \subseteq E$ of edges, we denote by $\chi^F \in \{0,1\}^E$ the incidence vector of F.

$$\sum_{T\in\mathcal{T}} \mu_T \cdot \chi^{E[T]} \leq x, \qquad \textit{and} \qquad \forall v \in V\colon \sum_{T\in\mathcal{T}\colon v\in V[T]} \mu_T = y_v.$$

As also mentioned in Footnote 4, this lemma gives rise to an immediate 2-approximation for PCTSP that follows the framework described above: Choosing a tree $T \in \mathcal{T}$ with probability μ_T, and performing parity correction by doubling the tree gives a tour of expected length at most $2c^\top x$, while the expected penalty incurred for vertices that are not visited can easily be seen to be $\pi^\top(1-y)$.

It is an intriguing question whether parity correction can be done at expected cost $c^\top x/2$. Such a result would immediately lead to a $3/2$-approximation algorithm for PCTSP, matching the guarantee of the Christofides-Serdyukov algorithm for TSP. While in the setting of classical TSP, we have $x/2 \in P^{\uparrow}_{Q\text{-join}}$ for any set Q of even cardinality (and can thus bound the cost of parity correction for *any* tree T by $c^\top x/2$), this is no longer true in the prize-collecting setting.

Given a tree T, we first prune it to obtain a tree T' (in a way we will shortly explain) and then perform parity correction. To analyze the cost of the parity correction, we construct a point $z \in P^{\uparrow}_{\text{odd}(T)\text{-join}}$ that is of the form

$$z = \alpha \cdot x + \beta \cdot \chi^{E[T']} \tag{2}$$

for some coefficients $\alpha, \beta \in \mathbb{R}_{\geq 0}$. (In fact, we will choose different coefficients β for different parts of the tree.) As the existence of cuts $S \subsetneq V$ for which both $x(\delta(S))$ and $\delta_{T'}(S)$ are small may require to choose α and β large, we preprocess both the solution (x, y) of the PCTSP LP relaxation that we use as well as the trees that we obtain from it through Lemma 1.

In our first preprocessing step, we get rid of cuts S for which $x(\delta(S))$ is very small. To this end, observe that our algorithm may always drop vertices $v \in V \setminus \{r\}$ for which y_v is very small. Concretely, if our tour does not visit a vertex v with $y_v \leq \delta$ for some $\delta \in [0, 1)$, we pay a penalty of π_v, which is at most a factor of $1/(1-\delta)$ larger than the fractional penalty $\pi_v(1 - y_v)$ occurring in the LP objective. Thus, if we aim for an α-approximation algorithm, we may safely choose $\delta = 1 - 1/\alpha$ and drop all vertices with $y_v \leq \delta$. Crucially for our analysis, we can also perform this dropping on the level of solutions of the PCTSP LP relaxation by using the so-called *splitting off* technique [18,27,28]. For a fixed vertex $v \in V$, splitting off allows to decrease the x-weight on two well-chosen edges $\{v, s\}$ and $\{v, t\}$ incident to v (and thereby also the value y_v) while increasing the weight on the edge $\{s, t\}$ by the same amount, without affecting feasibility for the PCTSP LP relaxation.[7] Note that such a *feasible splitting* at v does not increase the cost of the solution by the triangle inequality. A sequence of feasible splittings at v that result in $y_v = 0$ is called a *complete splitting*. Complete splittings always exist [18] and can be found in polynomial time through a polynomial number of minimum s-t cut computations by trying

[7] Generally, one can even guarantee that minimum s-t cut sizes are preserved for all $s, t \in V \setminus \{v\}$. Feasibility for the PCTSP LP relaxation is already maintained by preserving minimum r-u cut sizes for all $u \in V \setminus \{r, v\}$.

all candidate pairs of edges (see, e.g., [29,30] for more efficient procedures). Summarizing the above directly gives the following.

Theorem 2 (Splitting off). *Let (x^*, y^*) be a feasible solution of the PCTSP LP relaxation. Let $v \in V \setminus \{r\}$. There is a deterministic algorithm that computes in polynomial time a complete splitting at v, i.e., a sequence of feasible splittings at v resulting in a feasible solution (x, y) of the PCTSP LP relaxation with $y_v = 0$, as well as*

$$c^\top x \leq c^\top x^* \quad \text{and} \quad \forall u \in V \setminus \{v\}: \; y_u = y_u^*.$$

Our first preprocessing step then consists of repeatedly applying Theorem 2 to vertices $v \in V \setminus \{r\}$ with $y_v < \delta$ for some parameter $\delta \in [0, 1)$ that we fix later.

Our second preprocessing step affects the trees that we obtain from Lemma 1. While the first preprocessing step guarantees that there are no non-trivial cuts S with very small $x(\delta(S))$, we also want to eliminate cuts S with moderately small $x(\delta(S))$ for which T has only a single edge in $\delta(S)$, so that the combination z defined in (2) gets a significant contribution from at least one of x or $\chi^{E[T]}$ on every non-trivial cut. To this end, we use a pruning step defined as follows (also see Fig. 1).

Definition 1 (Core). *For a fixed solution (x, y) of the PCTSP LP relaxation, a tree T containing the root vertex, and a threshold γ, the* core *of T with respect to γ, denoted by* core(T, γ), *is the inclusion-wise minimal subtree of T that spans all vertices $v \in V[T]$ with $y_v \geq \gamma$.*

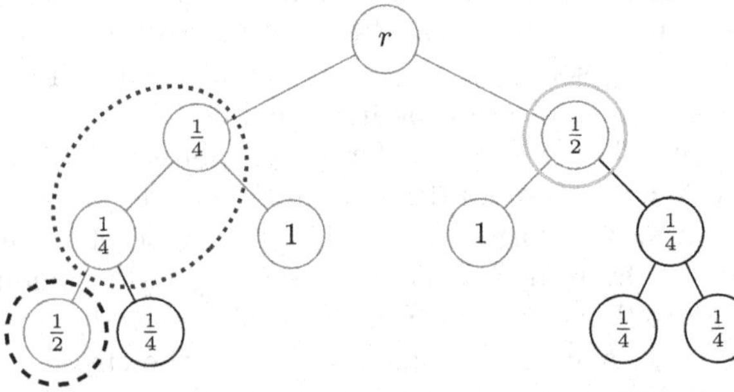

Fig. 1. The core $T' = \text{core}(T, 1/2)$ of the underlying tree T at threshold $\gamma = 1/2$ is highlighted in red, where y_v is shown for each node $v \in V[T]$. We emphasize the different situations that can occur in terms of cuts: For the dotted blue cut S, $x(\delta(S))$ may be small as it does not contain a vertex with $y_v \geq 1/2$, however there are at least three edges of T' in $\delta(S)$ that make up for this. The dashed black cut is a cut S with only one tree edge in $\delta(S)$, however it contains a vertex with $y_v \geq 1/2$ so $x(\delta(S))$ is large. Finally, for the solid green cut, $x(\delta(S))$ may be small and there are only two tree edges in $\delta(S)$, however for this cut, $\delta_{T'}(S)$ is even, so there is no corresponding constraint in the dominant of the odd(T')-join polytope. (Color figure online)

Indeed, if $T' = \mathrm{core}(T, \gamma)$, we know that for any non-empty cut $S \subsetneq V[T']$, we either have $x(\delta(S)) \geq 2\gamma$, or $|\delta_T(S)| > 1$. Additionally, the only relevant such cuts S in terms of parity correction on T' are those with $|S \cap \mathrm{odd}(T')|$ odd, which is well-known to be equivalent to $|\delta_{T'}(S)|$ being odd because

$$\begin{aligned} |S \cap \mathrm{odd}(T')| &\equiv \sum_{v \in S} \deg_{T'}(v) \\ &= 2 \cdot \left| E[T'] \cap \binom{S}{2} \right| + |\delta_{T'}(S)| \equiv |\delta_{T'}(S)| \pmod 2. \end{aligned} \tag{3}$$

Thus, for cuts S with $|S \cap \mathrm{odd}(T')|$ odd, $|\delta_{T'}(S)| > 1$ implies $|\delta_{T'}(S)| \geq 3$, thereby further boosting the load on $\delta(S)$ in z for this case (also see Fig. 1 for examples of the different types of cuts that may appear). Altogether, we are now ready to state our new algorithm for PCTSP, Algorithm 1.

Algorithm 1: Our new approximation algorithm for PCTSP

Input: *PCTSP instance* (G, r, c, π) *on* $G = (V, E)$, $\delta \in [0, 1)$.

1. Compute an optimal solution (x^*, y^*) of the PCTSP LP relaxation.
2. Perform complete splittings at all $v \in V$ with $y_v < \delta$ (see Theorem 2), resulting in a feasible solution (x, y) of the PCTSP LP relaxation.
3. Compute a set $\mathcal{T}$ of trees through Lemma 1 applied to (x, y).
4. Let
$$\mathcal{T}' = \bigcup_{\gamma \in \{y_v \,:\, v \in V\}} \{\mathrm{core}(T, \gamma) : T \in \mathcal{T}\}.$$

return *Best tour found by doing parity correction on all trees in* $\mathcal{T}'$.

We remark that Algorithm 1 can be implemented to run in polynomial time. Indeed, by Lemma 1 and Theorem 2, Steps 2 to 4 run in polynomial time. Moreover, we can compute an optimum solution to the PCTSP LP relaxation in polynomial time as the seperation problem can be solved by computing minimum r-v cut sizes for each $v \in V \setminus \{r\}$.

We show in the next section that there is a constant δ for which Algorithm 1 is a $(1+\sqrt{5})/2$-approximation algorithm. To go beyond that and prove Theorem 1, we will later allow an instance-specific choice of δ.

3 A $(1+\sqrt{5})/2$-Approximation Guarantee

In this section, we prove the following result that gives the golden ratio as approximation guarantee. This is slightly weaker than what Theorem 1 claims, but the proof is simple and illustrates our main ideas.

Theorem 3. *Algorithm 1 is an α-approximation algorithm for PCTSP with*

$$\alpha := \max\left\{\frac{5-2\delta}{3-\delta}, \frac{3-\delta}{2-\delta}, \frac{1}{1-\delta}\right\}.$$

In particular, for $\delta = 3-\sqrt{5}/2 \approx 0.382$, we get $\alpha = (1+\sqrt{5})/2 \approx 1.618$.

Throughout this section, we fix a solution (x, y) of the PCTSP LP relaxation that was obtained from an optimal solution (x^*, y^*) through complete splittings as in Step 2 of Algorithm 1, and we fix a set $\mathcal{T}$ of trees with weights $(\mu_T)_{T\in\mathcal{T}}$ that is obtained in Step 3, i.e., through Lemma 1 applied to (x, y). Moreover, we sample a random tree T from the set $\mathcal{T}'$ constructed in Step 4 of Algorithm 1 as follows: For a fixed value $\kappa \in [\delta, 1]$, sample a threshold $\gamma \in [\delta, \kappa]$ such that for any $t \in [\delta, \kappa]$, we have

$$\mathbb{P}[\gamma \leq t] = \frac{3-\delta-\kappa}{3-\delta-t}. \tag{4}$$

Independently, sample a tree $T_0 \in \mathcal{T}$ with marginals given by $(\mu_T)_{T\in\mathcal{T}}$. Then, define

$$T := \mathrm{core}(T_0, \gamma). \tag{5}$$

By definition of the core, it is clear that $T \in \mathcal{T}'$ even if $\gamma \notin \{y_v : v \in V\}$. To prove Theorem 3, we bound the expected cost of a tour constructed from T by parity correction. We remark that for proving Theorem 3, we only need $\kappa = 1$; we nonetheless proceed in this generality here to be able to reuse some of the following statements in a proof of Theorem 1. To start with, we bound the expected tour length.

Lemma 2. *Let $T = \mathrm{core}(T_0, \gamma)$ be a random tree generated as described in and above* (5)*, and let C be the cycle obtained through parity correction on T and shortcutting an Eulerian walk in the resulting graph. Then*

$$\mathbb{E}[c(E[C])] \leq \frac{7-2\delta-2\kappa}{3-\delta} \cdot c^\top x^*.$$

Proof. Let $\eta_1 > \ldots > \eta_k$ such that $\{\eta_1, \ldots, \eta_k\} = \{y_v : v \in V\}$. Define

$$E_i := \begin{cases} E[\mathrm{core}(T_0, 1)] & \text{for } i = 1 \\ E[\mathrm{core}(T_0, \eta_i)] \setminus E[\mathrm{core}(T_0, \eta_{i-1})] & \text{for } i \in \{2, \ldots, k\} \end{cases}.$$

We refer to Fig. 2 for an illustration of the sets E_i. In particular, this definition implies that

$$c(E[T]) = c\big(E[\mathrm{core}(T_0, \gamma)]\big) = \sum_{i\in[k]:\, \eta_i \geq \gamma} c(E_i). \tag{6}$$

To bound the cost of parity correction on T, we claim that

$$z := \frac{1}{3-\delta} \cdot x + \sum_{i\in[k]:\, \eta_i\geq\gamma} \left(1 - \frac{2\eta_i}{3-\delta}\right) \cdot \chi^{E_i}$$

lies in the dominant of the odd(T)-join polytope. This implies that a shortest odd(T)-join J has length

$$c(E[J])) \leq \frac{1}{3-\delta} \cdot c^\top x + \sum_{i \in [k]:\, \eta_i \geq \gamma} \left(1 - \frac{2\eta_i}{3-\delta}\right) \cdot c(E_i).$$

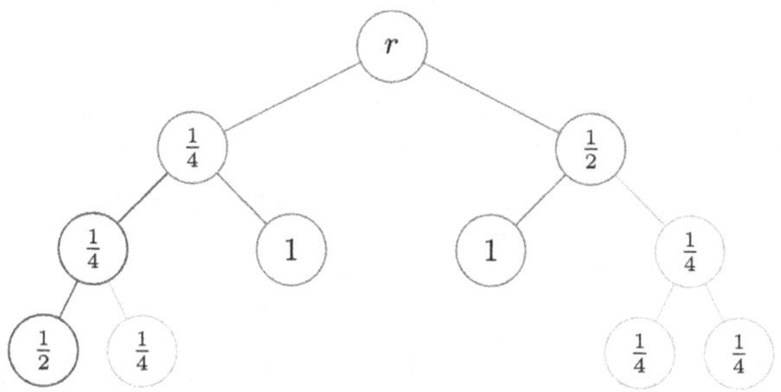

Fig. 2. Here, the variables η_i are given by the sequence 1, 1/2, 1/4. Edges in E_1 are drawn in red, those in E_2 in blue, and those in E_3 in green. We also color the nodes for intuition, however the sets E_i consist solely of edges. (Color figure online)

Combining this with (6) and taking expectations immediately gives the desired:

$$\begin{aligned}
\mathbb{E}[c(E[C])] &\leq \frac{1}{3-\delta} \cdot c^\top x + \sum_{T_0 \in \mathcal{T}} \mu_{T_0} \sum_{i=1}^{k} \mathbb{P}[\gamma \leq \eta_i] \cdot \left(2 - \frac{2\eta_i}{3-\delta}\right) \cdot c(E_i) \\
&\leq \frac{1}{3-\delta} \cdot c^\top x + \sum_{T_0 \in \mathcal{T}} \mu_{T_0} \sum_{i=1}^{k} \frac{6 - 2\delta - 2\kappa}{3-\delta} \cdot c(E_i) \qquad \text{(using (4))} \\
&= \frac{1}{3-\delta} \cdot c^\top x + \frac{6 - 2\delta - 2\kappa}{3-\delta} \cdot \sum_{T_0 \in \mathcal{T}} \mu_{T_0} c(E[T_0]) \\
&\leq \frac{7 - 2\delta - 2\kappa}{3-\delta} \cdot c^\top x. \qquad \text{(using Lemma 1)}
\end{aligned}$$

By the construction of (x, y) from (x^*, y^*) through splitting off, we have $c^\top x \leq c^\top x^*$, hence the above implies the desired. It thus remains to show that $z \in P^{\uparrow}_{\text{odd}(T)\text{-join}}$, i.e., that $z(\delta(S)) \geq 1$ for every $S \subseteq V$ with $|S \cap \text{odd}(T)|$ odd. As remarked in (3), $|S \cap \text{odd}(T)|$ is odd if and only if $|\delta_T(S)|$ is. If $|\delta_T(S)| \geq 3$,

$$\begin{aligned}
z(\delta(S)) &= \frac{1}{3-\delta} \cdot x(\delta(S)) + \sum_{i \in [k]:\, \eta_i \geq \gamma} \left(1 - \frac{2\eta_i}{3-\delta}\right) \cdot |E_i \cap \delta_T(S)| \\
&\geq \frac{2\delta}{3-\delta} + \left(1 - \frac{2}{3-\delta}\right) \cdot 3 = 1,
\end{aligned}$$

where we used that $x(\delta(S)) \geq 2\delta$ because $y_v \geq \delta$ for all $v \in S$, and that $\eta_i \leq 1$ for each $i \in [k]$. Otherwise, we have $|\delta_T(S)| = 1$. Let $i \in [k]$ such that $\delta_T(S) \subseteq E_i$. Then $x(\delta(S)) \geq 2\eta_i$, hence also in this case

$$z(\delta(S)) \geq \frac{2\eta_i}{3-\delta} + \left(1 - \frac{2\eta_i}{3-\delta}\right) = 1.$$

□

Next, we analyze the expected penalty incurred when starting with such a random tree T.

Lemma 3. *Let* $T = \mathrm{core}(T_0, \gamma)$ *be a random tree generated as described in and above* (5). *Then, for every* $v \in V$, *we have*

$$\mathbb{P}[v \in V[T]] \geq \begin{cases} 0 & \textit{if } y_v^* \in [0, \delta) \\ y_v^* \cdot \frac{3-\delta-\kappa}{3-\delta-y_v^*} & \textit{if } y_v^* \in [\delta, \kappa] \\ y_v^* & \textit{if } y_v^* \in (\kappa, 1] \end{cases}.$$

Proof. By construction, the solution (x, y) has the property that for all $v \in V$, either $y_v = 0$ or $y_v = y_v^* \geq \delta$. Consequently, no tree T_0 in the family $\mathcal{T}$ generated through Lemma 1 contains vertices $v \in V$ with $y_v^* < \delta$, and thus the same holds for T. Consequently, for such vertices $v \in V$, we get $\mathbb{P}[v \in V[T]] = 0$. For vertices $v \in V$ with $y_v = y_v^* \geq \delta$, we have $\mathbb{P}[v \in V[T_0]] = y_v^*$ by Lemma 1. Hence,

$$\mathbb{P}[v \in V[T]] \geq \mathbb{P}[v \in V[T_0]] \cdot \mathbb{P}[\gamma \leq y_v^*] = y_v^* \cdot \mathbb{P}[\gamma \leq y_v^*].$$

If $y_v^* > \kappa$, then $\mathbb{P}[\gamma \leq y_v^*] = 1$ and $\mathbb{P}[v \in V[T]] = y_v^*$. In the remaining case $y_v^* \in [\delta, \kappa]$, we use (4) to obtain the desired. □

Together, Lemmas 2 and 3 allow us to conclude Theorem 3.

Proof of Theorem 3. Let T be a random tree generated as described in and above (5), and let C be the cycle obtained through parity correction on T and shortcutting an Eulerian walk in the resulting graph.

Let $v \in V$. By Lemma 3, if $y_v^* < \delta$, then $\mathbb{P}[v \notin V[C]] = 1 \leq \frac{1}{1-\delta}(1 - y_v^*)$; if $y_v^* > \kappa$, then $\mathbb{P}[v \notin V[C]] \leq 1 - y_v^*$. If $\delta \leq y_v^* \leq \kappa$ then, again by Lemma 3,

$$\begin{aligned} \mathbb{P}[v \notin V[C]] &\leq 1 - y_v^* \cdot \frac{3-\delta-\kappa}{3-\delta-y_v^*} \\ &= \frac{(3-\delta)(1-y_v^*) - y_v^*(1-\kappa)}{3-\delta-y_v^*} \\ &\leq \frac{3-\delta}{3-\delta-\kappa} \cdot (1 - y_v^*). && \text{(because } y_v^* \leq \kappa \leq 1\text{)} \end{aligned}$$

Hence, together with Lemma 2 we get

$$\begin{aligned} &\mathbb{E}[c(E[C]) + \pi(V \setminus V[C])] \\ &\quad\leq \frac{7-2\delta-2\kappa}{3-\delta} \cdot c^\top x^* + \max\left\{\frac{3-\delta}{3-\delta-\kappa}, \frac{1}{1-\delta}\right\} \cdot \pi^\top(1 - y^*) \\ &\quad\leq \max\left\{\frac{7-2\delta-2\kappa}{3-\delta}, \frac{3-\delta}{3-\delta-\kappa}, \frac{1}{1-\delta}\right\} \cdot (c^\top x^* + \pi^\top(1-y^*)). \end{aligned} \tag{7}$$

Independently of the realization of the involved random variables, the cycle C is one that is generated in Algorithm 1. The maximum in (7) is minimized for $\kappa = 1$ and $\delta = (3-\sqrt{5})/2$, where it evaluates to $(1+\sqrt{5})/2$, thus giving the guarantee claimed in Theorem 3. $\square$

4 Getting Below 1.6

To improve upon the analysis given in Sect. 3, we exploit some remaining flexibility: Choosing not only γ but also δ randomly allows for balancing off costs better than before. The distribution used here is not best possible (though close, see Remark 1). For a deterministic algorithm, we show how sampling δ can be avoided by trying polynomially many instance-dependent values.

Proof of Theorem 1 *(Sketch).* For κ_0 and κ to be fixed later, sample $\delta \in [\kappa_0, \kappa]$ from a distribution with density $f(\delta) \propto (3-\delta)(\kappa-\delta)^{2.2}$ on $[\kappa_0, \kappa]$. Using this δ, sample a tree T as described in and above (5) (using the same κ as here), and let C be the cycle generated from parity correction on T and shortcutting an Eulerian walk in the resulting graph. By Lemma 2, the expected length of C is

$$\mathbb{E}[c(E[C])] \leq \underbrace{\int_{\kappa_0}^{\kappa} \frac{7-2\delta-2\kappa}{3-\delta} f(\delta)\,\mathrm{d}\delta}_{=:g(\kappa,\kappa_0)} \cdot c^\top x^* \ ,$$

where our choice of f allows to explicitly compute the integral defining $g(\kappa, \kappa_0)$. To bound the expected penalty, let $v \in V$. By Lemma 3, $\mathbb{P}[v \notin V[C]] = 1 \leq \frac{1}{1-\kappa_0} \cdot (1 - y_v^*)$ if $y_v^* < \kappa_0$, $\mathbb{P}[v \notin V[C]] = 1 - y_v^*$ if $y_v^* > \kappa$; for $y_v^* \in [\kappa_0, \kappa]$, we get

$$\mathbb{P}[v \notin V[C]] = 1 - y_v^* \cdot \int_{\kappa_0}^{y_v^*} \frac{3-\delta-\kappa}{3-\delta-y_v^*} f(\delta)\,\mathrm{d}\delta.$$

Our choice of f allows to bound the latter by a function $h_{y_v^*}(\kappa, \kappa_0)$ not involving integrals that can be evaluated exactly (for details, see [9]). We get

$$\mathbb{E}[\pi(V \setminus V[C])] \leq \underbrace{\max\left\{\frac{1}{1-\kappa_0}, \max_{y\in[\kappa_0,\kappa]} \frac{h_y(\kappa,\kappa_0)}{1-y}\right\}}_{=:h(\kappa,\kappa_0)} \cdot \pi^\top(1-y^*) \ . \tag{8}$$

Together, this gives a bound on the expected total cost of the form

$$\mathbb{E}[c(E[C]) + \pi(V \setminus V[C])] \leq \max\{g(\kappa,\kappa_0), h(\kappa,\kappa_0)\} \cdot \left(c^\top x^* + \pi^\top(1-y^*)\right).$$

The latter maximum evaluates to slightly below 1.599 for $\kappa_0 = 0.3724$ and $\kappa = 0.9971$, giving the desired guarantee. We remark that our explicit form of $h_y(\kappa, \kappa_0)$ allows to approximate the inner maximum in (8) up to a (quantifiable) minor error by evaluating the argument for each y in a sufficiently fine discretization of $[\kappa_0, \kappa]$ (for details, we again refer to [9]).

Finally, the choice of δ can be derandomized by trying the instance-specific values in the set $\{y_v^*: v \in V\}$ obtained from the optimal LP solution (x^*, y^*) that is used. Indeed, if $\delta \notin \{y_v^*: v \in V\}$, the bound on the expected cost of C in Lemma 2 improves (given $\kappa \geq 1/2$) by using the minimal δ' in $\{y_v^*: v \in V, y_v^* \geq \delta\}$, whereas the bound on the expected penalty cost does not change. □

Remark 1. Computational experiments based on discretizing a distribution over pairs (δ, κ) suggest that an analysis following the one in the above proof cannot achieve an approximation ratio of 1.59. We emphasize that this does not exclude that the actual approximation guarantee of Algorithm 1 is below 1.59.

References

1. Ahmadi, A., Gholami, I., Hajiaghayi, M., Jabbarzade, P., Mahdavi, M.: 2-approximation for prize-collecting Steiner forest. In: Proceedings of the 35th Annual ACM-SIAM Symposium on Discrete Algorithms (SODA 2024), pp. 669–693 (2024)
2. Archer, A., Bateni, M., Hajiaghayi, M., Karloff, H.: Improved approximation algorithms for prize-collecting Steiner tree and TSP. SIAM J. Comput. **40**(2), 309–332 (2011)
3. Ausiello, G., Bonifaci, V., Leonardi, S., Marchetti-Spaccamela, A.: Prize-collecting traveling salesman and related problems. In: Handbook of Approximation Algorithms and Metaheuristics, pp. 1–13. Chapman and Hall (2007)
4. Ausiello, G., Leonardi, S., Marchetti-Spaccamela, A.: On salesmen, repairmen, spiders, and other traveling agents. In: Bongiovanni, G., Petreschi, R., Gambosi, G. (eds.) CIAC 2000. LNCS, vol. 1767, pp. 1–16. Springer, Heidelberg (2000). https://doi.org/10.1007/3-540-46521-9_1
5. Balas, E.: The prize collecting traveling salesman problem. Networks **19**(6), 621–636 (1989)
6. Bang-Jensen, J., Frank, A., Jackson, B.: Preserving and increasing local edge-connectivity in mixed graphs. SIAM J. Discret. Math. **8**(2), 155–178 (1995)
7. Bateni, M., Chekuri, C., Ene, A., Hajiaghayi, M., Korula, N., Marx, D.: Prize-collecting Steiner problems on planar graphs. In: Proceedings of the 22nd Annual ACM-SIAM Symposium on Discrete Algorithms (SODA 2011), pp. 1028–1049 (2011)
8. Bienstock, D., Goemans, M.X., Simchi-Levi, D., Williamson, D.: A note on the prize collecting traveling salesman problem. Math. Program. **59**, 413–420 (1993)
9. Blauth, J., Klein, N., Nägele, M.: A better-than-1.6-approximation for prize-collecting TSP. arXiv: 2308.06254v2 [cs.DS] (2024)
10. Blauth, J., Nägele, M.: An improved approximation guarantee for Prize-Collecting TSP. In: Proceedings of the 55th Annual ACM Symposium on Theory of Computing (STOC 2023), pp. 1848–1861 (2023)
11. Blum, A., Chawla, S., Karger, D., Lane, T., Meyerson, A., Minkoff, M.: Approximation algorithms for orienteering and discounted-reward TSP. SIAM J. Comput. **37**(2), 653–670 (2007)
12. Chan, T.-H.H., Jiang, H., Jiang, S.H.-C.: A unified PTAS for prize collecting TSP and Steiner tree problem in doubling metrics. ACM Trans. Algorithms **16**(2), 1–23 (2020)

13. Chekuri, C., Korula, N., Pál, M.: Improved algorithms for orienteering and related problems. ACM Trans. Algorithms **8**(3), 1–27 (2012)
14. Christofides, N.: Worst-case analysis of a new heuristic for the travelling salesman problem. Oper. Res. Forum **3** (2022)
15. Christofides, N.: Worst-case analysis of a new heuristic for the travelling salesman problem. Technical report 388, Graduate School of Industrial Administration, Carnegie Mellon University (1976)
16. Dezfuli, S., Friggstad, Z., Post, I., Swamy, C.: Combinatorial algorithms for rooted prize-collecting walks and applications to orienteering and minimum-latency problems. In: Aardal, K., Sanità, L. (eds.) IPCO 2022. LNCS, vol. 13265, pp. 195–209. Springer, Cham (2022). https://doi.org/10.1007/978-3-031-06901-7_15
17. Edmonds, J., Johnson, E.: Matching, Euler tours and the Chinese postman. Math. Program. **5**, 88–124 (1973)
18. Frank, A.: On a theorem of Mader. Discret. Math. **101**(1), 49–57 (1992)
19. Garg, N.: Saving an Epsilon: a 2-approximation for the k-MST problem in graphs. In: Proceedings of the 37th Annual ACM Symposium on Theory of Computing (STOC 2005), pp. 396–402 (2005)
20. Goemans, M.X.: Combining approximation algorithms for the prize-collecting TSP. arXiv: 0910.0553 [cs.DS] (2009)
21. Goemans, M.X., Williamson, D.P.: A general approximation technique for constrained forest problems. SIAM J. Comput. **24**(2), 296–317 (1995)
22. Hajiaghayi, M.T., Jain, K.: The prize-collecting generalized Steiner tree problem via a new approach of primal-dual schema. In: Proceedings of the Seventeenth Annual ACM-SIAM Symposium on Discrete Algorithm (SODA 2006), pp. 631–640 (2006)
23. Karlin, A.R., Klein, N., Oveis Gharan, S.: A (slightly) improved approximation algorithm for metric TSP. In: Proceedings of the 53rd Annual ACM SIGACT Symposium on Theory of Computing (STOC 2021), pp. 32–45 (2021)
24. Karlin, A.R., Klein, N., Oveis Gharan, S.: A deterministic better-than-3/2 approximation algorithm for metric TSP. In: Del Pia, A., Kaibel, V. (eds.) IPCO 2023. LNCS, vol. 13904, pp. 261–274. Springer, Cham (2023). https://doi.org/10.1007/978-3-031-32726-1_19
25. Karpinski, M., Lampis, M., Schmied, R.: New inapproximability bounds for TSP. J. Comput. Syst. Sci. **81**(8), 1665–1677 (2015)
26. Könemann, J., Olver, N., Pashkovich, K., Ravi, R., Swamy, C., Vygen, J.: On the integrality gap of the prize-collecting Steiner forest LP. In: Approximation, Randomization, and Combinatorial Optimization. Algorithms and Techniques (APPROX/RANDOM 2017), pp. 17:1–17:13 (2017)
27. Lovász, L.: On some connectivity properties of Eulerian graphs. Acta Math. Acad. Scientiarum Hung. **28**(1), 129–138 (1976)
28. Mader, W.: A reduction method for edge-connectivity in graphs. Ann. Discrete Math. **3**, 145–164 (1978)
29. Nagamochi, H.: A fast edge-splitting algorithm in edge-weighted graphs. IEICE Trans. Fundam. Electron. Commun. Comput. Sci. **E89-A**(5), 1263–1268 (2006)
30. Nagamochi, H., Ibaraki, T.: Deterministic $\tilde{O}(nm)$ time edge-splitting in undirected graphs. J. Comb. Optim. **1**(6), 5–46 (1997)
31. Nguyen, V.H.: A primal-dual approximation algorithm for the asymmetric prize-collecting TSP. J. Comb. Optim. **25**, 265–278 (2013)
32. Paul, A., Freund, D., Ferber, A., Shmoys, D.B., Williamson, D.P.: Budgeted prize-collecting traveling salesman and minimum spanning tree problems. Math. Oper. Res. **45**(2), 576–590 (2020)

33. Paul, A., Freund, D., Ferber, A., Shmoys, D.B., Williamson, D.P.: Erratum to 'budgeted prize-collecting traveling salesman and minimum spanning tree problems'. Math. Oper. Res. **45**, 576–590 (2022)
34. Serdyukov, A.I.: O nekotorykh ekstremal'nykh obkhodakh v grafakh. Upravlyaemye sistemy **17**, 76–79 (1987)
35. Traub, V., Vygen, J., Zenklusen, R.: Reducing path TSP to TSP. SIAM J. Comput. **51**(3), STOC20-24–STOC20-53 (2022)
36. van Bevern, R., Slugina, V.A.: A historical note on the 3/2-approximation algorithm for the metric traveling salesman problem. Hist. Math. **53**, 118–127 (2020)

On Matrices over a Polynomial Ring with Restricted Subdeterminants

Marcel Celaya[1], Stefan Kuhlmann[2(✉)], and Robert Weismantel[2]

[1] School of Mathematics, Cardiff University, Cardiff, Wales, UK
[2] Department of Mathematics, ETH Zürich, Zürich, Switzerland
Stefan.Kuhlmann@math.ethz.ch

Abstract. This paper introduces a framework to study discrete optimization problems which are parametric in the following sense: their constraint matrices correspond to matrices over the ring $\mathbb{Z}[x]$ of polynomials in one variable. We investigate in particular matrices whose subdeterminants all lie in a fixed set $S \subseteq \mathbb{Z}[x]$. Such matrices, which we call *totally S-modular matrices*, are closed with respect to taking submatrices, so it is natural to look at minimally non-totally S-modular matrices which we call *forbidden minors for S*. Among other results, we prove that if S is finite, then the set of all determinants attained by a forbidden minor for S is also finite. Specializing to the integers, we subsequently obtain the following positive complexity results: the recognition problem for totally $\pm\{0, 1, a, a+1, 2a+1\}$-modular matrices with $a \in \mathbb{Z}\backslash\{-3, -2, 1, 2\}$ and the integer linear optimization problem for totally $\pm\{0, a, a+1, 2a+1\}$-modular matrices with $a \in \mathbb{Z}\backslash\{-2, 1\}$ can be solved in polynomial time.

1 Introduction

For a matrix $\boldsymbol{M}$, let $\Delta(\boldsymbol{M})$ denote the maximal absolute value of a subdeterminant of $\boldsymbol{M}$. Since decades it is well-known that the number $\Delta(\boldsymbol{M})$ plays a crucial role in understanding complexity questions related to integer programming problems whose associated constraint matrix is $\boldsymbol{M}$. Important examples include bounds on the diameter of polyhedra [7,10,21], questions about the proximity between optimal LP solutions and optimal integral solutions [3,8], bounds on the size of the minimal support of integer vectors in standard form programs [1,2,4,11], and running time functions for dynamic programming algorithms [12]. Largely unexplored however remain two fundamental algorithmic questions: Given a finite set $S \subseteq \mathbb{N}$ of allowed values for the subdeterminants in absolute value of the matrix $\boldsymbol{M}$:

1. What is the complexity of solving an ILP with associated constraint matrix $\boldsymbol{M}$ in dependence of $|S|$ and $\max_{s \in S} s$? (Optimization problem)
2. How can we efficiently verify whether a matrix $\boldsymbol{M}$ has all its subdeterminants in absolute value in S? (Recognition problem)

If $S = \{0, 1\}$, both questions have been answered. Indeed, a famous theorem of Hoffmann and Kruskal [15] and a decomposition theorem of Seymour for totally

J. Vygen and J. Byrka (Eds.): IPCO 2024, LNCS 14679, pp. 43–56, 2024.
https://doi.org/10.1007/978-3-031-59835-7_4

unimodular matrices [27] allow us to tackle both questions. We refer to [26] for further theory concerning totally unimodular matrices. If $S = \{0, 1, 2\}$, there exists a polynomial time algorithm to solve Question 1 [6]. In this generality, further optimization results are not known to us. There are other important results when one imposes additional restrictions on the constraint matrix. For instance, if one assumes that each row of the constraint matrix contains at most two non-zero entries, there is a polynomial time algorithm for the optimization problem [13]. A randomized polynomial time algorithm for the related integer feasibility question can be derived if the constraint matrix is $\{0, \Delta\}$-modular for $\Delta \leq 4$ [19,20]. There are also polynomial time algorithms for the optimization problem if the constraint matrix is $\{a, b, c\}$-modular [14] for $a, b, c \in \mathbb{N}$.

The main reason why there are only few general results on optimization and recognition problems is due to the lack of understanding integral matrices with bounded subdeterminants. The purpose of this paper is to develop a framework to investigate the structure of those matrices. Our point of departure is to consider matrices with entries being elements in $\mathbb{Z}[x]$, the ring of polynomials with integral coefficients in one indeterminate x. Moreover, we specify a set of polynomials $S \subseteq \mathbb{Z}[x]$ that corresponds to the allowed polynomials for the subdeterminants of those matrices. All other polynomials in $\mathbb{Z}[x] \backslash S$ are forbidden. Let us make this precise.

Definition 1. *Let $S \subseteq \mathbb{Z}[x]$ be finite. Let $1 \leq m, n$. A matrix $\boldsymbol{M} \in \mathbb{Z}[x]^{m \times n}$ is* totally S-modular *if every $k \times k$ subdeterminant of $\boldsymbol{M}$ is contained in S for $1 \leq k \leq \min\{m, n\}$. Let $2 \leq l$. The matrix $\boldsymbol{M} \in \mathbb{Z}[x]^{l \times l}$ is a* forbidden minor *for S if every $(l-1) \times (l-1)$ submatrix is totally S-modular but* $\det \boldsymbol{M} \notin S$. *By $F(S)$, we denote the set of all polynomials that arise as a determinant of some forbidden minor for S.*

The case $\mathbb{Z}$ can be recovered by restricting S to consist of constant polynomials. One advantage of operating in $\mathbb{Z}[x]$ is that we can extract a certain decomposition, Theorem 1, and simplify arguments due to arithmetic properties of $\mathbb{Z}[x]$. This allows us to make progress towards Question 1 and 2 if we evaluate the polynomials in $\boldsymbol{M}$ and S at integers. The disadvantage of our approach is that there are finitely many values of $a \in \mathbb{Z}$ for which our statements do not hold. For those values a, there exist a polynomial in $F(S)$ and a polynomial in S whose evaluations at a admit the same value. This implies that $F(S(a))$ cannot be the set of all polynomials that arise as a determinant of some forbidden minor for the set $S(a)$ of evaluations of all polynomials of S at a; cf. Lemma 3.

Let us remark that there is a related line of research. It involves understanding the matroids that admit a representation as a $\pm\{0, 1, 2, \ldots, \Delta\}$-modular matrix for $\Delta \geq 2$ [22,23]. We emphasize that there are crucial differences between this approach and the totally S-modular direction proposed in this work. For instance, here the notion of a forbidden minor depends on the concrete representation of the matrix.

We will later be able to make general statements about the structure of the set $F(S)$. However, when it comes to Questions 1 and 2, we need further restrictions to derive general statements.

1.1 The Smallest Non-trivial Cases

Given the requirements that $x \in S$ and S is as small as possible, it turns out that the first non-trivial cases are given by matrices with associated sets $S = \pm\{0, x, x+1, 2x+1\}$ and $S = \pm\{0, 1, x, x+1, 2x+1\}$. Let us next argue why this is the case. We assume that $0 \in S$. This assumption can be made without loss of generality because, if $0 \notin S$, we are quite restricted; see, for instance, [5]. So let $S = \pm\{0, x\}$. One can check that there is no invertible totally S-modular matrix in dimension larger than one. Hence, we may assume that S contains at least three different polynomials, i.e., $S = \pm\{0, x, y\}$ for $y \in \mathbb{Z}[x]$. The only ways of choosing y such that the set of totally S-modular matrices contains invertible instances in every dimension is either $y = 1$ or $y = x+1$. Let $S = \pm\{0, x, x+1\}$. After removing all-zero rows and columns, one can show that totally S-modular matrices $\boldsymbol{M}$ are submatrices of the following matrix up to a sign, row and column permutations, and duplicates:

$$\begin{pmatrix} x+1 & \dots & \dots & x+1 \\ \vdots & & \iddots & x \\ \vdots & \iddots & \iddots & \vdots \\ x+1 & x & \dots & x \end{pmatrix}.$$

In other words, the rows and columns of $\boldsymbol{M}$ can be totally ordered, or, equivalently, $\boldsymbol{M}$ excludes the submatrix

$$\begin{pmatrix} x+1 & x \\ x & x+1 \end{pmatrix}. \tag{1}$$

This implies that the matrix (1) is a forbidden minor for S with determinant $2x+1$. We therefore obtain a purely combinatorial description which does not take into account the knowledge about subdeterminants but still completely characterizes totally S-modular matrices. One of the advantages of this observation is that, after minor preprocessing, the recognition problem becomes surprisingly straightforward in large dimensions: Up to row and column permutations and multiplying with minus one, one can simply check whether the matrix has entries in $\{x, x+1\}$ and the forbidden submatrix in (1) does not appear. This leads to the natural question of what happens when we allow the matrix (1) to be present, i.e., we add $2x+1$ to the set S. This gives us $S = \pm\{0, x, x+1, 2x+1\}$. As a next step, one can add the choice $y = 1$ from above to our current set to obtain $S = \pm\{0, 1, x, x+1, 2x+1\}$.

1.2 Statements of Our Results

The statements are concerned with matrices that can be decomposed into $\boldsymbol{M} = \boldsymbol{T} + x \cdot \boldsymbol{u} \cdot \boldsymbol{v}^\top$ for integral $\boldsymbol{T}, \boldsymbol{u}$, and $\boldsymbol{v}$. To derive complexity results, we need to understand the matrix $\boldsymbol{T}$ for a given set S. The following decomposition result applies. Let $S(a)$ denote the set of evaluations of all polynomials of S at $a \in \mathbb{Z}$.

Theorem 1. *Let $S \subseteq \mathbb{Z}[x]$ be finite. Let $\boldsymbol{M} = \boldsymbol{T} + x \cdot \boldsymbol{u} \cdot \boldsymbol{v}^\top$ where $\boldsymbol{T}, \boldsymbol{u}$, and $\boldsymbol{v}$ are integral. If $\boldsymbol{M}$ is totally S-modular, then $\boldsymbol{T}$ is totally $S(0)$-modular.*

For instance, let $K_0 \subseteq \mathbb{N}$ be finite and $S = \pm\{kx-1, kx, kx+1 : k \in K_0\}$. Then Theorem 1 states that $\boldsymbol{T}$ is totally unimodular.

Theorem 1 can be significantly generalized. The assumption that $\boldsymbol{M} = \boldsymbol{T} + x \cdot \boldsymbol{u} \cdot \boldsymbol{v}^\top$ can be replaced by $\boldsymbol{M} = \boldsymbol{T} + x \cdot \boldsymbol{M}_1 + \ldots + x^l \cdot \boldsymbol{M}_l$ where $\boldsymbol{M}_1, \ldots, \boldsymbol{M}_l$ are arbitrary integral matrices for $l \in \mathbb{N}$. We refrain from proving this in detail. The reason is that we only use Theorem 1, and its evaluated form Corollary 1, to derive statements about optimization and recognition, but even then Theorem 1 itself is not enough. Further properties of the matrix $\boldsymbol{T}$ are required to tackle the following constraint parametric optimization problem over $\mathbb{Z}$ which is given in terms of a full-column-rank constraint matrix $\boldsymbol{M}(a) \in \mathbb{Z}^{m \times n}$ where every entry of $\boldsymbol{M}(a)$ is parametrized by a specific value $a \in \mathbb{Z}$:

$$\mathrm{ILP}(\boldsymbol{M}(a), \boldsymbol{b}, \boldsymbol{c}) : \max \boldsymbol{c}^\top \boldsymbol{x} \text{ s.t. } \boldsymbol{M}(a)\boldsymbol{x} \leq \boldsymbol{b},\ \boldsymbol{x} \in \mathbb{Z}^n$$

where $\boldsymbol{b}$ is integral. The following two results can be derived with the tools developed in this paper.

Theorem 2. *Let $a \in \mathbb{Z} \backslash \{-3, -2, 1, 2\}$ and $S(a) = \pm\{0, 1, a, a+1, 2a+1\}$. Given a matrix $\boldsymbol{M}(a) \in \mathbb{Z}^{m \times n}$, one can decide in polynomial time whether $\boldsymbol{M}(a)$ is totally $S(a)$-modular.*

Theorem 3. *Let $a \in \mathbb{Z} \backslash \{-2, 1\}$ and $S(a) = \pm\{0, a, a+1, 2a+1\}$. Let $\boldsymbol{M}(a) \in \mathbb{Z}^{m \times n}$ have full column rank and be totally $S(a)$-modular. Then one can solve $\mathrm{ILP}(\boldsymbol{M}(a), \boldsymbol{b}, \boldsymbol{c})$ for integral $\boldsymbol{b}$ and $\mathbf{c}$ in polynomial time.*

2 Tools

Throughout this paper we work with the lexicographical order of $\mathbb{Z}[x]$ which is defined by $s < t$ if and only if the leading coefficient of $t - s$ is positive. With respect to this ordering we further define the absolute value of $s \in \mathbb{Z}[x]$ to be

$$|s| = \begin{cases} s, & \text{if } s \geq 0, \\ -s, & \text{otherwise} \end{cases}.$$

Also, we interchangeably pass from polynomials $s \in \mathbb{Z}[x]$ to polynomial functions, denoted by $s(a)$, when evaluating polynomials. This can be done since the polynomial function is uniquely determined by the polynomial over $\mathbb{Z}[x]$; see [17, Chapter 4]. Let $I \subseteq [n]$ and $J \subseteq [n]$. We denote by $\boldsymbol{M}_{\backslash I, \backslash J}$ the submatrix of $\boldsymbol{M}$ without the rows indexed by I and columns indexed by J. In what follows, the notation $\boldsymbol{A}_{(k)} \subseteq \boldsymbol{M}$ is shorthand for specifying that $\boldsymbol{A}_{(k)}$ is a $k \times k$ submatrix of $\boldsymbol{M}$. Given a submatrix $\boldsymbol{A}$ of $\boldsymbol{M}$ and $i, j \in [n]$, we denote by $\boldsymbol{A}[i, j]$ the submatrix of $\boldsymbol{M}$ that contains $\boldsymbol{A}$ along with the extra row and column indexed by i and j respectively.

We now show two results that can be viewed as a generalization of the well-known fact that every matrix that is not totally unimodular and has entries in $\pm\{0,1\}$ contains a submatrix with determinant two in absolute value. For that purpose, we utilize a well-known determinant identity from linear algebra which is due to Sylvester [28] and commonly referred to as *Sylvester's determinant identity*. By convention, the determinant of an empty matrix is 1.

Lemma 1 (Sylvester's determinant identity). *Let $\boldsymbol{M} \in \mathbb{Z}[x]^{n\times n}$ for $n \geq 2$ and let $\boldsymbol{A}_{(k)} \subseteq \boldsymbol{M}$ for $k = 0,\dots,n$ with $\boldsymbol{A}_{(k)} = \boldsymbol{M}_{\setminus I,\setminus J}$ for the ordered sets $I = \{i_1,\dots,i_{n-k}\}$ and $J = \{j_1,\dots,j_{n-k}\}$. Then we get*

$$\det \boldsymbol{M} \cdot (\det \boldsymbol{A}_{(k)})^{n-1-k} = \det \begin{pmatrix} \det \boldsymbol{A}_{(k)}[i_1,j_1] & \dots & \det \boldsymbol{A}_{(k)}[i_1,j_{n-k}] \\ \vdots & \ddots & \vdots \\ \det \boldsymbol{A}_{(k)}[i_{n-k},j_1] & \dots & \det \boldsymbol{A}_{(k)}[i_{n-k},j_{n-k}] \end{pmatrix}.$$

For the most part, we work with the special case when $k = n-2$, which is also known as the *Desnanont-Jacobi identity.* In this case, we get the equation

$$\det \boldsymbol{M} \cdot \det \boldsymbol{A}_{(k)} = \det \begin{pmatrix} \det \boldsymbol{A}_{(k)}[i_1,j_1] & \det \boldsymbol{A}_{(k)}[i_1,j_2] \\ \det \boldsymbol{A}_{(k)}[i_2,j_1] & \det \boldsymbol{A}_{(k)}[i_2,j_2] \end{pmatrix} \tag{2}$$

for $I = \{i_1,i_2\}$ and $J = \{j_1,j_2\}$. This identity already implies our first bound:

Lemma 2. *Let $S \subseteq \mathbb{Z}[x]$ be finite. Then the set $F(S)$ of determinants attained by the forbidden minors of S is finite and*

$$\max_{x\in F(S)} |x| \leq 2 \cdot \max_{s\in S} s^2.$$

Proof. Select some forbidden minor $\boldsymbol{M}$ for S such that $\det \boldsymbol{M} \neq 0$. There exists an invertible submatrix $\boldsymbol{A}_{(n-2)} \subseteq \boldsymbol{M}$. By the Desnanot-Jacobi identity (2) applied to $\boldsymbol{A}_{(n-2)}$, we obtain that

$$\det \boldsymbol{M} = \frac{1}{\det \boldsymbol{A}_{(n-2)}} \cdot (s_1 s_2 - s_3 s_4),$$

where $s_i \in S$ for $i = 1,2,3,4$. Since $\det \boldsymbol{A}_{(n-2)} \in S$, the right hand side only attains finitely many values as S is finite. Hence, there are only finitely many values possible for $\det \boldsymbol{M}$. To obtain the inequality, we take absolute values on both sides, apply the triangle inequality, and observe that $|s_i| \leq \max_{s\in S} |s|$. The claim follows then from $1 \leq \left|\det \boldsymbol{A}_{(n-2)}\right|$. □

We immediately obtain that every forbidden minor for totally $\pm\{0,1,\dots,\Delta\}$-modular matrices over $\mathbb{Z}$ admits a determinant bounded by $2\Delta^2$ in absolute value. The bound in Lemma 2 is tight: Let $S \subseteq \mathbb{Z}[x]$ be finite such that $s \in S$ implies $-s \in S$ and $\{0\} \neq S$. Select $\tau = \max_{s\in S} |s|$. Then

$$\begin{pmatrix} \tau & \tau \\ -\tau & \tau \end{pmatrix}$$

is a forbidden minor for S and has determinant $2\tau^2$. One can strengthen the bound in Lemma 2 if the dimension is sufficiently large and $S \subseteq \mathbb{Z}$.

Theorem 4. *Let $S \subseteq \mathbb{Z}$ be finite. Let $\tau = \max_{s \in S} |s|$. Given a forbidden minor $\boldsymbol{M} \in \mathbb{Z}^{n \times n}$ for S of dimension $n \geq \lceil \log_2 \tau + 1 \rceil$, then*

$$|\det \boldsymbol{M}| \leq 2 \cdot \lceil \log_2 \tau + 1 \rceil \cdot \tau.$$

Proof. Suppose that the matrix $\boldsymbol{M}$ is invertible, otherwise we are done. Let $\Delta_k = \max \left\{ \left| \det \boldsymbol{A}_{(k)} \right| : \boldsymbol{A}_{(k)} \subseteq \boldsymbol{M} \right\}$ for all $k = 0, \dots, n$ and $\kappa = \lceil \log_2 \tau \rceil$. If $\kappa = 0$, we have $\tau = 1$ and, thus, $\Delta_{n-1}/\Delta_{n-2} \leq 1$ as $\Delta_{n-2} \geq 1$ and $\Delta_{n-1} \leq \tau = 1$. So the Desnanont-Jacobi identity (2) applied to some invertible $(n-2) \times (n-2)$ submatrix implies

$$|\det \boldsymbol{M}| \leq 2 \cdot \Delta_{n-1}/\Delta_{n-2} \cdot \Delta_{n-1} \leq 2 \cdot \Delta_{n-1} \leq 2 \cdot \tau.$$

For the remainder of the proof, we suppose that $1 \leq \kappa$. Observe that $\kappa + 1 \leq n$. Assume that $\Delta_{n-j}/\Delta_{n-j-1} > 2$ for all $j = 1, \dots, \kappa$. This yields

$$\prod_{j=1}^{\kappa} \Delta_{n-j}/\Delta_{n-j-1} > 2^{\kappa} \geq \tau.$$

However, we also have

$$\prod_{j=1}^{\kappa} \Delta_{n-j}/\Delta_{n-j-1} = \Delta_{n-1}/\Delta_{n-\kappa-1} \leq \tau$$

as $\Delta_{n-\kappa-1} \geq 1$ and $\Delta_{n-1} \leq \tau$, which is a contradiction. So we know there exists an index $l^* \in [\kappa]$ such that $\Delta_{n-l^*}/\Delta_{n-l^*-1} \leq 2$, where we have $n - l^* - 1 \geq 0$ by construction. Let $\boldsymbol{A} := \boldsymbol{A}_{(n-l^*-1)} \subseteq \boldsymbol{M}$ attain Δ_{n-l^*-1}. Then applying Sylvester's determinant identity, Lemma 1, to $\boldsymbol{A}$ gives us

$$\det \boldsymbol{M} \cdot (\det \boldsymbol{A})^{l^*} = \det \underbrace{\begin{pmatrix} \det \boldsymbol{A}[i_1, j_1] & \dots & \det \boldsymbol{A}[i_1, j_{l^*+1}] \\ \vdots & \ddots & \vdots \\ \det \boldsymbol{A}[i_{l^*+1}, j_1] & \dots & \det \boldsymbol{A}[i_{l^*+1}, j_{l^*+1}] \end{pmatrix}}_{=\boldsymbol{D}}$$

for suitable sets I and J. Dividing by $(\det \boldsymbol{A})^{l^*}$ and applying Laplace expansion to the first row on the right hand side yields

$$\det \boldsymbol{M} = \frac{1}{(\det \boldsymbol{A})^{l^*}} \cdot \sum_{k=1}^{l^*+1} (-1)^k \cdot \det \boldsymbol{A}[i_1, j_k] \cdot \det \boldsymbol{D}_{\setminus 1, \setminus k}.$$

Observe that $\det \boldsymbol{D}_{\setminus 1, \setminus k} = \det \boldsymbol{M}_{\setminus i_1, \setminus j_k} \cdot (\det \boldsymbol{A})^{l^*-1}$ by Lemma 1. Hence,

$$\det \boldsymbol{M} = \frac{1}{\det \boldsymbol{A}} \cdot \sum_{k=1}^{l^*+1} (-1)^k \cdot \det \boldsymbol{A}[i_1, j_k] \cdot \det \boldsymbol{M}_{\setminus i_1, \setminus j_k}.$$

Taking absolute values and using that $|\det \boldsymbol{A}| = \Delta_{n-l^*-1}$ gives

$$|\det \boldsymbol{M}| \leq (l^* + 1) \cdot \Delta_{n-l^*}/\Delta_{n-l^*-1} \cdot \Delta_{n-1} \leq (\kappa + 1) \cdot 2 \cdot \tau.$$

The claim follows from $\kappa = \lceil \log_2 \tau \rceil$. □

Let $S \subseteq \mathbb{Z}[x]$ be finite. If $\boldsymbol{M}$ is totally S-modular over $\mathbb{Z}[x]$, then $\boldsymbol{M}(a)$ is totally $S(a)$-modular for every $a \in \mathbb{Z}$. This raises the question of whether totally $S(a)$-modular matrices over $\mathbb{Z}$ for a fixed value $a \in \mathbb{Z}$ are also totally S-modular over $\mathbb{Z}[x]$, i.e., totally $S(a)$-modular for all $a \in \mathbb{Z}$. This is in general not true: Let $S = \pm\{0, x, x+1, 2x+1\} \subseteq \mathbb{Z}[x]$ and $a = 1$. We define the matrix

$$\boldsymbol{M} = \begin{pmatrix} x & x+1 & x & x \\ x+1 & x+1 & x+1 & x \\ x & x+1 & x+1 & x+1 \\ x & x & x+1 & x \end{pmatrix} \in \mathbb{Z}[x]^{4\times 4}.$$

One can check that $\boldsymbol{M}(1)$ is totally $S(1)$-modular. However, the matrix $\boldsymbol{M}$ satisfies $\det \boldsymbol{M} = 1 \notin S$. Nevertheless, if we evaluate at some $a \in \mathbb{Z}\backslash\{-2,-1,0,1\}$, we avoid this issue for this particular matrix since $\det \boldsymbol{M}(a) = 1 \notin S(a)$. One of our main results in this section states that this is a general phenomenon. Let

$$I(S) = \{a \in \mathbb{Z} : s(a) = f(a) \text{ for some } s \in S \text{ and } f \in F(S)\}$$

denote the set of integer valued intersections between the polynomial functions given by the elements in S and $F(S)$, the set of all polynomials that arise as a determinant of some forbidden minor for S.

Lemma 3. *Let $S \subseteq \mathbb{Z}[x]$ be finite and $a \in \mathbb{Z}\backslash I(S)$. Then $\boldsymbol{M}(a)$ is totally $S(a)$-modular if and only if $\boldsymbol{M}$ is totally S-modular over $\mathbb{Z}[x]$. Also, the matrix $\boldsymbol{M}(a)$ is a forbidden minor for $S(a)$ if and only if $\boldsymbol{M}$ is a forbidden minor for S. Furthermore, the set $I(S)$ is finite.*

Proof. Note that the statement about the forbidden minors follows directly from the first statement about totally S-modular matrices. Therefore, we only need to prove the first part of the statement. Recall that every totally S-modular matrix over $\mathbb{Z}[x]$ is totally $S(a)$-modular for every evaluation at $a \in \mathbb{Z}$. So it suffices to show the other direction.

A totally $S(a)$-modular matrix that is not the evaluation of a totally S-modular matrix over $\mathbb{Z}[x]$ contains a forbidden minor $\boldsymbol{M}$ for S by definition. This forbidden minor $\boldsymbol{M}$ satisfies $\det \boldsymbol{M}(a) \in S(a)$ and $\det \boldsymbol{M} \notin S$. In other words, let $f \in F(S)$ be the polynomial corresponding to $\det \boldsymbol{M}$, then $f \notin S$ but $f(a) = s(a)$ for some $s \in S$. Thus, we obtain $a \in I(S)$, a contradiction. So $\boldsymbol{M}$ does not contain a forbidden minor for S and is therefore totally S-modular. Since S and $F(S)$ are finite, see Lemma 2, we get that there are only finitely many of those intersections. □

We aim to make statements about matrices with entries being polynomials of degree at most one, that is, matrices given by $\boldsymbol{M} = \boldsymbol{T} + x \cdot \boldsymbol{R} \in \mathbb{Z}[x]^{n\times n}$ for integral $\boldsymbol{T}$ and $\boldsymbol{R}$. The following result relates the rank of $\boldsymbol{R}$ to the largest degree among the polynomials in a given set S. We write $\deg(s)$ for the degree of $s \in \mathbb{Z}[x]$.

Lemma 4. *Let $S \subseteq \mathbb{Z}[x]$ be finite. Let $\boldsymbol{M} = \boldsymbol{T} + x \cdot \boldsymbol{R} \in \mathbb{Z}[x]^{m\times n}$ be totally S-modular for integral $\boldsymbol{T}$ and $\boldsymbol{R}$. Then we have* $\operatorname{rank} \boldsymbol{R} \leq \max_{s\in S} \deg(s)$.

Proof. Let $r = \operatorname{rank} \boldsymbol{R}$ and $\boldsymbol{R}' \in \mathbb{Z}^{r\times r}$ be an invertible submatrix of $\boldsymbol{R}$ and $\boldsymbol{M}' = \boldsymbol{T}' + x \cdot \boldsymbol{R}'$ the corresponding submatrix of $\boldsymbol{M}$. We get

$$\det \boldsymbol{M}' = \det (\boldsymbol{T}' + x \cdot \boldsymbol{R}') = \det \boldsymbol{R}' \cdot \det \left(\boldsymbol{R}'^{-1}\boldsymbol{T}' + x \cdot \boldsymbol{I}\right),$$

where $\boldsymbol{I}$ denotes the unit matrix. Using the Leibniz formula for the determinant $\det(\boldsymbol{R}'^{-1}\boldsymbol{T}' + x \cdot \boldsymbol{I})$, we observe that there exists only one term with degree r, namely the term corresponding to the identity permutation, and no term with degree larger than r. Thus, the right hand side in the equation above is a polynomial of degree r. So $\det \boldsymbol{M}'$ is also a polynomial of degree r. Since the matrix $\boldsymbol{M}'$ is totally S-modular, we have that $r = \deg(\det \boldsymbol{M}') \leq \max_{s\in S} \deg(s)$. $\square$

So, if $\max_{s\in S} \deg(s) = 1$ and $\boldsymbol{R}$ is non-zero, then Lemma 4 implies that $\operatorname{rank} \boldsymbol{R} = 1$. Hence, we can express $\boldsymbol{R}$ as a rank-1 update and obtain $\boldsymbol{M} = \boldsymbol{T} + x \cdot \boldsymbol{u} \cdot \boldsymbol{v}^\top$ for some integral $\boldsymbol{u}$ and $\boldsymbol{v}$. Given such a matrix $\boldsymbol{M} = \boldsymbol{T} + x \cdot \boldsymbol{u} \cdot \boldsymbol{v}^\top$ without any restriction on its subdeterminants, it is possible to establish that subdeterminants of $\boldsymbol{M}$ are indeed polynomials of degree at most one. This is a consequence of the well-known matrix determinant lemma for rank-1 updates. A variant of this lemma is stated below.

Lemma 5 (Matrix determinant lemma). *Let $\boldsymbol{M} = \boldsymbol{T} + x \cdot \boldsymbol{u} \cdot \boldsymbol{v}^\top \in \mathbb{Z}[x]^{n\times n}$ for integral $\boldsymbol{T}$, $\boldsymbol{u}$, and $\boldsymbol{v}$. Then we have*

$$\det \boldsymbol{M} = \left(\det \boldsymbol{T} - \det \left(\boldsymbol{T} - \boldsymbol{u} \cdot \boldsymbol{v}^\top\right)\right) \cdot x + \det \boldsymbol{T}.$$

This is the essential ingredient to prove Theorem 1.

Proof (of Theorem 1). Let $\boldsymbol{M} = \boldsymbol{T} + x \cdot \boldsymbol{u} \cdot \boldsymbol{v}^\top$ be totally S-modular. Then Lemma 5 implies that $\det \boldsymbol{M} = \lambda_1 x + \lambda_0$ for $\lambda_0, \lambda_1 \in \mathbb{Z}$ and $\det \boldsymbol{T} = \lambda_0$. This holds for all such totally S-modular matrices. So the claim follows. $\square$

We will also use an implication of Theorem 1 for matrices over $\mathbb{Z}$.

Corollary 1. *Let $S \subseteq \mathbb{Z}[x]$ be finite and $a \in \mathbb{Z}\backslash I(S)$. If $\boldsymbol{M}(a) = \boldsymbol{T} + a \cdot \boldsymbol{u} \cdot \boldsymbol{v}^\top$ with integral $\boldsymbol{T}$, $\boldsymbol{u}$, and $\boldsymbol{v}$ is totally $S(a)$-modular, then the matrix $\boldsymbol{T}$ is totally $S(0)$-modular.*

Proof. Since $a \in \mathbb{Z}\backslash I(S)$, we know that every totally $S(a)$-modular matrix $\boldsymbol{M}(a) = \boldsymbol{T} + a \cdot \boldsymbol{u} \cdot \boldsymbol{v}^\top$ corresponds to a totally S-modular matrix $\boldsymbol{M} = \boldsymbol{T} + x \cdot \boldsymbol{u} \cdot \boldsymbol{v}^\top$ over $\mathbb{Z}[x]$ by Lemma 3. From Theorem 1 it follows that $\boldsymbol{T}$ is totally $S(0)$-modular. $\square$

3 Properties of $I(S)$

Following Lemma 3, we demonstrate how to calculate all intersections between the polynomial functions in a given set S and the polynomial functions in $F(S)$. We start by proving a lemma which holds under more general assumptions. For that purpose, we remark that the ring $\mathbb{Z}[x]$ is a unique factorization domain.

So every element in $\mathbb{Z}[x]$ admits a unique factorization into smaller irreducible elements up to multiplying with ± 1. Therefore, the notion of greatest common divisors carries naturally over from $\mathbb{Z}$ to $\mathbb{Z}[x]$. Analogously to $\mathbb{Z}$, elements in $\mathbb{Z}[x]$ are relatively prime if their greatest common divisor is 1. We call a matrix $\boldsymbol{M} \in \mathbb{Z}[x]^{m\times n}$ *totally unimodular* if every subdeterminant is contained in $\pm\{0,1\}$.

Lemma 6. *Let $S \subseteq \mathbb{Z}[x]$ be such that all non-zero $y, z \in S$ are relatively prime or $|y| = |z|$ and $2 \notin S$. Let $n \geq 3$ and $\boldsymbol{M} \in \mathbb{Z}[x]^{n\times n}$ be invertible and every $(n-1)\times(n-1)$ submatrix of $\boldsymbol{M}$ is totally S-modular. Then either $\boldsymbol{M}$ is totally unimodular or there exist invertible submatrices $\boldsymbol{A}_{(n-2)}, \tilde{\boldsymbol{A}}_{(n-2)} \subseteq \boldsymbol{M}$ such that*

$$\left|\det \boldsymbol{A}_{(n-2)}\right| \neq \left|\det \tilde{\boldsymbol{A}}_{(n-2)}\right|.$$

Proof. Let $n = 3$. The $(n-2)\times(n-2)$ submatrices of $\boldsymbol{M}$ are the entries of $\boldsymbol{M}$. Assume that every entry of $\boldsymbol{M}$ is in $\pm\{0,s\}$ for some $s \in S$. If $s = 1$, we get that $\boldsymbol{M} \in \pm\{0,1\}^{3\times 3}$. Since $2 \notin S$, the matrix $\boldsymbol{M}$ is totally unimodular, cf. [26, Theorem 19.3]. If we assume $s \neq 1$, we have $\det \boldsymbol{A}_{(2)} \in \pm\{s^2, 2s^2\}$ for an invertible submatrix $\boldsymbol{A}_{(2)} \subseteq \boldsymbol{M}$ as every entry is contained in $\pm\{0,s\}$. Since the non-zero elements in S are pairwise relatively prime and $s \in S$, we get $\det \boldsymbol{A}_{(2)} \notin S$, contradicting that $\boldsymbol{A}_{(2)}$ is totally S-modular.

So we suppose that $n \geq 4$. Let $\boldsymbol{M}$ not be totally unimodular. Again, we assume that every invertible $(n-2)\times(n-2)$ submatrix has determinant s or $-s$ for some $s \in S$. By the induction hypothesis applied to some invertible submatrix $\boldsymbol{A}_{(n-1)} \subseteq \boldsymbol{M}$, there exists an invertible submatrix $\boldsymbol{A}_{(n-3)} \subseteq \boldsymbol{A}_{(n-1)}$ with $\left|\det \boldsymbol{A}_{(n-3)}\right| \neq |s|$. We apply the Desnanont-Jacobi identity, (2), to $\boldsymbol{A}_{(n-3)} \subseteq \boldsymbol{A}_{(n-1)}$ and obtain that

$$\det \boldsymbol{A}_{(n-1)} = \frac{1}{\det \boldsymbol{A}_{(n-3)}}\left(s_1 s_2 - s_3 s_4\right) \in \pm\frac{1}{\det \boldsymbol{A}_{(n-3)}} \cdot \{s^2, 2\,s^2\}$$

with $s_i \in \pm\{0,s\}$ for $i = 1,2,3,4$. If $\left|\det \boldsymbol{A}_{(n-3)}\right| \neq 1$, then $\det \boldsymbol{A}_{(n-3)}$ does not divide s^2 or $2s^2$ as the non-zero elements in S are pairwise relatively prime and $2 \notin S$. However, this implies that $\det \boldsymbol{A}_{(n-1)} \notin \mathbb{Z}[x]$, a contradiction. If $\left|\det \boldsymbol{A}_{(n-3)}\right| = 1$, we get that $\det \boldsymbol{A}_{(n-1)} \in \pm\{s^2, 2s^2\}$. This gives us a contradiction since $\det \boldsymbol{A}_{(n-1)}$ and s are not relatively prime. □

We are in the position to determine the set of possible determinants given by the forbidden minors for a specific set S. We showcase this for $S = \pm\{0,1,x,x+1,2x+1\}$.

Lemma 7. *Let $S = \pm\{0,1,x,x+1,2x+1\}$ and D be the set of all determinants attained by a 2×2 forbidden minor for S. Then we have*

$$F(S) \subseteq \pm\{2, x-1, x+2, 2x, 2x+2, 3x+1, 3x+2, 4x+2\} \cup D.$$

Proof. Let $n \geq 3$ and $\boldsymbol{M}$ be a forbidden minor. There exists an invertible submatrix $\boldsymbol{A}_{(n-2)} \subseteq \boldsymbol{M}$. Applying the Desnanont-Jacobi identity (2), we get

$$\det \boldsymbol{M} \cdot \det \boldsymbol{A}_{(n-2)} = s_1 s_2 - s_3 s_4$$

with $s_i \in S$ for $i = 1, 2, 3, 4$. This equation needs to have a solution for $\det \boldsymbol{M}$ in $\mathbb{Z}[x]$. Suppose that $\det \boldsymbol{M}$ has degree larger than one. This implies $|\det \boldsymbol{A}_{(n-2)}| = 1$, an equality that has to hold for every invertible $(n-2)\times(n-2)$ subdeterminant. By Lemma 6, we deduce that $\boldsymbol{M}$ is totally unimodular and, thus, not a forbidden minor. So we assume $\det \boldsymbol{M}$ has degree at most one. Using suitable software such as SageMath, one can enumerate all feasible solutions to the Desnanont-Jacobi identity with degree at most one. This gives us the following feasible values for $\det \boldsymbol{M}$ up to a sign:

$$2S\backslash 0 \cup \{x-1, x+2, 2x-1, 2x+3, 3x+1, 3x+2, 4x, 4x+1, 4x+3, 4x+4\}.$$

However, it is possible to show that the values $\pm\{2x-1, 2x+3, 4x, 4x+1, 4x+3, 4x+4\}$ for $\det \boldsymbol{M}$ can only appear for a unique choice of $|\det \boldsymbol{A}_{(n-2)}|$. Therefore, if there exists a forbidden minor with such a determinant, every invertible $(n-2)\times(n-2)$ submatrix needs to have the same determinant in absolute value. This contradicts Lemma 6. Hence, we can exclude these values and obtain the determinants from the statement. □

Computing the intersections of the elements in $S = \pm\{0, 1, x, x+1, 2x+1\}$ and the set in Lemma 7 yields that $I(S) \subseteq \{-3, \ldots, 2\}$ in Corollary 1. We remark that it is possible to show that both sets are indeed equal.

Remark 1. For the set $S = \pm\{0, x, x+1, 2x+1\}$, one can show analogously to the proof of Lemma 7 that $F(S) \subseteq \pm\{1, 2x, 2x+2, 3x+1, 3x+2, 4x+2\} \cup D$ and $I(S) = \{-2, -1, 0, 1\}$, where D is the set of all determinants attained by a 2×2 forbidden minor for S.

4 Proofs of Theorems 2 and 3

For both cases, $S = \pm\{0, 1, x, x+1, 2x+1\}$ and $S = \pm\{0, x, x+1, 2x+1\}$, we recover that $S(a) = \pm\{0, 1\}$ if $a \in \{-1, 0\}$. Since one can optimize $\text{ILP}(\boldsymbol{M}(a), \boldsymbol{b}, \boldsymbol{c})$ in polynomial time for totally unimodular constraint matrices, see, for instance, [26, Chapter 19], and recognize totally unimodular matrices in polynomial time [26, Chapter 20], both theorems hold in this case. In the following, we prove Theorems 2 and 3 for the other values of a.

4.1 Proof of Theorem 2

Recall that $S = \pm\{0, 1, x, x+1, 2x+1\}$. By Lemma 4, every totally S-modular matrix $\boldsymbol{M}$ admits a decomposition into $\boldsymbol{M} = \boldsymbol{T} + x \cdot \boldsymbol{u} \cdot \boldsymbol{v}^\top$ for some suitable integral valued $\boldsymbol{T}$, $\boldsymbol{u}$, and $\boldsymbol{v}$. One can test in polynomial time whether such a decomposition exists. To finish the proof of Theorem 2, we use the complete characterization given below and the fact that one can recognize totally unimodular matrices in polynomial time; see [26, Chapter 20].

Lemma 8. *Let $\boldsymbol{T}$, $\boldsymbol{u}$, and $\boldsymbol{v}$ be integral. The matrix $\boldsymbol{M} = \boldsymbol{T} + x \cdot \boldsymbol{u} \cdot \boldsymbol{v}^\top$ is totally S-modular if and only if $\boldsymbol{T}$ and $\boldsymbol{T} - \boldsymbol{u} \cdot \boldsymbol{v}^\top$ are totally unimodular.*

Proof. We abbreviate $\bar{\boldsymbol{T}} = \boldsymbol{T} - \boldsymbol{u} \cdot \boldsymbol{v}^\top$. It suffices to show the statement for square matrices. So we assume without loss of generality that $m = n$. Let $\boldsymbol{T}$ and $\bar{\boldsymbol{T}}$ be totally unimodular. We have

$$\det \boldsymbol{M} = \left(\det \boldsymbol{T} - \det \bar{\boldsymbol{T}}\right) \cdot x + \det \boldsymbol{T} \tag{3}$$

by Lemma 5. So $\det \boldsymbol{M}$ is completely determined by $\det \boldsymbol{T}$ and $\det \bar{\boldsymbol{T}}$. As $\det \boldsymbol{T}$ and $\det \bar{\boldsymbol{T}}$ are both in $\pm\{0,1\}$, we obtain that $\det \boldsymbol{M} \in \pm\{0,1,x,x+1,2x+1\} = S$ by going through all feasible cases. The other direction follows directly by evaluating S, $\boldsymbol{M}$, and (3) at $x \in \{-1,0\}$. □

By Lemma 3 and $I(S) \subseteq \{-3,\dots,2\}$ from the previous section, we can now derive Theorem 2 by replacing S with $S(a)$ for $a \in \mathbb{Z}\backslash\{-3,-2,-1,0,1,2\}$.

4.2 Proof of Theorem 3

Following Remark 1, we fix some $a \in \mathbb{Z}\backslash\{-2,-1,0,1\}$. We use the decomposition $\boldsymbol{M}(a) = \boldsymbol{T} + a \cdot \boldsymbol{u} \cdot \boldsymbol{v}^\top$ for integral $\boldsymbol{u}, \boldsymbol{v}$ from Lemma 4 where $\boldsymbol{T}$ is totally unimodular by Corollary 1. By multiplying rows and columns with -1, we assume without loss of generality that $\boldsymbol{u}$ and $\boldsymbol{v}$ are non-negative. Observe that this already implies that the entries of $\boldsymbol{T}$ are in $\{0,1\}$ since the entries of $\boldsymbol{M}(a)$ are in $S(a)$. Let without loss of generality $\boldsymbol{M}(a)$ have no zero column or row. This implies that no entry of $\boldsymbol{u}$ and $\boldsymbol{v}$ equals zero. If there exists an entry of $\boldsymbol{M}(a)$ which is $2a+1$, then due to the rank-1 update the whole row and/or column containing this entry have entries $2a+1$. If there exists a row of $\boldsymbol{M}(a)$ whose entries are all $2a+1$, we can divide this row by $a/(2a+1)$ and round down the right hand side. So we can always assume that $\boldsymbol{u} = \boldsymbol{1}$. If there are multiple columns whose entries are all $2a+1$, we aggregate them. Hence, we assume that there is at most one column with entries $2a+1$. So we can select $\boldsymbol{v} \in \{1,2\}^n$ where at most one entry of $\boldsymbol{v}$ is 2. We suppose without loss of generality that $\boldsymbol{v}_1 \in \{1,2\}$ and $\boldsymbol{v}_{\backslash 1} = \boldsymbol{1}$. Let the first column of $\boldsymbol{M}(a)$ be $\boldsymbol{M}(a)_{\cdot,1}$. Notice that $\boldsymbol{M}(a)_{\cdot,\backslash 1}$ is a matrix with entries in $\{a, a+1\}$. We set $y = \boldsymbol{1}^\top \tilde{\boldsymbol{x}}$ for $\boldsymbol{x} = (\boldsymbol{x}_1, \tilde{\boldsymbol{x}})^\top$ to reformulate $\mathrm{ILP}(\boldsymbol{M}(a), \boldsymbol{b}, \boldsymbol{c})$ in the form of the mixed integer linear program

$$\max \boldsymbol{c}^\top \boldsymbol{x} \;\text{ s.t. }\; \boldsymbol{T}_{\backslash 1,\cdot} \tilde{\boldsymbol{x}} \leq \boldsymbol{b} - \boldsymbol{x}_1 \cdot \boldsymbol{M}(a)_{\cdot,1} - a \cdot \boldsymbol{u} \cdot y, \;\; \boldsymbol{1}^\top \tilde{\boldsymbol{x}} = y, \;\; \boldsymbol{x}_1, y \in \mathbb{Z}.$$

We next show that the constraint matrix, the transpose of $[\boldsymbol{T}_{\backslash 1,\cdot}|\boldsymbol{1}]$, is totally unimodular. If this is true, then a solution of the mixed integer linear program in two integer variables can be obtained in polynomial time using Lenstra's algorithm [18] or its improved successors [9,16,24]. It corresponds to a solution of $\mathrm{ILP}(\boldsymbol{M}(a), \boldsymbol{b}, \boldsymbol{c})$.

Consider $[\boldsymbol{T}_{\backslash 1,\cdot}|\boldsymbol{1}]$. The matrix $\boldsymbol{T}_{\backslash 1,\cdot}$ is totally unimodular since its a submatrix of $\boldsymbol{T}$. Let $\boldsymbol{T}'$ be an invertible submatrix of the constraint matrix that contains the row $\boldsymbol{1}$. By adding $a \cdot \boldsymbol{1}$ to every row of $\boldsymbol{T}'$ that is not the all-ones row, we obtain a new matrix $\boldsymbol{M}(a)'$ that contains a submatrix of $\boldsymbol{M}(a)_{\cdot,\backslash 1}$, whose entries are in $\{a, a+1\}$, and one all-ones row. Since each submatrix of $\boldsymbol{M}(a)$ is totally $S(a)$-modular, we obtain $\pm 1 = \det \boldsymbol{M}(a)' = \det \boldsymbol{T}'$ by Theorem 5 below. So $[\boldsymbol{T}_{\backslash 1,\cdot}|\boldsymbol{1}]$ is totally unimodular.

Theorem 5. *Let* $S = \pm\{0, x, x+1, 2x+1\} \subseteq \mathbb{Z}[x]$ *and* $n \geq 2$. *Let* $\boldsymbol{M} \in \{x, x+1\}^{n\times(n-1)}$ *be totally* S*-modular. Then* $\det[\boldsymbol{M}|\mathbf{1}] \in \pm\{0,1\}$.

Proof. For the purpose of deriving a contradiction, assume that the statement does not hold. We multiply the all-ones column with $x+1$. So we deduce that $\det[\boldsymbol{M}|(x+1)\cdot\mathbf{1}] \notin \pm\{0, x+1\}$ by assumption. Note that $\det[\boldsymbol{M}|(x+1)\cdot\mathbf{1}]$ is divisible by $x+1$. So we obtain that $\det[\boldsymbol{M}|(x+1)\cdot\mathbf{1}] \notin S$. Thus, the matrix $[\boldsymbol{M}|(x+1)\cdot\mathbf{1}]$ contains a forbidden minor for S. As $\boldsymbol{M}$ is totally S-modular, the forbidden minor contains the all-$(x+1)$'s column. For the remainder of the proof, we show that no forbidden minor with entries in $\{x, x+1\}$ contains an all-$(x+1)$ column or row, which finishes the proof of the theorem.

One can verify that for $n=2$ no forbidden minor with entries in $\{x, x+1\}$ exists. So we assume that $n \geq 3$. Suppose without loss of generality that the last row of $\boldsymbol{M}$, which we denote by $\boldsymbol{M}_{n,\cdot}$, has greatest common divisor larger than 1. We derive a contradiction.

We observe that $\gcd \boldsymbol{M}_{n,\cdot} \in \{x, x+1\}$. By Remark 1, we have $|\det \boldsymbol{M}| = 2\cdot\gcd \boldsymbol{M}_{n,\cdot}$ as $\det \boldsymbol{M}$ has to be divisible by $\gcd \boldsymbol{M}_{n,\cdot}$. Let $\boldsymbol{A}_{(n-1)} \subseteq \boldsymbol{M}$ be an invertible submatrix not containing the last row. Take an invertible submatrix $\boldsymbol{A}_{(n-2)} \subseteq \boldsymbol{A}_{(n-1)}$. The Desnanont-Jacobi identity, (2), applied to $\boldsymbol{A}_{(n-2)}$ yields

$$2\gcd \boldsymbol{M}_{n,\cdot} \det \boldsymbol{A}_{(n-2)} = \det \boldsymbol{M} \det \boldsymbol{A}_{(n-2)} = s_1 s_4 - s_2 s_3,$$

where $s_1, s_2 \in S$ and $s_3, s_4 \in \pm\{0, \gcd \boldsymbol{M}_{n,\cdot}\}$. This equation cannot be satisfied if $s_i = 0$ for some $i = 1,2,3,4$. So we assume that $s_i \neq 0$ for all $i = 1,2,3,4$. By division with $\gcd \boldsymbol{M}_{n,\cdot}$ and assuming that without loss of generality $s_3 = \gcd \boldsymbol{M}_{n,\cdot} = s_4$, we get $2\cdot\det \boldsymbol{A}_{(n-2)} = s_1 - s_2$ for $s_1, s_2 \in S\backslash 0$. The only solutions are given when $|\det \boldsymbol{A}_{(n-2)}| = s_1 = -s_2$ or $|\det \boldsymbol{A}_{(n-2)}| = -s_1 = s_2$. Hence, every invertible $(n-2)\times(n-2)$ submatrix of $\boldsymbol{A}_{(n-1)}$ has the same determinant as $\boldsymbol{A}_{(n-1)}$ in absolute value. If $n=3$, this implies that the $(n-2)\times(n-2)$ subdeterminants correspond to entries. As the entries of $\boldsymbol{M}$ are in $\{x, x+1\}$, this means that all the entries in $\boldsymbol{A}_{(n-1)}$ are the same which gives us $\det \boldsymbol{A}_{(n-1)} = 0$, a contradiction. So let $n \geq 4$. Take an invertible submatrix $\boldsymbol{A}_{(n-3)} \subseteq \boldsymbol{A}_{(n-1)}$. We apply the Desnanont-Jacobi identity, (2), to $\boldsymbol{A}_{(n-3)} \subseteq \boldsymbol{A}_{(n-1)}$ and get $\det \boldsymbol{A}_{(n-3)} \in \pm\{\det \boldsymbol{A}_{(n-1)}, 2\det \boldsymbol{A}_{(n-1)}\}$ which yields $|\det \boldsymbol{A}_{(n-3)}| = |\det \boldsymbol{A}_{(n-1)}|$. This holds for all invertible $(n-3)\times(n-3)$ submatrices of $\boldsymbol{A}_{(n-1)}$. By Lemma 6, this implies that $\boldsymbol{A}_{(n-1)}$ is totally unimodular. However, this contradicts that $\det \boldsymbol{A}_{(n-1)} \in S$. □

5 Finiteness of Forbidden Minors

Recall from the introduction that totally $\pm\{0, x, x+1\}$-modular matrices with entries in $\{x, x+1\}$ are completely characterized by excluding the submatrix (1). In light of the celebrated Robertson-Seymour theorem [25], one might ask whether there is always a finite list of forbidden minors for a finite set $S \subseteq \mathbb{Z}[x]$ and matrices with entries in $\{x, x+1\}$. This is not the case. An intriguing example with infinitely many forbidden minors is already given by the set $S =$

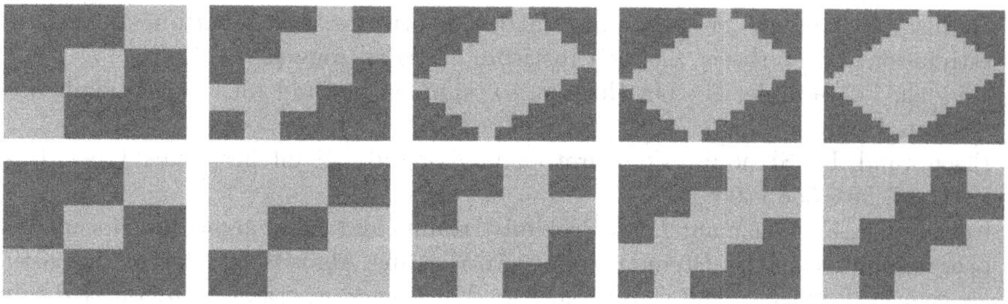

Fig. 1. The blue boxes depict the value x and the green boxes $x+1$ or vice versa. The first row of matrices corresponds to the first five elements of an infinite sequence of matrices that can be obtained by generalizing the existing pattern. It can be shown that those infinitely many matrices are forbidden minors for $\pm\{0,1,x,x+1,2x+1\}$. The matrices in the last row correspond to five forbidden minors for $\pm\{0,x,x+1,2x+1\}$. (Color figure online)

$\pm\{0,1,x,x+1,2x+1\}$; see Fig. 1 for an incomplete list. Interestingly, if one removes the 1 from S and passes to $\pm\{0,x,x+1,2x+1\}$, then we are only aware of finitely many forbidden minors; see Fig. 1. It is open whether this list is complete. If this is true, this might support the following conjecture for a finite set $S \subseteq \mathbb{Z}[x]$ such that $s \in S$ implies $-s \in S$: there exists a finite list of forbidden minors if and only if $1 \notin S$.

Acknowledgements. The authors are grateful to the reviewers for their valuable suggestions and comments.

References

1. Aliev, I., Averkov, G., de Loera, J.A., Oertel, T.: Sparse representation of vectors in lattices and semigroups. Math. Program. **192**, 519–546 (2022)
2. Aliev, I., De Loera, J.A., Eisenbrand, F., Oertel, T., Weismantel, R.: The support of integer optimal solutions. SIAM J. Optim. **28**(3), 2152–2157 (2018)
3. Aliev, I., Henk, M., Oertel, T.: Distances to lattice points in knapsack polyhedra. Math. Program. **182**, 175–198 (2020)
4. Aliev, I., De Loera, J.A., Oertel, T., O'Neill, C.: Sparse solutions of linear Diophantine equations. SIAM J. Appl. Algebra Geom. **1**, 239–253 (2017)
5. Artmann, S., Eisenbrand, F., Glanzer, C., Oertel, T., Vempala, S., Weismantel, R.: A note on non-degenerate integer programs with small sub-determinants. Oper. Res. Lett. **44**(5), 635–639 (2016)
6. Artmann, S., Weismantel, R., Zenklusen, R.: A strongly polynomial algorithm for bimodular integer linear programming. In: Proceedings of the 49th Annual ACM SIGACT Symposium on Theory of Computing, pp. 1206–1219 (2017)
7. Bonifas, N., Di Summa, M., Eisenbrand, F., Hähnle, N., Niemeier, M.: On subdeterminants and the diameter of polyhedra. Discret. Comput. Geom. **52**, 102–115 (2014)
8. Celaya, M., Kuhlmann, S., Paat, J., Weismantel, R.: Proximity and flatness bounds for linear integer optimization. Math. Oper. Res. (2023)

9. Dadush, D.: Integer programming, lattice algorithms, and deterministic volume estimation. Ph.D. thesis, Georgia Institute of Technology (2012)
10. Dadush, D., Hähnle, N.: On the shadow simplex method for curved polyhedra. Discrete Comput. Geom. **56**(4), 882–909 (2016)
11. Eisenbrand, F., Shmonin, G.: Carathéodory bounds for integer cones. Oper. Res. Lett. **34**, 564–568 (2006)
12. Eisenbrand, F., Weismantel, R.: Proximity results and faster algorithms for integer programming using the Steinitz lemma. ACM Trans. Algorithms **16**(1), 1–14 (2019)
13. Fiorini, S., Joret, G., Weltge, S., Yuditsky, Y.: Integer programs with bounded subdeterminants and two nonzeros per row. In: 2021 IEEE 62nd Annual Symposium on Foundations of Computer Science (FOCS), pp. 13–24 (2022)
14. Glanzer, C., Stallknecht, I., Weismantel, R.: Notes on {a, b, c}-modular matrices. Viet. J. Math. **50**(2), 469–485 (2022)
15. Hoffman, A.J., Kruskal, J.B.: Integral boundary points of convex polyhedra. In: Kuhn, H.W., Tucker, A.J. (eds.) Linear Inequalities and Related Systems, pp. 223–246 (1956)
16. Kannan, R.: Minkowski's convex body theorem and integer programming. Math. Oper. Res. **12**(3), 415–440 (1987)
17. Lang, S.: Algebra. Graduate Texts in Mathematics, Springer, New York (2005). https://doi.org/10.1007/978-1-4613-0041-0
18. Lenstra, H.W.: Integer programming with a fixed number of variables. Math. Oper. Res. **8**(4), 538–548 (1983)
19. Nägele, M., Nöbel, C., Santiago, R., Zenklusen, R.: Advances on strictly Δ-modular IPs. In: Del Pia, A., Kaibel, V. (eds.) IPCO 2023. LNCS, vol. 13904, pp. 393–407. Springer, Cham (2023). https://doi.org/10.1007/978-3-031-32726-1_28
20. Nägele, M., Santiago, R., Zenklusen, R.: Congruency-constrained TU problems beyond the bimodular case. In: Proceedings of the 2022 ACM-SIAM Symposium on Discrete Algorithms, SODA, pp. 2743–2790. SIAM (2022)
21. Narayanan, H., Shah, R., Srivastava, N.: A spectral approach to polytope diameter. arxiv.org/abs/2101.12198 (2021)
22. Oxley, J., Walsh, Z.: 2-modular matrices. SIAM J. Discret. Math. **36**(2), 1231–1248 (2022)
23. Paat, J., Stallknecht, I., Xu, L., Walsh, Z.: On the column number and forbidden submatrices for Δ-modular matrices. SIAM J. Discret. Math. **38**(1), 1–18 (2024)
24. Reis, V., Rothvoss, T.: The subspace flatness conjecture and faster integer programming (2023). https://arxiv.org/abs/2303.14605
25. Robertson, N., Seymour, P.D.: Graph minors. XX. Wagner's conjecture. J. Combin. Theory Ser. B **92**(2), 325–357 (2004). Special Issue Dedicated to Professor W.T. Tutte
26. Schrijver, A.: Theory of Linear and Integer Programming. Wiley, Hoboken (1986)
27. Seymour, P.D.: Decomposition of regular matroids. J. Combin. Theory Ser. B **28**(3), 305–359 (1980)
28. Sylvester, J.J.: On the relation between the minor determinants of linearly equivalent quadratic functions. London Edinburgh Dublin Philos. Mag. J. Sci. **1**(4), 295–305 (1851)

A First Order Method for Linear Programming Parameterized by Circuit Imbalance

Richard Cole[1], Christoph Hertrich[2(✉)], Yixin Tao[3], and László A. Végh[4,5]

[1] Courant Institute, New York University, New York, USA
[2] Université Libre de Bruxelles, Bruxelles, Belgium
christoph.hertrich@ulb.be
[3] Key Laboratory of Interdisciplinary Research of Computation and Economics, Ministry of Education, Shanghai University of Finance and Economics, Shanghai, China
[4] London School of Economics, London, UK
[5] Corvinus Institute for Advanced Studies, Corvinus University, Budapest, Hungary

Abstract. Various first order approaches have been proposed in the literature to solve Linear Programming (LP) problems, recently leading to practically efficient solvers for large-scale LPs. From a theoretical perspective, linear convergence rates have been established for first order LP algorithms, despite the fact that the underlying formulations are not strongly convex. However, the convergence rate typically depends on the Hoffman constant of a large matrix that contains the constraint matrix, as well as the right hand side, cost, and capacity vectors.

We introduce a first order approach for LP optimization with a convergence rate depending polynomially on the circuit imbalance measure, which is a geometric parameter of the constraint matrix, and depending logarithmically on the right hand side, capacity, and cost vectors. This provides much stronger convergence guarantees. For example, if the constraint matrix is totally unimodular, we obtain polynomial-time algorithms, whereas the convergence guarantees for approaches based on primal-dual formulations may have arbitrarily slow convergence rates for this class. Our approach is based on a fast gradient method due to Necoara, Nesterov, and Glineur (Math. Prog. 2019); this algorithm is called repeatedly in a framework that gradually fixes variables to the boundary. This technique is based on a new approximate version of Tardos's method, that was used to obtain a strongly polynomial algorithm for combinatorial LPs (Oper. Res. 1986).

Keywords: Linear Programming · First Order Methods · Hoffman Proximity · Circuit Imbalances

---

The full version is available at arXiv:2311.01959.

J. Vygen and J. Byrka (Eds.): IPCO 2024, LNCS 14679, pp. 57–70, 2024.
https://doi.org/10.1007/978-3-031-59835-7_5

1 Introduction

In this paper, we develop new first order algorithms for approximately solving the linear program

$$\begin{aligned} \min\ & \langle c, x\rangle \\ Ax &= b\,, \\ \mathbf{0} \le x &\le u\,, \end{aligned} \qquad (\mathrm{LP}(A,b,c,u))$$

where $A \in \mathbb{R}^{m\times n}$, $b \in \mathbb{R}^m$, $c, u \in \mathbb{R}^n$. We assume that $m \le n$. We use the notation $[\mathbf{0}, u] = \{x \in \mathbb{R}^n \mid \mathbf{0} \le x \le u\}$, and denote the feasible region as $\mathcal{P}_{A,b,u} := \{x \in \mathbb{R}^n \mid Ax = b\,, x \in [\mathbf{0}, u]\}$.

Linear programming (LP) is one of the most fundamental optimization problems with an immense range of applications in applied mathematics, operations research, computer science, and more. While Dantzig's Simplex method works well in practice, its running time may be exponential in the worst case. Breakthrough results in the 1970s and 1980s led to the development of the first polynomial time algorithms, the ellipsoid method [12] and interior point methods (IPMs) [11]. The Simplex method was one of the earliest computations implemented on a computer, and there are highly efficient LP solvers available, based on the Simplex and interior point methods.

Linear programming can also be seen as a special case of more general optimization models: it can be captured by various convex programs, saddle point problems, and linear complementarity problems. These connections led to the development of new LP algorithms being an important driving force in the development of optimization theory.

In this paper, we focus on first order methods (FOMs) for LP. The benefit of FOMs is cheap iteration complexity and efficient implementability for large-scale problems. In contrast to IPMs, they do not require careful initialization. FOMs are prevalent in optimization and machine learning, but they are not an obvious choice for LP for two reasons. First, the standard formulation has a complicated polyhedral feasible region, and therefore standard techniques are not directly applicable. Second, FOMs usually do not lead to polynomial running time guarantees: this is in contrast with IPMs that are polynomial and also efficient in practice.

Nevertheless, FOMs turn out to be practically efficient for large-scale LPs. In a recent paper Applegate et al. [1] use a restarted primal-dual hybrid gradient (PDHG) method based on a saddle point formulation. Their implementation outperforms the state-of-the art commercial Simplex and IPM solvers on standard benchmark instances, and is able to find high accuracy solutions to large-scale PageRank instances.

The number of iterations needed to find an ε-approximate solution in standard FOMs is typically $O(1/\varepsilon)$ or $O(1/\sqrt{\varepsilon})$. However, strong convexity properties can yield *linear convergence*, i.e., an $O(\log(1/\sqrt{\varepsilon}))$ dependence. No strongly convex formulation is known to capture LP. Despite this, there is a long line of work on FOMs that achieve linear convergence guarantees for LP, starting from

Eckstein and Bertsekas's alternating direction method from 1990 [5], followed by a variety of other techniques, e.g., [1,2,8,9,13,19,20].

Before discussing these approaches, let us specify the notion of approximate solutions. By a *δ-feasible* solution, we mean an $x \in [\mathbf{0}, u]$ with $\|Ax-b\|_1 \leq \delta\|A\|_1$. If LP$(A, b, c, u)$ is feasible, we let $\Phi(A, b, c, u)$ denote the optimum value. A *δ-optimal* solution satisfies $\langle c, x\rangle \leq \Phi(A, b, c, u)+\delta\|c\|_\infty$. Our goal will be to find a δ-feasible and δ-optimal solution for a required accuracy $\delta > 0$. Different papers may use different norms and normalizations in their accuracy requirement, but these can be easily translated to each other.

The above mentioned works are able to find δ-feasible and δ-optimal solutions in running times that depend polynomially on $\log(1/\delta)$, n, and $C(A, b, c, u)$, a constant depending on the problem input. In particular, Applegate, Hinder, Lu, and Lubin [2] give a running time bound $O(C \log(1/\delta))$ for restarted PDHG, where C is the Hoffman-constant associated with the primal-dual embedding of the LP. Recently, for the case when A is a totally unimodular matrix and there are no upper bounds u, Hinder [9] bounded the running time of restarted PDHG as $O(Hn^{2.5}\sqrt{\text{nzz}(A)}\log(Hm/\delta))$, where $\text{nzz}(A)$ is the number of nonzero entries of A, and b, c are integer vectors with $\|b\|_\infty, \|c\|_\infty \leq H$.

However, the constants involved in the running time bound are typically not polynomial in the binary encoding length of the input. In this paper, we give the first FOM-based algorithm with polynomial dependence on $\log(1/\delta)$, n, a constant $\bar{\kappa}(\mathcal{X}_A)$, and $\log\|b\|$, $\log\|c\|$, and $\log\|u\|$, as stated in Theorem 1. The constant $\bar{\kappa}(\mathcal{X}_A)$ is the *max circuit imbalance measure* defined and discussed below. In particular, it is upper bounded by the maximal subdeterminant $\Delta(A)$, but it is often much smaller than $\Delta(A)$. We have $\bar{\kappa}(\mathcal{X}_A) = 1$ if A is a totally unimodular matrix. Note that the running time depends polynomially on the logarithms of the capacity and cost vectors. In contrast, the bound in [9] only applies for the totally unimodular case and the running time is linear in $\|b\|_\infty + \|c\|_\infty$.

We also note that one may always rescale the matrix A to have $\|A\|_1 = 1$; however, such a rescaling may change $\bar{\kappa}(\mathcal{X}_A)$.

Theorem 1. *There is an FOM-based algorithm for LP(A, b, c, u) that obtains a solution x that is δ-feasible and δ-optimal, or concludes that no feasible solution exists, and whose runtime is dominated by the cost of performing* $O\big(n^{1.5}m^2\|A\|_1^2 \cdot \bar{\kappa}^3(\mathcal{X}_A)\log^3\left((\|u\|_1+\|b\|_1)nm \cdot \kappa(\mathcal{X}_A)\|A\|_1/\delta\right)\big)$ *gradient descent updates, where we assume* $\|A\|_1 \geq 1$. *Additionally, our algorithm returns a dual solution certifying approximate optimality of the solution in* $O\big(m\|A\|_2 \cdot \bar{\kappa}(\mathcal{X}_A)\cdot\log(n\|c\|_1/\delta)\big)$ *gradient descent updates.*

All proofs missing from this extended abstract can be found in the full version.

Hoffman Bounds and Quadratic Function Growth. The main underlying tool for proving linear convergence bounds is Hoffman-proximity theory, introduced by Hoffman in 1952 [10]. Let $A \in \mathbb{R}^{m\times n}$, let $\|.\|_\alpha$ be a norm in $\mathbb{R}^m$ and $\|.\|_\beta$ be a norm in $\mathbb{R}^n$. Then there exists a constant $\theta_{\alpha,\beta}(A)$ such that for any $x \in [\mathbf{0}, u]$, and any $b \in \mathbb{R}^m$, whenever $\mathcal{P}_{A,b,u}$ is nonempty, there exists an $\bar{x} \in \mathcal{P}_{A,b,u}$ such that $\|\bar{x} - x\|_\beta \le \theta_{\alpha,\beta}(A)\|Ax - b\|_\alpha$. To see how such bounds can lead to linear convergence, let us first focus on finding a feasible solution in $\mathcal{P}_{A,b,u}$. This can be formulated as a convex quadratic minimization problem:

$$\min \tfrac{1}{2}\|Ax - b\|^2 \quad \text{s.t.} \quad x \in [\mathbf{0}, u]\,. \tag{1}$$

This is a smooth objective function, but not strongly convex. Nevertheless, Hoffman-proximity guarantees that for any $x \in [\mathbf{0}, u]$ where $f(x) := \frac{1}{2}\|Ax - b\|^2$ is close to the optimum value, there exists *some* optimal solution $\bar{x}$ nearby. Nesterov, Necoara, and Glineur [13] introduce various relaxations of strong convexity, including the notion of μ_f-quadratic growth, and show that these weaker properties suffice for linear convergence. The R-FGM algorithm, a restart variant of fast gradient descent in [13] leads to the following bound for (1):

Theorem 2. *There is an algorithm* `Feasible`(A, b, u, δ), *which, on input* $A \in \mathbb{R}^{m\times n}$, $\|A\|_1 \ge 1$, $b \in \mathbb{R}^m$, $u \in \mathbb{R}^n$, *supposing the system* $Ax = b$, $x \in [\mathbf{0}, u]$ *is feasible, finds a* δ*-feasible solution using* $O\big(\|A\|_2 \cdot \theta_{2,2}(A) \cdot \log(m\|b\|_1/\delta)\big)$ *iterations of* `R-FGM`.

To solve $\mathrm{LP}(A, b, c, u)$, [13] in effect uses the standard reduction from optimization to feasibility by writing the primal and dual systems together. By strong duality, if $\mathrm{LP}(A, b, c, u)$ is feasible and bounded, then x is a primal and (π, w^+, w^-) is a dual optimal solution if and only if

$$Ax = b\,, \quad A^\top\pi + w^- - w^+ = c\,, \quad \langle c, x\rangle - \langle b, \pi\rangle + \langle u, w^+\rangle = 0\,, \quad x, w^-, w^+ \ge 0\,. \tag{2}$$

We can use R-FGM for this larger feasibility problem. However, the constraint matrix M now also includes the vectors c, b, and u. In particular, while $\theta_{2,2}(A) \le m$ for a TU matrix, $\theta_{2,2}(M)$ may be unbounded, as shown in the full version.

Other previous works obtain linear convergence bounds using different approaches, but share the above characteristics: their running time includes a constant term $C(A, b, c, u)$. For example, [5] and [19] use an alternating direction method based on an augmented Lagrangian, and [2] and [9] use restart PDHG. The convergence bounds depend not only on the Hoffman-constant of the system, but also linearly on the maximum possible norm of the primal and dual iterates seen during the algorithm.

Our Approach. We present an algorithm in the FOM family with polynomial dependence on $\log(1/\delta)$, n, m, $\log\|u\|$, $\log\|b\|$, $\log\|c\|$ and a constant $C(A)$ only dependent on A. Our algorithm repeatedly calls R-FGM on a potential function of the form

$$F_\tau(x) := \frac{1}{2}(\max\{0, \langle\hat{c}, x\rangle - \tau\})^2 + \frac{1}{2\|A\|_1^2}\|Ax - b\|_2^2\,, \tag{3}$$

for a suitably chosen parameter $\tau \in \mathbb{R}$, and a modified cost function $\hat{c}$. If we use $\hat{c} = c/\|c\|_\infty$, and τ is slightly below the optimum value, then one can show that a near-minimizer x of $F_\tau(x)$ is a near optimal primal solution to the original LP, and moreover, we can use the gradient $\nabla F_\tau(x)$ to construct a near-optimal dual solution to the problem.

Thus, one could find a δ-approximate and δ-optimal solution to $\mathrm{LP}(A, b, c, u)$ with $\log(1/\delta)$ dependence by doing a binary search over the possible values of τ, and running R-FGM for each guess. This already improves on the parameter dependence, however, it still involves a constant $C(A, c)$. One can formulate the minimization of $F_\tau(x)$ in the form (1); the constraint matrix also includes the vector $\hat{c}$. The resulting Hoffman constant can be arbitrarily worse than the one for the original system.

To overcome this issue, we instead define $\hat{c}$ as an ε-discretization of $c/\|c\|_\infty$. We show that the Hoffman constant remains bounded in terms of the Hoffman constant of the feasibility system and a suitably chosen $\varepsilon > 0$. Now, for the appropriate choice of τ, a near-minimizer of $F_\tau(x)$ only gives a crude approximation to the original LP: the error depends on the discretization parameter ε, and to keep the Hoffman constant under control we cannot choose ε very small. Nonetheless, the dual solution obtained from the gradient contains valuable information. For certain indices $i \in N$, using primal-dual slackness, one can conclude $x_i^* \approx 0$ or $x_i^* \approx u_i$ for an optimal solution x^* to the original LP. We fix all such x_i to 0 or u_i, respectively, and recurse. Even if we do not find any such x_i, we make progress by replacing our cost function by an equivalent reduced cost with the ℓ_∞ norm decreasing by at least a factor two.

To summarize: our overall algorithm has an outer loop that gradually fixes the variables to the upper and lower bounds, and repeatedly replaces the cost by a reduced cost. In the inner loop, we call R-FGM in a binary search framework that guesses the parameter τ. We note that while R-FGM is run on a number of systems, the total number of these systems is logarithmically bounded in $1/\delta$ and the input parameters. Moreover, besides the first order updates, we only perform simple arithmetic operations: based on the gradient, we eliminate a subset of variables and shift the cost function. On a high level, our algorithm is a repeatedly applied FOM, where after each run, we 'zoom in' to a 'critical' part of the problem based on what we learned from the previous iteration.

1.1 Circuit Imbalance Measures and Proximity

The key parameters for our algorithm are *circuit imbalance measures*. For a linear space $W \subseteq \mathbb{R}^n$, an elementary vector is a support minimal nonzero vector in W. A *circuit* in W is the support of some elementary vector; these are precisely the circuits in the associated linear matroid $\mathcal{M}(W)$. We let $\mathcal{F}(W) \subseteq W$ denote the set of elementary vectors in the space W. The subspaces $W = \{\mathbf{0}\}$ and $W = \mathbb{R}^N$ are called trivial subspaces; all other subspaces are nontrivial. We define the *fractional circuit imbalance measure* as

$$\kappa(W) := \max\left\{\left|\frac{g_j}{g_i}\right| : g \in \mathcal{F}(W), i, j \in \mathrm{supp}(g)\right\}$$

for nontrivial subspaces, and $\kappa(W) := 1$ for trivial subspaces.

Further, if W is a rational linear space, we let $\bar{\mathcal{F}}(W) \subseteq \mathcal{F}(W)$ denote the set of integer elementary vectors $g \in \mathbb{Z}^n \cap \mathcal{F}(W)$ such that the largest common divisor of the entries is 1. We define the *max circuit imbalance measure* as $\bar{\kappa}(W) := \max\{\|g\|_\infty : g \in \bar{\mathcal{F}}(W)\}$. Note that $\kappa(W) \leq \bar{\kappa}(W)$ but they may not be equal. For example, if the single elementary vector up to scaling is $(4, 7, 8)$, then $\kappa(W) = 2$ but $\bar{\kappa}(W) = 8$.

Let $A \in \mathbb{R}^{m\times n}$ be a matrix, and let $W = \ker(A)$ be the kernel of A. We let $\mathcal{F}(A)$, $\kappa(A)$, $\bar{\kappa}(A)$ denote $\mathcal{F}(W)$, $\kappa(W)$, $\bar{\kappa}(W)$, respectively, for the kernel space $W = \ker(A)$. We define the subspace $\mathcal{X}_A = \ker(A| - I_m)$; thus, $(v, -Av) \in \mathcal{X}_A$ for any $v \in \mathbb{R}^n$.

We refer the reader to the survey [6] for properties and applications of circuit imbalances, and just highlight some key facts. For a matrix $A \in \mathbb{Z}^{m\times n}$, let $\Delta(A)$ denote the largest absolute value of a subdeterminant of A. A matrix is is *totally unimodular (TU)* if $\Delta(A) = 1$. We note that $\kappa(W) = 1$ if and only if there exists a TU matrix $A \in \mathbb{Z}^{m\times n}$ such that $W = \ker(A)$. Further, it is easy to verify that for $A \in \mathbb{Z}^{m\times n}$, the inequality $\bar{\kappa}(A) \leq \Delta(A)$ holds. However, $\bar{\kappa}(A)$ can be arbitrarily smaller: $\bar{\kappa}(A) = 2$ for the node-edge incidence matrix of any undirected graph, whereas $\Delta(A)$ can be exponentially large. Our algorithm also uses the self-duality of κ: $\kappa(W) = \kappa(W^\perp)$ where $W^\perp$ denotes the orthogonal complement subspace (see [3]).

Circuit imbalances play two roles in our algorithm. First, they are used to bound the number of iterations of R-FGM by giving Hoffman-proximity bounds.

Lemma 1. *Let $A \in \mathbb{R}^{n\times m}$. Then $\theta_{1,\infty}(A) \leq \kappa(\mathcal{X}_A)$ and $\theta_{2,2}(A) \leq m \cdot \kappa(\mathcal{X}_A)$.*

These bounds can be tight for $\theta_{1,\infty}(A)$ and up to a factor $\sqrt{m}$ for $\theta_{2,2}(A)$, as discussed in the full version. To bound the number of iterations in R-FGM, we need a Hoffman bound—equivalently, a circuit imbalance bound—for a matrix representation of (3); this matrix B arises by adding an additional row containing $\hat{c}$ to the matrix A. For this, we need to use the max circuit imbalance measure $\bar{\kappa}(\mathcal{X}_A)$; we can show $\kappa(\mathcal{X}_B) \leq 2m \cdot \bar{\kappa}^2(\mathcal{X}_A)/\varepsilon$. Remember that the ε comes from $\hat{c}$, which is an ε-discretization of $c/\|c\|_\infty$.

The second role of $\kappa(\mathcal{X}_A)$ is for the variable fixing argument in the outer loop of the algorithm. The analysis relies on the following key proximity lemma, proved in the full version using conformal circuit decompositions.

Lemma 2. *Let $A \in \mathbb{R}^{m\times n}$ and let $z \in \mathbb{R}^n$ be cycle-free with respect to A, meaning that no vector $y \in \ker(A)$, $y \neq 0$ exists with $y_i z_i > 0$ whenever $y_i \neq 0$. Then*

$$\|z\|_\infty \leq \kappa(\mathcal{X}_A) \cdot \|Az\|_1 \quad \textit{and} \quad \|z\|_2 \leq m \cdot \kappa(\mathcal{X}_A) \cdot \|Az\|_2.$$

The inner loop returns a near optimal primal solution with respect to the rounded cost, as well as a near optimal dual solution derived from the gradient of the potential function. The above lemma can be used to argue about the existence of a nearby optimal solution. Based on this, we can infer that variables with a large positive or negative dual slack can be rounded to the lower or upper bounds. To make such an inference, the rounding accuracy ε needs to be calibrated to

$\kappa(\mathcal{X}_A)$. The larger $\kappa(\mathcal{X}_A)$ is, the more refined the rounding needed to obtain such guarantees.

Guessing the Condition Numbers. It is NP-hard to approximate $\kappa(A)$ within a factor $2^{\text{poly}(m)}$ for $A \in \mathbb{R}^{m\times n}$, see [3], using a result of Tunçel [17]. Still, our algorithm requires explicit bounds on the circuit imbalance measures both in the inner and outer loops. As discussed in the full version, this can be circumvented by a standard doubling guessing procedure: starting with the guess $\hat{\kappa} = 1$, we either succeed with the current guess or restart after doubling the guess. We will be able to detect failure since our algorithm—if run with the correct guess—returns a pair of approximately optimal primal and dual solutions.

1.2 Related Work

We recall that an LP algorithm is strongly polynomial if it only uses basic arithmetic operations $(+, -, \times, /)$ and comparisons, and the number of such operations is polynomial in the number of variables and constraints. Further, the algorithm must be in PSPACE, that is, the size of the numbers appearing in the computation must remain bounded in the input size. The existence of a strongly polynomial algorithm for LP is on Smale's list of main challenges for 21st century mathematics [14].

The variable fixing idea in our algorithm traces its roots to Tardos's strongly polynomial algorithm for minimum-cost circulations. The same idea was extended by Tardos [16] to obtain a $\text{poly}(n, \log \Delta(A))$ time algorithm for finding an exact solution to $\text{LP}(A, b, c, u)$ for an integer constraint matrix $A \in \mathbb{Z}^{m\times n}$ with largest subdeterminant $\Delta(A)$. This running time bound is strongly polynomial for 'combinatorial LPs', that is, LPs with all entries being integers of absolute value $\text{poly}(n)$.

We note that $\kappa(A) \le \kappa(\mathcal{X}_A) \le \bar{\kappa}(\mathcal{X}_A) \le \Delta(A)$ for an integer matrix A. Dadush et al. [4] strengthened Tardos' result by replacing $\Delta(A)$ by $\kappa(A)$, and removing all integrality-based arguments, and obtained a $\text{poly}(n, \log \kappa(A))$ running time bound. The algorithm is of black-box nature, and can use any LP solver; an exact optimal solution can be found by running nm LP-solvers to accuracy $\delta = 1/\text{poly}(n, \log \kappa(A))$.

Our algorithm uses variable fixing in a different manner, giving a robust extension to the approximate setting. Our end goal is not an exact optimal solution, but rather an approximate one (that could serve as an input for the black-box algorithm [4]). The approximate solution obtained from the FOM in the inner loop has weaker guarantees. Tardos [16] also uses subproblems with a similarly rounded cost function, but requires exact feasibility, which cannot be obtained from an FOM.

For this reason, we obtain weaker guarantees, and may fix variables to 0 that are small but positive in all optimal solutions. However, this is acceptable if we are only aiming for an approximate solution. On the positive side, we only need a logarithmic number of executions of the outer loop, in contrast to nm in [4,16].

This is because for us it is already sufficient progress to decrease the norm of the reduced cost, even if we cannot fix any variables.

A poly$(n, \log \kappa(A))$ running time for LP can also be achieved by a special class of 'combinatorial' interior point methods, called *Layered Least Squares (LLS) IPMs*. This class was introduced by Vavasis and Ye [18]. The parameter dependence was on the Dikin–Stewart–Todd condition measure $\bar{\chi}(A)$, but [3] observed that the two condition numbers are close to each other. Further, they gave a stronger LLS IPM with running time dependent on the optimal value $\kappa^*(A)$ of $\kappa(A)$ achievable by column rescaling. We refer the reader to the survey [6] for further results related to circuit imbalances and their uses in LP, including also diameter and circuit diameter bounds.

We also note that Fujishige et al. [7] recently gave a poly$(n, \kappa(A))$ algorithm for the minimum norm point problem (1) by combining FOMs and active set methods. Their algorithm terminates with an exact solution; on the other hand, it also uses projection steps that involve solving a system of linear equations. Thus, it is not an FOM; moreover, it is not applicable for optimization LP.

2 Proximity Tools and the Outer Loop of the Algorithm

We describe our algorithm under the simplifying assumption that the exact values of $\kappa(A)$ and $\bar{\kappa}(A)$ are known for the input matrix as well as all submatrices obtained by column deletions. As discussed in the Introduction, this assumption can be circumvented by repeated guessing. We also assume that LP(A, b, c, u) is feasible. In the full version, we show that this assumption can be removed using a two-stage approach similar to the Simplex algorithm.

Our optimization algorithm has two components: the outer loop and the inner loop. In the outer loop, we gradually reduce LP(A, b, c, u) by fixing some variables to their upper or lower bounds, and replacing the cost vector by an equivalent one of smaller norm.

2.1 The Proximity Tools

In the outer loop, our goal is to find a δ^{feas}-feasible and δ^{opt}-optimal solution to LP(A, b, c, u). We distinguish these two accuracy parameters for the sake of the recursive algorithm, where the required feasibility and optimality accuracies need to be changed differently in the recursive calls.

Primal-Dual Optimality and Cost Shifting. We use primal-dual arguments, making variable fixing decisions based on approximate complementarity conditions. The dual to LP(A, b, c, u) can be written as

$$\begin{aligned} \max\ & \langle b, \pi \rangle - \langle u, w^+ \rangle \\ A^\top \pi + w^- - w^+ &= c \\ w^-, w^+ &\geq 0. \end{aligned} \qquad (\text{Dual}(A, b, c, u))$$

Note that given $\pi \in \mathbb{R}^m$, the unique best choice of the variables w^- and w^+ is $w^- = \max\{c - A^\top \pi, 0\}$ and $w^+ = \max\{A^\top \pi - c, 0\}$. When we speak of a dual solution $\pi \in \mathbb{R}^m$, we mean its extension with these variables. Recall the primal-dual optimality conditions: $x^* \in \mathcal{P}_{A,b,u}$ and $\pi \in \mathbb{R}^m$ are optimal respectively to LP(A, b, c, u) and to Dual(A, b, c, u) if and only if the following holds:

$$\text{if } A_i^\top \pi < c_i \text{ then } x_i = 0, \text{ and if } A_i^\top \pi > c_i \text{ then } x_i = u_i \text{ for every } i \in [n]. \quad (4)$$

Approximate Complementarity and Proximity. Assume now that we have a pair of primal and dual solutions x and π that do not satisfy complementarity, but we have a quantitative bound on the violation. Namely, for a suitably chosen threshold $\sigma \geq 0$, let

$$\theta(x, \pi, \sigma) := \sum_{c_i - A_i^\top \pi > \sigma} x_i + \sum_{c_i - A_i^\top \pi < -\sigma} (u_i - x_i), \quad \text{and}$$
$$J(\pi, \sigma) := \left\{ i \in [n] : |c_i - A_i^\top \pi| > n \cdot \lceil \kappa(\mathcal{X}_A) \rceil \cdot \sigma \right\}.$$

Note that if x and π are primal and dual optimal, then the primal-dual complementarity constraints (4) imply $\theta(x, \pi, 0) = 0$. Let us assume that for some $\sigma > 0$, this quantity is still small. Note also that $J(\pi, \sigma)$ is the set of indices where the absolute value of the slack is much higher than the threshold σ. In particular, $\min\{x_i, u_i - x_i\} \leq \theta(x, \pi, \sigma)$ on these indices. Our key proximity result asserts that there exists an optimal solution that is close to the current solution on these indices. The proof is given in the full version, using Lemma 2.

Lemma 3. *Let $x \in \mathcal{P}_{A,b,u}$ be a feasible solution. Then there exists an optimal solution x^* for LP(A, b, c, u) such that*

$$|x_i - x_i^*| \leq \kappa(\mathcal{X}_A) \cdot \theta(x, \pi, \sigma)$$

for all $i \in J(\pi, \sigma)$.

Variable Fixing. Assume that from the inner loop of the algorithm we get $x \in [\mathbf{0}, u]$ and $\pi \in \mathbb{R}^m$ such that the feasibility violation $\|Ax - b\|_1$ and the complementarity violation $\theta(x, \pi, \sigma)$ are both tiny for the choice $\sigma := \|c\|_\infty / (4n\lceil \kappa(\mathcal{X}_A) \rceil)$. Note that for this choice, the threshold in the definition of $J(\pi, \sigma)$ becomes $\|c\|_\infty/4$. We partition $J(\pi, \sigma)$ into $J_1 := \left\{ i \in J(\pi, \sigma) \mid c_i - A_i^\top \pi < -\frac{\|c\|_\infty}{4} \right\}$ and $J_2 := \left\{ i \in J(\pi, \sigma) \mid c_i - A_i^\top \pi > \frac{\|c\|_\infty}{4} \right\}$. We apply Lemma 3 to the problem with the modified right hand side $b' = Ax$. By ensuring that $\theta(x, \pi, \sigma)$ is sufficiently small, we will see that there is an optimal solution x^* with $x_i^* \approx 0$ for $i \in J_1$ and $x_i^* \approx u_i$ for $i \in J_2$.

We fix these variables to the lower and upper bounds, respectively, and shift the cost function according to π. Thus, we specify the following new LP. Let $N := [n] \setminus (J_1 \cup J_2)$ and $\bar{b} := A_N x_N$.

$$\begin{aligned} \min \ & \langle c_N - A_N^\top \pi, z \rangle \\ A_N z &= \bar{b} \\ \mathbf{0}_N \leq z &\leq u_N . \end{aligned} \qquad (\text{LP}(A_N, \bar{b}, c_N - A_N^\top \pi, u_N))$$

We show the following result, which says the optimal solution of $\mathrm{LP}(A_N, \bar{b}, c_N - A_N^\top \pi, u_N)$ provides an approximately feasible and optimal solution to $\mathrm{LP}(A, b, c, u)$. The approximation is in terms of $\theta(x, \pi, \sigma)$ and $\|Ax - b\|_1$. Recall that $\Phi(A, b, c, u)$ denotes the optimal value, the value achieved by the solution to $\mathrm{LP}(A, b, c, u)$.

Theorem 3. *For $A \in \mathbb{R}^{m\times n}$, $b \in \mathbb{R}^m$, and $c, u \in \mathbb{R}^n$ such that $LP(A, b, c, u)$ is feasible, let $\sigma := \|c\|_\infty/(4n \cdot \lceil \kappa(\mathcal{X}_A)\rceil)$, and let $x \in [\mathbf{0}, u]$ and $\pi \in \mathbb{R}^m$ be a pair of (not necessarily feasible) primal and dual solutions. Then, $LP(A_N, \bar{b}, c_N - A_N^\top \pi, u_N)$ is feasible and, in addition satisfies the following:*

- *feasibility condition:* $\|b - \bar{b} - A_{J_2} u_{J_2}\|_1 \le \theta(x, \pi, \sigma) \cdot \|A\|_1 + \|Ax - b\|_1$.
- *optimality condition:* $\left|\Phi(A_N, \bar{b}, c_N - A_N^\top \pi, u_N) + \langle \bar{b}, \pi\rangle + \langle c_{J_2}, u_{J_2}\rangle - \Phi(A, b, c, u)\right|$ $\le \kappa(\mathcal{X}_A) \cdot \|c\|_1 \cdot \|Ax - b\|_1 + |J_1 \cup J_2| \cdot \kappa(\mathcal{X}_A) \cdot \|c\|_1 \cdot (2 + \kappa(\mathcal{X}_A)\|A\|_1) \cdot \theta(x, \pi, \sigma)$
- *cost reduction:* $\|c_N - A_N^\top \pi\|_\infty \le \|c\|_\infty/4$.

With this theorem, if one can find a pair (x, π) such that the right hand sides of the feasibility and optimality bounds are tiny, then $\mathrm{LP}(A, b, c, u)$ can be reduced to $\mathrm{LP}(A_N, \bar{b}, c_N - A_N^\top \pi, u_N)$ with a tiny loss on feasibility and optimality. Moreover, by replacing c with $c_N - A_N^\top \pi$, each iteration reduces the ℓ_∞-cost on the remaining variables by a factor 4. One can repeat this procedure and ultimately reduce the original problem to one with an extremely small objective function value and possibly with fewer variables. Solving this problem will give a good enough solution to the original $\mathrm{LP}(A, b, c, u)$, after restoring any variables fixed to the lower or upper bounds.

It is possible that both J_1 and J_2 are empty. This means that $\|c - A^\top \pi\|_\infty \le \|c\|_\infty/4$; we can simply recurse with the same b but improved cost function. Note that we could make progress more agressively by a preprocessing step that projects c to the kernel of A; this gets a cost vector of the form $c' = c - A^\top \pi$ with the smallest possible ℓ_2-norm—such a preprocessing is used in the strongly polynomial algorithms [4, 15, 16]. Setting a slightly smaller σ would then guarantee variable fixing in every iteration. However, the projection amounts to solving a system of linear equations that may be computationally more expensive. We instead proceed with lazier updates as above.

2.2 Description of the Outer Loop

Algorithm 1 takes as input (A, b, c, u) such that $\mathrm{LP}(A, b, c, u)$ is feasible, and accuracy parameters δ^{feas} and δ^{opt} such that $0 \le \delta^{\text{feas}} \cdot 8n\sqrt{m} \cdot \kappa(\mathcal{X}_A) \cdot \|A\|_1 \le \delta^{\text{opt}}$. The inner loop will give the subroutine $\texttt{GetPrimalDualPair}(A, b, c, u, \delta^{\text{feas}}, \delta^{\text{opt}})$, specified as follows.

Subroutine `GetPrimalDualPair`
Input: $A \in \mathbb{R}^{m\times n}$, $b \in \mathbb{R}^m$, $c, u \in \mathbb{R}^n$, $\delta^{\text{feas}}, \delta^{\text{opt}} > 0$ such that (A, b, c, u) is feasible, $\delta^{\text{opt}} \leq \|u\|_1$, and $\delta^{\text{feas}}\|A\|_1 \leq \delta^{\text{opt}}$.
Output: (x, π), $x \in [\mathbf{0}, u]$, $\pi \in \mathbb{R}^m$ such that

- The right-hand side of the feasibility bound in Theorem 3 is at most $\delta^{\text{feas}}\|A\|_1/n$.
- The right-hand side of the optimality bound in Theorem 3 is at most $\delta^{\text{opt}}\|c\|_\infty/n$.
- $\|\pi\|_\infty \leq 4n\sqrt{m}\cdot\kappa(\mathcal{X}_A)\cdot\|c\|_\infty$.

The algorithm uses a subroutine `Dual-Certificate`$(x, A, b, c, u, \delta^{\text{feas}}, \delta^{\text{opt}})$ that returns a suitable dual certificate of δ^{opt}-optimality of a δ^{feas}-feasible solution x, described in the full version. It uses the feasibility algorithm Theorem 2 on a dual system. The main correctness and running time statement is the following.

Theorem 4. *If LP(A, b, c, u) is feasible, then Algorithm 1 returns a δ^{feas}-feasible solution that is δ^{opt}-optimal along with a $2\delta^{\text{opt}}$-certificate. It makes at most $\log_2(n\|u\|_1/\delta^{\text{opt}})$ many recursive calls.*

Algorithm 1: `SolveLP`

Input : $A \in \mathbb{R}^{m\times n}$, $b \in \mathbb{R}^m$, $c, u \in \mathbb{R}^n$, $0 < \delta^{\text{feas}}(8n\sqrt{m}\cdot\kappa(\mathcal{X}_A)\|A\|_1) \leq \delta^{\text{opt}}$.
Output : A δ^{feas}-feasible and δ^{opt}-optimal solution to LP(A, b, c, u) along with a $2\delta^{\text{opt}}$-certificate

if $\delta^{\text{opt}} \geq \|u\|_1$ **then**
  $\bar{x} \leftarrow$ `Feasible`$(A, b, c, u, \delta^{\text{feas}})$ (See Theorem 2.) ;
  return $\bar{x}$
else
  $(x, \pi) \leftarrow$ `GetPrimalDualPair`$(A, b, c, u, \delta^{\text{feas}}, \delta^{\text{opt}})$;
  Define J_1, J_2, N, $\bar{b}$ as for Theorem 3 ;
  $c^{\text{new}} \leftarrow c - A^\top\pi$;
  $\lambda \leftarrow \frac{\|c_N\|_\infty}{2\|c_N^{\text{new}}\|_\infty}$;
  if $J_1 \cup J_2 = \emptyset$ **then**
    $x^{\text{out}} \leftarrow$ `SolveLP`$(A, b, c^{\text{new}}, u, \delta^{\text{feas}}, \lambda\delta^{\text{opt}})$
  else
    $x_{J_1}^{\text{out}} \leftarrow 0$; $x_{J_2}^{\text{out}} \leftarrow u_{J_2}$;
    $x_N^{\text{out}} \leftarrow$ `SolveLP`$(A_N, \bar{b}, c_N^{\text{new}}, u_N, \delta^{\text{feas}}\cdot|N|/n, \lambda\delta^{\text{opt}}\cdot|N|/n)$;
    $\bar{\pi} \leftarrow$ `Dual-Certificate`$(x, A, b, c, u, \delta^{\text{feas}}, \delta^{\text{opt}})$;
  return $(x^{\text{out}}, \bar{\pi})$

3 The Inner Loop of the Algorithm

Next, we describe the subroutine `GetPrimalDualPair`$(A, b, c, u, \delta^{\text{feas}}, \delta^{\text{opt}})$ in the inner loop. The required output is primal and dual vectors (x, π) satisfying the feasibility and optimality bounds in Theorem 3. In particular, we need to bound $\theta(x, \pi, \sigma) = \sum_{i:c_i+A_i^\top \pi>\sigma} x_i + \sum_{i:c_i+A_i^\top \pi<-\sigma}(u_i - x_i)$ and $\|Ax - b\|_1$ for $\sigma = \|c\|_\infty/(4n \cdot \lceil\kappa(\mathcal{X}_A)\rceil)$. We use a potential function $F_\tau(x)$ of the form (3) for a modified cost function $\hat{c}$.

As noted in the introduction, in order to keep the Hoffman-constant corresponding to $F_\tau(x)$ small, we define $\hat{c}$ as the discretization of $c/\|c\|_\infty$ into integer multiples of $\varepsilon = 1/(8n \cdot \lceil\kappa(\mathcal{X}_A)\rceil) = \sigma/(2\|c\|_\infty)$. More precisely, we round down the positive entries and round up the negative entries. Recalling that $1/\varepsilon$ is an integer, we have $\|\hat{c}\|_\infty = 1$.

For F_τ as in (3), let $F_\tau^\star := \min\{F_\tau(x) \mid x \in [\mathbf{0}, u]\}$ denote the optimum value. We say that $x \in [\mathbf{0}, u]$ is a *ζ-approximate minimizer* of F_τ if $F(x) \le F_\tau^\star + \zeta$. The following proposition is immediate.

Proposition 1. *$F_\tau^\star$ is a non-increasing continuous function of τ. If LP$(A, b, \hat{c}, u)$ is feasible, then $\Phi(A, b, \hat{c}, u)$ is the smallest value of τ such that $F_\tau^\star = 0$.*

The main driver of our algorithm is Necoara, Nesterov, and Glineur's R-FGM algorithm, applied to F_τ. We specify this subroutine as follows.

Subroutine R-FGM
Input: $A \in \mathbb{R}^{m \times n}$, $b \in \mathbb{R}^m$, $c, u \in \mathbb{R}^n$, $\tau \in \mathbb{R}$, $\zeta > 0$.
Output: A ζ-approximate minimizer $x \in [\mathbf{0}, u]$ of F_τ.

The purpose of `GetPrimalDualPair` (Algorithm 2) is to identify a value τ by binary search that is slightly below $\Phi(A, b, \hat{c}, u)$. It uses the following parameters to calibrate the accuracy of the binary search: $\overline{C} := 64n^2\sqrt{m} \cdot \kappa^2(\mathcal{X}_A) \cdot \|A\|_1$ and $\zeta := \left(\frac{\tilde{\delta}}{4\kappa^2(\mathcal{X}_A)n^4\overline{C}\sqrt{m}}\right)^2$. The correctness and performance statements are summarized as follows.

Theorem 5. *Assume LP(A, b, c, u) is feasible. Algorithm 2 makes $O(\log[\|u\|_1 nm \cdot \kappa(\mathcal{X}_A)/\tilde{\delta}])$ calls to* R-FGM*, and altogether these calls use $O(n^{1.5} m^2\|A\|_1^2 \cdot \bar{\kappa}^3(\mathcal{X}_A) \cdot \log^2[\|u\|_1 nm \cdot \kappa(\mathcal{X}_A)/\tilde{\delta}])$ iterations. On terminating, it outputs (x, π) satisfying:*

(i) $\theta(x, \pi, \sigma) \cdot \|A\|_1 + \|Ax - b\|_1 \le \delta^{\text{feas}}\|A\|_1/n$.
(ii) $\kappa(\mathcal{X}_A) \cdot \|c\|_1 \cdot \|Ax - b\|_1 + |J_1 \cup J_2| \cdot \kappa(\mathcal{X}_A) \cdot \|c\|_1 \cdot (2 + \kappa(\mathcal{X}_A)\|A\|_1) \cdot \theta(x, \pi, \sigma) \le \delta^{\text{opt}}\|c\|_\infty/n$.
(iii) $\|\pi\|_\infty \le 4n\sqrt{m} \cdot \kappa(\mathcal{X}_A) \cdot \|c\|_\infty$.

The proof, as well as the derivation of Theorem 1 from Theorems 4 and 5 are given in the full version. We now provide intuition why a suitable τ can be found by binary search. Given x, the dual solution π is defined based on the gradient of $F_\tau(x)$ as $\pi := \frac{\|c\|_\infty}{\|A\|_1^2 \alpha}(b - Ax)$ for $\alpha := \max\{0, \langle \hat{c}, x\rangle - \tau\}$ if $\alpha > 0$.

To bound the infeasibility $\|Ax - b\|_1$, we need to find a solution x where $F_\tau(x)$ is small, since $\|Ax - b\|_1 \le (2n\|A\|_1 F_\tau(x))^{1/2}$. Therefore, τ should not be much smaller than the optimum value $\Phi(A, b, c, u)$ of $\mathrm{LP}(A, b, c, u)$. The bound on $\theta(x, \pi, \sigma)$ can be shown by arguing that the improving directions of the gradient are small at an approximately optimal solution x: $x_i \approx 0$ if $\nabla_i F_\tau(x) \gg 0$ and $x_i \approx u_i$ if $\nabla_i F_\tau(x) \ll 0$, and that $|c_i - \hat{c}_i| \cdot \|c\|_\infty \le \|c\|_\infty \cdot \varepsilon = \sigma/2$. We also need that $\alpha > 0$ and is not too small. Based on these requirements, we can establish a narrow (but not too narrow) interval of τ where a sufficiently accurate approximate solution to $F_\tau(x)$ exists. Using Proposition 1, we can find a suitable τ by binary search.

Algorithm 2: `GetPrimalDualPair`

Input : $A \in \mathbb{R}^{m\times n}$, $b \in \mathbb{R}^m$, $c, u \in \mathbb{R}^n$, and $\delta^{\text{feas}}, \delta^{\text{opt}} > 0$ such that (A, b, c, u) is feasible, and $\delta^{\text{feas}}\|A\|_1 \le \delta^{\text{opt}} < \|u\|_1$

. **Output** : $x \in [\mathbf{0}, u]$ and $\pi \in \mathbb{R}^n$.

$\tau^+ \leftarrow \|u\|_1$ and $\tau^- \leftarrow -\|u\|_1 - 2\overline{C}\sqrt{\zeta}$;

Repeat

  $\tau \leftarrow \frac{\tau^+ + \tau^-}{2}$;

  $x \leftarrow$ `R-FGM`$(A, b, c, u, \tau, \zeta)$;

  if $F_\tau(x) > 2\overline{C}^2\zeta$ **then** $\tau^- \leftarrow \tau$;

  if $F_\tau(x) < \overline{C}^2\zeta$ **then** $\tau^+ \leftarrow \tau$;

  if $F_\tau(x) \in [\overline{C}^2\zeta, 2\overline{C}^2\zeta]$ **then**

    $\alpha \leftarrow \max\{0, \langle \hat{c}, x\rangle - \tau\}$;

    $\pi \leftarrow \frac{\|c\|_\infty}{\|A\|_1^2\alpha}(b - Ax)$;

    $w^+ \leftarrow \max\{A^\top\pi - c, 0\}$; $w^- \leftarrow \max\{c - A^\top\pi, 0\}$;

    return (x, π)

Acknowledgements. This work was supported by the European Research Council (ERC) under the European Union's Horizon 2020 research and innovation programme (grant agreement no. ScaleOpt–757481; for C. Hertrich additionally via grant agreement no. ForEFront–615640). Y. Tao also acknowledges Grant 2023110522 from SUFE, National Key R&D Program of China (2023YFA1009500), NSFC grant 61932002. Part of the work was done while L. Végh was visiting the Corvinus Institute for Advanced Studies, Corvinus University, Budapest, Hungary, and while C. Hertrich was affiliated with London School of Economics, UK, and with Goethe-Universität Frankfurt, Germany.

References

1. Applegate, D., et al.: Practical large-scale linear programming using primal-dual hybrid gradient. Adv. Neural. Inf. Process. Syst. **34**, 20243–20257 (2021)
2. Applegate, D., Hinder, O., Lu, H., Lubin, M.: Faster first-order primal-dual methods for linear programming using restarts and sharpness. Math. Program. **201**(1), 133–184 (2023)
3. Dadush, D., Huiberts, S., Natura, B., Végh, L.A.: A scaling-invariant algorithm for linear programming whose running time depends only on the constraint matrix. Math. Program. (2023). (in press)
4. Dadush, D., Natura, B., Végh, L.A.: Revisiting Tardos's framework for linear programming: faster exact solutions using approximate solvers. In: Proceedings of the 61st Annual IEEE Symposium on Foundations of Computer Science (FOCS), pp. 931–942 (2020)
5. Eckstein, J., Bertsekas, D.P., et al.: An alternating direction method for linear programming. Technical report LIDS-P-1967 (1990)
6. Ekbatani, F., Natura, B., Végh, L.A.: Circuit imbalance measures and linear programming. In: Surveys in Combinatorics 2022. London Mathematical Society Lecture Note Series, pp. 64–114. Cambridge University Press (2022)
7. Fujishige, S., Kitahara, T., Végh, L.A.: An update-and-stabilize framework for the minimum-norm-point problem. In: Del Pia, A., Kaibel, V. (eds.) IPCO 2023. LNCS, vol. 13904, pp. 142–156. Springer, Cham (2023). https://doi.org/10.1007/978-3-031-32726-1_11
8. Gilpin, A., Pena, J., Sandholm, T.: First-order algorithm with convergence for-equilibrium in two-person zero-sum games. Math. Program. **133**(1–2), 279–298 (2012)
9. Hinder, O.: Worst-case analysis of restarted primal-dual hybrid gradient on totally unimodular linear programs. arXiv preprint arXiv:2309.03988 (2023)
10. Hoffman, A.J.: On approximate solutions of systems of linear inequalities. J. Res. Natl. Bur. Stand. **49**(4), 263–265 (1952)
11. Karmarkar, N.: A new polynomial-time algorithm for linear programming. In: Proceedings of the 16th Annual ACM Symposium on Theory of Computing (STOC), pp. 302–311 (1984)
12. Khachiyan, L.G.: A polynomial algorithm in linear programming. In: Doklady Academii Nauk SSSR, vol. 244, pp. 1093–1096 (1979)
13. Necoara, I., Nesterov, Y., Glineur, F.: Linear convergence of first order methods for non-strongly convex optimization. Math. Program. **175**, 69–107 (2019)
14. Smale, S.: Mathematical problems for the next century. Math. Intell. **20**, 7–15 (1998)
15. Tardos, É.: A strongly polynomial minimum cost circulation algorithm. Combinatorica **5**(3), 247–255 (1985)
16. Tardos, É.: A strongly polynomial algorithm to solve combinatorial linear programs. Oper. Res. **34**, 250–256 (1986)
17. Tunçel, L.: Approximating the complexity measure of Vavasis-Ye algorithm is NP-hard. Math. Program. **86**(1), 219–223 (1999)
18. Vavasis, S.A., Ye, Y.: A primal-dual interior point method whose running time depends only on the constraint matrix. Math. Program. **74**(1), 79–120 (1996)
19. Wang, S., Shroff, N.: A new alternating direction method for linear programming. Adv. Neural Inf. Process. Syst. **30** (2017)
20. Yang, T., Lin, Q.: RSG: beating subgradient method without smoothness and strong convexity. J. Mach. Learn. Res. **19**(1), 236–268 (2018)

Approximately Packing Dijoins via Nowhere-Zero Flows

Gérard Cornuéjols, Siyue Liu(✉), and R. Ravi

Carnegie Mellon University, Pittsburgh, USA
{gc0v,siyueliu,ravi}@andrew.cmu.edu

Abstract. In a digraph, a dicut is a cut where all the arcs cross in one direction. A dijoin is a subset of arcs that intersects each dicut. Woodall conjectured in 1976 that in every digraph, the minimum size of a dicut equals to the maximum number of disjoint dijoins. However, prior to our work, it was not even known whether at least 3 disjoint dijoins exist in an arbitrary digraph whose minimum dicut size is sufficiently large. By building connections with nowhere-zero (circular) k-flows, we prove that every digraph with minimum dicut size τ contains $\frac{\tau}{k}$ disjoint dijoins if the underlying undirected graph admits a nowhere-zero (circular) k-flow. The existence of nowhere-zero 6-flows in 2-edge-connected graphs (Seymour 1981) directly leads to the existence of $\frac{\tau}{6}$ disjoint dijoins in a digraph with minimum dicut size τ, which can be found in polynomial time as well. The existence of nowhere-zero circular $\frac{2p+1}{p}$-flows in $6p$-edge-connected graphs (Lovász et al. 2013) directly leads to the existence of $\frac{\tau p}{2p+1}$ disjoint dijoins in a digraph with minimum dicut size τ whose underlying undirected graph is $6p$-edge-connected.

Keywords: Woodall's conjecture · Nowhere-zero flow · Approximation algorithm

1 Introduction

Dicuts and Dijoins. Given a digraph $D = (V, A)$ and a subset U of its vertices with $U \neq \emptyset, V$, denote by $\delta^+_D(U)$ and $\delta^-_D(U)$ the arcs leaving and entering U, respectively. The cut induced by U is $\delta_D(U) := \delta^+_D(U) \cup \delta^-_D(U)$. We omit the subscript D if the context is clear. For an arc subset $B \subseteq A$, $\delta^+_B(U) := \delta^+_D(U) \cap B$. A *dicut* is an arc subset of the form $\delta^+(U)$ such that $\delta^-(U) = \emptyset$. A *dijoin* is a subset $J \subseteq A$ that intersects every dicut at least once. More generally, we will also work with the notion of a τ-*dijoin*, which is a subset $J \subseteq A$ that intersects every dicut at least τ times. If D is a weighted digraph with arc weights $w : A \to \mathbb{Z}_+$, we say that D can *pack* k dijoins if there exist k dijoins $J_1, ..., J_k$ such that no arc e is contained in more than $w(e)$ of these k dijoins. In this case, we say that $J_1, ..., J_k$ is a *packing* of D under weight w. In particular, when the digraph is *unweighted*, i.e., $w(e) = 1$ for every $e \in A$, D packs k dijoins if and only if D contains k (arc-)disjoint dijoins. The *value* of the packing is the number k of dijoins in the packing. Edmonds and Giles [8] conjectured the following.

Conjecture 1 (Edmonds-Giles). Let $D = (V, A)$ be a digraph with arc weights $w \in \{0, 1\}^A$. If the minimum weight of a dicut is τ, then D can pack τ dijoins.

J. Vygen and J. Byrka (Eds.): IPCO 2024, LNCS 14679, pp. 71–84, 2024.
https://doi.org/10.1007/978-3-031-59835-7_6

We can assume without loss of generality that $w \in \{0, 1\}^A$ because we can always replace an arc e with integer weight $w(e) > 1$ by $w(e)$ parallel arcs of weight 1. Note that the weight 0 arcs cannot be removed because they, together with the weight 1 arcs, determine the dicuts. The above conjecture was disproved by Schrijver [20]. However, the following unweighted version of the Edmonds-Giles conjecture, proposed by Woodall [26], is still open.

Conjecture 2 (Woodall). In every digraph, the minimum size of a dicut equals the maximum number of disjoint dijoins.

Several weakenings of Woodall's conjecture have been made in the literature. It has been conjectured that there exists some integer $\tau \geq 3$ such that every digraph with minimum dicut size at least τ contains 3 disjoint dijoins [6]. Shepherd and Vetta [23] raised the following question. Let $f(\tau)$ be the maximum value such that every weighted digraph whose dicuts all have weight at least τ, can pack $f(\tau)$ dijoins. They conjectured that $f(\tau)$ is of order $\Omega(\tau)$. In Sect. 2, we give an affirmative answer to this conjecture in the *unweighted* case. The main results in this paper are the following approximate versions of Woodall's conjecture.

Theorem 1. *Every digraph $D = (V, A)$ with minimum dicut size τ contains $\lfloor \frac{\tau}{6} \rfloor$ disjoint dijoins, and such dijoins can be found in polynomial time.*

Given a digraph $D = (V, A)$, the *underlying undirected graph* is the graph with vertex set V and edge set obtained by replacing each arc $(u, v) \in A$ with an undirected edge (u, v). To exclude the cases $\tau = 0$ and $\tau = 1$, when Woodall's conjecture holds trivially, we assume $\tau \geq 2$ throughout the paper, which implies that the underlying undirected graph is 2-edge-connected.

Theorem 2. *Let p be a positive integer. Every digraph $D = (V, A)$ with minimum dicut size τ and with the property that its underlying undirected graph is $6p$-edge-connected contains $\left\lfloor \frac{\tau p}{2p+1} \right\rfloor$ disjoint dijoins.*

Nowhere-Zero Circular Flows. Let $G = (V, E)$ be an *undirected* graph and let $k \geq 2$ be an integer. Tutte [25] introduced the notion of a *nowhere-zero k-flow* of G, which is an orientation E^+ and $f : E^+ \to \{1, 2, ..., k-1\}$ such that $\sum_{e \in \delta^+_{E^+}(v)} f(e) = \sum_{e \in \delta^-_{E^+}(v)} f(e)$ for every vertex $v \in V$. Goddyn et al. [10] extended the definition to allowing k to take fractional values. Let p, q be two integers such that $0 < p \leq q$. A *nowhere-zero circular $\frac{p+q}{p}$-flow* of G is an orientation E^+ and $f : E^+ \to \{1, 1 + \frac{1}{p}, ..., \frac{q}{p}\}$, such that $\sum_{e \in \delta^+_{E^+}(v)} f(e) = \sum_{e \in \delta^-_{E^+}(v)} f(e)$. When $p = 1$ we recover Tutte's notion.

Both theorems above are consequences of the following main theorem we prove.

Theorem 3. *For a digraph $D = (V, A)$ with minimum dicut size τ, if the underlying undirected graph admits a nowhere-zero circular k-flow, where $k \geq 2$ is a rational, then D contains $\lfloor \frac{\tau}{k} \rfloor$ disjoint dijoins.*

The first ingredient of our approach to proving the above results is reducing the problem of packing dijoins in a digraph to that of packing strongly connected digraphs. This reduction is not new and it was already explored by Shepherd and Vetta [23].

Augment the input digraph D by adding reverse arcs for all input arcs and assigning weights τ to the original arcs and 1 to the newly added reverse arcs. Denote the augmented digraph by $\vec{G}$ with weight w^D. Define a *τ-strongly-connected digraph (τ-SCD)* to be a weighted digraph such that the arcs leaving every cut have weight at least τ. Note that a 1-SCD is a *strongly connected digraph (SCD)*. It is not hard to see that for a digraph D with minimum dicut size τ, the augmented digraph $\vec{G}$ with weight w^D is τ-strongly-connected. One can then show that packing $\tau' \leq \tau$ dijoins in the original digraph D is equivalent to decomposing the augmented weighted digraph $\vec{G}$ into τ' strongly connected digraphs (Proposition 2).

We then draw a connection to nowhere-zero flows. There is a rich literature on the existence of nowhere zero k-flows from which we will use two important results. Seymour [22] showed that there always exists a nowhere-zero 6-flow in 2-edge-connected graphs. Younger [27] gave a polynomial time algorithm to construct a nowhere-zero 6-flow in 2-edge-connected graphs.

Theorem 4 ([22,27]). *Every 2-edge-connected graph admits a nowhere-zero 6-flow which can be found in polynomial time.*

Lovász et al. [17] proved the following existence result for nowhere-zero circular flows under stronger connectivity requirements.

Theorem 5 ([17]). *Let p be a positive integer. Every $6p$-edge-connected graph admits a nowhere-zero circular $\frac{2p+1}{p}$-flow.*

Returning to dijoins and the augmented digraph $\vec{G}$, we need to decompose this augmented digraph into some $\tau' \leq \tau$ disjoint strongly connected digraphs. In general, decomposing a digraph into strongly connected digraphs is a notoriously hard problem. It is not known whether there exists an integer τ such that every τ-strongly-connected digraph can be decomposed into 2 disjoint strongly connected digraphs [4]. To get around this difficulty, we reduce our goal to finding *two* disjoint subdigraphs of $\vec{G}$, each of which can be decomposed into τ' in or out r-arborescences for some fixed root r. The idea of pairing up in- and out-arborescences was already used successfully by Shepherd and Vetta [23] to find a half-integral packing of dijoins of value $\frac{\tau}{2}$. Here, we crucially argue (in Theorem 6) that if the underlying undirected graph of D admits a nowhere-zero k-flow, then the digraph $\vec{G}$ with weight w^D can be decomposed into two disjoint $\lfloor \frac{\tau}{k} \rfloor$-SCD's. (Note that we do not prove this for any arbitrary τ-SCD.) Using Edmonds' disjoint arborescences theorem [7], we can now extract $\lfloor \frac{\tau}{k} \rfloor$ disjoint in r-arborescences from the first and the same number of out r-arborescences from the second. Pairing them up gives us the final set of $\lfloor \frac{\tau}{k} \rfloor$ strongly connected digraphs. Our results then follow from the prior theorems about the existence of nowhere-zero flows.

Strongly Connected Orientations. In Sect. 3, we give equivalent forms of Woodall's conjecture and of the Edmonds-Giles conjecture, respectively, in terms of packing strongly connected orientations, which are of independent interest. Given an undirected graph $G = (V, E)$, let $\vec{G} = (V, E^+ \cup E^-)$ be a digraph obtained from making two copies of each edge $e \in E$ and directing them oppositely, one arc being denoted by $e^+ \in E^+$ and the other by $e^- \in E^-$. A *τ-strongly connected orientation (τ-SCO)* of G is a multi-subset of arcs from $E^+ \cup E^-$ picking exactly τ many of e^+ and e^- (possibly with repetitions) for each e such that at least τ arcs leave every cut. In particular, a *strongly connected orientation (SCO)* of G is a 1-SCO of G. In other words, a τ-SCO is an integral vector in the polyhedron

$$P_0^\tau := \big\{x \in \mathbb{R}^{E^+\cup E^-} \,\big|\, x_{e^+} \geq 0,\ x_{e^-} \geq 0,\ \forall e \in E,\quad x_{e^+} + x_{e^-} = \tau,\ \forall e \in E,\quad x(\delta^+(U)) \geq \tau,\ \forall U \subsetneq V, U \neq \emptyset\big\}. \tag{1}$$

One may ask whether a τ-SCO can always be decomposed into τ disjoint SCO's. This is not the case. Indeed we prove in Theorem 9 that this question is equivalent the Edmonds-Giles conjecture.

In contrast, we define x to be a *nowhere-zero τ-SCO* if it is a τ-SCO and $x_e \geq 1$ for every arc e. In other words, a nowhere-zero τ-SCO is an integral vector in the polyhedron

$$P_1^\tau := \big\{x \in \mathbb{R}^{E^+\cup E^-} \,\big|\, x_{e^+} \geq 1,\ x_{e^-} \geq 1,\ \forall e \in E,\quad x_{e^+} + x_{e^-} = \tau,\ \forall e \in E,\quad x(\delta^+(U)) \geq \tau,\ \forall U \subsetneq V, U \neq \emptyset\big\}. \tag{2}$$

In Theorem 10, we prove that Woodall's conjecture is true if and only if for every undirected graph G, a *nowhere-zero* τ-SCO can be decomposed into τ disjoint SCO's.

Related Work

Shepherd and Vetta [23] raised the question of approximately packing dijoins. They also introduced the idea of adding reverse arcs to make the digraph τ-strongly-connected, then packing strongly connected subdigraphs, and finally pairing up in- and out-arborescences. Yet, this approach itself only gives a half integral packing of value $\frac{\tau}{2}$ in a digraph with minimum dicut size τ. It is conjectured by Király [15] that every digraph with minimum dicut size τ contains two disjoint $\lfloor\frac{\tau}{2}\rfloor$-dijoins, see also [1]. One might notice that if this conjecture is true, together with the approach of combining in and out r-arborescences, one can show that there exist $\lfloor\frac{\tau}{2}\rfloor$ disjoint dijoins in a digraph with minimum dicut size τ. Abdi et al. [2] proved that every digraph can be decomposed into a dijoin and a $(\tau - 1)$-dijoin. Abdi et al. [1] further showed that a digraph with minimum dicut size τ can be decomposed into a k-dijoin and a $(\tau - k)$-dijoin for every integer $k \in \{1, ..., \tau - 1\}$ under the condition that the underlying undirected graph is τ-edge-connected. Mészáros [18] proved that when the underlying undirected graph is $(q-1, 1)$-partition-connected for some prime power q, the digraph can be decomposed into q disjoint dijoins. However, none of these approaches tell us how to decompose a digraph with minimum dicut size τ into a large number of disjoint dijoins without connectivity requirements. We also refer to the papers that view the problem from the perspective of reorienting the directions of a subset of arcs to make the graph strongly connected, such as [1,5,19]. For the context of nowhere-zero k-flow, we refer interested readers to [11–13,17,22,24,27] and the excellent survey by Jaeger [14]. Finally, Schrijver's unpublished notes [19] reformulate Woodall's conjecture into the problem of partitioning the arcs of the digraph into strengthenings. A *strengthening* is an arc set $J \subseteq A$ which, on flipping the orientation of the arcs in J, makes the digraph strongly connected. This inspired the reformulations in Theorem 9 and Theorem 10.

2 An Approximate Packing of Dijoins

In this section we prove our main result, Theorem 3. We begin by observing that the existence of a constant valued nowhere-zero circular flow implies that there is a nearly balanced orientation in the sense that, for each cut, the number of arcs entering it differs by a constant factor from the number of arcs leaving it. This is already pointed out in different places (e.g. see in [9, 10, 24]). We summarize this key fact in the following lemma. Since we will reuse this fact we also give the proof here. In a digraph $D = (V, A)$, denote by e^{-1} the reverse of arc $e \in A$, and by B^{-1} the arcs obtained by reversing the directions of the arcs in $B \subseteq A$.

Lemma 1. *Let $G = (V, E)$ be an undirected graph that admits a nowhere-zero circular k-flow E^+ and $f : E^+ \to [1, k-1]$, where $k \geq 2$ is a rational number. Let $E^- = (E^+)^{-1}$. Then, for every $U \subsetneq V, U \neq \emptyset$,*

$$\frac{1}{k}|\delta_G(U)| \leq |\delta^+_{E^+}(U)| \leq \frac{k-1}{k}|\delta_G(U)|,$$
$$\frac{1}{k}|\delta_G(U)| \leq |\delta^+_{E^-}(U)| \leq \frac{k-1}{k}|\delta_G(U)|.$$

Proof. By flow conservation, $f(\delta^+_{E^+}(U)) = f(\delta^-_{E^+}(U)), \forall U \subsetneq V, U \neq \emptyset$. Thus, one has $1 \cdot |\delta^+_{E^+}(U)| \leq f(\delta^+_{E^+}(U)) = f(\delta^-_{E^+}(U)) \leq (k-1) \cdot |\delta^-_{E^+}(U)|$. Similarly, one also has $1 \cdot |\delta^-_{E^+}(U)| \leq f(\delta^-_{E^+}(U)) = f(\delta^+_{E^+}(U)) \leq (k-1) \cdot |\delta^+_{E^+}(U)|$. It follows from the equality $|\delta_G(U)| = |\delta^+_{E^+}(U)| + |\delta^-_{E^+}(U)|$ that $\frac{1}{k}|\delta_G(U)| \leq |\delta^+_{E^+}(U)| \leq \frac{k-1}{k}|\delta_G(U)|$ and $\frac{1}{k}|\delta_G(U)| \leq |\delta^-_{E^+}(U)| \leq \frac{k-1}{k}|\delta_G(U)|$. By noticing that $|\delta^+_{E^-}(U)| = |\delta^-_{E^+}(U)|$, the inequality $\frac{1}{k}|\delta_G(U)| \leq |\delta^+_{E^-}(U)| \leq \frac{k-1}{k}|\delta_G(U)|$ holds. □

Let $D = (V, A)$ be a digraph. By Lemma 1, both the subdigraph consisting of the arcs that are in the same orientation as the nowhere-zero circular flow and its complement intersect every dicut in a large proportion of its size. This gives us a way to decompose the digraph into two k-dijoins with a large k (an example is in Fig. 1). Recall that a k-dijoin is an arc set that intersects each dicut at least k times.

Proposition 1. *For a digraph $D = (V, A)$ with minimum dicut size τ, if the underlying undirected graph admits a nowhere-zero circular k-flow for some rational number $k \geq 2$, then D contains two disjoint $\lfloor \frac{\tau}{k} \rfloor$-dijoins.*

Proof. Let E^+ and $f : E^+ \to [1, k-1]$ be a nowhere-zero circular k-flow of the underlying undirected graph G of D. By Lemma 1, $\frac{1}{k}|\delta_G(U)| \leq |\delta^+_{E^+}(U)| \leq \frac{k-1}{k}|\delta_G(U)|$ for every $U \subsetneq V, U \neq \emptyset$. Take $J = A \cap E^+$ to be the arcs that have the same directions in A and E^+. Then, for a dicut $\delta^+_D(U)$ such that $\delta^-_D(U) = \emptyset$, we have $|J \cap \delta^+_D(U)| = |\delta^+_{E^+}(U)| \geq \frac{1}{k}|\delta_G(U)| = \frac{1}{k}|\delta^+_D(U)| \geq \frac{\tau}{k}$ and $|(A \setminus J) \cap \delta^+_D(U)| = |\delta^+_D(U)| - |\delta^+_{E^+}(U)| \geq |\delta^+_D(U)| - \frac{k-1}{k}|\delta_G(U)| = |\delta^+_D(U)| - \frac{k-1}{k}|\delta^+_D(U)| \geq \frac{\tau}{k}$. Thus, both J and $A \setminus J$ are $\lfloor \frac{\tau}{k} \rfloor$-dijoins. □

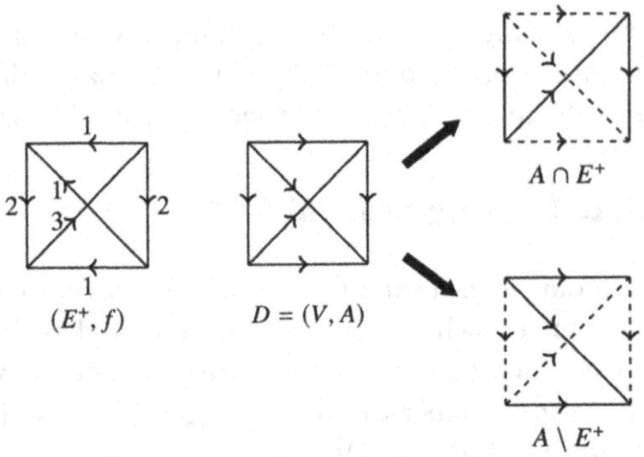

Fig. 1. (E^+, f) is a nowhere-zero 4-flow of $G = K_4$. $D = (V, A)$, whose underlying undirected graph is G, can be decomposed into a dijoin $A \cap E^+$ and a 2-dijoin $A \setminus E^+$.

In a digraph D with minimum dicut size τ, although Proposition 1 suggests that D can be decomposed into two digraphs, each being a $\lfloor \frac{\tau}{k} \rfloor$-dijoin, there is no guarantee that the new digraphs have minimum dicut size at least $\lfloor \frac{\tau}{k} \rfloor$. This is because a non-dicut in D may become a dicut when we delete arcs, which can potentially have very small size. This is a general difficulty with inductive proofs for decomposing a digraph into dijoins.

The key observation here is that, by switching to the setting of strongly connected digraphs, we can bypass this issue. Given a digraph $D = (V, A)$ with minimum dicut size τ, let G be the underlying undirected graph of D and $\vec{G} = (V, E^+ \cup E^-)$ be the digraph obtained by copying each edge of G twice and directing them oppositely. For convenience, we let $E^+ = A$ and $E^- = A^{-1}$. Define the *weights associated with* D to be $w^D \in \mathbb{Z}^{E^+ \cup E^-}$ such that $w^D_{e^+} = \tau, \forall e^+ \in E^+$ and $w^D_{e^-} = 1, \forall e^- \in E^-$. It is easy to see that $\vec{G}$ with weight w^D is τ-SCD. Indeed, for every $U \subsetneq V, U \neq \emptyset$ such that $\delta^+_D(U) \neq \emptyset$, there exists some arc $e^+ \in E^+$ such that $e^+ \in \delta^+_{\vec{G}}(U)$, and thus $w^D(\delta^+_{\vec{G}}(U)) \geq w^D_{e^+} = \tau$. Otherwise, $\delta^+_D(U) = \emptyset$ which means $\delta^-_D(U)$ is a dicut. Therefore, $w^D(\delta^+_{\vec{G}}(U)) = w^D(\delta^+_{E^-}(U)) = |\delta^-_D(U)| \geq \tau$. This means that the augmented digraph $\vec{G}$ with weight w^D is τ-strongly-connected. We first reformulate the problem of packing dijoins in D into a problem of packing strongly connected digraphs in $\vec{G}$ under weight w^D. We then prove a decomposition result into two $\lfloor \tau/k \rfloor$-strongly-connected digraphs with the help of nowhere-zero circular k-flows. The following reformulation has essentially been stated and used in [23]. We include its proof here.

Proposition 2. *For an integer $k \leq \tau$, the digraph D contains k disjoint dijoins if and only if $\vec{G}$ with weight w^D can pack k strongly connected digraphs.*

Proof. Let $F_1, \ldots, F_k$ be k strongly connected digraphs of G that is a packing of $\vec{G}$ under weight w^D. Define $J_i := \{e^+ \in E^+ \mid \chi_{F_i}(e^-) = 1, e^- \in E^-\}$. We claim each J_i is a dijoin of D. Suppose not. Then there exists some dicut $\delta^-_D(U)$ such that $J_i \cap \delta^-_D(U) = \emptyset$. This implies $F_i \cap \delta^+_{\vec{G}}(U) = \emptyset$, contradicting the fact that F_i is a strongly connected digraph of $\vec{G}$. Moreover, since $w^D_{e^-} = 1$, at most one of $F_1, ..., F_k$ uses e^-, $\forall e^- \in E^-$. Thus, at most one of $J_1, ..., J_k$ uses $e^+, \forall e^+ \in E^+ = A$. Therefore, $J_1, ..., J_k$ are disjoint dijoins of D.

Conversely, let $J_1, ..., J_k$ be k disjoint dijoins in D. W.l.o.g. we can assume each J_i is a minimal dijoin, that is, J_i is not contained in another dijoin. It is shown by Frank (see e.g. in [16], Chap. 6) that each minimal dijoin is a strengthening, i.e., $(A \setminus J_i) \cup J_i^{-1}$ is a strongly connected digraph. Let $F_i := (A \setminus J_i) \cup J_i^{-1}$, $\forall i$. The same argument as for the other direction applies to argue that $F_1, ..., F_k$ is a valid packing of strongly connected digraphs in $\vec{G}$ under weight w^D. □

Theorem 6. *Let $D = (V, A)$ be a digraph with minimum dicut size τ. If the underlying undirected graph admits a nowhere-zero circular k-flow for some rational number $k \geq 2$, then the weight w^D associated with D contains two disjoint $\lfloor \frac{\tau}{k} \rfloor$-SCD's.*

Proof. Let E^+ and $f : E^+ \to \{1, ..., k-1\}$ be a nowhere-zero k-flow of G. Let E^- be obtained by reversing the arcs of E^+. Let G be the underlying undirected graph of D and $\vec{G} = (V, E^+ \cup E^-)$. Construct $x \in \mathbb{Z}^{E^+ \cup E^-}$ as follows.

$$x_e = \begin{cases} \lceil \frac{\tau}{2} \rceil, & e \in A \cap E^+ \\ \lfloor \frac{\tau}{2} \rfloor, & e \in A \cap E^- \\ 1, & e \in A^{-1} \cap E^+ \\ 0, & e \in A^{-1} \cap E^- \end{cases}, \quad \text{and equivalently} \quad (w^D - x)_e = \begin{cases} \lfloor \frac{\tau}{2} \rfloor, & e \in A \cap E^+ \\ \lceil \frac{\tau}{2} \rceil, & e \in A \cap E^- \\ 0, & e \in A^{-1} \cap E^+ \\ 1, & e \in A^{-1} \cap E^- \end{cases}.$$

We prove that both x and $(w^D - x)$ are $\lfloor \frac{\tau}{k} \rfloor$-SCD's. We discuss two cases.

If $\delta_D(U)$ is a dicut, then $|\delta_G(U)| \geq \tau$. Since $\tau \geq 2$, we have $x_e \geq 1$, $\forall e \in \delta^+_{E^+}(U)$. Therefore, $x(\delta^+_{\vec{G}}(U)) \geq x(\delta^+_{E^+}(U)) \geq |\delta^+_{E^+}(U)| \geq \frac{1}{k}|\delta_G(U)| \geq \frac{\tau}{k}$, where the third inequality follows from Lemma 1. On the other hand, $(w^D - x)_e \geq 1$, $\forall e \in \delta^+_{E^-}(U)$. Therefore, $(w^D - x)(\delta^+_{\vec{G}}(U)) \geq (w^D - x)(\delta^+_{E^-}(U)) \geq |\delta^+_{E^-}(U)| \geq \frac{1}{k}|\delta_G(U)| \geq \frac{\tau}{k}$.

If $\delta_D(U)$ is not a dicut, then $\delta^+_{\vec{G}}(U) \cap A \neq \emptyset$. Therefore, $x(\delta^+_{\vec{G}}(U)) \geq x(\delta^+_{\vec{G}}(U) \cap A) \geq \lfloor \frac{\tau}{2} \rfloor \geq \lfloor \frac{\tau}{k} \rfloor$ since $k \geq 2$. Also, $(w^D - x)(\delta^+_{\vec{G}}(U)) \geq (w^D - x)(\delta^+_{\vec{G}}(U) \cap A) \geq \lfloor \frac{\tau}{2} \rfloor \geq \lfloor \frac{\tau}{k} \rfloor$ since $k \geq 2$. Therefore, both x and $(w^D - x)$ are $\lfloor \frac{\tau}{k} \rfloor$-SCD's. □

Proof of Theorem 3

From Proposition 2, given a digraph D, we can reduce the problem of packing dijoins of D to that of packing strongly connected digraphs of the augmented digraph $\vec{G}$ with weight w^D which is τ-strongly-connected. To achieve the goal, we recall a classical theorem about decomposing digraphs into arborescences. In a digraph $D = (V, A)$ with a fixed root r, an *out (in) r-arborescence* is a directed spanning tree such that each vertex in $V \setminus \{r\}$ has exactly one arc entering (leaving) it. If the root is not fixed it is called an *out (in) arborescence*. Edmonds' disjoint arborescences theorem [7] states that when fixing a root r, every *rooted-τ-connected* digraph, i.e., $|\delta^+_D(U)| \geq \tau, \forall U \subsetneq V, r \in U$, can be decomposed into τ disjoint out r-arborescences. Furthermore, this decomposition can be done in polynomial time.

Theorem 7 ([7]). *Given a digraph D and a root r, if D is rooted-τ-connected, then D contains τ disjoint out r-arborescences, and such r-arborescences can be found in polynomial time.*

A τ-strongly-connected digraph is in particular rooted-τ-connected. Therefore, fixing a root $r \in V$, a τ-strongly-connected digraph contains τ disjoint out r-arborescences. If we reverse the directions of the arcs and apply Theorem 7, we see that a τ-strongly-connected digraph also contains τ disjoint in r-arborescences.

Therefore, we can decompose digraph $\vec{G}$ with weight w^D into τ in r-arborescences, or into τ out r-arborescences. Pairing each in r-arborescence with an out r-arborescence, we obtain τ strongly connected digraphs. However, each arc can be used in both in and out r-arborescences. Shepherd and Vetta [23] use this idea to obtain a half integral packing of dijoins of value $\frac{\tau}{2}$. Yet, finding disjoint in and out arborescences together is quite challenging. It is open whether there exists τ such that a τ-strongly-connected digraph can even pack one in-arborescence and one out-arborescence [3].

Theorem 6 paves the way to approximately packing disjoint in and out arborescences in our instances. Fixing a root r, if we are able to decompose the graph into two τ'-strongly-connected graphs and thereby find τ' disjoint in r-arborescences in the first graph and τ' disjoint out r-arborescences in the second graph, then we can combine them to get a strongly connected digraph.

Proof of Theorem 3. By Proposition 2, it suffices to prove that w^D can pack $\lfloor \frac{\tau}{k} \rfloor$ strongly connected digraphs. By Theorem 6, $\vec{G}$ with weight w^D can be decomposed into weighted digraphs J_1 and J_2 such that each of them is $\lfloor \frac{\tau}{k} \rfloor$-strongly-connected. Fixing an arbitrary root r, since a $\lfloor \frac{\tau}{k} \rfloor$-strongly-connected digraph is in particular rooted-$\lfloor \frac{\tau}{k} \rfloor$-connected, by Theorem 7, J_1 can be decomposed into $\lfloor \frac{\tau}{k} \rfloor$ disjoint out r-arborescences $S_1, \ldots, S_{\lfloor \frac{\tau}{k} \rfloor}$. Similarly, J_2 can be decomposed into $\lfloor \frac{\tau}{k} \rfloor$ disjoint in r-arborescences $T_1, \ldots, T_{\lfloor \frac{\tau}{k} \rfloor}$. Let $F_i := S_i \cup T_i$, for $i = 1, \ldots, \lfloor \frac{\tau}{k} \rfloor$. Each F_i is a strongly connected digraph. This is because every out r-cut $\delta^+_{\vec{G}}(U), r \in U$ is covered by S_i and every in r-cut $\delta^+_{\vec{G}}(U), r \notin U$ is covered by T_i and thus every cut $\delta^+_{\vec{G}}(U)$ is covered by F_i. Therefore, $F_1, \ldots, F_{\lfloor \frac{\tau}{k} \rfloor}$ forms a packing of strongly connected digraphs under weight w^D. □

Theorem 1 now follows by combining Theorem 3 and Theorem 4 and noting that the underlying undirected graph of a digraph with minimum dicut size $\tau \geq 2$ is 2-edge-connected. By Theorem 4, the nowhere-zero 6-flow can be found in polynomial time, and thus the decomposition described in Theorem 6 can be done in polynomial time. Moreover, further decomposing J_1 and J_2 into in and out r-arborescences can also be done in polynomial time due to Theorem 7. Thus in the end we can find $\lfloor \frac{\tau}{6} \rfloor$ disjoint dijoins in polynomial time. Theorem 2 now follows by combining Theorem 3 and Theorem 5. However, as far as we know there is no constructive version of Theorem 5, which means Theorem 2 cannot be made algorithmic directly.

3 A Reformulation of Woodall's Conjecture in Terms of Strongly Connected Orientations

In this section, we discuss the relation between packing dijoins, strongly connected orientations and strongly connected digraphs. We also discuss another reformulation of Woodall's conjecture in terms of strongly connected orientations.

Given an undirected graph $G = (V, E)$, let $\vec{G} = (V, E^+ \cup E^-)$ be the digraph obtained by copying each edge of G twice and orienting them in opposite directions. Denote by $\chi_F \in \{0, 1\}^E$ the characteristic vector of F. Let

$$SCO(G) := \{x \in \{0,1\}^{E^+ \cup E^-} | x = \chi_O \text{ for some strongly connected orientation } O \text{ of } G\}. \tag{3}$$

Recall that strongly connected orientations (SCO's) are $0, 1$ vectors in the polyhedron P_0^1 defined in (1). Recall that given a digraph $D = (V, A)$, a strengthening is a subset $J \subseteq A$ such that by flipping the orientation of the arcs in J the digraph becomes strongly connected [16]. Note that a strengthening is necessarily a dijoin.

Schrijver observed the following reformulation of Woodall's conjecture in terms of strengthenings in his unpublished note ([19], Sect. 2).

Theorem 8 ([19]). *Woodall's conjecture is true if and only if, in every digraph with minimum dicut size τ, the arcs can be partitioned into τ strengthenings.*

Another way to look at $SCO(G)$ is to fix a direction E^+ and view it as a lift of the set of strengthenings of $G^+ = (V, E^+)$. Indeed, given a strengthening $J \subseteq E^+$, $(E^+ \setminus J) \cup J^{-1}$ is a strongly connected orientation of G. Conversely, given a strongly connected orientation $O \subseteq E^+ \cup E^-$, $E^+ \setminus O$ is a strengthening of G^+. The characteristic vectors of the strengthenings of G^+ are the $0, 1$ vectors in the following polyhedron:

$$\{x \in \mathbb{R}^{E^+} | \, 0 \le x_{e^+} \le 1, \ \forall e \in E,$$
$$x(\delta^-_{G^+}(U)) - x(\delta^+_{G^+}(U)) \le |\delta^-_{G^+}(U)| - 1, \forall U \subsetneq V, U \neq \emptyset\}.$$

Due to the Edmonds-Giles submodular flow theorem [8], this is an integral polytope and thus it describes the convex hull of the set of strengthenings of G^+, see also [21]. Note that P_0^1 is a linear transformation of the above polyhedron and thus it is also integral, which means $\text{conv}(SCO(G)) = P_0^1$.

Recall that a τ-SCO is an integral vector in P_0^τ defined in (1). A nowhere-zero τ-SCO is an integral vector in P_1^τ defined in (2). A τ-SCO cannot always be integrally decomposed into τ SCO's, however, due to the following equivalence.

Theorem 9. *The Edmonds-Giles Conjecture 1 is true if and only if for every undirected graph G and integer $\tau > 0$, every τ-SCO can be decomposed into τ SCO's.*

The following counterexample to the Edmonds-Giles Conjecture 1 discovered by Schrijver [20] can be translated to disprove the statement that every τ-SCO can be decomposed into τ SCO's. Let $x \in \mathbb{Z}^{E^+ \cup E^-}$ be defined by $x_e = 1$ if e is solid, $x_e = 2$ if e is dashed, and $x_e = 0$ for the reverse of the dashed arcs (which we do not draw here). The vector x is a 2-SCO but it cannot be decomposed into 2 strongly connected orientations. (See Fig. 2).

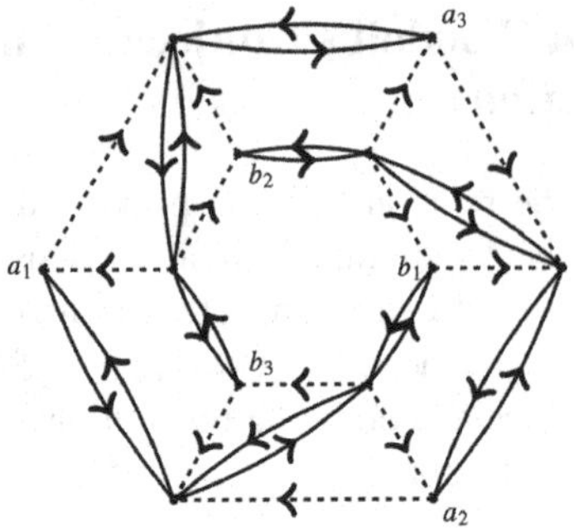

Fig. 2. The solid arcs with weight 1 and dashed arcs with weight 2 cannot be decomposed into 2 SCO's O_1, O_2. Assume for a contradiction that O_1, O_2 exist. The dashed arcs have their orientation fixed in both O_i. Three paths consisting of solid arcs in between a_i, b_i have to be directed paths in both O_i, otherwise there is a trivial dicut along the paths in some O_i. Both O_i need to enter the inner hexagon from the outer hexagon, which means each O_i should have at least one directed path oriented as $a_i \to b_i$. Thus, one O_i has exactly one directed path oriented as $b_i \to a_i$ and two oriented as $a_i \to b_i$. Assume O_1 has orientation $b_1 \to a_1, a_2 \to b_2$ and $a_3 \to b_3$. This leaves no arc to go from the left half to the right half of the graph, a contradiction to O_1 being an SCO.

However, slightly revising the statement, we obtain an equivalent form of Woodall's conjecture 2, which is still open.

Theorem 10. *Woodall's Conjecture 2 is true if and only if for every undirected graph G and integer $\tau > 0$, every nowhere-zero τ-SCO can be decomposed into τ SCO's.*

We will first prove Theorem 10 and modify the proof to prove Theorem 9. Our proof of Theorem 10 is inspired by Schrijver's Theorem 8. Schrijver's reformulation essentially covers the special case when $\bar{w}^D \in \mathbb{Z}^{E^+ \cup E^-}$ with $\bar{w}^D_{e^+} = \tau - 1, \forall e^+ \in E^+$ and $\bar{w}^D_{e^-} = 1, \forall e^- \in E^-$ in Theorem 10. One can also easily verify that $\bar{w}^D$ is a nowhere-zero τ-SCO. We generalize the weights to be any nowhere-zero τ-SCO of D, and thus give a stronger consequence of Woodall's conjecture. By allowing the entries of a τ-SCO to take 0 values, we give an equivalent statement of the Edmonds-Giles conjecture in Theorem 9, showing a contrast between the two conjectures.

Proof of Theorem 10. We first prove the "if" direction. Let $D = (V, A)$ be a digraph (e.g. Figure 3-(1)) whose underlying undirected graph is $G = (V, E)$. Let τ be the size of a minimum dicut of D. We assume $\tau \geq 2$ w.l.o.g. and this implies that the size of minimum cut of D is also greater than or equal to 2. By making two copies of each edge of G and orienting them oppositely, we obtain $\vec{G} = (V, E^+ \cup E^-)$. For convenience we will assume that e^+ and e^- are defined according to their direction in D, i.e., $e = (u, v) \in A$ iff $e^+ = (u, v)$ and $e^- = (v, u)$. In other words, $E^+ = A$ and $E^- = A^{-1}$. Take $x \in \mathbb{Z}^{E^+ \cup E^-}$ such that $x_{e^+} = \tau - 1$, $x_{e^-} = 1$ for every $e \in E$ (as shown in Fig. 3-(2)). We claim that $x \in P_1^{\tau}$. The only nontrivial constraint to prove is $x(\delta^+_{\vec{G}}(U)) \geq \tau$ for every $U \subsetneq V, U \neq \emptyset$. If $\delta^-_D(U)$ is a dicut such that $\delta^+_D(U) = \emptyset$, then $x(\delta^+_{\vec{G}}(U)) = x(\delta^+_{E^-}(U)) = |\delta^-_D(U)| \geq \tau$. Otherwise, $\delta^+_D(U) \neq \emptyset$

and thus $\delta^+_{\vec{G}}(U)$ contains at least one arc in E^+. Moreover, since $|\delta_D(U)| \geq 2$, one has $x(\delta^+_{\vec{G}}(U)) = x(\delta^+_{E^+}(U)) + x(\delta^+_{E^-}(U)) \geq (\tau - 1) + 1 = \tau$. Thus, $x \in P^\tau_1$. By the assumption, $x = \sum_{i=1}^{\tau} \chi_{O_i}$ where each O_i is a strongly connected orientation. Take $J_i = \{e^+ \in E^+ \mid \chi_{O_i}(e^-) = 1, e^- \in E^-\}$. Note that $(A \setminus J_i) \cup (J_i^{-1}) = O_i$. Therefore, $(A \setminus J_i) \cup (J_i^{-1})$ is strongly connected, which means J_i is a strengthening of D, and thus a dijoin of D. Since $x_{e^-} = 1$ for each $e \in E$, J_i's are disjoint. Thus we get τ disjoint dijoins of D. We now prove the "only if" direction. Given an undirected graph $G = (V, E)$, consider the corresponding directed graph $\vec{G} = (V, E^+ \cup E^-)$ with each edge of E copied and oppositely oriented. For an integral $x \in P^\tau_1$, (e.g. Figure 3-(2)) construct a new digraph D from G in the following way. For each edge $e = (u, v) \in E$ where $e^+ = (u, v)$ and $e^- = (v, u)$, add a node w_e, add $x_{e^+} \geq 1$ arcs from u to w_e and $x_{e^-} \geq 1$ arcs from v to w_e, and delete e (as shown in Fig. 3-(3)). We claim that the size of a minimum dicut of D is τ. Every vertex w_e induces a dicut $\delta^-_D(\{w_e\})$ of size $x_{e+}+x_{e^-} = \tau$. Thus, we only need to show that the size of every dicut of D is at least τ. Given U such that $\delta^-_D(U) = \emptyset$, if there exists $e = (u, v) \in E$ such that $u, v \in U$ but $w_e \notin U$, then $|\delta^+_D(U)| \geq |\delta^-_D(\{w_e\})| \geq \tau$. Thus, we may assume w.l.o.g. that for every $e = (u, v) \in E$ such that $u, v \in U$, we also have $w_e \in U$. Since $\delta^-_D(U) = \emptyset$, for every $e = (u, v) \in E$ such that $u, v \notin U$, we also have $w_e \notin U$. Moreover, for every $e = (u, v) \in E$ such that $u \in U$, $v \notin U$, since there is at least an arc from v to w_e but $\delta^-_D(U) = \emptyset$, we infer that $w_e \notin U$. Thus, $\delta^+_D(U) = \{uw_e \mid e = (u, v) \in E, u \in U, v \notin U\}$. Thus, by the way we construct D, $|\delta^+_D(U)| \geq x(\delta^+_{\vec{G}}(U)) \geq \tau$. Therefore, D has minimum dicut size τ. By Woodall's conjecture, there exists τ disjoint dijoins $J_1, ..., J_\tau$ in D. In particular, each dijoin intersects dicut $\delta^-_D(\{w_e\})$ exactly once since $|\delta^-_D(\{w_e\})| = \tau$. Let O_i be an orientation defined by $O_i := \{e^+ \mid uw_e \in J_i\} \cup \{e^- \mid vw_e \in J_i\}$. Note that O_i is indeed an orientation since exactly one of uw_e and vw_e is in J_i, for every $e \in E$. We claim that each O_i is a strongly connected orientation of G. Assume not. Then there exists $U \subseteq V$, such that $\delta^+_{\vec{G}}(U) \cap O_i = \emptyset$. Let $U' := U \cup \{w_e \mid e = (u, v) \in E, u, v \in U\}$. It is easy to see that U' is a dicut of D such that $\delta^-_D(U') = \emptyset$. It follows from $\delta^+_{\vec{G}}(U) \cap O_i = \emptyset$

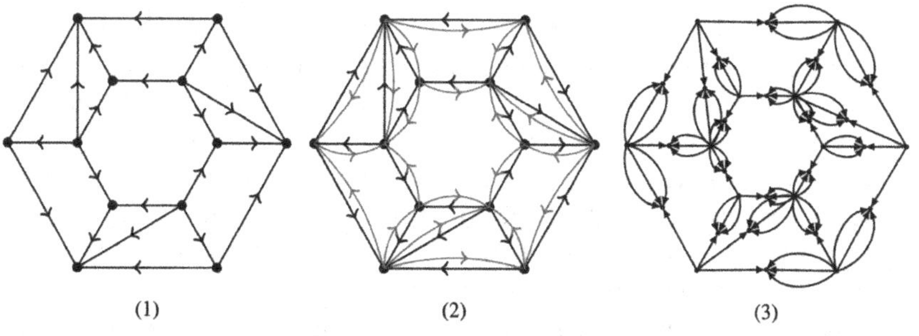

Fig. 3. (1) is a digraph with minimum dicut size 4. In (2) the weights of black arcs are 3 and the weights of gray arcs are 1. This figure illustrates how to convert from a digraph D (1) to a weighted digraph $\vec{G}$ (2) in the first part of the proof of Theorem 10 and how to convert from a weighted digraph $\vec{G}$ (2) to a digraph D (3) in the second part of the proof of Theorem 10.

that $\delta^+_D(U') \cap J_i = \emptyset$, a contradiction to J_i being a dijoin of D. Moreover, by the way we construct D, for each $e^+ = (u, v)$, $\sum_{i=1}^{\tau} \chi_{O_i}(e^+) = |\{J_i \mid uw_e \in J_i\}| = x_{e^+}$. For each $e^- = (v, u)$, $\sum_{i=1}^{\tau} \chi_{O_i}(e^-) = |\{J_i \mid vw_e \in J_i\}| = x_{e^-}$. Therefore, $\sum_{i=1}^{\tau} \chi_{O_i} = x$. This ends the proof of this direction. □

To prove Theorem 9, we need a structural lemma.

Lemma 2. *Let $D = (V, A)$ be a digraph with weight $w \in \{0, 1\}^A$ and assume that the minimum weight of a dicut is $\tau \geq 2$. Let $e \in A$ be some arc such that $w_e = 1$. If there exists a cut $\delta_D(U)$ such that $\delta^+_D(U) = \{e\}$ and $w(\delta^-_D(U)) = 0$, then e is not contained in any minimum dicut of D.*

Proof. Suppose not. Then there exists a dicut $\delta^-_D(W)$ such that $w(\delta^-_D(W)) = \tau$ and $e \in \delta^-_D(W)$. Let D' be obtained from D by deleting e. Then $\delta^-_{D'}(U)$ becomes a dicut of D'. Therefore, $\delta^-_{D'}(U \cap W)$ and $\delta^-_{D'}(U \cup W)$ are both dicuts of D'. However, since e leaves U and enters W, e goes from $U \setminus W$ to $W \setminus U$. Thus, $e \notin \delta_D(U \cap W)$ and $e \notin \delta_D(U \cup W)$. Therefore, both $\delta^-_D(U \cap W)$ and $\delta^-_D(U \cup W)$ are dicuts of D. Moreover, $w(\delta^-_D(U \cap W)) + w(\delta^-_D(U \cup W)) = w(\delta^-_D(U)) + w(\delta^-_D(W)) - 1 = \tau - 1$. It follows that $w(\delta^-_D(U \cap W)) \leq \tau - 1$ and $w(\delta^-_D(U \cup W)) \leq \tau - 1$. Notice that either $U \cap W \neq \emptyset$ or $U \cup W \neq V$. Otherwise, e is a bridge of D, contradicting $\tau \geq 2$. Therefore, either $\delta^-_D(U \cap W)$ or $\delta^-_D(U \cup W)$ violates the assumption that the size of a minimum dicut is τ, a contradiction. □

Proof of Theorem 9. We modify the proof of Theorem 10 to prove Theorem 9. We first prove the "if" direction. Let $D = (V, A)$ be a digraph with weights $w \in \{0, 1\}^A$ and minimum dicut $\tau \geq 2$. We can assume there is no arc $e \in A$ with weight 1 such that there exists a cut $\delta_D(U)$ such that $\delta^+_D(U) = \{e\}$ and $w(\delta^-_D(U)) = 0$. For otherwise, by Lemma 2, e is not contained in any minimum dicut, which means we can set the weight of e to be 0 without decreasing the size of a minimum dicut. Any packing of τ dijoins in the new graph will be a valid packing of τ dijoins of the old graph.

Let G be the underlying undirected graph of D and $\vec{G} = (V, E^+ \cup E^-)$ be defined as before such that $E^+ = A$ and $E^- = A^{-1}$. Define $x \in \mathbb{Z}^{E^+ \cup E^-}$ as follows. For weight 1 arcs $e^+ \in A$, we define $x_{e^+} = \tau - 1$ and $x_{e^-} = 1$ as before. For the weight 0 arcs $e^+ \in A$, we define $x_{e^+} = \tau$ and $x_{e^-} = 0$. To argue that $x(\delta^+_{\vec{G}}(U)) \geq \tau$ for every $U \subsetneq V, U \neq \emptyset$, if $\delta^-_D(U)$ is a dicut it follows in the same way as in the proof of Theorem 10. Therefore, without loss of generality, we assume there exists at least one arc $e^+ \in E^+$ in $\delta^+_D(U)$. If there exists such an arc with $w(e^+) = 0$, then $x(\delta^+_{\vec{G}}(U)) \geq x_{e^+} \geq \tau$. Otherwise, all the arcs in $\delta^+_D(U)$ have weight 1. If there exist at least 2 arcs of weight 1 in $\delta_D(U)$, we follow the same argument as in the earlier proof. The only case left is when $\delta^+_D(U)$ is a single arc of weight 1 and all the arc in $\delta^-_D(U)$ has weight 0, which has been excluded in the beginning. Therefore, we have proved $x(\delta^+_{\vec{G}}(U)) \geq \tau$ for every $U \subsetneq V, U \neq \emptyset$, which implies $x \in P_0^{\tau}$. By the assumption, $x = \sum_{i=1}^{\tau} \chi_{O_i}$ where each O_i is a strongly connected orientation. We define the dijoins in the same way as the other proof. Note that the dijoins are disjoint and never use weight 0 arcs. Therefore, we find τ dijoins that form a valid packing of graph D with weight w.

Next, we prove the "only if" direction. Given an undirected graph $G = (V, E)$, the corresponding $\vec{G} = (V, E^+ \cup E^-)$, and an integral $x \in P_0^{\tau}$, we construct weighted digraph

D as follows. For an edge $e = (u, v)$ such that $x_{e^+}, x_{e^-} \geq 1$, we construct node w_e and arcs uw_e, vw_e in the same way as in the proof of Theorem 10. For an edge $e^+ = (u, v)$ such that $x_{e^+} = \tau$, $x_{e^-} = 0$, we add node w_e, add τ arcs of weight 1 from u to w_e and add a weight 0 arc from v to w_e. Similarly, for $e^+ = (u, v)$ with $x_{e^+} = 0$, $x_{e^-} = \tau$, we add node w_e, add a weight 0 arc from u to w_e and τ arcs of weight 1 from v to w_e. The same argument applies to see the minimum dicut size of D is τ. By Edmonds-Giles' conjecture, we can find τ disjoint dijoins in the weighted digraph D. As before, we can find τ strongly connected orientations accordingly that sum up to x. □

4 Conclusions and Discussions

We showed that every digraph with minimum dicut size τ can pack $\lfloor \frac{\tau}{6} \rfloor$ dijoins, or $\left\lfloor \frac{\tau p}{2p+1} \right\rfloor$ dijoins when the digraph is $6p$-edge-connected. The existence of nowhere-zero circular k-flow for a smaller k (< 6) when special structures are imposed on the underlying undirected graphs would lead to a better ratio, i.e., $\lfloor \frac{\tau}{k} \rfloor$, approximate packing of dijoins for those digraphs. The limitation of this approach is that we cannot hope that nowhere-zero 2-flows always exist because this is equivalent to the graph being Eulerian. Thus, bringing the number up to $\lfloor \frac{\tau}{2} \rfloor$ disjoint dijoins would be challenging using this approach. However, it is necessary for Woodall's conjecture to be true that every digraph with minimum dicut size τ contains two disjoint $\lfloor \frac{\tau}{2} \rfloor$-dijoins. Therefore, new ideas are needed to prove or disprove whether such a decomposition exists.

The careful reader may have noticed that the approach only works for the unweighted case. Yet, by a slight modification of the argument, it extends to the weighted case when the underlying undirected graph of the weight-1 arcs is 2-edge-connected. In this case, we can find a nowhere-zero k-flow on the weight-1 arcs and construct the decomposition of weight-1 arcs the same way as in Theorem 6. However, unfortunately, in general, the underlying graph of weight-1 arcs may have bridges or be disconnected, in which case the above argument does not work. Studying a proper analogue of nowhere-zero flows in mixed graphs could be helpful in resolving the question in weighted digraphs.

Acknowledgements. The first author is supported by the U.S. Office of Naval Research under award number N00014-22-1-2528, the second author is supported by a Balas Ph.D. Fellowship from the Tepper School, Carnegie Mellon University and the third author is supported by the U.S. Office of Naval Research under award number N00014-21-1-2243 and the Air Force Office of Scientific Research under award number FA9550-20-1-0080. We thank Ahmad Abdi, Olha Silina and Michael Zlatin for invaluable discussions, and the referees for their detailed suggestions.

References

1. Abdi, A., Cornuéjols, G., Zambelli, G.: Arc connectivity and submodular flows in digraphs (2023). https://www.andrew.cmu.edu/user/gc0v/webpub/connectedflip.pdf
2. Abdi, A., Cornuéjols, G., Zlatin, M.: On packing dijoins in digraphs and weighted digraphs. SIAM J. Discret. Math. **37**(4), 2417–2461 (2023)
3. Bang-Jensen, J., Kriesell, M.: Disjoint sub (di) graphs in digraphs. Electron. Notes Discrete Math. **34**, 179–183 (2009)

4. Bang-Jensen, J., Yeo, A.: Decomposing k-arc-strong tournaments into strong spanning subdigraphs. Combinatorica **24**, 331–349 (2004)
5. Chudnovsky, M., Edwards, K., Kim, R., Scott, A., Seymour, P.: Disjoint dijoins. J. Combin. Theory Ser. B **120**, 18–35 (2016)
6. Devos, M.: Woodall's conjecture. http://www.openproblemgarden.org/op/woodalls_conjecture
7. Edmonds, J.: Edge-disjoint branchings. Combin. Algorithms 91–96 (1973)
8. Edmonds, J., Giles, R.: A min-max relation for submodular functions on graphs. Ann. Discrete Math. **1**, 185–204 (1977)
9. Goddyn, L.A.: Some open problems I like. https://www.sfu.ca/~goddyn/Problems/problems.html
10. Goddyn, L.A., Tarsi, M., Zhang, C.-Q.: On (k, d)-colorings and fractional nowhere-zero flows. J. Graph Theory **28**(3), 155–161 (1998)
11. Jaeger, F.: On nowhere-zero flows in multigraphs. In: Proceedings, Fifth British Combinatorial Conference, Aberdeen, pp. 373–378 (1975)
12. Jaeger, F.: Flows and generalized coloring theorems in graphs. J. Combin. Theory Ser. B **26**(2), 205–216 (1979)
13. Jaeger, F.: On circular flows in graphs. In: Finite and Infinite Sets, pp. 391–402. Elsevier (1984)
14. Jaeger, F.: Nowhere-zero flow problems. Sel. Top. Graph Theory **3**, 71–95 (1988)
15. Király, T.: A result on crossing families of odd sets (2007). https://egres.elte.hu/tr/egres-07-10.pdf
16. Lovász, L.: Combinatorial Problems and Exercises, vol. 361. American Mathematical Society (2007)
17. Lovász, L.M., Thomassen, C., Wu, Y., Zhang, C.-Q.: Nowhere-zero 3-flows and modulo k-orientations. J. Combin. Theory Ser. B **103**(5), 587–598 (2013)
18. Mészáros, A.: A note on disjoint dijoins. Combinatorica **38**(6), 1485–1488 (2018)
19. Schrijver, A.: Observations on Woodall's conjecture. https://homepages.cwi.nl/~lex/files/woodall.pdf
20. Schrijver, A.: A counterexample to a conjecture of Edmonds and Giles. Discret. Math. **32**, 213–214 (1980)
21. Schrijver, A.: Total dual integrality from directed graphs, crossing families, and sub-and supermodular functions. In: Progress in Combinatorial Optimization, pp. 315–361. Elsevier (1984)
22. Seymour, P.D.: Nowhere-zero 6-flows. J. Combin. Theory Ser. B **30**(2), 130–135 (1981)
23. Shepherd, B., Vetta, A.: Visualizing, finding and packing dijoins. Graph Theory Combin. Optim. 219–254 (2005)
24. Thomassen, C.: The weak 3-flow conjecture and the weak circular flow conjecture. J. Combin. Theory Ser. B **102**(2), 521–529 (2012)
25. Tutte, W.T.: A contribution to the theory of chromatic polynomials. Can. J. Math. **6**, 80–91 (1954)
26. Woodall, D.R.: Menger and König systems. In: Theory and Applications of Graphs: Proceedings, Michigan, 11–15 May 1976, pp. 620–635 (1978)
27. Younger, D.H.: Integer flows. J. Graph Theory **7**(3), 349–357 (1983)

Capacitated Facility Location with Outliers and Uniform Facility Costs

Rajni Dabas[1], Naveen Garg[2](✉), and Neelima Gupta[1]

[1] Department of Computer Science, University of Delhi, New Delhi, India
{rajni,ngupta}@cs.du.ac.in
[2] Indian Institute of Technology Delhi, New Delhi, India
naveen@cse.iitd.ac.in

Abstract. We consider the capacitated facility location problem with outliers when facility costs are uniform. Our main result is the first constant factor approximation for this problem. We give a local search algorithm that requires only 2 operations and is a $6.372+\epsilon$ approximation. In developing this result we also improve the approximation guarantee of the capacitated facility location problem with uniform facility costs. Our local search algorithm is extremely simple to analyze and is a $3.732+\epsilon$ approximation thus improving on the 4-approximation of Kao [12].

Keywords: Approximation Algorithms · Facility Location · Outliers · Local Search

1 Introduction

Facility location and its many variants are well-studied NP-hard problems in the operations research and theoretical computer science communities. In the classical facility location problem, we are given a set of clients and (potential) facilities in a metric space. Every facility has an opening cost (*facility cost*) and for every client, facility pair we are given the cost of serving the client by the facility (*service cost*). The objective is to open a subset of facilities so that the cost of opening these facilities and serving all clients from one of the open facilities is minimized. The distance between them determines the cost of serving a client by a facility. When the underlying distances do not form a metric, facility location and its many variants are at least as hard to approximate as set-cover. Hence, in this paper we will be crucially using the fact that the distances (and hence the service costs) form a metric.

Several useful extensions of the facility location problem have been considered by researchers and one of these is the capacitated facility location (CFL), where we are also given a bound on the maximum number of clients that a facility can serve. Hard capacity constraints in facility location and related problems are notoriously difficult to handle. For example, the standard LP for the capacitated

Part of this work was done when the second author was at the University of Warwick on a Royal Society Wolfson visiting Fellowship.

J. Vygen and J. Byrka (Eds.): IPCO 2024, LNCS 14679, pp. 85–98, 2024.
https://doi.org/10.1007/978-3-031-59835-7_7

facility location problem (CFL) is known to have an unbounded integrality gap even when the capacities are uniform. In this paper, we consider the facility opening costs to be uniform, i.e., every facility has the same opening cost. Levi et al. [14] obtained a 5-approximation for CFL with uniform facility costs, Aardal et al. [1] improved this guarantee to $4.562 + \epsilon$ and this was further improved to a 4-approximation by Kao [12]. Our first result is a very simple 2-operation local search algorithm that is a $3.732 + \epsilon$ approximation for CFL with uniform facility costs. The analysis of our algorithm is equally simple and this not only allows us to argue a sharper approximation but also lends itself to extension to facility location with outliers.

In facility location problems serving a few distant clients can affect the cost of the solution disproportionately. In such scenarios, leaving these clients unserved can improve the cost of the solution significantly. To capture this, Charikar et al. [6] introduced the notion of outliers. Outliers are the clients that can be left unserved in our solution. In the facility location with outliers (FLO) problem we are additionally given a positive integer L that specifies the maximum number of clients (outliers) that can be left unserved. The objective now is to open a subset of facilities and identify a set of L outliers such that the cost of opening these facilities and serving the remaining clients from these open facilities is minimized. Charikar et al. [6] showed that the natural LP has an unbounded integrality gap for facility location with outliers and gave a $(3+\epsilon)$ approximation for the problem.

In the capacitated version of facility location with outliers (CFLO) we additionally want to satisfy the capacity constraints on the facilities. The hard constraints of capacities and number of outliers make the CFLO problem very challenging and no constant approximation is known for this problem. Our second and main result is the first constant factor approximation for CFLO assuming uniform facility opening costs. We use local search to obtain a $6.372 + \epsilon$ approximation.

Our local search algorithms for both CFL and CFLO require only 2 operations one of which is an `add` operation to add a facility not already part of the solution. For CFL, our second operation `open` opens a facility, t, and closes a subset of facilities, S'. The best such set - under certain restrictions - can be found by solving a knapsack problem and this operation has also been key in prior work on capacitated facility location. For the outlier version, we modify the open operation to open 2 facilities, t_1, t_2 and close a subset S' of facilities. We impose novel restrictions - clients served by the facilities in S' are either served by t_1 or made outliers and t_2 serves the right outliers from the current solution - on the operation to allow us to find the best S' in polynomial time and this involves solving a 2-dimensional knapsack problem.

As with most local search algorithms, identifying the correct set of operations is only a small part of our contribution. Most of the effort goes into putting together a suitable set of inequalities that would lead to a bound on the quality of the locally optimum solution. The outliers pose a challenge here and our key contribution is to define a suitable bijection between the outliers in the locally optimum solution and the global optimum. Unlike CFL where the choice of inequalities is such that each facility of the optimum solution is opened only a bounded number of times, the presence of outliers does not permit us this luxury. In fact, the inequalities we use to bound the facility cost of the locally optimum solution might require some facilities to be opened many times. However, since facility costs are uniform we can amortize this and argue that on average each facility of the optimum solution is opened only a small number of times.

The remainder of the paper is organized as follows. After discussing some related results in Sect. 2 we present our result for CFL with uniform facility costs in Sect. 3. This also serves as a starting point for our discussion on CFL with outliers in Sect. 4. In Sect. 5 we analyze the cost of our locally optimum solution for CFLO. Due to space constraints, some proofs (marked with a †) will only be available in the full version of the paper.

2 Preliminaries and Prior Work

Let $(\mathcal{P}, c)$ be a metric space where $\mathcal{P}$ is a finite sets of points and $c : \mathcal{P} \times \mathcal{P} \to \mathbb{R}^+$ is a distance function satisfying triangle inequality and symmetry.

Definition 1 (Capacitated Facility Location (CFL)). *We are given a set, $X \subseteq \mathcal{P}$, of n clients and a set, $F \subseteq \mathcal{P}$ of m facility locations. A facility $i \in F$ has a facility opening cost $f_i = f$ and a capacity $u(i) \in \mathbb{N}$. The objective is to find,*

- *a set $S \subseteq F$ of facilities to open and,*
- *an assignment $\sigma : X \to S$ respecting the capacities, i.e., for each facility $i \in S$, $|\sigma^{-1}(i)| \leq u(i)$*

such that the total cost, $f|S| + \sum_{j \in X} c(j, \sigma(j))$, is minimized.

Given a set S of facilities, the assignment of clients to facilities can be determined by solving a minimum cost assignment problem. Hence the set of open facilities completely determines the solution and so we use S to denote both the solution and the set of open facilities in the solution. For a solution S, let $C(S) = C_f(S) + C_s(S)$ denote the cost of solution S where $C_s(S)$ and $C_f(S)$ denote the service cost and facility cost of solution S respectively. Let S be a locally optimum solution and S^* an optimal solution for a given instance of CFL. Further let $\sigma : X \to S$ denote the assignment of clients in S and $\sigma^* : X \to S^*$ denote the assignment of clients in S^*.

Let $X(s)$ be the set of clients served by facility s in S then $X(A) = \cup_{s \in A} X(s)$ for $A \subseteq S$ denote the set of all clients served by facilities in A in solution S. Similarly, let $X^*(t)$ be the set of clients served by facility t in S^* then $X^*(A) = \cup_{t \in A} X^*(t)$ for $A \subseteq S^*$ denote the set of all clients served by facilities in A in solution S^*.

Chudak and Williamson [8] and Pál et. al. [15] construct a directed bipartite graph $G = (F \cup X, E)$ with F, X as the two sides of the vertex set. If $j \in X(i)$, edge (i, j) is added to E and give it a length $c(i, j)$. Similarly, if $j \in X^*(i)$, edge (j, i) is added to E and give it a length $c(i, j)$. Every vertex $j \in X$ has one incoming and one outgoing edge. The edges of E are then decomposed into cycles and maximal paths. Note that the number of maximal paths starting from $i \in S$ equals $|X(i)| - |X^*(i)|$ and the number of maximal paths ending at $i \in S^*$ equals $|X^*(i)| - |X(i)|$. By the triangle inequality, a maximal path from $s \in S$ to $t \in S^*$ has length at least $c(s, t)$. Further, the total length of all maximal paths is at most the total length of all edges in E which equals $C_s(S) + C_s(S^*)$.

Pál et al. [15] next formulate a transhipment problem where a facility $i \in S$ is a supply node with supply $|X(i)|$ and a facility $i \in S^*$ is a demand node with demand $|X^*(i)|$. Note that a facility $i \in S \cap S^*$ is both a supply node and a demand node. The cost of shipping 1 unit of flow from $s \in S$ to $t \in S^*$ equals $c(s, t)$. Since $c(i, i) = 0$, for $i \in S \cap S^*$ any solution for this transhipment problem would ship $\min\{|X(i)|, |X^*(i)|\}$ flow from the supply node i to the demand node i. This fact, together with the maximal paths in the decomposition of edges in E, defines a solution to this transhipment problem of cost at most $C_s(S) + C_s(S^*)$.

Let x be an optimum solution to the transhipment problem and let $x(s, t)$ be the amount of flow shipped from $s \in S$ to $t \in S^*$. Pál et al. [15] next construct an undirected *exchange graph*, $H = (V', E')$ where V' has a vertex for each facility in S and a vertex for each facility in S^*; thus a facility in $S \cap S^*$ corresponds to two vertices in V'. Define $E' = \{(s, t) | x(s, t) > 0\}$. The exchange graph H and the optimum solution x to the transhipment problem have the following properties.

1. H is acyclic and this follows from the fact that x is an optimum solution to the transhipment problem [15].
2. If $i \in S \cap S^*$ then H has an edge between the two vertices in V' corresponding to i and one of these vertices is a leaf of H.
3. $\forall s \in S$, $\sum_{t \in S^*} x(s, t) = |X(s)| \leq u(s)$.
4. $\forall t \in S^*$, $\sum_{s \in S} x(s, t) = |X^*(t)| \leq u(t)$.
5. $\sum_{s \in S, t \in S^*} x(s, t) c(s, t) \leq C_s(S) + C_s(S^*)$.

We remark that the above description of [15] is not entirely accurate. In particular, the exchange graph constructed by [15] has a vertex for each facility in $S \setminus S^*$ and a vertex for each facility in S^*. Our more symmetric construction of the exchange graph helps us in extending it to handle outliers.

Definition 2 (Capacitated Facility Location with Outliers (CFLO)). *In addition to the input for the capacitated facility location problem, we are given a bound, $L \leq n$, on the number of outliers permitted in a feasible solution. The objective is to find,*

- *a set $S \subseteq F$ of facilities,*
- *a set $X' \subseteq X$ of outliers of size at most L,*
- *an assignment $\sigma : (X \setminus X') \rightarrow S$ respecting the capacities, i.e., for each facility $i \in S$, $|\sigma^{-1}(i)| \leq u(i)$*

such that the cost, $f|S| + \sum_{j \in X \setminus X'} c(j, \sigma(j))$, *is minimized.*

Let O be the set of outliers in solution S. Once the set of open facilities S is fixed, the outliers and the assignment of clients to facilities are easily determined by solving a minimum cost flow problem as follows. We set up a bipartite graph with vertex sets X and $S \cup \{o\}$; the vertex o will allow us to identify outliers. An edge from client j to facility $i \in S$ has cost $c(j,i)$ and capacity 1 while an edge from client j to vertex o has cost 0 and capacity 1. There is a zero-cost edge from every vertex $i \in S$ to a sink, t, of capacity $u(s)$ and a zero-cost edge from vertex o to t of capacity L. Similarly, there is an edge from a source, s, to all vertices in X of cost 0 and capacity 1. A min-cost flow that routes $|X|$ units from s to t gives the best assignment of clients to facilities and identifies the L outliers. Hence there is no ambiguity in using S to denote both the set of open facilities and the solution.

Other Related Work: There is a vast amount of literature on facility location problems. The natural LP for capacitated facility location has an unbounded integrality gap and An et al. [2] strengthened it by adding network flow inequalities to establish a constant integrality gap. The integrality gap was recently improved to 9.0927 by Kao [12] and the best approximation algorithm known uses local search and is a 5-approximation [4].

In the k-median problem the objective is to open k facilities such that the cost of serving clients from the open facilities is minimized. The first constant approximation for k-median with outliers was obtained by Chen [7]. Krishnaswamy et al. [13] used iterative rounding to obtain a 7.081-approximation and this was further improved to $(6.994 + \epsilon)$-approximation by Gupta et al. [11].

Friggstad et al.[10] gave a PTAS for facility location with outliers with uniform facility costs on doubling and shortest path metrics. They also obtain a PTAS on these metrics and a $(3 + \epsilon)$ approximation on general metrics, for k-median with outliers but their algorithm opens $(1 + \epsilon)k$ facilities. Dabas and Gupta [9] considered CFLO with uniform capacities and gave an $O(1/\epsilon^2)$-approximation which violates outliers and capacities by a $(1 + \epsilon)$ factor.

3 Capacitated Facility Location with Uniform Facility Costs

Our algorithm does local search: we start with an arbitrary feasible solution (set of open facilities of total capacity at least n) and keep performing local search steps till they improve the cost of the solution. Let S be the solution at any step in this algorithm.

3.1 Local Search Operations

add(t): For $t \notin S$, if $C(S \cup \{t\}) < C(S)$ then $S \leftarrow S \cup \{t\}$. A facility t which is not in the current solution S is added to S if its addition improves the cost of the solution.

We define open(t, S') as an operation which opens a facility $t \in F$ and closes $S' \subseteq S$. If $t \in S$ then the operation is defined only if S' contains t; in this case, the operation closes facilities in $S' \setminus \{t\}$. In determining the cost of this operation we assume all clients in $X(S')$ are reassigned to t. Thus the reduction in $C(S)$ if this operation is performed is $\sum_{s \in S'} \sum_{j \in X(s)} c(j,s) - c(j,t) + f(|S'|-1)$.

open(t): This is the same as the operation open(t, S') for a subset $S' \subseteq S$ for which the cost of the operation is minimum. Given t, the problem of finding the optimal such S' can be formulated as a knapsack problem and solved in polynomial time (Lemma 1). The operation $S \leftarrow S \setminus S' \cup \{t\}$, $S' \subseteq S$, is performed only if it improves the cost of the solution S.

Lemma 1 **([15], Lemma 3.2).** *Given t, one can, in polynomial time, find a set S' that minimizes the cost of open(t, S') among all subsets $S' \subseteq S$.*

The algorithm stops if neither of the two operations improves the cost of the solution. Note that if a facility in S does not serve any client we close that facility. The solution S at the end of the algorithm is a locally optimum solution. The number of improvement steps of this local search algorithm can be made polynomial in the input size and $1/\epsilon$ by requiring that a local improvement be made only if it improves the cost of the solution by a $(1+\epsilon)$-factor.

3.2 Analysis

The service cost of the solution S can be bounded by the add operation.

Lemma 2 **([3], Lemma 4.1).** $C_s(S) \leq C_s(S^*) + C_f(S^*) = C(S^*)$.

We need a suitable set of inequalities to bound the facility opening cost of the locally optimum solution S. We construct a bipartite graph G and formulate a transhipment problem similar to Pál et al. [15] as described in Sect. 2. Let x be the optimum solution of the transhipment problem and H be the exchange graph defined in Sect. 2.

Every connected component of H is a tree and contains an edge of E' and hence a vertex in S^*. We root each tree in H at an arbitrary vertex in S^*. Let $\mathcal{L}$ be the set of facilities in S that are leaves in H. For $t \in S^*$, let $C(t)$ be the set of children of t and let $C_{\mathcal{L}}(t)$ be the set $C(t) \cap \mathcal{L}$. Further, let $F^* = \{t \in S^* : |C_{\mathcal{L}}(t)| \geq 1\}$. Thus, for $t \in F^*, C_{\mathcal{L}}(t) \neq \phi$.

Since our solution is locally optimal, if we open $t \in F^*$ and close facilities in $C_{\mathcal{L}}(t)$, we do not improve the cost of the solution. We capture this fact by writing an inequality which says that the cost of open($t, C_{\mathcal{L}}(t)$) is non-negative for every $t \in F^*$. (Refer Fig. 3.2). Note that if $t \in S$ then by Property 2 of the exchange graph H, $t \in C_{\mathcal{L}}(t)$ and hence the operation open($t, C_{\mathcal{L}}(t)$) is well-defined.

1. Since $s \in C_{\mathcal{L}}(t)$ is a leaf of H, $x(s,t) = |X(s)|$. Due to property 4 of the exchange graph H, the number of clients assigned to t in operation open($t, C_{\mathcal{L}}(t)$), $t \in F^*$ is $\sum_{s \in C_{\mathcal{L}}(t)} |X(s)| = \sum_{s \in C_{\mathcal{L}}(t)} x(s,t) \leq \sum_{s \in S} x(s,t) \leq u(t)$. Hence the capacity of t is not violated.
2. The increase in service cost over all operations open($t, C_{\mathcal{L}}(t)$), $t \in F^*$, is bounded by $\sum_{t \in F^*} \sum_{s \in C_{\mathcal{L}}(t)} x(s,t)c(s,t) \leq \sum_{t \in S^*} \sum_{s \in S} x(s,t)c(s,t) = C_s(S) + C_s(S^*)$ due to property 5 of the exchange graph H.

3. The decrease in facility opening costs over all operations open$(t, C_{\mathcal{L}}(t))$, $t \in F^*$, is $f \cdot (|\mathcal{L}| - |F^*|)$.

Since S is a locally optimum solution, using points 2, 3 above, we get

$$f \cdot |\mathcal{L}| \leq C_s(S) + C_s(S^*) + f \cdot |F^*|. \tag{1}$$

Lemma 3. *The number of facilities in $S \setminus \mathcal{L}$ is bounded by $|S^*|$.*

Proof. Let T be a tree in H rooted at an arbitrary vertex in S^*. With every facility in S which is not a leaf of T, we associate a child in S^*. A facility in S^* can be associated with at most one facility of S in this manner. Summing over all trees in H then proves the lemma.

Combining Eq. 1 with Lemma 3 and then applying Lemma 2 yields,

$$C_f(S) \leq 2C_f(S^*) + C_s(S) + C_s(S^*) \leq 2C_s(S^*) + 3C_f(S^*). \tag{2}$$

Having bounded both the facility cost and the services cost of our solution, we now employ a scaling technique introduced by Charikar et al. [5]. This involves scaling the facility costs by a factor γ and then running the local search algorithm. The service cost of the optimum solution is unchanged but the facility cost scales by γ. Lemma 2 can now be written as

$$C_s(S) \leq C_s(S^*) + \gamma C_f(S^*) \tag{3}$$

and inequality 2 corresponds to

$$\gamma C_f(S) \leq 2C_s(S^*) + 3\gamma C_f(S^*). \tag{4}$$

Thus $C(S) = C_s(S) + C_f(S) \leq (1 + \frac{2}{\gamma})C_s(S^*) + (\gamma + 3)C_f(S^*)$. For $\gamma = \sqrt{3} - 1$, we get $C(S) \leq (\sqrt{3} + 2)(C_s(S^*) + C_f(S^*)) = (\sqrt{3} + 2)C(S^*)$ which implies a 3.732 approximation.

Theorem 1. *The local search procedure with 'add' and 'open' operations yields a locally optimal solution that is a (3.732 $+\epsilon$) -approximation to the optimum solution of the capacitated facility location problem with uniform facility opening costs.*

4 The Algorithm for CFLO with Uniform Facility Costs

We start with a feasible solution S and perform the following operations whenever they improve the cost of the solution.

add(t): The operation is the same as defined in Sect. 3.1.

We define multiSwap(t_1, t_2, S') as an operation which opens facilities $t_1, t_2 \in F$ (t_1, t_2 may be identical) and closes $S' \subseteq S$. As in the open operation in Sect. 3.2 if $t_1 \in S$ then multiSwap(t_1, t_2, S') is defined only if S' contains t_1; in this case, the operation closes facilities in $S' \setminus \{t\}$. In determining the cost of this operation we assume that

1. clients in $X(S \setminus S')$ continue to be served by the facility that served them in S,
2. clients in $X(S')$ are either served by t_1 or are made outliers, and
3. clients in O are either served by t_2 or remain outliers.

Thus the reduction in $C_s(S)$ if multiSwap(t_1, t_2, S') is performed equals the service cost of clients in $X(S')$ minus the cost of servicing clients in $X(S')$ by t_1 if they are not made outliers minus the cost of servicing clients in O by t_2 if they are no longer outliers.

multiSwap(t_1, t_2)**:** This is the same as the operation multiSwap(t_1, t_2, S') for a set $S' \subseteq S$ which minimizes the cost of the operation. Given t_1, t_2 determining the optimal S' can be formulated as a 2-dimensional knapsack problem and solved in polynomial time (Lemma 4). The operation $S \leftarrow S \setminus S' \cup \{t_1, t_2\}, t_1, t_2 \in F, S' \subseteq S$, is performed only if it improves the cost of the solution S.

Lemma 4 (†). *Given t_1, t_2, one can, in polynomial time, find a set S' that minimizes the cost for multiSwap(t_1, t_2, S') among all subsets $S' \subseteq S$.*

The algorithm stops when neither of the two operations improves the cost of the solution. Note that if a facility in S does not serve any client we close that facility. The solution S at the end of the algorithm is a locally optimum solution. The number of steps of the algorithm can be made polynomial by making a small sacrifice on the quality of the approximation as in Sect. 3.1.

5 Bounding the Locally Optimum Solution

Let $S \subseteq F$ be a locally optimal solution and $S^* \subseteq F$ an optimal solution for a given instance of CFLO. Let O and O^* represent the set of outliers in solutions S and S^* respectively. We will assume $O \cap O^* = \phi$. This is without loss of generality as S is also a locally optimum solution for an instance with clients $X \setminus (O \cap O^*)$ and number of outliers $L - |O^* \cap O|$. Recall that, for a facility $s \in S$, $X(s)$ represents the set of clients served by s in the locally optimum solution and for a facility, $t \in S^*$, $X^*(t)$ represents the set of clients served by t in the optimum solution. Let $\sigma : (X \setminus O) \rightarrow S$ be the assignment of clients in S and let $\sigma^* : (X \setminus O^*) \rightarrow S^*$ be the assignment of clients in S^*.

The add operation can be used to bound the service cost of the solution as stated in Lemma 5.

Lemma 5 (†). $C_s(S) \leq C_s(S^*) + C_f(S^*) = C(S^*)$.

We can use the same approach for bounding the number of facilities in S as we followed in the case of capacitated facility location. However, when we close a subset $S' \subseteq S$ of facilities, some clients in $X(S')$ might be outliers in S^* (i.e., they belong to O^*) and we will not be able to assign these clients to the facility we open. We get around this by letting these clients be outliers in the new solution. Since the number of outliers needs to remain bounded by L, we serve some clients in O by opening another facility from S^*. To ensure facilities are not opened too often we will carefully choose a mapping from clients in O^* to O.

5.1 Modifying H to Handle Outliers

We modify the graph G constructed in Sect. 3.2 by introducing two vertices o, o^* which correspond to outliers in O and O^* respectively. If client j is in O then we add edge (o, j) to G and the edge (j, o^*) if j is in O^*; these edges are not assigned any cost. Once again we decompose edges of G into cycles and maximal paths and ignore all cycles from further consideration.

For a facility, $s \in S$, consider the maximal paths from s to o^* in this decomposition and let $O^*(s) \subseteq O^*$ be the set of clients preceding o^* on these paths. Similarly, let $O(t) \subseteq O$ be the set of clients following o on the maximal paths from o to $t \in S^*$.

The decomposition of edges of G into maximal paths gives us paths from a facility $s \in S$ to a facility $t \in S^*$. The number of such paths starting from $s \in S$ is exactly $|X(s)| - |X^*(s)| - |O^*(s)|$. Exactly $|X^*(t)| - |X(t)| - |O(t)|$ of such paths end at $t \in S^*$. Motivated by this observation we formulate a transhipment problem which has a supply of $|X(s)| - |O^*(s)|$ at node $s \in S$ and a demand of $|X^*(t)| - |O(t)|$ at node $t \in S^*$. Note that nodes in S may have zero supply and nodes in S^* zero demand. The cost of shipping one unit of flow from $s \in S$ to $t \in S^*$ is $c(s,t)$. As before, it is no loss of generality to assume that in any solution to the transhipment problem for $s \in S \cap S^*$, $\min(|X(s)| - |O^*(s)|, |X^*(s)| - |O(s)|)$ flow will be shipped from the supply node s to the demand node s at zero cost. Let x be an optimum solution to the transhipment problem.

Lemma 6 (†).

$$\sum_{s \in S, t \in S^*} x(s,t)c(s,t) + \sum_{t \in S^*} \sum_{j \in O(t)} c(j,t) + \sum_{s \in S} \sum_{j \in O^*(s)} c(s,j) \leq C_s(S) + C_s(S^*).$$

We use the optimum solution x of the transhipment problem to construct an undirected exchange graph $H = (V', E')$ as in Sect. 3.2. x, H have the following properties.

1. H is acyclic and this follows from the fact that x is an optimum solution to the transhipment problem [15].
2. If $s \in S \cap S^*$ then H has an edge between the two vertices in V' corresponding to s and at least one of these vertices is a leaf of H.
3. $\forall s \in S$, $\sum_{t \in S^*} x(s,t) = |X(s)| - |O^*(s)|$.
4. $\forall t \in S^*$, $\sum_{s \in S} x(s,t) = |X^*(t)| - |O(t)|$.

As in Sect. 3.2, we root each component of H at a facility in S^* if one exists. Unlike Sect. 3.2, we might have some isolated facilities in S in the forest H; these are the facilities with zero supply. Let s be an isolated facility in S and $t \in S^*$ be the root of some tree in H. We add edge (s,t) to H and assign $x(s,t) = 0$.

5.2 Constructing $\kappa : O^* \to O$

Let $\mathcal{L}$ be the facilities in S that are leaves in H. Let $F^* \subseteq S^*$ be the facilities that are the parent of some facility in $\mathcal{L}$. For $t \in S^*$, let $C_{\mathcal{L}}(t)$ be the children of t in $\mathcal{L}$.

As in Sect. 3, to bound the number of facilities in $\mathcal{L}$ we can write inequalities for operations which open a facility $t \in F^*$ and close facilities in $C_{\mathcal{L}}(t)$. However, note that t may not have sufficient capacity to serve all clients served by facilities in $C_{\mathcal{L}}(t)$ since some of these clients could correspond to outliers in S^*. To overcome this difficulty we define a bijection, κ, from O^* to O which helps us identify a facility in S^* to open to serve some outliers in S to account for outliers in S^* served by $C_{\mathcal{L}}(t)$.

Let $S^* = \{t_1, t_2, \ldots, t_{|S^*|}\}$. We construct the mapping in 2 steps using Algorithm 1.

Step 1: We consider facilities of S^* in increasing order of the indices and let t_i be the facility under consideration. If $t_i \in F^*$ then consider the facilities in $C_{\mathcal{L}}(t_i)$. For a facility $s \in C_{\mathcal{L}}(t_i)$ we map clients in $O^*(s)$ to clients in $O(t_i)$. If this is not possible since no unmapped clients are remaining in $O(t_i)$ we save the remaining clients in $O^*(s)$ in an array A; such clients shall be mapped in Step 2. It is important that we consider all clients in $O^*(s)$ before moving on to the next facility in $C_{\mathcal{L}}(t_i)$. After this step, for all $t_i \in F^*$, either all clients in $O(t_i)$ or all clients in $\bigcup_{s \in C_{\mathcal{L}}(t_i)} O^*(s)$ are assigned under the mapping κ.

Step 2: We consider facilities in S^* in the same order as in Step 1. If t_i is the facility under consideration and a client in $O(t_i)$ is unmapped, we map a client saved in the array A to such a client in $O(t_i)$.

The mapping constructed at the end of the above steps is extended to a bijection from O^* to O by assigning the unmapped clients arbitrarily.

5.3 Constructing the Bipartite Graph, Q

We construct a bipartite graph Q, with vertex sets $U = \{u_i, 1 \leq i \leq |S^*|\}$ and $W = \{w_i, 1 \leq i \leq |S^*|\}$; vertices u_i, w_i correspond to facility $t_i \in S^*$. For a facility $s \in \mathcal{L}$, we add an edge (u_i, w_j) to Q if t_i is the parent of s in H and some client in $O^*(s)$ is mapped to a client in $O(t_j)$. The edge is given a label s, i.e., $L(u_i, w_j) = \{s\}$. All edges with the same endpoints are combined into one edge and assigned a label which is the union of labels on these edges.
Edges of Q that arose due to the mapping in Step 1 are called *type 1* edges and those that arose due to the mapping in Step 2 are called *type 2* edges. Note that,

P1: The edge $(u_i, w_j), 1 \leq i, j \leq |S^*|$ is type 1 if $i = j$ and type 2 if $i \neq j$.
P2: For $1 \leq i \leq S^*$, at least one of u_i, w_i has no type 2 edge incident on it.
P3: At most one type 1 edge is incident to any vertex in Q.

Lemma 7. *Q is a forest.*

Algorithm 1: Mapping κ

```
   Output: κ : O* → O
1  for j = 1 to |O*| do
2  |  κ(j) = Null
   // Step 1
3  count = 0
4  for i = 1 to |S*| do
5  |  for s ∈ C_L(t_i) do
6  |  |  for j ∈ O*(s) do
7  |  |  |  if ∃ j' ∈ O(t_i) and κ^{-1}(j') == Null then
8  |  |  |  |  κ(j) = j'
9  |  |  |  else
10 |  |  |  |  A[count++] = j // A[] stores clients not mapped in Step
   |  |  |  |    1
   // Step 2
11 count = 0
12 for i = 1 to |S*| do
13 |  if ∃ j' ∈ O(t_i) and κ^{-1}(j') == Null then
14 |  |  κ(A[count++]) = j
15 return κ
```

Proof. Two edges (u_i, w_j) and (u_k, w_l) are *crossing* if $i < k$ and $l < j$. We begin with a couple of claims.

Claim. No pair of type 2 edges are crossing.

Proof. For contradiction assume that Q has edges (u_i, w_j) and (u_k, w_l) where $i < k$ and $l < j$. The existence of type 2 edges incident at u_i, u_k implies that after step 1 not all clients in $\cup_{s \in C_{\mathcal{L}}(t_i)} O^*(s), \cup_{s \in C_{\mathcal{L}}(t_k)} O^*(s)$ are mapped by κ. Since $i < k$, t_i is considered before t_k in step 1 and unmapped clients in $\cup_{s \in C_{\mathcal{L}}(t_i)} O^*(s)$ appear before unmapped clients in $\cup_{s \in C_{\mathcal{L}}(t_k)} O^*(s)$ in array A.

In step 2, since $l < j$, t_l is considered before t_j. In Lines 13-14 of Algorithm 1, unmapped clients in $\cup_{s \in C_{\mathcal{L}}(t_i)} O^*(s)$ are mapped to unmapped clients in $O(t_l)$ until either all clients in $\cup_{s \in C_{\mathcal{L}}(t_i)} O^*(s)$ or all clients in $O(t_l)$ are mapped. In the former case, Q cannot contain (u_i, w_j) while in the latter case, it cannot contain edge (u_k, w_l).

Claim. Type 2 edges form a forest in Q.

Proof. For contradiction assume that C is a cycle formed by type 2 edges and let (u_{i_1}, w_{j_1}) and (w_{j_1}, u_{i_2}) be adjacent edges on C. It is no loss of generality to assume $i_1 < i_2$. Let (u_{i_2}, w_{j_2}) be the next edge on C. Since it does not cross edge (u_{i_1}, w_{j_1}), it follows that $j_2 > j_1$. Continuing this argument we observe that the vertices of U, and W on the cycle are monotonically increasing sequences. This yields a contradiction.

To argue that Q is a forest we start with the forest of type 2 edges in Q and add type 1 edges one by one maintaining the invariant that the graph is a forest.

Let Q_i be the subgraph at Step i and suppose we add the type 1 edge (u_j, w_j). If this creates a cycle then there must exist a path in Q_i between u_j and w_j. This implies both these vertices have type 2 edges incident at them which contradicts property **P2**.

5.4 Bounding the Number of Facilities

Let $\mathcal{L}_0$ be the facilities in $\mathcal{L}$ that do not label any edge, $\mathcal{L}_1$ the facilities that label exactly one edge, and $\mathcal{L}_{>1}$ the facilities that label more than one edge in Q. We first bound the number of facilities in the set $\mathcal{L}_{>1}$.

Lemma 8. $|\mathcal{L}_{>1}| \leq |S^*|$.

Proof. Let $s \in \mathcal{L}_{>1}$ be a facility that labels multiple edges in Q. If $s \in C_{\mathcal{L}}(t_i)$ then these edges have u_i as a common endpoint.

Suppose s labels the type 1 edge (u_i, w_i). The other edges incident to u_i are type 2 and by property **P2** w_i does not have any type 2 edge incident to it. We associate s with $w_i \in W$.

Now assume s does not label any type 1 edge. Let $(u_i, w_j), (u_i, w_{j+1}), \ldots, (u_i, w_k)$ be edges incident at u_i which are labeled s. We associate s with w_j. If w_j is associated with another $s' \in \mathcal{L}_{>1}$, then there exists l such that edge (u_l, w_j) has a label s'. By our argument above $l \neq j$ and hence (u_l, w_j) is type 2. By interchanging roles of l, i we can assume $l > i$. The edge (u_i, w_{j+1}) which is type 2 crosses another type 2 edge (u_l, w_j) yielding a contradiction.

We conclude that a node in W is associated with at most one facility in $\mathcal{L}_{>1}$. Since $|W| = |S^*|$ the lemma follows.

We now bound $|\mathcal{L} \setminus \mathcal{L}_{>1}|$. Note that for $s \in \mathcal{L}_0$, $O^*(s) = \phi$. Consider a facility $s \in \mathcal{L}_0$ and let $t_i \in S^*$ be its parent in H. We add the label s to edge (u_i, w_i) and thus a facility which was earlier in $\mathcal{L}_0$ is now included in $\mathcal{L}_1$.

To bound the number of facilities in $\mathcal{L}_1$ we perform a multiSwap operation for some edges in Q. In particular, for an edge (u_i, w_j) in Q, if $L(u_i, w_j) \cap \mathcal{L}_1 \neq \phi$, we write the inequality corresponding to multiSwap(t_i, t_j, S') where $S' = (L(u_i, w_j) \cap \mathcal{L}_1)$. In writing this inequality we assume a certain reassignment of the clients. Recall S' are the facilities we close and let $s \in S'$.

T1 Clients in $X(s)$ which are not on a maximal path from s to o^* are assigned to t_i while the remaining clients in $X(s)$ are made outliers.
T2 Clients in O which are mapped to clients in $O^*(s)$ are served by t_j.

We now argue that reassignment of clients does not violate capacities. Note that,

1. In multiSwap(t_i, t_j, S'), the total number of clients assigned to facility t_i is $\sum_{s \in S'}(|X(s)| - |O^*(s)|) = \sum_{s \in S'} x(s, t_i) \leq \sum_{s \in S} x(s, t_i) \leq |X^*(t_i)| - |O(t_i)| \leq u(t_i)$ and these inequalities follow from the properties of x. If $t_i \in S$ then an additional $|X(t_i)| - |O^*(t_i)| = x(t_i, t_i)$ clients are assigned to t_i. The total number of clients assigned to t_i is still bounded by $\sum_{s \in S} x(s, t_i)$.

2. The total number of clients assigned to t_j is $\sum_{s\in S'}|O^*(s)|$ which by our mapping κ and choice of S' is at most $|O(t_j)|$. This in turn is at most $u(t_j)$. If $t_j \in S$ then the total number of clients assigned to t_j is $|X(t_j)|+|O(t_j)|$ and this is at most $u(t_j)$. This follows from the fact that the number of maximal paths that end at t_j in the decomposition of graph G is at most $u(t_j)-|X(t_j)|$ and this is only more than the number of maximal paths from o to t_j which is $|O(t_j)|$.
3. If $t_i = t_j$ then from the above argument the total number of clients assigned to t_i is at most $|X^*(t_i)| - |O(t_i)|$ which together with the observation that at most $|O(t_j)| = |O(t_i)|$ outliers are assigned to t_j, implies that the capacity of t_i is not violated.

Lemma 9. *The total increase in service cost due to the multiSwap operations performed is at most* $C_s(S) + C_s(S^*)$.

Proof. We perform a multiSwap operation for some edges of the forest Q. The total increase in service cost due to assignments of type **T1** for all these operations is bounded by $\sum_{t\in F^*}\sum_{s\in C_{\mathcal{L}}(t)} x(s,t)c(s,t) \le \sum_{t\in S^*}\sum_{s\in S} x(s,t)c(s,t)$. The total change in service cost due to assignments of type **T2** is bounded by $\sum_{t\in S^*}\sum_{j\in O(t)} c(t,j)$. By Lemma 6 it follows that the total increase in service costs due to **T1**, **T2** is at most $C_s(S) + C_s(S^*)$.

Note that every facility in $\mathcal{L}\setminus\mathcal{L}_{>1}$ is closed in one of the multiSwap operations defined above. Since none of these multiSwap operations leads to an improvement in the total cost, $f\cdot|\mathcal{L}\setminus\mathcal{L}_{>1}|$ is at most the total cost of facilities opened and the total increase in service cost in these multiSwap operations.

The forest Q has $2|S^*|$ vertices. However, u_i has an edge incident to it only if $t_i \in F^*$ and hence the number of edges in Q is at most $|F^*|+|S^*|$ edges. Of these edges exactly $|F^*|$ edges are type 1 and hence at most $|S^*|$ are type 2.

For each type 1 edge (u_i, w_i), we write an inequality corresponding to the operation multiSwap(t_i, t_i). Hence the number of facilities opened due to multiSwap operations corresponding to type 1 edges is at most $|F^*|$. The number of facilities opened due to multiSwap operation corresponding to type 2 edges is at most $2|S^*|$. Therefore, the total number of facilities opened over all mutliSwap operations is at most $3|S^*|$.

This together with the Lemma 9 implies that $f\cdot|\mathcal{L}\setminus\mathcal{L}_{>1}| \le 3f|S^*|+C_s(S)+C_s(S^*)$. By Lemma 8, $f\cdot|\mathcal{L}_{>1}| \le f\cdot|S^*|$ and by Lemma 3, $f\cdot|S\setminus\mathcal{L}| \le f\cdot|S^*|$. Combining, with $f\cdot|\mathcal{L}\setminus\mathcal{L}_{>1}| \le 3f|S^*|+C_s(S)+C_s(S^*)$ we get

$$C_f(S) = f\cdot|S| \le 5C_f(S^*) + C_s(S) + C_s(S^*). \tag{5}$$

5.5 Putting Things Together

We are now ready to combine the bounds on the service and facility costs of the locally optimum solution, S.

$$\begin{aligned} C(S) &= C_s(S) + C_f(S) \\ &\le C_s(S) + 5C_f(S^*) + C_s(S) + C_s(S^*) \end{aligned}$$

$$= 2C_s(S) + 5C_f(S^*) + C_s(S^*)$$
$$\leq 3C_s(S^*) + 7C_f(S^*).$$

where the first inequality follows from inequality 5 and the last inequality follows from Lemma 5. Using the scaling idea from Sect. 3, we obtain a 6.372 approximation.

Theorem 2. *The local search procedure with 'add' and 'multiSwap' operations yields a locally optimal solution that is a* $(6.372 + \epsilon)$*-approximation to the optimum solution of the capacitated facility location with outliers problem when facility opening costs are uniform.*

References

1. Aardal, K., van den Berg, P.L., Gijswijt, D., Li, S.: Approximation algorithms for hard capacitated k-facility location problems. EJOR **242**(2), 358–368 (2015)
2. An, H.C., Singh, M., Svensson, O.: LP-based algorithms for capacitated facility location. In: FOCS, 2014, pp. 256–265 (2014)
3. Arya, V., Garg, N., Khandekar, R., Meyerson, A., Munagala, K., Pandit, V.: Local search heuristics for k-median and facility location problems. SIAM J. Comput. **33**(3), 544–562 (2004)
4. Bansal, M., Garg, N., Gupta, N.: A 5-approximation for capacitated facility location. In: ESA, pp. 33–144 (2012)
5. Charikar, M., Guha, S.: Improved combinatorial algorithms for facility location problems. SIAM J. Comput. **34**(4), 803–824 (2005)
6. Charikar, M., Khuller, S., Mount, D.M., Narasimhan, G.: Algorithms for facility location problems with outliers. In: SODA, pp. 642–651 (2001)
7. Chen, K.: A constant factor approximation algorithm for k-median clustering with outliers. In: SODA, pp. 826–835 (2008)
8. Chudak, F.A., Williamson, D.P.: Improved approximation algorithms for capacitated facility location problems. In: IPCO, pp. 99–113 (1999)
9. Dabas, R., Gupta, N.: Capacitated facility location with outliers/penalties. In: Zhang, Y., Miao, D., Mohring, R. (eds.) Computing and Combinatorics. Lecture Notes in Computer Science, vol. 13595, pp. 549–560. Springer, Cham (2022). https://doi.org/10.1007/978-3-031-22105-7_49
10. Friggstad, Z., Khodamoradi, K., Rezapour, M., Salavatipour, M.R.: Approximation schemes for clustering with outliers. ACM Trans. Algorithms **15**(2), 1–26 (2019)
11. Gupta, A., Moseley, B., Zhou, R.: Structural iterative rounding for generalized k-median problems. In: 48th International Colloquium on Automata, Languages, and Programming, ICALP 2021. LIPIcs, vol. 198, pp. 1–18 (2021)
12. Kao, M.: Improved LP-based approximation algorithms for facility location with hard capacities. CoRR **abs/2102.06613** (2021)
13. Krishnaswamy, R., Li, S., Sandeep, S.: Constant approximation for k-median and k-means with outliers via iterative rounding. In: STOC, pp. 646-659 (2018)
14. Levi, R., Shmoys, D.B., Swamy, C.: LP-based approximation algorithms for capacitated facility location. J. Math. Program. **131**(1–2), 365–379 (2012)
15. Pál, M., Tardos, E., Wexler, T.: Facility location with nonuniform hard capacities. In: Proceedings of the 42nd IEEE Symposium on Foundations of Computer Science (FOCS), Las Vegas, Nevada, USA, pp. 329–338 (2001)

Integer Points in Arbitrary Convex Cones: The Case of the PSD and SOC Cones

Jesús A. De Loera[1], Brittney Marsters[1], Luze Xu[1(✉)], and Shixuan Zhang[2]

[1] University of California, Davis, CA 95616, USA
{jadeloera,bmmarsters,lzxu}@ucdavis.edu
[2] Texas A&M University, College Station, TX 77843, USA
shixuan.zhang@tamu.edu

Abstract. We investigate the semigroup of integer points inside a convex cone. We extend classical results in integer linear programming to integer conic programming. We show that the semigroup associated with nonpolyhedral cones can sometimes have a notion of finite generating set. We show this is true for the cone of positive semidefinite matrices (PSD) and the second-order cone (SOC). Both cones have a finite generating set of integer points, similar in spirit to Hilbert bases, under the action of a finitely generated group. We also extend notions of total dual integrality, Gomory-Chvátal closure, and Carathéodory rank to integer points in arbitrary cones.

Keywords: Integer Points · Convex Cones · Semigroups · Hilbert bases · Conic Programming · Positive Semidefinite Cone · Second-Order Cone

1 Introduction

A semigroup S is a subset of $\mathbb{Z}^N$ that contains $\mathbf{0}$ and is closed under addition. Given a convex cone $C \subseteq \mathbb{R}^N$, the integer points $S_C := C \cap \mathbb{Z}^N$ form a semigroup which we will call the *conical semigroup* of C. In particular, given any compact convex body $K \subseteq \mathbb{R}^n$, the integer points $\operatorname{cone}(K \times \{1\}) \cap \mathbb{Z}^{n+1}$ form a conical semigroup. Conical semigroups appear not just in optimization [1,6], but also

Due to space limitations, we omit several proofs. These can be found at https://www.math.ucdavis.edu/~deloera/IPCO2024.pdf.

This research is partially based upon work supported by the National Science Foundation under Grant No. DMS-1929284 while the first, third, and fourth authors were in residence at the Institute for Computational and Experimental Research in Mathematics in Providence, RI, during the Discrete Optimization program. We thank Kurt Anstreicher, Renata Sotirov, Pablo Parrilo, Chiara Meroni, and Bento Natura for relevant comments.

J. Vygen and J. Byrka (Eds.): IPCO 2024, LNCS 14679, pp. 99–112, 2024.
https://doi.org/10.1007/978-3-031-59835-7_8

in algebra and number theory [2,3]. Given a convex cone $C \subseteq \mathbb{R}^N$ for $N \geq 1$, we say a subset $B \subseteq S_C$ is an *integral generating set* of S_C if for any $s \in S_C$ there exist $b_1, \ldots, b_m \in B$ and $c_1, \ldots, c_m \in \mathbb{Z}_{\geq 0}$ such that $s = \sum_{i=1}^m c_i b_i$, for some $m \geq 1$. Furthermore, we call B a *conical Hilbert basis* if B is an inclusion-minimal integral generating set.

When the defining cone C is rational, polyhedral and pointed, there is abundant literature on the topic. It is well-known that we have a unique finite Hilbert basis in this case [11,20]. Historically, Hilbert bases have been fundamental in the theory and algorithms of combinatorial optimization. For example, determining if a rational system $A\mathbf{x} \leq \mathbf{b}$ is totally dual integral (TDI) is equivalent to checking if, for every face F of the polyhedron $P := \{\mathbf{x} : A\mathbf{x} \leq \mathbf{b}\}$, the rows of A which are active in a face F form a Hilbert basis for cone(F) [20].

It is natural to ask, what properties transfer from rational polyhedral cones to arbitrary convex cones? For instance, *do we preserve any notion of finiteness in generating sets for semigroups when we relax the polyhedral condition and instead consider general conical semigroups? Are there Hilbert bases for general cones?* This paper discusses finite generation for conical semigroups and extends the polyhedral cone theory of Hilbert bases to nonpolyhedral convex cones. Our main results will pertain to the semigroups arising from the cone of positive semidefinite matrices and the second-order cone. Both cones play a key role in modern optimization [4,5]. We also discuss some applications of our nonpolyhedral point of view.

In what follows, we denote $\mathrm{GL}(N, \mathbb{Z}) := \{U \in \mathbb{Z}^{N \times N} : |\det(U)| = 1\}$. Here is a new notion of finite generation for conical semigroups.

Definition 1. *Given a conical semigroup $S_C \subset \mathbb{Z}^N$, we call it (R, G)-finitely generated if there is a finite subset $R \subseteq S_C$ and a finitely generated subgroup $G \subseteq \mathrm{GL}(N, \mathbb{Z})$ acting on C linearly such that*

1. *both the cone C and the semigroup S_C are invariant under the group action, i.e., $G \cdot C = C$ and $G \cdot S_C = S_C$, and*
2. *every element $s \in S_C$ can be represented as*

$$s = \sum_{i \in K} \lambda_i g_i \cdot r_i$$

for $r_i \in R$, $g_i \in G$, and $\lambda_i \in \mathbb{Z}_{\geq 0}$, and where K is a finite index set.

Note that when C is a (pointed) rational polyhedral cone, then the conical semigroup $S_C = C \cap \mathbb{Z}^N$ is (R, G)-finitely generated by R, its Hilbert basis, and G, the trivial group $\{I_N\}$. Similarly, note that if S_C is an (R, G)-finitely generated semigroup, then $\cup_{r \in R} G \cdot r$ is an integral generating set of S_C, which is a superset of a conical Hilbert basis. We call R the set of *roots* of S_C, and $\cup_{r \in R} G \cdot r$ the set of generators for S_C.

While a nonpolyhedral cone cannot be finitely generated in the usual sense, using a possibly infinite (finitely generated) group G allows us to extend our understanding beyond the polyhedral case. Because the possibly infinite generators for S_C can be obtained by group action G on a finite set R and G is finitely generated, this allows for the possibility of algorithmic methods. The well-known Krein-Milman theorem states that any point in a closed pointed cone C can be generated by extreme rays, denoted by $\operatorname{ext}(C)$ [4]. When we restrict to the conical semigroup S_C and nonnegative integer combinations, the primitive integer points on the extreme rays of C must be contained in the set of generators of S_C, where an integer point $x = (x_1, \ldots, x_N) \in \mathbb{Z}^N$ is *primitive* if $\gcd(x_1, \ldots, x_N) = 1$. We call the integer points of S_C on the extreme rays of C *extreme points*, denoted by $\operatorname{ext}(S_C) := \{y \in S_C : y \in \operatorname{ext}(C) \cap \mathbb{Z}^N\}$. However, as in the polyhedral case, the generators will often include extra nonextreme boundary points or even interior points. We provide the following definition of *sporadic points* that cannot have an extreme point subtracted from them and still remain within the cone.

Definition 2. *We say a point $x \in S_C = C \cap \mathbb{Z}^N$ is sporadic if there does not exist $y \in \operatorname{ext}(S_C)$ such that $x - y \in S_C$.*

If $x \in S_C$ is sporadic, then x cannot be written as an integer conical combination of extreme points (even though it can be written as a real combination of them). From the definition of sporadic points, we know that all points $x \in S_C$ can be written as an integer conical combination of primitive extreme points and one sporadic point. To show that a semigroup is (R, G)-finitely generated, it is sufficient to show that the set of primitive extreme points and sporadic points are finite or can be obtained from a finitely generated group G that acts on a finite set of roots, R.

The two convex cones of interest in this work are positive semidefinite cone (PSD) and second-order cone (SOC). In Sects. 2 and 3 of this paper, we will present the following two main results pertaining to integer points in the PSD cone $\mathcal{S}_+^n(\mathbb{Z})$, and those in the SOC $\mathrm{SOC}(n) \cap \mathbb{Z}^n$.

Theorem 1. *The conical semigroup of the cone of $n \times n$ positive semidefinite matrices, $\mathcal{S}_+^n(\mathbb{Z})$, is (R, G)-finitely generated by $G \cong \mathrm{GL}(n, \mathbb{Z})$ where G acts on $X \in \mathcal{S}_+^n(\mathbb{Z})$ by $X \mapsto UXU^\mathsf{T}$ for each $U \in \mathrm{GL}(n, \mathbb{Z})$, and by R, the union of any single rank-one matrix and a finite subset of the sporadic points. Moreover,*

1. *If $n \leq 5$, then there are no sporadic points. Thus, $R = \{\mathbf{e}_1\mathbf{e}_1^\mathsf{T}\}$, where $\mathbf{e}_1$ is the first unit vector.*
2. *If $n = 6$, then $R = \{\mathbf{e}_1\mathbf{e}_1^\mathsf{T}, M\}$, where M is a single sporadic point defined in Sect. 2 Proposition 5.*

Theorem 2. *For dimension $3 \leq n \leq 10$, the conical semigroup $\mathrm{SOC}(n) \cap \mathbb{Z}^n$ is (R, G)-finitely generated with G and R defined in Sect. 3.*

We say that two matrices X_1, X_2 are unimodularly equivalent if $X_2 = U \cdot X_1 = UX_1U^\mathsf{T}$ for some $U \in \mathrm{GL}(n, \mathbb{Z})$. It is easy to see that it defines an equivalence relation for all integer PSD matrices. Note that the equivalence class of $\mathbf{e}_1\mathbf{e}_1^\mathsf{T}$ are all rank-1 integer matrix $\mathbf{x}\mathbf{x}^\mathsf{T}$ for some primitive integer vector $\mathbf{x} \in \mathbb{Z}^n$. An interpretation of Theorem 1 is that for dimension $n \leq 5$, every integer PSD matrix can be represented as the sum of rank-1 matrices $\mathbf{x}\mathbf{x}^\mathsf{T}$ for some primitive integer vector $\mathbf{x} \in \mathbb{Z}^n$. However, the same result fails for dimension $n = 6$. In this case, we will have that every integer PSD matrix can be represented as the sum of rank-1 matrices and one sporadic matrix Y, which is unimodularly equivalent to M (this matrix was first found by [19]). In general, every integer PSD matrix can be represented as the sum of rank-1 matrices and one sporadic matrix, which is unimodularly equivalent to a matrix in the finite set R. Regarding prior work that inspired us, we mention [16] that contains a similar rank-1 decomposition structure for PSD $\{0,1\}$ matrices: a PSD $\{0,1\}$ matrix $X \in \mathcal{S}_+^n(\mathbb{Z}) \cap \{0,1\}^{n\times n}$ satisfies $X = \sum_{i\in K} \mathbf{x}_i \mathbf{x}_i^\mathsf{T}$ for $\mathbf{x}_i \in \{0,1\}^n$, where K is a finite index set. Similarly, [18] extended the results to PSD $\{0,\pm 1\}$ matrices: a PSD $\{0,\pm 1\}$ matrix $X \in \mathcal{S}_+^n(\mathbb{Z}) \cap \{0,\pm 1\}^{n\times n}$ satisfies $X = \sum_{i\in K} \mathbf{x}_i \mathbf{x}_i^\mathsf{T}$ for $\mathbf{x}_i \in \{0,\pm 1\}^n$, where K is a finite index set. Our results extend to all integer positive semidefinite matrices. For the second-order cone, we extended the construction of the Barning-Hall tree in [8] for the primitive extreme points (or Pythagorean tuples) to classify the sporadic points.

While it might be tempting to believe that these results hint that all conical semigroups are (R, G)-finitely generated for some finite set R and some subgroup $G \subseteq \mathrm{GL}(N,\mathbb{Z})$, we conjecture the contrary:

Conjecture 1. *There exists a conical semigroup S_C that is not (R, G)-finitely generated for any choice of R and G.*

What is the significance of these results beyond their connections to classical geometry of numbers, lattices, and number theory? (See e.g., [14]). We motivate our interest about conical semigroups with two applications in optimization. In what follows, we assume that our cone $C \subset \mathbb{R}^N$ is full-dimensional.

The first application regards the notion of *Chvátal-Gomory cuts* which is useful in the branch-and-cut methods for integer programming. How much of this can be extended to conic integer programming? Given a linear map $\mathcal{A} : \mathbb{R}^m \to \mathbb{R}^N$ and $\mathbf{c} \in \mathbb{R}^N$, we define a *linear conical inequality* (LCI) system as

$$\mathrm{LCI}_C(\mathbf{c}, \mathcal{A}) := \{\mathbf{x} \in \mathbb{R}^m : \mathbf{c} - \mathcal{A}(\mathbf{x}) \in C\}$$

where $\mathbf{c} \in \mathbb{Z}^N$ and $\mathcal{A}(\mathbb{Z}^m) \subseteq \mathbb{Z}^N$. When C is the cone of positive semidefinite matrices in $\mathcal{S}^n(\mathbb{R})$, then $N = \binom{n+1}{2}$ and $\mathcal{A}(\mathbf{x}) = \sum_{i=1}^m x_i A_i$ for some matrices $A_1, \ldots, A_m \in \mathcal{S}^n(\mathbb{Z})$. This is known as a *linear matrix inequality* and defines a *spectrahedron*. An important concept for LCI is called total dual integrality (TDI), which has been well-known for polyhedral cones C [12,13] and recently extended to spectrahedral cones [7,17]. We use C^* to denote the dual cone of C, $\mathcal{A}^*$ to denote the adjoint linear map of $\mathcal{A}$, and give a definition for general cones here.

Definition 3. *An LCI system $\mathbf{c} - \mathcal{A}(\mathbf{x}) \in C$ is totally dual integral, if for any $\mathbf{b} \in \mathbb{Z}^m$, the dual optimization problem*

$$\min \quad y(\mathbf{c}) \quad \text{s.t.} \quad \mathcal{A}^*(y) = \mathbf{b},\ y \in C^*,$$

whenever feasible, has an integer optimal solution $y^* \in C^* \cap \mathbb{Z}^N$.

To approximate the convex hull of $Z := \mathrm{LCI}_C(\mathbf{c}, \mathcal{A}) \cap \mathbb{Z}^m$, a commonly used approach (quite similar to its polyhedral version) is to add *Chvátal-Gomory* (CG) cuts, which are defined as follows [17]. If $\mathbf{u} \in \mathbb{Z}^m$ is an integral vector and $v \in \mathbb{R}$ a real number such that the linear inequality $\mathbf{u}^\mathsf{T}\mathbf{x} \leq v$ is *valid* for all $x \in \mathrm{LCI}_C(c, \mathcal{A})$, then the inequality $\mathbf{u}^\mathsf{T}\mathbf{x} \leq \lfloor v \rfloor$ is valid for all $\mathbf{x} \in Z$ and called a CG cut. There are possibly infinitely many CG cuts so we define the (elementary) CG closure as

$$\text{CG-cl}(Z) := \bigcap_{\substack{(\mathbf{u},v)\in\mathbb{Z}^m\times\mathbb{R}:\\ S\subseteq\{\mathbf{x}:\mathbf{u}^\mathsf{T}\mathbf{x}\leq v\}}} \left\{\mathbf{x} \in \mathbb{R}^m : \mathbf{u}^\mathsf{T}\mathbf{x} \leq \lfloor v \rfloor\right\}. \tag{1}$$

Now take any linear function $w \in C^*$ such that $w(\mathbb{Z}^N) \subseteq \mathbb{Z}$. Then, a CG cut can be generated by

$$w \circ \mathcal{A}(\mathbf{x}) \leq \lfloor w(\mathbf{c}) \rfloor,$$

as, by definition, $w \circ \mathcal{A}(\mathbb{Z}^m) \in \mathbb{Z}$. Conversely, if the conical semigroup $S_{C^*} := C^* \cap \mathbb{Z}^N$ is (R, G)-finitely generated, then we can get all CG cuts through R and G for our TDI LCI system. This is one of the nice consequences of this property.

Theorem 3. *Suppose $C \subset \mathbb{R}^N$ is a full-dimensional convex cone such that $S_{C^*} := C^* \cap \mathbb{Z}^N$ is (R, G)-finitely generated, and $\mathrm{LCI}_C(\mathbf{c}, \mathcal{A})$ is TDI. Then the CG closure for $Z := \mathrm{LCI}_C(\mathbf{c}, \mathcal{A}) \cap \mathbb{Z}^m$ can be described by*

$$\textit{CG-cl}(Z) = \left\{\mathbf{x} \in \mathbb{R}^m : (g \cdot r)^\mathsf{T}\mathcal{A}(\mathbf{x}) \leq \lfloor (g \cdot r)^\mathsf{T}\mathbf{c} \rfloor, \quad \forall r \in R, g \in G\right\}.$$

The final application has to do with classical notions of integer rank [10]. Just like the notion of (real) rank of a linear system allows us to bound the number of nonzero entries in a solution of a linear system, we want to know how many elements are needed to decompose any element of a conical semigroup as a linear combination of generators with nonnegative integer coefficients. Suppose that our conical semigroup $S_C = C \cap \mathbb{Z}^N$ has an integer generating set B. For any element $s \in S_C$, there exist integer generators $b_1, \ldots, b_m \in B$ and $\lambda_1, \ldots, \lambda_m \in \mathbb{Z}_{\geq 1}$ such that $s = \sum_{i=1}^m \lambda_i b_i$, for some $m \geq 1$. The minimum number m needed in the sum is called the *integer Carathéodory rank* (ICR) of s, and the maximum number over all $s \in S_C$ is the ICR of the conical semigroup S_C or the cone C. We show an upper bound on the ICR that depends only on the dimension N. The proof is almost identical to the popular polyhedral result in [10,22] but we must use the extreme point characterization of semi-infinite linear optimization [9] to allow infinite generating sets.

Theorem 4. *Let $C \subset \mathbb{R}^N$ be an arbitrary pointed convex cone and $S_C := C \cap \mathbb{Z}^N$. Then $\mathrm{ICR}(S_C) \leq 2N - 2$.*

2 The Positive Semidefinite (PSD) Cone

Let $\mathcal{S}^n(\mathbb{Z})$ (resp. $\mathcal{S}^n(\mathbb{R})$) denote the set of $n \times n$ symmetric matrices of integer (resp. real) entries. For a matrix $X \in \mathcal{S}^n(\mathbb{Z})$, we say that X is PSD (denoted as

$X \succeq 0$) if and only if it is so when regarded as a real matrix $X \in \mathcal{S}^n(\mathbb{R})$. We denote $\mathcal{S}^n_+(\mathbb{Z})$ as the set of integer PSD matrices.

The group $\mathrm{GL}(n, \mathbb{Z})$ embeds into $\mathrm{GL}(N, \mathbb{Z})$ as follows. Given a matrix $U \in \mathrm{GL}(n, \mathbb{Z})$ and any $X \in \mathcal{S}^n(\mathbb{Z})$, we define the action $U \cdot X := UXU^\mathsf{T}$. This action is a linear map and takes integer points in $\mathbb{Z}^N$ to integer points, and thus can be represented by the multiplication with a matrix in $\mathrm{GL}(N, \mathbb{Z})$. It is well-known that this group $\mathrm{GL}(n, \mathbb{Z})$ is finitely generated [23]. For the convenience of discussion, we still use the matrix $U \in \mathrm{GL}(n, \mathbb{Z})$ to denote this matrix multiplication in the subgroup of $\mathrm{GL}(N, \mathbb{Z})$.

2.1 Lemmas for $n \le 5$ and $n = 6$

The following *integer rank-1 decomposition* for PSD integer matrices is studied in [19]. We recast their arguments with a modern geometric perspective, and use it to extend the notion of (R, G)-finite generation to the PSD cone.

Lemma 1. *If $n \le 5$, then for any $X \in \mathcal{S}^n_+(\mathbb{Z})$, we can find a finite index set K and vectors $\mathbf{x}_i \in \mathbb{Z}^n$, $i \in K$ such that*

$$X = \sum_{i \in K} \mathbf{x}_i \mathbf{x}_i^\mathsf{T}. \tag{2}$$

To restate Definition 2 in the PSD case, we say an integer matrix $X \in \mathcal{S}^n(\mathbb{Z})$ is *sporadic* if there does not exist $\mathbf{x} \in \mathbb{Z}^n \setminus \{\mathbf{0}\}$ such that $X - \mathbf{x}\mathbf{x}^\mathsf{T} \succeq 0$. Lemma 1 is equivalent to the fact that there is no sporadic point in $\mathcal{S}^n_+(\mathbb{Z})$ when $n \le 5$.

Proposition 1. *There is no sporadic point in $\mathcal{S}^n_+(\mathbb{Z})$ if and only if every positive semidefinite integer matrix in $\mathcal{S}^n_+(\mathbb{Z})$ has an integer rank-1 decomposition.*

For any matrix $X \in \mathcal{S}^n(\mathbb{R})$, we can define a convex set $C(X) := \{\mathbf{x} \in \mathbb{R}^n : X - \mathbf{x}\mathbf{x}^\mathsf{T} \succeq 0\}$. Since $X - \mathbf{x}\mathbf{x}^\mathsf{T} \succeq 0$ if and only if, for any $\mathbf{v} \in \mathbb{R}^n$, $|\mathbf{v}^\mathsf{T}\mathbf{x}|^2 \le \mathbf{v}^\mathsf{T} X \mathbf{v}$, we see that $C(X)$ is a compact convex set that is symmetric about the origin. This provides another equivalent formulation of the integer rank-1 decomposition.

Proposition 2. *For $X \in \mathcal{S}^n_+(\mathbb{Z})$, X is sporadic if and only if $C(X) \cap \mathbb{Z}^n = \{\mathbf{0}\}$.*

This provides a geometric perspective to our problem. Note that the set $C(X)$ is a (possibly degenerate) ellipsoid because

$$X \succeq \mathbf{x}\mathbf{x}^\mathsf{T} \iff \begin{bmatrix} 1 & \mathbf{x}^\mathsf{T} \\ \mathbf{x} & X \end{bmatrix} \succeq 0 \iff \mathbf{x}^\mathsf{T} X^\dagger \mathbf{x} \le 1, (I - XX^\dagger)\mathbf{x} = 0,$$

by the positive semidefiniteness of Schur complements, where $X^\dagger$ denotes the pseudoinverse of X. In the case where $\det(X) > 0$,

$$C(X) = \{\mathbf{x} \in \mathbb{R}^n : \mathbf{x}^\mathsf{T} \mathrm{adj}(X) \mathbf{x} \le \det(X)\} \text{ with } \mathrm{vol}(C(X)) = V_n \sqrt{\det(X)}$$

where $\mathrm{adj}(X)$ is the adjugate of X satisfying $\mathrm{adj}(X) = \det(X)X^{-1}$ and $V_n := \pi^{n/2}/\Gamma(\frac{n}{2}+1)$ is the volume of the unit n-ball. Note that $C(X)$ is not necessarily full-dimensional if X is rank deficient. The degenerate case for rank 1 is characterized by the following proposition.

Proposition 3. *Suppose* $X \in \mathcal{S}_+^n(\mathbb{Z})$, *and* $\mathrm{rank}(X) = 1$, *then* $X = \lambda \mathbf{x}\mathbf{x}^\mathsf{T}$, *where* $\mathbf{x} \in \mathbb{Z}^n$ *and* $\lambda \in \mathbb{Z}_{\geq 1}$.

From Proposition 3, we can directly prove the case for $n = 2$ using Minkowski's first Theorem (for example, see [11]): if $\mathrm{vol}(C(X)) > 4$, then there is a nonzero integral point.

Proposition 4. *Lemma 1 holds for* $n = 2$.

In the general degenerate case, we can reduce the problem to one involving full-rank matrices of some lower dimension.

Lemma 2. *Let* $X \in \mathcal{S}^n(\mathbb{Z})$. *If* $r = \mathrm{rank}(X) < n$, *then* X *is unimodularly equivalent to*

$$\begin{bmatrix} \mathbf{0} & \mathbf{0} \\ \mathbf{0} & \hat{X} \end{bmatrix},$$

for some $\hat{X} \in \mathbb{Z}^{r \times r}$, $\mathrm{rank}(\hat{X}) = r$.

To prove Lemma 1, we need to use a more sophisticated method based on the *Hermite constant* [21]

$$\gamma_n := \left(\max_{A \succ 0} \frac{\lambda_1(A)}{(\det(A))^{\frac{1}{n}}} \right)^n, \text{ where } \lambda_1(A) = \min_{\mathbf{x} \in \mathbb{Z}^n \setminus \{\mathbf{0}\}} (\mathbf{x}^\mathsf{T} A \mathbf{x}).$$

Remark 1. Hermite gives a bound $\gamma_n \leq (\frac{4}{3})^{\frac{n(n-1)}{2}}$. The exact value of γ_n is only known for $n \leq 8$ and $n = 24$.

n	2	3	4	5	6	7	8	24
γ_n	$\frac{4}{3}$	2	2	8	$\frac{64}{3}$	64	256	4^{24}

Proof (for Lemma 1). The case $n = 1$ follows from Proposition 3. We will show that $C(X) \cap \mathbb{Z}^n \neq \{\mathbf{0}\}$ for $2 \leq n \leq 5$, where $C(X) = \{\mathbf{x} \in \mathbb{R}^n : \mathbf{x}^\mathsf{T}\mathrm{adj}(X)\mathbf{x} \leq \det(X)\}$. By the definition of the Hermite constant, we have

$$\min_{\mathbf{x} \in \mathbb{Z}^n \setminus \{\mathbf{0}\}} (\mathbf{x}^\mathsf{T}\mathrm{adj}(X)\mathbf{x}) \leq (\gamma_n \det(\mathrm{adj}(X)))^{\frac{1}{n}} = (\gamma_n (\det(X))^{n-1})^{\frac{1}{n}}.$$

For $n = 2, 3, 4, 5$, we have $\frac{n^n}{(n-1)^{n-1}} > \gamma_n$. By taking the derivative with respect to $\det(X)$, we know that $\frac{(\det(X)+1)^n}{(\det(X))^{n-1}} \geq \frac{n^n}{(n-1)^{n-1}}$. Thus, $\gamma_n < \frac{(\det(X)+1)^n}{\det(X)^{n-1}}$. Therefore,

$$\min_{\mathbf{x} \in \mathbb{Z}^n \setminus \{\mathbf{0}\}} (\mathbf{x}^\mathsf{T}\mathrm{adj}(X)\mathbf{x}) \leq (\gamma_n \det(\mathrm{adj}(X)))^{\frac{1}{n}} = (\gamma_n \det(X)^{n-1})^{\frac{1}{n}} < \det(X) + 1.$$

Because $x^\mathsf{T}\text{adj}(X)x, \det(X) \in \mathbb{Z}$ for any $x \in \mathbb{Z}^n$, we have

$$\min_{\mathbf{x}\in\mathbb{Z}^n\setminus\{\mathbf{0}\}} (\mathbf{x}^\mathsf{T}\text{adj}(X)\mathbf{x}) \leq \det(X),$$

which implies that $C(X) \cap (\mathbb{Z}^n \setminus \{\mathbf{0}\}) \neq \emptyset$. Lemma 1 now follows from Propositions 2 and 1, and Lemma 2. □

The argument used to prove Lemma 1 fails for $n \geq 6$, but it implies that the determinant of the sporadic matrices is bounded by a constant only dependant on n. For example, in the case of $n = 6$, the argument only fails when $3 \leq \det(X) \leq 14$; for $n = 7$, it only fails when $2 \leq \det(X) \leq 56$, and for $n = 8$, it only fails when $1 \leq \det(X) \leq 247$. We summarize this observation in the following corollary.

Corollary 1. *If $X \in \mathcal{S}^n_+(\mathbb{Z})$ is sporadic, then* $\det(X) < \gamma_n$.

A sporadic matrix for $n = 6$ was initially found in [19].

Proposition 5. *In $n = 6$, the matrix M is sporadic, i.e., $C(M) \cap \mathbb{Z}^n = \{\mathbf{0}\}$.*

$$M = \begin{bmatrix} 2 & 0 & 1 & 1 & 1 & 1 \\ 0 & 2 & 0 & 1 & 1 & 1 \\ 1 & 0 & 2 & 1 & 1 & 1 \\ 1 & 1 & 1 & 2 & 1 & 1 \\ 1 & 1 & 1 & 1 & 2 & 1 \\ 1 & 1 & 1 & 1 & 1 & 2 \end{bmatrix} \quad \textit{with } \det(M) = 3.$$

Moreover, in [15], it is shown that for $n = 6$, M is the unique sporadic matrix under unimodular equivalence. Using this fact, we have the following Lemma.

Lemma 3. *If $n = 6$, then for any $X \in \mathcal{S}^n_+(\mathbb{Z})$,*

$$X = \sum_{i\in K} \mathbf{x}_i\mathbf{x}_i^\mathsf{T} + Y$$

for $\mathbf{x}_i \in \mathbb{Z}^n$ and Y unimodularly equivalent to M, where K is a finite index set.

2.2 Proof of Theorem 1

Proof. We know that the primitive extreme points are generated from the group $\text{GL}(n, \mathbb{Z})$ that acts on $\{\mathbf{e}_1\mathbf{e}_1^\mathsf{T}\}$. The finiteness of the index set K follows from similar argument in Proposition 1. We only need to prove that the sporadic points are generated from the group $\text{GL}(n, \mathbb{Z})$ on a finite set R.

Corollary 1 shows that for any sporadic matrix X, $\det(X) < \gamma_n$. By [21, Theorem 2.4], there exists a constant $\alpha_n > 0$ depending only on n, such that for any positive definite matrix $X \in \mathcal{S}^n(\mathbb{Z})$, there is a unimodularly equivalent matrix X' of X with diagonal entries satisfy

$$\prod_{i=1}^{n} X'_{ii} \leq \alpha_n \det(X') = \alpha_n \det(X) < \alpha_n\gamma_n.$$

Because $X' \in \mathcal{S}^n(\mathbb{Z})$ is positive definite, $X'_{ii} \geq 1$ and thus is bounded from above. From this we see that there are only finitely many possibilities for such X' because each off-diagonal entry must satisfy $|X'_{ij}|^2 \leq X'_{ii}X'_{jj}$ for any $1 \leq i, j \leq n$.

The two special cases $n \leq 5$ and $n = 6$ follow from Lemma 1 and 3. □

3 The Second-Order Cone (SOC)

In this section, we will let T_n be the conical semigroup $\mathrm{SOC}(n) \cap \mathbb{Z}^n$ where

$$\mathrm{SOC}(n) := \left\{ \mathbf{x} \in \mathbb{R}^n : 0 \leq \sqrt{x_1^2 + \cdots + x_{n-1}^2} \leq x_n \right\}.$$

Additionally, for $\mathbf{a}, \mathbf{b} \in \mathbb{R}^n$, consider the bilinear form

$$\langle \mathbf{a}, \mathbf{b} \rangle := a_1 b_1 + a_2 b_2 + \cdots + a_{n-1} b_{n-1} - a_n b_n.$$

In this quadratic space, the reflection in vector $\mathbf{w}$ is defined as $\mathbf{x} \to \mathbf{x} - 2\frac{\langle \mathbf{x}, \mathbf{w} \rangle}{\langle \mathbf{w}, \mathbf{w} \rangle}\mathbf{w}$.

Definition 4. *Let $P_{i,j}$ be the permutation matrix that swaps the i-th and j-th columns and define Q_k be the matrix determined by*

$$(Q_k)_{i,j} = \begin{cases} -1 & \text{if } i = j = k \\ 1 & \text{if } i = j \neq k \\ 0 & \text{if } i \neq j. \end{cases}$$

For $n = 3$, let A_3 denote the matrix associated with the reflection in the vector $(1, 1, 1)$. For $4 \leq n \leq 10$, let A_n denote the matrix associated with the reflection in the vector $(1, 1, 1, 0, \ldots, 0, 1)$ also associated to this bilinear form:

$$A_3 = \begin{pmatrix} -1 & -2 & 2 \\ -2 & -1 & 2 \\ -2 & -2 & 3 \end{pmatrix}, \quad A_n = \left(\begin{array}{ccc|c|c} 0 & -1 & -1 & & 1 \\ -1 & 0 & -1 & \mathbf{0} & 1 \\ -1 & -1 & 0 & & 1 \\ \hline & \mathbf{0} & & I_{n-4} & \mathbf{0} \\ \hline -1 & -1 & -1 & \mathbf{0} & 2 \end{array}\right)$$

We define the matrix $A_n^+ = Q_1 Q_2 \ldots Q_{n-1} A_n$. Note that A_n is unimodular.

Elements $\mathbf{s} \in T_n$ such that $\langle \mathbf{s}, \mathbf{s} \rangle = \mathbf{0}$ belong to the boundary of T_n, and we will denote the set of these points as ∂T_n. In number theory, these points are called Pythagorean tuples. In [8], they proved that the set of primitive Pythagorean tuples, denoted as $\mathrm{ext}^p(T_n)$, is generated by finitely many matrices acting on a finite set R for $3 \leq n \leq 10$.

Lemma 4 (Theorem 1 in [8]). *For $3 \leq n \leq 10$, $\mathrm{ext}^p(T_n) = \cup_{r \in R} G \cdot r$, where the group*

$$G = \langle A_n, Q_1, \ldots, Q_{n-1}, P_{1,2}, P_{1,3}, \ldots P_{1,n-1} \rangle$$

and the sets

1. $R = \{(1,0,\ldots,0,1)^\mathsf{T}\}$ *for* $3 \le n < 10$,
2. $R = \{(1,0,0,0,0,0,0,0,0,1)^\mathsf{T}, (1,1,1,1,1,1,1,1,1,3)^\mathsf{T}\}$ *for* $n = 10$,

where G acts on R by left multiplication.

Note that G maps ∂T_n to ∂T_n.

We will begin this section by discussing the structural properties of the sporadic points of T_n. Then we use the structural properties of Pythagorean tuples and sporadic points to prove Theorem 2.

3.1 Sporadic Points of $\mathrm{SOC}(n) \cap \mathbb{Z}^n$

In this section, we will begin by restating the definition of sporadic points in the case of SOC(n) and offer two partial characterizations of sporadic elements of SOC(n).

Definition 5. *We call a point* $\mathbf{s} \in T_n$ *sporadic if there is no integral point* $\mathbf{p}$ *such that* $\langle \mathbf{p}, \mathbf{p} \rangle = 0$ *and* $\mathbf{s} - \mathbf{p} \in T_n$.

Just as the group G takes elements of ∂T_n to ∂T_n, the group G will take sporadic elements to sporadic elements. This closure ensures that our action by G on the semigroup T_n is well-defined.

Lemma 5. *Let* $\mathbf{s} \in T_n$.

1. *Then,* $A_n^+\mathbf{s}$ *and* $(A_n^+)^{-1}\mathbf{s}$ *are both in* T_n.
2. *If* $\mathbf{s}$ *is sporadic, then* $A_n^+\mathbf{s}$ *and* $(A_n^+)^{-1}\mathbf{s}$ *are both sporadic.*

Next, we will provide some Lemmas about the properties of sporadic points necessary to prove Theorem 2. For a detailed exposition of the technical proofs, please refer to the extended version. Lemmas 6 and 7 show that sporadic points are close to the boundary, ∂T_n.

Lemma 6. *Suppose* $\mathbf{s} \in T_n$ *is a primitive sporadic with nonnegative entries such that* $s_n > 1$ *and* $s_i \neq 0$ *for some* $i \in [n-1]$. *Then,*

$$s_n = \left\lceil \sqrt{s_1^2 + s_2^2 + \cdots + s_{n-1}^2} \right\rceil$$

Lemma 7. *Let* $\mathbf{s} \in T_n$. *If* $\langle \mathbf{s}, \mathbf{s} \rangle = -1$, *then* $\mathbf{s}$ *is sporadic. When* $3 \le n < 7$, *we also have the converse: if* $\mathbf{s}$ *is sporadic, then* $\langle \mathbf{s}, \mathbf{s} \rangle = -1$.

Inspired by the structure of Pythagorean tuples, we analyze the set of sporadic points that remain at the same height in T_n after multiplication by $(A_n^+)^{-1}$. Let $(p)_n$ denotes the nth coordinate of p,

Lemma 8. *Let* $n \le 10$. *Suppose* $\mathbf{s} \in T_n$ *is a primitive sporadic such that* $s_1 \geq \cdots \geq s_{n-1} \geq 0$ *and* $s_n > 1$. *The following list of tuples are the only such* $\mathbf{s}$ *where* $((A_n^+)^{-1}\mathbf{s})_n = s_n$.

- *For* $n = 7$, *we have the following tuple:* $(1, 1, 1, 1, 1, 1, 3)$.
- *For* $n = 8$, *we have the following tuples:* $(1, 1, 1, 1, 1, 1, 1, 3)$, $(1, 1, 1, 1, 1, 1, 0, 3)$.
- *For* $n = 9$, *we have the following tuples:*

$$(1,1,1,1,1,1,1,1,3), (1,1,1,1,1,1,1,0,3), (1,1,1,1,1,1,0,0,3), (2,2,2,2,2,2,2,1,6).$$

- *For* $n = 10$, *we have the following tuples:*

$$(1,1,1,1,1,1,1,1,0,3), \quad (1,1,1,1,1,1,1,0,0,3), \quad (1,1,1,1,1,1,0,0,0,3),$$
$$(2,2,2,2,2,2,2,2,1,6), \quad (2,2,2,2,2,2,2,1,0,6).$$

Then we show that besides the points listed in Lemma 8, every other sporadic points will reduce to a strictly lower height after multiplication by $(A_n^+)^{-1}$.

Lemma 9. *Let* $\mathbf{s} \in T_n$ *be sporadic with nonnegative entries such that* $s_1 \geq s_2 \geq \cdots \geq s_{n-1}$, $s_1 \geq 1$ *and* $3 \leq n \leq 10$. *For* s *not listed in Lemma 8,*

$$((A_n^+)^{-1}\mathbf{s})_n < (\mathbf{s})_n.$$

Proof. This is equivalent to showing that $-s_1 - s_2 - s_3 + 2s_n < s_n$, or rather $s_n < s_1 + s_2 + s_3$. The case of $n = 3$ reduces to the inequality $s_3 < s_1 + s_2$. By Lemma 6, we have that

$$\begin{aligned} s_n &= \left\lceil \sqrt{s_1^2 + s_2^2 + \cdots + s_{n-1}^2} \right\rceil \\ &\leq \left\lceil \sqrt{s_1^2 + s_2^2 + s_3^2 + 2s_1s_2 + 2s_1s_3 + 2s_2s_3} \right\rceil \qquad (3) \\ &= \left\lceil \sqrt{(s_1 + s_2 + s_3)^2} \right\rceil = s_1 + s_2 + s_3, \end{aligned}$$

where the inequality follows from the order $s_1s_2 \geq s_1s_3 \geq s_2s_3 \geq s_4^2 \geq ... \geq s_{n-1}^2$ and $n \leq 10$. As s is not one of the tuples listed in Lemma 8, the inequality (3) can be made strict. Therefore, $s_n < s_1 + s_2 + s_3$, which implies that $(A_n^+)^{-1}\mathbf{s}$ sits at a strictly lower height in the cone than $\mathbf{s}$. □

3.2 Proof of Theorem 2

We now present a complete formulation of Theorem 2 followed by its proof.

Theorem 5. *For dimension* $3 \leq n \leq 10$, *the conical semigroup* $\mathrm{SOC}(n) \cap \mathbb{Z}^n$ *is* (R, G)*-finitely generated by*

$$G = \langle A_n^+, Q_1, \ldots, Q_{n-1}, P_{1,2}, P_{1,3}, \ldots P_{1,n-1} \rangle$$

and a finite set R. *More specifically,*

1. *If* $3 \leq n \leq 6$, *then* $R = \{(1, 0, \ldots, 0, 1)^\mathsf{T}, (0, \ldots, 0, 1)^\mathsf{T}\}$.

2. *If* $n = 7$, *then*

$$R = \left\{(1,0,0,0,0,0,1)^\mathsf{T}, (0,0,0,0,0,0,1)^\mathsf{T}, (1,1,1,1,1,1,3)^\mathsf{T}\right\}.$$

3. *If* $n = 8$, *then*

$$R = \left\{(1,0,0,0,0,0,0,1)^\mathsf{T}, (0,0,0,0,0,0,0,1)^\mathsf{T}, (1,1,1,1,1,1,1,3)^\mathsf{T}, (1,1,1,1,1,1,0,3)^\mathsf{T}\right\}.$$

4. *If* $n = 9$, *then*

$$R = \left\{(1,0,0,0,0,0,0,0,1)^\mathsf{T}, (0,0,0,0,0,0,0,0,1)^\mathsf{T}, (1,1,1,1,1,1,1,1,3)^\mathsf{T},\right.$$
$$\left.(1,1,1,1,1,1,1,0,3)^\mathsf{T}, (1,1,1,1,1,1,0,0,3)^\mathsf{T}, (2,2,2,2,2,2,2,1,6)^\mathsf{T}\right\}.$$

5. *If* $n = 10$, *then*

$$R = \left\{(1,0,0,0,0,0,0,0,0,1)^\mathsf{T}, (1,1,1,1,1,1,1,1,1,3)^\mathsf{T}, (0,0,0,0,0,0,0,0,0,1)^\mathsf{T},\right.$$
$$(1,1,1,1,1,1,1,1,0,3)^\mathsf{T}, (1,1,1,1,1,1,1,0,0,3)^\mathsf{T}, (1,1,1,1,1,1,0,0,0,3)^\mathsf{T},$$
$$\left.(2,2,2,2,2,2,2,2,1,6)^\mathsf{T}, (2,2,2,2,2,2,2,1,0,6)^\mathsf{T}\right\}.$$

Proof. This follows directly from Lemma 9 and that fact that $(0,0,\ldots,0,1)$ is the sporadic of minimal height in this cone. Let $\mathbf{s} \in T_n$. If $\mathbf{s}$ is not sporadic, we can represent it as

$$\mathbf{s} = \lambda_1\mathbf{p}_1 + \lambda_2\mathbf{p}_2 + \cdots + \lambda_k\mathbf{p}_k + \lambda\mathbf{p} \tag{4}$$

where $\lambda, \lambda_i \in \mathbb{Z}_{\geq 0}$, each $\mathbf{p}_i$ is a primitive Pythagorean tuple and $\mathbf{p}$ is sporadic. By Lemma 4, each $\mathbf{p}_i$ can be decomposed as $\mathbf{p}_i = G_i(1,0,\ldots,0,1)^\mathsf{T}$ when $3 \leq n < 10$ or $\mathbf{p}_i = G_i(1,0,\ldots,0,1)^\mathsf{T} + \tilde{G}_i(1,1,\ldots,1,3)^\mathsf{T}$ when $n = 10$ where each $G_i, \tilde{G}_i \in G$. It remains to consider the sporadic $\mathbf{p}$. Given any primitive sporadic tuple $\mathbf{s}$, we can recover an element of R as follows:

1. Multiply $\mathbf{p}$ by the appropriate permutation matrices $P_{i,j}$ and sign changing matrices Q_j so that $\mathbf{p}$ has nonnegative entries and $p_1 \geq \cdots \geq p_{n-1}$. Call this resulting vector $\mathbf{p}'$.
2. Multiply $\mathbf{p}'$ by $(A_n^+)^{-1}$ and repeat step 1 as necessary. By Lemma 9, the height of the resulting vector will be strictly lower then that of the vector we started with or the resulting vector will belong to R.
3. Repeat step 2 until the resulting vector $\mathbf{r}$ belongs to R. By Lemma 8, the only possibilities for the resulting vector belong to R.

This process gives the equality $\mathbf{r} = G_1 \ldots G_k\mathbf{p}$. If we let $G' = G_1 \ldots G_k$, then we have $(G')^{-1}\mathbf{r} = \mathbf{p}$. Therefore, for $3 \leq n \leq 10$, the conical semigroup $\mathrm{SOC}(n)$ is (R, G)-finitely generated by the claimed R and G. □

When $n = 9$, the primitive sporadic point $(2, 2, 2, 2, 2, 2, 2, 1, 6)$ can be written as the sum of two sporadic points with smaller heights:

$$(2, 2, 2, 2, 2, 2, 2, 1, 6) = (1, 1, 1, 1, 1, 1, 1, 0, 3) + (1, 1, 1, 1, 1, 1, 0, 0, 3).$$

We can similarly decompose $(2, 2, 2, 2, 2, 2, 2, 2, 1, 6)$ and $(2, 2, 2, 2, 2, 2, 2, 1, 0, 6)$ for $n = 10$. In this sense, these sporadic points fail to be minimal. Thus, if we remove them from the set of roots R, our semigroup S remains (R, G)-finitely generated. However, when we remove these point from our root sets, our decomposition in equality (4) requires modification and we must allow for multiple sporadic points in the expression.

Remark 2. Lastly, it is worth noting that inequality (3) would fail in dimensions larger than 10. Thus, this line of argumentation would fail to produce results for $n > 10$.

References

1. Aliev, I., De Loera, J.A., Eisenbrand, F., Oertel, T., Weismantel, R.: The support of integer optimal solutions. SIAM J. Optim. **28**(3), 2152–2157 (2018)
2. Aliev, I., Averkov, G., De Loera, J.A., Oertel, T.: Sparse representation of vectors in lattices and semigroups. Math. Program. **192**(1–2), 519–546 (2022)
3. Aliev, I., De Loera, J.A., Oertel, T., O'Neill, C.: Sparse solutions of linear Diophantine equations. SIAM J. Appl. Algebra Geom. **1**(1), 239–253 (2017)
4. Barvinok, A.: A Course in Convexity, vol. 54. American Mathematical Society, Providence (2002)
5. Ben-Tal, A., Nemirovski, A.: Lectures on modern convex optimization: analysis, algorithms, and engineering applications. SIAM (2001)
6. Berndt, S., Brinkop, H., Jansen, K., Mnich, M., Stamm, T.: New support size bounds for integer programming, applied to MakeSpan minimization on uniformly related machines (2023). ArXiv:2305.08432
7. de Carli Silva, M.K., Tuncel, L.: A notion of total dual integrality for convex, semidefinite, and extended formulations. SIAM J. Discret. Math. **34**(1), 470–496 (2020)
8. Cass, D., Arpaia, P.J.: Matrix generation of Pythagorean n-tuples. Proc. Am. Math. Soc. **109**(1), 1–7 (1990)
9. Charnes, A., Cooper, W.W., Kortanek, K.: Duality in semi-infinite programs and some works of Haar and Carathéodory. Manage. Sci. **9**(2), 209–228 (1963)
10. Cook, W., Fonlupt, J., Schrijver, A.: An integer analogue of Caratheodory's theorem. J. Comb. Theory, Ser. B **40**(1), 63–70 (1986)
11. De Loera, J., Hemmecke, R., Köppe, M.: Algebraic and Geometric Ideas in the Theory of Discrete Optimization. MOS-SIAM Series on Optimization, Society for Industrial and Applied Mathematics (2013)
12. Edmonds, J., Giles, R.: Total dual integrality of linear inequality systems. In: Progress in Combinatorial Optimization, pp. 117–129. Elsevier (1984)
13. Giles, F.R., Pulleyblank, W.R.: Total dual integrality and integer polyhedra. Linear Algebra Appl. **25**, 191–196 (1979)
14. Kaveh, K., Khovanskii, A.G.: Newton-Okounkov bodies, semigroups of integral points, graded algebras and intersection theory. Ann. Math. (2) **176**(2), 925–978 (2012)

15. Ko, C.: On the decomposition of quadratic forms in six variables (dedicated to professor LJ Mordell on his fiftieth birthday). Acta Arith **3**(1), 64–78 (1939)
16. Letchford, A.N., Sørensen, M.M.: Binary positive semidefinite matrices and associated integer polytopes. Math. Program. **131**(1–2), 253–271 (2012)
17. de Meijer, F., Sotirov, R.: The Chátal-Gomory procedure for integer SDPs with applications in combinatorial optimization. Math. Program. (2024)
18. de Meijer, F., Sotirov, R.: On integrality in semidefinite programming for discrete optimization. SIAM J. Optim. **34**(1), 1071–1096 (2024)
19. Mordell, L.J.: The representation of a definite quadratic form as a sum of two others. Ann. Math. **38**(4), 751 (1937)
20. Schrijver, A.: Theory of Linear and Integer Programming. Wiley Series in Discrete Mathematics & Optimization. Wiley, Hoboken (1998)
21. Schürmann, A.: Computational Geometry of Positive Definite Quadratic Forms: Polyhedral Reduction Theories, Algorithms, and Applications, vol. 48. American Mathematical Society, Providence (2009)
22. Sebö, A.: Hilbert bases, Caratheodory's theorem and combinatorial optimization. In: Proceedings of the 1st Integer Programming and Combinatorial Optimization Conference, pp. 431–455 (1990)
23. Trott, S.M.: A pair of generators for the unimodular group. Can. Math. Bull. **5**(3), 245–252 (1962)

The Extension Complexity of Polytopes with Bounded Integral Slack Matrices

Sally Dong(✉) and Thomas Rothvoss

University of Washington, Seattle, USA
{sallyqd,rothvoss}@uw.edu

Abstract. We show that any bounded integral function $f : A \times B \mapsto \{0, 1, \ldots, \Delta\}$ with rank r has deterministic communication complexity $\Delta^{O(\Delta)} \cdot \sqrt{r} \cdot \log r$, where the rank of f is defined to be the rank of the $A \times B$ matrix whose entries are the function values. As a corollary, we show that any n-dimensional polytope that admits a slack matrix with entries from $\{0, 1, \ldots, \Delta\}$ has extension complexity at most $\exp(\Delta^{O(\Delta)} \cdot \sqrt{n} \cdot \log n)$.

Keywords: Polytope · Slack matrix · Extension complexity

1 Introduction

In classical communication complexity, two players, Alice and Bob, are given a Boolean function $f : A \times B \mapsto \{0, 1\}$, as well as separate inputs $a \in A$ and $b \in B$, and wish to compute $f(a, b)$ while minimizing the total amount of communication. Alice and Bob have unlimited resources for pre-computations and agree on a deterministic *communication protocol* to compute f before receiving their respective inputs. The *length* of their protocol is defined to be the maximum number of bits exchanged over all possible inputs. The *deterministic communication complexity* of f, denoted by $CC^{\det}(f)$, is the minimum length of a protocol to compute f.

A major open problem in communication complexity is the *log-rank conjecture* proposed by Lovász and Saks [6], which asks if $CC^{\det}(f) \leq (\log r)^{O(1)}$ for all Boolean functions f of rank r, where the *rank* of a two-party function f on $A \times B$ is defined to be the rank of the matrix $\mathbf{M} \in \mathbb{R}^{A \times B}$ with $\mathbf{M}_{a,b} = f(a, b)$ for all $(a, b) \in A \times B$. The best upper bound currently known is due to Lovett [7], who showed $CC^{\det}(f) \leq O(\sqrt{r} \log r)$ using discrepancy theory techniques.

In this work, we obtain similar deterministic communication complexity bounds for a larger class of functions:

Theorem 1 (Main result, communication complexity). *Let $f : A \times B \mapsto \{0, 1, \ldots, \Delta\}$ be a bounded integral function of rank r. Then there exists a deterministic communication protocol to compute f with length at most $\Delta^{O(\Delta)} \cdot \sqrt{r} \cdot \log r$ bits.*

The function f can be directly viewed as a non-negative matrix $\mathbf{M} \in \{0, \ldots, \Delta\}^{A \times B}$ of rank r. We use the matrix representation exclusively in the

J. Vygen and J. Byrka (Eds.): IPCO 2024, LNCS 14679, pp. 113–123, 2024.
https://doi.org/10.1007/978-3-031-59835-7_9

remainder of this paper. Let us adopt the convention that a *rectangle* in $\mathbf{M}$ is a (non-contiguous) submatrix $\mathbf{M}[A', B']$ indexed by some $A' \subseteq A$ and $B' \subseteq B$. A rectangle is *monochromatic* with color i if all entries in the rectangle have value i.

The *non-negative rank* of a non-negative matrix $\mathbf{M}$, denoted by $\text{rank}_+(\mathbf{M})$, is defined as the minimum r such that $\mathbf{M}$ can be written as the sum of r non-negative rank-1 matrices, or equivalently, as $\mathbf{M} = \mathbf{UV}$ for non-negative matrices $\mathbf{U} \in \mathbb{R}_{\geq 0}^{A \times r}$ and $\mathbf{V} \in \mathbb{R}_{\geq 0}^{r \times B}$. It is straightforward to see $\text{rank}_+(\mathbf{M}) \leq 2^{CC^{\text{det}}(\mathbf{M})}$ (c.f. Rao and Yehudayoff [9], Chap. 1, Theorem 1.6): The *protocol tree* to compute $\mathbf{M}$ has at most $2^{CC^{\text{det}}(\mathbf{M})}$ leaves, each corresponding to a monochromatic rectangle of $\mathbf{M}$. These rectangles are disjoint over all leaves, and their union is $\mathbf{M}$. Since a monochromatic rectangle is a non-negative matrix of rank 0 or 1, we conclude that $\mathbf{M}$ can be written as a sum of at most $2^{CC^{\text{det}}(\mathbf{M})}$ rank-1 non-negative matrices. The *positive semidefinite rank* of $\mathbf{M}$, denoted by $\text{rank}_{\text{psd}}(\mathbf{M})$, generalizes non-negative rank and has an analogous relationship to quantum communication complexity [3]. It is defined as the minimum r such that there are positive semidefinite matrices $\mathbf{U}_1, \ldots \mathbf{U}_A$ and $\mathbf{V}_1, \ldots, \mathbf{V}_B$ of dimension $r \times r$ satisfying $\mathbf{M}_{i,j} = \text{Tr}(\mathbf{U}_i \mathbf{V}_j)$ for all i, j; trivially, $\text{rank}_{\text{psd}}(\mathbf{M}) \leq \text{rank}_+(\mathbf{M})$. Barvinok [1] showed that if $\mathbf{M}$ has at most k distinct entries, as is the setting studied in this paper, then $\text{rank}_{\text{psd}}(\mathbf{M}) \leq \binom{k-1+\text{rank}(\mathbf{M})}{k-1}$.

Non-negative rank brings us to a beautiful connection with *extension complexity*, introduced in the seminal work of Yannakakis [12] in the context of writing combinatorial optimization problems as linear programs. The *extension complexity* of a polytope $\mathcal{P}$, denoted $(\mathcal{P})$, is defined as the minimum number of facets of some higher dimensional polytope $\mathcal{Q}$ (its *extended formulation*) such that there exists a linear projection of $\mathcal{Q}$ to $\mathcal{P}$.

A foundational theorem from [12] states that $(\mathcal{P}) = \text{rank}_+(\mathbf{S})$, where $\mathbf{S}$ is the *slack matrix* of $\mathcal{P}$, defined as follows: Suppose $\mathcal{P}$ has facets $\mathcal{F}$ and vertices $\mathcal{V}$. Then $\mathbf{S}$ is a non-negative $\mathcal{F} \times \mathcal{V}$ matrix, where the (f, v)-entry indexed by facet $f \in \mathcal{F}$ defined by the halfspace $a^\top x \leq b$ and vertex $v \in \mathcal{V}$ has value $\mathbf{S}_{f,v} = b - a^\top v$. (A facet may be defined by many equivalent halfspaces, and therefore the slack matrix is not unique; the result holds for all valid slack matrices.) In fact, [12] showed that a factorization of $\mathbf{S}$ with respect to non-negative rank gives an extended formulation of $\mathcal{P}$ and vice versa. For a comprehensive preliminary survey, see [2]. A number of breakthrough results in extended complexity in recent years emerged from lower-bounding the non-negative rank of the slack matrix for specific polytopes, such as the TSP, Cut, and Stable-Set polytopes by Fiorini-Massar-Pokutta-Tiwary-De Wolf [4], and the Perfect-Matching polytope by Rothvoss [11].

Connecting extension complexity to deterministic communication complexity via non-negative rank, we have:

Corollary 1 (Main result, extension complexity). *Let $\mathcal{P}$ be a n-dimensional polytope that admits slack matrix $\mathbf{S}$. Suppose the entries of $\mathbf{S}$ are integral and bounded by Δ. Then the extension complexity of $\mathcal{P}$ is at most $\exp(\Delta^{O(\Delta)} \cdot \sqrt{n} \cdot \log n)$.*

Proof. Since $\mathcal{P}$ is n-dimensional, we know $\mathbf{S}$ has rank at most n, as there are at most n linearly independent vertices of $\mathcal{P}$, and the slack of a vertex with respect to all the facets is a linear function. We combine the theorem of [12] and Theorem 1 for the overall conclusion.

Our proof follows the approach discussed in the note of Rothvoss [10] simplifying Lovett's result. The following lemma shows that there are indeed concrete polytopes for which our result applies:

Lemma 1. *Suppose there are polytopes* $\mathcal{P} \subseteq \mathcal{Q} \subseteq \mathbb{R}^n$ *where* $\mathcal{P} = \operatorname{conv}\{\boldsymbol{x}_1, \ldots, \boldsymbol{x}_v\}$ *and* $\mathcal{Q} = \{\boldsymbol{x} \in \mathbb{R}^n : \mathbf{A}\boldsymbol{x} \le \boldsymbol{b}\}$ *with* $\mathbf{A} \in \mathbb{R}^{f \times n}$. *Suppose the partial slack matrix* $\mathbf{S} \in \mathbb{R}^{f \times v}$ *with* $\mathbf{S}_{i,j} = \boldsymbol{b}_i - \mathbf{A}_i \boldsymbol{x}_j$ *for* $i \in [f], j \in [v]$ *is integral and bounded by* Δ. *Then there exists a polytope* $\mathcal{K}$ *with extension complexity at most* $\exp(\Delta^{O(\Delta)} \cdot \sqrt{n} \cdot \log n)$ *so that* $\mathcal{P} \subseteq \mathcal{K} \subseteq \mathcal{Q}$.

Proof. From above, we have $\operatorname{rank}_+(\mathbf{S}) \le \exp(\Delta^{O(\Delta)} \cdot \sqrt{n} \cdot \log n)$. Moreover, it is well-known that $s \overset{\text{def}}{=} \operatorname{rank}_+(\mathbf{S})$ is the extension complexity of some polytope $\mathcal{K}$ such that $\mathcal{P} \subseteq \mathcal{K} \subseteq \mathcal{Q}$. (It is in fact the minimum extension complexity over all such sandwiched polytopes.) The conclusion follows.

For completeness, we show the latter fact: Suppose $\mathbf{S} = \mathbf{U}\mathbf{V}$ is a non-negative factorization of $\mathbf{S}$ with $\mathbf{U} \in \mathbb{R}^{f \times s}_{\ge 0}$ and $\mathbf{V} \in \mathbb{R}^{s \times v}_{\ge 0}$. Let $\mathcal{K}^{\text{lift}} \overset{\text{def}}{=} \{(\boldsymbol{x}, \boldsymbol{y}) \in \mathbb{R}^{n+s} : \mathbf{A}\boldsymbol{x} + \mathbf{U}\boldsymbol{y} = \boldsymbol{b}, \boldsymbol{y} \ge \mathbf{0}\}$, and let $\mathcal{K}$ be the projection of $\mathcal{K}^{\text{lift}}$ onto the first n coordinates. It is immediately clear that $\mathcal{K} \subseteq \mathcal{Q}$, and $(\mathcal{K}) \le s$ by definition. For each $j \in [v]$, the point $\boldsymbol{x}_j$ satisfies $\mathbf{A}\boldsymbol{x}_j + \mathbf{S}^j = \boldsymbol{b}$, where $\mathbf{S}^j$ is the j-th column of $\mathbf{S}$ given by $\mathbf{U}\mathbf{V}\boldsymbol{e}_j$. It follows that $(\boldsymbol{x}_j, \mathbf{V}\boldsymbol{e}_j) \in \mathcal{K}^{\text{lift}}$, so $\boldsymbol{x}_j \in \mathcal{K}$. As this holds for each $\boldsymbol{x}_1, \ldots, \boldsymbol{x}_v$, we conclude $\mathcal{P} \subseteq \mathcal{K}$.

We give a direct example in a combinatorial optimization setting: Consider the k-Set-Packing problem, where we are given a collection of n sets $S_1, \ldots, S_n \subseteq [N]$ with $N \gg n$, and want to find a maximum subcollection such that each element $j \in [N]$ is contained in at most k sets. The k-Set-Packing polytope $\mathcal{P}$ is is the convex hull of all feasible subcollections of sets, given by

$$\mathcal{P} = \operatorname{conv}\Big\{x \in \{0,1\}^n : \sum_{i: j \in S_i} x_i \le k \;\; \forall j \in [N]\Big\}.$$

Its natural LP relaxation $\mathcal{Q}$ is

$$\mathcal{Q} = \Big\{x \in [0,1]^n : \sum_{i: j \in S_i} x_i \le k \;\; \forall j \in [N]\Big\}.$$

In the regime where $N \gg n$, a priori, the extension complexity of $\mathcal{P}$ and $\mathcal{Q}$ could be as large as N. But interestingly, let $\mathbf{S}$ be the partial slack matrix with respect to $\mathcal{P}$ and $\mathcal{Q}$ as defined in Lemma 1. Then $\mathbf{S}$ contains integral values in $\{0, \ldots, k\}$, and so we conclude there exists a sandwiched polytope $\mathcal{P} \subseteq \mathcal{K} \subseteq \mathcal{Q}$ with $(\mathcal{K}) \le \exp(k^{O(k)} \cdot \sqrt{n} \cdot \log n)$.

2 Communication Protocol

In this section, we give a deterministic communication protocol for a bounded integral matrix $\mathbf{S}$ assuming it has the crucial property that any submatrix contains a large monochromatic rectangle. The protocol is based on the protocol from Nisan and Widgerson [8] that is expanded on by Lovett [7].

Lemma 2. *Let* $0 < \delta < 1$. *Let* $\mathbf{M} \in \{0, 1, \ldots, \Delta\}^{A \times B}$ *be a bounded integral matrix with rank* r, *and suppose for any submatrix* $\mathbf{S} \stackrel{\text{def}}{=} \mathbf{M}[A', B']$ *where* $A' \subseteq A$ *and* $B' \subseteq B$, *there is a monochromatic rectangle in* $\mathbf{S}$ *of size* $\geq \exp(-\delta(r))|A'||B'|$ *for some function* δ *of* r. *Then,*

$$CC^{\det}(\mathbf{M}) \leq \Theta\left(\log \Delta + \log^2 r + \sum_{i=0}^{\log r} \delta(r/2^i)\right).$$

We begin by proving two helper lemmas relating to $\mathbf{M}$ and the rank of its submatrices, which we subsequently use to bound the communication complexity.

Lemma 3. *Let* $\mathbf{A}, \mathbf{B}, \mathbf{C}, \mathbf{R}$ *be matrices of the appropriate dimensions, and let* $\mathbf{R}$ *have rank 0 or 1. Then*

$$\operatorname{rank}\begin{pmatrix}\mathbf{R}\\ \mathbf{B}\end{pmatrix} + \operatorname{rank}\begin{pmatrix}\mathbf{R} & \mathbf{A}\end{pmatrix} \leq \operatorname{rank}\begin{pmatrix}\mathbf{R} & \mathbf{A}\\ \mathbf{B} & \mathbf{C}\end{pmatrix} + 3.$$

Proof. We use a sequence of elementary rank properties:

$$\begin{aligned}\operatorname{rank}\begin{pmatrix}\mathbf{R} & \mathbf{A}\\ \mathbf{B} & \mathbf{C}\end{pmatrix} + 1 \geq \operatorname{rank}\begin{pmatrix}\mathbf{0} & \mathbf{A}\\ \mathbf{B} & \mathbf{C}\end{pmatrix} &\geq \operatorname{rank}(\mathbf{A}) + \operatorname{rank}(\mathbf{B})\\ &\geq \operatorname{rank}\begin{pmatrix}\mathbf{R} & \mathbf{A}\end{pmatrix} + \operatorname{rank}\begin{pmatrix}\mathbf{R}\\ \mathbf{B}\end{pmatrix} - 2.\end{aligned}$$

Lemma 4. *Suppose the rank of* $\mathbf{M} \in \{0, 1, \ldots, \Delta\}^{A \times B}$ *is* r. *Then* $\mathbf{M}$ *contains at most* $(\Delta + 1)^r$ *different rows and columns.*

Proof. We show the argument for rows: Let $\mathbf{U} \in \{0, \ldots, \Delta\}^{A \times r}$ denote a submatrix of $\mathbf{M}$ consisting of r linearly independent columns of $\mathbf{M}$. Clearly $\mathbf{U}$ has at most $(\Delta+1)^r$ different rows. If row i and row j of $\mathbf{U}$ are identical, then row i and j of $\mathbf{M}$ are also identical, since by definition of rank, the columns of $\mathbf{M}$ are obtained by taking linear combinations of columns of $\mathbf{U}$.

Now, we can design a communication protocol to compute $\mathbf{M}$ using standard techniques. Without loss of generality, we may assume $\mathbf{M}$ does not contain identical rows or columns, so that the conclusion of Lemma 4 can be applied.

Proof of Lemma 2. Suppose Alice has input $a \in A$ and Bob has $b \in B$. Then Alice and Bob compute $\mathbf{M}_{a,b}$ by recursively reducing the matrix $\mathbf{M}$ to a smaller submatrix in one of two ways: they communicate the bit 0 which guarantees a decrease in rank in the resulting submatrix, and the bit 1 which guarantees a decrease in size.

Let $\mathbf{S}$ denote the submatrix to be considered at a recursive iteration. Alice and Bob begin with $\mathbf{S} \stackrel{\text{def}}{=} \mathbf{M}$. During a recursive iteration, they first write $\mathbf{S}$ in the form

$$\mathbf{S} = \begin{pmatrix} \mathbf{R} & \mathbf{A} \\ \mathbf{B} & \mathbf{C} \end{pmatrix},$$

where $\mathbf{R}$ denote the large monochromatic rectangle in $\mathbf{S}$ that is guaranteed to exist, with $|\mathbf{R}| \geq \exp(-\delta(r))|\mathbf{S}|$. By Lemma 3, either $\operatorname{rank}\begin{pmatrix}\mathbf{R} & \mathbf{A}\end{pmatrix} \leq \frac{1}{2}\operatorname{rank}(\mathbf{S}) + \frac{3}{2}$, or $\operatorname{rank}\begin{pmatrix}\mathbf{R} & \mathbf{B}\end{pmatrix} \leq \frac{1}{2}\operatorname{rank}(\mathbf{S}) + \frac{3}{2}$.

In the first case, Alice communicates. If her input row a is in the upper submatrix of $\mathbf{S}$, Alice sends the bit 0, and both Alice and Bob update $\mathbf{S} = \begin{pmatrix}\mathbf{R} & \mathbf{A}\end{pmatrix}$. On the other hand, if row a is in the lower submatrix of $\mathbf{S}$, Alice sends the bit 1, and they both update $\mathbf{S} = \begin{pmatrix}\mathbf{B} & \mathbf{C}\end{pmatrix}$. In the second case, Bob communicates. If his input column b is in the left submatrix of $\mathbf{S}$, Bob sends 0, and they update $\mathbf{S} = \begin{pmatrix}\mathbf{R} \\ \mathbf{B}\end{pmatrix}$. Otherwise, Bob sends 1, and they update $\mathbf{S} = \begin{pmatrix}\mathbf{A} \\ \mathbf{C}\end{pmatrix}$. If $\mathbf{S}$ has size 1, Alice can simply output the entry of $\mathbf{S}$, which is precisely $\mathbf{M}_{a,b}$.

Consider the communication protocol up until the rank of $\mathbf{S}$ is halved (i.e., when the first 0 bit is communicated): The protocol tree has at most $\Theta\left(\frac{-\log|\mathbf{S}|}{\log(1-\exp(-\delta(r)))}\right) = \Theta\left(\frac{r\log(\Delta+1)}{\exp(-\delta(r))}\right)$-many leaves at this point, which is the max number of 1 bits that could have been communicated. Standard balancing techniques (c.f. [9, Chapter 1, Theorem 1.7]) then allow us to balance the protocol tree; combined with the bound $|\mathbf{S}| \leq (\Delta+1)^{2r}$ by Lemma 4, we conclude that there exists a protocol of length $O(\log r + \log\log(\Delta+1) + \delta(r))$.

Next, consider the phase where the protocol continues until the rank drops from $r/2$ to $r/4$. Note that we may assume the submatrix at the start of this phase has unique rows and columns, as Alice and Bob has unlimited computation at the start. Then this phase can be simulated by a protocol of length $O(\log(r/2) + \log\log(\Delta+1) + \delta(r/2))$. We proceed in the same fashion where at phase i, the rank drops from $r/2^i$ to $r/2^{i+1}$. Summing over $i = 0, \ldots, \log r$, we get that the total protocol length is at most

$$\sum_{i=0}^{\log r} \left(\log\left(\frac{r}{2^i}\right) + \log\log(\Delta+1) + \delta\left(\frac{r}{2^i}\right)\right)$$
$$\leq \log^2 r + \log r \log\log(\Delta+1) + \sum_{i=0}^{\log r} \delta\left(\frac{r}{2^i}\right).$$

In the base case, $\mathbf{S}$ has rank 1 or 2. Suppose it has rank 2, and let $\mathbf{S} = \boldsymbol{u}\boldsymbol{v}^\top + \boldsymbol{u}'\boldsymbol{v}'^\top$ be a factorization, where $\boldsymbol{u}$ and $\boldsymbol{u}'$ are two linearly independent columns of $\mathbf{S}$. Then Alice sends $\boldsymbol{u}_a$ and $\boldsymbol{u}'_a$, which Bob uses to compute $\boldsymbol{u}_a\boldsymbol{v}_b + \boldsymbol{u}'_a\boldsymbol{v}'_b = \mathbf{S}_{a,b} = \mathbf{M}_{a,b}$. Since $\boldsymbol{u}$ and $\boldsymbol{u}'$ are columns of $\mathbf{S}$, their entries take values from $\{0, 1, \ldots, \Delta\}$, so Alice communicates $O(\log\Delta)$ bits in total. Finally we can omit the term $\log r \log\log(\Delta+1)$ as it is dominated by $\log^2 r + \log\Delta$.

3 Finding Large Monochromatic Rectangles

In this section, we show that the assumption for applying the communication protocol from Sect. 2 does indeed hold. That is, any bounded integral matrix contains a sufficiently large monochromatic rectangle.

We first reduce the problem of finding large monochromatic rectangles to finding *almost-monochromatic* rectangles. We say a rectangle is $(1-\varepsilon)$-*monochromatic* if at least a $(1-\varepsilon)$-fraction of its entries have the same value.

Lemma 5. *Suppose* $\mathbf{M} \in \mathbb{R}^{A \times B}$ *has rank* $r \geq 1$, *and is* $(1-\frac{1}{16r})$-*monochromatic with color* α. *Then* $\mathbf{M}$ *contains a monochromatic rectangle of size* $\geq \frac{|A||B|}{8}$.

Proof. Let us call a column of $\mathbf{M}$ *bad* if it contains at least $\frac{1}{8r}|A|$-many non-α entries. By Markov's inequality, at most half the columns are bad. Let $B' \subseteq B$ be the remaining good columns, where each contains at least $(1-\frac{1}{8r})|A|$-many α entries. Let $B'' \subseteq B'$ be a maximal set of linearly independent good columns. We know $|B''| \leq r$ as the rank of $\mathbf{M}$ is r.

Let $\mathbf{U} = \mathbf{M}[A, B'']$. Since each column in B'' contains at most $\frac{1}{8r}|A|$-many non-α entries, there are at most $r \cdot \frac{1}{8r}|A|$ rows of $\mathbf{U}$ that contain non-α entries. Let A' denote the $|A| - \frac{1}{8}|A| \geq \frac{1}{2}|A|$ rows of $\mathbf{U}$ that contain only α entries. So $\mathbf{M}[A', B'']$ contains only α entries.

Let $\mathbf{T} = \mathbf{M}[A', B']$. Since the columns in B' are linear combinations of columns in B'', each column of $\mathbf{T}$ must be of the form $\beta\mathbf{1}$ for some β. Finally, we know $|\mathbf{T}| = |A'||B'| \geq \frac{1}{4}|A||B|$, and there are at most $\frac{1}{16r}|A||B|$ non-α entries in $\mathbf{M}$ in total, so at least half the columns of $\mathbf{T}$ must have value α.

Now, it remains to show that we can find large almost-monochromatic rectangles.

Lemma 6. *Let* $\mathbf{M} \in \{0, 1, \ldots, \Delta\}^{A \times B}$ *be a bounded integral matrix of rank* r. *Then* $\mathbf{M}$ *contains a* $(1-\frac{1}{16r})$-*monochromatic rectangle of size at least* $|A||B| \cdot \exp(-\Delta^{O(\Delta)} \cdot \sqrt{r} \cdot \log r)$.

To prove Lemma 6, we first define a distribution $\mathcal{D}$ over the rectangles of $\mathbf{M}$, and then use the probabilistic method with respect to $\mathcal{D}$ to show that there exists a large enough almost-monochromatic rectangle. We begin with the technical ingredients:

Definition 1 (Factorization norm). *For a matrix* $\mathbf{M} \in \mathbb{R}^{A \times B}$, *define its* γ_2-*norm as*

$$\gamma_2(\mathbf{M}) \stackrel{\text{def}}{=} \min\{R \geq 0 : \textit{there are families of vectors } \{\boldsymbol{u}_a\}_{a \in A}, \{\boldsymbol{v}_b\}_{b \in B} \textit{ so that } \mathbf{M}_{a,b} = \langle \boldsymbol{u}_a, \boldsymbol{v}_b \rangle \textit{ and } \|\boldsymbol{u}_a\|_2 \|\boldsymbol{v}_b\|_2 \leq R \textit{ for all } a, b\}.$$

In other words, $\gamma_2(\mathbf{M})$ gives the Euclidean length needed to factor the matrix $\mathbf{M}$. Note that by rescaling the $\boldsymbol{u}_a$'s and $\boldsymbol{v}_b$'s, it is straightforward to guarantee $\|\boldsymbol{u}_a\|_2 \leq \sqrt{R}$ and $\|\boldsymbol{v}_b\|_2 \leq \sqrt{R}$ for all a, b in the definition. Here $\boldsymbol{u}_a, \boldsymbol{u}_b$ are vectors of any dimension (of course one may choose $\boldsymbol{u}_a, \boldsymbol{u}_b$ to have dimension rank(**M**)). The terminology γ_2-norm or factorization norm is indeed justified as γ_2 is a norm on the space of real matrices.

The following lemma is well-known:

Lemma 7 (Lemma 4.2,[5]). *Any matrix* $\mathbf{M} \in \mathbb{R}^{A\times B}$ *satisfies* $\gamma_2(\mathbf{M}) \leq \|\mathbf{M}\|_\infty \cdot \sqrt{\text{rank}(\mathbf{M})}$.

It will be convenient for us to factor **M** with vectors of the same Euclidean length, which comes at the expense of the dimension:

Lemma 8. *For any matrix* $\mathbf{M} \in \mathbb{R}^{A\times B}$ *and* $s \geq \sqrt{\gamma_2(\mathbf{M})}$*, there are vectors* $\{\boldsymbol{u}_a\}_{a\in A}, \{\boldsymbol{v}_b\}_{b\in B}$ *such that* $\mathbf{M}_{a,b} = \langle \boldsymbol{u}_a, \boldsymbol{v}_b\rangle$ *and* $\|\boldsymbol{u}_a\|_2 = \|\boldsymbol{v}_b\|_2 = s$ *for all* $a \in A$ *and* $b \in B$.

Proof. Construct vectors $\boldsymbol{u}_a, \boldsymbol{v}_b$ each with 2-norm $\|\boldsymbol{u}_a\|_2, \|\boldsymbol{v}_b\|_2 \leq s$ and $\mathbf{M}_{a,b} = \langle \boldsymbol{u}_a, \boldsymbol{v}_b\rangle$ using Lemma 7. Then add $|A| + |B|$ new coordinates, where each $\boldsymbol{u}_a$ and $\boldsymbol{v}_b$ receives a "private" coordinate. Set the private coordinate of $\boldsymbol{u}_a$ to $\sqrt{s^2 - \|\boldsymbol{u}_a\|_2^2}$, and similarly for $\boldsymbol{v}_b$.

We denote $N^n(0,1)$ as the n-dimensional standard Gaussian. Let $S^{n-1} \stackrel{\text{def}}{=} \{x \in \mathbb{R}^n : \|x\|_2 = 1\}$ be the n-dimensional unit sphere. The following argument is usually called hyperplane rounding in the context of approximation algorithms:

Lemma 9 (Sheppard's formula). *Any vectors* $\boldsymbol{u}, \boldsymbol{v} \in S^{n-1}$ *with* $\langle \boldsymbol{u}, \boldsymbol{v}\rangle = \alpha$ *satisfy*

$$\Pr_{\boldsymbol{g}\sim N^n(0,1)}[\langle \boldsymbol{g}, \boldsymbol{u}\rangle \geq 0 \text{ and } \langle \boldsymbol{g}, \boldsymbol{v}\rangle \geq 0] = h(\alpha) \stackrel{\text{def}}{=} \frac{1}{2}\left(1 - \frac{\arccos(\alpha)}{\pi}\right).$$

More generally, any vectors $\boldsymbol{u}, \boldsymbol{v} \in \mathbb{R}^n \setminus \{\mathbf{0}\}$ *satisfy*

$$\Pr_{\boldsymbol{g}\sim N^n(0,1)}[\langle \boldsymbol{g}, \boldsymbol{u}\rangle \geq 0 \text{ and } \langle \boldsymbol{g}, \boldsymbol{v}\rangle \geq 0] = h\left(\frac{\langle \boldsymbol{u}, \boldsymbol{v}\rangle}{\|\boldsymbol{u}\|_2\|\boldsymbol{v}\|_2}\right).$$

We can now define a suitable distribution over the rectangles of **M** using the above tools. In particular, we want the probability that a rectangle contains an entry to be a function of the entry value.

Lemma 10. *Let* $\mathbf{M} \in \{0, 1, \ldots, \Delta\}^{A\times B}$ *be a bounded integral matrix of rank* r*. Then for any* k*, there is a distribution* $\mathcal{D}_k$ *over the rectangles of* **M** *so that for all* $a \in A$ *and* $b \in B$*,*

$$\Pr_{\mathbf{R}\sim\mathcal{D}_k}[(a,b) \in \mathbf{R}] = \left(h\left(\frac{\mathbf{M}_{a,b}}{\Delta\sqrt{r}}\right)\right)^k,$$

where h *is the function defined in Lemma 9.*

Proof. We use Lemma 8 to factor **M**, which gives vectors $\boldsymbol{u}_a, \boldsymbol{v}_b$ for all $a \in A$ and $b \in B$, such that $\mathbf{M}_{a,b} = \langle \boldsymbol{u}_a, \boldsymbol{v}_b\rangle$, and $\|\boldsymbol{u}_a\|_2 = \|\boldsymbol{v}_b\|_2 = \Delta^{1/2}r^{1/4}$. Then we sample independent Gaussians $\boldsymbol{g}_1, \ldots, \boldsymbol{g}_k \sim N^n(0,1)$ and set

$$\mathbf{R}_i \stackrel{\text{def}}{=} \{a \in A : \langle \boldsymbol{u}_a, \boldsymbol{g}_i \rangle \geq 0\} \times \{b \in B : \langle \boldsymbol{v}_b, \boldsymbol{g}_i \rangle \geq 0\},$$

and then $\mathbf{R} \stackrel{\text{def}}{=} \mathbf{R}_1 \cap \cdots \cap \mathbf{R}_k$. Note that $\mathbf{R}$ is indeed a rectangle. For each i and each $(a, b) \in A \times B$ we have

$$\Pr[(a, b) \in \mathbf{R}_i] = h\left(\frac{\langle \boldsymbol{u}_a, \boldsymbol{v}_b \rangle}{\|\boldsymbol{u}_a\|_2 \|\boldsymbol{v}_b\|_2}\right) = h\left(\frac{\mathbf{M}_{a,b}}{\Delta\sqrt{r}}\right).$$

The overall expression follows by independence of the k rectangles.

Finally, we use the above distribution to show the existence of large almost-monochromatic rectangles.

Proof of Lemma 6. Let $\mathcal{E}_0 \,\dot\cup \cdots \dot\cup\, \mathcal{E}_\Delta$ be the partition of the entries $A \times B$ based on entry values, so that $\mathcal{E}_j \stackrel{\text{def}}{=} \{(a, b) \in A \times B : \mathbf{M}_{a,b} = j\}$. Let $m_j \stackrel{\text{def}}{=} (64r\Delta)^{(8\Delta)^j}$ for each $j = 0, \ldots, \Delta$, and let i be the index such that $m_i \cdot |\mathcal{E}_i|$ is maximized. From this, we also get

$$|\mathbf{M}| = |A| \cdot |B| = \sum_{j=0}^{\Delta} |\mathcal{E}_j| \leq \sum_{j=0}^{\Delta} \frac{m_i}{m_j} |\mathcal{E}_i| \leq m_i |\mathcal{E}_i|. \tag{3.1}$$

For notational convenience, recall $h(\alpha) \stackrel{\text{def}}{=} \frac{1}{2}\left(1 - \frac{\arccos(\alpha)}{\pi}\right)$, and let $c(j) \stackrel{\text{def}}{=} h\left(\frac{j}{\Delta\sqrt{r}}\right)$. We observe that on $[0, 1]$, the function h is convex, monotone increasing, lowerbounded by $h(0) = 1/4$, upperbounded by $h(1) = 1/2$, and $h'(\alpha) = \frac{1}{2\pi\sqrt{1-\alpha^2}} \geq \frac{1}{2\pi}$. Additionally, the following claim about c will be useful for our calculations later:

For $1 \leq j \leq \Delta$, we have

$$\frac{c(j)}{c(j-1)} \geq \frac{c(j-1) + c'(j-1)}{c(j-1)} \geq 1 + \frac{1}{2\pi\Delta\sqrt{r} \cdot c(j-1)} \geq 1 + \frac{4}{3\pi\Delta\sqrt{r}}. \tag{3.2}$$

where we used the fact that $c(j-1) \leq 3/8$. Furthermore, we have

$$\frac{c(\Delta)}{c(0)} \leq \frac{c(0) + \Delta \cdot c'(\Delta)}{c(0)} = 1 + \frac{4\Delta}{\Delta\sqrt{r} \cdot 2\pi\sqrt{1 - 1/r}} \leq 1 + \frac{4}{\pi\sqrt{r}}. \tag{3.3}$$

Next, let $\mathcal{D}_k$ be the distribution from Lemma 10, and generate $\mathbf{R} \sim \mathcal{D}_k$ for some choice of k to be determined. We will show that there exists a k such that $\mathbf{R}$ is expected to be $(1 - \frac{1}{16r})$ monochromatic with color i, and is sufficiently large. Specifically, the number of i-entries in $\mathbf{R}$ is greater than the number all other entries in $\mathbf{R}$ by a factor of $16r$ in expectation, and moreover, this difference is sufficiently large, which in turn means $\mathbf{R}$ is sufficiently large.

$$\mathbb{E}_{\mathbf{R}\sim\mathcal{D}_k}\left[|\mathcal{E}_i \cap \mathbf{R}| - 16r\sum_{j\neq i}|\mathcal{E}_j \cap \mathbf{R}|\right]$$

$$= |\mathcal{E}_i|\cdot c(i)^k - 16r\sum_{j\neq i} c(j)^k|\mathcal{E}_j| \qquad \text{(by Lemma 10)}$$

$$\geq |\mathcal{E}_i|\cdot c(i)^k\left(1 - 16r\sum_{j\neq i}\frac{c(j)^k m_i}{c(i)^k m_j}\right)$$

Suppose $\frac{c(j)^k m_i}{c(i)^k m_j} \leq \frac{1}{64r\Delta}$ for each $j \neq i$, then we can conclude

$$\geq |\mathcal{E}_i|\cdot 4^{-k}(1-\frac{1}{4}) \qquad \text{(Since } c(i)\geq \tfrac{1}{4}\text{)}$$

$$\geq \frac{|A||B|}{m_i}\frac{1}{2\cdot 4^k}. \qquad \text{(by Lemma 3.1)}$$

Claim There exists a choice of k such that $\frac{c(j)^k m_i}{c(i)^k m_j} \leq \frac{1}{64r\Delta}$ for all $j \neq i$.

Proof We consider two cases:

(1) $0 \leq j < i$: In this case we can bound $\frac{c(j)^k m_i}{c(i)^k m_j} \leq \left(\frac{c(i-1)}{c(i)}\right)^k \frac{m_i}{m_0}$. To get our claim, it suffices to choose k to satisfy

$$\left(\frac{c(i-1)}{c(i)}\right)^k \frac{m_i}{m_0} \leq \frac{1}{64r\Delta}$$

$$\Leftarrow \qquad k \geq \frac{((8\Delta)^i+1)\log(64r\Delta)}{\log\frac{c(i)}{c(i-1)}}.$$

Using the lower bound from Eq. (3.2), along with $\log(1+x) \geq x/2$ for $x \leq 1$, we conclude it suffices to choose k to satisfy

$$k \geq ((8\Delta)^i+1)\log(64r\Delta)\cdot\frac{3}{2}\pi\Delta\sqrt{r}. \tag{3.4}$$

(2) $i < j \leq \Delta$: In this case we can bound $\frac{c(j)^k m_i}{c(i)^k m_j} \leq \left(\frac{c(\Delta)}{c(0)}\right)^k \frac{m_i}{m_{i+1}}$. To get our claim, it suffices to choose k to satisfy

$$\left(\frac{c(\Delta)}{c(0)}\right)^k \frac{m_i}{m_{i+1}} \leq \frac{1}{64r\Delta}$$

$$\Leftarrow \qquad k \leq \frac{((8\Delta)^{i+1}-(8\Delta)^i-1)\log(64r\Delta)}{\log\frac{c(\Delta)}{c(0)}}.$$

Using the upper bound from Eq. (3.3), we conclude it suffices to choose k to satisfy

$$k \leq ((8\Delta)^{i+1}-(8\Delta)^i-1)\log(64r\Delta)\cdot\frac{\pi\sqrt{r}}{4}. \tag{3.5}$$

To choose k to simultaneously satisfy the two cases, we first verify that the lower and upper bound for k in Eq. (3.4) and Eq. (3.5) are consistent. Indeed when $i \neq 0$, we have

$$\frac{3}{2}((8\Delta)^i + 1)\Delta \leq \frac{1}{4}((8\Delta)^{i+1} - (8\Delta)^i - 1),$$

so we may choose k to be equal to the lower bound. If $i = 0$, then the lower bound from Eq. (3.4) does not apply, so we choose k to be equal to the upper bound.

For any established choice of k, we always have $k \leq (8\Delta)^{\Delta} \log(64r\Delta)\pi\Delta\sqrt{r}$. Moreover, we have $\log m_i \leq (8\Delta)^{\Delta} \log(64r\Delta)$. We conclude that

$$\begin{aligned}\mathbb{E}_{\mathbf{R}\sim\mathcal{D}_k}\left[|\mathcal{E}_i \cap \mathbf{R}| - 16r\sum_{j\neq i}|\mathcal{E}_j \cap \mathbf{R}|\right] &\geq \frac{|A||B|}{m_i}\frac{1}{2\cdot 4^k} \\ &\geq \exp(-\Delta^{O(\Delta)}\cdot\sqrt{r}\cdot\log r)|A||B|.\end{aligned}$$

Then any $\mathbf{R}$ attaining this expectation will simultaneously satisfy $|\mathbf{R}| \geq \exp(-\Delta^{O(\Delta)}\cdot\sqrt{r}\cdot\log r)|A||B|$ and be $(1-\frac{1}{16r})$-monochromatic.

4 Proof of Theorem 1

By Lemma 6, we know any submatrix $\mathbf{S}$ of $\mathbf{M}$ contains a $(1-\frac{1}{16r})$-monochromatic rectangle of size $\exp(-\Delta^{O(\Delta)}\cdot\sqrt{r}\cdot\log r)|\mathbf{S}|$. Therefore, by Lemma 5, $\mathbf{S}$ contains a monochromatic rectangle of size $\frac{1}{8}\exp(-\Delta^{O(\Delta)}\cdot\sqrt{r}\cdot\log r)|\mathbf{S}|$. We substitute $\delta(r) = \Delta^{O(\Delta)}\sqrt{r}\log r$ in Lemma 2 to get

$$CC^{\det}(\mathbf{M}) \leq \Theta\left(\log\Delta + \log^2 r + \sum_{i=0}^{\log r}\delta(r/2^i)\right).$$

The summation simplifies as follows:

$$\begin{aligned}\sum_{i=0}^{\log r}\delta(r/2^i) &= \sum_{i=0}^{\log r}\Delta^{O(\Delta)}\sqrt{r}\cdot 2^{-i/2}\cdot\log\left(\frac{r}{2^i}\right) \\ &\leq \Delta^{O(\Delta)}\sqrt{r}\log r\cdot\sum_{i=0}^{\log r}2^{-i/2},\end{aligned}$$

where the sum converges. We ignore the lower order terms in the communication complexity expression to conclude

$$CC^{\det}(\mathbf{M}) \leq \Theta\left(\Delta^{O(\Delta)}\sqrt{r}\log r\right).$$

□

Acknowledgments. We thank Sam Fiorini for helpful discussions, and the anonymous reviewers for helpful feedback.

Disclosure of Interests. The authors have no competing interests.

References

1. Barvinok, A.: Approximations of convex bodies by polytopes and by projections of spectrahedra. arXiv preprint: arXiv:1204.0471 (2012)
2. Conforti, M., Cornuéjols, G., Zambelli, G.: Extended formulations in combinatorial optimization. 4OR **8**(1), 1–48 (2010)
3. Fawzi, H., Gouveia, J., Parrilo, P.A., Robinson, R.Z., Thomas, R.R.: Positive semidefinite rank. Math. Program. **153**, 133–177 (2015)
4. Fiorini, S., Massar, S., Pokutta, S., Tiwary, H.R., De Wolf, R.: Exponential lower bounds for polytopes in combinatorial optimization. J. ACM (JACM) **62**(2), 1–23 (2015)
5. Linial, N., Mendelson, S., Schechtman, G., Shraibman, A.: Complexity measures of sign matrices. Combinatorica **27**, 439–463 (2007)
6. Lovász, L., Saks, M.: Lattices, mobius functions and communications complexity. In: 29th Annual Symposium on Foundations of Computer Science, pp. 81–90. IEEE Computer Society (1988)
7. Lovett, S.: Communication is bounded by root of rank. J. (JACM) **63**(1), 1–9 (2016)
8. Nisan, N., Wigderson, A.: On rank vs. communication complexity. Combinatorica **15**(4), 557–565 (1995)
9. Rao, A., Yehudayoff, A.: Communication Complexity and Applications. Cambridge University Press, Cambridge (2020)
10. Rothvoß, T.: A direct proof for Lovett's bound on the communication complexity of low rank matrices. arXiv preprint: arXiv:1409.6366 (2014)
11. Rothvoß, T.: The matching polytope has exponential extension complexity. J. ACM (JACM) **64**(6), 1–19 (2017)
12. Yannakakis, M.: Expressing combinatorial optimization problems by linear programs. In: Proceedings of the Twentieth Annual ACM Symposium on Theory of Computing, pp. 223–228 (1988)

Assortment Optimization with Visibility Constraints

Théo Barré[1], Omar El Housni[2(✉)], and Andrea Lodi[2]

[1] IEOR, UC Berkeley, Berkeley, USA
theo_barre@berkeley.edu
[2] ORIE, Cornell Tech, Cornell University, New York, USA
{oe46,al748}@cornell.edu

Abstract. Motivated by applications in e-retail and online advertising, we study the problem of assortment optimization under visibility constraints, referred to as APV. We are given a universe of substitutable products and a stream of T customers. The objective is to determine the optimal assortment of products to offer to each customer in order to maximize the total expected revenue, subject to the constraint that each product is required to be shown to a minimum number of customers. The minimum display requirement for each product is given exogenously and we refer to these constraints as *visibility constraints*. We assume that customer choices follow a Multinomial Logit model. We provide a characterization of the structure of the optimal assortments and present an efficient polynomial time algorithm for solving APV. To accomplish this, we introduce a novel function called the "expanded revenue" of an assortment and establish its supermodularity. Our algorithm takes advantage of this structural property. Additionally, we demonstrate that APV can be formulated as a compact linear program. Finally, we propose a novel, *fair* strategy for pricing the revenue loss due to the enforcement of visibility constraints.

Keywords: Assortment Optimization · Supermodularity · Algorithm Design · Multinomial Logit model · Visibility Constraints

1 Introduction

Assortment optimization is a crucial aspect of decision making in many industries such as e-retail and online advertising. The goal of assortment optimization is to select a subset of available products to offer to customers in order to maximize a specific objective function, such as maximizing revenue, profit, or market share. For example, e-retailers seek to strategically select which products to display to a customer in order to maximize the expected revenue. Online advertisers strategically select the most effective combination of advertisements to maximize user engagement and desired outcomes, such as click-through rates. The assortment choice is crucial due to the substitution effect, where a product's sales depend not only on its intrinsic value but also on the alternatives presented to

J. Vygen and J. Byrka (Eds.): IPCO 2024, LNCS 14679, pp. 124–138, 2024.
https://doi.org/10.1007/978-3-031-59835-7_10

the customer. For example, offering a high-quality, high-priced product alongside a comparable product at a significantly lower price may result in poor sales for the higher-priced product, leading to an unsatisfactory platform revenue. This highlights the importance of carefully selecting assortments.

Traditionally, assortment optimization frameworks often overlook a crucial element in contemporary e-commerce: product visibility. In today's complex business landscape, where companies adhere to Service-Level Agreements (SLAs) with suppliers and prioritize sponsored product promotion, product visibility within an assortment is pivotal. SLAs often define conditions for product representation, ensuring equitable visibility for each supplier's products on the platform. Moreover, the concept of sponsored products has gained traction, with brands willing to pay for prominent display and increased visibility. While these strategies influence consumer behavior, solely focusing on products visibility without considering broader assortment optimization can lead to an imbalanced product mix, resulting in reduced customer satisfaction and overall revenue.

In this paper, we introduce the notion of *visibility constraints* in the context of Assortment Optimization. The purpose is to enforce a minimum display of each product, i.e., each product has to be shown at least a certain number of times in the displayed assortments. This constraint is modeling both Service-Level Agreements and sponsored products. It can also capture the settings where the platform would like to ensure some fairness notion among vendors by ensuring that each product is given a "fair" chance, i.e., it is shown at least to a certain number of customers. Specifically, we are given a universe of substitutable products and a stream of T customers. For each customer, we have to offer an assortment from the universe of products. The customer decides to purchase one of these products, or to leave without purchasing any product (no-purchase option). We assume that the choice of the customer is governed by a Multinomial Logit (MNL) choice model. We enforce the constraint that each product in the universe has to be shown a minimum number of times among the T assortments offered. The minimum display requirement for each product is given exogenously. Our objective is to maximize the total expected revenue from the T customers. We refer to this combinatorial optimization problem as *Assortment optimization Problem with Visibility* constraints, briefly denoted as APV.

A first natural question concerns the complexity of APV. In fact, without visibility constraints, the problem reduces to the classical unconstrained assortment optimization under MNL, for which we know that the optimal assortment is revenue-ordered [19] and therefore can be solved in polynomial time. However, by enforcing the visibility constraints, we might have to include certain products in the assortment that cannibalize the sales of other more profitable products because of the substitution effect. Consequently, determining the optimal assortments for APV is not immediately clear. A subsequent question emerging from the visibility problem is the task of quantifying the revenue loss incurred by enforcing visibility constraints compared to the relaxed unconstrained problem. This challenge compels us to develop a pricing strategy that appropriately apportions the loss to different vendors based on the impact of their product on the overall revenue. Such a scenario frequently occurs within the framework of SLAs. Typically, a contract between a platform and a vendor includes a clause that guarantees a certain level of visibility to the vendor's product. In return for this visibility, the vendor compensates the platform with a fee.

1.1 Our Contributions

Our main technical contribution is to design a polynomial time algorithm for APV. In order to achieve that, we introduce the notion of expanded revenue and expanded set of an assortment and leverage their structural properties to characterize the structure of an APV optimal solution, and consequently design an efficient algorithm to compute it. A subsequent contribution is to quantify the revenue loss caused by the visibility constraints compared to the unconstrained assortment optimization problem and design a strategy to share this loss among the different vendors of the products for which visibility constraints have been enforced. Our contributions are summarized as follows.

(a) Expanded revenue: monotonicity and supermodularity[1]. In Sect. 1, we introduce the expanded revenue function. Given a universe of products $\mathcal{N}$ and an assortment $A \subseteq \mathcal{N}$. The expanded revenue of assortment A is defined as the maximum revenue of any assortment in $\mathcal{N}$ that contains A. The expanded set is the assortment that achieves this maximum revenue. We show that the expanded revenue function is closely related to the APV objective function in the case of a single customer. In Lemma 1, we provide a linear time algorithm to compute the expanded revenue of any assortment. Then, we prove in Lemma 2 that the expanded revenue function is monotone, i.e., the expanded revenue of an assortment decreases as the assortment gets larger. Finally, in Lemma 3, we prove the main theoretical result on which our final algorithm relies: the supermodularity of the expanded revenue function.

(b) Polynomial time algorithm and LP formulation. Building on the previous properties of the expanded revenue, we identify in Theorem 1 a very simple nested structure for an APV optimal solution, and devise a polynomial time algorithm to efficiently solve the problem. Additionally, we demonstrate in Theorem 2 that APV can be formulated as a compact linear program.

(c) Price of visibility. The introduction of visibility constraints results in a reduction in the total expected revenue compared to the unconstrained version of the assortment problem. We aim to evaluate this revenue loss and propose a fair strategy for distributing it among vendors based on their respective contributions to the loss. Building on the LP formulation of APV, we devise in Sect. 4, a pricing strategy as follows: For each product that negatively impacts the overall revenue, we charge the vendor a fraction of the loss proportional to the ratio between the negative contribution of the product and the sum of the negative contributions of all products. We demonstrate that this strategy satisfies natural fairness properties and exhibits favorable computational tractability.

[1] A function $f : \Omega \rightarrow \mathbb{R}$ is supermodular $\iff \forall A, B \in \Omega,\ f(A \cup B) + f(A \cap B) \geq f(A) + f(B)$.

1.2 Related Literature

Assortment optimization under the Multinomial Logit model is a well-established problem in scientific literature. Initially introduced by [12], with subsequent works by [14] and [10], the MNL choice model has gained popularity for modeling customer choices due to its simplicity in computing the choice probabilities, its predictive power and its computational tractability compared to more complex choice models. It has been extensively used in various research works such as [7,9,13,18,19], to mention a few. The MNL model proves particularly useful in assortment optimization, as demonstrated by [19], who showed that under the MNL model, the optimal assortment in the unconstrained setting is revenue-ordered. This means that it contains all products whose revenues exceed a certain threshold, simplifying the optimization problem by avoiding the consideration of exponentially numerous potential subsets. Moreover, Gallego et al. [8] give a linear programming formulation for the unconstrained assortment problem under MNL. Rusmevichientong et al. [16] solved the version of the problem with a cardinality constraint, proving it is still solvable in polynomial time, and Desir et al. [6] studied more general capacity constraints, showing it is NP-hard to solve in the general case. Sumida et al. [18] and Davis et al. [4] studied totally unimodular constraint structures for the assortment and showed that the resulting problem can be reformulated as a linear program. However, when considering mixtures of MNL models (MMNL), the assortment optimization problem becomes NP-hard even in the unconstrained setting with two classes of customers as shown in Rusmevichientong et al. [17].

To the best of our knowledge, this paper is the first to study assortment optimization with visibility constraints. The topic of visibility in assortment planning has barely been covered: Chen et al. [2] studied visibility under a fairness approach, trying to enforce similar visibility for products with similar characteristics, while Wang et al. [20] studied a version of the assortment optimization problem in which they can increase the attractiveness of some products through an advertising budget. In addition, the topic of assortment optimization for a stream of customers is more often studied from an online perspective, where decisions are made sequentially such as in Davis et al. [5] and [3]. In contrast, we study a static version of the problem, where we plan the entirety of our assortments in advance. Other versions of the problem, such as in [11], consider a flow of customers with randomized preferences, to which we offer a common assortment. Finally, in revenue management, pricing problems are often considered in the sense of optimizing the selling price of each product, as for example in [21], [15] and [1], while, for our problem, selling prices are fixed and we study instead how to price the loss generated by enforcing visibility of each product.

2 Model Formulation

The MNL Choice Model. Let $\mathcal{N} := \{1, \ldots, n\}$ be a universe of substitutable products at our disposal. Each product $i \in \mathcal{N}$ has a price $p_i \geq 0$. Without loss of generality, we order the products by non-increasing prices, i.e., $p_1 \geq p_2 \geq \ldots \geq p_n$. An assortment of products or an offer set, is simply a subset of products $S \subseteq \mathcal{N}$. Additionally, the option of not selecting any product is symbolically represented as product 0, and referred to it as the no-purchase option.

We assume that customers make choices according to a Multinomial Logit model. Under this model, each product $i \in \mathcal{N}$ is associated with a preference weight $v_i > 0$. Note that v_i captures the attractiveness of product i, meaning a high preference weight indicates a high popularity. Without loss of generality, we use the standard convention that the no-purchase preference weight is normalized to $v_0 = 1$. We use the notation $V(S) := \sum_{i \in S} v_i$, which is the total weight of a subset $S \subseteq \mathcal{N}$. Under the MNL model, if we offer an assortment $S \subseteq \mathcal{N}$, the customer chooses product i with probability $\phi(i, S) := \frac{v_i}{1+V(S)}$. We refer to $\phi(i, S)$ as the choice probability of product i given assortment S. Alternatively, the customer may decide to not purchase any product, which happens with the complementary probability $\phi(0, S) := \frac{1}{1+V(S)}$. Let $R(S)$ be the expected revenue we get from a customer if we offer assortment S. In particular, we have

$$R(S) := \sum_{i \in S} p_i \phi(i, S) = \frac{\sum_{i \in S} p_i v_i}{1 + \sum_{i \in S} v_i}.$$

Assortment Optimization with Visibility Constraints. We are presented with a stream of T customers. Each customer t will be offered an assortment S_t. Customers make choices according to the same MNL model, i.e., a customer decides to purchase product i from assortment S_t with a probability $\phi(i, S_t)$, or they may choose the no-purchase option with probability $\phi(0, S_t)$. The expected revenue we obtain from customer t is $R(S_t)$. To ensure visibility, we impose constraints that require each product $i \in \mathcal{N}$ to be shown to at least ℓ_i customers. Note that the parameters ℓ_i are exogenous and satisfy $\ell_i \in [\![0, T]\!]$ for all $i \in \mathcal{N}$. Our objective is to determine the assortment S_t to offer to each customer t in order to maximize the total expected revenue while (strictly) satisfying the visibility constraints. We refer to this problem as the *Assortment optimization Problem with Visibility constraints* (APV). It can be formulated as follows:

$$\begin{aligned} \max_{S_1,\ldots,S_T \subseteq \mathcal{N}} \quad & \sum_{t=1}^{T} R(S_t) \\ s.t. \quad & \sum_{t=1}^{T} \mathbb{1}(i \in S_t) \geq \ell_i, \quad \forall i \in \mathcal{N}. \end{aligned} \tag{APV}$$

3 Polynomial Time Algorithm for APV

The primary contribution of this paper is the development of a polynomial time algorithm for APV. To achieve this, we introduce in Sect. 3.1 the concepts of the "Expanded Revenue" and "Expanded Set" of an assortment. In Sect. 3.2, we present a polynomial time algorithm to compute the expanded set and expanded revenue and demonstrate the monotonicity and supermodularity of the expanded revenue function. Leveraging these properties, we characterize the structure of an optimal solution for APV and present an algorithm that compute it in $O(nT)$ time (Sect. 3.3). Finally, in Sect. 3.4, we demonstrate that APV can be formulated as a compact linear program.

3.1 Expanded Revenue and Expanded Set

We begin our analysis by examining APV in the context of a single customer. In this particular scenario, the visibility constraints are given such that either $\ell_i = 0$ or $\ell_i = 1$. Let A denote the subset of all products where $\ell_i = 1$. Consequently, APV is transformed into the problem of identifying the assortment that maximizes revenue while including A. This particular problem will serve as the building block for our analysis, as it lays the foundation for understanding the general case involving T customers. Thus, it leads us to introduce the subsequent definitions that will aid us in our analysis.

Definition 1 (Expanded revenue). *Let $A \subseteq \mathcal{N}$. The expanded revenue of A, denoted as $\overline{R}(A)$, is defined as the maximum expected revenue achieved by any assortment in $\mathcal{N}$ that contains A. In particular, it is given by*

$$\overline{R}(A) := \max_{S \subseteq \mathcal{N},\, A \subseteq S} R(S). \tag{1}$$

The optimal solution of the maximization problem in (1) is referred to as the expanded set of A. In case multiple optimal solutions exist, we break ties by selecting the optimal assortment with the largest cardinality. Lemma 1 will show that the expanded set is well defined. Formally, we have the following definition.

Definition 2 (Expanded set). *The expanded set of A, denoted as $\overline{A}$, is defined as the assortment within $\mathcal{N}$ that maximizes the expected revenue among all assortments containing A. If multiple assortments achieve the same maximum expected revenue, $\overline{A}$ is selected as the assortment with the largest cardinality. Mathematically, $\overline{A}$ is given by*

$$\overline{A} := \underset{S \subseteq \mathcal{N},\, A \subseteq S}{\arg\max} \left\{ |S| \ : \ R(S) = \overline{R}(A) \right\}.$$

Problem (1) can be viewed as equivalent to APV when considering a single customer scenario ($T = 1$) and defining A as the set of products that need to be shown once, i.e., $A = \{i \in \mathcal{N} : \ell_i = 1\}$. Thus, $\overline{A}$ represents the optimal assortment, with the largest cardinality, for the problem.

In our analysis, we consider $\overline{R}$ as a set function that takes an assortment $A \subseteq \mathcal{N}$ as input and returns $\overline{R}(A)$. Note that $\overline{R}(A) = R(\overline{A})$. In the subsequent section, we delve into examining various properties of this function, as well as properties associated with the expanded set.

3.2 Properties of the Expanded Revenue

In this section, we first show that we can compute the expanded revenue and the expanded set of a given assortment in polynomial time (Lemma 1. Then, we show that the expanded revenue is a non-increasing function and the expanded set is a non-decreasing function (Lemma 2) Finally, we show that the expanded revenue function is supermodular (Lemma 3) which is the most fundamental for our algorithm design later in the paper.

Computing the Expanded Revenue and Expanded Set. Recall w.l.o.g. that $p_1 \geq \ldots \geq p_n$. We define an assortment S to be price-ordered if $S = \{1, \ldots, k\}$ for some $1 \leq k \leq n$. Essentially, a price-ordered assortment prioritizes products with high prices. It is worth noting there are only n possible price-ordered assortments. Consider an assortment $A \subseteq \mathcal{N}$, and its expanded set $\overline{A}$. In the following lemma, we demonstrate that $\overline{A}$ is the union of A and a price-ordered assortment. Since there are only n possible price-ordered assortments, it is sufficient to compute the expected revenue of the assortments $A \cup \{1, \ldots, k\}$ for each $k \in \{1, \ldots, n\}$. The expanded set corresponds to the assortment with the highest expected revenue. In the case of multiple assortments with the same maximum revenue, we break ties by selecting the one with the largest cardinality. Thus, the expanded set $\overline{A}$ can be computed in linear time, specifically $O(n)$. The expanded revenue is simply $\overline{R}(A) = R(\overline{A})$. The proof of Lemma 1 leverages some structural properties of the revenue function under MNL.

Lemma 1. *For any $A \subseteq \mathcal{N}$, the expanded set of A is given by $\overline{A} = A \cup \{i \in \mathcal{N} : p_i \geq \overline{R}(A)\}$. Furthermore, $\overline{R}(A)$ and $\overline{A}$ can be computed in time $O(n)$.*

Proof. By definition, we have $A \subseteq \overline{A}$. Hence, we can write $\overline{A} = A \cup B$, where $B \subseteq \mathcal{N} \setminus A$. We will prove that

$$B = \{i \in \mathcal{N} \setminus A \ : \ p_i \geq \overline{R}(A)\}.$$

Assume that there exists $i \in B$ such that $p_i < \overline{R}(A) = R(A \cup B)$. It is known that, under the MNL model, when we add a product j to an assortment S, the revenue of this assortment increases if and only if $p_j \geq R(S)$. For completeness, we provide the statement and the proof of this known result in the full version of this work. Using this lemma implies that removing i from B would strictly increase the expected revenue $R(A)$, which contradicts the optimality of $A \cup B$. Assume there exists $i \in \mathcal{N} \setminus A$ such that $p_i \geq \overline{R}(A)$ but $i \notin B$. Adding i to B would increase the revenue. If this increase is strict, it contradicts the optimality of $A \cup B$. If the revenue stays the same, it contradicts the definition of $\overline{A} = A \cup B$ as the optimal solution with maximum cardinality. Thus,

$$\overline{A} = A \cup \{i \in \mathcal{N} \setminus A, p_i \geq \overline{R}(A)\}.$$

Finally, $\overline{A}$ can be computed in time $O(n)$. Indeed, we just have to start from A, and add elements by decreasing price. At each iteration, we can compute the new revenue from the previous one in constant time if we store the current numerator and denominator, since we only need to add $p_i v_i$ to the former and v_i to the latter when we reach element i. Finally, we just have to pick the highest revenue set among the n sets computed. □

Lemma 2 (Monotonicity). *If $A \subseteq B \subseteq \mathcal{N}$, then $\overline{A} \subseteq \overline{B}$ and $\overline{R}(A) \geq \overline{R}(B)$.*

Proof. For $A \subseteq B \subseteq \mathcal{N}$, we have $\{S \subseteq \mathcal{N} : B \subseteq S\} \subseteq \{S \subseteq \mathcal{N} : A \subseteq S\}$. So every feasible solution for $\max_{S\subseteq\mathcal{N},\, B\subseteq S} R(S)$ is a feasible solution for $\max_{S\subseteq\mathcal{N},\, A\subseteq S} R(S)$. Therefore, $\overline{R}(A) \geq \overline{R}(B)$. It follows that $\{i \in \mathcal{N}, p_i \geq \overline{R}(A)\} \subseteq \{i \in \mathcal{N}, p_i \geq \overline{R}(B)\}$, and therefore $\overline{A} = A \cup \{i \in \mathcal{N}, p_i \geq \overline{R}(A)\} \subseteq B \cup \{i \in \mathcal{N}, p_i \geq \overline{R}(B)\} = \overline{B}$. □

Lemma 3 (Supermodularity). *The expanded revenue function $\overline{R}$ is supermodular, i.e.,*

$$\forall A, B \subseteq \mathcal{N},\ \ \overline{R}(A \cup B) + \overline{R}(A \cap B) \geq \overline{R}(A) + \overline{R}(B).$$

Proof. First, we provide an intermediate computation: $\forall S, B \subseteq \mathcal{N}$,

$$\begin{aligned} R(S) - R(S \cup B) &= R(S) - R(S \cup (B \setminus S)) \\ &= R(S) - \frac{\sum_{i\in S\cup(B\setminus S)} p_i v_i}{1 + V(S) + V(B\setminus S)} \\ &= \frac{R(S)(1+V(S)) + R(S)V(B\setminus S) - \sum_{i\in B\cup S} p_i v_i}{1+V(S\cup B)} \\ &= \frac{\sum_{i\in S} p_i v_i + R(S)V(B\setminus S) - \sum_{i\in S} p_i v_i - \sum_{i\in B\setminus S} p_i v_i}{1+V(S\cup B)} \\ &= \frac{\sum_{i\in B\setminus S}(R(S)-p_i)v_i}{1+V(S\cup B)}. \end{aligned} \tag{2}$$

Now, we show that $\forall A, B \subseteq \mathcal{N}, R(\overline{A} \cup \overline{B}) + \overline{R}(A \cap B) \geq \overline{R}(A) + \overline{R}(B)$. This is equivalent to $R(\overline{A \cap B}) - R(\overline{B}) \geq R(\overline{A}) - R(\overline{A} \cup \overline{B})$, which can be rewritten as

$$\frac{\sum_{i\in\overline{B}\setminus\overline{A\cap B}}(R(\overline{A\cap B}) - p_i)v_i}{1+V(\overline{B})} \geq \frac{\sum_{i\in\overline{A}\cup\overline{B}\setminus\overline{A}}(R(\overline{A}) - p_i)v_i}{1+V(\overline{A}\cup\overline{B})}, \tag{3}$$

using Eq. (2), and the fact that $A \cap B \subseteq B \implies \overline{A \cap B} \subseteq \overline{B}$ by Lemma 2. We know that $\forall i \in \overline{B} \setminus \overline{A \cap B}, p_i < R(\overline{A \cap B})$. Indeed,

$$\overline{A\cap B} = (A \cap B) \cup \{i \in \mathcal{N}, p_i \geq R(\overline{A\cap B})\} \supseteq \{i \in \mathcal{N}, p_i \geq R(\overline{A\cap B})\}.$$

For the same reason, $\forall i \in \overline{A} \cup \overline{B} \setminus \overline{A},\ p_i < R(\overline{A})$. Then, we show that $\overline{A} \cup \overline{B} \setminus \overline{A} \subseteq \overline{B} \setminus \overline{A \cap B}$. Indeed, $\overline{A} \cup \overline{B} \setminus \overline{A} \subseteq \overline{B}$, and then $A \cap B \subseteq A \implies \overline{A \cap B} \subseteq \overline{A}$ by Lemma 2. So since $\overline{A} \cup \overline{B} \setminus \overline{A}$ contains no element of $\overline{A}$, it also contains no element of $\overline{A \cap B}$. Thus, $\overline{A} \cup \overline{B} \setminus \overline{A} \subseteq \overline{B} \setminus \overline{A \cap B}$. In addition, Lemma 2 gives $R(\overline{A \cap B}) \geq R(\overline{A})$. Therefore,

$$0 < \sum_{i\in\overline{A}\cup\overline{B}\setminus\overline{A}} (R(\overline{A}) - p_i)v_i \leq \sum_{i\in\overline{B}\setminus\overline{A\cap B}} (R(\overline{A\cap B}) - p_i)v_i, \tag{4}$$

because for each term in the left sum, there is a different term in the right sum that is superior or equal to it, and the additional terms in the right sum are all positive. Finally, we have

$$V(\overline{B}) = \sum_{i \in \overline{B}} v_i \leq \sum_{i \in \overline{A} \cup \overline{B}} v_i = V(\overline{A} \cup \overline{B}), \tag{5}$$

Combining (4) and (5) give the desired inequality (3), which is equivalent to

$$\forall A, B \subseteq \mathcal{N}, R(\overline{A} \cup \overline{B}) + \overline{R}(A \cap B) \geq \overline{R}(A) + \overline{R}(B).$$

Then, as $A \subseteq \overline{A}$ and $B \subseteq \overline{B}$, we have $A \cup B \subseteq \overline{A} \cup \overline{B}$, and therefore by Lemma 2, $\overline{R}(A \cup B) \geq \overline{R}(\overline{A} \cup \overline{B}) \geq R(\overline{A} \cup \overline{B})$. Therefore, $R(\overline{A} \cup \overline{B}) + \overline{R}(A \cap B) \geq \overline{R}(A) + \overline{R}(B)$ which implies $\overline{R}(A \cup B) + \overline{R}(A \cap B) \geq \overline{R}(A) + \overline{R}(B)$. □

3.3 Optimal Solution for APV

In this section, we present the main technical result in this paper. In particular, we characterize the structure of an optimal solution of APV. Our characterization relies on the supermodularity property of the expanded revenue function. Moreover, we show that we can compute such a solution in $O(n + T)$, which gives us a polynomial time algorithm to solve APV.

Optimal Solution. Consider an instance of APV. Recall that for all $i \in \mathcal{N}$, ℓ_i is the lower bound on the minimum number of customers for which we must offer product i. For $t \in \{0, 1, \ldots, T\}$, we define the following sets

$$L_t = \{i \in \mathcal{N}, \ell_i = t\}. \tag{6}$$

Our candidate solution for APV is given by

$$S_t^* = \overline{\bigcup_{t \leq u \leq T} L_u}, \quad \forall t \in \{1, \ldots, T\}. \tag{7}$$

Note that $(L_t)_{0 \leq t \leq T}$ is a partition of $\mathcal{N}$. Moreover, since $\overline{\bigcup_{t+1 \leq u \leq T} L_u} \subseteq \overline{\bigcup_{t \leq u \leq T} L_u}$, the monotonicity property in Lemma 2 implies that $S_{t+1}^* \subseteq S_t^*$ for any $t = 0, \ldots, T-1$. Therefore, our solution has nested structure, i.e., $S_T^* \subseteq S_{T-1}^* \cdots \subseteq S_1^*$. In the following, we prove that the assortments given by (7) are optimal for APV. Moreover, they can be computed in polynomial time. Indeed, Lemma 1 shows that each of them can be computed in time $O(n)$, so the entire solution can be computed in time $O(nT)$. We can further improve the running time to $O(n + T)$ as shown below.

Theorem 1. *The sequence of assortments $(S_t^*)_{1 \leq t \leq T}$ defined in (7) is optimal for APV. Moreover, such a solution can be computed in $O(n + T)$ time.*

Proof. We reason by recurrence using the supermodularity property. For $T = 1$, the problem we want to solve is exactly $\max_{S \subseteq \mathcal{N} \text{ s.t. } L_1 \subseteq S} R(S) = \overline{R}(L_1)$, and by definition we know that $S_1^* := \overline{L_1}$ is an optimal solution.

Let us now take an instance of problem APV with $T \geq 2$, and assume that the result is valid for $T-1$. For simplicity, we use the notation $R_T(S_1, \ldots, S_T) := \sum_{t=1}^T R(S_t)$. Let $\hat{S}_1, \ldots, \hat{S}_T$ be an optimal solution of our problem and let

$$R_T^* := \max_{S_1, \ldots, S_T} R_T(S_1, \ldots, S_T) = \sum_{t=1}^T R(\hat{S}_t).$$

We notice that $\forall t \in [\![1, T]\!], R(\hat{S}_t) = \overline{R}(\hat{S}_t)$. In fact, by definition of expanded revenue, we have $R(\hat{S}_t) \leq \overline{R}(\hat{S}_t)$. Moreover, if the inequality was strict, we could complement $\hat{S}_t$ by adding to it the elements in $\overline{\hat{S}_t} \backslash \hat{S}_t$. This would strictly increase the objective value of APV while still being a feasible solution, which contradicts the optimality of $\hat{S}_1, \ldots, \hat{S}_T$. Hence, $R(\hat{S}_t) = \overline{R}(\hat{S}_t)$ for all $t \in [\![1, T]\!]$.

We now iteratively build a new feasible solution for APV. We initialize $S_t \leftarrow \hat{S}_t$ for each $t \in [\![1, T]\!]$. Then, for t going from 2 to T, we apply the following operations: $S_1 \leftarrow S_1 \cup \hat{S}_t$, $S_t \leftarrow S_1 \cap \hat{S}_t$. Finally, we take $\overline{S_1}, \ldots, \overline{S_T}$ as our new solution. We will refer to this algorithmic process as Algo.

At each iteration of Algo, the solution $S_1, \ldots, S_T$ remains feasible for problem APV. Indeed, if A and B are two assortments, then $A \cup B$ and $A \cap B$ contain the same products and with the same number of occurrences as A and B. This is because products that were only in A are now only in $A \cup B$ but not in $A \cap B$. Similarly, products that were only in B are now only in $A \cup B$ but not in $A \cap B$. Finally, products that were in both A and B are in $A \cup B$ and $A \cap B$.

Consider step t of Algo where the current solution is $S_1, \ldots, S_T$. From Lemma 3, the supermodularity of the expanded revenue function gives

$$\overline{R}(S_1 \cup \hat{S}_t) + \overline{R}(S_1 \cap \hat{S}_t) \geq \overline{R}(S_1) + \overline{R}(\hat{S}_t).$$

Moreover, $\overline{R}(S_u)$ remains the same as in step $t-1$ for all $u \notin \{1, t\}$. Therefore, the sum of the expanded revenue of the assortments at each step of Algo increases. Hence, by induction, the final assortments $S_1, S_2, \ldots, S_T$ that we obtain at the end verify $\sum_{t=1}^T \overline{R}(S_t) \geq \sum_{t=1}^T \overline{R}(\hat{S}_t) = R_T^*$. Moreover, $R_T(\overline{S_1}, \ldots, \overline{S_T}) = \sum_{t=1}^T \overline{R}(S_t)$. Therefore, $\overline{S_1}, \ldots, \overline{S_T}$ is also optimal for APV. Note that at the end of Algo we have $S_1 = \bigcup_{1 \leq t \leq T} \hat{S}_t$. Because of the visibility constraints, we have

$$\bigcup_{1 \leq t \leq T} L_t \subseteq \bigcup_{1 \leq t \leq T} \hat{S}_t = S_1.$$

Finally, let $Z_1 = S_1 \setminus L_0 = \bigcup_{1 \leq t \leq T} L_t$ and $Z_t = S_t$ for all $t = 2, \ldots, T$. By monotonicity of the expanded revenue, Lemma 2 gives $\overline{R}(Z_1) \geq \overline{R}(S_1)$, i.e., $R(\overline{Z_1}) \geq R(\overline{S_1})$, which implies $R_T(\overline{Z_1}, \ldots, \overline{Z_T}) \geq R_T^*$. Hence, $\overline{Z_1}, \ldots, \overline{Z_T}$ is also an optimal solution of APV. Note that $\sum_{t=2}^T \mathbb{1}(i \in Z_t) \geq \ell_i - 1 \; \forall i \in \mathcal{N}$ since all elements that need to appear at least once are already in $Z_1 \subseteq \overline{Z_1}$. Hence, we can optimize

$$R^*_{T-1} := \max_{S_2,\ldots,S_T} \sum_{t=2}^{T} R(S_t),$$

independently of $\overline{Z_1}$ and under the constraints $\sum_{t=2}^{T} \mathbb{1}(i \in Z_t) \geq \ell_i - 1\ \forall i \in \mathcal{N}$, which is exactly our initial problem but with $T-1$ customers. By recurrence hypothesis, $(\overline{\bigcup_{t \leq u \leq T} L_u})_{2 \leq t \leq T}$ is an optimal solution for R^*_{T-1}. Moreover, $\overline{Z_1} = \overline{\bigcup_{1 \leq t \leq T} L_t}$. Therefore, $(\overline{\bigcup_{t \leq u \leq T} L_u})_{1 \leq t \leq T}$ is an optimal solution for APV.

The running time of the algorithm for APV can be improved from $O(nT)$ to $O(n+T)$. In fact, we proceed by induction. We start by computing S^*_T. Then, when computing S^*_t, we do not need to directly compute $\overline{\bigcup_{t \leq u \leq T} L_u}$. Instead, since $S^*_t \supseteq S^*_{t+1}$, we just need to check the products $\{i \in \mathcal{N} \backslash (S^*_{t+1}) : p_i \geq \overline{R}(S^*_t)\}$ as in Lemma 1, which could be finished in $O(1 + |S^*_t| - |S^*_{t+1}|)$. Therefore, the running time is $O(1 + |S^*_T|) + \sum_{t=1}^{T-1} O(1 + |S^*_t| - |S^*_{t+1}|) = O(|S^*_1| + T) = O(n+T)$. □

3.4 Linear Program for APV

Consider the classical assortment problem, i.e., that without any constraints, under MNL model for a single customer

$$\max_{S \subseteq \mathcal{N}} \quad R(S). \tag{AP}$$

It is known that AP can be formulated as the following LP (Gallego et al. [8]),

$$\max_{S \subseteq \mathcal{N}} R(S) = \max_{(\alpha_i)_{0 \leq i \leq n}} \left\{ \sum_{i=1}^{n} p_i \alpha_i \quad s.t. \quad \forall i \in \mathcal{N}, 0 \leq \frac{\alpha_i}{v_i} \leq \alpha_0, \quad \sum_{i=0}^{n} \alpha_i = 1 \right\}.$$

Motivated by the structure of the above LP and the structure of our optimal solution of APV given in Equation (7), we propose the following linear formulation for APV.

Theorem 2 (LP for APV). APV *is equivalent to the following linear program:*

$$\begin{aligned}
&\max_{(\alpha_i^t)_{1 \leq t \leq T, 0 \leq i \leq n}} && \sum_{i=1}^{n} p_i \sum_{t=1}^{T} \alpha_i^t \\
&\quad s.t. && \forall t \in [\![1, T]\!], \ \sum_{i=0}^{n} \alpha_i^t = 1, \\
& && \forall i \in \mathcal{N}, \ \ \forall t \in [\![1, \ell_i]\!], \ \ \alpha_i^t = v_i \alpha_0^t, \\
& && \forall i \in \mathcal{N}, \ \ \forall t \in [\![\ell_i + 1, T]\!], \ \ 0 \leq \alpha_i^t \leq v_i \alpha_0^t.
\end{aligned}$$

Proof. Let $(S_1^*, \ldots, S_T^*)$ be the optimal solution of APV given in (7), i.e., $S_t^* = L_t \cup \ldots \cup L_T$, for all $t = 1, \ldots, T$. We define the solution

$$\alpha_0^t = \frac{1}{1+V(S_t^*)}, \quad \text{and} \quad \forall i \in \mathcal{N}, \quad \alpha_i^t = \mathbb{1}(\{i \in S_t^*\})\frac{v_i}{1+V(S_t^*)}.$$

For any $i \in \mathcal{N}$ and $t \in [\![1, \ell_i]\!]$, we have $L_i \subseteq S_t^*$. Hence, our solution verify the second line of constraints of the LP. The first and third line of constraints are verified by definition of the α_i^t. Therefore our solution is feasible for the LP and it easy to verify that the objective function of our solution is the same as the optimal objective value of APV. This proves that the optimal value of the LP is superior or equal to the optimal value of APV.

Now, let $(\alpha_i^t)_{\substack{1 \le t \le T \\ 0 \le i \le n}}$ be an optimal solution to the LP. We define $S_t = \{i \in \mathcal{N} : \alpha_i = v_i \alpha_0\}$. We have $T(n+1)$ variables in the LP, so at least $T(n+1)$ constraints are active. The $T + \sum_{i=1}^n \ell_i$ equality constraints will always be verified, which means that out of the $2\sum_{i=1}^n (T - \ell_i)$ remaining inequality constraints, at least half of them are tight. Each lower bound inequality is incompatible with the upper bound inequality for each i, so we have $\forall i \in [\![1, n]\!], \alpha_i^t \in \{0, v_i \alpha_0^t\}$. Therefore, the two objective functions have the same value. Then, because at least the ℓ_i first sets contain product i because of constraint $\forall i \in \mathcal{N}, \forall t \in [\![1, \ell_i]\!], \frac{\alpha_i^t}{v_i} = \alpha_0^t$, the solution $(S_1, \ldots, S_T)$ created is feasible for APV. So the optimal value of the LP is inferior or equal to the optimal value of APV. This proves that the two problems have the same optimal value, and each solution of the LP allows us the reconstruct a solution to APV. □

4 Price of Visibility

In this section, we explore a scenario where each product within our universe is associated with a specific vendor. As mentioned earlier, vendors can impose visibility constraints on their products within the platform. These constraints can be established through service level agreements or product sponsorships. Enforcing these constraints may result in a decrease in the platform's revenue. In what follows, we present an instance to show that enforcing the visibility constraints can imply a gap that is arbitrary large as compared to the unconstrained setting (without visibility constraints).

Example. Consider an instance of APV with two products $n = 2$ and T customers. Let $p_1 = 1, v_1 = 1, \ell_1 = 0$ and $p_2 = 0, v_2 = M, \ell_2 = T$. We consider the setting where M is large. To compute the optimal assortment for AP. As mentioned earlier, it sufficient to evaluate the revenue of the price-ordered assortments and choose the one with the highest revenue. We have $R(\{1\}) = \frac{1}{2}$ and $R(\{1,2\}) = \frac{1}{2+M} < R(\{1\})$. Therefore, the optimal assortment is $S^* = \{1\}$ and $R(S^*) = 1$. On the other hand, let $(S_1, S_2, \ldots, S_T)$ a feasible solution for APV. Because, $\ell_2 = T$, we have to include product 2 in every assortment S_t. Adding product 1 to an assortment only increases the revenue because $R(\{1,2\}) = \frac{1}{2+M} > R(\{2\}) = 0$. Therefore, it is optimal to offer product 1 and 2 in every assortment

S_t in an optimal solution of APV. Therefore, $S_t^* = \{1, 2\}$ for all $t = 1, \ldots, T$. Now, consider the ratio

$$\frac{TR(S^*)}{\sum_{t=1}^{T} R(S_t^*)} = \frac{T}{T\frac{1}{2+M}} = M + 2$$

that goes to infinity as M increases. Hence, the gap can be arbitrarily large.

To address this issue, the platform can implement a fee structure based on the vendors' contributions to the revenue loss. We introduce a novel, *fair* method to distribute the loss among different products in proportion to their contribution to the overall loss.

Consider the unconstrained assortment optimization AP and let S^* be its optimal solution, which is known to be price-ordered and can be computed efficiently. In the absence of visibility constraints and with T customers, it is optimal to offer assortment S^* to each customer. Consequently, the total expected revenue in the unconstrained setting can be expressed as $TR(S^*)$. As the unconstrained problem serves as a relaxation of APV, it possesses a higher objective function. Let $(S_1^*, S_2^*, \ldots, S_T^*)$ be an optimal solution for APV. We denote the revenue loss due to the visibility constraints as

$$\Delta := TR(S^*) - \sum_{t=1}^{T} R(S_t^*). \tag{8}$$

A First Naive Approach. One approach is to allocate the loss based solely on the parameters ℓ_i. In this case, the proportion of the loss assigned to the vendor of product i would be determined by $\frac{\ell_i}{\sum_{j=1}^{n} \ell_j}$. However, this distribution would not be equitable in the sense that we should not impose any charges on a product that already belongs to the optimal set S^*, even if $\ell_i > 0$. Moreover, this allocation fails to consider the impact of each product on the overall loss. For example, if there is a product with exceptionally high preference weight v_i but a significantly lower price p_i, while other products have higher prices and lower preference weights, enforcing the visibility of the first product would drive revenue down to zero, whereas the others would have a lesser impact. In this scenario, the former product should be responsible for covering almost the entire revenue loss.

Our Approach. Let $S_1^*, \ldots, S_T^*$ be an optimal solution to APV. First, observe that $R(S_t^*) = \frac{\sum_{i \in S_t^*} p_i v_i}{1 + \sum_{i \in S_t^*} v_i}$ implies that

$$R(S_t^*) = \sum_{i \in S_t^*} (p_i - R(S_t^*)) v_i$$

for every set S_t^*. This decomposition of the revenue gives us which products drive the revenue down (the ones with $p_i < R(S_t^*)$) and which products increase the revenue ($p_i > R(S_t^*)$). It also shows that the contribution of product i to the revenue is proportional to the difference between the price of product i and the actual revenue, as well as proportional to the preference weight v_i. We can then rewrite the total revenue of the assortments $(S_1^*, S_2^*, \ldots, S_T^*)$ as

$$\sum_{t=1}^{T} R(S_t^*) = \sum_{t=1}^{T} \sum_{i \in S_t^*} (p_i - R(S_t^*))v_i = \sum_{i=1}^{n} \sum_{t=1}^{T} \mathbb{1}(i \in S_t^*)(p_i - R(S_t^*))v_i.$$

Therefore, we view the contribution of product $i \in \mathcal{N}$ to the total revenue as $\sum_{t=1}^{T} \mathbb{1}(i \in S_t^*)(p_i - R(S_t^*))v_i$.

Pricing the Loss. For each product $i \in \mathcal{N}$, we propose to charge its vendor the fraction of the loss corresponding to the negative contribution of the product, divided by the sum of the negative contributions of all the products:

$$\Gamma_i := \frac{\left[\sum_{t=1}^{T} \mathbb{1}(i \in S_t^*)(p_i - R(S_t^*))v_i\right]^-}{\sum_{j=1}^{n} \left[\sum_{t=1}^{T} \mathbb{1}(j \in S_t^*)(p_j - R(S_t^*))v_j\right]^-} \cdot \Delta \tag{9}$$

where $x^- = max(-x, 0)$ is the negative part of x. This pricing strategy verifies several important properties. First, it ensures a fair distribution of fees among products. Products with the same price and weight incur the same fee under similar visibility constraints. Additionally, this pricing strategy is monotone in the sense that the fee of a product increases as its visibility parameter grows. Finally, the pricing strategy is interpretable and can be computed efficiently, making it practical for real-world applications. We provide a detailed discussion of all these properties in the full version of this paper.

References

1. Alptekinoğlu, A., Semple, J.H.: The exponomial choice model: a new alternative for assortment and price optimization. Oper. Res. **64**(1), 79–93 (2016)
2. Chen, Q., Golrezaei, N., Susan, F., Baskoro, E.: Fair assortment planning. arXiv preprint: arXiv:2208.07341 (2022)
3. Cheung, W.C., Simchi-Levi, D.: Thompson sampling for online personalized assortment optimization problems with multinomial logit choice models. Available at SSRN 3075658 (2017)
4. Davis, J., Gallego, G., Topaloglu, H.: Assortment planning under the multinomial logit model with totally unimodular constraint structures (2013)
5. Davis, J.M., Topaloglu, H., Williamson, D.P.: Assortment optimization over time. Oper. Res. Lett. **43**(6), 608–611 (2015)
6. Désir, A., Goyal, V., Zhang, J.: Capacitated assortment optimization: hardness and approximation. Oper. Res. **70**(2), 893–904 (2022)
7. El Housni, O., Topaloglu, H.: Joint assortment optimization and customization under a mixture of multinomial logit models: on the value of personalized assortments. Oper. Res. **71**, 1197–1215 (2022)
8. Gallego, G., Ratliff, R., Shebalov, S.: A general attraction model and sales-based linear program for network revenue management under customer choice. Oper. Res. **63**(1), 212–232 (2015)

9. Gao, P., et al.: Assortment optimization and pricing under the multinomial logit model with impatient customers: sequential recommendation and selection. Oper. Res. **69**(5), 1509–1532 (2021)
10. Hausman, J., McFadden, D.: Specification tests for the multinomial logit model. Econometrica **52**(5), 1219–40 (1984)
11. Li, Z.: A single-period assortment optimization model. Prod. Oper. Manag. **16**, 369–380 (2009)
12. Luce, R.D.: Individual Choice Behavior. John Wiley, Hoboken (1959)
13. Mahajan, S., Van Ryzin, G.: Stocking retail assortments under dynamic consumer substitution. Oper. Res. **49**(3), 334–351 (2001)
14. McFadden, D.: Conditional logit analysis of qualitative choice behavior. Institute of Urban and Regional Development, University of California (1973)
15. Miao, S., Chao, X.: Dynamic joint assortment and pricing optimization with demand learning. Manuf. Serv. Oper. Manag. **23**(2), 525–545 (2021)
16. Rusmevichientong, P., Shen, Z.J.M., Shmoys, D.B.: Dynamic assortment optimization with a multinomial logit choice model and capacity constraint. Oper. Res. **58**(6), 1666–1680 (2010)
17. Rusmevichientong, P., Shmoys, D., Tong, C., Topaloglu, H.: Assortment optimization under the multinomial logit model with random choice parameters. Prod. Oper. Manag. **23**(11), 2023–2039 (2014)
18. Sumida, M., Gallego, G., Rusmevichientong, P., Topaloglu, H., Davis, J.: Revenue-utility tradeoff in assortment optimization under the multinomial logit model with totally unimodular constraints. Manage. Sci. **67**(5), 2845–2869 (2021)
19. Talluri, K., Van Ryzin, G.: Revenue management under a general discrete choice model of consumer behavior. Manage. Sci. **50**(1), 15–33 (2004)
20. Wang, C., Wang, Y., Tang, S.: When advertising meets assortment planning: joint advertising and assortment optimization under multinomial logit model. Available at SSRN 3908616 (2021)
21. Wang, R.: Capacitated assortment and price optimization under the multinomial logit model. Oper. Res. Lett. **40**(6), 492–497 (2012)

Adaptivity Gaps in Two-Sided Assortment Optimization

Omar El Housni[1(✉)], Alfredo Torrico[2], and Ulyssee Hennebelle[3]

[1] Cornell Tech, Cornell University, New York, USA
oe46@cornell.edu
[2] CDSES, Cornell University, Ithaca, USA
alfredo.torrico@cornell.edu
[3] Ecole Polytechnique, Palaiseau, France
ulysse.hennebelle@polytechnique.edu

Abstract. We study a two-sided assortment optimization framework to address the challenge of choice congestion faced by matching platforms. The goal is to decide the assortments to offer to agents in order to maximize the expected number of matches. We identify several classes of policies that the platforms can use in their design. Our main goal is to measure the value that one class of policies have over another one. For this, we define the adaptivity gap as the worst-case ratio between the optimal value achieved by two different policy classes. First, we show that the adaptivity gap between the class of policies that statically show assortments to one-side first and the class of policies that adaptively show assortments to one-side first is exactly $1 - 1/e$. Second, we show that the adaptivity gap between the latter class of policies and the fully adaptive class of policies that show assortments to agents one by one is exactly $1/2$. Finally, we observe that the worst policies are those who simultaneously show assortments to all the agents, in fact, we show that the adaptivity gap with respect to one-sided policies can be arbitrarily small. These results showcase the benefit of each class of policies and, in particular, demonstrate that the optimal value of the best class of adaptive policies is a constant multiplicative factor away from those of one-sided policies.

Keywords: Matching markets · Assortment optimization · Adaptivity gap

1 Introduction

Two-sided online platforms have transformed the way we move around cities, how we connect and interact with others and even have facilitated how we outsource tasks and search for jobs. Classic examples include freelancing platforms like Taskrabbit and Upwork, dating platforms such as Bumble and Tinder, accommodation companies like Airbnb and Vrbo, and ride-sharing apps such as Blablacar. These matching markets are generally two-sided in nature whose participants' interaction mainly depends on the specific application. For example, some platforms are designed with one-sided interactions where one side initiates a request,

J. Vygen and J. Byrka (Eds.): IPCO 2024, LNCS 14679, pp. 139–153, 2024.
https://doi.org/10.1007/978-3-031-59835-7_11

e.g., someone needs to design a website, while the responding side is searching to serve requests, e.g., someone looking for web design jobs. Other platforms, instead, allow for bi-directional interactions where both sides can search the market and initiate an interaction, for example, two people liking their profiles in a dating app. These matching markets do not only account for the way in which their participants interact, but also for their preferences. In general, each agent in the platform has preferences over the opposite side, for instance, the years of experience in web design or the height of a partner. These preferences may lead to *choice congestion* which occurs when the most popular participants concentrate more requests than they can handle, resulting in market inefficiencies. In summary, both the platform's design and the preferences of the participants will ultimately affect the outcome of the market. These challenges have been widely studied in the literature either for general matching markets or for some of the applications above, see e.g. [4,9,15,20,21,24,27,28,30].

Two-sided assortment optimization has recently gained attention as an algorithmic tool to reduce choice congestion [4,7,28,29,36]. Broadly speaking, a two-sided platform first elicits their participants' preferences and then carefully craft the *assortment* of alternatives that will be visible to each of them. Then, the participants search over their options and either choose one of them or leave the platform. A match is realized when two opposite participants select each other. Therefore, the role of the platform is to design these assortments and the way in which they will be presented to the participants, which will ultimately affect how they interact in the platform. A natural trade-off arises between displaying relevant options to each of the participants, as an incentive to not leave the platform, and reducing choice congestion. Carefully balancing these two will result in maximizing the number of matches in the platform, and consequently, in maximizing revenue.

Our main focus is to study the two-sided assortment optimization problem from a policy perspective. By accounting for the agents' preferences, a policy determines the assortments and the order in which they are presented to each of the agents in the platform. Our goal is to measure the benefit that a two-sided platform can obtain by implementing certain classes of policies over others. For this, we define a two-sided assortment optimization framework and distinguish different policy classes. We then compare the optimal matching outcome between different policies, and show how valuable policies that *adapt* to the agents' choices are when compared to those that do not adapt.

1.1 Our Results and Contributions

In this work, we introduce a general two-sided assortment optimization framework on a bipartite graph composed by customers and suppliers. Our goal is to address the choice congestion challenge faced by a matching platform that wants to maximize the total expected number of matches where a match occurs when two individuals mutually select each other. The matching process is determined by the agents' preferences, the assortments of alternatives that the platform constructs and the order in which these menus are presented to the agents.

Given this, feasible solutions to the two-sided assortment optimization problem corresponds to policies. To simply put, in each step, a policy selects a group of agents, design their assortments and then observe their choices; the policy continues with the remaining individuals until no more agents left. Within this broad definition of a policy we identify several classes. First, we define those *fully static* policies that simultaneously present assortments to all of the agents in the platform and we denote by (Fully Static) the two-sided assortment optimization problem restricted to this class. On the other extreme, we classify those *fully adaptive* policies that select one agent in each step, so they are able to adapt to the choices of each agent; we denote by (Fully Adaptive) the corresponding optimization problem. Finally, we can classify those *one-sided* policies which first show assortments to all of the agents on one side, and then continue with the opposite side. As before, we identify *one-sided static* and *one-sided adaptive* policies, and we denote by (One-sided Static) and (One-sided Adaptive) the corresponding optimization problems, respectively.

Given these policy classes, we define the *adaptivity gap* as the worst-case ratio between the optimal expected number of matches achieved by two different policy classes. First, we study the one-sided policies and our main result is:

Theorem 1. *The adaptivity gap between* (One-sided Static) *and* (One-sided Adaptive) *is* $1 - \frac{1}{e}$.

We prove this result in Sect. 3. Next, we focus on the fully adaptive policies.

Theorem 2. *The adaptivity gap between* (One-sided Adaptive) *and* (Fully Adaptive) *is* $1/2$.

We prove this result in Sect. 4. Finally, we measure the value of the fully static policies. For this, we compute the adaptivity gap between (Fully Static) and (One-sided Static). In Proposition 1, we show that the ratio between the optimal value of (Fully Static) and the optimal value of (One-sided Static) is $\mathcal{O}(1/n)$, where n is the number of agents on one side of the bipartition. We summarize our results in Fig. 1.

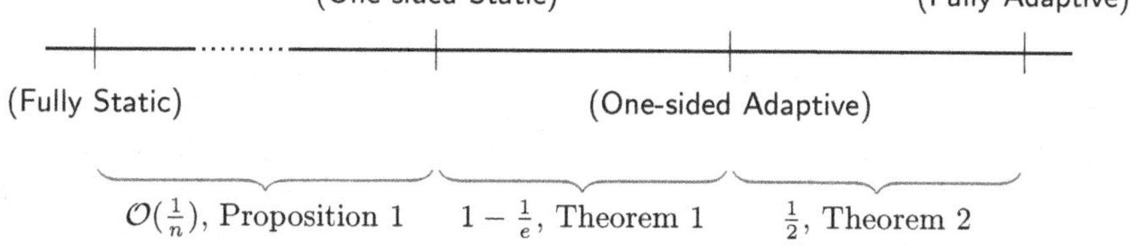

Fig. 1. The results in this paper are in blue. The adaptivity gaps, presented below each bracket, correspond to the ratio between the smaller and larger optimal values. (Color figure online)

1.2 Related Literature

In the following, we present the most relevant streams of literature. First, the one-sided assortment optimization framework has been extensively studied in the revenue management literature. In particular, the single-customer assortment optimization problem was introduced in [35], and, since then, numerous offline and dynamic models, in the unconstrained and constrained settings, have been considered. The offline problem has been studied under various choice models such as the Logit-based [12,32,35], the Rank-based [3] the Markov chain-based [10] and general choice models [8]. Several of these optimization problems are NP-hard, e.g., [3,32]. The setting with capacity constraints has also been studied under several choice models, see e.g. [11,13,31]. For a survey on assortment optimization, we refer to [25]. The online version of the problem has been studied under several settings and arrival models, see e.g. [14,16–18,26].

One of the first works to consider an assortment optimization problem in a two-sided setting was [7]. The authors consider a two-sided market with customers and suppliers where the goal of the platform is to simultaneously show an assortment of suppliers to each customer such that maximizes the expected number of matches. The authors show that the problem is strongly NP-hard and provide an algorithm that guarantees a constant approximation factor. Their results were later improved in [36] who show a $1 - 1/e$ approximation factor for a more general setting. For the same problem, parameterized guarantees are given in [2]. Other two-sided assortment optimization models have been considered for dating markets [28,29], labor markets [4] and markets with endogenous prices [33].

There is an extensive literature on market congestion starting with the seminal work of [30]. Since then, the literature has studied several market interventions to reduce congestion such as matchmaking strategies [34], signaling competition levels [9], market recommendations [22] and ranking [15], limiting the visibility [5,19] and choice [23], controlling who initiates contact [6,24].

2 Model

2.1 Problem Formulation

Consider a two-sided platform represented by a bipartite graph where, on one side, we have a set $\mathcal{C}$ with n customers and, on the opposite side, we have a set $\mathcal{S}$ with m suppliers. The entire set of agents in the platform will be denoted as $\mathcal{A} = \mathcal{C} \cup \mathcal{S}$. In this platform, customers and suppliers are looking to match with each other, so we assume that each agent has preferences over the agents on the opposite side. In the following, we formalize the agents' preferences and the platform's objective.

The preferences of each customer $i \in \mathcal{C}$ are captured by a discrete choice model $\phi_i : \mathcal{S} \cup \{0\} \times 2^{\mathcal{S}} \rightarrow [0, 1]$ where $\{0\}$ represents the outside option (e.g., a different platform). For any assortment of suppliers $S \subseteq \mathcal{S}$ presented to customer i, the value $\phi_i(j, S)$ corresponds to the probability that customer i chooses option $j \in \mathcal{S} \cup \{0\}$, where these values are such that $\sum_{j \in S \cup \{0\}} \phi_i(j, S) = 1$ and $\phi_i(j, S) = 0$ for any $j \notin S \cup \{0\}$. In other words, for a given assortment, the customer probabilistically chooses either the outside option or one alternative from the assortment. We denote the demand function of customer i as $f_i : 2^{\mathcal{S}} \rightarrow [0, 1]$ that, for any subset $S \subseteq \mathcal{S}$, returns $f_i(S) = \sum_{j \in S} \phi_i(j, S)$, i.e., the probability that customer i chooses someone from S. Similarly, the preferences of each supplier $j \in \mathcal{S}$ are determined by a choice model $\phi_j : \mathcal{C} \cup \{0\} \times 2^{\mathcal{C}} \rightarrow [0, 1]$ whose demand function will be denoted as f_j. We emphasize that agents can have heterogeneous preferences, i.e., we do not assume the same choice model for each agent. Throughout this work, we consider the following standard assumption which is satisfied by the wide class of random-utility choice models [8] such as the well-known Logit-based [32,35] and Markov-chain [10,13] choice models.

Assumption 1. *The demand function of any agent $a \in \mathcal{A}$ is a monotone submodular set function.*[1]

As we mentioned before, the goal of the platform is to facilitate the matching between both sides. To achieve this, the platform aims to clear the market by showing an assortment to each agent (possibly in an adaptive way) such that the total expected number of matches is maximized, where a match between a customer and a supplier is realized when both mutually select each other. Formally, the design of the assortments and the order in which they are presented to the agents are determined by a *policy*. A feasible policy π consists of a sequence of actions such that, in each step, π decides to *process* a subset of the agents, specifically: (1) it selects a subset $A \subset \mathcal{A}$ and (2) it presents a *feasible* assortment to each agent $a \in A$.[2] Then, the policy observes the choices made by each agent in A[3] and continue to the next step with the remaining agents $\mathcal{A} \setminus A$, if any. The policy stops when no agents remain to be processed. To simply put, a policy corresponds to a decision tree where each node indicates the action (agents to be processed and the corresponding assortments) and each outgoing arc corresponds to one of the possible outcomes (choices made by the processed agents). We denote by Π the space of all feasible policies and, for a policy $\pi \in \Pi$, let M_π be the random variable that indicates the total number of matches obtained under π. The platform then is interested in solving the following two-sided assortment optimization problem:

$$\max\Big\{\mathbb{E}[\mathsf{M}_\pi] : \ \pi \in \Pi\Big\} \tag{1}$$

where the expectation is taken over the agents' choices and the possible randomized actions made by the policy. Note that the optimal value of Problem (1) largely depends on how much freedom to *adapt* to the agents' choices the platform has. Intuitively, if the platform restrict Π to policies that process all of

[1] A non-negative set function $f : 2^{\mathcal{E}} \rightarrow \mathbb{R}_+$ defined over a set of elements $\mathcal{E}$ is *monotone* if for every pair of subsets $E \subseteq E' \subseteq \mathcal{E}$, we have $f(E) \leq f(E')$. Function f is *submodular* if for every $e \in \mathcal{E}$ and $E \subseteq E' \subseteq \mathcal{E} \setminus \{e\}$, we have $f(E \cup \{e\}) - f(E) \geq f(E' \cup \{e\}) - f(E')$.

[2] We do not restrict the model to any particular assortment feasibility family.

[3] The agents' choices are done independently, irrevocably and simultaneously.

the agents in one step, then the optimal value would be different than if Π is constrained to policies that process agents one by one. In this work, our main goal is to measure the *gap* that exists between optimal values when the problem is restricted to policies in different classes. By doing this, we aim to quantify how valuable for matching platforms is to focus on policies that *adapt* to the agents' preferences and choices.

2.2 Policy Classes and Problems of Interest

Policies can be largely classified on how much past information (agents' choices) can be used to design the assortments in future steps or how many agents can be processed in each step. On the one hand, we define $\Pi_{(\mathsf{FS})}$ as the class of *fully static* policies that process all the agents at the same time. Note that these policies are not able to adapt to the agents choices since the assortments are simultaneously designed. On the other hand, we define $\Pi_{(\mathsf{FA})}$ as the class of *fully adaptive* policies that process agents one by one. These policies are allowed to process different types of agents in consecutive iterations. Moreover, in each iteration, these policies are allowed to use the information obtained from the choices made by previous processed agents. We note that $\Pi_{(\mathsf{FA})} \supset \Pi_{(\mathsf{FS})}$ since any policy in $\Pi_{(\mathsf{FS})}$ can be simulated by a policy in $\Pi_{(\mathsf{FA})}$.[4]

Note that the classes above can be further distinguished by which side is entirely processed first; we call these policies *one-sided.* In this context, we also consider two policy classes. First, we define $\Pi_{(\mathsf{OS})}$ as the class of *one-sided static* policies which either: (1) simultaneously process all the customers first, it observes their choices and then it processes all the suppliers; or (2) simultaneously process all the suppliers first, it observes their choices and then it processes all the customers. Second, we define $\Pi_{(\mathsf{OA})}$ as the class of *one-sided adaptive* policies which either: (1) process all the customers one by one (observing their choices), and then all the suppliers; or (2) process all the suppliers one by one, and then all the customers. As we noted with the fully adaptive and static policies, we have that $\Pi_{(\mathsf{OA})} \supset \Pi_{(\mathsf{OS})}$. In fact, all the policy classes defined above are nested in the following way:

Observation 1. *Our policy classes verify* $\Pi_{(\mathsf{FS})} \subset \Pi_{(\mathsf{OS})} \subset \Pi_{(\mathsf{OA})} \subset \Pi_{(\mathsf{FA})}$.

Given the definitions above, we now define our main problems of interest. On one extreme, we have Problem (1) restricted to $\Pi_{(\mathsf{FS})}$, i.e., to the class of policies that simultaneously process all the agents in $\mathcal{A}$.

Definition 1. *We define the* fully static *problem and its optimal objective value as*

$$\mathrm{OPT}_{(\mathsf{FS})} = \max\Big\{\mathbb{E}[\mathsf{M}_\pi] : \ \pi \in \Pi_{(\mathsf{FS})}\Big\}. \qquad \text{(Fully Static)}$$

[4] Note the we are slightly abusing the word adaptive here, since in this simulation, agents are processed one by one but the assortments were chosen in advance.

On the other extreme, we have Problem (1) restricted to $\Pi_{(\mathsf{FA})}$, i.e., to the class of policies that process agents in $\mathcal{A}$ one by one.

Definition 2. *We define the* fully adaptive *problem and its optimal objective value as*

$$\mathrm{OPT}_{(\mathsf{FA})} = \max\left\{\mathbb{E}[\mathsf{M}_\pi] : \pi \in \Pi_{(\mathsf{FA})}\right\}. \qquad \text{(Fully Adaptive)}$$

We observe that (Fully Adaptive) can be solved via a dynamic programming formulation with exponentially many states and variables, which we include in the full version of the paper. Also, from Observation 1, we note that $\mathrm{OPT}_{(\mathsf{FA})}$ is at least $\mathrm{OPT}_{(\mathsf{FS})}$, since the optimal solution of the latter can be used as a feasible policy in the former. Moreover, any other general policy for Problem (1) in Π has an objective value that lies between $\mathrm{OPT}_{(\mathsf{FA})}$ and $\mathrm{OPT}_{(\mathsf{FS})}$.

Let us now focus on the problems defined by one-sided policies. On the one hand, we have Problem (1) restricted to $\Pi_{(\mathsf{OS})}$, i.e., to the class of policies that simultaneously process one entire side first.

Definition 3. *We define the* one-sided static *problem and its optimal objective value as*

$$\mathrm{OPT}_{(\mathsf{OS})} = \max\left\{\mathbb{E}[\mathsf{M}_\pi] : \pi \in \Pi_{(\mathsf{OS})}\right\}. \qquad \text{(One-sided Static)}$$

Note that we can further restrict (One-sided Static) to one specific side, for example, [7] introduced the problem in which customers are all simultaneously processed first. Formally, we define ($\mathcal{C}$-One-sided Static) and ($\mathcal{S}$-One-sided Static) as Problem (1) restricted to $\Pi^{\mathcal{C}}_{(\mathsf{OS})}$ and $\Pi^{\mathcal{S}}_{(\mathsf{OS})}$, respectively, where the superscript represents the initial side. We denote by $\mathrm{OPT}_{(\mathcal{C}\text{-}\mathsf{OS})}$ and $\mathrm{OPT}_{(\mathcal{S}\text{-}\mathsf{OS})}$ their respective optimal objective values. Note that $\mathrm{OPT}_{(\mathsf{OS})} = \max\left\{\mathrm{OPT}_{(\mathcal{C}\text{-}\mathsf{OS})}, \mathrm{OPT}_{(\mathcal{S}\text{-}\mathsf{OS})}\right\}$.

Finally, on the other hand, we have Problem (1) restricted to $\Pi_{(\mathsf{OA})}$, i.e., to the class of policies that process first an entire side one by one.

Definition 4. *We define the* one-sided adaptive *problem and its optimal objective value as*

$$\mathrm{OPT}_{(\mathsf{OA})} = \max\left\{\mathbb{E}[\mathsf{M}_\pi] : \pi \in \Pi_{(\mathsf{OA})}\right\}. \qquad \text{(One-sided Adaptive)}$$

Similarly to the one-sided static setting, we can further restrict (One-sided Adaptive) to a specific side. Formally, we define ($\mathcal{C}$-One-sided Adaptive) and ($\mathcal{S}$-One-sided Adaptive) as Problem (1) restricted to $\Pi^{\mathcal{C}}_{(\mathsf{OA})}$ and $\Pi^{\mathcal{S}}_{(\mathsf{OA})}$, respectively. We denote by $\mathrm{OPT}_{(\mathcal{C}\text{-}\mathsf{OA})}$ and $\mathrm{OPT}_{(\mathcal{S}\text{-}\mathsf{OA})}$ their respective optimal objective values. We analogously observe that $\mathrm{OPT}_{(\mathsf{OA})} = \max\left\{\mathrm{OPT}_{(\mathcal{C}\text{-}\mathsf{OA})}, \mathrm{OPT}_{(\mathcal{S}\text{-}\mathsf{OA})}\right\}$.

As we noted with the fully static and fully adaptive problem, $\mathrm{OPT}_{(\mathsf{OA})}$ is at least $\mathrm{OPT}_{(\mathsf{OS})}$ since any optimal solution in the former problem can be transformed into a feasible policy for the latter problem. In fact, these two values lie in between $\mathrm{OPT}_{(\mathsf{FS})}$ and $\mathrm{OPT}_{(\mathsf{FA})}$ as shown in Fig. 1.

2.3 The Adaptivity Gap

Note that a policy that process certain agents earlier than others will output a better outcome than a policy that shows assortments to everyone at once. This simply happens because, in the former type of policy, the platform is able to "collect" more information and, subsequently, to adapt to the agents' choices. In this section, we define our main metric, the *adaptivity gap*, which aims to measure the benefit (in terms of optimal expected number of matches) that one policy class has over another one. Formally, we define the adaptivity gap as:

Definition 5. *For a given pair of policy classes $\Pi', \Pi'' \subseteq \Pi$ such that $\Pi' \subseteq \Pi''$ we define the adaptivity gap between them as:*

$$\textsc{Gap}(\Pi', \Pi'') = \min_{\text{all instances } I} \left\{ \frac{\mathrm{OPT}_{\Pi'}(I)}{\mathrm{OPT}_{\Pi''}(I)} \right\},$$

where an instance I is determined by $\mathcal{C}$, $\mathcal{S}$ and the choice models ϕ and $\mathrm{OPT}(I)$ *indicates the dependence of the optimal value on I. To minimize notation, in the remainder of the paper, we do not write this dependence when it is understood from context.*

We emphasize that in the definition above, $\Pi' \subseteq \Pi''$ means that any policy in Π' can be simulated by some policy in Π''. Note that since $\Pi' \subseteq \Pi''$, then we have that $\textsc{Gap}(\Pi', \Pi'') \in [0, 1]$, where 1 means that the amount of information used by policies in Π'' does not result in a better matching outcome when compared to policies in Π'. Our main goal is to analyze the gap between four main classes of policies: $\Pi_{\mathsf{(FS)}}$, $\Pi_{\mathsf{(OS)}}$, $\Pi_{\mathsf{(OA)}}$ and $\Pi_{\mathsf{(FA)}}$. Since these classes are nested, we are interested in three adaptivity gaps, which are shown in blue brackets in Fig. 1. We now present our first result.

Proposition 1. *There exists an instance such that* $\mathrm{OPT}_{\mathsf{(FS)}} = \mathcal{O}\left(\frac{1}{n}\right) \cdot \mathrm{OPT}_{\mathsf{(OS)}}$.

The proof of Proposition 1 can be found in the full version of this work. To simply put, the result in Proposition 1 imply that fully static policies, which process all the agents simultaneously, can be arbitrarily bad as compared to one-sided static policies. In particular, the adaptivity gap between (One-sided Static) and (Fully Static) can be arbitrarily small as n gets large.

3 Gap Between (One-sided Static)and (One-sided Adaptive)

In this section, we focus on the proof of Theorem 1 which we separate in two parts: In Sect. 3.1, we show Lemma 1 that states that the gap is always at least $1 - 1/e$. Then, in Sect. 3.2, we construct an instance for which the gap is asymptotically close to $1 - 1/e$ (Lemma 2).

Lemma 1. *For every instance, we have that* $\mathrm{OPT}_{\mathsf{(OS)}} \geq \left(1 - \frac{1}{e}\right) \cdot \mathrm{OPT}_{\mathsf{(OA)}}$.

Lemma 2. *There exists an instance such that* $\mathrm{OPT}_{\mathsf{(OS)}} = \left(1 - \frac{1}{e} + \mathcal{O}\left(\frac{\log n}{n}\right)\right) \cdot \mathrm{OPT}_{\mathsf{(OA)}}$. *In particular, the adaptivity gap tends to $1 - 1/e$ as n goes to ∞.*

3.1 Proof of Lemma 1

In the following, we fix the initiating side to $\mathcal{C}$. We will show that for any instance, we have

$$\mathrm{OPT}_{(\mathcal{C}\text{-OS})} \geq \left(1 - \frac{1}{e}\right) \cdot \mathrm{OPT}_{(\mathcal{C}\text{-OA})}. \tag{2}$$

The same result can be analogously obtained when $\mathcal{S}$ is the initiating side. To prove Inequality (2), we introduce an appropriate LP relaxation which upper bounds $\mathrm{OPT}_{(\mathcal{C}\text{-OA})}$. Specifically, we consider the following formulation:

$$\begin{aligned}
\max \quad & \sum_{j \in \mathcal{S}} \sum_{C \subseteq \mathcal{C}} f_j(C) \cdot \lambda_{j,C} && && (3)\\
s.t. \quad & \sum_{C \subseteq \mathcal{C}} \lambda_{j,C} = 1, && \text{for all } j \in \mathcal{S} \\
& \sum_{C: C \ni i} \lambda_{j,C} = \sum_{S: S \ni j} \tau_{i,S} \cdot \phi_i(j, S), && \text{for all } i \in \mathcal{C},\ j \in \mathcal{S} \\
& \sum_{S \subseteq \mathcal{S}} \tau_{i,S} = 1, && \text{for all } i \in \mathcal{C}, \\
& \lambda_{j,C},\ \tau_{i,S} \geq 0 && \text{for all } j \in \mathcal{S},\ C \subseteq \mathcal{C},\ i \in \mathcal{C},\ S \subseteq \mathcal{S}.
\end{aligned}$$

We can interpret $\lambda_{j,C}$ as the probability that supplier j gets chosen by exactly customers in C and $\tau_{i,S}$ as the probability that we show assortment S to customer i. Note that both C and S can potentially be $\emptyset$. The first and third constraints are distribution constraints. The second constraint corresponds to the probability that a customer i is in the "backlog" of supplier j, i.e., the probability that i sees and chooses j. We note that a closely related relaxation was studied for the i.i.d. online arrival model introduced in [4], however, our relaxation is more general in that it applies to the adaptive setting that we consider in this paper and to the online model studied in [4] under the random order arrival setting. Problem (3) will play a crucial role in our proof and, in fact, it is an upper bound of ($\mathcal{C}$-One-sided Adaptive):

Lemma 3. *Problem* (3) *is a relaxation of* ($\mathcal{C}$-One-sided Adaptive).

The proof of Lemma 3 is in the full version of this work. Now we focus on ($\mathcal{C}$-One-sided Static) which coincides with the setting introduced in [7]. In this problem, a policy can be simply viewed as simultaneously selecting a family of assortments $S_1, \ldots, S_n$ where subset S_i will be shown to customers $i \in \mathcal{C}$. As noted in [36], we can consider a randomized solution in which for each customer $i \in \mathcal{C}$, we have a distribution τ_i over assortments and the objective value in ($\mathcal{C}$-One-sided Static) corresponds to

$$\sum_{j \in \mathcal{S}} \sum_{C \subseteq \mathcal{C}} f_j(C) \cdot \prod_{i \in C} \left(\sum_{\substack{S \subseteq \mathcal{S}: \\ S \ni j}} \tau_{i,S} \cdot \phi_i(j, S) \right) \prod_{i \in \mathcal{C} \setminus C} \left(1 - \sum_{\substack{S \subseteq \mathcal{S}: \\ S \ni j}} \tau_{i,S} \cdot \phi_i(j, S) \right), \tag{4}$$

In fact, the following distribution, for all $j \in \mathcal{C},\ C \subseteq \mathcal{C}$

$$\lambda^{\mathsf{ind}}_{j,C} = \prod_{i \in C} \left(\sum_{S \subseteq \mathcal{S}: S \ni j} \tau_{i,S} \cdot \phi_i(j,S) \right) \prod_{i \in \mathcal{C} \setminus C} \left(1 - \sum_{S \subseteq \mathcal{S}: S \ni j} \tau_{i,S} \cdot \phi_i(j,S) \right), \quad (5)$$

is a feasible solution in Problem (3); due to the page limit, we omit the proof of feasibility which can be easily checked by the reader. A distribution of this form is referred to as the *independent distribution* and we will denote it as λ^{ind}. We are now ready to finalize the proof of Lemma 1. [1] defines the *correlation gap* between the "worst-case" distribution and the independent distribution. In particular, for any monotone submodular objective function, [1] show the following:

Lemma 4 ([1]). *For any monotone submodular function g, we have that*

$$\mathbb{E}_{A \sim D^{\mathsf{ind}}}[g(A)] \geq \left(1 - \frac{1}{e}\right) \cdot \mathbb{E}_{A \sim D}[g(A)],$$

where D^{ind} is the independent distribution and D is any distribution over subsets with the same marginals.

We conclude the proof of Lemma 1 as follows: let $\tau^\star$ and $\lambda^\star$ be an optimal solution of Problem (3). Therefore, by Lemma 3 we know that $\mathrm{OPT}_{(\mathcal{C}\text{-OA})}$ is upper bounded by the optimal value of Problem (3) which is determined by $\lambda^\star$. On the other hand, we construct the independent distribution λ^{ind} as in Eq. (5) with marginal values $\sum_{S \subseteq \mathcal{S}: S \ni j} \tau^\star_{i,S} \cdot \phi_i(j,S)$. Note that $\lambda^\star$ and λ^{ind} have the same marginal values. Then, we obtain the following bound for the ratio between the optimal values of ($\mathcal{C}$-One-sided Static) and ($\mathcal{C}$-One-sided Adaptive),

$$\frac{\mathrm{OPT}_{(\mathcal{C}\text{-OS})}}{\mathrm{OPT}_{(\mathcal{C}\text{-OA})}} \geq \frac{\sum_{j \in \mathcal{S}} \sum_{C \subseteq \mathcal{C}} f_j(C) \cdot \lambda^{\mathsf{ind}}_{j,C}}{\sum_{j \in \mathcal{S}} \sum_{C \subseteq \mathcal{C}} f_j(C) \cdot \lambda^\star_{j,C}} \geq 1 - \frac{1}{e},$$

where the first inequality follows by noting that $\tau^\star$ can be used as a randomized feasible solution in ($\mathcal{C}$-One-sided Static) whose objective value is equal to Expression (4) evaluated in $\tau^\star$ which is exactly the expected value with respect to the independent distribution λ^{ind}. The last inequality is due to Lemma 4. Finally, an analogous construction and analysis can be done to show that $\mathrm{OPT}_{(\mathcal{S}\text{-OS})} \geq (1 - \frac{1}{e}) \cdot \mathrm{OPT}_{(\mathcal{S}\text{-OA})}$. □

3.2 Proof of Lemma 2

Consider an instance with n customers and n suppliers. On the one hand, all customers have the same choice model, such that for any $i \in \mathcal{C}$, any subset of suppliers $S \subseteq \mathcal{S}$, $\phi_i(j,S) = 1/|S|$ for all $j \in S$ and $\phi(0,S) = 0$. For any $k \in [n]$, define the sequence $\beta_k = k(1 - e^{-\frac{1}{k}})$ and $\beta_0 = 0$. Note that $0 \leq \beta_k \leq 1$. On the other hand, all suppliers have the same choice model: for any $j \in \mathcal{S}$, any subset of customer $C \subseteq \mathcal{C}$ with $|C| = k$, we consider $\phi_j(i, C) = \beta_k/k$ for all $i \in C$ and $\phi(0, C) = 1 - \beta_k$. It is easy to see that all these choice models satisfy Assumption 1.

First let us compute the optimal solution of (One-sided Static). For this, consider ($\mathcal{S}$-One-sided Static) and denote by M_1 the number of matches obtained under a family of assortments $C_1, \dots, C_n \subseteq \mathcal{C}$ where for each $j \in S$, C_j is the feasible assortment offered to supplier j with $|C_j| = k_j$. Since the outside option probability in the choice model of customers is always 0, then the number of matches is equal to the number of customers who are chosen after suppliers are processed, i.e., $\mathsf{M}_1 = \sum_{i=1}^{n} \mathbb{1}_{\{i \text{ is chosen}\}}$. Therefore, $\mathbb{E}[\mathsf{M}_1] = n - \sum_{i=1}^{n} \mathbb{P}(i \text{ is not chosen})$. We have for any $i \in \mathcal{C}$,

$$\frac{1}{n}\sum_{i=1}^{n} \mathbb{P}(i \text{ is not chosen}) = \frac{1}{n}\sum_{i=1}^{n} \prod_{\substack{j \in \mathcal{S}:\\ i \in C_j}} \left(1 - \frac{\beta_{k_j}}{k_j}\right) = \frac{1}{n}\sum_{i=1}^{n} \exp\left(- \sum_{\substack{j \in \mathcal{S}:\\ i \in C_j}} \frac{1}{k_j}\right),$$

where the first equality is because $\mathbb{P}(i \text{ is not chosen})$ is equal to the product of the probabilities that each j did not choose i, for all $j \in \mathcal{S}$ that sees i. The following equality is due to the definition of β_{k_j}. Finally, by Jensen's inequality, we can prove that the last expression is at least e^{-1} which implies that $\mathbb{E}[\mathsf{M}_1] \leq n - ne^{-1}$. Now, consider the ($\mathcal{C}$-One-sided Static) problem and denote by M_2 the number of matches obtained by showing a family of assortments $S_1, \dots, S_n \subseteq \mathcal{S}$ where for each $i \in \mathcal{C}$, S_i is a feasible assortment offered to customer i. After customers make their choices, for each $j \in \mathcal{S}$, let $\alpha_j \in \{0, 1, \dots, n\}$ be the random number of customers who choose supplier j. Since the outside option probability in the customers' choice models is 0, then $\sum_{j=1}^{n} \alpha_j = n$. Therefore, conditioned on $\alpha_1, \dots, \alpha_n$:

$$\mathbb{E}[\mathsf{M}_2 | \alpha_1, \dots, \alpha_n] = \sum_{j=1}^{n} \beta_{\alpha_j} = \sum_{j=1}^{n} \alpha_j (1 - e^{-\frac{1}{\alpha_j}}) = n - \sum_{j=1}^{n} \alpha_j e^{-\frac{1}{\alpha_j}},$$

where in the first equality we use that j sees α_j customers and the demand equals β_{α_j}. Thus, by applying Jensen's inequality, we get $\mathbb{E}[\mathsf{M}_2 | \alpha_1, \dots, \alpha_n] \leq n - ne^{-1}$. Therefore, by taking expectation over $\alpha_1, \dots, \alpha_n$ we conclude $\mathrm{OPT}_{(\mathsf{OS})} \leq n(1 - 1/e)$. To show that is equal to $n(1 - 1/e)$, we give a feasible solution. Consider the following solution for ($\mathcal{C}$-One-sided Static): each customer is offered exactly and exclusively one supplier. Then, each customer will choose the supplier with probability 1, and the supplier will choose back that customer with probability $\beta_1 = 1 - 1/e$. Hence, the expected number of matches is $n(1 - 1/e)$ which implies that $\mathrm{OPT}_{(\mathsf{OS})} = n(1 - 1/e)$.

For the (One-sided Adaptive) problem consider the following feasible policy: We start processing suppliers in any random order one by one. For each supplier, we offer all customers who are not chosen yet. In particular, the first supplier is offered all customers in $\mathcal{C}$ and makes a choice. If the supplier chooses a customer i, we update the list of customers who are not chosen yet to $\mathcal{C} \setminus \{i\}$ and proceed to the next supplier. Recall that all customers and suppliers in our instance are similar. Let $\mathsf{M}_{p,q}$ be the number of matches under our policy when we have a sub-instance with p customers and q suppliers. By conditioning on the choice of first supplier, we get $\mathbb{E}[\mathsf{M}_{n,n}] = (1+\mathbb{E}[\mathsf{M}_{n-1,n-1}])\cdot\beta_n+\mathbb{E}[\mathsf{M}_{n,n-1}]\cdot(1-\beta_n)$. It is clear that $\mathbb{E}[\mathsf{M}_{n,n-1}] \geq \mathbb{E}[\mathsf{M}_{n-1,n-1}]$ because adding one customer can only make the matching better. Therefore, $\mathbb{E}[\mathsf{M}_{n,n}] \geq \beta_n + \mathbb{E}[\mathsf{M}_{n-1,n-1}]$. Hence, by induction, we get $\mathrm{OPT}_{(\mathsf{OA})} \geq \mathbb{E}[\mathsf{M}_{n,n}] \geq \sum_{k=1}^{n} \beta_k$. Note that $\beta_k = k(1-e^{-1/k}) \geq k(\frac{1}{k} - \frac{1}{2k^2}) = 1-\frac{1}{2k}$. Therefore, $\mathrm{OPT}_{(\mathsf{OA})} \geq n-\frac{1}{2}\sum_{k=1}^{n}\frac{1}{k}$, i.e., $\mathrm{OPT}_{(\mathsf{OA})} = n-\mathcal{O}(\log n)$. We conclude that $\mathrm{OPT}_{(\mathsf{OS})}/\mathrm{OPT}_{(\mathsf{OA})} = (1-1/e+\mathcal{O}(\log n/n))$. □

4 Gap Between (One-sided Adaptive) and (Fully Adaptive)

Now, we focus on the proof of Theorem 2 which we separate in two parts: In Sect. 4.1, we show Lemma 5 and we defer the proof of Lemma 6 to the full version of this work.

Lemma 5. *For every instance, we have that* $\mathrm{OPT}_{(\mathsf{OA})} \geq \frac{1}{2}\cdot\mathrm{OPT}_{(\mathsf{FA})}$.

Lemma 6. *There exists an instance such that* $\mathrm{OPT}_{(\mathsf{OA})} = \left(\frac{1}{2}+\frac{1}{2(n-1)}\right)\cdot \mathrm{OPT}_{(\mathsf{FA})}$. *In particular, the adaptivity gap tends to* $1/2$ *as* n *goes to* ∞.

4.1 Proof of Lemma 5

Consider π^* an optimal policy for (Fully Adaptive). For each agent $a \in \mathcal{A}$, we denote as X_a the random variable which takes value 1 if a is matched in π^* and 0 otherwise. We denote Y_a the random variable which takes value 1 if a is matched to an opposite agent that was processed before a in π^* and 0 otherwise. We denote Z_a the random variable which takes value 1 if a is matched to an opposite agent that was processed after a in π^* and 0 otherwise. Clearly, we have $X_a = Y_a + Z_a$. Given this notation, observe that $\sum_{i\in\mathcal{C}} Y_i = \sum_{j\in\mathcal{S}} Z_j$ and $\sum_{j\in\mathcal{S}} Y_j = \sum_{i\in\mathcal{C}} Z_i$. Let M_{π^*} be the number of matches obtained by π^* which is given by

$$\mathsf{M}_{\pi^*} = \sum_{i\in\mathcal{C}} X_i = \sum_{i\in\mathcal{C}} Y_i + \sum_{i\in\mathcal{C}} Z_i = \sum_{j\in\mathcal{S}} Z_j + \sum_{i\in\mathcal{C}} Z_i.$$

Observe that $\sum_{j\in\mathcal{S}} Z_j$ represents the total number of matches that occur between a subset of suppliers and a subset of customers, where all the suppliers in the subset are processed by π^* before any of the customers in the other subset. Let us define a feasible one sided adaptive policy $\pi_{\mathcal{S}}$ that process suppliers first. We will couple this policy with π^*. In particular, $\pi_{\mathcal{S}}$ offers exactly the same assortment as π^* whenever π^* processes a supplier. If π^* processes a customer before finishing all suppliers at some time, then $\pi_{\mathcal{S}}$ does not process any agent at that time but just simulate the choice of the customer so that π^* and $\pi_{\mathcal{S}}$ follow the same decision tree. After processing all suppliers, $\pi_{\mathcal{S}}$ offers to all customers who were chosen the same assortments as in π^*. Observe that the number of matches obtained by $\pi_{\mathcal{S}}$ denoted by $\mathsf{M}_{\mathcal{S}}$, in this joint probability space, is at least $\sum_{j\in\mathcal{S}} Z_j$. Therefore,

$$\mathrm{OPT}_{(\mathcal{S}\text{-OA})} \geq \mathbb{E}[\mathsf{M}_{\mathcal{S}}] \geq \mathbb{E}\left[\sum_{j\in\mathcal{S}} Z_j\right].$$

Similarly, we have $\mathrm{OPT}_{(\mathcal{C}\text{-OA})} \geq \mathbb{E}[\sum_{i\in\mathcal{C}} Z_i]$. Therefore, we conclude that

$$\begin{aligned}\mathrm{OPT}_{(\mathsf{OA})} = \max\{\mathrm{OPT}_{(\mathcal{C}\text{-OA})}, \mathrm{OPT}_{(\mathcal{S}\text{-OA})}\} &\geq \frac{1}{2}(\mathrm{OPT}_{(\mathcal{C}\text{-OA})} + \mathrm{OPT}_{(\mathcal{S}\text{-OA})}) \\ &\geq \frac{1}{2}\mathbb{E}\left[\sum_{i\in\mathcal{C}} Z_i\right] + \frac{1}{2}\mathbb{E}\left[\sum_{j\in\mathcal{S}} Z_j\right] \\ &= \frac{1}{2}\mathbb{E}[\mathsf{M}_{\pi^*}] = \frac{1}{2}\mathrm{OPT}_{(\mathsf{FA})}.\end{aligned}$$

□

References

1. Agrawal, S., Ding, Y., Saberi, A., Ye, Y.: Correlation robust stochastic optimization. In: Proceedings of the Twenty-First Annual ACM-SIAM Symposium on Discrete Algorithms, pp. 1087–1096. SIAM (2010)
2. Ahmed, A., Sohoni, M.G., Bandi, C.: Parameterized approximations for the two-sided assortment optimization. Oper. Res. Lett. **50**(4), 399–406 (2022)
3. Aouad, A., Farias, V., Levi, R., Segev, D.: The approximability of assortment optimization under ranking preferences. Oper. Res. **66**(6), 1661–1669 (2018)
4. Aouad, A., Saban, D.: Online assortment optimization for two-sided matching platforms. Manag. Sci. **69**(4), 2069–2087 (2023)
5. Arnosti, N., Johari, R., Kanoria, Y.: Managing congestion in matching markets. Manuf. Serv. Oper. Manag. **23**(3), 620–636 (2021)
6. Ashlagi, I., Braverman, M., Kanoria, Y., Shi, P.: Clearing matching markets efficiently: informative signals and match recommendations. Manag. Sci. **66**(5), 2163–2193 (2020)
7. Ashlagi, I., Krishnaswamy, A.K., Makhijani, R., Saban, D., Shiragur, K.: Technical note-assortment planning for two-sided sequential matching markets. Oper. Res. **70**(5), 2784–2803 (2022)
8. Berbeglia, G., Joret, G.: Assortment optimisation under a general discrete choice model: a tight analysis of revenue-ordered assortments. Algorithmica **82**(4), 681–720 (2020)
9. Besbes, O., Fonseca, Y., Lobel, I., Zheng, F.: Signaling competition in two-sided markets. Available at SSRN (2023)
10. Blanchet, J., Gallego, G., Goyal, V.: A Markov chain approximation to choice modeling. Oper. Res. **64**(4), 886–905 (2016)
11. Davis, J., Gallego, G., Topaloglu, H.: Assortment planning under the multinomial logit model with totally unimodular constraint structures. Work in Progress (2013)
12. Davis, J.M., Gallego, G., Topaloglu, H.: Assortment optimization under variants of the nested logit model. Oper. Res. **62**(2), 250–273 (2014)

13. Désir, A., Goyal, V., Segev, D., Ye, C.: Constrained assortment optimization under the Markov chain-based choice model. Manag. Sci. **66**(2), 698–721 (2020)
14. Feng, Y., Niazadeh, R., Saberi, A.: Near-optimal Bayesian online assortment of reusable resources. In: Proceedings of the 23rd ACM Conference on Economics and Computation, pp. 964–965 (2022)
15. Fradkin, A.: Search frictions and the design of online marketplaces. Technical report, Working Paper, MIT (2015)
16. Golrezaei, N., Nazerzadeh, H., Rusmevichientong, P.: Real-time optimization of personalized assortments. Manag. Sci. **60**(6), 1532–1551 (2014)
17. Gong, X.Y., Goyal, V., Iyengar, G.N., Simchi-Levi, D., Udwani, R., Wang, S.: Online assortment optimization with reusable resources. Manag. Sci. **68**(7), 4772–4785 (2022)
18. Goyal, V., Iyengar, G., Udwani, R.: Asymptotically optimal competitive ratio for online allocation of reusable resources. arXiv preprint arXiv:2002.02430 (2020)
19. Halaburda, H., Jan Piskorski, M., Yıldırım, P.: Competing by restricting choice: the case of matching platforms. Manag. Sci. **64**(8), 3574–3594 (2018)
20. Hitsch, G.J., Hortaçsu, A., Ariely, D.: Matching and sorting in online dating. Am. Econ. Rev. **100**(1), 130–63 (2010)
21. Horton, J.J.: Online labor markets. In: Saberi, A. (ed.) WINE 2010. LNCS, vol. 6484, pp. 515–522. Springer, Heidelberg (2010). https://doi.org/10.1007/978-3-642-17572-5_45
22. Horton, J.J.: The effects of algorithmic labor market recommendations: evidence from a field experiment. J. Law Econ. **35**(2), 345–385 (2017)
23. Immorlica, N., Lucier, B., Manshadi, V., Wei, A.: Designing approximately optimal search on matching platforms. In: Proceedings of the 22nd ACM Conference on Economics and Computation, pp. 632–633 (2021)
24. Kanoria, Y., Saban, D.: Facilitating the search for partners on matching platforms. Manag. Sci. **67**(10), 5990–6029 (2021)
25. Kök, A.G., Fisher, M.L., Vaidyanathan, R.: Assortment planning: review of literature and industry practice. In: Agrawal, N., Smith, S. (eds.) Retail Supply Chain Management. ISOR, vol. 122, pp. 99–153. Springer, Boston (2008). https://doi.org/10.1007/978-0-387-78902-6_6
26. Ma, W., Simchi-Levi, D.: Algorithms for online matching, assortment, and pricing with tight weight-dependent competitive ratios. Oper. Res. **68**(6), 1787–1803 (2020)
27. Manshadi, V., Rodilitz, S.: Online policies for efficient volunteer crowdsourcing. Manag. Sci. **68**(9), 6572–6590 (2022)
28. Rios, I., Saban, D., Zheng, F.: Improving match rates in dating markets through assortment optimization. Manuf. Serv. Oper. Manag. **25**(4), 1304–1323 (2023)
29. Rios, I., Torrico, A.: Platform design in matching markets: a two-sided assortment optimization approach. arXiv preprint arXiv:2308.02584 (2023)
30. Rochet, J.C., Tirole, J.: Platform competition in two-sided markets. J. Eur. Econ. Assoc. **1**(4), 990–1029 (2003)
31. Rusmevichientong, P., Shen, Z.J.M., Shmoys, D.B.: Dynamic assortment optimization with a multinomial logit choice model and capacity constraint. Oper. Res. **58**(6), 1666–1680 (2010)
32. Rusmevichientong, P., Shmoys, D., Tong, C., Topaloglu, H.: Assortment optimization under the multinomial logit model with random choice parameters. Prod. Oper. Manag. **23**(11), 2023–2039 (2014)
33. Shi, P.: Optimal match recommendations in two-sided marketplaces with endogenous prices. Available at SSRN 4034950 (2022)

34. Shi, P.: Optimal matchmaking strategy in two-sided marketplaces. Manag. Sci. **69**(3), 1323–1340 (2023)
35. Talluri, K., Van Ryzin, G.: Revenue management under a general discrete choice model of consumer behavior. Manag. Sci. **50**(1), 15–33 (2004)
36. Torrico, A., Carvalho, M., Lodi, A.: Multi-agent assortment optimization in sequential matching markets. arXiv preprint arXiv:2006.04313 (2023)

Two-Stage Stochastic Stable Matching

Yuri Faenza(✉), Ayoub Foussoul(✉), and Chengyue He(✉)

IEOR, Columbia University, New York, USA
{yf2414,af3209,ch3480}@columbia.edu

Abstract. We introduce and study a two-stage stochastic stable matching problem between students and schools. A decision maker chooses a stable matching in a marriage instance; then, after some agents enter or leave the market following a probability distribution $\mathcal{D}$, chooses a stable matching in the new instance. The goal is, roughly speaking, to maximize the expected quality of the matchings across the two stages and minimize the expected students' discontent for being downgraded to a less preferred school in the second-stage. We consider both the case when $\mathcal{D}$ is given explicitly and when it is accessed via a sampling oracle. In the former case, we give a polynomial time algorithm. In the latter case, we show that, unless P = NP, no algorithm can find the optimal value or the optimal solution of the problem in polynomial-time. On the positive side, we give a pseudopolynomial algorithm that computes a solution of arbitrarily small additive error. Our techniques include the use of a newly defined poset of stable pairs, which may be of independent interest.

Keywords: Stable Matching · Two-stage stochastic programming · Poset

1 Introduction

Stability is a fundamental concept in matching markets problems when we do not only wish to optimize a global objective function, but we also care that the output solution is fair at the level of individual agents. Since its introduction in the seminal work by Gale and Shapley [7], stability has been employed in many real-world applications, including matching medical residents to hospitals, assigning students to schools, and matching organ donors to recipients. An instance $(A, B, \succ)$ of the marriage model (see, e.g., [8,12]) consists of a two-sided market of students A and schools B, and for every $a \in A$ (resp., $b \in B$), a strict order $\succ_a$ (resp., $\succ_b$) over $B^+ = B \cup \{\emptyset\}$ (resp., $A^+ = A \cup \{\emptyset\}$), where $\emptyset$ denotes the outside option. An element $v \in A \cup B$ is called an *agent*. In the stable matching problem, the goal is to find a matching[1] M that is *stable*: a matching with no *blocking pairs* or *blocking agents*. More precisely, let $M(v) \in A \cup B \cup \{\emptyset\}$

[1] Throughout the paper, a matching in a marriage instance $(A, B, \succ)$ refers to a subset of pairs of $A^+ \times B^+$ where every agent $v \in A \cup B$ is paired with either a unique agent from the other side or with the outside option $\emptyset$.

J. Vygen and J. Byrka (Eds.): IPCO 2024, LNCS 14679, pp. 154–167, 2024.
https://doi.org/10.1007/978-3-031-59835-7_12

$$\begin{array}{l|l}
a^1\colon \boxed{b^1} \succ_{a^1} b^2 \succ_{a^1} b^3 \succ_{a^1} \underline{b^4} \succ_{a^1} b^5 \succ_{a^1} \emptyset & b^1\colon a^4 \succ_{b^1} a^5 \succ_{b^1} a^3 \succ_{b^1} a^2 \succ_{b^1} a^1 \succ_{b^1} \emptyset \\
a^2\colon \boxed{b^2} \succ_{a^2} b^1 \succ_{a^2} b^4 \succ_{a^2} \underline{b^3} \succ_{a^2} b^5 \succ_{a^2} \emptyset & b^2\colon a^3 \succ_{b^2} a^5 \succ_{b^2} a^4 \succ_{b^2} a^1 \succ_{b^2} a^2 \succ_{b^2} \emptyset \\
a^3\colon \boxed{b^3} \succ_{a^3} b^4 \succ_{a^3} b^1 \succ_{a^3} \underline{b^2} \succ_{a^3} b^5 \succ_{a^3} \emptyset & b^3\colon a^2 \succ_{b^3} a^5 \succ_{b^3} a^1 \succ_{b^3} a^4 \succ_{b^2} a^3 \succ_{b^3} \emptyset \\
a^4\colon \boxed{b^4} \succ_{a^4} b^3 \succ_{a^4} b^2 \succ_{a^4} \underline{b^1} \succ_{a^3} b^5 \succ_{a^4} \emptyset & b^4\colon a^1 \succ_{b^4} a^5 \succ_{b^4} a^2 \succ_{b^4} a^3 \succ_{b^4} a^4 \succ_{b^4} \emptyset \\
a^5\colon \boxed{\underline{b^5}} \succ_{a^5} b^3 \succ_{a^5} b^2 \succ_{a^5} b^1 \succ_{a^5} b^4 \succ_{a^5} \emptyset & b^5\colon a^5 \succ_{b^5} a^1 \succ_{b^5} a^2 \succ_{b^5} a^3 \succ_{b^5} a^4 \succ_{b^5} \emptyset
\end{array}$$

Fig. 1. *Example adapted from* [6, *Example 8.1*]. *Consider the first-stage instance* I_1 *given above, where* $a^1, \dots, a^5$ *are the students. Then* $M_0 = \{a^1b^1, a^2b^2, a^3b^3, a^4b^4, a^5b^5\}$ *and* $\underline{M} = \{a^1b^4, a^2b^3, a^3b^2, a^4b^1, a^5b^5\}$ *are two stable matchings of* I_1*, with* M_0 *being student-optimal (the partner of each student* a *in* M_0 *and* $\underline{M}$ *is, respectively, boxed and underlined in* a*'s preference list). If* b^5 *leaves the market in the second stage, the only stable matching is* $M' = \underline{M} \setminus \{a^5b^5\}$*, hence,* $\underline{M}$ *minimizes the number of students downgraded in the second stage.*

denote the partner of an agent $v \in A \cup B$ in M, then a pair $ab \in A \times B$ is called a blocking pair if both a and b prefer each other to their partner in M, that is, $b \succ_a M(a)$ and $a \succ_b M(b)$, and an agent $v \in A \cup B$ is called a blocking agent if v prefers the outside option to its partner in M, that is, $\emptyset \succ_v M(v)$.

While the classical stable matching problem assumes static and fully known input, in many applications the input changes over time as new agents enter or leave the market. For instance, each year, and after an assignment of the students to San Francisco public middle schools has been decided, around 20% of the students who were allotted a seat choose not to use it, mostly to join private schools instead [3]. Also, new students move to the city and need to be allotted a seat. On the other side of the market, the schools may have unforeseen budget cuts or expansions, leading to a change in the number of available seats.

In all of the above scenarios, a second-stage reallocation of at least part of the seats is required. For instance, while a subset of students might commit to their assigned seats in the first round of matching (and hence leave the market along with their allotted seats), the original assignment needs to be adjusted for the rest of the students after new agents arrive (or old agents depart) in order to maintain stability. It is therefore important that the initial stable matching, besides being of good quality, is adaptable to a changing environment with small adjustments so that it leads to a good quality second round matching and to small dissatisfaction of the students present in both rounds from being downgraded to a less preferred school. It is easy to find examples (see Fig. 1) where, in order to achieve such goal, one might want to go beyond the student optimal stable matching[2], which is the obvious solution that is almost always employed (see, e.g., [1,11]).

Motivated by these considerations, in this paper we introduce and study a two-stage stochastic stable matching problem, where agents enter and leave the market between the two stages. The goal of the decision maker is, roughly

[2] The student optimal matching is a stable matching that all students prefer to every other stable matching.

speaking, to maximize the expected quality of the matchings across the two stages and minimize the expected students' discontent from being downgraded to a less preferred school when going from the first to the second stage. Below, we make these ideas formal.

1.1 The Model

An instance of the two-stage stochastic stable matching problem is as follows.

First-Stage Instance: a marriage instance $I_1 = (A, B, \succ)$ of agents present in the first stage.

Second-Stage Instance: a distribution $\mathcal{D}$ over subsets of $A \cup B$, from which the agents present in the second stage are sampled[3].

Cost: cost functions $c_1, c_2 : A^+ \times B^+ \to \mathbb{Q}$ and a penalty coefficient $\lambda \in \mathbb{Q}_+$. Let S denote the set of agents sampled from $\mathcal{D}$ in the second-stage, and $I_2 = (A \cap S, B \cap S, \succ)$ denote the corresponding second-stage instance[4]. Given a first- (resp., second-) stage matching M_1 (resp., M_2), we measure the quality of M_1 (resp., M_2) by $c(M_1) = \sum_{ab \in M_1} c_1(ab)$ (resp., $c(M_2) = \sum_{ab \in M_2} c_2(ab)$), and measure the dissatisfaction of students for being moved to a less preferred school between M_1 and M_2 by

$$d_S(M_1, M_2) = \lambda \sum_{a \in A \cap S} [R_a(M_2(a)) - R_a(M_1(a))]^+. \tag{1}$$

Here, $R_a : B^+ \to \mathbb{N}$ is the *rank function* of a, that is, $R_a(b) = i$ iff b is the i-th most preferred choice of a (among the schools B and the outside position) and $[x]^+ = \max\{0, x\}$. The coefficient λ is the per unit of rank change dissatisfaction of a student from switching to a school of higher rank between M_1 and M_2. This is a natural measure of dissatisfaction where students are unhappy to be downgraded to a less preferred school in the second-stage, such that the worse the school is, the more unsatisfied they are.

Objective Function: The goal is to solve the two-stage stochastic problem

$$\min_{M_1 \in \mathcal{M}_{I_1}} c_1(M_1) + \mathbb{E}_{S \sim \mathcal{D}} \left[\min_{M_2 \in \mathcal{M}_{I_2}} c_2(M_2) + d_S(M_1, M_2) \right], \tag{2STO}$$

where $I_1 = (A, B, \succ)$ is the first-stage marriage instance, S is the set of agents of the second-stage sampled from $\mathcal{D}$, $I_2 = (A \cap S, B \cap S, \succ)$ is the second-stage instance, and $\mathcal{M}_I$ denotes the set of stable matchings of a marriage instance I.

In (2STO), we therefore wish to select a first-stage matching M_1 such that the cost of M_1 plus the expected cost that we have to pay in the second-stage

[3] This setting subsumes the more general setting where agents enter and leave the market (see Remark 1).

[4] Given subsets $A' \subset A$ and $B' \subset B$, we use $(A', B', \succ)$ to denote the marriage instance $(A', B', \succ')$ where $\succ'$ is the restriction of the collection of orders $\succ$ to agents $A' \cup B'$.

is minimized. The second-stage cost is given by the cost of the second-stage matching plus the total dissatisfaction of the students for being downgraded to a less preferred school between the first- and second-stage. We note that once a first-stage matching M_1 is fixed and for every fixed second-stage scenario S, the second-stage problem

$$\min_{M_2 \in \mathcal{M}_{I_2}} c_2(M_2) + d_S(M_1, M_2)$$

is a minimum weight stable matching problem and hence can be solved efficiently (see, e.g., [8]).

Remark 1 (Generalizations). We remark that our model captures the more general setting where agents can also enter the market in the second stage. In particular, for every entering student a in an instance $\mathcal{I}$, one can consider an instance $\mathcal{I}'$ with an extra dummy school b' such that a and b' are the most preferred partners of each other. Hence, a, b' are matched to each other in each stable matching of the first stage. Then a entering the market in the second stage is equivalent to b' leaving the market. It is easy to see that c_2 can be adjusted so as to discount the dissatisfaction of a in the new model and obtain the original objective function. An entering school is modelled analogously.

Also, our algorithm and analysis extend seamlessly to capture the following generalization of the dissatisfaction function (1): given $\lambda \in \mathbb{Q}_+$, and a set of non-negative scores for the pairs $w : A^+ \times B^+ \to \mathbb{Q}_+$, with the property that $w(ab) \geq w(ab')$ if $b \succ_a b'$, define

$$d_S(M_1, M_2) = \lambda \sum_{a \in A \cap S} [w(aM_2(a)) - w(aM_1(a))]^+.$$

This more general version of the dissatisfaction function captures, for example, the setting when the per unit of rank change dissatisfaction from changing school is student-dependent (in which case $w(ab) = \alpha_a \cdot R_a(b)$ for some set of non-negative weights $\{\alpha_a\}_{a \in A}$) or where a student does not care if the change happens between, say, their top 5 schools (in which case $w(ab_1) = w(ab_2) = \cdots = w(ab_5)$, where $b_1, \ldots, b_5$ are the top 5 schools in a's list) but cares if it happens between one of their top 5 schools and the rest. For simplicity of exposition, we restrict the presentation to the unweighted case, and briefly discuss the changes needed for the general case when we present our algorithm (Sect. 2.2).

1.2 Our Results and Techniques

Explicit Second-Stage Distribution. Assume first that the second-stage distribution is given explicitly by the list of all possible second-stage scenarios Θ and their respective probabilities of occurrence $\{p_S\}_{S \in \Theta}$. Let $\mathsf{N}^{\mathsf{exp}}(\mathcal{I})$ denote the input size[5] of instance $\mathcal{I}$ of (2STO) in this setting. Our first main result is a polynomial time algorithm for (2STO) in this input model.

[5] For $x \in \mathbb{Q}$, let $\|x\|$ denote the encoding length of parameter x. Then $\mathsf{N}^{\mathsf{exp}}(\mathcal{I}) = \sum_{S \in \Theta} \|p_S\| + \sum_{ab}(\|c_1(ab)\| + \|c_2(ab)\|) + \|\lambda\| + |A \cup B|^2$.

Theorem 1. *There exists an algorithm such that given any instance $\mathcal{I}$ of (2STO) where the second-stage distribution is given explicitly solves the the problem in time polynomial in the input size* $\mathsf{N}^{\mathsf{exp}}(\mathcal{I})$.

To prove Theorem 1, we show that the two-stage problem (2STO) can be polynomially reduced to a minimum $s-t$ cut problem in a carefully constructed directed graph. In our construction, we introduce and use a new poset defined over stable pairs of a marriage instance, that is, the student-school pairs that are contained in some stable matching. We call this poset the *poset of stable pairs* (see Sect. 2.1). This poset might be of independent interest in approaching stable matching optimization problems[6].

Implicit Second-Stage Distribution. Next, we consider the more general model where the second-stage distribution $\mathcal{D}$ is given by a sampling oracle. Let $\mathsf{N}^{\mathsf{imp}}(\mathcal{I})$ denote the input size[7] of instance $\mathcal{I}$ of (2STO) in this setting. We show the following hardness result for the two-stage problem (2STO) under this more general model.

Theorem 2. *Unless P=NP, there exists no algorithm that given any instance $\mathcal{I}$ of (2STO) where the second-stage distribution is specified implicitly by a sampling oracle, solves[8] the problem in time and number of calls to the sampling oracle that is polynomial in the input size* $\mathsf{N}^{\mathsf{imp}}(\mathcal{I})$. *This hardness result holds even if the cost parameters λ and $\{c_1(ab)\}_{ab}, \{c_2(ab)\}_{ab}$ are in $\{0, 1\}$.*

Our proof of Theorem 2 relies on a reduction from the problem of counting the number of vertex covers of an undirected graph, which is #P-Hard [9].

On the positive side, we give an arbitrary good additive approximation to (2STO) when the second-stage distribution is specified implicitly. Our algorithm runs in time pseudopolynomial in the input size. In particular, let $M_1^{\mathcal{I}}$ be the optimal solution of an instance $\mathcal{I}$ of (2STO) and let $\mathsf{val}_{\mathcal{I}}(M_1)$ denote the objective value of a first-stage stable matching M_1. We show the following result.

Theorem 3. *There exists an algorithm that, given an instance $\mathcal{I}$ of (2STO) where the second-stage distribution is specified implicitly by a sampling oracle, and two parameters $\epsilon > 0$ and $\alpha \in (0, 1)$, gives a first-stage stable matching solution M_1 such that,*

$$\mathbb{P}\left(\mathsf{val}_{\mathcal{I}}(M_1) \leq \mathsf{val}_{\mathcal{I}}(M_1^{\mathcal{I}}) + \epsilon\right) \geq 1 - \alpha.$$

---

[6] Our construction could also be done starting from the poset of rotations [8]; however, our poset (defined over pairs) gives more intuitive constructions and a characterization of stable families of pairs as antichains of the poset, leading to natural reformulations for other problems (e.g., the maximum cardinality of a family of pairs that is stable in two marriage instances can be reformulated using our poset as the maximum independent set in the union of two perfect graphs, studied in [4]).

[7] $\mathsf{N}^{\mathsf{imp}}(\mathcal{I}) = \sum_{ab}(\|c_1(ab)\| + \|c_2(ab)\|) + \|\lambda\| + |A \cup B|^2$.

[8] In the sense that it gives at least one of the two, the optimal value or the optimal solution. Note that the hardness of finding the optimal value does not exclude the possibility of finding the optimal solution in polynomial time.

The algorithm runs in time polynomial in $\mathsf{N}^{\mathsf{imp}}(\mathcal{I})$, $\max_{ab} |c_2(ab)|$, λ, $1/\epsilon$ *and* $\log(1/\alpha)$.

The algorithm from Theorem 3 runs in polynomial time when λ, c_2 are polynomially bounded (recall that, already in this case, by Theorem 2, no polynomial-time algorithm exists unless P = NP). Our algorithm employs the widely used Sample Average Approximation method (e.g. [5,10,13]) to approximate an instance of (2STO) with an implicitly specified second-stage distribution by an instance of explicitly specified second-stage distribution. Then we show that a relatively small number of samples (calls to the sampling oracle) is enough to get a good additive approximation with high probability. Whether there exists and FPRAS for (2STO) in this model is an interesting open question.

Numerical Experiments. We illustrate our method on randomly generated instances. We compare the performance of five first-stage matchings: The optimal solution of our problem denoted by M^*, the one-side optimal matchings, the matching minimizing $c_1(\cdot)$, and finally the best stable matching in hindsight denoted by M^{off}, which is the first-stage stable matching that would have been optimal to pick if one knew the realization of the second-stage distribution. Our results suggest that for a wide range of values of the penalty coefficient λ, our solution M^* is a strictly better choice to pick in the first stage and that it has a performance that is not very far from the performance of the best matching in hindsight M^{off}.

Note. Due to space limitation, all proofs are deferred to the full version of the paper.

1.3 Notation and Definitions

Consider a marriage instance $I = (A, B, \succ)$. We say that a pair $ab \in A \times B$ (resp., family of pairs) is *stable* if it is contained in some stable matching of I. An agent is stable if it is not matched to the outside option in some stable matching of I[9]. Let $\mathcal{S}(I)$ and $\mathcal{F}(I)$ denote the set of stable pairs and stable families of I, respectively. For a student a, let $\mathcal{S}_a(I)$ denote the set of stable pairs containing student a. Let $A^{\mathsf{st}}(I)$ and $B^{\mathsf{st}}(I)$ denote the set of stable students and schools of I respectively. The collection of orders $(\succ_a)_{a\in A}$ induce a natural partial order $>_A$ over the set of pairs $A \times B^+$, such that $ab >_A a'b'$ iff $a = a'$ and $b \succ_a b'$. Given a non-empty finite set Y and a total order $>$ over Y, we denote by $\max_> Y$ (resp., $\min_> Y$) the maximum (resp., minimum) of Y with respect to $>$.

2 Explicit Second-Stage Distribution

We consider in this section an instance of (2STO) given by a first-stage marriage instance $(A, B, \succ)$, a second-stage distribution $\mathcal{D}$ given explicitly by the list of all

9 We recall that if an agent is not matched to the outside option in a stable matching then they will not be matched to the outside option in all stable matchings (see, e.g., [8]).

possible second-stage scenarios Θ and their probabilities of occurrence $\{p_S\}_{S\in\Theta}$, and finally costs c_1, c_2, and λ. The two-stage stochastic problem (2STO) can be written as follows,

$$\min_{M_1\in\mathcal{M}_{I_1}} c_1(M_1) + \sum_{S\in\Theta} p_S \left(\min_{M_2^S\in\mathcal{M}_{I_2^S}} c_2(M_2^S) + d_S(M_1, M_2^S) \right),$$

or equivalently,

$$\min_{\substack{M_1\in\mathcal{M}_{I_1}\\ \{M_2^S\in\mathcal{M}_{I_2^S}\}_{S\in\Theta}}} c_1(M_1) + \sum_{S\in\Theta} p_S c_2(M_2^S) + p_S d_S(M_1, M_2^S). \qquad \text{(EXP-2STO)}$$

where $I_2^S = (A\cap S, B\cap S, \succ)$ is the second-stage instance under scenario S. We now give a polynomial time algorithm to solve (EXP-2STO). Our algorithm uses a new partial order that we introduce next.

2.1 Poset of Stable Pairs

In this section, we introduce the *poset of stable pairs*. This is a partial order that we define over the set of stable pairs, and which characterizes the stable families of a marriage instance.

Definition 1. *(Poset of stable pairs) Let $I = (A, B, \succ)$ be a marriage instance. The* poset of stable pairs *of I denoted by $(\mathcal{S}(I), >)$ is the set of stable pairs $\mathcal{S}(I)$ over which we define the partial order $>$ as follows: for every $ab, a'b' \in \mathcal{S}(I)$, we have $a'b' < ab$ if and only if for every stable matching M such that $ab \in M$ it holds that $M(a') \succ_{a'} b'$. We write $a'b' \leq ab$ if either $ab = a'b'$ or $a'b' < ab$.*

In particular, a stable pair $ab \in \mathcal{S}(I)$ is greater than a stable pair $a'b' \in \mathcal{S}(I)$ if whenever the larger pair ab appears in some stable matching M, the student of the smaller pair (i.e., a') is matched in M to a partner they strictly prefer to b'. Note that in particular, $ab' < ab$ when $ab' <_A ab$ or equivalently $b' \prec_a b$. In the reminder of the paper, we use the order $<_A$ when comparing (non-necessarily stable) pairs involving the same student, we use the order $<$ when comparing stable pairs (potentially involving different students), and we use the order $\prec_v$ when comparing agents (and the outside option) from the opposite side of v. The next lemma shows that $(\mathcal{S}(I), >)$ is, as claimed in Definition 1, a poset, and can be constructed efficiently.

Lemma 1. *Let I be a marriage instance. Then, $(\mathcal{S}(I), >)$ is a partially ordered set and can be constructed in time polynomial in the number of agents.*

The following lemma gives a characterization of the stable families of a marriage instance by the means of its poset of stable pairs.

Lemma 2. *Let I be a marriage instance. Then $F \in \mathcal{F}(I)$ if and only if F is an antichain of $(\mathcal{S}(I), >)$.*

We now leverage Lemma 2 to construct a capacitated directed graph with vertices $\mathcal{S}(I)$ in which cuts with finite capacity are in a one-to-one correspondence with the maximal antichains of $(\mathcal{S}(I), >)$, hence with stable matchings. We begin with the definition of the *Next-to-Smallest Dominating Stable Pair.*

Definition 2. *(Next-to-Smallest Dominating Stable Pair) Let $I = (A, B, \succ)$ be a marriage instance. Let $ab \in \mathcal{S}(I)$, $a' \in A^{\text{st}}(I)$ with $a' \neq a$, and $\mathcal{S}_{a'}(I) = \{a'b'_1, a'b'_2, \ldots, a'b'_{n_{a'}}\}$ with $b'_1 \prec_{a'} \cdots \prec_{a'} b'_{n_{a'}}$. Suppose ab is dominated by an element of $\mathcal{S}_{a'}(I)$, that is, there exists $i \in [n_{a'}]$ such that $ab < a'b'_i$, and let $i^* = \min\{i \in [n_{a'}] \mid ab < a'b'_i\}$. We define the* Next-to-Smallest Dominating Stable Pair *of ab in $\mathcal{S}_{a'}(I)$ as $\sigma(ab, a') = a'b'_{i^*-1}$.*

In order for $\sigma(ab, a')$ to be well-defined in the above definition, b'_{i^*-1} needs to be well-defined (i.e., $i^* \geq 2$). The following lemma shows this is always the case.

Lemma 3. *Let $I = (A, B, \succ)$ be a marriage instance. Let $ab \in \mathcal{S}(I)$ and $a' \in A^{\text{st}}(I)$ with $a' \neq a$. Let $\mathcal{S}_{a'}(I) = \{a'b'_1, a'b'_2, \ldots, a'b'_{n_{a'}}\}$ with $b'_1 \prec_{a'} \cdots \prec_{a'} b'_{n_{a'}}$. If ab is dominated by an element of $\mathcal{S}_{a'}(I)$ then $\min\{i \in [n_{a'}] \mid ab < a'b'_i\} \geq 2$.*

We are now ready to define the *Cut Graph* of a marriage instance. This is a capacitated direct graph whose finite cuts are in a one-to-one correspondence with the stable matchings of the instance.

Definition 3. *(Cut Graph) Consider a marriage instance $I = (A, B, \succ)$. The* Cut Graph *of I is a capacitated directed graph denoted by $\mathrm{G}_{\text{cut}}(I)$. The vertices of $\mathrm{G}_{\text{cut}}(I)$ are the stable pairs $\mathcal{S}(I)$, a source vertex s and a sink vertex t. The arcs of $\mathrm{G}_{\text{cut}}(I)$ are as follows: (i) for every $a \in A^{\text{st}}(I)$, let $\mathcal{S}_a(I) = \{ab_1, ab_2, \ldots, ab_{n_a}\}$ with $b_1 \prec_a \cdots \prec_a b_{n_a}$. Add an arc from ab_i to ab_{i+1} for every $i \in [n_a - 1]$ and an arc from s to ab_{n_a}. (ii) for every $ab \in \mathcal{S}(I)$ and $a' \in A^{\text{st}}(I)$ such that $a' \neq a$, if ab is dominated by an element of $\mathcal{S}_{a'}(I)$, add an arc from ab to $\sigma(ab, a')$. All arcs have infinite capacity.*

Note that t is isolated in the cut graph. The following lemma shows that there is a one-to-one correspondence between the $s - t$ cuts of finite capacity of $\mathrm{G}_{\text{cut}}(I)$ and the stable matchings of I.

Lemma 4. *Let $I = (A, B, \succ)$ be a marriage instance. The following is a bijection between $s - t$ cuts of finite capacity in $\mathrm{G}_{\text{cut}}(I)$ and stable matchings in I:*

(a) Let $(C, \overline{C})$ be a finite capacity $s - t$ cut of $\mathrm{G}_{\text{cut}}(I)$. Then, for every $a \in A^{\text{st}}(I)$, $\mathcal{S}_a(I) \cap C \neq \emptyset$. Moreover, the matching M where every agent $a \in A^{\text{st}}(I)$ is matched to its least preferred pair in C, i.e., the pair $\min_{<_A} \mathcal{S}_a(I) \cap C$, and every other non-stable agent $v \notin A^{\text{st}}(I) \cup B^{\text{st}}(I)$ is matched to the outside option is a stable matching.

(b) Let M be a stable matching of I. Let

$$C = \{s\} \cup \bigcup_{a \in A^{\text{st}}(I)} \{ab \in \mathcal{S}_a(I) \mid aM(a) \leq_A ab\},$$

which is the union of $\{s\}$ and the set of all stable pairs that are greater than $aM(a)$ for every stable agent $a \in A^{\text{st}}(I)$. Then $(C, \overline{C})$ is an $s - t$ cut of finite capacity of $\mathrm{G}_{\text{cut}}(I)$.

2.2 Algorithm

We now present our algorithm for (EXP-2STO). First of all, note that adding a positive constant $c \in \mathbb{Q}_+$ to the cost functions c_1 and c_2 changes the objective value by exactly $c|A^{\mathrm{st}}(I_1)|+c\sum_{S\in\Theta} p_S|A^{\mathrm{st}}(I_2^S)|$ which is a constant and does not depend on the chosen solution $M_1, \{M_2^S\}_{S\in\Theta}$. Hence, without loss of generality, we suppose that the cost functions c_1 and c_2 are non-negative. From now on, when necessary, we add a superscript indicating the instance at hand when referring to an order. For example, we use $>^{I_1}$, to refer to the poset of stable pairs of the first-stage instance I_1. We construct a capacitated directed graph $\mathcal{N}(\mathcal{I})$ and show that (EXP-2STO) is equivalent to the minimum $s-t$ cut problem over $\mathcal{N}(\mathcal{I})$.

Consider the cut graphs $\mathrm{G_{cut}}(I_1)$ and $\{\mathrm{G_{cut}}(I_2^S)\}_{S\in\Theta}$ constructed following Definition 3 for the first-stage instance I_1 and second-stage instances $\{I_2^S\}_{S\in\Theta}$ respectively. In order to distinguish the vertices of these graphs, we use the notation $[v]$ to denote vertex v when it belongs to $\mathrm{G_{cut}}(I_1)$ and use $[v]_S$ to denote v when it belongs to $\mathrm{G_{cut}}(I_2^S)$. For example, $[s]$ denotes the vertex s of $\mathrm{G_{cut}}(I_1)$ and $[s]_S$ denotes the vertex s of $\mathrm{G_{cut}}(I_2^S)$.

High-Level Idea. Our directed graph $\mathcal{N}(\mathcal{I})$ is constructed in three steps.

(i) *Bijection between finite capacity cuts and solutions* $(M_1, \{M_2^S\}_{S\in\Theta})$. We consider the cut graphs $\mathrm{G_{cut}}(I_1)$ and $\{\mathrm{G_{cut}}(I_2^S)\}_{S\in\Theta}$ to which we add a super source vertex s (resp., super sink vertex t) that we link to $[s], \{[s]\}_{S\in\Theta}$ (resp., $[t], \{[t]_S\}_{S\in\Theta}$) with infinite capacity arcs in such a way that any $s-t$ cut of finite capacity always includes (resp., precludes) the vertices $[s], \{[s]\}_{S\in\Theta}$ (resp., $[t], \{[t]_S\}_{S\in\Theta}$). This ensures that any $s-t$ cut of finite capacity of $\mathcal{N}(\mathcal{I})$ corresponds to a solution $(M_1, \{M_2^S\}_{S\in\Theta})$, where M_1 (resp., M_2^S) is a first- (resp., second-)stage stable matching. We next add arcs of finite positive capacity to capture the cost of the solution $(M_1, \{M_2^S\}_{S\in\mathcal{S}})$.

(ii) *Cost* $\lambda \cdot \sum_{S\in\mathcal{S}} p_S d_S(M_1, M_2^S)$. This is the most complex part of the construction, and the one for which the new poset comes in handy. It is achieved by adding edges between graphs $\mathrm{G_{cut}}(I_1)$ and $\mathrm{G_{cut}}(I_2^S)$ for all $S \in \Theta$. In particular, for every $S \in \Theta$ we add edges between the vertices of $\mathrm{G_{cut}}(I_1)$ and those of $\mathrm{G_{cut}}(I_2^S)$ such that an $s-t$ cut of finite capacity is traversed by a subset of these edges of total capacity $\lambda p_S d_S(M_1, M_2^S)$, where M_1, M_2^S are the stable matchings corresponding to the cut. To do so, for every $a \in A \cap S$, we would like to add an arc of capacity λp_S from the copy of a pair ab in $\mathrm{G_{cut}}(I_2^S)$ to the copy of the pair ab in $\mathrm{G_{cut}}(I_1)$ for every b such that $b \succeq_a \emptyset$. This would ensure that an $s-t$ cut of finite capacity that takes from $\mathcal{S}_a(I_1)$ the pairs that a weakly prefers to ab_1 and from $\mathcal{S}_a(I_2^S)$ the pairs that weakly prefers to ab_2 (in which case a will be matched to b_1 and b_2 in the corresponding first and second stage stable matchings respectively) is traversed by exactly $[R_a(b_2) - R_a(b_1)]^+$ many pairs of capacity λp_S. However, ab is not necessarily stable in I_1 (resp., I_2^S) and hence $[ab]$

(resp., $[ab]_S$) may not belong to the cut graph $\mathrm{G}_{\mathrm{cut}}(I_1)$ (resp., $\mathrm{G}_{\mathrm{cut}}(I_2^S)$). When ab is not stable in a cut graph, our arcs connect instead, roughly speaking, the closest stable pair ab' to ab such that $b' \prec_a b$.

(iii) *Cost* $c_1(M_1) + \sum_{S \in \mathcal{S}} p_S c_2(M_2^S)$. These costs are captured using arcs between nodes of $\mathrm{G}_{\mathrm{cut}}(I_1)$ (resp., nodes of $\mathrm{G}_{\mathrm{cut}}(I_2^S)$ for $S \in \Theta$) such that an $s - t$ cut of finite capacity of $\mathcal{N}(\mathcal{I})$ corresponding to the first- (resp., second-)stage stable matching M_1 (resp., M_2^S) is traversed by a subset of arcs of total capacity $c_1(M_1) + \eta$ (resp., $p_S c_2(M_2^S) + \eta$) for some constant η.

See Fig. 2 for an example.

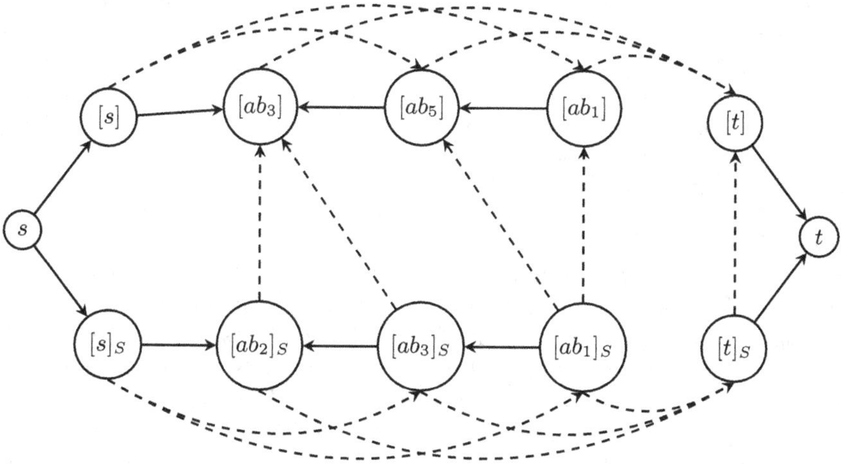

Fig. 2. An example of the construction of $\mathcal{N}(\mathcal{I})$. Fix $a \in A \cap S$ and assume a has a preference list $b_2 \succ_a b_3 \succ_a b_5 \succ_a b_1 \succ_a \emptyset$. Suppose that $\mathcal{S}_a(I_1) = \{ab_3, ab_5, ab_1\}$ and $\mathcal{S}_a(I_2^S) = \{ab_2, ab_3, ab_1\}$ for some scenario $S \in \Theta$. The figure depicts the subgraph of $\mathcal{N}(\mathcal{I})$ induced by these stable pairs. The solid arcs have capacity ∞ while the dashed arcs have finite capacity. Suppose the solution gives a cut $(C, \bar{C})$ such that $\{s, [s], [ab_3], [s]_S, [ab_2]_S, [ab_3]_S, [ab_1]_S\} \subset C$ and $\{[ab_5], [ab_1], [t], [t]_S, t\} \subset \bar{C}$, such cut will correspond to matchings such that $M_1(a) = b_3$ and $M_2^S(a) = b_1$, and in this case the rank change is 2, which is consistent with the fact that exactly two arcs from $\mathrm{G}_{\mathrm{cut}}(I_2^S)$ to $\mathrm{G}_{\mathrm{cut}}(I_1)$ traverse this cut.

Formal Definition. Formally, our directed graph $\mathcal{N}(\mathcal{I})$ consists of the disjoint union of the cut graphs $\mathrm{G}_{\mathrm{cut}}(I_1)$ and $\{\mathrm{G}_{\mathrm{cut}}(I_2^S)\}_{S \in \Theta}$ to which we add a super source vertex s and super sink vertex t and the following arcs,

(i) An arc from s to $[s]$, an arc from s to $[s]_S$ for all $S \in \Theta$, an arc from $[t]$ to t and an arc from $[t]_S$ to t for all $S \in \Theta$, all of infinite capacity.

(ii) Consider a second-stage scenario $S \in \Theta$ and let $a \in A \cap S$ be an agent present both in the first- and second-stage. Let $b \in B^+$ such that $b \succeq_a \emptyset$ and let $\Pi(ab) = \{ab' \in \mathcal{S}_a(I_1) \mid ab' \leq_A ab\}$ and $\Pi_S(ab) = \{ab' \in \mathcal{S}_a(I_2^S) \mid ab' \leq_A ab\}$ define,

$$\Psi(ab) = \begin{cases} [s] & \text{if } a \notin A^{\text{st}}(I_1) \\ [t] & \text{else if } \Pi(ab) = \emptyset \\ [\max_{<_A} \Pi(ab)] & \text{otherwise} \end{cases}, \quad \Psi_S(ab) = \begin{cases} [s]_S & \text{if } a \notin A^{\text{st}}(I_2^S) \\ [t]_S & \text{else if } \Pi_S(ab) = \emptyset \\ [\max_{<_A} \Pi_S(ab)]_S & \text{otherwise} \end{cases};$$

add an arc of capacity λp_S[10] from $\Psi_S(ab)$ to $\Psi(ab)$.

(iii) Let $a \in A^{\text{st}}(I_1)$ and let $\mathcal{S}_a(I_1) = \{ab_1, ab_2, \ldots, ab_{n_a}\}$ with $b_1 \prec_a \cdots \prec_a b_{n_a}$. Add an arc from $[s]$ to $[ab_i]$ of capacity $c_1(ab_{i+1})$ for every $i \in [1, n_a - 1]$. Add an arc from $[ab_i]$ to $[t]$ of capacity $c_1(ab_i)$ for every $i \in [1, n_a]$. Similarly, for every $S \in \Theta$, let $a \in A^{\text{st}}(I_2^S)$ and let $\mathcal{S}_a(I_2^S) = \{ab_1, ab_2, \ldots, ab_{n_a}\}$ with $b_1 \prec_a \cdots \prec_a b_{n_a}$, add an arc arc from $[s]_S$ to $[ab_i]_S$ of capacity $p_S c_2(ab_{i+1})$ for every $i \in [1, n_a - 1]$ and an arc from $[ab_i]_S$ to $[t]_S$ of capacity $p_S c_2(ab_i)$ for every $i \in [1, n_a]$.

The following lemma shows that solving (EXP-2STO) is equivalent to solving the minimum $s-t$ cut problem over $\mathcal{N}(\mathcal{I})$. Our first main result Theorem 1 follows immediately from Lemma 5.

Lemma 5. *Let $(C, \overline{C})$ be a minimum $s-t$ cut in $\mathcal{N}(\mathcal{I})$. Let $[C]$ (resp., $[C]_S$) denote the pairs corresponding to the vertices of C from the first-stage (resp., scenario S of the second-stage) cut graph. Let M_1 be the matching where every stable agent in the first stage $a \in A^{\text{st}}(I_1)$ is matched to its least preferred stable pair in $[C]$, i.e., the pair $\min_{<_A} \mathcal{S}_a(I_1) \cap [C]$ (which exists as $\mathcal{S}_a(I_1) \cap [C] \neq \emptyset$ by Lemma 4), and every other non-stable agent $v \notin A^{\text{st}}(I_1) \cup B^{\text{st}}(I_1)$ is matched to the outside option. Similarly, for every scenario $S \in \Theta$, let M_2^S be the matching where every stable agent in the second stage $a \in A^{\text{st}}(I_2^S)$ is matched to its least preferred stable pair in $[C]_S$, i.e., the pair $\min_{<_A} \mathcal{S}_a(I_2^S) \cap [C]_S$ (which exists as $\mathcal{S}_a(I_2^S) \cap [C]_S \neq \emptyset$ by Lemma 4), and every other non-stable agent $v \notin A^{\text{st}}(I_2^S) \cup B^{\text{st}}(I_2^S)$ is matched to the outside option. Then M_1 and $\{M_2^S\}_{S \in \Theta}$ are stable matchings and an optimal solution of (EXP-2STO).*

3 Sampling Oracle Model

We consider in this section the more general model where the second-stage distribution $\mathcal{D}$ of an instance $\mathcal{I}$ is given by a sampling oracle. We show an hardness result for the problem under this more general input model and give an arbitrary good additive approximation.

---

[10] For the generalized version of the dissatisfaction function in Remark 1, if b is not the most preferred school of a, let $b' = \min_{\succ_a} \{b'' | b'' \succ_a b\}$, then use here capacity $(w(ab) - w(ab')) \cdot \lambda p_S \geq 0$ instead.

3.1 Computational Complexity

Theorem 2 follows from Lemma 6 below and the fact that the counting problem of vertex covers is #P-Hard [9], hence a polynomial time algorithm for this problem would imply that #P=FP, which in turns implies that P=NP (see, e.g., [2]).

Lemma 6. *Suppose that there exists an algorithm $\mathcal{A}$ such that given any instance $\mathcal{I}$ of (2STO), algorithm $\mathcal{A}$ solves the two-stage problem over $\mathcal{I}$ (i.e., finds either the optimal solution or the optimal value) in time and number of calls to the sampling oracle that is polynomial in the input size $\mathsf{N}^{\mathsf{imp}}(\mathcal{I})$. Then $\mathcal{A}$ can be used to count the number of vertex covers in any undirected graph $G(V,E)$ in time polynomial in $|V|$.*

3.2 Algorithm

We give a near-optimal additive approximation to the two-stage problem (2STO). Our algorithm is sampling based and runs in time pseudopolynomial in the input.

Our starting point is to approximate the expected value in the two-stage stochastic problem (2STO) by a sample average over N i.i.d. samples. This method, known in the stochastic programming literature as the *Sample Average Approximation* (SAA) method has been widely used to approximate two-stage stochastic combinatorial problems [5,10,13]. In particular, let $S^1,\dots,S^N$ be N i.i.d. samples drawn from the distribution $\mathcal{D}$. We replace the expected value in the two-stage stochastic problem (2STO) with the sample average taken over $S^1,\dots,S^N$. We get the following sample average minimization problem,

$$\min_{M_1\in\mathcal{M}_{I_1}} c_1(M_1) + \frac{1}{N}\sum_{j=1}^{N} \min_{M_2^j\in\mathcal{M}_{I_2^j}} c_2(M_2^j) + d_{S^j}(M_1, M_2^j). \tag{SAA}$$

Note that (SAA) is an instance of (EXP-2STO) where $\Theta = \{S^1,\dots,S^N\}$ and $p_S = \frac{1}{N}$ for all $S\in\Theta$, and can therefore be solved exactly using our algorithm for explicitly specified distributions, in time polynomial in N and the other input parameters. It remains to show that, for any $\epsilon > 0$, a small number of samples N is enough to get an ϵ additive approximation of (SAA) with high probability.

Let $\hat{M}_1$ be an optimal solution of the sample average problem (SAA). The following lemma bounds the (additive) quality of the first-stage stable matching $\hat{M}_1$ in the original problem (SAA) as a function of the number of samples N.

Lemma 7. *Let $\alpha\in(0,1)$. Then with probability at least $1-\alpha$ it holds that,*

$$\mathsf{val}_{\mathcal{I}}(\hat{M}_1) \le \mathsf{val}_{\mathcal{I}}(M_1^{\mathcal{I}}) + |A|(\max_{ab}|c_2(ab)| + \lambda|B|)\sqrt{\frac{\max\{|A|,|B|\}\log(3.88/\alpha)}{N}}.$$

Our third result Theorem 3 follows immediately from Lemma 7. In particular, for any $\epsilon > 0$, a number of samples

$$N = \left(|A|(\max_{ab} |c_2(ab)| + \lambda|B|) \right)^2 \frac{\max\{|A|, |B|\} \log(3.88/\alpha)}{\epsilon^2}$$

is enough to get an ϵ additive approximation with probability at least $1 - \alpha$.

4 Numerical Experiments

Experimental Setup. We consider an instance of (2STO) with $n = 50$ students and schools of uniformly randomly generated preferences. We suppose that each agent leaves the market in the second-stage with probability $p = 0.25$ independently of the other agents. In terms of costs, we set the cost of a pair ab, $c_1(ab)$ and $c_2(ab)$, to be the average rank between the two sides[11], and solve the problem for different values of the penalty coefficient λ. We compare the performance of five first-stage matchings: the optimal matching computed by our algorithm denoted by M^*, the one-side optimal matchings M_0 and M_z that are the common practice that is usually used, the matching $\hat{M}$ maximizing c_1, and the optimal matching in hindsight M^{off}, that is, the first-stage stable matching that would have been optimal to take if one knew the realization of the second-stage distribution.

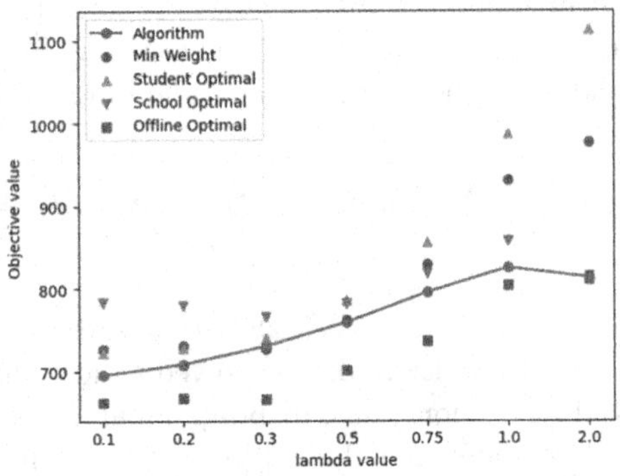

Fig. 3. Simulation results on an instance with 50 students and 50 schools.

Results. The results are given in Fig. 3. We observe that for a wide range of λ, the optimal stable matching is different from M_0, M_z, and $\hat{M}$. Moreover, the optimal value $\mathsf{val}(M^*)$ is relatively close to the optimal value in hindsight $\mathsf{val}(M^{\text{off}})$. Observe that for large values of λ (in which case we are more in favor of minimizing the students downgrades in the second-stage than the quality of

[11] This is known as an *egalitarian* stable matching (see, e.g., [8]), since it balances between the utilities of the two sides of the market.

the matching), the optimal matching becomes M_z. Our results suggest that for a wide range of λ, it is better to choose M^* over choosing M_0, M_z, or $\hat{M}$ in the first-stage. They also suggest that M^* will not be a lot worse than what we could have done in hindsight.

Choosing $\boldsymbol{\lambda}$. The coefficient λ measures how much we favor the minimization of changes between the two stages to the maximization of the quality of the matchings. In practice, the appropriate λ is problem-dependent and a good coefficient λ can be learned from a few applications of the method. In particular, given some measure of performance (e.g. surveys where students/faculty express how satisfied they are with the matching process), one can choose, for example, λ that maximizes the performance on average.

Acknowledgments. The authors thank the IPCO referees for their useful comments. Yuri Faenza and Chengyue He acknowledge the support of the NSF Grant 2046146 *CAREER: An Algorithmic Theory of Matching Markets.*

References

1. Abdulkadiroğlu, A., Pathak, P.A., Roth, A.E.: The New York City high school match. Am. Econ. Rev. **95**(2), 364–367 (2005)
2. Arora, S., Barak, B.: Computational Complexity: A Modern Approach. Cambridge University Press, Cambridge (2009)
3. Ashlagi, I., Graur, A., Lo, I., Mentzer, K.: Overbooking with priority-respecting reassignment. In: Presentation at the 3rd ACM Conference on Equity and Access in Algorithms, Mechanisms, and Optimization (EAAMO'23) (2023)
4. Chakaravarthy, V.T., Pandit, V., Roy, S., Sabharwal, Y.: Finding independent sets in unions of perfect graphs. In: IARCS Annual Conference on Foundations of Software Technology and Theoretical Computer Science (FSTTCS 2010), pp. 251–259 (2010)
5. Charikar, M., Chekuri, C., Pál, M.: Sampling bounds for stochastic optimization. In: International Workshop on Approximation Algorithms for Combinatorial Optimization, pp. 257–269 (2005)
6. Faenza, Y., Zhang, X.: Legal assignments and fast EADAM with consent via classic theory of stable matchings. Oper. Res. **70**(3), 1873–1890 (2022)
7. Gale, D., Shapley, L.S.: College admissions and the stability of marriage. Am. Math. Mon. **69**(1), 9–15 (1962)
8. Gusfield, D., Irving, R.W.: The Stable Marriage Problem: Structure and Algorithms. MIT Press, Cambridge (1989)
9. Provan, J.S., Ball, M.O.: The complexity of counting cuts and of computing the probability that a graph is connected. SIAM J. Comput. **12**(4), 777–788 (1983)
10. Ravi, R., Sinha, A.: Hedging uncertainty: approximation algorithms for stochastic optimization problems. Math. Program. **108**, 97–114 (2006)
11. Roth, A.E., Peranson, E.: The redesign of the matching market for American physicians: some engineering aspects of economic design. Am. Econ. Rev. **89**(4), 748–780 (1999)
12. Roth, A.E., Sotomayor, M.: Two-sided matching. Handb. Game Theory Econ. Appl. **1**, 485–541 (1992)
13. Swamy, C., Shmoys, D.B.: Sampling-based approximation algorithms for multi-stage stochastic optimization. In: 46th Annual IEEE Symposium on Foundations of Computer Science (FOCS 2005), pp. 357–366 (2005)

Von Neumann-Morgenstern Stability and Internal Closedness in Matching Theory

Yuri Faenza[1], Clifford S. Stein[1], and Jia Wan[2(✉)]

[1] Columbia University, New York, NY 10027, USA
yf2414@columbia.edu, cliff@ieor.columbia.edu
[2] Massachusetts Institute of Technology, Cambridge, MA 02139, USA
jiawan@mit.edu

Abstract. Gale and Shapley's stability criterion enjoys a rich mathematical structure, which propelled its application in various settings. Although immensely popular, the approach by Gale and Shapley cannot encompass all the different features that arise in applications, motivating the search for alternative solution concepts. We investigate alternatives that rely on the concept of internal stability, a notion introduced for abstract games by von Neumann and Morgenstern and motivated by the need of finding a set of mutually compatible solutions. The set of stable matchings is internally stable (IS). However, the class of IS sets is much richer, for an IS set of matchings may also include unstable matchings and/or exclude stable ones. In this paper, we focus on two families of IS sets of matchings: *von Neumann-Morgenstern* (vNM) stable and *internally closed*. We study algorithmic questions around those concepts in both the marriage and the roommate models. One of our results imply that, in the marriage model, internally closed sets are an alternative to stable matchings that is as tractable as stable matchings themselves, a fairly rare occurrence in the area. Both our positive and negative results rely on new structural insights and extensions of classical algebraic structures associated with sets of matchings, which we believe to be of independent interest.

Keywords: Stable matching · rotation · poset · distributive lattice · vNM stability

1 Introduction

The *marriage model* was introduced by Gale and Shapley in their classical work [12] to address the problem of fairly allocating college seats to students. In (a slight generalization of) their setting, we are given a two-sided matching market, with each agent listing a subset of the agents from the opposite side of the market in a strict preference order. The goal is to find a matching that respects a fairness property called *stability*. Stable matchings enjoy a rich

J. Vygen and J. Byrka (Eds.): IPCO 2024, LNCS 14679, pp. 168–181, 2024.
https://doi.org/10.1007/978-3-031-59835-7_13

mathematical structure that has been leveraged on to design various algorithms (see, e.g., [15,23,25]) to assign students to schools, doctors to hospitals, workers to firms, partners in online dating, agents in ride-sharing platforms, and more, see, e.g., [1,16,24,30]. Although immensely popular, stability is not the right notion for some applications, where we may want, e.g., a matching of larger size or more favorable to one side of the market. Concepts alternative to stability, such as popularity [8,9,13,17,19,21] and Pareto-optimality [2,26] have therefore become an important area of research. However, such alternative concepts often lack many of the attractive structural and algorithmic properties of stable matchings. Much research has therefore been devoted to defining alternative solution concepts that are computationally tractable [3,5,11,23].

Internal Stability, von Neumann-Morgenstern Stability, and Internal Closedness. In this paper, we study two solution concepts alternative to stability, in both the marriage and the roommate (i.e., non-bipartite) model. These concepts rely on the fundamental game theory notion of *internal stability* [28]. A set of matchings $\mathcal{M}'$ of a (marriage or roommate) instance is *internally stable* if there are no matchings $M, M' \in \mathcal{M}'$ such that an edge of M blocks M', i.e., there is no pair of agents matched in M that strictly prefer each other to their respective partners in M'. Note that while stability is a property of a matching, internal stability is a property of a set of matchings, and that the set of stable matchings is internally stable. As discussed by von Neumann and Morgenstern [28], a family of internally stable solutions $\mathcal{M}'$ to a game (a family of matchings, in our case) can be thought of as the family of "standard behaviours" within an organization: solutions in $\mathcal{M}'$ are all and only those that are compatible with predefined rules. Also, as argued in [7,28], one can question why other solutions are excluded from being members of $\mathcal{M}'$. Hence, in order for an internally stable set $\mathcal{M}'$ to be deemed an acceptable standard, it is required that $\mathcal{M}'$ satisfies further conditions. Typically, one requires $\mathcal{M}'$ to be *externally stable*: for every matching $M \notin \mathcal{M}'$, there is a matching $M' \in \mathcal{M}'$ containing an edge that blocks M. A set that is both internally and externally stable is called *von Neumann-Morgenstern (vNM)* stable. Von Neumann and Morgenstern proposed vNM stability as the main solution concept for cooperative games [28]. Shubik [27] reports more than 100 works that investigate this concept until 1973 (see also [22]). Later research in game theory shifted the focus from vNM stability to the core of games, which enjoys stronger properties, and is often easier to work with, than vNM stable sets [6]. However, vNM stable sets in the marriage model have been shown to enjoy strong algorithmic and structural properties [6,10,29].

To define internal closedness, we relax the external stability condition to inclusionwise maximality. A set is *internally closed* if it is an inclusionwise maximal set of internally stable matchings. Our definition is motivated by two considerations. First, a vNM stable set may not always exist in a roommate instance, while an internally closed set of matchings always exists. Second, a central planner may want a specific set of internally stable matchings $\mathcal{M}'$ to be part of the family of feasible solutions. Consider for instance the problem the planner faces when given a family of internally stable matchings $\mathcal{M}'$, representing the currently accepted matchings. The planner's goal is to look for a more comprehensive set of solutions, that are possibly an improvement over the status quo, but are also compatible with it, i.e., matchings that neither block nor are blocked by matchings in $\mathcal{M}'$. Formally, can we efficiently obtain an internally closed set $\mathcal{M}''$ so that $\mathcal{M}'' \supseteq \mathcal{M}'$? Clearly, $\mathcal{M}''$ always exists, as one can start from $\mathcal{M}'$ and iteratively enlarge it as to achieve inclusionwise maximality while preserving internal stability. As we will see, our analysis of internally closed sets of matching also leads to an algorithmic understanding of vNM stable sets.

1.1 Overview of Contributions and Techniques

Motivated by the discussion above, in this paper we investigate structural and algorithmic properties of vNM stable and internally closed sets of matching. Our results have implications for the theory of stable matchings more generally. We present our contributions from the end, i.e., from their algorithmic implications.

Algorithmic and Complexity Results. We show that, in a marriage instance, one can find an internally closed set of matchings containing any given internally stable set of matchings in polynomial time (Theorem 1). Conversely, in a roommate instance, even deciding if a set of matchings is internally closed, or vNM stable, is co-NP-hard, and the problem of finding a vNM stable set of matchings is also co-NP-hard (Theorem 6).

From Matchings to Edges. The algorithmic statements from the previous paragraph glossed over the complexity issue of representing the input and the output to our problems, since a family of internally stable matchings may have size exponential in the number of agents. We bypass this concern by showing that every internally closed set of matchings coincides with the set $\mathcal{S}'$ of stable matchings in a *subinstance* of the original instance, i.e., an instance obtained from it by removing certain entries from preference lists. Hence, denoting an instance by a pair $(G, >)$ (with G being the graph with agents as nodes and matchable pairs as edges, and $>$ the agents' preferences), our input is given by a set $E_0 \subseteq E(G)$ implicitly describing the set $\mathcal{S}'$ of stable matchings in $(G[E_0], >)$. Similarly, an internally closed or vNM stable set of matchings is described compactly by $E_C \subseteq E(G)$. See Sect. 2 for details. This fact allows us to work with (polynomially-sized) sets of edges, rather than (possibly exponentially-sized) sets of matchings. It also implies that any question on an internally closed set of matchings, once the set E_C had been determined, reduces to the analogous question on stable matchings, for which algorithms are often known.

Our next step lies in understanding how to enlarge the input set of edges E_0, in order to obtain the set E_C. The challenge here lies in the fact that adding *any single edge* to E_0 may not lead to a strictly larger internally stable set of matchings, while adding certain edges may prevent the possibility of finding an internally closed set of matching. Hence, any algorithm will need to iteratively add *sets of edges* rather than single edges, possibly leading to a search space that is exponentially large. However, (extensions of) classical algebraic properties of stable matchings come to our rescue.

T =

Agent	Preference List
1	A B C E
2	B A C D
3	C D A E
4	D C A B
5	E C
A	3 4 2 1
B	4 1 2
C	5 1 2 4 3
D	2 3 4
E	1 5 3

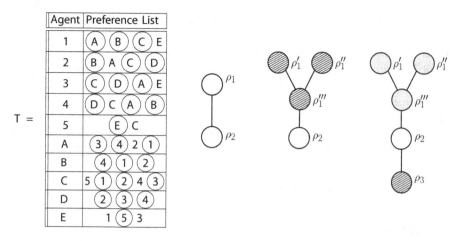

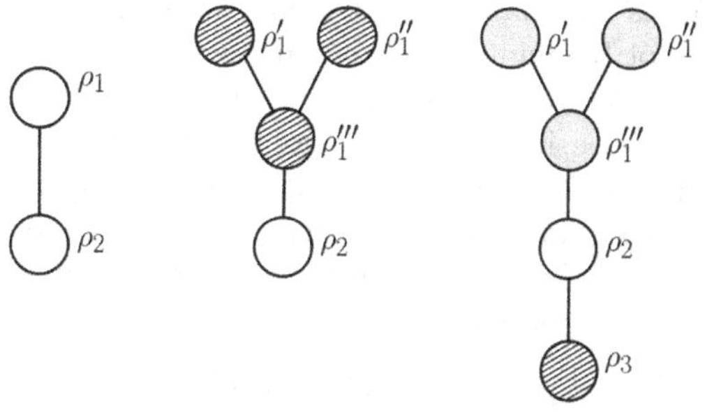

Fig. 1. Left: A marriage instance described by its preference table T with, circled, the entries of subtable T'. The set $\mathcal{S}'$ of stable matchings of T' is internally stable but not internally closed: matching $\{1B, 2A, 3D, 4C, 5E\}$ neither blocks nor is blocked by any matching from $\mathcal{S}'$. Right: We iteratively expand the poset of rotations of $\mathcal{S}'$ first by dissecting ρ_1 into $\rho_1', \rho_1'', \rho_1'''$ (leading to the poset of rotations of the stable matchings of $T \setminus \{1E, 5C, 3E\}$) and then by vertically expanding the poset via ρ_3 (poset associated to $T \setminus \{3E\}$). The set of stable matchings of $T \setminus \{3E\}$ is internally closed: no further enlargement of the poset is possible.

From Edges to Rotations: The Marriage Case. We first focus on the marriage case, investigated in Sect. 3. Since vNM stable sets in this model are well understood [6,10,29], we consider only internally closed sets of matchings. A classical result gives a bijection between the set of stable matchings of a marriage instance and the family of closed sets of the associated poset of *rotations* $(R, \sqsupseteq)$ [20]. Rotations are certain cycles in the marriage instance (see Sect. 3). We first give a characterization of internally closed sets of matchings that relies on a certain "maximality" property of the poset of rotations. Roughly speaking, a family of matchings is internally closed if and only if, in the poset of rotations associated to it, no rotation can be *dissected*, i.e., replaced with a poset of new rotations, and we cannot *vertically expand* the poset by adding rotations that are maximal or minimal wrt $\sqsupseteq$ (see Theorem 4). On the way to this result, we introduce the novel concept of *generalized rotations*, that may be useful for other questions in the area. Our characterization leads to a polynomial-time algorithm that iteratively enlarges a set of internally stable matchings by trying to dissect rotations or vertically expand the associated poset (see Fig. 1 for an example). This algorithms shows that an internally closed set of matchings containing a given internally stable set of matchings can be found in polynomial time (see Theorem 1).

From Edges to Rotations: The Roommate Case. We investigate the roommate case in Sect. 4. The set of stable matchings in the roommate case does not seem to have a relevant lattice structure, and may be empty. However, when it is non-empty, it can be described via a poset of *singular* and *non-singular* rotations (first introduced in [14], defined differently from the marriage case). Our main structural contribution here is to show that the set of stable matchings is internally closed if and only if the poset of rotations associated to it cannot be augmented via what we call a *stitched rotation* (see Definition 6 and Theorem 5). We then show how to construct, from any instance ϕ of 3-SAT, a roommate instance that has a stitched rotation if and only if ϕ is satisfiable, thus proving the claimed hardness result. Internally closed sets give us a new tool to study vNM stability, also leading to a proof of the co-NP-hardness of deciding if a set of matchings is vNM stable, and of finding a vNM stable set (see Theorem 6).

Note. All proofs are deferred to the journal version of the paper.

2 Preliminaries

Basic Notation. For $n \in \mathbb{N}$, we let $[n] = \{1, \ldots, n\}$ and $[n]_0 = \{0\} \cup [n]$. On top of the classical graph representation for marriage and roommate instances, we represent instances via preference tables (see, e.g., [15]). A *(roommate) instance* is therefore described as a set A of agents and, for each $z \in A$, a *preference list* consisting of a subset $A(z) \subseteq A \backslash \{z\}$ and a strict ordering (i.e., without ties) of elements from this list. For each $z \in A$, $A(z)$ contains therefore all and only the agents that are acceptable to z. The collection of all preference lists is then represented by the *(preference) table* $T(A, >)$ (or T in short), where $>$ collects, for $z \in A$, the strict ordering within the preference list of agent z, denoted as $>_z$. We assume that preferences are symmetric, i.e., z_1 is on z_0's preference list if and only if z_0 is on z_1's. See Fig. 1 and Fig. 2 for examples.

Preferences. For $z, z_1, z_2 \in A$ with $z_1, z_2 \in A(z)$, we say that z *strictly prefers* z_1 to z_2 if $z_1 >_z z_2$. We say that z *(weakly) prefers* z_1 to z_2, and write $z_1 \geq_z z_2$, if $z_1 >_z z_2$ or $z_1 = z_2$. For $z \in A$, we extend $>_z$ to $\emptyset$ by letting $z_1 >_z \emptyset$ for all $z_1 \in A(z)$. That is, all agents strictly prefer being matched to some agent in their preference list than being left unmatched. Because of the symmetry assumption, we say that $z_0z_1 \in T$ if z_1 is on z_0's preference list and call z_0z_1 an *edge (of T)*. Let $r_T(z_0, z_1)$ denote the *ranking* (i.e., the position, counting from left to right) of z_1 in the preference list of z_0 in preference table T: $r_T(z_0, z_1) < r_T(z_0, z_2)$ if and only if agent z_0 strictly prefers z_1 to z_2 within preference table T. We let $r_T(z, \emptyset) = +\infty$. For a preference table T and agent z, let $f_T(z)$, $s_T(z)$, $\ell_T(z)$ denote the first, second and last agent on z's preference list.

Consistency, Subtables. Two roommate instances T, T' have *consistent* preference lists if, for any pair $z_0z_1, z_0z_2 \in T$ and $z_0z_1, z_0z_2 \in T'$, we have $r_T(z_0, z_1) < r_T(z_0, z_2)$ if and only if $r_{T'}(z_0, z_1) < r_{T'}(z_0, z_2)$. We write $T' \subseteq T$ when T', T have the following properties: (a) $z_0z_1 \in T$ for all $z_0z_1 \in T'$ and (b) T, T' have consistent preference lists. In this case, we call T' a *subtable* of T. For subtables T_1, T_2 of T, we let $T_1 \cup T_2$ be the subtable of T that contains all edges that are in T_1, T_2, or both.

Matching Basics. Fix a roommate instance $T(A, >)$. A *matching* M of T is a collection of disjoint pairs of agents from A, with the property that if $zz' \in M$, then z appears in z' preference list. For $z_0 \in A$, we let $M(z_0)$ be the partner of z_0 in matching M. If $z_0 z_1 \notin M$ for every $z_1 \in A$, we write $M(z_0) = \emptyset$. If $M(z_0) \neq \emptyset$, we say that z_0 is *matched (in M)*. For a matching $M \subseteq T$, we say that $ab \in T$ is a *blocking pair* for M (and M is blocked by ab) if $b >_a M(a)$ and $a >_b M(b)$. M is *stable* if it is not blocked by any pair in T, and *unstable* if it is blocked by some pair in T. We say that a matching M' *blocks* a matching M if M' contains a blocking pair ab for M. Any matching M can be interpreted as a preference table T, where $z_0 z_1 \in T$ if and only if $z_0 z_1 \in M$. For a roommate instance T, we let $\mathcal{M}(T)$ denote the set of matchings of T, and $\mathcal{S}(T)$ denote the set of stable matchings of T. If $xy \in T$ is contained in some stable matching of T, then it is called a *stable edge* or *stable pair.* Let $E_S(T)$ denote the subtable of T containing all and only the stable edges. If $E_S(T) = T$, then T is called a *stable table.* When T is clear from the context, we abbreviate $\mathcal{M}(T), \mathcal{S}(T), r_T(\cdot,\cdot), E_S(T), \ldots$ by $\mathcal{M}$, $\mathcal{S}$, $r(\cdot,\cdot), E_S, \ldots$. We say an instance T is *solvable* if it admits at least one stable matching.

Internal Stability and Related Concepts. Let T be a roommate instance. We say that a set of matchings $\mathcal{M}' \subseteq \mathcal{M}(T)$ is *internally stable* if given any two matchings $M, M' \in \mathcal{M}'$, M does not block M' and M' does not block M. For an internally stable set of matchings $\mathcal{M}' \subseteq \mathcal{M}(T)$, we define its closure $\overline{\mathcal{M}'} = \{M \in \mathcal{M}(T) : \{M\} \cup \mathcal{M}' \text{ is internally stable}\}$. Note that $\overline{\mathcal{M}'}$ may not be internally stable. If $\overline{\mathcal{M}'} = \mathcal{M}'$, we say that $\mathcal{M}'$ is *internally closed.* Note that internally closed sets of matchings are exactly the inclusionwise maximal internally stable sets. For an internally stable set $\mathcal{M}'' \supseteq \mathcal{M}'$, we say that $\mathcal{M}''$ is an *internal closure* of $\mathcal{M}'$ if $\mathcal{M}''$ is internally closed. Clearly, every internally stable set of matching admits an internal closure, which may not be unique. The following lemma gives more basic structural results.

Lemma 1. *Let T be an instance of the roommate problem, $\mathcal{M}' \subseteq \mathcal{M}(T)$, and $\widetilde{T} = \cup\{M | M \in \mathcal{M}'\}$. (a) $\mathcal{M}'$ is internally stable if and only if $\mathcal{M}' \subseteq \mathcal{S}(\widetilde{T})$; (b) If $\mathcal{M}'$ is internally closed, then $\mathcal{M}' = \mathcal{S}(\widetilde{T})$.*

By Lemma 1, part (b), we can succinctly represent any internally closed set of matchings $\mathcal{M}'$ via $\widetilde{T} \subseteq T$ such that $\mathcal{M}' = \mathcal{S}(\widetilde{T})$. A set $\mathcal{M}' \subseteq \mathcal{M}(T)$ is called *externally stable* (in T) if for each $M \in \mathcal{M}(T) \setminus \mathcal{M}'$, there exists $M' \in \mathcal{M}'$ that blocks M. A set that is both internally and externally stable is called *vNM stable* (in T). Note that any vNM stable set $\mathcal{M}'$ is necessarily internally closed, therefore by Lemma 1, part (b), we have $\mathcal{M}' = \mathcal{S}(\widetilde{T})$, where $\widetilde{T} = \cup\{M : M \in \mathcal{M}'\}$.

The Problems of Interest. We now present the algorithmic questions investigated in this paper. Marriage instances are a special case of roommate instances (see Sect. 3 for a definition). As previously discussed, the input to the first three problems below consists of a subtable $\widetilde{T} \subseteq T$, which implicitly describes the associated set of internally stable matchings $\mathcal{S}(\widetilde{T})$, see Lemma 1.

Find an internal closure of a given internally stable set, Marriage Case($\widetilde{T}, T$) (IStoIC-MC)

Given: A marriage instance T and a stable table $\widetilde{T}$ such that $\widetilde{T} \subseteq T$.
Find: $T' \subseteq T$ such that $\mathcal{S}(T')$ is an internal closure of $\mathcal{S}(\widetilde{T})$.

Check Internal Closedness($\widetilde{T}, T$) (CIC)

Given: A roommate instance T and a stable table $\widetilde{T}$ such that $\widetilde{T} \subseteq T$.
Decide: If $\mathcal{S}(\widetilde{T})$ is internally closed.

Check vNM Stability($\widetilde{T}, T$) (CvNMS)

Given: A roommate instance T and a stable table $\widetilde{T}$ such that $\widetilde{T} \subseteq T$.
Decide: If $\mathcal{S}(\widetilde{T})$ is vNM stable.

Find a vNM Stable Set(T) (FvNMS)

Given: A solvable roommate instance T.
Find: $T' \subseteq T$ such that $\mathcal{S}(T')$ is a vNM stable set, or conclude that no vNM stable set exists.

3 Internally Closed Sets: The Marriage Case

In a *marriage* instance [12], the set of agents A can be partitioned into two disjoint sets $X = \{x_1, x_2, \ldots, x_p\}$ and $Y = \{y_1, y_2, \cdots, y_\ell\}$, where the preference list of any agent in X consists only of a subset of agents in Y, and vice versa. All marriage instances are solvable [12]. An agent $x \in X$ is called an *X-agent*, and similarly an agent $y \in Y$ is called a *Y-agent.* In this section, we give a characterization of internally closed sets of matchings in the marriage case that relies on a generalization of the classical concept of rotations (see Theorem 4). We then use the characterization to show the following result.

Theorem 1. *Given a marriage instance with n agents, IStoIC-MC can be solved in time $O(n^4)$.*

Throughout this section, fix a marriage instance T.

3.1 Algebraic Structures Associated to Stable Matchings

Partial Order. We start by discussing known features of the poset of stable matchings obtained via an associated partial order (the poset is in fact a distributive lattice, but we will not use this fact explicitly in the exposition). For an extensive treatment of this topic, see [15]. Define the following domination relationship between matchings:

$$M \succeq_X M' \text{ if, for every } x \in X,\ M(x) \geq_x M'(x).$$

If, in addition, there exists at least one agent $x \in X$ such that $M(x) >_x M'(x)$ (or, equivalently, if $M' \neq M$), then we write $M \succ_X M'$. We symmetrically define the relations $M \succeq_Y M'$ and $M \succ_Y M'$. It is well-known that, for $Z \in \{X, Y\}$, there exists a matching $M_Z^T \in \mathcal{S}$ (or simply M_Z) such that $M_Z^T \succeq_Z M'$ for all $M' \in \mathcal{S}$. M_Z^T is called *Z-optimal.* Note that M_Z^T is, by definition, stable.

Classical Rotations and Properties. We sum up here definitions and known facts about the classical concept of rotations (first introduced in [20]).

Definition 1. *Let $M \in \mathcal{S}(T)$. Following [15], given distinct X-agents $x_0, \ldots, x_{r-1} \in X$ and Y-agents $y_0, \ldots, y_{r-1} \in Y$, we call the finite ordered list of ordered pairs*

$$\rho = (x_0, y_0), (x_1, y_1), \ldots, (x_{r-1}, y_{r-1}) \tag{1}$$

*a (*classical*)* X-rotation[1] *exposed in M if, for every $i \in [r-1]_0$:*

a) $x_i y_i \in M$; b) $x_i >_{y_{i+1}} x_{i+1}$; c) $y_i >_{x_i} y_{i+1}$;
d) $M(y) >_y x_i$ for all $y \in Y$ such that $x_i y \in T$ and $y_i >_{x_i} y >_{x_i} y_{i+1}$.

We abuse notation and write $x \in \rho$ (resp., $y \in \rho$) if $(x, y) \in \rho$ for some y (resp., for some x), and similarly we write $x \notin \rho$ (resp., $y \notin \rho$) if such y (resp., x) does not exist. The *elimination* of an X-rotation ρ exposed in a stable matching M maps M to $M' := M/\rho$ where $M(x) = M'(x)$ for $x \in X \backslash \rho$ and $M'(x_i) = y_{i+1}$ for all $i \in [r-1]_0$. Note that $M \succ_X M'$ and M' differs from M by a cyclic shift of each X-agent in ρ to the partner in M of the next X-agent in ρ. Rotations can be used to describe the set of stable matchings, as shown next.

Theorem 2. *There is exactly one set of X-rotations $R_X = \{\rho_1, \rho_2, \ldots, \rho_h\}$ such that, for $i \in [h]$, ρ_i is exposed in $((M_X/\rho_1)/\rho_2)\ldots/\rho_{i-1}$, and $M_Y = M_X/R_X = (((M_X/\rho_1)/\rho_2)/\ldots)/\rho_h$. Moreover, R_X is exactly the set of all X-rotations exposed in some stable matching of T.*

Extending the definition from the previous theorem, for $R = \{\rho_1, \rho_2, \ldots, \rho_k\} \subseteq R_X$, such that, for $i \in [k]$, ρ_i is exposed in $((M/\rho_1)/\rho_2)\ldots/\rho_{i-1}$, we let $M/R := (((M/\rho_1)/\rho_2)/\ldots)/\rho_k$. Define the poset $(R_X, \sqsupseteq)$ as follows: $\rho \sqsupseteq \rho'$ if

[1] We often omit "X-" and call ρ simply a rotation. Note that an X-rotation (and a generalized X-rotation, defined later) with r elements is equivalent up to a constant shift (modulo r) of all indices of its pairs. Hence, we will always assume that indices in entries of a (generalized) X-rotation are taken modulo r.

for any sequence of X-rotation eliminations $M_X/\rho_1/\dots/\rho_k$ with $\rho = \rho_k$, we have $\rho' \in \{\rho_0, \rho_1, \dots, \rho_k\}$. If moreover $\rho' \neq \rho$, we write $\rho \sqsupset \rho'$. When we want to stress the instance T we use to build the (po)set of X-rotations, we denote it by $R_X(T)$.

A set $R \subseteq R_X$ is called *closed* if $\rho \in R, \rho' \in R_X : \rho \sqsupseteq \rho' \Rightarrow \rho' \in R$. For $\rho \in R_X$, we let $R(\rho) = \{\rho' \in R_X : \rho \sqsupset \rho'\}$. Note that $R(\rho)$ is closed and does not include ρ. The following extension of Theorem 2 can be seen as a specialization of Birkhoff's representation theorem [4] to the lattice of stable matchings.

Theorem 3. *The following map defines a bijection between closed sets of rotations and stable matchings:* $R \subseteq R_X$, *R closed* $\rightarrow M_X/R$.

Generalized X-Rotations. We now introduce an extension of the classical concept of X-rotation, which we call *generalized X-rotation*, and define its associated digraph.

Definition 2. *Let $M \in \mathcal{M}(T)$. Given distinct X-agents $x_0, \dots, x_{r-1}$ and Y-agents $y_0, \dots, y_{r-1}$, we call the ordered set of ordered pairs*

$$\rho^g = (x_0, y_0), (x_1, y_1), \dots, (x_{r-1}, y_{r-1}) \tag{2}$$

a generalized X-rotation exposed in M *if, for every $i \in [r-1]_0$, it satisfies properties a)-b)-c) from Definition 1 (but not necessarily d). Often, we omit "X-" when clear from the context.*

Note that a classical X-rotation exposed in a stable matching M is also a generalized rotation exposed in M. We again write $x \in \rho^g$ (resp., $y \in \rho^g$) if $(x, y) \in \rho^g$ for some y (resp., for some x). The *elimination* of a generalized X-rotation, or of a set of generalized X-rotations, is defined analogously to the classical rotation case. For a (generalized) X-rotation ρ^g exposed at a matching M, we still have $M \succ_X M/\rho^g$. For a generalized X-rotation ρ^g as in (2), we let $E(\rho^g) = \{x_i y_i\}_{i \in [r-1]_0} \cup \{x_i y_{i+1}\}_{i \in [r-1]_0}$, and also interpret $E(\rho^g)$ as a subtable of T. Similarly, we interpret ρ^g as a subtable with edges $x_i y_i$ for $i \in [r-1]_0$.

The Generalized Rotation Digraphs. We define the following generalized X-rotation digraph for a matching M of T, denoted as $D_X(M,T)$ or simply $D_X(M)$ when T is clear from the context. The set of nodes is given by $X \cup Y$. For any agents $x \in X, y \in Y$, add arc (x, y) if $x >_y M(y)$ and $M(x) >_x y$; and add (y, x) if $M(y) = x$. Note that the outdegree of each X-agent can be larger than 1, but the outdegree of every Y-agent is at most 1, see Fig. 2 and Example 1. The next lemma follows directly from the definition of $D_X(M)$ and it is similar to a known statement for classical rotations, see, e.g., [15,20].

Lemma 2. *Let $M \in \mathcal{M}(T)$. $x_0 \rightarrow y_1 \rightarrow x_1 \rightarrow \dots \rightarrow y_0 \rightarrow x_0$ is a cycle in $D_X(M,T)$ if and only if $\rho^g = (x_0, y_0), (x_1, y_1), \dots, (x_{r-1}, y_{r-1})$ is a generalized X-rotation exposed in M. We say that ρ^g and the cycle* correspond *to each other.*

Classical and generalized Y-rotations are defined similarly to classical and generalized X-rotations, with the role of agents in X and Y swapped. By symmetry, all definitions and properties carry over.

Example 1. Consider the instance T in Fig. 2, left, and its stable matching $M = \{x_1y_4, x_2y_2, x_3y_3, x_4y_1\}$. The generalized rotation digraph $D_X(M)$ is given in Fig. 2, center. $P = x_1 \to y_3 \to x_3 \to y_2 \to x_2 \to y_1 \to x_4 \to y_4 \to x_1$ is a cycle in $D_X(M)$. Cycle P corresponds to a generalized X-rotation $\rho = (x_1, y_4), (x_3, y_3), (x_2, y_2), (x_4, y_1)$, which is exposed in M (wrt table T). One can check that ρ satisfies properties a)-b)-c) but not d) from Definition 1. On the other hand, $\rho_1 = (x_1, y_4), (x_4, y_1)$ satisfies properties a)-b)-c)-d) from Definition 1, hence it is an X-rotation exposed in M, corresponding to cycle $x_1 \to y_1 \to x_4 \to y_4 \to x_1$ of $D_X(M)$.

$T =$

Agent	Preference list
x_1	$y_4\ y_1\ y_3$
x_2	$y_2\ y_1$
x_3	$y_3\ y_2$
x_4	$y_1\ y_4$
y_1	$x_2\ x_1\ x_4$
y_2	$x_3\ x_2$
y_3	$x_1\ x_3$
y_4	$x_4\ x_1$

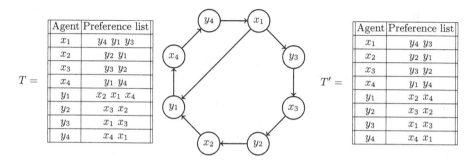

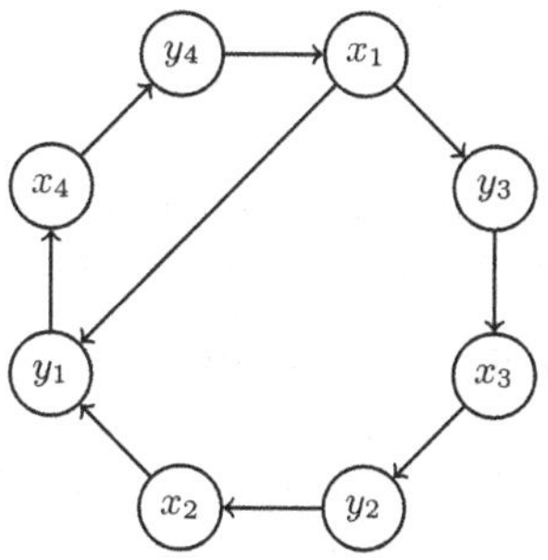

$T' =$

Agent	Preference list
x_1	$y_4\ y_3$
x_2	$y_2\ y_1$
x_3	$y_3\ y_2$
x_4	$y_1\ y_4$
y_1	$x_2\ x_4$
y_2	$x_3\ x_2$
y_3	$x_1\ x_3$
y_4	$x_4\ x_1$

Fig. 2. Illustrations from Example 1. From left to right: T, $D_X(M)$, T'.

Dissection of a Rotation. A new, key concept for proving our characterization of internally closed sets is that of dissecting set for a rotation.

Dissecting a Rotation(T, T', ρ) (DR)

Given: Marriage instances $T' \subseteq T$ with T' stable, and $\rho \in R_X(T')$.
Find: A set $R = \{\rho_1, \rho_2, \dots, \rho_k\}$ satisfying a) $R \subseteq R_X(T^*) \setminus R_X(T')$ with $T^* = T' \cup_{j=1}^k E(\rho_j)$; b) $M_X^{T'}/R(\rho)/\rho = M_X^{T'}/R(\rho)/R$; or output $\{\rho\}$ if R as above does not exist.

If DR(T, T', ρ) outputs a set $R \neq \{\rho\}$, then R is called a *dissecting set* for (T, T', ρ). Note that a dissecting set has at least two elements.

Example 2. Consider again the instance from Example 1. T has 3 inclusionwise maximal matchings: M, $M_1 = \{x_1y_1, x_2y_2, x_3y_3, x_4y_4\}$, $M_2 = \{x_1y_3, x_2y_1, x_3y_2, x_4y_4\}$. Let $T' = M \cup E(\rho)$. M is the X-optimal matching within T' and ρ is the only (classical) X-rotation exposed in M in T'. Note that we have $M_2 = M/\rho$. Let $T^* = T = T' \cup E(\rho_1) \cup E(\rho_2)$, with ρ_1 as in Example 1 and $\rho_2 = (x_1, y_1), (x_3, y_3), (x_2, y_2)$. M is also the X-optimal matching within T. Note that $\rho_1 \in R_X(T^*)$ is exposed in M, $\rho_2 \in R_X(T^*)$ is exposed in M/ρ_1, and $M_2 = M/\rho_1/\rho_2$. Yet, $\rho_1, \rho_2 \notin R_X(T')$. Hence, $\{\rho_1, \rho_2\}$ is a dissecting set for (T, T', ρ). When enlarging T' to T^*, the set of stable matchings becomes larger, and ρ is dissected into ρ_1 and ρ_2, increasing the size of the poset of rotations.

In Example 2, the poset of rotations in T^* is an "expansion" of the poset in T'. We next show that whenever a rotation is dissected, a similar phenomenon happens.

Lemma 3 (Internal expansion of the rotation poset via rotation dissection). *Let $T' \subseteq T$ with T' stable. Let $\rho \in R_X(T')$ and $R = \{\rho_1, \ldots, \rho_k\}$ be a dissecting set for (T, T', ρ). Then a) $T^* = T' \cup_{j=1}^k \rho_j$ is a stable table and b) $R_X(T^*) = (R_X(T') \setminus \{\rho\}) \cup \{\rho_1, \ldots, \rho_k\}$.*

Generalized Rotations Exposed in the Y-Optimal Stable Matching. We next show that an internally stable set of matchings can be "vertically expanded" by adding to a stable table the edges from a generalized X- (resp., Y-)rotation exposed in the Y- (resp., X-)optimal stable matching. The next lemma is stated for X-rotations, but by symmetry extends to Y-rotations.

Lemma 4 (Vertical expansion of the rotation poset). *Let T be a marriage instance and $T' \subseteq T$, with T' stable. Let $M_Y = M_Y^{T'}$ and suppose ρ^g is a generalized X-rotation corresponding to a cycle of $D_X(M_Y, T)$. Let $M^* = M_Y/\rho^g$ and $\overline{T} = T' \cup E(\rho^g) = T' \cup M^*$. Then a) $M^* \notin \mathcal{S}(T')$, b) the set $\{M^*\} \cup \mathcal{S}(T')$ is internally stable, c) $\overline{T}$ is a stable table, d) M^* is the Y-optimal stable matching in $\mathcal{S}(\overline{T})$, and e) $R_X(\overline{T}) = R_X(T') \cup \{\rho^g\}$.*

3.2 Characterization of Internally Closed Sets of Matchings

We now have all ingredients to state a characterization of internally closed sets of matchings based on generalized rotations and on rotation dissections.

Theorem 4. *Let T be a marriage instance and $\mathcal{M}' \subseteq \mathcal{M}(T)$. $\mathcal{M}'$ is internally closed if and only if $\mathcal{M}' = \mathcal{S}(T')$ for a stable subtable $T' \subseteq T$ such that:*

a) (no rotation can be dissected) DR(T, T', ρ) returns $\{\rho\}$ for all $\rho \in R_X(T')$.
b) (no vertical expansion) $D_X(M_Y^{T'}, T)$ and $D_Y(M_X^{T'}, T)$ have no cycles.

We can now give a sketch of the proof of how to solve (IStoIC-MC) and prove Theorem 1. Starting from input $\widetilde{T}$, we iteratively check if conditions a) and b) from Theorem 4 are verified and, if not, rely on Lemma 3 and Lemma 4 to create internally stable tables strictly containing $\widetilde{T}$. While Lemma 3 does not explain how to enlarge the current subtable of T and the rotation poset, we can perform all operations efficiently by using properties of the stable matchings lattice and of (generalized) rotations.

4 Internally Closed and vNM Stable Sets: The Roommate Case

In this section, we deal with roommate instances. Our major structural contribution is an extension of the concept of rotations which we call *stitched rotations.* We show that the set of stable matchings of a solvable roommate instance is internally closed if and only if a stitched rotation does not exist (Theorem 5). The complexity of finding stitched rotations then allow us to deduce our hardness results, see Theorem 6.

Rotations. A major difference between the marriage and the roommate problem is the absence of a (known) relevant lattice structure in the latter. This change calls for a different concept of rotation (first defined in [18]) for analyzing the problem. Throughout the section, assume that T is a solvable roommate instance.

Definition 3. *Let $T' \subseteq T$. A sequence $\rho = (x_0, y_0), (x_1, y_1), ..., (x_{r-1}, y_{r-1})$ is called a* rotation exposed *in T' if $y_i = f_{T'}(x_i)$ and $y_{i+1} = s_{T'}(x_i)$ for all $i \in [r-1]_0$, where indices are taken modulo r. T'/ρ denotes the table obtained from T' by deleting all pairs $y_i z$ such that y_i strictly prefers x_{i-1} to z. We refer to this process as to the* elimination *of ρ in T. Recall that $f_{T'}(x)$ gives the first preference of x within table T', and $s_{T'}(x)$ gives the second preference.*

Let $\mathcal{Z}(T)$ denote the set of all rotations exposed in some table obtained from T by iteratively eliminating exposed rotations. Throughout the rest of the section, we fix a rotation $\rho = (x_0, y_0), (x_1, y_1), ..., (x_{r-1}, y_{r-1}) \in \mathcal{Z}(T)$ and again we also interpret ρ as a table with entries $x_i y_i$ for $i \in [r-1]_0$. Rotations can be further classified into two categories: *singular* and *non-singular.*

Definition 4. *ρ is called a* non-singular *rotation if $\overline{\rho} \in \mathcal{Z}(T)$, where*

$$\overline{\rho} = (y_1, x_0), (y_2, x_1), ..., (y_i, x_{i-1}), ..., (y_0, x_{r-1}), \tag{3}$$

and it is called singular *otherwise. The subset of $\mathcal{Z}(T)$ containing all singular (resp., non-singular) rotations is denoted by $\mathcal{Z}_s(T)$ (resp., $\mathcal{Z}_{ns}(T)$).*

Antipodal Edges and Stitched Rotations. Let T^* be the subtable of T containing all and only its stable edges, i.e., $T^* = E_S(T)$. We first argue that to "expand" $\mathcal{S}(T^*)$ to a strictly larger internally stable set of matchings, i.e., to find a stable table $T' \supsetneq T^*$ such that $\mathcal{S}(T') \supsetneq \mathcal{S}(T^*)$, any edge $e \in T' \backslash T^*$ must satisfy what we call the *antipodal* condition, defined below. We then introduce stitched rotations, a new object that allows us to assemble antipodal edges and expand the set of stable matchings to a larger internally stable set.

Definition 5. *We say that an edge $e = xy \in T \backslash T^*$ satisfies the antipodal condition (wrt T^*) if exactly one of the following is true:*

$$y >_x f_{T^*}(x) \, and \, x <_y \ell_{T^*}(y), \quad or \quad x >_y f_{T^*}(y) \, and \, y <_x \ell_{T^*}(y).$$

Lemma 5. *Let $e \in T \backslash T^*$. Assume that e is not an antipodal edge. Then every $M \in \mathcal{M}(T)$ with $e \in M$ is blocked by some edge from T^*.*

Definition 6. *Suppose $\rho' \in \mathcal{Z}(T^* \cup \rho')$ is exposed in $T^* \cup E(\rho')$ and, for all $i \in [r-1]_0$, $x_i y_i$ satisfies the antipodal condition with respect to T^*. If $\rho' \in \mathcal{Z}_{ns}(T^* \cup \rho')$, we call ρ' a* stitched rotation *with respect to T^*.*

The next theorem shows that the stable subtable T^* of a solvable instance is internally closed if and only if there is no stitched rotation w.r.t. T^*.

Theorem 5 (Expansion of internally stable sets via stitched rotations). *1. Let $M \in \mathcal{M}(T) \backslash \mathcal{S}(T^*)$ and assume that $\{M\} \cup \mathcal{S}(T^*)$ is internally stable. Then there exists a stitched rotation w.r.t. T^*. 2. Conversely, if ρ' is a stitched rotation w.r.t. T^*, there exists $M \in \mathcal{M}(T) \backslash \mathcal{S}(T^*)$ with $\rho' \subseteq M$ so that $\{M\} \cup \mathcal{S}(T^*)$ is internally stable.*

Hardness Results. For each instance ϕ of 3-SAT, we create an instance T with stable subtable $\widetilde{T}$ such that $\mathcal{S}(T) = \mathcal{S}(\widetilde{T})$. Moreover, there exists a satisfiable assignment for ϕ if and only if there exists a rotation that is stitched w.r.t. $\widetilde{T}$. Together with Theorem 5, this reduction and variations of it lead to the following (the details of these constructions are postponed to the journal version).

Theorem 6. *`CIC`, `CvNMS`, and `FvNMS` are co-NP-hard.*

Acknowledgments. Yuri Faenza acknowledges the support of the NSF Grant 2046146 *CAREER: An Algorithmic Theory of Matching Markets.* Cliff Stein is supported in part by NSF grant CCF-2218677, ONR grant ONR-13533312, and by the Wai T. Chang Chair in Industrial Engineering and Operations Research.

References

1. Abdulkadiroğlu, A., Sönmez, T.: School choice: a mechanism design approach. Am. Econ. Rev. **93**(3), 729–747 (2003)
2. Abraham, D.J., Cechlárová, K., Manlove, D.F., Mehlhorn, K.: Pareto optimality in house allocation problems. In: Fleischer, R., Trippen, G. (eds.) ISAAC 2004. LNCS, vol. 3341, pp. 3–15. Springer, Heidelberg (2004). https://doi.org/10.1007/978-3-540-30551-4_3
3. Abraham, D.J., Levavi, A., Manlove, D.F., O'Malley, G.: The stable roommates problem with globally-ranked pairs. In: Deng, X., Graham, F.C. (eds.) WINE 2007. LNCS, vol. 4858, pp. 431–444. Springer, Heidelberg (2007). https://doi.org/10.1007/978-3-540-77105-0_48
4. Birkhoff, G.: Rings of sets. Duke Math. J. **3**(3), 443–454 (1937)
5. Cseh, Á., Faenza, Y., Kavitha, T., Powers, V.: Understanding popular matchings via stable matchings. SIAM J. Discrete Math. **36**(1), 188–213 (2022)
6. Ehlers, L.: Von Neumann-Morgenstern stable sets in matching problems. Journal of Economic Theory **134**(1), 537–547 (2007)
7. Ehlers, L., Morrill, T.: (Il) legal assignments in school choice. Rev. Econ. Stud. **87**(4), 1837–1875 (2020)

8. Faenza, Y., Kavitha, T.: Quasi-popular matchings, optimality, and extended formulations. Math. Oper. Res. **47**(1), 427–457 (2022)
9. Faenza, Y., Kavitha, T., Powers, V., Zhang, X.: Popular matchings and limits to tractability. In: Proceedings of the Thirtieth Annual ACM-SIAM Symposium on Discrete Algorithms, pp. 2790–2809. SIAM (2019)
10. Faenza, Y., Zhang, X.: Legal assignments and fast EADAM with consent via classical theory of stable matchings. Oper. Res. **70**(3), 1873–1890 (2022)
11. Faenza, Y., Zhang, X.: Affinely representable lattices, stable matchings, and choice functions. Math. Program. **197**(2), 721–760 (2023)
12. Gale, D., Shapley, L.S.: College admissions and the stability of marriage. Am. Math. Mon. **69**(1), 9–15 (1962)
13. Gärdenfors, P.: Match making: assignments based on bilateral preferences. Behav. Sci. **20**(3), 166–173 (1975)
14. Gusfield, D.: The structure of the stable roommate problem: efficient representation and enumeration of all stable assignments. SIAM J. Comput. **17**(4), 742–769 (1988)
15. Gusfield, D., Irving, R.W.: The Stable Marriage Problem: Structure and Algorithms. MIT Press, Cambridge (1989)
16. Hitsch, G.J., Hortaçsu, A., Ariely, D.: Matching and sorting in online dating. Am. Econ. Rev. **100**(1), 130–63 (2010)
17. Huang, C.C., Kavitha, T.: Popularity, mixed matchings, and self-duality. Math. Oper. Res. **46**(2), 405–427 (2021)
18. Irving, R.W.: An efficient algorithm for the "stable roommates" problem. J. Algorithms **6**(4), 577–595 (1985)
19. Irving, R.W., Kavitha, T., Mehlhorn, K., Michail, D., Paluch, K.E.: Rank-maximal matchings. ACM Trans. Algorithms (TALG) **2**(4), 602–610 (2006)
20. Irving, R.W., Leather, P.: The complexity of counting stable marriages. SIAM J. Comput. **15**(3), 655–667 (1986)
21. Kavitha, T.: A size-popularity tradeoff in the stable marriage problem. SIAM J. Comput. **43**(1), 52–71 (2014)
22. Lucas, W.F.: von Neumann-Morgenstern stable sets. In: Handbook of Game Theory with Economic Applications, vol. 1, pp. 543–590 (1992)
23. Manlove, D.: Algorithmics of Matching Under Preferences, vol. 2. World Scientific (2013)
24. Roth, A.E.: Stability and polarization of interests in job matching. Econometrica J. Econometric Soc. **52**(1), 47–57 (1984)
25. Roth, A.E., Sotomayor, M.: Two-sided matching. In: Handbook of Game Theory with Economic Applications, vol. 1, pp. 485–541 (1992)
26. Shapley, L., Scarf, H.: On cores and indivisibility. J. Math. Econ. **1**(1), 23–37 (1974)
27. Shubik, M.: Game Theory in the Social Sciences: Concepts and Solutions (1982)
28. Von Neumann, J., Morgenstern, O.: Theory of Games and Economic Behavior. Princeton University Press, Princeton (2007)
29. Wako, J.: A polynomial-time algorithm to find von Neumann-Morgenstern stable matchings in marriage games. Algorithmica **58**(1), 188–220 (2010)
30. Wang, X., Agatz, N., Erera, A.: Stable matching for dynamic ride-sharing systems. Transp. Sci. **52**(4), 850–867 (2018)

Fully-Dynamic Load Balancing

Ayoub Foussoul[1(✉)], Vineet Goyal[1], and Amit Kumar[2]

[1] IEOR, Columbia University, New York, USA
{af3209,vg2277}@columbia.edu
[2] Department of Computer Science and Engineering, IIT Delhi, New Delhi 110016, India
amit.kumar@cse.iitd.ac.in

Abstract. We study the classical load balancing problem in a fully dynamic setting where jobs both arrive and depart. Each job can only be assigned to a subset of machines and can be reassigned at any time step. The goal is to maintain a near-optimal maximum load at all time steps with a small total number of reassignments. We consider the setting where the degree of the jobs (number of machines they can be assigned to) is bounded. This is motivated from natural settings where jobs can only be locally assigned to a small number of machines (e.g., bike sharing [10], map-reduce settings [21]) and generalizes the classical EDGE-ORIENTATION problem. We give a constant competitive algorithm with amortized constant number of reassignments. We also consider the generalizations of our problem to arbitrary reassignment costs and arbitrary job sizes. The generalizations require different techniques and we give a different randomized algorithm for these.

1 Introduction

We consider the classical online load balancing problem in the fully dynamic setting where jobs both arrive and depart in an online manner. More specifically, we are given a set of m machines. At each time t, a new job may arrive or an existing job may terminate or depart. Note that the arrival or departure time of a job is only known when the corresponding event happens. When a job j arrives, it also specifies a subset $S(j) \subseteq [m]$ of machines to which it can be assigned. An algorithm assigns each arriving job j to a machine in $S(j)$. The load on a machine at a time t is defined as the number of jobs assigned to it at time t. Our goal is to minimize the maximum load on any machine. The competitive ratio of an algorithm is defined as the maximum over all inputs and over all time steps t of the ratio of the maximum load on a machine at time t and the maximum load of an optimal assignment at time t (that minimizes the maximum load in the assignment restricted to the jobs alive at time t in the input). We refer to this problem as DYNLOADBALANCE.

Observe that once a job is assigned to a machine, it cannot be reassigned to another machine. In this model, Azar et al. [2] showed that the competitive ratio of any randomized algorithm is $\Omega(\sqrt{m})$. Azar et al. [3] gave a matching upper bound. There has also been work on models that allow jobs to be reassigned –

J. Vygen and J. Byrka (Eds.): IPCO 2024, LNCS 14679, pp. 182–195, 2024.
https://doi.org/10.1007/978-3-031-59835-7_14

the main challenge here is to understand the trade-off between the extent of reassignment and competitive ratio. Phillips and Westbrook [18] gave an algorithm that, for every constant $\rho \geq 1$, has competitive ratio $^{12 \log m}/_{\rho}$, while performing at most ρT reassignments of jobs over T arrivals and departures, i.e., at most ρ reassignment amortized over each arrival or departure of a job. Their algorithm does no reassignments unless a job departs from the system and hence does not have a competitive ratio better than $\Omega(\log m)$ [4]. In the special case when the machines are identical (every job can be assigned to every machine) Andrews et al. [1] gave an algorithm with competitive ratio 3.5981 performing at most $6.8285 \cdot T$ reassignments over a sequence of T arrivals and departures. In the more general setting of unrelated machines, Bhattacharya et al. [5] gave a $(2+\epsilon)$-competitive deterministic algorithm with $O_\epsilon((\log T \log d) \cdot (\mathsf{OptR} + T))$ total reassignments, where d denotes the maximum degree of a job (i.e. maximum number of machines to which a job can be assigned), and OptR denotes the minimum number of reassignments required by any off-line algorithm to maintain an optimal assignment at all time steps over T job arrivals and departures. They also gave a randomized algorithm maintaining an $O(\frac{\log \log T}{\log \log \log T})$-competitive assignment in expectation with $O(\log d \cdot (\mathsf{OptR} + T))$ expected reassignments.

In the so-called "arrival only" model, where jobs do not depart after arrival, Azar et al. [4] gave an $O(\log m)$-competitive algorithm and show that this is the best possible ratio that can be achieved by any randomized algorithm. When reassignments are allowed, Gupta et al. [14] show that one can maintain a 2-competitive assignment with $O(T)$ reassignments over T job arrivals. We note that the algorithm of [14] can be shown to be 2-competitive with a total of $O(T \log T)$ reassignments when jobs depart as well, and that this is the best possible recourse bound this algorithm can achieve (refer to the full version for more details on this). For identical machines, the best competitive ratio is 1.92 [9] with a lower bound of 1.88 [19]. When reassignments are allowed, Sanders et al. [20] show that a competitive ratio of 1.5 can be achieved using an amortized 2 reassignments per job. In the more general setting of unrelated machines, Krishnaswamy et al. [16] gave an $(2+\epsilon)$-competitive algorithm with $O_\epsilon(T \log T)$ total reassignments.

Another interesting special case is the EdgeOrientation problem. In this problem, edges over a given set of vertices arrive (or depart) online. Whenever an edge arrives, the algorithm directs the edge, and the goal is to minimize the maximum in-degree of a vertex. This is a special case of DynLoadBalance where each edge corresponds to a job that can be assigned to only two machines, namely the ones corresponding to the two end-points of this edge. If reassignments are not allowed Grove et al. [11] show a logarithmic lower bound of the competitive ratio. When reassignments are allowed, Brodal and Fagerberg [7] gave a constant competitive algorithm with $O(\mathsf{OptR}+T)$ reassignments, where OptR denotes the minimum number of reassignments required by any off-line algorithm to maintain an optimal assignment at all time steps over T job arrivals and departures. However no such result is known when the degree of each job j, i.e., the size of the set $S(j)$, is bounded by a constant (larger than 2). This is the main focus of this paper.

Many other optimization problems have been considered in a fully dynamic setting with reassignments including bin packing [8], set cover [13], graph coloring [6] and Steiner tree [12,15,17].

In this paper, we consider the DynLoadBalance problem with reassignments. In particular, our algorithm maintain a constant competitive assignment with $O(\mathsf{OptR} + T)$ reassignments over T arrivals and departures when the jobs are of bounded degree. Here OptR denotes the minimum number of reassignments required to maintain an optimal assignment off-line. As noted above, this generalizes the known results for the EdgeOrientation problem. Our study is motivated by natural settings where machines can be spread over a geographically large area, and hence each job can be locally assigned to a small number of machines only (e.g., bike sharing [10], map-reduce settings [21]).

1.1 Our Results and Techniques

We now give details of our contributions. We consider the DynLoadBalance problem with reassignments where the degree of each job is bounded by d. There are two parameters of interest: (i) the maximum load on any machine, and (ii) the number of reassignments. Given an instance $\mathcal{I}$, let $\mathsf{OptL}_t(\mathcal{I})$ be the optimal maximum load on any machine when the set of jobs is given by the jobs alive at time t. Let $\mathsf{OptR}(\mathcal{I})$ denote the minimum number of reassignments done by any off-line algorithm on the instance $\mathcal{I}$ that maintains maximum load $\mathsf{OptL}_t(\mathcal{I})$ at each time t. Our first result is as follows:

Theorem 1. *There exists an algorithm that, for any instance $\mathcal{I}$ of* DynLoadBalance *with job degrees bounded by d, is $O(d^2)$-competitive with a total number of reassignments $O(d^2(\mathsf{OptR}(\mathcal{I}) + T))$.*

Our algorithm maintains a weight for each job-machine pair (j, i), where $i \in S(j)$. The weights of these pairs corresponding to job j shall be a permutation of $\{1, 2, \ldots, \mathsf{deg}(j)\}$, where $\mathsf{deg}(j) := |S(j)|$ denotes the degree of job j. Now, consider an (off-line) sequence of assignments $(\sigma_t^*)_{t\in[T]}$ that maintains maximum load $\mathsf{OptL}_t(\mathcal{I})$ at time t and has $\mathsf{OptR}(\mathcal{I})$ job reassignments. The weight of the pair (j, i) at time t shall represent "how confident we are" that job j is assigned to machine i by σ_t^*, with smaller weight signifying higher confidence. The algorithm uses the current weights to assign jobs to machines (using a suitable low weight assignment) until it becomes clear that some of our weights are wrong (i.e., when the load on a subset of machines becomes high). In this case, the jobs on high load machines are reassigned and their corresponding weights are updated. The analysis of this algorithm relies on a novel potential function that intuitively represents "how correct our weights are". The bound follows from linking the gain in potential (intuitively the increase in correctness of our weights) to the number of reassignments.

Next, we consider the generalization of the DynLoadBalance problem to the setting where jobs can have varying reassignment costs, i.e., reassigning a job $j \in J$ incurs a costs c_j. We refer to this problem as GCDynLoadBalance.

Given an instance $\mathcal{I}$ of GCDYNLOADBALANCE, we can define $\mathsf{OptL}_t(\mathcal{I})$ in an analogous manner: it is the optimal maximum load on a machine when the set of jobs is given by the jobs alive at time t. Similarly, let $\mathsf{OptR}(\mathcal{I})$ denote the minimum reassignment cost incurred by any off-line algorithm on the instance $\mathcal{I}$ that maintains maximum load $\mathsf{OptL}_t(\mathcal{I})$ at each time t. We show the following:

Theorem 2. *Given an instance $\mathcal{I}$ of* GCDYNLOADBALANCE *with job degrees bounded by d, there exists a randomized $O(d)$-competitive algorithm with total expected reassignment cost $O(d(\mathsf{OptR}(\mathcal{I}) + \sum_{j\in J} c_j))$.*

Our techniques for DYNLOADBALANCE do not extend to arbitrary reassignment costs. In particular, consider an off-line sequence of assignments $(\sigma_t^*)_{t\in[T]}$ that maintains load $\mathsf{OptL}_t(\mathcal{I})$ at time t with total reassignment cost $\mathsf{OptR}(\mathcal{I})$. As in the unit cost setting, suppose we maintain a weight for each job-machine pair (j, i) representing "how confident we are" that job j is assigned to machine i in σ^*. Suppose the load on a subset of machines becomes high. Consequently, the jobs on these machines will need to be reassigned and the weights of corresponding pairs need to be updated. Let B be the set of jobs on these machines whose assignment is the same as in the optimal sequence σ^*. Updating the edge weights corresponding to these jobs decreases the correctness of our weights. Let G be the set of jobs assigned to a different machine in σ^*. Updating the edge weights for such jobs increases the correctness of our weights. Now it might happen that the jobs in B have much larger cost than those in G, and in such a setting, the net gain (in correctness of our weights) of the algorithm by such weight updates may be negative.

Instead, we develop a different randomized approach for GCDYNLOADBALANCE. In particular, instead of reassigning all the jobs on high load machines, we will reassign one job at a time until no machine has high load. More precisely, fix a time step t and suppose there exists a machine i of load larger than $\Theta(d) \cdot \mathsf{OptL}_t(\mathcal{I})$. We select one of the jobs j assigned to i where jobs with higher costs are selected with smaller probability and reassign j to one of the machines in $S(j)$ chosen uniformly at random. We repeat this random process until no more machines have load larger than $\Theta(d) \cdot \mathsf{OptL}_t(\mathcal{I})$. Intuitively, since the optimal assignments $(\sigma_t^*)_t$ have a maximum load $\mathsf{OptL}_t(\mathcal{I})$, it must be the case that a large fraction of the jobs on machine i are assigned to a machine different from i in σ_t^*. Therefore, there is a high probability of picking one of these jobs and changing its assignment and hence getting closer to the optimal assignment σ_t^*. We consider a potential function that represents "how close we are from the optimal assignment σ^*" and link the expected increase in potential during the random reassignment process to the number of reassignments made in the process, allowing us to prove our claimed bounds.

We note that our randomized algorithm achieves better bounds than our deterministic algorithm (Theorem 1) for uniform reassignment costs. However, the bound on the number of reassignments holds only in expectation in our randomized algorithm as opposed to worst case in the deterministic one.

Finally, we consider the generalization of the DYNLOADBALANCE problem to arbitrary job sizes where each job j has size p_j. The load on a machine i is now given by the sum of the sizes of the jobs assigned to i. We refer to this problem as GSDynLoadBalance. Given an instance $\mathcal{I}$ of GSDynLoadBalance, define $\mathsf{OptL}_t(\mathcal{I})$ and $\mathsf{OptR}(\mathcal{I})$ in an analogous manner. We get the following result:

Theorem 3. *Given an instance $\mathcal{I}$ of* GSDynLoadBalance *with job degrees bounded by d, there exists a randomized $O(d)$-competitive algorithm with total expected number of reassignment $O(d(\mathsf{OptR}(\mathcal{I}) + T))$.*

Our algorithm for GSDynLoadBalance is similar to that for GCDynLoadBalance, though some more technical details are needed in the analysis. In particular, given a high load machine i, we now pick and reassign a random job j from machine i with probability proportional to its size. As the total size of the jobs assigned to machine i in an optimal solution σ^* is relatively small, this implies a high probability of picking and reassigning a job j which is not assigned to i in σ^* and hence getting closer to σ^*.

It is worth noting that if we allow an off-line algorithm to maintain a maximum load slightly higher than the optimal value $\mathsf{OptL}_t(\mathcal{I})$ at each time t, then the number/cost of reassignments can decrease significantly (refer to the full version for more details on this). Let $\mathsf{OptR}(\alpha, \mathcal{I})$ denote the minimum number/cost of reassignments performed by an off-line algorithm that maintains maximum load at most $\alpha \cdot \mathsf{OptL}_t(\mathcal{I})$ at each time t where $\alpha \geq 1$. Then our algorithms and analysis can be extended seamlessly to give an $O(\alpha d^2)$-competitive algorithm with $O(d^2(\mathsf{OptR}(\alpha, \mathcal{I}) + T))$ reassignments for the unit costs/sizes case, $O(\alpha d)$-competitive with expected reassignment cost $O(d(\mathsf{OptR}(\alpha, \mathcal{I}) + \sum_j c_j))$ for arbitrary costs, and $O(\alpha d)$-competitive with $O(d(\mathsf{OptR}(\alpha, \mathcal{I}) + T))$ expected reassignments for arbitrary sizes.

2 Preliminaries

We define an instance $\mathcal{I}$ of the DynLoadBalance problem. We are given a set of m machines and a set J of n jobs arriving and departing online over a time horizon of length T. Note that T here denotes the total number of arrivals and departures and hence $[T] = [2n]$ (for a positive integer a, let $[a]$ denote the set $\{1, 2, \ldots, a\}$). Each job j has an arrival time r_j and a departure time d_j which are revealed on arrival and departure of the job respectively. Further, each job j specifies, on arrival, a subset $S(j)$ of machines to which it can be assigned. Let J_t denote the set of jobs present at the end of time step $t \in [T]$, i.e., $J_t := \{j \in J : t \in [r_j, d_j)\}$.

At every time $t \in [T]$, the algorithm maintains an assignment $\sigma_t : J_t \to [m]$ of the jobs in J_t to the set of machines, where $\sigma_t(j) \in S(j)$ for each job j. Note that the assignment of a particular job may change over time. There are two parameters of interest: the total *reassignment cost* and the *maximum load* on a machine at each time t.

For each time t, and an assignment $\psi : J_t \to [m]$, define $\mathsf{Load}_i(\psi)$, the load on a machine i, as the number of jobs assigned to machine i, i.e., $|\{j : \psi(j) = i\}|$.

Let

$$\mathsf{Load}(\psi) := \max_{i\in[m]} \mathsf{Load}_i(\psi).$$

Finally, let

$$\mathsf{OptL}_t(\mathcal{I})$$

denote the minimum over all assignments $\psi : J_t \to [m]$ of $\mathsf{Load}(\psi)$. We say that an assignment $\psi : J_t \to [m]$ (resp., sequence of assignments $(\psi_t : J_t \to [m])_{t\in[T]}$) is *$\alpha$-approximate* if $\mathsf{Load}(\psi) \le \alpha \cdot \mathsf{OptL}_t(\mathcal{I})$ (resp., $\mathsf{Load}(\psi_t) \le \alpha \cdot \mathsf{OptL}_t(\mathcal{I})$ for all $t \in [T]$). Given a sequence of assignments $(\psi_t)_{t\in[T]}$, the total number of reassignments performed by the sequence is given by

$$\sum_{t=1}^{T-1} \sum_{j\in J_t\cap J_{t+1}} \mathbf{1}_{\psi_t(j)\neq\psi_{t+1}(j)}.$$

Let $(\sigma_t^*)_{t\in[T]}$ denote a sequence of optimal (1-approximate) assignments minimizing the number of reassignments among all sequences of optimal assignments. Let

$$\mathsf{OptR}(\mathcal{I})$$

denote the number of reassignments performed by $(\sigma_t^*)_{t\in[T]}$.

Objective. Let $\mathcal{I}$ be an instance of DynLoadBalance. Given two parameters $\beta \ge 1$ and $\gamma \ge 1$, we say that a solution $(\psi_t)_{t\in[T]}$ to $\mathcal{I}$ is (β,γ)-competitive if the following two conditions hold:

(i) $\mathsf{Load}(\psi_t) \le \beta \cdot \mathsf{OptL}_t(\mathcal{I})$ for all $t \in [T]$.
(ii) The solution $(\psi_t)_{t\in[T]}$ perform at most $\gamma(\mathsf{OptR}(\mathcal{I}) + T)$ reassignments.

In other words, a (β,γ)-competitive solution maintains a β-approximate assignment online and performs a number of reassignments $\gamma\mathsf{OptR}(\mathcal{I}) + \gamma T$ which is within a multiplicative γ compared to the minimum number of reassignments required off-line $\mathsf{OptR}(\mathcal{I})$ plus an additive γT which is, in amortized time, γ reassignments per job.

3 Our Algorithm

We now give our algorithm for the DynLoadBalance problem. We refer to this algorithm as LearnWeights. Let $d = \max_{j\in J} \mathsf{deg}(j)$, where $\mathsf{deg}(j) := |S(j)|$ denotes the degree of job j. We assume without loss of generality that $d \ge 2$ (otherwise the problem is trivial).

3.1 Algorithm LearnWeights

Let $\beta := 2d^2$. At every time t, our algorithm maintains a weighted bipartite graph $H^{(t)}$. The vertex set is given by $J_t \cup [m]$, i.e., the left side contains all the active jobs at time t and the right side has one vertex for every machine.

There is an edge (j, i), where $j \in J_t, i \in [m]$ iff $i \in S(j)$. Every edge e has a weight $w_e^{(t)}$. We shall maintain the invariant that at any time t and for any job $j \in J_t$, the set of weights incident to j at time t is equal to the set $[\mathsf{deg}(j)]$, i.e., if we consider the edges incident to j in some order, then the sequence of the corresponding edge weights is a permutation of $[\mathsf{deg}(j)]$. We now specify details of the algorithm when a job arrives or departs. Unless specified otherwise, we shall assume that $w_e^{(t)} = w_e^{(t-1)}$ (resp. $\sigma_t(j) = \sigma_{t-1}(j)$) for an edge e (resp. job j) which appears in both $H^{(t)}$ and $H^{(t-1)}$.

Arrival: when a new job j arrives at a time t, before assigning job j, call procedure REASSIGN(t) described in Algorithm 1 (this procedure makes sure that the load on any machine does not exceed $\beta \cdot \mathsf{OptL}_t(\mathcal{I})$), then assign (arbitrarily) the weights $1, 2, \ldots, \mathsf{deg}(j)$ to the edges incident with j in $H^{(t)}$, and finally assign the job j to the machine i such that $w_{(j,i)}^{(t)} = 1$.

Departure: when a job j departs at a time t, simply remove it in the graph $H^{(t)}$, then call procedure REASSIGN(t) (we need to call the procedure after departure of a job as well because $\mathsf{OptL}_t(\mathcal{I})$ may decrease when jobs depart).

We now describe the reassignment procedure which makes sure that the load on any machine is at most $\beta \cdot \mathsf{OptL}_t(\mathcal{I})$.

Reassignment: Note that procedure REASSIGN(t) is called before a new job is assigned and after a departing job is deleted. Hence, the set of jobs considered by the procedure is always $J_{t-1} \cap J_t$. In procedure REASSIGN(t), we first check if there is a machine i for which the current load is more than $\beta \cdot \mathsf{OptL}_t(\mathcal{I})$ (line 1.1). If so, we need to reassign jobs to machines such that this condition is satisfied for all machines. Consider the subset of machines on which the load is more than $(\beta - 1) \cdot \mathsf{OptL}_t(\mathcal{I})$ and let J_t^{full} be the set of jobs assigned to these machines (line 1.2). We compute a minimum weight assignment ψ_t^{full} of jobs in J_t^{full} to the machines in $[m]$ such that each machine receives at most $\mathsf{OptL}_t(\mathcal{I})$ jobs. Such an assignment clearly exists by definition of $\mathsf{OptL}_t(\mathcal{I})$ (line 1.3). Observe that we use the current weights $w^{(t-1)}$ for this computation. The final assignment σ_t of jobs is given by ψ_t^{full} for jobs in J_t^{full} and remains unchanged for rest of the jobs (line 1.5). Finally, we update the weights of the edges as follows. We maintain the invariant that if a job j is assigned to a machine i, then the weight of the corresponding edge is 1. Thus, if a job $j \in J_t^{\mathsf{full}}$ is assigned to a machine i_j by ψ_t^{full} and $w := w_{(j,i_j)}^{(t-1)} > 1$, then we need to update the weights of the edges incident with j (line 1.9). Order the edges incident with j according to their weights: then the first edge e_1 gets weight $\mathsf{deg}(j)$, whereas the edge e_w at position w (i.e., (j, i_j)) gets weight 1. The weights of all the edges between e_1 and e_w remains unchanged, and we decrease the weights of rest of the edges by 1. This ensures that the weights assigned to these edges are distinct and lie in the range $[\mathsf{deg}(j)]$. The weights remain unchanged for the rest of the edges. This completes the description of the procedure.

Algorithm 1: Procedure REASSIGN(t)

if *there is a machine with load larger than* $\beta \cdot \mathsf{OptL}_t(\mathcal{I})$ **then**

  Let $J_t^{\mathsf{full}} \subset J_t \cap J_{t-1}$ denote the set of jobs assigned to machines of load at least $(\beta - 1)\mathsf{OptL}_t(\mathcal{I})$.

  Find a minimum weight (w.r.t. weights $w^{(t-1)}$) assignment ψ_t^{full} of jobs in J_t^{full} to $[m]$ such that each machine $i \in [m]$ is assigned at most $\mathsf{OptL}_t(\mathcal{I})$ jobs from J_t^{full}.

  for *each job* $j \in J_t^{\mathsf{full}}$ **do**

    Set the assignment $\sigma_t(j)$, of job j, as follows:

$$\sigma_t(j) \leftarrow \psi_t^{\mathsf{full}}(j)$$

  for *each job* $j \in J_t^{\mathsf{full}}$ **do**

    Let i_j denote $\psi_t^{\mathsf{full}}(j)$.

    if $w := w_{(j,i_j)}^{(t-1)} > 1$ **then**

      Set the weights $w_{(j,i)}^{(t)}, i \in [\mathsf{deg}(j)]$, as follows:

$$w_{(j,i)}^{(t)} \leftarrow \begin{cases} \mathsf{deg}(j) & \text{if } w_{(j,i)}^{(t-1)} = 1 \\ 1 & \text{if } i = i_j \\ w_{(j,i)}^{(t-1)} & \text{if } 2 \leq w_{(j,i)}^{(t-1)} < w \\ w_{(j,i)}^{(t-1)} - 1 & \text{if } w < w_{(j,i)}^{(t-1)} \leq \mathsf{deg}(j) \end{cases}$$

The following is a brief description of the high level idea behind our algorithm. The formal analysis is given in Sect. 3.2.

High Level Idea. Suppose for simplicity that the optimal off-line sequence $(\sigma_t^*)_{t\in[T]}$ is static, that is $\mathsf{OptR}(\mathcal{I}) = 0$ and that all the jobs have degree d. Hence, each job j is assigned to a fixed machine i_j^* in σ^* when it is alive.

Our algorithm maintains at each time $t \in [T]$ and for each job j a set of weights $\{w_{(j,i)}^{(t)}\}_{i\in S(j)}$ for its adjacent edges. The weight $w_{(j,i)}^{(t)}$ shall represents "how confident we are" that $i = i_j^*$ with smaller weight signifying higher confidence. We define the following potential function,

$$\Phi_t = -\sum_{j\in J_t} w_{(j,i_j^*)}^{(t)}$$

that measures "how correct our weights are", such that higher potential signifies that smaller weights are put on the optimal assignments (j, i_j^*).

The idea behind our algorithm is as follows: we follow the current weights until a subset of machines accumulate a high load ($\sim (\beta - 1) \cdot \mathsf{OptL}_t(\mathcal{I})$) in which case, on the one hand, the jobs J_t^{full} on these machines must be reassigned in order to keep a bounded maximum load. On the other hand, because the maximum load in σ_t^* is at most $\mathsf{OptL}_t(\mathcal{I})$, it must be the case that at least a fraction $\frac{(\beta-1)\cdot\mathsf{OptL}_t(\mathcal{I})-\mathsf{OptL}_t(\mathcal{I})}{(\beta-1)\cdot\mathsf{OptL}_t(\mathcal{I})} = (\beta-2)/(\beta-1)$ of the jobs J_t^{full} are such that $\sigma_t(j) \neq i_j^*$. Now, suppose we increase the weight of the edge $(j, \sigma_t(j))$ and decrease the weights of the edges (j, i) for $i \neq \sigma_t(j)$ for all jobs $j \in J_t^{\mathsf{full}}$. This would decrease the weight of edge (j, i_j^*) for at least a fraction $(\beta-2)/(\beta-1)$ of the jobs J_t^{full} and increase it for at most a fraction $1/(\beta-1)$. With a large enough β, this would increase the potential (i.e., increase the correctness of our weights). Hence, whenever a reassignment cost is paid, a gain in the correctness of our weights is made. Our bounds follow from linking the gain in potential to the number of reassigned jobs whenever a reassignment occurs.

However, the weight updates described above are not always possible if one wants to maintain the same set of weights all along. Instead, we use a minimum weight assignment based update (line 1.9) that we show to have a similar effect on the potential (refer to the full version for a discussion of our choice of weights and their updates).

3.2 Analysis

In this section, we give the formal analysis of algorithm LEARNWEIGHTS. The first observation is that the maximum load on any machine at any time t is at most $\beta \cdot \mathsf{OptL}_t(\mathcal{I})$.

Lemma 1. *At any time t, the load on any machine is at most $(2d^2+1)\cdot\mathsf{OptL}_t(\mathcal{I})$.*

We now bound the number of reassignments made by the algorithm. We need to define a suitable potential function here. Recall the definition of $(\sigma_t^*)_{t\in T}$ – it is the sequence minimizing the total number of reassignments among all the sequences of assignments that maintain load at most $\mathsf{OptL}_t(\mathcal{I})$ on any machine at any time t (i.e., with total number of reassignments $\mathsf{OptR}(\mathcal{I})$). For every $t \in [T]$, define the following potential

$$\Phi_t = -\sum_{j\in J_t} w^{(t)}_{(j,\sigma_t^*(j))}.$$

By convention $J_t^{\mathsf{full}} = \emptyset$ when no reassignments are done at step t, $J_0 = \emptyset$ and $\Phi_0 = 0$. For every $t \in [T]$, the change in potential $(\Phi_t - \Phi_{t-1})$ between the time steps $t-1$ and t can be decomposed into four parts,

$$\left[\sum_{j\in J_{t-1}\setminus J_t} w^{(t-1)}_{(j,\sigma_{t-1}^*(j))}\right] + \left[-\sum_{j\in J_t\setminus J_{t-1}} w^{(t)}_{(j,\sigma_t^*(j))}\right]$$
$$+\left[\sum_{j\in J_t\cap J_{t-1}} w^{(t-1)}_{(j,\sigma_{t-1}^*(j))} - w^{(t-1)}_{(j,\sigma_t^*(j))}\right] + \left[\sum_{j\in J_t\cap J_{t-1}} w^{(t-1)}_{(j,\sigma_t^*(j))} - w^{(t)}_{(j,\sigma_t^*(j))}\right]$$

The first term arises if a job departs at time t, and the second occurs when a new job arrives at time t. The third term measures the change in potential when switching from assignment σ_{t-1}^* to σ_t^* (under the weights $w^{(t-1)}$) and the fourth one captures the potential change when the weight changes from $w^{(t-1)}$ to $w^{(t)}$ during the REASSIGN(t) step. Let $\Delta\Phi_t^{\mathsf{del}}, \Delta\Phi_t^{\mathsf{add}}, \Delta\Phi_t^{\mathsf{off}}$ and $\Delta\Phi_t^{\mathsf{wc}}$ denote these terms respectively. Let

$$\mathsf{OptR}_t(\mathcal{I}) = \sum_{j \in J_{t-1} \cap J_t} \mathbf{1}_{\{\sigma_{t-1}^*(j) \neq \sigma_t^*(j)\}}$$

denote the number of reassignments between σ_{t-1}^* and σ_t^*. The following lemma bounds each of these four terms.

Lemma 2. *For every $t \in [T]$, we have $\Delta\Phi_t^{\mathsf{del}} \geq 0$, $\Delta\Phi_t^{\mathsf{add}} \geq -d$, $\Delta\Phi_t^{\mathsf{off}} \geq -d \cdot \mathsf{OptR}_t(\mathcal{I})$ and finally $\Delta\Phi_t^{\mathsf{wc}} \geq \frac{d^2-1}{(2d^2-1)(d+1)}|J_t^{\mathsf{full}}|$.*

In the proof of Lemma 2, we divide the jobs J_t^{full} into three groups: good jobs that raise the potential after the weight updates, bad jobs that decrease the potential, and neutral jobs that leave the potential unchanged. We then use the fact that we reassign the jobs J_t^{full} to a minimum weight assignment to show that the increase in potential from the good jobs dominates the decrease from bad jobs leading to a net increase in potential of at least $\Theta(1/d)|J_t^{\mathsf{full}}|$. The following corollary of Lemma 2 bounds the total number of reassignments performed by the algorithm.

Corollary 1. *The total number of reassignments performed by* LEARNWEIGHTS *is at most*

$$\frac{d(d+1)(2d^2-1)}{d^2-1} \cdot (\mathsf{OptR}(\mathcal{I}) + T).$$

Our first main result Theorem 1 follows from Lemma 1 and corollary 1.

4 Arbitrary Reassignment Costs

We consider in this section the generalization of our problem to arbitrary reassignment costs GCDYNLOADBALANCE, i.e., the reassignment of a job $j \in J$ costs $c_j > 0$. As before, for every instance $\mathcal{I}$ of GCDYNLOADBALANCE and time step t, $\mathsf{OptL}_t(\mathcal{I})$ denotes the optimum load at time t. We define the total reassignment cost of a sequence of assignments $(\psi_t)_{t\in[T]}$ as

$$\sum_{t=1}^{T-1} \sum_{j \in J_t \cap J_{t+1}} c_j \cdot \mathbf{1}_{\psi_t(j) \neq \psi_{t+1}(j)}.$$

As before, we denote by $(\sigma_t^*)_{t\in[T]}$ a sequence of assignments which maintains a maximum load $\mathsf{OptL}_t(\mathcal{I})$ at every time step and minimizes the total reassignment cost. Let $\mathsf{OptR}(\mathcal{I})$ denote the total reassignment cost of $(\sigma_t^*)_{t\in[T]}$.

We extend the definition of a (β, γ)-competitive solution to this settings as follows: Given an instance $\mathcal{I}$ of GCDYNLOADBALANCE we say that a solution $(\psi_t)_{t\in[T]}$ is (β, γ)-competitive if

(i) $\mathsf{Load}(\psi_t) \leq \beta \cdot \mathsf{OptL}_t(\mathcal{I})$ for all $t \in [T]$.
(ii) The total reassignment cost of the solution $(\psi_t)_{t\in[T]}$ is at most $\gamma(\mathsf{OptR}(\mathcal{I}) + \sum_{j\in J} c_j)$.

Note that $\sum_{j\in J} c_j$ is the cost of reassigning each job $j \in J$ once.

4.1 Algorithm GCLearnAssignment

Let $\beta = (2d + 1)$. Our algorithm works as follows.

Arrival: when a new job j arrives at a time t, before assigning job j, call procedure GCReassign(t) described in Algorithm 2 (this procedure ensures that the load on any machine is at most $\beta \cdot \mathsf{OptL}_t(\mathcal{I})$), then assign the job j to a machine $i \in S(j)$ arbitrarily.

Departure: when job j departs at time t, simply remove it from our assignment, then call procedure GCReassign(t) (we need to call the procedure after a job departure because $\mathsf{OptL}_t(\mathcal{I})$ may decrease when jobs depart).

We now describe procedure GCReassign(t) which ensures that the load on any machine is at most $\beta \cdot \mathsf{OptL}_t(\mathcal{I})$.

Reassignments: In procedure GCReassign(t), we first arbitrarily select (e.g., the one with the smallest index) a machine i on which the load is larger than $\beta \cdot \mathsf{OptL}_t(\mathcal{I})$ (line 2.2). Next, we randomly select a job j from the set of jobs assigned to machine i (line 2.4) – the probability of choosing a job is inversely proportional to its reassignment cost. Now we choose a machine $i' \in S(j)$ uniformly at random (line 2.5). The job j gets reassigned to machine i' (line 2.6). This random reassignment process is repeated until the maximum load on any machine is at most $\beta \cdot \mathsf{OptL}_t(\mathcal{I})$.

Algorithm 2: Procedure GCReassign(t)

2.1 **while** *there is a machine with load larger than* $\beta \cdot \mathsf{OptL}_t(\mathcal{I})$ **do**
2.2 Select a machine i of load larger than $\beta \cdot \mathsf{OptL}_t(\mathcal{I})$.
2.3 Let $S(i)$ denote the set of jobs assigned to i in the current assignment.
2.4 Select job $j \in S(i)$ with probability $\frac{1/c_j}{\sum_{\ell\in S(i)} 1/c_\ell}$.
2.5 Select a random machine $i' \in S(j)$, where machine l is selected with probability $1/\mathsf{deg}(j)$.
2.6 Reassign job j to machine i'.

The following is a brief description of the high level idea behind our algorithm. The formal analysis is given in Sect. 4.2.

High Level Idea. We define a potential function that represents "how close our current assignment is from the optimal assignment σ^*" and show that each single reassignment in the procedure GCReassign(t) is such that the expected increase in the potential is at least a fraction of the expected cost of the reassignment. Intuitively, since the optimal assignments $(\sigma_t^*)_{t\in[T]}$ have maximum load

$\mathsf{OptL}_t(\mathcal{I})$, and machine i from which a job is selected for reassignment has load at least $\beta \cdot \mathsf{OptL}_t(\mathcal{I})$, it must be the case that a large fraction of the jobs on machine i are assigned to a different machine in σ^*. Hence, there is a high probability of picking one of these jobs and changing its assignment and hence getting closer to the optimal assignment σ_t^*. Finally, we use a Wald type identity to show that the total expected increase in potential during the whole GCREASSIGN(t) procedure (of random length) is at least a fraction of to the total reassignment cost paid by the procedure.

4.2 Analysis

In this section, we give the formal analysis of algorithm GCLEARNASSIGNMENT. Assuming the reassignment procedure terminates in finite time almost surely (which we show later in Lemma 5), the maximum load at any time step t is bounded as follows:

Lemma 3. *At any time t, the load on any machine is at most $(2d+2) \cdot \mathsf{OptL}_t(\mathcal{I})$.*

We now bound the expected reassignment cost of our algorithm. Fix a time step $t \in [T]$. For every assignment $\psi_t : J_t \to [m]$, we define the *weighted potential of ψ_t* as

$$\Phi_t(\psi_t) = -\sum_{j \in J_t} c_j \mathbf{1}_{\{\psi_t(j) \neq \sigma_t^*(j)\}}.$$

Recall that σ_t is the assignment maintained by the algorithm at time t (which is the assignment after any deletion, assignment and reassignments). By convention, let $\Phi_0(\sigma_0) = 0$. For every $t \in [T]$, the change in potential, $(\Phi_t(\sigma_t) - \Phi_{t-1}(\sigma_{t-1}))$ between two consecutive time steps $t-1$ and t can be decomposed into four parts:

$$\left[\sum_{j \in J_{t-1} \setminus J_t} c_j \mathbf{1}_{\{\sigma_{t-1}(j) \neq \sigma_{t-1}^*(j)\}}\right] + \left[-\sum_{j \in J_t \setminus J_{t-1}} c_j \mathbf{1}_{\{\sigma_t(j) \neq \sigma_t^*(j)\}}\right]$$
$$+ \left[\sum_{j \in J_{t-1} \cap J_t} c_j \mathbf{1}_{\{\sigma_{t-1}(j) \neq \sigma_{t-1}^*(j)\}} - \sum_{j \in J_{t-1} \cap J_t} c_j \mathbf{1}_{\{\sigma_{t-1}(j) \neq \sigma_t^*(j)\}}\right]$$
$$+ \left[\sum_{j \in J_{t-1} \cap J_t} c_j \mathbf{1}_{\{\sigma_{t-1}(j) \neq \sigma_t^*(j)\}} - \sum_{j \in J_{t-1} \cap J_t} c_j \mathbf{1}_{\{\sigma_t(j) \neq \sigma_t^*(j)\}}\right]$$

The first term arises if a job departs. The second term occurs when a job arrives. The third term is a change in potential due to the off-line reassignment between σ_{t-1}^* and σ_t^* and the fourth term measures the change in potential when switching from assignment σ_{t-1} to σ_t. We denote these terms by $\Delta\Phi_t^{\mathsf{del}}$, $\Delta\Phi_t^{\mathsf{add}}$, $\Delta\Phi_t^{\mathsf{off}}$ and $\Delta\Phi_t^{\mathsf{ass}}$ respectively. Let C_t denote the total reassignment cost due to procedure GCREASSIGN(t) and let

$$\mathsf{OptR}_t(\mathcal{I}) = \sum_{j \in J_{t-1} \cap J_t} c_j \mathbf{1}_{\{\sigma_{t-1}^*(j) \neq \sigma_t^*(j)\}}$$

denote the reassignment cost paid by the off-line solution between $t-1$ and t. The following lemma bounds each of the four terms.

Lemma 4. *For every $t \in [T]$, it holds that $\Delta\Phi_t^{\mathsf{del}} \geq 0$, $\Delta\Phi_t^{\mathsf{add}} \geq -\sum_{j \in J_t \setminus J_{t-1}} c_j$, $\Delta\Phi_t^{\mathsf{off}} \geq -\mathsf{OptR}_t(\mathcal{I})$ and $\mathbb{E}(\Delta\Phi_t^{\mathsf{ass}}) \geq 1/(2d+1) \cdot \mathbb{E}(C_t)$.*

In the proof of Lemma 4, we show that the expected change in potential at every iteration of the reassignment procedure GCReassign(t) is at least a fraction $1/(2d+1)$ of the expected cost paid at that iteration. We then use a Wald type identity to link the total expected change in potential to the total cost paid during the whole reassignment procedure (of random length). The following corollary of Lemma 4 bounds the total expected reassignment cost made by the algorithm.

Corollary 2. *The total expected reassignment cost made by algorithm* GCLearnAssignment *is at most* $(4d+2)\left(\mathsf{OptR}(\mathcal{I}) + \sum_{j \in J} c_j\right)$.

It follows from the above corollary that our algorithm is finite almost surely.

Lemma 5. *Algorithm* GCLearnAssignment *is finite almost surely.*

Our second main result Theorem 2 follows from Lemma 3 and corollary 2.

5 Arbitrary Job Sizes

We consider in this section the generalization of our problem to arbitrary job sizes denoted by GSDynLoadBalance and where each job j has size $p_j > 0$. The load on a machine is now given by the sum of the sizes of the jobs assigned to the machine. Our algorithm for GSDynLoadBalance that we denote by GSLearnAssignment is similar to our algorithm for the arbitrary costs setting with extra technical details needed in the analysis. In particular, given a machine i with high load, we now pick and reassign a random job j from machine i with probability proportional to its size. As the total size of the jobs assigned to machine i in an optimal solution σ^* is relatively small, this implies a high probability of picking and reassigning a job j which is not assigned to i in σ^* and hence getting closer to σ^*. Our algorithm and analysis for this setting are deferred to the full version of the paper.

References

1. Andrews, M., Goemans, M.X., Zhang, L.: Improved bounds for on-line load balancing. Algorithmica **23**, 278–301 (1999)
2. Azar, Y., Broder, A.Z., Karlin, A.R.: On-line load balancing. Theoret. Comput. Sci. **130**(1), 73–84 (1994)
3. Azar, Y., Kalyanasundaram, B., Plotkin, S., Pruhs, K.R., Waarts, O.: On-line load balancing of temporary tasks. J. Algorithms **22**(1), 93–110 (1997)

4. Azar, Y., Naor, J., Rom, R.: The competitiveness of on-line assignments. J. Algorithms **18**(2), 221–237 (1995)
5. Bhattacharya, S., Buchbinder, N., Levin, R., Saranurak, T.: Chasing positive bodies. In: 2023 IEEE 64th Annual Symposium on Foundations of Computer Science (FOCS), pp. 1694–1714 (2023)
6. Bosek, B., Zych-Pawlewicz, A.: Dynamic coloring of unit interval graphs with limited recourse budget. In: 30th Annual European Symposium on Algorithms (ESA 2022), vol. 244, pp. 1–14 (2022)
7. Brodal, G.S., Fagerberg, R.: Dynamic representations of sparse graphs. In: Workshop on Algorithms and Data Structures, pp. 342–351 (1999)
8. Feldkord, B., et al.: Fully-dynamic bin packing with little repacking. In: 45th International Colloquium on Automata, Languages, and Programming (ICALP 2018), vol. 107, pp. 1–24 (2018)
9. Fleischer, R., Wahl, M.: On-line scheduling revisited. J. Sched. **3**(6), 343–353 (2000)
10. Fricker, C., Gast, N.: Incentives and redistribution in homogeneous bike-sharing systems with stations of finite capacity. Euro J. Transp. Logist. **5**, 261–291 (2016)
11. Grove, E.F., Kao, M.Y., Krishnan, P., Vitter, J.S.: Online perfect matching and mobile computing. In: Algorithms and Data Structures: 4th International Workshop, WADS'95, pp. 194–205 (1995)
12. Gu, A., Gupta, A., Kumar, A.: The power of deferral: maintaining a constant-competitive steiner tree online. In: Proceedings of the Forty-Fifth Annual ACM Symposium on Theory of Computing, pp. 525–534 (2013)
13. Gupta, A., Krishnaswamy, R., Kumar, A., Panigrahi, D.: Online and dynamic algorithms for set cover. In: Proceedings of the 49th Annual ACM SIGACT Symposium on Theory of Computing, pp. 537–550 (2017)
14. Gupta, A., Kumar, A., Stein, C.: Maintaining assignments online: matching, scheduling, and flows. In: Proceedings of the Twenty-Fifth Annual ACM-SIAM Symposium on Discrete Algorithms, SODA, pp. 468–479 (2014)
15. Imase, M., Waxman, B.M.: Dynamic Steiner tree problem. SIAM J. Discret. Math. **4**(3), 369–384 (1991)
16. Krishnaswamy, R., Li, S., Suriyanarayana, V.: Online unrelated-machine load balancing and generalized flow with recourse. In: Proceedings of the 55th Annual ACM Symposium on Theory of Computing, pp. 775–788 (2023)
17. Megow, N., Skutella, M., Verschae, J., Wiese, A.: The power of recourse for online MST and TSP. SIAM J. Comput. **45**(3), 859–880 (2016)
18. Phillips, S., Westbrook, J.: On-line load balancing and network flow. Algorithmica **21**, 245–261 (1998)
19. Rudin, J.F., III., Chandrasekaran, R.: Improved bounds for the online scheduling problem. SIAM J. Comput. **32**(3), 717–735 (2003)
20. Sanders, P., Sivadasan, N., Skutella, M.: Online scheduling with bounded migration. Math. Oper. Res. **34**(2), 481–498 (2009)
21. Wang, W., Zhu, K., Ying, L., Tan, J., Zhang, L.: MapTask scheduling in MapReduce with data locality: throughput and heavy-traffic optimality. IEEE/ACM Trans. Networking **24**(1), 190–203 (2014)

Pairwise-Independent Contention Resolution

Anupam Gupta[1], Jinqiao Hu[2], Gregory Kehne[3], and Roie Levin[4](✉)

[1] New York University, New York, NY, USA
anupam.g@nyu.edu
[2] Peking University, Beijing, China
cppascalinux@gmail.com
[3] University of Texas at Austin, Austin, TX, USA
gregorykehne@gmail.com
[4] Rutgers University, New Brunswick, NJ, USA
roie.levin@rutgers.edu

Abstract. We study online contention resolution schemes (OCRSs) and prophet inequalities for non-product distributions. Specifically, when the active set is sampled according to a *pairwise-independent* (PI) distribution, we show a $(1-o_k(1))$-selectable OCRS for uniform matroids of rank k, and $\Omega(1)$-selectable OCRSs for laminar, graphic, cographic, transversal, and regular matroids. These imply prophet inequalities with the same ratios when the set of values is drawn according to a PI distribution. Our results complement recent work of Dughmi et al. [14] showing that no $\omega(1/k)$-selectable OCRS exists in the PI setting for general matroids of rank k.

Keywords: Online Algorithms · Prophet Inequalities · Contention Resolution

1 Introduction

Consider the *prophet inequality* problem: a sequence of independent positive real-valued random variables $\mathbf{X} = \langle X_1, X_2, \ldots, X_n \rangle$ are revealed one by one. Upon seeing X_i the algorithm must decide whether to select or discard the index i; these decisions are irrevocable. The goal is to choose some subset S of the indices $\{1, 2, \ldots, n\}$ to maximize $\mathbb{E}[\sum_{i\in S} X_i]$, subject to the set S belonging to a well-behaved family $\mathcal{I} \subseteq 2^{[n]}$. The goal is to get a value close to $\mathbb{E}[\max_{S\in\mathcal{I}} \sum_{i\in S} X_i]$, the value that a clairvoyant "prophet" could obtain in expectation. This problem originally arose in optimal stopping theory, where the case of $\mathcal{I}$ being the set of all singletons was considered [23]: more recently, the search for good prophet inequalities has become a cornerstone of stochastic optimization and online decision making, with the focus being on generalizing to broad classes of downward-closed sets $\mathcal{I}$ [16,22,28], considering additional assumptions on the order in which

A. Gupta—Supported in part by NSF awards CCF-1955785 and CCF-2006953.
R. Levin—Work was done while the author was a Fulbright Israel Postdoctoral Fellow.

J. Vygen and J. Byrka (Eds.): IPCO 2024, LNCS 14679, pp. 196–209, 2024.
https://doi.org/10.1007/978-3-031-59835-7_15

these random variables are revealed [1,2,15,17], obtaining optimal approximation guarantees [11,24], and competing with nonlinear objectives [18,29].

One important and interesting direction is to reduce the requirement of independence between the random variables: *what if the r.v.s are correlated?* The case of negative correlations is benign [26,27], but general correlations present significant hurdles—even for the single-item case, it is impossible to get value much better than the trivial $\mathbb{E}[\max_i X_i]/n$ value obtained by random guessing [20]. As another example, the model with *linear correlations*, where $\mathbf{X} = A\mathbf{Y}$ for some independent random variables $\mathbf{Y} \in \mathbb{R}_+^d$ and known non-negative matrix $A \in \mathbb{R}_+^{d \times n}$, also poses difficulties in the single-item case [21].

Given these impossibilities, Caragiannis et al. [7] gave single-item prophet inequalities in the setting of *weak correlations*: specifically, for the setting of *pairwise-independent distributions.* As the name suggests, these are distributions that look like product distributions when restricted to any two random variables. While pairwise-independent distributions have long been studied in other contexts [25], they have received less attention in the context of stochastic optimization. Caragiannis et al. [7] give both algorithms and some limitations for $\mathbf{X}$ exhibiting pairwise independence. They also considered related pricing and bipartite matching problems.

We ask the question: can we extend the prophet inequalities known for richer classes of constraint families $\mathcal{I}$ to the pairwise-independent case? In particular,

Which matroids admit good pairwise-independent prophet inequalities?

Specifically, we investigate the analogous questions for *(online) contention resolution schemes (OCRSs)* [16], another central concept in online decision making, and a close relative of prophet inequalities. In an OCRS, a random subset of a ground set is marked active. Elements are sequentially revealed to be active or inactive, and the OCRS must decide irrevocably on arrival whether to select each active element, subject to the constraint that the selected element set belongs to a constraint family $\mathcal{I}$. The goal is to ensure that each element, conditioned on being active, is picked with high probability. It is intuitive from the definitions (and formalized by Feldman et al. [16]) that good OCRSs imply good prophet inequalities (see also [24]).

1.1 Our Results

Our first result is for the k-uniform matroid, where the algorithm can pick up to k items: we achieve a $(1 - o_k(1))$-factor of the expected optimal value.

Theorem 1 (Uniform Matroid PI Prophets). *There is an algorithm in the prophet model for k-uniform matroids that achieves expected value at least $(1 - O(k^{-1/5}))$ of the expected optimal value.*

We prove this by giving a $(1 - O(k^{-1/5}))$-*selectable* online contention resolution scheme for k-uniform matroids, even when the underlying generative process is only pairwise-independent. Feldman et al. [16] showed that selectable OCRSs immediately lead to prophet inequalities (against an almighty adversary) in the fully independent case, and we observe that their proofs translate to pairwise-independent distributions as well. Along the way we also show a $(1 - O(k^{-1/3}))$ (offline) CRS for the pairwise-independent k-uniform matroid case.

We then show $\Omega(1)$-selectable OCRSs for sevaral classes of matroids, again via pairwise-independent OCRSs.

Theorem 2 (Other Matroids PI Prophets). *There exist $\Omega(1)$-selectable OCRSs for laminar (Sect. 3), graphic (Sect. 4), cographic (full version), transversal (full version), and regular (Sect. 5) matroids. These immediate imply $\Omega(1)$ prophet inequalities for these matroids against almighty adversaries.*

Finally, we consider the single-item case in greater detail in the full version. For this single-item case the current best result from Caragiannis et al. [7] uses a multiple-threshold algorithm to achieve a $(\sqrt{2} - 1)$-prophet inequality; however, this bound is worse than the $1/2$-prophet inequality known for fully independent distributions. We show that no (non-adaptive) multiple-threshold algorithm (i.e., one that prescribes a sequence of thresholds τ_i up-front, and picks the first index i such that $X_i \geq \tau_i$) can beat $2(\sqrt{5} - 2) \approx 0.472$, suggesting that if $1/2$ is at all possible it will require adaptive algorithms.

Theorem 3 (Upper Bound for Multiple Thresholds). *Any multiple-threshold algorithm for the single-item PI prophet inequality is at most* 0.472-*competitive.*

In the full version, we also give a *single-sample* single-item PI prophet inequality.

Theorem 4 (Single-Sample Prophet Inequality). *There is an algorithm that draws a single sample from the underlying pairwise-independent distribution $\langle \widetilde{X}_1, \ldots, \widetilde{X}_n \rangle \sim \mathcal{D}$ on $\mathbb{R}^n_+$, and then faced with a second sample $\langle X_1, \ldots, X_n \rangle \sim \mathcal{D}$ (independent from $\langle \widetilde{X}_1, \ldots, \widetilde{X}_n \rangle$), picks a single item i from $X_1, \ldots, X_n$ with expected value at least $\Omega(1) \cdot \mathbb{E}_{\mathbf{X} \sim \mathcal{D}}[\max_i X_i]$.*

1.2 Related Work

In independent and concurrent work, Dughmi et al. [14] also study the pairwise-independent versions of prophet inequalities and (online) contention resolution schemes. This work can be considered complementary to ours: they show that for arbitrary linear matroids, nothing better than $O(1/r)$ factors can be achieved for pairwise-independent versions of OCRSs, and nothing better than $O(1/(\log r))$ factors can be achieved for pairwise-independent versions of matroid prophet inequalities (where r is the rank of the matroid). They also obtain $\Omega(1)$-selectable OCRSs for uniform, graphical, and bounded degree transversal matroids by observing that these have the α-partition property (see [5]), reducing to the single-item setting. Another motivation for our work is the famous matroid secretary problem, since the latter is known to be equivalent to the existence of good OCRSs for *arbitrary* distributions that admit $\Omega(1)$-balanced CRSs against a *random-order* adversary [13].

The original single-item prophet inequality for product distributions was proven by Krengel and Sucheston [23]. There is a vast literature on variants and extensions of prophet inequalities, which we cannot survey here for lack of space. Contention resolution schemes were introduced by Chekuri et al. [10] in the context of constrained submodular function maximization, and these were generalized by Feldman et al. [16] to the online setting in order to give prophet inequalities for richer constraint families.

Limited-independence versions of prophet inequalities were studied from the early days e.g. by Hill and Kertz and Rinott and Samuel-Cahn [20,27]. Many stochastic optimization problems have been studied recently in correlation-robust settings, e.g., by Bateni et al., Chawla et al., Immorlica et al. [6,9,21]; pairwise-independent prophet inequalities were introduced by Caragiannis et al. [7].

There is a line of work on single-sample prophet inequalities in the i.i.d. setting [3,4,8,19,30]. This is the first such study for pairwise-independent distributions.

1.3 Preliminaries

We provide several essential definitions here, and a more complete preliminaries section in the full version. We assume the reader is familiar with the basics of matroid theory, and refer to Schrijver [31] for definitions. For a matroid $M = (E, \mathcal{I})$ the matroid polytope is defined to be $\mathcal{P}_M := \{\mathbf{x} \in \mathbb{R}^E : \mathbf{x} \geq \mathbf{0},\ \mathbf{x}(S) \leq \text{rank}(S)\ \forall S \subseteq E\}$. For polytope $\mathcal{P}$ and scalar $b \in \mathbb{R}$, define $b\mathcal{P} := \{b\mathbf{x} : \mathbf{x} \in \mathcal{P}\}$.

We focus on pairwise independent versions of *contention resolution schemes* (CRSs), in both offline and online settings. Our setting entails a set $R \subseteq E$ drawn from a distribution $\mathcal{D}$ with marginal probabilities given by some $\mathbf{x} \in b\mathcal{P}_M$, and the goal is to select items $I \subseteq R$, $I \in \mathcal{I}$, such that $\Pr[i \in I \mid i \in R] \geq c$ for all i. An algorithm which does this is a (b, c)*-balanced* CRS.

For *online* contention resolution schemes (OCRSs) the items arrive one-at-a-time; a scheme must decide whether to include each arriving element into its independent set I or irrevocably reject it [16]. Generally the events $[i \in R]$ are taken to be independent, so that $\mathbf{x}$ determines $\mathcal{D}$. For a *pairwise-independent* (PI) OCRS these events are only pairwise independent under $\mathcal{D}$.

The Almighty Adversary. The almighty adversary knows everything. It first sees the realization of $R \sim \mathcal{D}$, as well as all randomness the algorithm will use. It then adversarially orders R. To describe PI-OCRS's with guarantees against the almighty adversary, we adopt ideas from Feldman et al. [16], and restrict our attention to a schemes which coincide with their *greedy* OCRSs:

Definition 1 (**(b, c)-selectable PI-OCRS**). *Let $\mathcal{P} \subseteq [0,1]^n$ be some convex polytope. We call a (randomized) algorithm $\pi : 2^{[n]} \to 2^{[n]}$ a* (b, c)-selectable PI-OCRS *if it satisfies the following:*

1. *Algorithm π precommits to some feasible set family $\mathcal{F} \subseteq \mathcal{I}$, and then adds each arriving i to I only if the resulting set is in $\mathcal{F}$.*
2. *For any $\mathbf{x} \in b\mathcal{P}$, any distribution $\mathcal{D}$ PI-consistent with $\mathbf{x}$ and any $i \in [n]$, let $\mathcal{F}$ be the feasible set family defined by π. Let R be sampled according to $\mathcal{D}$, then*
$$\Pr_{R \sim \mathcal{D}}[I \cup \{i\} \in \mathcal{F} \quad \forall I \subseteq R, I \in \mathcal{F} \mid i \in R] \geq c. \tag{1.1}$$
Here the probability is over R and internal randomness of π in defining $\mathcal{F}$.

Notice the definition here is slightly different from [16], as we need to condition on the event $i \in R$. This is due to our limited independence over events $i \in R$. For the mutually independent case, one can prove that $\Pr_{R\sim\mathcal{D}}[I \cup \{i\} \in \mathcal{F} \quad \forall I \subseteq R, I \in \mathcal{F} \mid i \in R] = \Pr_{R\sim\mathcal{D}}[I \cup \{i\} \in \mathcal{F} \quad \forall I \subseteq R, I \in \mathcal{F}]$, but this may not hold in the pairwise-independent case.

A (b, c)-selectable PI-OCRS implies a $(1, bc)$-selectable PI-OCRS (or for short bc-selectable PI-OCRS), and gives guarantees against an almighty adversary. For details see the full version.

The Offline Adversary and Prophet Inequalities. The offline adversary does not know the randomness of π and must choose an arrival order for $[n]$ before $R \sim \mathcal{D}$ is sampled. PI-OCRSs that are (b, c)-*balanced* are effective against the offline adversary, and since the offline adversary is weaker than the almighty adversary, a (b, c)-selectable PI-OCRS is always (b, c)-balanced. Once again, a (b, c)-balanced PI-OCRS may be converted to a $(1, bc)$-balanced PI-OCRS (or a bc-balanced PI-OCRS for short) via independent subsampling of R.

Feldman et al. [16] showed connections between OCRSs and prophet inequalities. In the full version, we formally establish this connection in the pairwise-independent setting through the formulation of a *PI matroid prophet game*, and we demonstrate that balanced PI-OCRSs are enough to give prophet inequalities.

As an upshot, we show that our results imply matroid prophet inequalities for the pairwise-independent setting; for each class of matroids, any c-balanced PI-OCRS yields a c-competitive prophet inequality for values drawn from a pairwise-independent distribution. This generalizes the single-item PI prophet inequality of Caragiannis et al. [7] in the setting where the gambler knows the joint distribution as well as the marginals.

2 Uniform Matroids

Recall that the independent sets of a *uniform matroid* $M = (E, \mathcal{I})$ of rank k are all subsets of E of size at most k; hence our goal is to pick some set of size at most k. Identifying E with $[n]$, the corresponding matroid polytope is $\mathcal{P}_M := \{\mathbf{x} \in [0,1]^n : \sum_{i=1}^n x_i \leq k\}$. Our main results for uniform matroids are the following, which imply Theorem 1.

Theorem 5 (Uniform Matroids). *For uniform matroids of rank k, there is*

(i) a $(1 - O(k^{-1/3}))$-balanced PI-CRS, and
(ii) a $(1 - O(k^{-1/5}))$-selectable PI-OCRS.

A simple greedy PI-OCRS follows by choosing the feasible set family $\mathcal{F} = \mathcal{I}$, i.e. selecting R as the resulting set if $|R| \leq k$. However, conditioning on $i \in R$, pairwise independence only guarantees the marginals of the events $j \in R$ (and they might have arbitrary correlation), so we can only use Markov's inequality to bound $\Pr[|R| \leq k \mid i \in R]$. This analysis only gives a $(b, 1-b)$-selectable PI-OCRS for k-uniform matroid. (For details see the full version.)

Hence instead of conditioning on some $i \in R$ and using Markov's inequality, we consider all items together and use Chebyshev's inequality to bound $\Pr[|R| \geq k, i \in R]$. The following lemma is key for both our PI-CRS and PI-OCRS.

Lemma 1. *Let $M = (E, \mathcal{I})$ be a k-uniform matroid, where E is identified as $[n]$. Given $\mathbf{x} \in (1-\delta)\mathcal{P}_M$ and a distribution $\mathcal{D}$ of subsets of E that is PI-consistent with $\mathbf{x}$, let $R \subseteq E$ be the random set sampled according to $\mathcal{D}$. Then*

$$\sum_{i=1}^{n} \Pr[|R| \geq k, i \in R] \leq \frac{1-\delta^2}{\delta^2}.$$

Proof. The left-hand side can be written as

$$\sum_{i=1}^{n} \Pr[|R| \geq k, i \in R] = \sum_{i=1}^{n} \sum_{t=k}^{n} \sum_{\substack{S: \\ |S|=t}} \mathbb{1}[i \in S] \Pr[R = S] = \sum_{t=k}^{n} \sum_{\substack{S: \\ |S|=t}} \Pr[R = S]|S|$$
$$= \sum_{t=k}^{n} t \cdot \Pr[|R| = t] = k \Pr[|R| \geq k] + \sum_{t=k+1}^{n} \Pr[|R| \geq t].$$

We now bound the two parts separately using Chebyshev's inequality. Let $X_i := \mathbb{1}[i \in R]$ be the indicator for i being active, and let $X = \sum_{i \in E} X_i$. Since X_i are pairwise independent, $\text{Var}[X] = \sum_i \text{Var}[X_i] \leq \sum_i \mathbb{E}[X_i^2] = \sum_i \mathbb{E}[X_i] = \mathbb{E}[X]$. For the first part, we have

$$k \cdot \Pr[|R| \geq k] = k \cdot \Pr[X \geq k] \leq k \cdot \frac{\text{Var}[X]}{(k - \mathbb{E}[X])^2} \qquad \text{(Chebyshev's ineq.)}$$
$$\leq k \cdot \frac{1-\delta}{\delta^2 k} = \frac{1-\delta}{\delta^2}. \qquad (2.1)$$

For the second part,

$$\sum_{t=k+1}^{n} \Pr[|R| \geq t] = \sum_{t=k+1}^{n} \Pr[X \geq t] \leq \sum_{t=k+1}^{n} \frac{\mathrm{Var}[X]}{(t - \mathbb{E}[X])^2} \quad \text{(Chebyshev's ineq.)}$$
$$\leq \sum_{t=k+1}^{n} \frac{(1-\delta)k}{(t-(1-\delta)k)^2} \leq (1-\delta)k \cdot \sum_{t\geq 1} \frac{1}{(\delta k + t)^2}$$
$$\leq (1-\delta)k \cdot \frac{1}{\delta k} = \frac{1-\delta}{\delta}, \tag{2.2}$$

where we used the inequality

$$\sum_{j\geq 1} \frac{1}{(x+j)^2} \leq \sum_{j\geq 1} \frac{1}{(x+j-1)(x+j)} = \sum_{j\geq 1} \left(\frac{1}{x+j-1} - \frac{1}{x+j} \right) = \frac{1}{x}.$$

Summing up the (2.1) and (2.2) finishes the proof. □

Using this lemma, we can bound $\min_i \Pr[|R| \geq k \mid i \in R]$ and obtain a $(1-O(k^{-1/3}))$-balanced PI-CRS (in the same way that [10, Lemma 4.13] implies a $(b, 1-b)$-CRS in the i.i.d. setting). The details are deferred to the full version.

2.1 A $(1 - O(k^{-1/5}))$-Selectable PI-OCRS for Uniform Matroids

Our PI-CRS has to consider the elements in a specific order, and therefore it does not work in the online setting where the items come in adversarial order. The key idea for our PI-OCRS is to separate "good" items and "bad" items, and control each part separately. Let us assume R is sampled according to some distribution $\mathcal{D}$ PI-consistent with $\mathbf{x}$, and that $\mathbf{x}$ is on a face of $(1-\varepsilon)\mathcal{P}_M$, i.e.

$$\sum_{i=1}^{n} \Pr[i \in R] = (1-\varepsilon)k. \tag{2.3}$$

We will choose the value of ε later. For some other constants $r, b \in (0,1)$ define an item i to be *good* if $\Pr[|R| > \lfloor (1-r\varepsilon)k \rfloor \mid i \in R] \leq b$. Let E_g denote the set of good items, and $E_b := E \backslash E_g$ the remaining *bad* items. Our algorithm keeps two buckets, one for the good items and one for the bad, such that

(i) the good bucket has a capacity of $\lfloor (1-r\varepsilon)k \rfloor$, and
(ii) the bad bucket has a capacity of $\lceil r\varepsilon k \rceil$.

When an item arrives, we put it into the corresponding bucket as long as that bucket is not yet full. Finally, we take the union of the items in the two buckets as the output of our OCRS. This algorithm is indeed a greedy PI-OCRS with the feasible set family $\mathcal{F} = \{I \in \mathcal{I} : |I \cap E_g| \leq \lfloor (1-r\varepsilon)k \rfloor, |I \cap E_b| \leq \lceil r\varepsilon k \rceil\}$.

We show that for any item i, $\Pr[I \cup \{i\} \in \mathcal{F} \;\; \forall I \in \mathcal{F}, I \subseteq R \mid i \in R] \geq 1 - o(1)$. First, for a good item i, by definition

$$\Pr[I \cup \{i\} \in \mathcal{F} \ \ \forall I \in \mathcal{F},\ I \subseteq R \mid i \in R] = 1 - \Pr[|R \cap E_g| > \lfloor (1 - r\varepsilon)k \rfloor \mid i \in R] \geq 1 - \Pr[|R| > \lfloor (1 - r\varepsilon)k \rfloor \mid i \in R] \geq 1 - b.$$

Next, for a bad item i, we can use Markov's inequality conditioning on $i \in R$:

$$\Pr[I \cup \{i\} \in \mathcal{F} \ \ \forall I \in \mathcal{F}, I \subseteq R \mid i \in R] = 1 - \Pr[|R \cap E_b| > \lceil r\varepsilon k \rceil \mid i \in R]$$
$$\geq 1 - \frac{\sum_{j \in E_b} \Pr[j \in R \mid i \in R]}{r\varepsilon k} = 1 - \frac{\sum_{j \in E_b} \Pr[j \in R]}{r\varepsilon k}, \tag{2.4}$$

where we use Markov's inequality, and the last step uses pairwise independence of events $i \in R$. We now need to bound $\sum_{j \in E_b} \Pr[j \in R]$. If we define ε' as $1 - \varepsilon' = \frac{1-\varepsilon}{1-r\varepsilon}$, then we have

$$\begin{aligned}
\sum_{j \in E_b} \Pr[j \in R] &= \sum_{j \in E_b} \frac{\Pr[|R| \geq \lfloor (1 - r\varepsilon)k \rfloor, j \in R]}{\Pr[|R| \geq \lfloor (1 - r\varepsilon)k \rfloor \mid j \in R]} \\
&\leq \sum_{j \in E_b} \frac{\Pr[|R| \geq \lfloor (1 - r\varepsilon)k \rfloor, j \in R]}{b} \qquad \text{(since } j \text{ is bad)} \\
&\overset{(\star)}{\leq} \frac{(1 - (\varepsilon')^2)/(\varepsilon')^2}{b} \leq \frac{1}{(1-r)^2 \varepsilon^2 b},
\end{aligned}$$

where $(\star)$ uses Lemma 1. Substituting back into (2.4), $\Pr[I \cup \{i\} \in \mathcal{F} \ \ \forall I \in \mathcal{F}, I \subseteq R \mid i \in R] \geq 1 - ((1-r)^2 r\varepsilon^3 bk)^{-1}$.

To balance the good and bad items, we set $b = ((1-r)^2 r\varepsilon^3 bk)^{-1} = ((1-r)^2 r\varepsilon^3 k)^{-1/2}$. If we set $r = 1/3$, then we have an $(1-\varepsilon, 1-(\frac{4}{27}\varepsilon^3 k)^{-1/2})$-selectable PI-OCRS. Finally, if we set $\varepsilon = k^{-1/5}$, since a (b, c)-selectable PI-OCRS implies a (bc)-selectable PI-OCRS, we have a $(1 - O(k^{-1/5}))$-selectable PI-OCRS.

3 Laminar Matroids

In this section we give an $\Omega(1)$-selectable PI-OCRS for laminar matroids. A laminar matroid is defined by a laminar family $\mathcal{A}$ of subsets of E, and a capacity function $c : \mathcal{A} \to \mathbb{Z}$; a set $S \subseteq E$ is independent if $|S \cap A| \leq c(A)$ for all $A \in \mathcal{A}$.

The outline of the algorithm is as follows: we construct a new capacity function c' by rounding down $c(A)$ to powers of two; satisfying these more stringent constraints loses only a factor of two. Then we run greedy PI-OCRSs for uniform matroids from Sect. 2.1 *independently* for each capacity constraint $c'(A)$, $A \in \mathcal{A}$. Finally, we output the intersection of these feasible sets. For our analysis, we apply a union bound on probability of an item being discarded by some greedy PI-OCRS; this is a geometric series by our choice of c'.

As the first step, we define $c'(A)$ to be the largest power of 2 smaller than $c(A)$, for each $A \in \mathcal{A}$. (For simplicity we assume that $E \in \mathcal{A}$.) Moreover, if sets $A, B \in \mathcal{A}$ with $A \subseteq B$ and $c'(A) \geq c'(B)$, then we can discard A from the collection. In conclusion, the final constraints satisfy:

1. The new laminar family is $\mathcal{A}' \subseteq \mathcal{A}$.
2. For any $A \in \mathcal{A}'$, $c'(A)$ is power of 2, and $c(A)/2 < c'(A) \leq c(A)$.
3. (Strict Monotonicity) For any $A, B \in \mathcal{A}'$ with $A \subsetneq B$, we have $c'(A) < c'(B)$.

Let M' denote the laminar matroid defined by the new set of constraints. We can check that any c-selectable PI-OCRS for M' is a $(1/2, c)$-selectable PI-OCRS for M. Hence, it suffices to give a $\Omega(1)$-selectable PI-OCRS for M'.

Now we run greedy OCRSs for uniform matroids to get a $(1/25, 1/2.661)$-selectable PI-OCRS: for a set A with capacity $c'(A)$, from Sect. 2 we have both a $(1-b, b)$-selectable PI-OCRS and a $(1-b, 1-(\frac{4}{27}b^3 c'(A))^{-1/2})$-selectable PI-OCRS: the former is better for small capacities, whereas the latter is better for larger capacities. Setting a threshold of $t = 13$ and choosing $b = 24/25$, we use the former when $c'(A) < 2^t$, else we use the latter. Now a union bound over the various sets containing an element gives us the result: the crucial fact is that we get a contribution of $t(1-b)$ from the first smallest scales and a geometric sum giving $O(2^{-t/2} b^{-3/2})$ from the larger ones. The details appear in the full version.

4 Graphic Matroids

Recall that graphic matroids correspond to forests (acyclic subgraphs) of a given (multi)graph. For these matroids we show the following.

Theorem 6. *For $b \in (0, 1/2)$, there is a $(b, 1-2b)$-selectable PI-OCRS scheme for graphic matroids.*

Let $M = (E, \mathcal{I})$ be a graphic matroid defined on (multi)graph $G = (V, E)$. Let $\mathcal{D}$ be any distribution over 2^E that is PI-consistent with some $\mathbf{x} \in b\mathcal{P}_M$, and R sampled according to $\mathcal{D}$. We follow the construction of OCRS of Feldman et al. [16]. Our goal is to construct a chain of sets: $\varnothing = E_l \subsetneq E_{l-1} \subsetneq \cdots \subsetneq E_0 = E$ where for any $i \in \{0 \cdots l-1\}$ and any $e \in E_i \backslash E_{i+1}$,

$$\Pr[e \in \operatorname*{span}_{M/E_{i+1}} (((R \cap (E_i \backslash E_{i+1}))\backslash e) \mid e \in R] \leq 2b. \tag{4.1}$$

We can now define the feasible set for our greedy PI-OCRS as $\mathcal{F} = \{I \subseteq E : \forall i, I \cap (E_i \backslash E_{i+1}) \in \mathcal{I}(M/E_{i+1})\}$. By definition of contraction, $\mathcal{F} \subseteq \mathcal{I}(M)$. To check selectability, for an edge e in $E_i \backslash E_{i+1}$, we have $\Pr[I \cup \{e\} \in \mathcal{F} \quad \forall I \in \mathcal{F}, I \subseteq R \mid e \in R] = \Pr[e \notin \operatorname{span}_{M/E_{i+1}}(((R \cap (E_i \backslash E_{i+1}))\backslash e) \mid e \in R] \geq 1 - 2b$ (using (4.1)). Therefore this is a $(b, 1-2b)$-selectable PI-OCRS. It remains to show how to construct such a chain. We use the following recursive procedure:

1. Initialize $E_0 = E, i = 0$.
2. Set $S = \varnothing$.
3. While there exists $e \in E_i \backslash S$ such that $\Pr[e \in \operatorname{span}_{M/S}((R \cap (E_i \backslash S))\backslash e) \mid e \in R] > 2b$, add e into S.
4. $i \leftarrow i+1$, set $E_i = S$.
5. If $E_i \neq \varnothing$, goto step 2; otherwise set $l = i$ and terminate.

Inequality (4.1) is automatically satisfied by this procedure. It remains to show that the process always terminates, i.e. that step 3 always leaves at least one element unidentified, and hence $E_{i+1} \subsetneq E_i$. We start with the following claim.

Claim 1. *If $u_0 \in V$ satisfies $\sum_{e \in \mathcal{E}(u_0)} x_e \leq 2b$, then in the above procedure generating E_1 from E, we have that for all $e \in \mathcal{E}(u_0)$, $e \notin S$.*

Proof. We prove our claim using induction. For any edge $e \in \mathcal{E}(u_0) \cap R$, $e \in \text{span}(R \backslash \{e\})$ implies the existence of a circuit $C \subseteq R$ containing e. By the definition of circuits, C must contain some edge $e' \in \mathcal{E}(u_0) \backslash \{e\}$. By the pairwise independence of events $e \in R$, we have $\Pr[e \in \text{span}(R \backslash \{e\}) \mid e \in R] \leq \Pr[\exists e' \in \mathcal{E}(u_0) \backslash \{e\}, e' \in R \mid e \in R] \leq \sum_{e \in \mathcal{E}(u_0)} x_e \leq 2b$.

Therefore we do not add any $e \in \mathcal{E}(u_0)$ into S in the first iteration. Suppose no $e \in \mathcal{E}(u_0)$ has been added to S during the first i iterations, then before the $(i+1)^{th}$ iteration starts, u_0 has not been merged with any other vertex in the contracted graph G/S, so $\mathcal{E}(u_0)$ in G/S is the same as the original graph G. Thus $\sum_{e \in \mathcal{E}(u_0)} x_e \leq 2b$ still holds for u_0 in G/S, and by the same argument as the first iteration, no $e \in \mathcal{E}(u_0)$ will be added to S in the $(i+1)^{th}$ iteration. □

Since $\mathbf{x} \in b\mathcal{P}_M$, we have $\sum_{e \in E} x_e \leq b(n-1)$, which implies $\sum_{u \in V} \sum_{e \in \mathcal{E}(u)} x_e \leq 2b(n-1)$. By averaging, there exists a vertex $u_0 \in V$ such that $\sum_{e \in \mathcal{E}(u_0)} x_e \leq 2b(n-1)/n \leq 2b$, and by Claim 1, $\mathcal{E}(u_0) \cap E_1 = \varnothing$. Assuming no isolated vertex in V, $E_1 \subsetneq E_0$. Similarly, for any i, since $M|_{E_i}$ is also a graphic matroid and $\mathbf{x}|_{E_i} \in b\mathcal{P}_{M|_{E_i}}$, the same argument holds for it. Therefore $E_{i+1} \subsetneq E_i$ always holds, which finishes our proof of termination for our construction.

5 Regular Matroids

We now give a $\Omega(1)$-competitive PI-OCRS for regular matroids. We use the regular matroid decomposition theorem of Seymour [32] and its modification by Dinitz and Kortsarz [12], which decomposes any regular matroid into 1-sums, 2-sums, and 3-sums of graphic matroids, cographic matroids, and a specific 10-element matroid R_{10}. (These matroids are called the *basic* matroids of the decomposition). We now define 1,2,3-sums, and argue that it suffices to run a PI-OCRS for each of the basic matroids and to output the union of their outputs.

Definition 2 (Binary Matroid Sums [12,32]**).** *Given two matroids $M_1 = (E_1, \mathcal{I}_1)$ and $M_2 = (E_2, \mathcal{I}_2)$, the* matroid sum *$M$ defined on the ground set $E(M_1) \Delta E(M_2)$ is as follows. The set C is a cycle in M iff it can be written as $C_1 \Delta C_2$, where C_1 and C_2 are cycles of M_1 and M_2. respectively. Furthermore,*

1. *If $E_1 \cap E_2 = \varnothing$, then M is called 1-sum of M_1 and M_2.*
2. *If $|E_1 \cap E_2| = 1$, then we call M the 2-sum of M_1 and M_2.*
3. *If $|E_1 \cap E_2| = 3$, let $Z = E_1 \cap E_2$. If Z is a circuit of both M_1 and M_2, then we call M the 3-sum of M_1 and M_2.*

(The i-sum is denoted $M_1 \oplus_i M_2$.) Our definition differs from [12,32] as we have dropped some conditions on the sizes of M_1 and M_2 that we do not need. A $\{1, 2, 3\}$-*decomposition* of a matroid $\widetilde{M}$ is a set of matroids $\mathcal{M}$ called the *basic matroids*, together with a rooted binary tree T in which $\widetilde{M}$ is the root and the leaves are the elements of $\mathcal{M}$. Every internal vertex in the tree is either the 1-, 2-, or 3-sum of its children. Seymour's decomposition theorem for regular matroids [32] says that every regular matroid $\widetilde{M}$ has a (poly-time computable) $\{1, 2, 3\}$-decomposition with all basic matroids being graphic, cographic or R_{10}.

The Dinitz-Kortsarz Modification. Dinitz and Kortsarz [12] modified Seymour's decomposition to give an $O(1)$-competitive algorithm for the regular-matroid secretary problem, as follows. Given a $\{1, 2, 3\}$-decomposition T for binary matroid $\widetilde{M}$ with basic matroids $\mathcal{M}$, we define Z_M, the *sum-set* of a non-leaf vertex M in T, to be the intersection of the ground sets of its children (the sum-set is thus not in the ground set of M). A sum-set Z_M for internal vertex M is either the empty set (if M is the 1-sum of its children), a single element (for 2-sums), or three elements that form a circuit in its children (for 3-sums). A $\{1, 2, 3\}$-decomposition is *good* if for every sum-set Z_M of size 3 associated with internal vertex $M = M_1 \oplus_3 M_2$, the set Z_M is contained in the ground set of a single basic matroid below M_1, and in the ground set of a single basic matroid below M_2. For a given $\{1, 2, 3\}$-decomposition of a matroid $\widetilde{M}$ with basic matroids $\mathcal{M}$, define the *conflict graph* G_T to be the graph on $\mathcal{M}$ where basic matroids M_1 and M_2 share an edge if their ground sets intersect. [12] show that if T is a good $\{1, 2, 3\}$-decomposition of $\widetilde{M}$, then G_T is a forest. We can root each tree in such a forest arbitrarily, and define the parent $p(M)$ of each non-root matroid $M \in \mathcal{M}$. Let A_M be the sum-set for the edge between matroid M and its parent, i.e., $A_M = E(M) \cap E(p(M))$.

Theorem 7 (Theorem 3.8 of [12]). *There is a good $\{1, 2, 3\}$-decomposition T for any binary matroid $\widetilde{M}$ with basic matroids $\mathcal{M}$ such that (a) each matroid $M \in \mathcal{M}$ has no circuits of size 2 consisting of an element of A_M and an element of $E(\widetilde{M})$, and (b) every basic matroid $M \in \mathcal{M}$ can be obtained from some $M' \in \widetilde{\mathcal{M}}$ by deleting elements and adding parallel elements.*

Dinitz and Kortsarz showed that a good $\{1, 2, 3\}$-decomposition for a matroid $\widetilde{M}$ can be used to construct independent sets for $\widetilde{M}$ as follows. Below, $\cdot|_S$ denotes restriction to the set S.

Lemma 2 (Lemma 4.4 of [12]). *Let T be a good $\{1, 2, 3\}$-decomposition for $\widetilde{M}$ with basic matroids $\mathcal{M}$. For each $M \in \mathcal{M}$, let I_M be an independent set of $(M/A_M)|_{(E(M)\cap E(\widetilde{M}))}$. Then $I = \bigcup_{M\in\mathcal{M}} I_M$ is independent in $\widetilde{M}$.*

Our Algorithm. Given the input matroid $\widetilde{M}$, our idea is to take a good decomposition T and run a PI-OCRS for $(M/A_M)|_{(E(M)\cap E(\widetilde{M}))}$ for each vertex M in the conflict graph G_T. Then we need to glue the pieces together using Lemma 2. One technical obstacle is that the input to an OCRS is a feasible point in the matroid polytope, so to use the framework of [12] we need to convert it into a feasible solution to the polytopes of the (modified) basic matroids. Our insight is captured by the following lemma.

Lemma 3. *Let T be a good $\{1,2,3\}$-decomposition of regular matroid $\widetilde{M}$ with basic matroids $\mathcal{M}$, and let vector $\mathbf{x} \in \frac{1}{3}\mathcal{P}_{\widetilde{M}}$. Then for every basic matroid $M \in \mathcal{M}$, if $\widehat{M} := (M/A_M)|_{(E(M)\cap E(\widetilde{M}))}$, then $\mathbf{x}|_{\widehat{M}} \in \mathcal{P}_{\widehat{M}}$.*

Proof. Fix a set $S \subseteq E(M)\cap E(\widetilde{M})$. We will show that $\text{rank}_{\widehat{M}}(S) \geq \frac{1}{3}\text{rank}_{\widetilde{M}}(S)$, from which the claim follows.

Case 1: $A_M = \{z\}$. For any maximal independent set $I \subset S$ according to M, there always exists $a \in I$ such that $(I \cup \{z\})\backslash\{a\}$ is independent in M, therefore $\text{rank}_{\widehat{M}}(S) \geq \text{rank}_{\widetilde{M}}(S) - 1$. Also since no element in S is parallel to z, for any non-empty S we have $\text{rank}_{M/A_M}(S) \geq 1$, and we can conclude that $\text{rank}_{M/A_M}(S) \geq \frac{1}{3}\text{rank}_{\widetilde{M}}(S)$.

Case 2: A_M is some 3-cycle $\{z_1, z_2, z_3\}$. For any maximal independent set $I \subset S$ according to M where $|I| \geq 3$, there always exists $a, b \in I$ such that $(I \cup \{z_1, z_2\})\backslash\{a, b\}$ is independent in M. Therefore $\text{rank}_{M_2/Z}(S) \geq \text{rank}_M(S) - 2$. We claim that there does not exist e in $E(M)\backslash A_M$ such that $e \in \text{span}(A_M)$.

Suppose for contradiction such an e exists. Then there is some circuit in $A_M \cup e$ containing e. Since there are no parallel elements, this circuit have size 3. Without loss of generality, assume this circuit is $C = \{z_1, z_2, e\}$. Since A_M is a circuit, by definition of binary matroids, the set $C\Delta A_M = \{z_3, e\}$ is a cycle, and thus e is parallel to z_3, a contradiction. Therefore for any non-empty S, we have that $\text{rank}_{M/A_M}(S) \geq 1$, and we conclude that $\text{rank}_{M/A_M}(S) \geq \frac{1}{3}\text{rank}_{\widetilde{M}}(S)$. □

We conclude the main theorem of the section (see the full version for a proof).

Theorem 8 (Regular Matroids). *There is a $(1/3, 1/12)$-selectable PI-OCRS for regular matroids.*

References

1. Adamczyk, M., Wlodarczyk, M.: Random order contention resolution schemes. In: 59th IEEE Annual Symposium on Foundations of Computer Science, FOCS 2018, Paris, France, 7–9 October 2018, pp. 790–801. IEEE Computer Society (2018)
2. Arsenis, M., Drosis, O., Kleinberg, R.: Constrained-order prophet inequalities. In: Proceedings of the 2021 ACM-SIAM Symposium on Discrete Algorithms, SODA 2021, Virtual Conference, 10–13 January 2021, pp. 2034–2046. SIAM (2021)
3. Azar, P.D., Kleinberg, R., Weinberg, S.M.: Prophet inequalities with limited information. In: Proceedings of the Twenty-Fifth Annual ACM-SIAM Symposium on Discrete Algorithms, SODA 2014, Portland, Oregon, USA, 5–7 January 2014, pp. 1358–1377. SIAM (2014)
4. Azar, P.D., Kleinberg, R., Weinberg, S.M.: Prior independent mechanisms via prophet inequalities with limited information. Games Econ. Behav. **118**, 511–532 (2019)

5. Babaioff, M., Dinitz, M., Gupta, A., Immorlica, N., Talwar, K.: Secretary problems: weights and discounts. In: Proceedings of the Twentieth Annual ACM-SIAM Symposium on Discrete Algorithms, SODA 2009, New York, NY, USA, 4–6 January 2009, pp. 1245–1254. SIAM (2009)
6. Bateni, M., Dehghani, S., Hajiaghayi, M., Seddighin, S.: Revenue maximization for selling multiple correlated items. In: Bansal, N., Finocchi, I. (eds.) Algorithms - ESA 2015. LNCS, vol. 9294, pp. 95–105. Springer, Heidelberg (2015). https://doi.org/10.1007/978-3-662-48350-3_9
7. Caragiannis, I., Gravin, N., Lu, P., Wang, Z.: Relaxing the independence assumption in sequential posted pricing, prophet inequality, and random bipartite matching. In: Feldman, M., Fu, H., Talgam-Cohen, I. (eds.) WINE 2021. LNCS, vol. 13112, pp. 131–148. Springer, Cham (2021). https://doi.org/10.1007/978-3-030-94676-0_8
8. Caramanis, C., Faw, M., Papadigenopoulos, O., Pountourakis, E.: Single-sample prophet inequalities revisited. CoRR abs/2103.13089 (2021)
9. Chawla, S., Gergatsouli, E., McMahan, J., Tzamos, C.: Approximating Pandora's box with correlations. In: Approximation, Randomization, and Combinatorial Optimization. Algorithms and Techniques, APPROX/RANDOM 2023. LIPIcs, Atlanta, Georgia, USA, 11–13 September 2023, vol. 275, pp. 26:1–26:24. Schloss Dagstuhl - Leibniz-Zentrum für Informatik (2023)
10. Chekuri, C., Vondrák, J., Zenklusen, R.: Submodular function maximization via the multilinear relaxation and contention resolution schemes. SIAM J. Comput. **43**(6), 1831–1879 (2014)
11. Correa, J.R., Foncea, P., Hoeksma, R., Oosterwijk, T., Vredeveld, T.: Posted price mechanisms and optimal threshold strategies for random arrivals. Math. Oper. Res. **46**(4), 1452–1478 (2021)
12. Dinitz, M., Kortsarz, G.: Matroid secretary for regular and decomposable matroids. SIAM J. Comput. **43**(5), 1807–1830 (2014)
13. Dughmi, S.: Matroid secretary is equivalent to contention resolution. In: 13th Innovations in Theoretical Computer Science Conference, ITCS 2022. LIPIcs, Berkeley, CA, USA, 31 January–3 February 2022, vol. 215, pp. 58:1–58:23. Schloss Dagstuhl - Leibniz-Zentrum für Informatik (2022)
14. Dughmi, S., Kalayci, Y.H., Patel, N.: Limitations of stochastic selection problems with pairwise independent priors. CoRR abs/2310.05240 (2023)
15. Esfandiari, H., Hajiaghayi, M., Liaghat, V., Monemizadeh, M.: Prophet secretary. SIAM J. Discrete Math. **31**(3), 1685–1701 (2017)
16. Feldman, M., Svensson, O., Zenklusen, R.: Online contention resolution schemes. In: Proceedings of the Twenty-Seventh Annual ACM-SIAM Symposium on Discrete Algorithms, pp. 1014–1033. SIAM (2016)
17. Giambartolomei, G., Mallmann-Trenn, F., Saona, R.: Prophet inequalities: separating random order from order selection. CoRR abs/2304.04024 (2023)
18. Goyal, V., Humair, S., Papadigenopoulos, O., Zeevi, A.: MNL-Prophet: sequential assortment selection under uncertainty. CoRR abs/2308.05207 (2023)
19. Gupta, A., Kehne, G., Levin, R.: Set covering with our eyes wide shut. In: Proceedings of the 2024 Annual ACM-SIAM Symposium on Discrete Algorithms (SODA), pp. 4530–4553. SIAM (2024)
20. Hill, T.P., Kertz, R.P.: A survey of prophet inequalities in optimal stopping theory. In: Strategies for Sequential Search and Selection in real time, Amherst, MA (1990). Contemp. Math. **125**, 191–207. American Mathematical Society, Providence (1992)

21. Immorlica, N., Singla, S., Waggoner, B.: Prophet inequalities with linear correlations and augmentations. In: EC 2020: The 21st ACM Conference on Economics and Computation, Virtual Event, Hungary, 13–17 July 2020, pp. 159–185. ACM (2020)
22. Kleinberg, R., Weinberg, S.M.: Matroid prophet inequalities and applications to multi-dimensional mechanism design. Games Econom. Behav. **113**, 97–115 (2019)
23. Krengel, U., Sucheston, L.: On semiamarts, amarts, and processes with finite value. In: Probability on Banach Spaces. Advanced Probability Related Topics, vol. 4, pp. 197–266. Dekker, New York (1978)
24. Lee, E., Singla, S.: Optimal online contention resolution schemes via ex-ante prophet inequalities. In: 26th Annual European Symposium on Algorithms, ESA 2018. LIPIcs, Helsinki, Finland, 20–22 August 2018, vol. 112, pp. 57:1–57:14. Schloss Dagstuhl - Leibniz-Zentrum für Informatik (2018)
25. Luby, M., Wigderson, A.: Pairwise independence and derandomization. Found. Trends Theor. Comput. Sci. **1**(4), 237–301 (2005)
26. Rinott, Y., Samuel-Cahn, E.: Comparisons of optimal stopping values and prophet inequalities for negatively dependent random variables. Ann. Statist. **15**(4), 1482–1490 (1987)
27. Rinott, Y., Samuel-Cahn, E.: Optimal stopping values and prophet inequalities for some dependent random variables. In: Stochastic Inequalities, Seattle, WA (1991). IMS Lecture Notes Monograph Series, vol. 22, pp. 343–358. Institute of Mathematical Statistics, Hayward (1992)
28. Rubinstein, A.: Beyond matroids: secretary problem and prophet inequality with general constraints. In: Proceedings of the 48th Annual ACM SIGACT Symposium on Theory of Computing, STOC 2016, Cambridge, MA, USA, 18–21 June 2016, pp. 324–332. ACM (2016)
29. Rubinstein, A., Singla, S.: Combinatorial prophet inequalities. In: Proceedings of the Twenty-Eighth Annual ACM-SIAM Symposium on Discrete Algorithms, SODA 2017, Barcelona, Spain, Hotel Porta Fira, 16–19 January 2017, pp. 1671–1687. SIAM (2017)
30. Rubinstein, A., Wang, J.Z., Weinberg, S.M.: Optimal single-choice prophet inequalities from samples. In: 11th Innovations in Theoretical Computer Science Conference, ITCS 2020. LIPIcs, Seattle, Washington, USA, 12–14 January 2020, vol. 151, pp. 60:1–60:10. Schloss Dagstuhl - Leibniz-Zentrum für Informatik (2020)
31. Schrijver, A.: Combinatorial Optimization: Polyhedra and Efficiency. Algorithms and Combinatorics, vol. 24. Springer, Berlin (2003)
32. Seymour, P.: Decomposition of regular matroids. A source book in matroid theory, p. 339 (1986)

An FPTAS for Connectivity Interdiction

Chien-Chung Huang[1], Nidia Obscura Acosta[2(✉)], and Sorrachai Yingchareonthawornchai[3]

[1] Department of Computer Science, École Normale Supérieure, Paris, France
cchuang@di.ens.fr
[2] Department of Computer Science, Aalto University, Espoo, Finland
nidia.obscuraacosta@aalto.fi
[3] The Hebrew University of Jerusalem, Jerusalem, Israel
sorrachai.cp@gmail.com

Abstract. In the connectivity interdiction problem, we are asked to find a global graph cut and remove a subset of edges under a budget constraint, so that the total weight of the remaining edges in this cut is minimized. This problem easily includes the knapsack problem as a special case, hence it is NP-hard. For this problem, Zenklusen [Zen14] designed a polynomial-time approximation scheme (PTAS), and for the special case of unit edge costs, exact algorithms. He posed the question of whether a fully polynomial-time approximation scheme (FPTAS) is possible for the general case. We give an affirmative answer. For the special case of unit edge costs, we also give faster exact and approximation algorithms.

Our main technical contribution is to establish a connection with an intermediate graph cut problem, called the *normalized* min-cut, which, roughly speaking, penalizes the edge weights of the remaining edges more severely, when more edges are taken out for free.

1 Introduction

We study the *connectivity interdiction problem* [Zen14]. In this problem, we are given an instance $(G = (V, E), w, c, b)$, where G is an undirected multi-graph which has weights $w : E \to \mathbb{Z}_{>0}$ and costs $c : E \to \mathbb{Z}_{>0}$ associated with its edges. A subset of edges $F \subseteq E$ is feasible if $c(F) = \sum_{e \in F} c(e)$ is no more than the budget $b \in \mathbb{Z}_{>0}$. The objective is to remove a feasible set of edges so that in the remaining graph $G' = (V, E \backslash F)$, the minimum weight of a global cut is minimized. Connectivity interdiction arises as a natural optimization problem in graphs. The potential applications include drug delivery intersection [Woo93], nuclear smuggling interdiction [MPS07], security of electric grids under terrorist attacks [SWB04], hospital infection control [Ass87], the vulnerability of the network infrastructure [MM09], and strategic military planning in the battlefield [GMT71].

We note that connectivity interdiction can also be regarded as the *b-free min-cut problem*, that is, we aim at finding a vertex set $S \subsetneq V$ and a feasible subset of edges $F \subseteq \delta(S)$ (where $\delta(S)$ is the set of edges with exactly one endpoint in S), so that $\sum_{e \in F} c(e) \leq b$ and $\sum_{e \in \delta(S) \backslash F} w(e)$ is minimized.

J. Vygen and J. Byrka (Eds.): IPCO 2024, LNCS 14679, pp. 210–223, 2024.
https://doi.org/10.1007/978-3-031-59835-7_16

It is easy to see that even in a graph as simple as two vertices connected by parallel edges, the b-free min-cut problem already contains the knapsack problem as a special case, and hence is NP-hard. A very natural special case is that of unit edge costs, that is when $c(e) = 1$ for all $e \in E$. This case is known to be in P.

In the rest of this paper, we assume that $n = |V|$ and $m = |E|$. For convenience, we define a cut as the set of edges instead of the set of vertices. That is, we say that a set of edges C is a cut if $C = \delta(S)$ for some $S \subsetneq V$. We denote the weight of a cut C, as $w(C) = \sum_{e \in C} w(e)$: the total sum of the weight of its edges. Throughout the paper, the notation $\tilde{O}(\cdot)$ hides poly-logarithmic factors in n and m. Zenklusen [Zen14] has shown the following two results for the connectivity problem.

- A PTAS for arbitrary edge costs.
- For the unit edge costs, an $\tilde{O}(m^2 n^4)$ deterministic exact algorithm and an $\tilde{O}(mn^4)$ randomized exact algorithm[1].

In [Zen14] Zenklusen has posed the open question whether it is possible to obtain an FPTAS with arbitrary edge costs.

1.1 Our Results and Techniques

We give an affirmative answer to Zenklusen's question, showing an FPTAS for arbitrary edge costs. Let $(G = (V, E), w, c, b)$ be an instance of the b-free min-cut problem.

Theorem 1 (FPTAS). *With arbitrary edge costs, there is a fully polynomial time approximation scheme for the b-free min-cut problem: given $\delta > 0$, in time $O(\frac{m^2 n^4}{\delta} \log(n w_{max} b))$, where $w_{\max} = \max_{e \in E} w(e)$, we can find an $(1+\delta)$-approximate solution.*

Remark 1. Furthermore, we show that when the value $W = \max w_i$ is polynomial on the input size, the connectivity interdiction problem admits a pseudo-polynomial time algorithm.

For the special case of unit edge costs, we also improve on the running time of Zenklusen. We design three algorithms that offer varying trade-offs between the running time and the accuracy.

Theorem 2. *With unit edge costs and for positive accuracy parameter $\varepsilon < 1/100$, we can*

1. *find an exact solution in $\tilde{O}(m + n^4 b)$ time;*
2. *or find an $(1+\varepsilon)$-approximate solution in $\tilde{O}(\frac{m}{\varepsilon} + n^3 b)$ time;*
3. *or find an $(2+\varepsilon)$-approximate solution in $\tilde{O}(\frac{m}{\varepsilon} + n^2 b)$ time.*

All these algorithms are randomized and succeed with high probability.

[1] $\tilde{O}(\cdot)$ hides poly-logarithmic factors.

Our Proof Method. Let positive $\varepsilon < 1/100$ be an accuracy parameter. Let $\alpha_1 = 2+2\varepsilon, \alpha_2 = 2-4\varepsilon, \alpha_3 = 1.5-2\varepsilon$. We say that a cut C is an α-approximate cut if the weight of C is at most α factor times the value of a global min-cut. The key approach is to show the following.

Theorem 3. *We can re-define a weight function $\tilde{w}^{(i)}$ where $i \in \{1,2,3\}$ in the same graph G so that at least one α_i-approximate min-cut C w.r.t. $\tilde{w}^{(i)}$ contains a feasible edge-set $F \subseteq C$ such that $w(C \backslash F)$ is β_i-approximation to the b-free min-cut problem where $\beta_1 = 1, \beta_2 = 1+O(\varepsilon), \beta_3 = 2+O(\varepsilon)$.*

To design an FPTAS for general connectivity interdiction, we use the weight function[2] $\tilde{w}^{(1)}$ as stated in Theorem 3 for which one of the α_1-approximate cut C (w.r.t. the weight function $\tilde{w}^{(1)}$) has a feasible set $F \subseteq C$ such that $w(C \backslash F)$ is an optimal solution to the interdiction problem. Then, we enumerate all α_1-approximate cuts in $\tilde{O}(n^{\lfloor 2\alpha_1 \rfloor}) = \tilde{O}(n^4)$ time using Nagamochi et al. [NNI97] cut enumeration algorithm. Given a cut, computing the optimal feasible edge-set corresponds to a knapsack minimization problem, where an FPTAS is known [GL80].

For the case of unit edge costs, we could use any $\tilde{w}^{(i)}$ where $i \in \{1,2,3\}$. By Theorem 3, either exact, $1+O(\varepsilon)$-, $2+O(\varepsilon)$-approximate solutions for connectivity interdiction can be found in the enumeration of α_i-approximate cuts where $i = 1,2,3$, respectively. By adapting the tree packing theorem [Kar00], we can show that the number of approximate cuts is $O(n^4), O(n^3), O(n^2)$, respectively. Given a cut, the optimal feasible edge set corresponds to the b heaviest edges in the cut w.r.t. w (since the cost is unitary).

Here as our goal is to have really fast algorithms; it would be inefficient to "explicitly" list all the α_i-approximate cuts and then sum up the edge weights after removing the heaviest b edges. To do so, we design a data structure that can output the value $w(C \backslash F)$ quickly. Such a data structure uses several ideas from a recent work of Chekuri and Quanrud [CQ17] (cf. [Kar00]). They introduced the data structures for enumerating and evaluating the weights of $O(n^2)$ many $(1+\epsilon)$-approximate cuts. We extend their framework to enumerate and evaluate the weights of $O(n^3)$ and $O(n^4)$ approximate cuts. To evaluate $w(C \backslash F) = w(C) - w(F)$ quickly, we use priority queues on top of their data structures to evaluate $w(F)$ in $\tilde{O}(b)$ time. This explains the terms $\tilde{O}(n^4 b), \tilde{O}(n^3 b), \tilde{O}(n^2 b)$, respectively in Theorem 2.

Normalized Min-Cut. The proof of Theorem 3 hinges on an intermediate problem that we call *the normalized min-cut problem*, defined as follows.

Definition 1 (Normalized min-cut). *Given (G,w,c,b) (originally an instance of a b-free min-cut problem), here the objective is to find a cut C, and a subset of its edges $F \subset C$ satisfying $0 \le c(F) \le b$, so that $\frac{w(C \backslash F)}{\delta_{c(F)}}$ is minimized where $\delta_i := b - i + 1$.*

---

[2] We will discuss how to compute $w^{(1)}$ efficiently later.

Intuitively, the cost of a cut is "normalized" according to the total cost of edges that are taken out: the less costly the subset F taken out, the more heavily the rest of the edges $C \backslash F$ would be normalized.

In fact, originally, the normalized min-cut problem arises in a different (but related) context. In a recent paper of Chalermsook et el. [CHN+22] on survivable network design, the same problem was first introduced (under a different name "minimum normalized free cut") to deal with a certain boxing constraint in the LPs. There, a special case of unit-edge costs is actually *solved as a technical necessity.* To obtain an FPTAS in this paper, we emphasize that we do not need to solve the normalized min-cut problem per se, but rather we use its optimal solution as a certificate in the analysis of the weight function $\tilde{w}^{(i)}$ in Theorem 3. We introduce another notation to facilitate our discussion.

Definition 2. *For each $0 \leq i \leq b$, let $\lambda_i = \min_{C, F \subseteq C, c(F)=i} w(C \backslash F)$. (In case that there is no C and $F \subseteq C$ so that $c(F) = i$, λ_i is understood to be infinity.)*

Furthermore, let OPT and OPT^N denote the value of the optimal solution for the b-free min-cut and the normalized min-cut, repsectively. Then by definition,

$$\mathsf{OPT} = \min_{0 \leq i \leq b} \lambda_i, \quad \mathsf{OPT}^N = \min\{\frac{\lambda_0}{b+1}, \frac{\lambda_1}{b}, \ldots, \frac{\lambda_i}{b-i+1}, \ldots, \frac{\lambda_{b-1}}{2}, \frac{\lambda_b}{1}\}.$$

Note that when c is unit edge costs, $\mathsf{OPT} = \lambda_b$, but for arbitrary costs this might not be true.

For both b-free min-cut and normalized min-cut, a solution involves a cut C and its feasible set of edges $F \subseteq C$. In our presentation, we often write them as a pair (C, F).

We now explain how the normalized min-cut problem comes into the picture. To define the intermediate weight function, we *truncate* the weights as follows.

Definition 3. *Let τ be a parameter. Define a global min-cut problem $(G = (V, E), \tilde{w}_\tau)$ by truncating the weights w as follows:*

$$\tilde{w}_\tau(e) = \begin{cases} \tau \cdot c(e) & \text{if } w(e) \geq \tau \cdot c(e), \\ w(e) & \text{otherwise.} \end{cases}$$

An edge $e \in E$ is heavy *if $w(e) \geq \tau \cdot c(e)$. Otherwise, it is* light.

Assuming that we are given an estimate τ of OPT^N (we will explain later how such guesses can be made and how to bound the number of such guesses). The following relations of the various optimal cuts (of different problems) is crucial.

Theorem 4 (Main).

Let (C^N, F^N) be the solution that realizes OPT^N in the normalized min-cut problem, i.e., $\mathsf{OPT}^N = \frac{w(C^N \backslash F^N)}{b-c(F^N)+1}$. Similarly, let (C^*, F^*) be a solution that realizes OPT in the b-free min-cut problem, i.e., $w(C^* \backslash F^*) = \mathsf{OPT}$.

Suppose that $\mathsf{OPT}^N(1-\varepsilon) \leq \tau \leq \mathsf{OPT}^N$ where $0 < \varepsilon \leq 1/100$. Consider the instance $(G, \tilde{w}_\tau)$ of the global min-cut problem, and denote $C^{\min}$ as an optimal solution.

Finally, let $\alpha_1 = 2 + 2\varepsilon, \alpha_2 = 2 - 4\varepsilon$, and $\alpha_3 = 1.5 - 2\varepsilon$. Then the following facts hold.

(i) $\tilde{w}_\tau(C^*) \leq \alpha_1 \cdot \tilde{w}_\tau(C^{\min})$. *That is, C^* is α_1-approximate min-cut w.r.t. $\tilde{w}_\tau$.*
(ii) If $\tilde{w}_\tau(C^*) > \alpha_2 \cdot \tilde{w}_\tau(C^{\min})$, then $\tilde{w}_\tau(C^N) \leq \alpha_2 \cdot \tilde{w}_\tau(C^{\min})$ and (C^N, F^N) is an $(1 + O(\varepsilon))$-approximate solution to the b-free min-cut problem.
(iii) If $\tilde{w}_\tau(C^*) > \alpha_3 \cdot \tilde{w}_\tau(C^{\min})$, then $\tilde{w}_\tau(C^N) \leq \alpha_3 \cdot \tilde{w}_\tau(C^{\min})$ and (C^N, F^N) is an $(2 + O(\varepsilon))$-approximate solution to the b-free min-cut problem.

Theorem 4 immediately implies Theorem 3, and gives us a trade-off in terms of the number of cuts to enumerate and the quality of the solutions from the normalized min-cut problem. Namely, the number of approximate cuts w.r.t. $\tilde{w}$ to enumerate is $O(n^{\lfloor 2\alpha_1 \rfloor}), O(n^{\lfloor 2\alpha_2 \rfloor}), O(n^{\lfloor 2\alpha_3 \rfloor})$, which is $O(n^4), O(n^3), O(n^2)$, respectively, in order to obtain exact, $(1 + O(\varepsilon))$ and $(2 + O(\varepsilon))$ approximate solutions respectively. To remove the assumption, $\mathsf{OPT}^N(1 - \varepsilon) \leq \tau \leq \mathsf{OPT}^N$, we guess the value of τ by the geometric series $(1 + \varepsilon)^i$ in an appropriate range.

Other Related Work. For many optimization problems, one can introduce a corresponding interdiction problem. See [SS19] for a recent survey on the interdiction problem in general. Roughly speaking, in the interdiction problem, there is a certain "interdictor", who tries to worsen the optimal value of the optimization problem as much as possible. The present work focuses on the global cut problem. But in fact, a large variety of classical optimization problems have been studied from the point of view of interdiction. Examples include: shortest paths [MMG89, SBNK95, BGV89, BSR20], network design [ASG13, DXT+12], minimum spanning tree [Zen15], graph matching [Zen10], clustering [BTV10], and network flows [CZ17].

1.2 Organization

In Sect. 2, we establish the connections between the normalized min-cut and the b-free min-cut problems, which allows us to prove our main technical result, Theorem 4. In Sect. 3, we show our FPTAS for general edge costs, and in Sect. 4 we show our algorithms for unit edge costs. Most of the omitted proofs can be found in the full version of the paper.

2 The Normalized Min-Cut Problem

We prove Theorem 4 in this section.

Proof Outline. We first show the two structural properties of the optimal solution (C^N, F^N) to the normalized min-cut problem: (1) every $e \in F^N$ is heavy, i.e., every edge in the optimal free-edge set is heavy, and (2) C^N is an $(1 + 2\varepsilon)$-approximate cut in the new weight function $\tilde{w}_\tau$ (Definition 3). Next, we prove that if C^* is not α-approximate cut in $\tilde{w}_\tau$, then we can establish a good lower bound on OPT in the original weight function w. This allows us to show that the optimal solution to the normalized min-cut (C^N, F^N) is a good approximate solution to the b-free min-cut problem. Depending on the parameter α, we obtain different guarantees as shown in Theorem 4. We now formalize the proofs.

Structures of (C^N, F^N). The following fact will be useful.

Lemma 1. *Suppose that* OPT^N *is realized by* (C^N, F^N)*, namely,* $\mathsf{OPT}^N = \frac{w(C^N \backslash F^N)}{\delta_{c(F)}}$. *Then every edge* $e \in F^N$ *is a heavy edge.*

Proof. Suppose, for a contradiction, that $e \in F^N$ is a light edge. Then $w(e) < \tau c(e)$. Let $F' = F^N \backslash \{e\}$. Then

$$\frac{w(C^N \backslash F')}{\delta_c(F')} = \frac{w(C^N \backslash F^N) + w(e)}{\delta_{c(F^N)} + c(e)} < \frac{w(C^N \backslash F^N) + \tau c(e)}{\delta_{c(F^N)} + c(e)}$$

$$\leq \frac{w(C^N \backslash F^N) + (\frac{w(C^N \backslash F^N)}{\delta_{c(F^N)}})c(e)}{\delta_{c(F^N)} + c(e)} = \frac{w(C^N \backslash F^N)}{\delta_{c(F^N)}}$$

where in the second inequality we use the fact that $\tau \leq \mathsf{OPT}^N$. As $c(F') \leq c(F^N) \leq b$, we have obtained a contradiction, since the pair (C^N, F') reaches a value smaller than OPT^N.

A main subroutine of our algorithms is to enumerate all α-approximate cuts in the global cut instance $(G, \tilde{w}_\tau)$. We will first establish a lower bound on the optimal value of the global cut in the latter.

Lemma 2. *Given any cut* C *in* $(G, \tilde{w}_\tau)$*, then* $\tilde{w}_\tau(C) \geq \tau(1 + b)$.

Proof. Let $F \subseteq C$ be the set of heavy edges of C. Suppose that $c(F) \geq b + 1$. Then $\tilde{w}_\tau(C) \geq \tilde{w}_\tau(F) \geq \tau c(F) \geq \tau(1 + b)$. On the other hand, suppose that $c(F) = i < b + 1$. Then

$$\tilde{w}_\tau(C) = \tilde{w}_\tau(F) + \tilde{w}_\tau(C \backslash F) = i\tau + w(C \backslash F) \geq i\tau + \mathsf{OPT}^N \delta_i \geq i\tau + \tau\delta_i = \tau(1+b),$$

where in the second inequality we use the fact that $w(C \backslash F) \geq \lambda_i \geq \mathsf{OPT}^N \delta_i$.

Lemma 3. *Suppose that* OPT^N *is realized by* (C^N, F^N)*. Furthermore,* C *is the optimal global cut in the instance* $(G, \tilde{w})$*. Then*

$$\frac{\tilde{w}_\tau(C^N)}{\tilde{w}_\tau(C)} < 1 + 2\varepsilon,$$

assuming that $\varepsilon < 1/2$.

Proof. By Lemma 2, we already know that $\tilde{w}_\tau(C) \geq \tau(1 + b)$. It remains to upper bound $\tilde{w}_\tau(C^N)$. By Lemma 1, all edges in F^N are heavy. Thus $\tilde{w}_\tau(F^N) = \tau c(F^N)$. Furthermore, as $\tilde{w}_\tau(C^N \backslash F^N) \leq w(C^N \backslash F^N) = \delta_{c(F^N)} \mathsf{OPT}^N$. Therefore

$$\tilde{w}_\tau(C^N) \leq \tau c(F^N) + \delta_{c(F^N)}\mathsf{OPT}^N \leq \tau c(F^N) + \frac{\tau}{1-\varepsilon}\delta_{c(F^N)}$$

$$\leq \frac{\tau}{1-\varepsilon}(c(F^N) + \delta_{c(F^N)}) = \frac{\tau}{1-\varepsilon}(1+b).$$

In conclusion, we have $\frac{\tilde{w}_\tau(C^N)}{\tilde{w}_\tau(C)} < 1/(1-\varepsilon) < 1+2\varepsilon$.

Lower Bound on OPT. The next lemma states that if after enumerating all α-approximate cuts in $(G, \tilde{w}_\tau)$, we still cannot find an optimal solution in the original b-free min-cut instance, we can then establish a lower bound on OPT.

Lemma 4. *Let C^* denote the optimal solution in the b-free min-cut instance (G, w, c, b). Namely, there exists $F^* \subseteq C^*$ so that $c(F^*) \leq b$ and $w(C^* \backslash F^*) =$ OPT. Then either C^* is an α-approximate cut in the instance $(G, \tilde{w}_\tau)$, for some $\alpha \geq 1$, or $\mathsf{OPT} \geq \tau(\alpha + (\alpha-1)b)$.*

Proof. Suppose that C^* is not an α-approximate cut in $(G, \tilde{w}_\tau)$. Let $\tilde{C}$ denote the optimal global cut according to the weight $\tilde{w}_\tau$. Then

$$\alpha \leq \frac{\tilde{w}_\tau(C^*)}{\tilde{w}_\tau(\tilde{C})} \leq \frac{\tilde{w}_\tau(F^*) + \tilde{w}_\tau(C^* \backslash F^*)}{\tau(1+b)} \leq \frac{c(F^*)\tau + w(C^* \backslash F^*)}{\tau(1+b)} \leq \frac{b\tau + \mathsf{OPT}}{\tau(1+b)},$$

where the second inequality uses Lemma 2.

Corollary 1. *Suppose that OPT^N and OPT are realized by (C^N, F^N) and (C^*, F^*) respectively, namely, $\mathsf{OPT}^N = \frac{w(C^N \backslash F^N)}{\delta_{c(F^N)}}$, and $w(C^* \backslash F^*) = \mathsf{OPT}$. Furthermore, C^* is not an α-approximate cut in $(G, \tilde{w}_\tau)$ for some $\alpha \geq 1$. Then*

$$\frac{w(C^N \backslash F^N)}{w(C^* \backslash F^*)} \leq \frac{1+b}{(1-\varepsilon)(\alpha + (\alpha-1)b)}.$$

Proof.

$$\frac{w(C^N \backslash F^N)}{w(C^* \backslash F^*)} = \frac{\mathsf{OPT}^N \cdot \delta_{c(F^N)}}{\mathsf{OPT}} \leq \frac{\mathsf{OPT}^N \cdot \delta_{c(F^N)}}{\tau(\alpha + (\alpha-1)b)}$$

$$\leq \frac{\frac{\tau}{1-\varepsilon}(1+b-c(F^N))}{\tau(\alpha+(\alpha-1)b)} \leq \frac{1+b}{(1-\varepsilon)(\alpha+(\alpha-1)b)},$$

where the first inequality uses Lemma 4.

Finishing the Proof. We are now ready to prove Theorem 4. For each item of the statement, we need to set an appropriate value of α. We next show that by setting α to be suitably large, the optimal solution in b-free min-cut must be an α-approximate cut in $(G, \tilde{w}_\tau)$—this leads to an FPTAS for non-uniform cost and an exact algorithm for uniform costs.

Theorem 5. *Suppose that $\alpha = 2 + 2\varepsilon$, with $0 < \varepsilon \leq 1$. Then the optimal cut C^* that realizes* OPT *must be α-approximate cut in $(G, \tilde{w}_\tau)$.*

Proof. For a contradiction, suppose that C^* is not α-approximate in $(G, \tilde{w}_\tau)$. Then by Corollary 1,

$$\frac{w(C^N \backslash F^N)}{w(C^* \backslash F^*)} \leq \frac{1+b}{(1-\varepsilon)(\alpha + (\alpha - 1)b)} < 1,$$

where the last inequality can be easily verified. Then we reach a contradiction that $C^N \backslash F^N$ is an even cheaper solution than $C^* \backslash F^*$.

Proof (Proof of Theorem 4).

We first prove part 1. Let $\alpha = 2 + 2\varepsilon$ with $\varepsilon = 0.01$. Theorem 5 already gives us that the optimal cut C^* which realizes OPT must be a $2 + 2\varepsilon$-approximate cut in $(G, \tilde{w}_\tau)$ as desired.

We next proceed to prove Parts 2 & 3. First note that by Lemma 3, $\tilde{w}_\tau(C^N) \leq \alpha \cdot \tilde{w}_\tau(C^{\min})$ for $\alpha = 2 - 4\varepsilon$ and $\alpha = 1.5 - 2\varepsilon$.

Let $\alpha = 2 - 4\varepsilon$, note that contrary to part 1, we cannot ensure that the cut C^* which obtains OPT will be a $2 - 4\varepsilon$-approx cut in $(G, \tilde{w}_\tau)$. However, we know that by Corollary 1 the cut C^N achieving the optimal value in the normalized mincut problem is an approximation to the optimum b-free min cut as follows:

$$\frac{w(C^N \backslash F^N)}{w(C^* \backslash F^*)} \leq \frac{1+b}{(1-\varepsilon)(\alpha + (\alpha - 1)b)} \leq \frac{1+b}{(1-\varepsilon)((2-4\varepsilon) + (1-4\varepsilon)b)}$$

$$\leq \frac{1}{(1-\varepsilon)(1-4\varepsilon)} \leq 1 + 8\varepsilon.$$

For $\alpha = \frac{3}{2} - 2\varepsilon$ we have:

$$\frac{w(C^N \backslash F^N)}{w(C^* \backslash F^*)} \leq \frac{1+b}{(1-\varepsilon)((\frac{3}{2} - 2\varepsilon) + (\frac{1}{2} - 2\varepsilon)b)} \leq \frac{1}{(1-\varepsilon)(\frac{1}{2} - 2\varepsilon)} \leq 2 + 13\varepsilon.$$

3 An FPTAS for General Connectivity Interdiction

We explain how to obtain an FPTAS with arbitrary edge costs, proving Theorem 1.

Theorem 6. *Let (C^*, F^*) be an optimal b-free min-cut of the instance $(G = (V, E), w, c, b)$. There is a weight function $\tilde{w}^* : E \to \mathbb{R}_{\geq 0}$ such that C^* is a (2.01)-approximate min-cut w.r.t. $\tilde{w}^*$. Furthermore, there is an efficient algorithm that outputs a collection $\mathcal{W}$ of weight functions such that (1) $\mathcal{W}$ contains such a function $\tilde{w}^*$ and (2) $|\mathcal{W}| = O(\log(m \cdot w_{\max} \cdot b))$.*

Theorem 6 follows from Theorem 4(i) by guessing the appropriate value of τ in the range of $(1+\varepsilon)^i$ for some fixed ε. For each guess τ, we define the weight function as in Definition 3 according to τ. We defer the proof to the end of the section.

Proposition 1. *Given a cut C, suppose that $F \subseteq C$ is the optimal feasible edge-set, then we can compute, in $O(m^2/\delta)$ time for $\delta > 0$, a feasible edge-set F' so that $c(F') \leq b$ and $w(C \backslash F') \leq (1+\delta) w(C \backslash F)$.*

Proposition 1 follows from a simple reduction to knapsack minimization problem for which an FPTAS is known [GL80].

Furthermore, in [Isl09] a pseudo-polynomial time algorithm running in time $O(n^2 W)$ for the knapsack minimization is shown (where $W = \max w_i$). As we will see in Remark 2 this allows us to get a pseudo-polynomial time algorithm for the general case of connectivity interdiction. We will prove Proposition 1 at the end of the section.

We are now ready to state the FPTAS. We try all weight functions from Theorem 6. For each weight function, we enumerate all 2.1-approximate cuts w.r.t. $\tilde{w}_\tau$. For each cut C, we compute $(1+\delta)$-approximate feasible edge-set $F \subseteq C$ using Proposition 1. Finally, we output the best pair of a cut and its free-edge set found so far. The algorithm is summarized in Algorithm 1.

Algorithm 1: FPTAS for general connectivity interdiction

Input: b-free min-cut instance (G, c, w, b) and accuracy parameter $\delta > 0$.
Output: (C', F'), where C' is a cut in G and F' is its free edge-set $F' \subseteq C'$.
foreach $\tilde{w}_\tau \in \mathcal{W}$ *in the set of weight functions in Theorem 6* **do**

1. Enumerate $C_1, \ldots, C_\ell$, all 2.1-approximate cuts with respect to $\tilde{w}_\tau$.
2. For all $i \leq \ell$, Compute F_i an $(1+\delta)$-approximate feasible edge-set of C_i as described in Proposition 1.

Return the cut and its free edge-set (C', F'), such that $w(C' \setminus F') \leq w(C \setminus F)$ for all pairs (C, F) computed in the previous step.

Running Time. We use Nagamochi et al. [NNI97] cut enumeration algorithm along with the fact that the number of α-approximate min-cuts is $O(n^{\lfloor 2\alpha \rfloor}) = O(n^4)$ [Kar00]. We can compute $(1+\delta)$-approximate feasible edge-set in $O(\frac{1}{\delta} \cdot m^2)$ time using Proposition 1. By Theorem 6, we repeat for $|\mathcal{W}| = O(\log(n \cdot w_{\max} \cdot b))$ iterations. Therefore, the algorithm runs in $O(\frac{m^2 n^4}{\delta} \log(n w_{max} b))$ time.

Correctness. By Theorem 6, there is $\tilde{w}^* \in \mathcal{W}$ such that C^* is an $(2+2\varepsilon)$-min-cut w.r.t. $\tilde{w}^*$ for $\varepsilon < 1/100$. Since we enumerate all 2.1-approximate min-cuts in w^*, we are bound to encounter C^* at some point in the enumeration. By Proposition 1, we can compute a set $F \subseteq C^*$ so that $w(C^* \backslash F) \leq (1+\delta) w(C^* \backslash F^*) \leq (1+\delta)\mathsf{OPT}^*$. As a result, the solution (C', F') returned by the algorithm must be an $(1+\delta)$-approximate solution to the b-free min-cut problem.

Remark 2. Note that in order to obtain a pseudo-polynomial time algorithm for the general connectivity interdiction problem when the value $W = \max w_i$ is polynomial in the input size, we replace step 2 in Algorithm 1 with the pseudo-polynomial time algorithm from [Isl09].

We now prove Theorem 6 and Proposition 1.

Proof (Proof of Theorem 6). Observe that by the definition of OPT^N, it follows easily that[3]

$$\frac{1}{b+1} \leq \mathsf{OPT}^N \leq \lambda_0.$$

Trivially, $\lambda_0 \leq m w_{\max}$. Thus $\mathsf{OPT}^N \in [\frac{1}{b+1}, m w_{\max}]$. We define the collection of weight functions as follows. Let $\mathcal{W} = \{\}$ and let $\varepsilon \leq 1/100$. For all $j = 0, 1, \cdots$ so that $\frac{(1+\varepsilon)^j}{b+1} \leq m w_{\max}$, set $\tau = \frac{(1+\varepsilon)^j}{b+1}$, and define $\tilde{w}_\tau$ according to τ (as in Definition 3) and add it into $\mathcal{W}$. The number of weight functions is $|\mathcal{W}| = O(\log m w_{\max} b)$; furthermore, one of the τs must satisfy $\mathsf{OPT}^N(1-\varepsilon) \leq \tau \leq \mathsf{OPT}^N$ for $\varepsilon \leq 1/100$. Therefore, Theorem 4(i) applies to that weight function.

Proof (Proof of Proposition 1). Finding the optimal set F^* reduces to the following knapsack minimization problem.

> (Knapsack Minimization) Find a subset $A^* \subseteq C^*$, which respects $c(A^*) \geq (\sum_{e \in C^*} c(e)) - b$, so that $w(A^*)$ is minimized.

$C^* \backslash F^*$ is exactly the optimal solution A^* in the above knapsack problem. It is known [GL80] that, in $O(m^2/\delta)$ time for $\delta > 0$, one can find a solution A so that $w(A) \geq (\sum_{e \in E} c(e)) - b$ and $w(A) \geq (1+\delta) w(A^*)$. Setting A^* to be $C^* \backslash F$ gives the desired proof.

4 Fast Algorithms for Unit Edge Costs

We prove Theorem 2 in this section. We assume that the given instance (G, c, w, b) has unit edge costs $c: E \to 1$. Notice that in this setting, once a cut C is chosen, the best feasible set of edges $F \subseteq C$ to be removed is simply the b heaviest edges according to w. We will call these edges the *free set* (of C).

[3] We can assume that there is no cut C such that $w(C) \leq b$. Therefore, all of the λ_is have value at least 1.

4.1 The Schematic Algorithm

We now describe the algorithm. The inputs consist of an b-free mincut instance (G, c, w, b) and two positive parameters $\varepsilon < 1/100$ and $\alpha \in \{\alpha_1, \alpha_2, \alpha_3\}$ where $\alpha_1 := 2 + 2\epsilon, \alpha_2 := 2 - 4\epsilon, \alpha_3 := 1.5 - 2\epsilon$. The parameter α controls the trade-off between the approximation ratio and running time. First, we compute an estimate τ such that $\mathsf{OPT}^N(1-\varepsilon) \leq \tau \leq \mathsf{OPT}^N$. Then, we define the weight function $\tilde{w}_\tau$ according to Definition 3. We enumerate all α-approximate cuts, and for each cut C we compute $w(C \backslash F)$ where F is the free set of C. Finally, we return the cut C' and its free set F' with the smallest value of $w(C' \backslash F')$ so far. We summarize the algorithm in Algorithm 2.

Analysis. We next describe the analysis of Algorithm 2.

Theorem 7 *Let (C', F') be the solution returned by Algorithm 2 on the b-free mincut instance (G, c, w, b) with parameters $\varepsilon < 1/100$ and $\alpha \in \{\alpha_1, \alpha_2, \alpha_3\}$.*

1. *If $\alpha = \alpha_1$, and $\varepsilon = 0.001$, then (C', F') is optimal and it can be implemented to run in $\tilde{O}(m + n^4 b)$ time.*
2. *If $\alpha = \alpha_2$, then (C', F') is $(1+O(\varepsilon))$-approximation and it can be implemented to run in $\tilde{O}(\frac{1}{\varepsilon} \cdot m + n^3 b)$ time.*
3. *If $\alpha = \alpha_3$, then (C', F') is $(2+O(\varepsilon))$-approximation and it can be implemented to run in $\tilde{O}(\frac{1}{\varepsilon} \cdot m + n^2 b)$ time.*

Every implementation is randomized and succeeds with high probability.

Theorem 2 follows immediately from Theorem 7. For implementation and running time analysis, we will discuss the necessary tools in Sect. 4.2. Here we just focus on proving the correctness.

Algorithm 2: The algorithm for the unit edge costs

Input: A b-free mincut instance (G, c, w, b) and two positive parameters $\varepsilon < 1/100$ and $\alpha \in \{\alpha_1, \alpha_2, \alpha_3\}$ where $\alpha_1 := 2 + 2\epsilon, \alpha_2 := 2 - 4\epsilon, \alpha_3 := 1.5 - 2\epsilon$.

Output: (C', F'), where C' is a cut in G and F' is its free edge-set $F' \subset C'$.

1. Let τ be an estimate of OPT^N such that $\mathsf{OPT}^N(1-\varepsilon) \leq \tau \leq \mathsf{OPT}^N$.
2. Let $\mathcal{C}_{\tilde{w}_\tau}$ be the set of α-approximate min-cuts in the instance $(G, \tilde{w}_\tau)$ where $\tilde{w}_\tau$ is defined in Definition 3, which depends on the value τ.
3. For each cut $C \in \mathcal{C}_{\tilde{w}_\tau}$, compute $w(C \setminus F)$ where F is the free set of C.
4. Return the cut and its free set (C', F') such that $w(C' \setminus F') \leq w(C \setminus F)$ for all pairs (C, F) computed in the previous step.

Correctness. We now prove the correctness of the algorithm in Theorem 7. Let (C^*, F^*) be an optimal solution to b-free min-cut problem, and (C^N, F^N) be an optimal solution to the normalized min-cut problem. In each case of $\alpha \in \{\alpha_1, \alpha_2, \alpha_3\}$, we apply Theorem 4 using $\tilde{w}_\tau$ as defined in Definition 3, where $\mathsf{OPT}^N(1-\epsilon) \leq \tau \leq \mathsf{OPT}^N$.

1. If $\alpha = \alpha_1$ and $\epsilon = 0.001$, then $\mathcal{C}_{\tilde{w}_\tau}$ contains C^*, and thus F^* was computed in step 3 of the algorithm. Therefore, at the end we have $w(C' \backslash F') \leq w(C^* \backslash F^*)$.
2. If $\alpha \in \{\alpha_2, \alpha_3\}$, then we consider two different cases depending on $\tilde{w}_\tau(C^*)$. If $\tilde{w}_\tau(C^*) \leq \alpha \tilde{w}_\tau(C^{\min})$, where $C^{\min}$ is global min-cut w.r.t. $\tilde{w}_\tau$, then $\mathcal{C}_{\tilde{w}_\tau}$ contains C^*, and then we are done by the same argument in as the previous case. Otherwise, by Theorem 4, $\tilde{w}_\tau(C^N)$ is α-approximate cut in $\tilde{w}_\tau$. Thus, $C^N \in \mathcal{C}_{\tilde{w}_\tau}$. In this case, F^N was also computed in step 3 of the algorithm. Therefore, if $\alpha = \alpha_2$, then $w(C' \backslash F') \leq w(C^N \backslash F^N) \leq (1 + O(\varepsilon))w(C^* \backslash F^*)$, where the last inequality follows from Theorem 4(ii). If $\alpha = \alpha_3$, then $w(C' \backslash F') \leq w(C^N \backslash F^N) \leq (2 + O(\varepsilon))w(C^* \backslash F^*)$, where the last inequality follows from Theorem 4(iii).

4.2 Fast Implementation

For the fast implementation of Algorithm 2, we need an efficient enumeration of α-approximate cuts and the efficient evaluation of weights of cuts when their b heaviest edges are removed. We achieve the enumeration via the adaptation of the tree packing theorem by Karger [Kar00] for 3 and 4-respecting cuts.

For the evaluation of cuts with respect to the original weight w when their b-heaviest edges are removed, we generalize a cut oracle data structure from [CQ17] that *efficiently* computes the weight of 1 and 2-respecting cuts, to also handle 3 and 4-respecting cuts and to handle the computation of their b heaviest edges via the addition of a max-heap data structure to the construction.

In this section, we will give a brief sketch of these tools and provide the full proofs in the full version of the paper.

Definition 4 (r-respecting cut). *Let $G = (V, E)$ be a graph and let T be a spanning tree of G. We say that a cut C in G r-respects a spanning tree T of G, if $|C \cap E(T)| = r$. Note that each set of r tree edges induces a unique cut in G.*

Karger [Kar00] (see also the discussion [CQ17]) gave a randomized algorithm to compute a collection of $O(\log n)$ spanning trees such that with high probability every α-approximate min-cut r-respects some spanning tree in the collection.

Theorem 8 (Tree Packing Theorem [Kar00] (Restated)).
Let $G = (V, E, w)$ be an undirected weighted graph. There is an algorithm that takes G and a parameter α (where $\alpha \in \{\alpha_1, \alpha_2, \alpha_3\}$, and $\varepsilon < 1/100$) as inputs and outputs a collection $\mathcal{T}$ of $O(\log n)$ spanning trees (the constant factor in $O(\cdot)$ depends on α) satisfying the following with high probability:

- *If $\alpha = \alpha_1$, then every α_1-approx. min-cut 4-respects a spanning tree in $\mathcal{T}$,*
- *If $\alpha = \alpha_2$, then every α_2-approx. min-cut 3-respects a spanning tree in $\mathcal{T}$,*
- *If $\alpha = \alpha_3$, then every α_3-approx. min-cut 2-respects a spanning tree in $\mathcal{T}$.*

Our approximate cut enumeration task now reduces to the enumeration of all 2, 3 and 4-respecting cuts of $\mathcal{T}$, and hence the enumeration of all $O(n^2)$, $O(n^3)$ or $O(n^4)$ possible combinations of 2, 3 or 4 edges from $\mathcal{T}$, respectively (see the full version of the paper for full proofs).

Denote by U a set of edges of tree T in the enumeration, and C be the corresponding cut induced by U in G. We denote U as $[[C]]_T$ to represent the encoding of C via the tree T. By Theorem 8, every α-approximate min-cut r-respects some tree in the tree packing $\mathcal{T}$ for some $r \leq 4$ with high probability. Then, every α-approximate min-cut is listed in the enumeration with high probability.

However, during the enumeration, we need to compute the value $w(C \backslash F) = w(C) - w(F)$ where F is the set of b heaviest edges in C. Naively, this takes $O(m)$ time per iteration. To speed up this computation, we preprocess each tree in the tree packing so that for any cut $[[C]]_T$ represented by the tree, we can compute $w(C \backslash F)$ in $\tilde{O}(b)$ time. This data structure is formalized in the following:

Lemma 5 (Cut Oracle). *There is a data structure $\mathcal{O}$ that given a weighted graph $G = (V, E, w)$, and a spanning tree T of G, supports the following operations, where $[[C]]_T$ is the set of at most 4 tree edges of T inducing C.*

- $\mathcal{O}$.WEIGHT($[[C]]_T$)*: Return $w(C)$ in $\tilde{O}(1)$ time.*
- $\mathcal{O}$.KFREEWEIGHT($[[C]]_T, k$)*: Return $w(C \backslash F)$ where $F \subseteq C$ is the set of k heaviest edges in C in $\tilde{O}(k)$ time.*
- $\mathcal{O}$.CUT($[[C]]_T$)*: returns the cut C in $O(m)$ time.*

The data structure can be constructed in $\tilde{O}(m)$ time.

For the proof of Lemma 5 and the final details of the fast implementation and running time see the full version of the paper.

Acknowledgements. This research was supported by the European Research Council (ERC) under the European Union's Horizon 2020 research and innovation programme under grant agreement No 759557, ERC Starting grant CODY 101039914, ISF grant 3011005535, Academy of Finland grant 335715, the group CASINO/ENS chair on algorithmics and machine learning, and also by the grants ANR-19-CE48-0016 and ANR-18-CE40-0025-01.

References

[ASG13] Addis, B., Di Summa, M., Grosso, A.: Identifying critical nodes in undirected graphs: complexity results and polynomial algorithms for the case of bounded treewidth. Discrete Appl. Math. **161**(16–17), 2349–2360 (2013)

[Ass87] Assimakopoulos, N.: A network interdiction model for hospital infection control. Comput. Biol. Med. **17**(6), 413–422 (1987)

[BGV89] Ball, M.O., Golden, B.L., Vohra, R.V.: Finding the most vital arcs in a network. Oper. Res. Lett. **8**(2), 73–76 (1989)

[BSR20] Busan, S., Schäfer, L.E., Ruzika, S.: The two player shortest path network interdiction problem. CoRR, abs/2004.08338 (2020)

[BTV10] Bazgan, C., Toubaline, S., Vanderpooten, D.: Complexity of determining the most vital elements for the 1-median and 1-center location problems. In: Wu, W., Daescu, O. (eds.) COCOA 2010. LNCS, vol. 6508, pp. 237–251. Springer, Heidelberg (2010). https://doi.org/10.1007/978-3-642-17458-2_20

[CHN+22] Chalermsook, P., Huang, C.-C., Nanongkai, D., Saranurak, T., Sukprasert, P., Yingchareonthawornchai, S.: Approximating k-edge-connected spanning subgraphs via a near-linear time LP solver. In: ICALP. LIPIcs, vol. 229, pp. 37:1–37:20. Schloss Dagstuhl - Leibniz-Zentrum für Informatik (2022)

[CQ17] Chekuri, C., Quanrud, K.: Approximating the held-karp bound for metric TSP in nearly-linear time. In: FOCS, pp. 789–800. IEEE Computer Society (2017)

[CZ17] Chestnut, S.R., Zenklusen, R.: Hardness and approximation for network flow interdiction. Networks **69**(4), 378–387 (2017)

[DXT+12] Dinh, T.N., Xuan, Y., Thai, M.T., Pardalos, P.M., Znati, T.: On new approaches of assessing network vulnerability: hardness and approximation. IEEE/ACM Trans. Netw. **20**(2), 609–619 (2012)

[GL80] Gens, G., Levner, E.: Complexity of approximation algorithms for combinatorial problems: a survey. SIGACT News **12**(3), 52–65 (1980)

[GMT71] Ghare, P.M., Montgomery, D.C., Turner, W.C.: Optimal interdiction policy for a flow network. Naval Res. Logist. Q. **18**(1), 37–45 (1971)

[Isl09] Islam, M.T.: Approximation algorithms for minimum knapsack problem. Ph.D. thesis, University of Lethbridge, Department of Mathematics and Computer Science, Lethbridge (2009)

[Kar00] Karger, D.R.: Minimum cuts in near-linear time. J. ACM **47**(1), 46–76 (2000)

[MM09] Matisziw, T.C., Murray, A.T.: Modeling s-t path availability to support disaster vulnerability assessment of network infrastructure. Comput. Oper. Res. **36**(1), 16–26 (2009)

[MMG89] Malik, K., Mittal, A.K., Gupta, S.K.: The k most vital arcs in the shortest path problem. Oper. Res. Lett. **8**(4), 223–227 (1989)

[MPS07] Morton, D.P., Pan, F., Saeger, K.J.: Models for nuclear smuggling interdiction. IIE Trans. **39**(1), 3–14 (2007)

[NNI97] Nagamochi, H., Nishimura, K., Ibaraki, T.: Computing all small cuts in an undirected network. SIAM J. Discrete Math. **10**(3), 469–481 (1997)

[SBNK95] Schieber, B., Bar-Noy, A., Khuller, S.: The complexity of finding most vital arcs and nodes. University of Maryland at College Park, USA (1995)

[SS19] Smith, J.C., Song, Y.: A survey of network interdiction models and algorithms. Eur. J. Oper. Res. **283**(3), 797–811 (2019)

[SWB04] Salmeron, J., Wood, K., Baldick, R.: Analysis of electric grid security under terrorist threat. IEEE Trans. Power Syst. **19**(2), 905–912 (2004)

[Woo93] Wood, R.K.: Deterministic network interdiction. Math. Comput. Model. **17**(2), 1–18 (1993)

[Zen10] Zenklusen, R.: Matching interdiction. Discrete Appl. Math. **158**(15), 1676–1690 (2010)

[Zen14] Zenklusen, R.: Connectivity interdiction. Oper. Res. Lett. **42**(6), 450–454 (2014)

[Zen15] Zenklusen, R.: An O(1)-approximation for minimum spanning tree interdiction. In: 2015 IEEE 56th Annual Symposium on Foundations of Computer Science, pp. 709–728 (2015)

Tight Lower Bounds for Block-Structured Integer Programs

Christoph Hunkenschröder[1], Kim-Manuel Klein[2], Martin Koutecký[3], Alexandra Lassota[4](✉), and Asaf Levin[5]

[1] Institut für Mathematik (formerly), TU Berlin, Berlin, Germany
hunkenschroeder@tu-berlin.de

[2] Institute for Theoretical Computer Science, University of Lübeck, Lübeck, Germany
kimmanuel.klein@uni-luebeck.de

[3] Computer Science Institute, Charles University, Prague, Czech Republic
koutecky@iuuk.mff.cuni.cz

[4] Eindhoven University of Technology, Eindhoven, The Netherlands
a.a.lassota@tue.nl

[5] Faculty of Data and Decisions Sciences, The Technion, Haifa, Israel
levinas@ie.technion.ac.il

Abstract. We study fundamental block-structured integer programs called tree-fold and multi-stage IPs. Tree-fold IPs admit a constraint matrix with independent blocks linked together by few constraints in a recursive pattern; and transposing their constraint matrix yields multi-stage IPs. The state-of-the-art algorithms to solve these IPs have an exponential gap in their running times, making it natural to ask whether this gap is inherent. We answer this question affirmative. Assuming the Exponential Time Hypothesis, we prove lower bounds showing that the exponential difference is necessary, and that the known algorithms are near optimal. Moreover, we prove unconditional lower bounds on the norms of the Graver basis, a fundamental building block of all known algorithms to solve these IPs. This shows that none of the current approaches can be improved beyond this bound.

Keywords: integer programming · n-fold · tree-fold · multi-stage · (unconditional) lower bounds · ETH · subset sum

C. Hunkenschröder acknowledges funding by Einstein Foundation Berlin. K.-M. Klein was supported by DFG project KL 3408/1-1. A. Lassota was partially supported by the Swiss National Science Foundation (SNSF) within the project *Complexity of Integer Programming (207365)*. M. Koutecký was partially supported by the Charles University project UNCE 24/SCI/008 and by the project 22-22997S of GA ČR. A. Levin is partially supported by ISF – Israel Science Foundation grant number 1467/22.
A full version of this paper can be found in [14].

J. Vygen and J. Byrka (Eds.): IPCO 2024, LNCS 14679, pp. 224–237, 2024.
https://doi.org/10.1007/978-3-031-59835-7_17

1 Introduction

In the past years, there has been tremendous progress in the algorithmic theory and in the applications of *block-structured integer programming*. An *integer program (IP)* in standard form is the problem $\min\{c^\intercal x : Ax = b, l \leq x \leq u, x \in \mathbb{Z}^n\}$. We deal with the setting where the constraint matrix A exhibits a certain block structure. One of the most prominent block-structure are *n-fold* integer programs (n-fold IPs), in which the constraint matrix A decomposes into a block-diagonal matrix after the first few rows are deleted. In other words, n-fold IPs are constructed from independent IPs of small dimensions that are linked by a few constraints. The generalization, in which the diagonal blocks themselves have an n-fold structure recursively, is called *tree-fold* IPs. The transpose of a tree-fold matrix yields another class of highly relevant constraint matrices called *multi-stage* matrix. We formally define those next.

Definition 1 (Tree-fold and multi-stage matrices). *Any matrix $A \in \mathbb{Z}^{m \times n}$ is a tree-fold matrix with one level and level size m. A matrix A is a tree-fold matrix with $\tau \geq 2$ levels and level sizes $\sigma = (\sigma_1, \ldots, \sigma_\tau) \in \mathbb{Z}^\tau_{\geq 1}$ if deleting the first σ_1 rows of A decomposes the matrix into a block-diagonal matrix, where each block is a tree-fold matrix with $\tau - 1$ levels and level sizes $(\sigma_2, \ldots, \sigma_\tau)$.*

If $A^\intercal$ is a tree-fold matrix with τ levels and level sizes σ, then A is called a multi-stage matrix *with τ stages and stage sizes σ.*

For a schematic picture, see Fig. 1

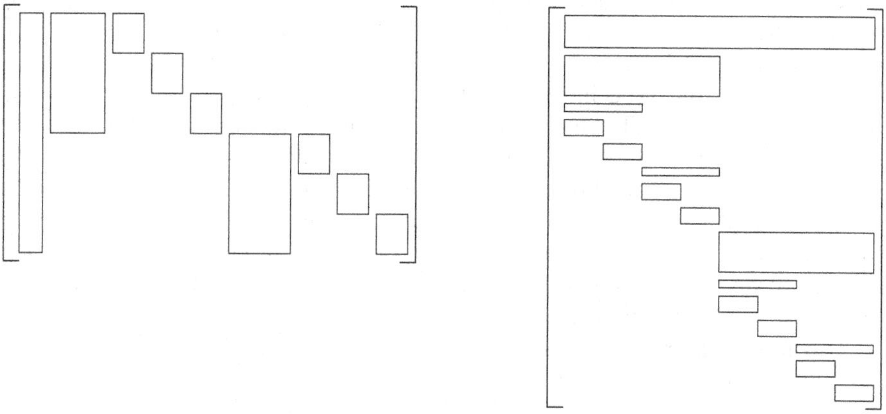

Fig. 1. On the left, a schematic multi-stage with three levels is presented. On the right, a schematic tree-fold with 4 layers is pictured. All entries within a rectangle can be non-zero, all entries outside of the rectangles must be zero.

Definition 2 (n-fold and 2-stage stochastic matrices). *The special case of a tree-fold matrix with two levels is called n-fold. Respectively, a multi-stage matrix with two stages is called 2-stage stochastic matrix.*

For a picture, see Fig. 2

$$\begin{bmatrix} A_1 & D_1 & & & \\ A_2 & & D_2 & & \\ \vdots & & & \ddots & \\ A_n & & & & D_n \end{bmatrix} \qquad \begin{bmatrix} C_1 & C_2 & \dots & C_n \\ D_1 & & & \\ & D_2 & & \\ & & \ddots & \\ & & & D_n \end{bmatrix}$$

Fig. 2. On the left, a 2-stage stochastic matrix with blocks A_i and D_i, $i \in [n]$ is presented. On the right, an n-fold matrix with blocks C_i and D_i, $i \in [n]$ is pictured. All entries not belonging to a block are zero.

The study of n-fold IPs was initiated in [30]. A milestone was the fixed-parameter tractable algorithm by Hemmecke et al. [12] whose running time depends polynomially on the dimension n, and exponentially only on the sizes of the small blocks. Faster and more generally applicable algorithms were subsequently developed, including strongly polynomial algorithms, near-linear (in n) time algorithms [4,8,9,18,28,30], and a new result where entries in the global part can be large [6]. At the same time, n-fold IPs found many applications, for instance in scheduling problems, stringology, graph problems, and computational social choice, see e.g. [3,11,13,16,17,22–26], solving long-standing open problems.

Multi-stage IPs and their special case for $\tau = 2$ (called 2-stage stochastic IPs) have been studied even longer than n-fold IPs, going back to the work of Aschenbrenner and Hemmecke [2]. They are commonly used in stochastic programming and often used in practice to model uncertainty of decision making over time [1,7,19,29]. The first known upper bounds on the complexity of solving multi-stage IPs had a huge and non-explicit dependence on τ and σ in their running time. The upper bound was subsequently improved to have an exponential tower of height τ with $\|\sigma\|_1$ appearing at the top (times a polynomial in the encoding length of the input I), and only very recently to have a triple-exponential (in $\|\sigma\|_1$) running time (times $|I|^{O(1)}$) [5,9,20,21]. In [6], a new FPT time algorithm to decide feasibility of a 2-stage stochastic IP is presented that can also handle large entries in the global part (i.e., the largest entry in the global part is not a parameter).

Intriguingly, the algorithms for tree-fold IPs have a running time that depends only double-exponentially on $\|\sigma\|_1$ [9]. A natural response to seeing this exponential gap is to ask whether it is inherent, and whether multi-stage integer programming is indeed harder than tree-fold IPs despite their similar nature of constraint matrices. This question was partially answered when Jansen et al. [15] showed that, assuming the Exponential Time Hypothesis (ETH), 2-stage stochastic IPs require a double-exponential running time in $\|\sigma\|_1$. This contrasts with known single-exponential algorithms w.r.t. $\|\sigma\|_1$ for n-fold IPs. Also for tree-fold IPs, a double-exponential lower bound w.r.t. τ is known [27], although this is stated in terms of the parameter *tree-depth*, which is linked to the largest number of non-zeroes in any column (see Sect. 6), still leaving the exact dependence on the number of levels τ open.

We settle the complexity of all aforementioned block-structured integer programs, answer the question whether the exponential gap is necessary affirmatively, and (nearly) close the gaps between algorithms and lower bounds:

1. (Theorem 1) We show an ETH-based lower bound for multi-stage IPs that is triple-exponential in the number of levels τ when the level sizes σ are constant, and recovers the existing double-exponential lower bound (in $\|\sigma\|_1$) [15] as a special case. This lower bound is comparable to the running time of the currently fastest algorithm [21].
2. (Theorem 2) We show an ETH-based lower bound for tree-fold IPs which recovers as a special case the result of [27], and is comparable to the running time of the currently best algorithm [9]. Our bound shows more accurately how the running time depends on τ.
3. (Corollary 3) A particularly interesting consequence of Theorem 2 is a lower bound of roughly $2^{\sigma_1 \sigma_2}$ for the special class of n-fold IPs.

The core technical idea behind the lower bounds in this paper relates bi- and tri-diagonal matrices to block-structured matrices, see Sect. 2.

4. (Lemma 1 and Corollary 1) Every bi-/tri-diagonal matrix can be reordered to obtain a multi-stage (Lemma 1) and a tree-fold (Corollary 1) matrix.

We believe this result is of independent interest, as it provides a new tool for solving combinatorial problems: Formulating any problem as a matrix with constant bandwidth is enough to be able to apply the tree-fold or multi-stage integer programming algorithms to solve it efficiently.

Since the hard instances we construct have bi- or tri-diagonal structure, we are able to obtain the required lower bounds. For the multi-stage IPs lower bound, we combine this idea with splitting one complicated constraint carefully into several sparse constraints with only few variables, similarly to the well-known reduction from a 3-SAT formula to a 3-SAT formula where each variable only appears constantly often. This is done in Sect. 3. Section 4 continues with the lower bound for tree-fold IPs.

The central concept to all the aforementioned algorithms is the *Graver basis* $\mathcal{G}(A)$ of the constraint matrix A, or in case of the new result [6], of some reduced constraint matrix A' with block-structure and small entries, and a closer examination shows that the main factor driving the complexity of those algorithms are the ℓ_∞- and ℓ_1-norms of elements of $\mathcal{G}(A)$ ($\mathcal{G}(A')$). Improved bounds on those norms would immediately lead to improved algorithms, contradicting ETH. We show that the ETH cannot be violated in this way by giving *unconditional* lower bounds on the norms of elements of $\mathcal{G}(A)$ for block-structured matrices.

We demonstrate these norm lower bounds on the matrices used in the proofs of Theorems 1 and 2, as described in Sect. 5. Our unconditional lower bounds for Graver basis elements traces back to the same matrix hardness was proven for in [20]. Extending these results to multi-stage IPs and advancing to tree-fold and n-fold IPs though required the here presented concept of rearranging bi- and tri-diagonal matrices. In Sect. 6, we briefly express our work in terms of the parameter *tree-depth* that is also commonly used to describe block-structured integer programs.

This paper partially builts on an arXiv preprint [9]. The present paper provides stronger and novel results. Specifically, it introduces bi- and tri-diagonal matrices formally as crucial components for the hardness proofs. This approach enables us to reframe the results for tree-fold IPs in terms of *levels* and the maximum number of rows in a level, rather than stating them solely on tree-depth as in both [9,27], which is the product of these parameters. This refined perspective allows for a more nuanced analysis of n-fold IPs which derives at its (near tight) lower bound, which was unattainable with previous methods. In the context of multi-stage IPs, our investigation picks up from where the proof in [15] for 2-stage stochastic IPs concluded. Notably, the proof for 2-stage stochastic IPs did not involve or observe any potential rearrangement of the stages; the first stage comprised only one variable, and the 2-stage stochastic structure emerged naturally from the underlying problem.

Due to space restrictions, some proofs are omitted. A proof sketch is given in this case. For a full version, see [14].

2 About Bi-Diagonal and Tri-Diagonal Matrices

This section is devoted to bi- and tri-diagonal matrices, i.e., matrices in which all non-zero entries are on two or three consecutive diagonals, respectively.

Definition 3 (Bi-diagonal, Tri-diagonal Matrix). *A matrix $\mathcal{A} \in \mathbb{Z}^{m \times n}$ is* bi-diagonal *if $a_{i,j} = 0$ for $i \notin \{j, j+1\}$. A matrix $\mathcal{A} \in \mathbb{Z}^{m \times n}$ is* tri-diagonal *if $a_{i,j} = 0$ for $i \notin \{j, j+1, j+2\}$.*

We show that any bi- or tri-diagonal matrix A can be viewed as a multi-stage or tree-fold matrix with the desired parameters. While the next lemma requires quite specific matrix dimensions, note that once a matrix is bi- or tri-diagonal, we can always add zero rows or columns, and it remains bi- or tri-diagonal.

Lemma 1. *Let $\tau \geq 1$, $\sigma \in \mathbb{Z}^{\tau}_{\geq 1}$, and define $S := \prod_{i=1}^{\tau}(\sigma_i + 1)$.*

i) Let $\mathcal{A} \in \mathbb{Z}^{S \times S-1}$ be bi-diagonal. Then $\mathcal{A}$ is a multi-stage matrix with τ stages and stage sizes σ, up to column permutations.
ii) Let $\mathcal{A} \in \mathbb{Z}^{2S \times 2S-2}$ be tri-diagonal. Then $\mathcal{A}$ is a multi-stage matrix with τ stages and stage sizes 2σ, up to column permutations.

Proof Sketch. We prove both claims by an induction on the number of stages τ. The claims trivially hold for $\tau = 1$ as any matrix can be interpreted as a multi-stage matrix of just one stage.

i) As for bi-diagonal matrices, note that if we delete a column i, the matrix separates into two independent bi-diagonal matrices, one from column 1 to column $i-1$, and the second from columns $(i+1)$ until $(S-1)$ (and the corresponding non-zero rows respectively).
We use this idea of splitting the matrix into independent matrices as follows: for $\tau > 1$, we permute σ_1 equidistant columns to the front. This gives us a 2-stage stochastic matrix with σ_1 columns in the first stage and a second stage that are exactly the independent matrices. For each of these independent matrices, we now need to find a re-arrangement into a multi-stage matrix with $\tau - 1$ stages and sizes $\sigma_2, \ldots, \sigma_\tau$, which is possible due to the induction hypothesis.

ii) The only adaption needed for the second claim is to shift two consecutive columns for each column chosen in i) to the front to split the matrix into independent matrices.

Proof. i): The proof is by induction on τ, and the base case for $\tau = 1$ is trivial. For $\tau > 1$, we briefly lay out the idea before providing the formal construction. Observe that if we delete any column, we can partition the remaining matrix into two blocks $\binom{A_1}{0}$, $\binom{0}{A_2}$ whose columns are orthogonal to each other as depicted below for column j:

$$\mathcal{A} = \left(\begin{array}{c|c|c} & 0 & \\ A_1 & \vdots & 0 \\ & a_{j,j} & \\ \hline & a_{j+1,j} & \\ 0 & \vdots & A_2 \\ & 0 & \end{array}\right).$$

Consequently, permuting σ_1 equidistant columns (that is, we take every ℓth column for an appropriate ℓ defined below) to the front, we obtain a 2-stage stochastic matrix with stage sizes (σ_1, S'), where $S' = \prod_{i=2}^{\tau}(\sigma_i + 1)$ and whose $\sigma_1 + 1$ diagonal blocks are again bi-diagonal, allowing us to induct on them.

Formally, let $A^{(0)} \in \mathbb{Z}^{S \times \sigma_1}$ be the matrix comprising the columns of $\mathcal{A}$ with index in $L := \{jS' : \ j = 1, \ldots, \sigma_1\}$. For $k = 1, \ldots, \sigma_1 + 1$, let

$$A^{(k)} \in \mathbb{Z}^{S' \times S' - 1}$$

denote the matrix where the (i,j)-th entry corresponds to the $((k-1)S' + i,\ (k-1)S' + j)$-th entry of $\mathcal{A}$.

The following 2-stage stochastic matrix arises from $\mathcal{A}$ by permuting A_0 to the front:

$$\left(\begin{array}{c|ccc} & \boxed{A^{(1)}} & & \\ A^{(0)} & & \ddots & \\ & & & \boxed{A^{(\sigma_1+1)}} \end{array}\right).$$

We have

$$A^{(k)}_{i,j} = \mathcal{A}_{(k-1)S'+i,(k-1)S'+j} = 0$$

whenever $i \notin \{j, j+1\}$, hence each matrix $A^{(k)}$ is again bi-diagonal. By induction, each $A^{(k)}$ is a multi-stage matrix with $\tau - 1$ stages and stage sizes $(\sigma_2, \ldots, \sigma_\tau)$, after a suitable permutation.

ii): The proof follows the same argument as for the bi-diagonal case, the only difference being that we have to delete two columns in order to split the matrix into independent blocks.

For $\tau = 1$, there is nothing to show. For $\tau \geq 2$, define

$$L := \bigcup_{j=1}^{\sigma_1} \{2jS' - 1, 2jS'\}$$

and let $A^{(0)} \in \mathbb{Z}^{2S \times 2\sigma_1}$ comprise the columns with indices in L. For $k = 1, \ldots, \sigma_1 + 1$, let the matrix $A^{(k)} \in \mathbb{Z}^{2S' \times 2S' - 2}$ arise from $\mathcal{A}$ by restricting to row indices $2(k-1)S' + 1, \ldots, 2kS'$ and column indices $2(k-1)S' + 1, \ldots, 2kS' - 2$. Again, the matrices $A^{(k)}$ are tri-diagonal, and applying the induction step on them yields the claim. □

Clearly, the result can be extended to any band-width on the diagonal. Also, the number of columns can be chosen individually even within a specific stage. However, the simpler version above suffices for our purposes.

By considering $\mathcal{A}^\intercal$, we obtain the following corollary.

Corollary 1. *Let $\tau \geq 1$, $\sigma \in \mathbb{Z}^\tau_{\geq 1}$, and define $S := \prod_{i=1}^{\tau} (\sigma_i + 1)$.*

i) Let $\mathcal{A}^\intercal \in \mathbb{Z}^{S \times S-1}$ be bi-diagonal. Then $\mathcal{A}$ is a tree-fold matrix with τ levels and level sizes σ, up to row permutations.
ii) Let $\mathcal{A}^\intercal \in \mathbb{Z}^{2S \times 2S-2}$ be tri-diagonal. Then $\mathcal{A}$ is a tree-fold matrix with τ levels and level sizes 2σ, up to row permutations.

We close this section with a brief remark. If $\mathcal{A}$ itself is bi-diagonal, we can add a zero column in the front. This way we obtain a matrix $\tilde{A}$ for which $\tilde{A}^\intercal$ is bi-diagonal. Similarly, we can add two columns in the tri-diagonal case.

3 Multi-stage Integer Programming

This section presents our main result regarding the hardness of multi-stage IPs. In particular, we reduce 3-SAT to multi-stage IPs, proving the following:

Theorem 1 (A lower bound for multi-stage IPs). *For every fixed number of stages $\tau \geq 1$ and stage sizes $\sigma \in \mathbb{Z}^\tau_{\geq 1}$, there is no algorithm solving every instance I of multi-stage integer programming in time $2^{2^{S^{o(1)}}} |I|^{O(1)}$, where $S = \prod_{i=1}^{\tau} (\sigma_i + 1)$ and $|I|$ is the encoding length of I, unless the ETH fails.*

By considering multi-stage IPs with constant stage sizes, we immediately get that every algorithm has to have a triple exponential dependency on τ when parametrized by the number of stages τ and the largest value of any coefficient of

A. Thus, we cannot increase the dependency on the other parameters to decrease the dependency on τ. Specifically, we rule out any algorithm solving multi-stage IPs in time less than triple exponential in τ for the parameters $\tau, \sigma, \Delta, \|c\|_\infty$, $\|b\|_\infty, \|\ell\|_\infty$. This proves tightness of the triple-exponential complexity w.r.t. τ of the current state-of-the-art algorithm of Klein and Reuter [21].

Corollary 2. *There is a family of multi-stage integer programming instances with $\tau \geq 1$ stages, constant stage sizes, and entries of constant value which, assuming the ETH, cannot be solved in time $2^{2^{2^{o(\tau)}}} |I|^{O(1)}$ where $|I|$ is the encoding length of the respective instance I.*

Proof Sketch (of Theorem 1). The proof starts where the proof for the double exponential lower bound in [15] ends. Specifically, we are given a 2-stage stochastic matrix which is, under ETH, the double exponentially hard instance for 2-stage stochastic IPs.

The blocks of the second stage, that is, the independent diagonal matrices are each *nearly* a diagonal matrices with an extra row with (possibly) just non-zero entries. Note that this row corresponds to a summation of scaled summands, i.e., it correspond to $az_1 + bz_2 + cz_3 + dz_4 + \ldots$ for row entries $a, b, \ldots$ and variables $z_1, z_2, \ldots$ respectively.

This sum is equal to $s_2 + cz_3 + dz_4$ with $az_1 + bz_2 = s_2$. This trick is already used in [9], and allows us by repeated application to split the full row into an equivalent bi-diagonal matrix. Combing this with the remaining entries in the column and some re-arrangement of the rows, we get a tri-diagonal matrix for each second stage block of the original 2-stage stochastic IP. This is the desired form to apply Lemma 1 to each of the second stage matrices giving us a multi-stage IP with the desired dimensions w.r.t. the ETH to proof the theorem.

4 Tree-Fold Integer Programming

In this section, we consider tree-fold IPs, whose constraint matrices are the transpose of multi-stage matrices. Our results can be viewed as a refinement of [27, Theorem 4]; while [27] only considers a single parameter (namely the *tree-depth*, discussed in Section 5), we take more aspects of the structure of the matrix into account. For example, as one case of our lower bound, we obtain the currently best known lower bound for the special class of n-fold IPs, i.e., tree-fold IPs with only two levels.

We reduce from the Subset Sum problem. There, we are given numbers $a_1, \ldots, a_n, b \in \mathbb{Z}_{\geq 0}$, and the task is to decide whether there exists a vector $x \in \{0,1\}^n$ satisfying $\sum_{i=1}^n a_i x_i = b$. Since all integers are non-negative, we can compare each a_i to b beforehand, and henceforth assume $0 \leq a_i < b$ for all i.

Lemma 2 ([27] **Lemma 12**). *Unless the ETH fails, there is no algorithm for* Subset Sum *that solves every instance $a_1, \ldots, a_n, b$ in time $2^{o(n+\log(b))}$.*

Theorem 2 (A lower bound for tree-fold IPs). *Assuming the ETH, for every fixed $\tau \geq 1$ and $\sigma \in \mathbb{Z}_{\geq 1}^{\tau}$, there is no algorithm solving every tree-fold IP with τ levels and level sizes σ in time $2^{o(\prod_{i=1}^{\tau}(\sigma_i+1))}|I|^{O(1)}$.*

As a corollary, we obtain the following lower bound for n-fold IPs.

Corollary 3 (A lower bound for n-fold IPs). *Assuming the ETH, there is no algorithm solving every n-fold IP with level sizes (σ_1, σ_2) in time $2^{o(\sigma_1\sigma_2)}|I|^{O(1)}$.*

Proof Sketch (of Theorem 2). We start with an instance of the SUBSET SUM problem. The goal is to first model this problem as an appropriate n-fold IP. Then, observing that the second level block-diagonal matrices are bi-diagonal allows us to apply Lemma 1, yielding the desired algorithm.

Let us thus focus on sketching how to obtain the n-fold matrix. We start with the straight-forward interpretation of SUBSET SUM as an integer program, that is, $\{\sum_{i=1}^{n} a_i x_i = b, x \in \{0,1\}^n\}$. We could directly apply the standard encoding as in Theorem 1 to lower the size of the entries, but this would yield $\sigma_1 = 1$. We could assign random rows to the first level to obtain larger values for σ_1, but this is arguably not a clean reduction for arbitrary values of σ_1. Instead, we use a doubly encoding of the large entries: First, we encode them to the base of Δ with $\Delta^{\sigma_1} \geq b > \Delta^{\sigma_1 - 1}$. This is done similar as in the standard trick, see e.g. Theorem 1, however, we use the transposed construction. The same arguments apply. Then, the standard encoding is used to obtain the desired small entries, and the statement immediately follows.

Proof (of Theorem 2). For $\tau = 1$, we have arbitrary IPs, and the statement follows by [27].

Let $a_1, \ldots, a_n, b \in \mathbb{Z}_{\geq 1}$ be a SUBSET SUM instance. Choose a constant $1 \leq \sigma_1 \leq \lceil \log_2(b) \rceil$ and let $\Delta \in \mathbb{Z}_{\geq 2}$ s.t. $\Delta^{\sigma_1} \geq b > \Delta^{\sigma_1 - 1}$, with $\Delta = b$ if $\sigma_1 = 1$. In both cases, we have

$$\sigma_1 \log_2(\Delta) \leq 2\log_2(b) \leq 2\sigma_1 \log_2(\Delta). \tag{1}$$

Let $\lambda = (1, \Delta, \Delta^2, \ldots, \Delta^{\sigma_1 - 1})^\intercal$ and let $c_i \in \{0, \ldots, \Delta - 1\}^{\sigma_1}$ be the Δ-encoding of a_i, i.e., $a_i = \lambda^\intercal c_i$. Observe that $a^\intercal x = b$ has a solution if and only if the system

$$(C|D)\begin{pmatrix} x \\ y \end{pmatrix} := \left(\begin{array}{cccc|ccccc} \vdots & \vdots & & \vdots & \Delta & & & & \\ & & & & -1 & \Delta & & & \\ c_1 & c_2 & \cdots & c_n & & -1 & \ddots & & \\ \vdots & \vdots & & \vdots & & & \ddots & \Delta & \\ & & & & & & & -1 & \end{array}\right) \begin{pmatrix} x \\ y \end{pmatrix} = \begin{pmatrix} b \\ 0 \\ 0 \\ \vdots \\ 0 \end{pmatrix}$$

has a solution $\binom{x}{y}$. (The "if"-direction is observing $\lambda^\intercal(C, D) = (a^\intercal, 0)$; the "only-if"-direction follows by observing that a solution x fixes the values of y.)

Let $t := \lceil \log_2 \Delta \rceil$ and fix a column c_k of C. We can express each entry of c_k in its binary representation. This is, let $z^\intercal = (1, 2, 4, \ldots, 2^{t-1})^\intercal \in \mathbb{Z}^t$ and let

$C_k \in \{0,1\}^{\sigma_1 \times t}$ be the unique matrix subject to $C_k z = c_k$. Similarly, for the k-th column d_k, there is a unique matrix $D_k \in \{0,1\}^{\sigma_1 \times t}$ subject to $d_k = D_k z$.

Again, let the encoding matrix be the matrix

$$E_t := \begin{pmatrix} 2 & -1 & & & \\ & 2 & -1 & & \\ & & \ddots & \ddots & \\ & & & 2 & -1 \end{pmatrix} \in \mathbb{Z}^{(t-1)\times t}, \tag{2}$$

and observe that the system $E_t x = 0$, $0 \le x \le 2^t$ only has the two solutions $\{0, z\}$. Therefore, the SUBSET SUM instance has a solution if and only if the system

$$\mathcal{A}\begin{pmatrix} x \\ y \end{pmatrix} := \begin{pmatrix} C_1 & \cdots & C_n & D_1 & \cdots & D_{s-1} \\ E_t & & & & & \\ & E_t & & & & \\ & & \ddots & & & \\ & & & & & E_t \end{pmatrix} \begin{pmatrix} x \\ y \end{pmatrix} = \begin{pmatrix} b \\ 0 \\ \vdots \\ 0 \end{pmatrix} \tag{n-fold IP}$$

has a solution satisfying $0 \le x \le 2^t$.

So far, the constructed IP is an n-fold IP with $\sigma_1 \in \Theta(\frac{\log_2(b)}{\log_2(\Delta)})$ rows in the top blocks, $t \in \Theta(\log(\Delta))$ columns per block, and $\hat{\sigma}_2 = t-1$ rows in the diagonal blocks. We have $n + \sigma_1$ blocks in total, and $\|\mathcal{A}\|_\infty \le 2$, hence the size of the constructed instance is polynomial in the size of the SUBSET SUM instance. The first claim follows already: If we could solve it in time

$$2^{o(\sigma_1 \hat{\sigma}_2)} \le 2^{o(\sigma_1 \log_2(\Delta))} \overset{(1)}{\le} 2^{o(\log_2(b))},$$

this would contradict Lemma 2, finishing the proof for $\tau = 2$ (Corollary 3).

To prove the statement for $\tau \ge 3$, we continue the construction. Choose τ s.t. $2 \le \tau - 1 \le \lceil \log_2(t) \rceil$ and $s \ge 1$ s.t. $(s+1)^{\tau-1} \ge t > s^{\tau-1}$. Furthermore, let $\ell \ge 0$ be s.t. $(s+1)^{\tau-1-\ell} s^\ell \ge t > (s+1)^{\tau-\ell-2} s^{\ell+1}$. Observe that if $s = 1$, then $\ell = 0$, since $\tau - 1 \le \lceil \log_2(t) \rceil$. Let $(\sigma_2, \dots, \sigma_\tau) \in \{s-1, s\}^{\tau-1}$ have $\tau - 1 - \ell$ entries s and ℓ entries $s-1$. This careful construction of σ allows us to estimate

$$\prod_{i=2}^{\tau} (\sigma_i + 1) = (s+1)^{\tau-1-\ell} s^\ell = \frac{s+1}{s} (s+1)^{\tau-\ell-2} s^{\ell+1} < 2t. \tag{3}$$

Since $E_t^\intercal$ is bi-diagonal, we can extend E_t with zeros to a tree-fold matrix with $\tau - 1$ levels and level sizes $(\sigma_2, \dots, \sigma_\tau)$ due to Corollary 1. In total, the matrix $\mathcal{A}$ is a tree-fold matrix with τ levels and level sizes σ, and its encoding length did not change. If there is an algorithm solving every such tree-fold IP in time

$$2^{o((\sigma_1+1)\prod_{i=2}^{\tau}(\sigma_i+1))} \overset{(3)}{\le} 2^{o(2\sigma_1 \cdot 2t)} \le 2^{o(8\sigma_1 \log_2(\Delta))} \overset{(1)}{\le} 2^{o(16 \log_2(b))},$$

this would again contradict Lemma 2. □

5 Lower Bounds on the Graver Basis

Definition 4. *Let $A \in \mathbb{Z}^{m \times n}$. The Graver basis $\mathcal{G}(A)$ of A is the set of all vectors $z \in \ker(A) \cap \mathbb{Z}^n \backslash (0, \ldots, 0)$ that cannot be written as $z = x + y$ with $x, y \in \ker(A) \cap \mathbb{Z}^n$ satisfying $x_i y_i \geq 0$ for all i. For any norm K, we define $g_K(A) = \max_{g \in \mathcal{G}(A)} \|g\|_K$.*

Our previous results state that, assuming ETH, there is no algorithm for multi-stage or tree-fold IPs that solves *every* instance within a certain time threshold. We now show further evidence orthogonal to the ETH. All known algorithms for block-structured IPs have complexities which are at least $g_\infty(A)$ or $g_1(A)$ for multi-stage IPs or tree-fold IPs, respectively. Thus, if there is no fundamentally different algorithm for those problems, lower bounding $g_\infty(A)$ and $g_1(A)$ lower bounds the complexity of those problems. We show that the instances we construct in the hardness proofs, in particular, the encoding matrix E_t, indeed have large $g_\infty(A)$ and $g_1(A)$, respectively.

Lemma 3. *Let $t \geq 2$, $\Delta \in \mathbb{Z}_{\geq 2}$. The encoding matrix*

$$E_t(\Delta) := \begin{pmatrix} \Delta & -1 & & & \\ & \Delta & -1 & & \\ & & \ddots & \ddots & \\ & & & \Delta & -1 \end{pmatrix} \in \mathbb{Z}^{(t-1) \times t}$$

satisfies $g_\infty(E_t(\Delta)) \geq \Delta^{t-1}$ and $g_1(E_t(\Delta)) \geq \frac{\Delta^t - 1}{\Delta - 1}$.

Proof. The matrix $E_t(\Delta)$ has full row rank. Therefore, its kernel has rank $t - (t-1) = 1$. Since $z^\intercal := (1, \Delta, \ldots, \Delta^{t-1}) \in \ker(E_t)$, every element in $\ker(E_t) \cap \mathbb{Z}^t$ is an integer multiple of z. Thus, $\{z, -z\}$ is the Graver basis of $E_t(\Delta)$. □

[10, Lemma 2] states that $g_1(A) \leq (2m\|A\|_\infty + 1)^m$ for any matrix $A \in \mathbb{Z}^{m \times n}$, so E_t is almost the worst case. As an immediate consequence of Lemma 1, we get:

Theorem 3. *Let $\tau \in \mathbb{Z}_{\geq 1}$, $\sigma \in \mathbb{Z}^\tau_{\geq 1}$, and define $S := \prod_{i=1}^{\tau} (\sigma_i + 1)$.*

1. *There is a multi-stage matrix A with τ stages and stage sizes σ that satisfies $g_\infty(A) \geq \|A\|_\infty^{S-2}$.*
2. *There is a tree-fold matrix A with τ levels and level sizes σ that satisfies $g_1(A) \geq \|A\|_\infty^{S-1}$.*

Proof. For the first claim, observe that the matrix $E_{S-1}(\Delta)$ satisfies $g_\infty(E_{S-1}(\Delta)) \geq \Delta^{S-2}$. If A arises from $E_{S-1}(\Delta)$ by adding a zero row in the top and the bottom, we immediately obtain $g_\infty(A) \geq \Delta^{S-2}$, and can apply Lemma 1.

For the second claim, we can apply Corollary 1 to the matrix $E_S(\Delta)$. □

6 Beyond Block Structure: Tree-Depth

There is the more general notion of primal and dual *tree-depth* of A to capture the above classes of block-structured IPs. This section provides a brief introduction and states our results in terms of these parameters.

The *primal graph* G of A has the columns of A as a vertex set, and an edge between two columns a, b if they share a non-zero entry, i.e., there is an index i with $a_i b_i \neq 0$. A *td-decomposition* of G is an arborescence T with $V(T) = V(G)$ s.t. for any edge $\{u, v\} \in E(G)$ there either is a (u, v)-path in T or a (v, u)-path. A td-decomposition of G with minimum height is a minimum td-decomposition. The primal tree-depth $\text{td}_P(A)$ of A is the maximum number of vertices on any path in a miminum td-decomposition of G. The *dual graph* and tree-depth are the primal graph and tree-depth of $A^\intercal$.

If A is a multi-stage matrix with τ stages and stage sizes σ, we can construct a primal td-decomposition: Start with a path through the first σ_1 columns. Since the rest of the matrix decomposes into blocks, there are no edges between any two blocks, and any edge inducing a path either is from one of the first σ_1 columns to a block, or within a block. Thus, we can recurse on the blocks and append another path of σ_2 columns to column σ_1. This way, we obtain a td-decomposition of height $\sum_{i=1}^{\tau} \sigma_i - 1$, and get:

Lemma 4.

1. *Let A be multi-stage with τ stages and stage sizes σ. Then* $\text{td}_P(A) \leq \sum_{i=1}^{\tau} \sigma_i$.
2. *Let A be tree-fold with τ levels and level sizes σ. Then* $\text{td}_D(A) \leq \sum_{i=1}^{\tau} \sigma_i$.

Using the constructions of Theorems 1 and 2, we have $\sigma \leq 2 \cdot \mathbf{1}$ and hence, $\text{td}_P(\mathcal{A}) \leq 2\tau$, $\text{td}_D(\mathcal{A}) \leq 2\tau$ respectively. We obtain the following corollary. While the second point was already proven in [27], the first point is a new consequence.

Corollary 4. *Assuming the ETH, there is no algorithm solving every IP in time* $2^{2^{2^{o(\text{td}_P(A))}}} |I|^{O(1)}$, *nor in time* $2^{o(2^{\text{td}_D(A)})} |I|^{O(1)}$.

References

1. Albareda-Sambola, M., van der Vlerk, M.H., Fernández, E.: Exact solutions to a class of stochastic generalized assignment problems. Eur. J. Oper. Res. **173**(2), 465–487 (2006)
2. Aschenbrenner, M., Hemmecke, R.: Finiteness theorems in stochastic integer programming. Found. Comput. Math. **7**(2), 183–227 (2007)
3. Chen, L., Marx, D., Ye, D., Zhang, G.: Parameterized and approximation results for scheduling with a low rank processing time matrix. In: STACS. LIPIcs, vol. 66, pp. 22:1–22:14. Schloss Dagstuhl - Leibniz-Zentrum für Informatik (2017)
4. Cslovjecsek, J., Eisenbrand, F., Hunkenschröder, C., Rohwedder, L., Weismantel, R.: Block-structured integer and linear programming in strongly polynomial and near linear time. In: SODA, pp. 1666–1681. SIAM (2021)

5. Cslovjecsek, J., Eisenbrand, F., Pilipczuk, M., Venzin, M., Weismantel, R.: Efficient sequential and parallel algorithms for multistage stochastic integer programming using proximity. In: ESA. LIPIcs, vol. 204, pp. 33:1–33:14. Schloss Dagstuhl - Leibniz-Zentrum für Informatik (2021)
6. Cslovjecsek, J., Koutecký, M., Lassota, A., Pilipczuk, M., Polak, A.: Parameterized algorithms for block-structured integer programs with large entries. In: SODA 2024 (2024). https://arxiv.org/abs/2311.01890
7. Dempster, M.A.H., Fisher, M.L., Jansen, L., Lageweg, B.J., Lenstra, J.K., Rinnooy Kan, A.H.G.: Analysis of heuristics for stochastic programming: results for hierarchical scheduling problems. Math. Oper. Res. **8**(4), 525–537 (1983)
8. Eisenbrand, F., Hunkenschröder, C., Klein, K.-M.: Faster algorithms for integer programs with block structure. In: ICALP. LIPIcs, vol. 107, pp. 49:1–49:13. Schloss Dagstuhl - Leibniz-Zentrum für Informatik (2018)
9. Eisenbrand, F., Hunkenschröder, C., Klein, K.-M., Koutecký, M., Levin, A., Onn, S.: An algorithmic theory of integer programming. arXiv preprint arXiv:1904.01361 (2019)
10. Eisenbrand, F., Weismantel, R.: Proximity results and faster algorithms for integer programming using the Steinitz lemma. ACM Trans. Algorithms (TALG) **16**(1), 1–14 (2019)
11. Gavenciak, T., Koutecký, M., Knop, D.: Integer programming in parameterized complexity: five miniatures. Discrete Optim. **44**(Part), 100596 (2022)
12. Hemmecke, R., Onn, S., Romanchuk, L.: n-fold integer programming in cubic time. Math. Program. **137**(1), 325–341 (2013)
13. Hemmecke, R., Onn, S., Weismantel, R.: n-fold integer programming and nonlinear multi-transshipment. Optim. Lett. **5**(1), 13–25 (2011)
14. Hunkenschröder, C., Klein, K.-M., Koutecký, M., Lassota, A., Levin, A.: Tight lower bounds for block-structured integer programs (2024). https://arxiv.org/abs/2402.17290
15. Jansen, K., Klein, K.-M., Lassota, A.: The double exponential runtime is tight for 2-stage stochastic ILPs. In: Singh, M., Williamson, D.P. (eds.) IPCO 2021. LNCS, vol. 12707, pp. 297–310. Springer, Cham (2021). https://doi.org/10.1007/978-3-030-73879-2_21
16. Jansen, K., Klein, K.-M., Maack, M., Rau, M.: Empowering the configuration-IP - new PTAS results for scheduling with setups times. In: ITCS. LIPIcs, vol. 124, pp. 44:1–44:19. Schloss Dagstuhl - Leibniz-Zentrum fuer Informatik (2019)
17. Jansen, K., Lassota, A., Maack, M., Pikies, T.: Total completion time minimization for scheduling with incompatibility cliques. In: ICAPS, pp. 192–200. AAAI Press (2021)
18. Jansen, K., Lassota, A., Rohwedder, L.: Near-linear time algorithm for n-fold ILPs via color coding. SIAM J. Discrete Math. **34**(4), 2282–2299 (2020)
19. Kall, P., Wallace, S.W.: Stochastic Programming. Springer, Heidelberg (1994)
20. Klein, K.-M.: About the complexity of two-stage stochastic IPs. Math. Program. **192**(1), 319–337 (2022)
21. Klein, K.-M., Reuter, J.: Collapsing the tower - on the complexity of multistage stochastic IPs (2022)
22. Knop, D., Koutecký, M.: Scheduling meets n-fold integer programming. J. Sched. **21**(5), 493–503 (2018)
23. Knop, D., Koutecký, M.: Scheduling kernels via configuration LP. In: ESA. LIPIcs, vol. 244, pp. 73:1–73:15. Schloss Dagstuhl - Leibniz-Zentrum für Informatik (2022)
24. Knop, D., Koutecký, M., Levin, A., Mnich, M., Onn, S.: Parameterized complexity of configuration integer programs. Oper. Res. Lett. **49**(6), 908–913 (2021)

25. Knop, D., Koutecký, M., Mnich, M.: Combinatorial n-fold integer programming and applications. Math. Program. **184**(1), 1–34 (2020)
26. Knop, D., Koutecký, M., Mnich, M.: Voting and bribing in single-exponential time. ACM Trans. Econ. Comput. **8**(3), 12:1–12:28 (2020)
27. Knop, D., Pilipczuk, M., Wrochna, M.: Tight complexity lower bounds for integer linear programming with few constraints. ACM Trans. Comput. Theory **12**(3), 1–19 (2020). https://doi.org/10.1145/3397484
28. Koutecký, M., Levin, A., Onn, S.: A parameterized strongly polynomial algorithm for block structured integer programs. In: 45th International Colloquium on Automata, Languages, and Programming. Leibniz International Proceedings in Informatics (LIPIcs), Germany, vol. 107, pp. 85:1–85:14. Schloss Dagstuhl–Leibniz-Zentrum fuer Informatik (2018)
29. Laporte, G., Louveaux, F.V., Mercure, H.: A priori optimization of the probabilistic traveling salesman problem. Oper. Res. **42**(3), 543–549 (1994)
30. De Loera, J.A., Hemmecke, R., Onn, S., Weismantel, R.: N-fold integer programming. Discrete Optim. **5**(2), 231–241 (2008)

A Lower Bound for the Max Entropy Algorithm for TSP

Billy Jin[1], Nathan Klein[2], and David P. Williamson[1](✉)

[1] Cornell University, Ithaca, NY, USA
{bzj3,davidpwilliamson}@cornell.edu
[2] Institute for Advanced Study, Princeton, NJ, USA
nklein@ias.edu

Abstract. One of the most famous conjectures in combinatorial optimization is the four-thirds conjecture, which states that the integrality gap of the subtour LP relaxation of the TSP is equal to $\frac{4}{3}$. For 40 years, the best known upper bound was 1.5, due to Wolsey [Wol80]. Recently, Karlin, Klein, and Oveis Gharan [KKO22] showed that the max entropy algorithm for the TSP gives an improved bound of $1.5 - 10^{-36}$. In this paper, we show that the approximation ratio of the max entropy algorithm is at least 1.375, even for graphic TSP. Thus the max entropy algorithm does not appear to be the algorithm that will ultimately resolve the four-thirds conjecture in the affirmative, should that be possible.

1 Introduction

In the traveling salesman problem (TSP), we are given a set of n cities and the costs c_{ij} of traveling from city i to city j for all i, j. The goal of the problem is to find the cheapest tour that visits each city exactly once and returns to its starting point. An instance of the TSP is called *symmetric* if $c_{ij} = c_{ji}$ for all i, j; it is *asymmetric* otherwise. Costs obey the *triangle inequality* (or are *metric*) if $c_{ij} \leq c_{ik} + c_{kj}$ for all i, j, k. All instances we consider will be symmetric and obey the triangle inequality. We treat the problem input as a complete graph $G = (V, E)$, where V is the set of cities, and $c_e = c_{ij}$ for edge $e = \{i, j\}$.

In the mid-1970s, Christofides [Chr76] and Serdyukov [Ser78] each gave a $\frac{3}{2}$-approximation algorithm for the symmetric TSP with triangle inequality. The algorithm computes a minimum-cost spanning tree, and then finds a minimum-cost perfect matching on the odd degree vertices of the tree to compute a connected Eulerian subgraph. Because the edge costs satisfy the triangle inequality, any Eulerian tour of this Eulerian subgraph can be "shortcut" to a tour of no greater cost. Until very recently, this was the best approximation factor known for the symmetric TSP with triangle inequality, although over the last decade substantial progress was made for many special cases and variants of the problem. For example, in *graphic TSP*, the input to the problem is an unweighted connected graph, and the cost of traveling between any two nodes is the number of edges in the shortest path between the two nodes. A sequence of papers led to a 1.4-approximation algorithm for this problem due to Sebő and Vygen [SV14].

J. Vygen and J. Byrka (Eds.): IPCO 2024, LNCS 14679, pp. 238–251, 2024.
https://doi.org/10.1007/978-3-031-59835-7_18

In the past decade, a variation on the Christofides-Serdyukov algorithm has been considered. Its starting point is a well-known linear programming relaxation of the TSP introduced by Dantzig, Fulkerson, and Johnson [DFJ54], sometimes called the *Subtour LP* or the *Held-Karp bound* [HK71]. It is not difficult to show that for any optimal solution x^* of this LP relaxation, $\frac{n-1}{n}x^*$ is a feasible point in the spanning tree polytope. The spanning tree polytope is known to have integer extreme points, and so $\frac{n-1}{n}x^*$ can be decomposed into a convex combination of spanning trees, and the cost of this convex combination is a lower bound on the cost of an optimal tour. In particular, the convex combination can be viewed a distribution over spanning trees such that the expected cost of a spanning tree sampled from this distribution is a lower bound on the cost of an optimal tour. The variation of Christofides-Serdyukov algorithm considered is one that samples a random spanning tree from a distribution on spanning trees given by the convex combination, and then finds a minimum-cost perfect matching on the odd vertices of the tree. This idea was introduced in work of Asadpour et al. [Asa+17] (in the context of the asymmetric TSP) and Oveis Gharan, Saberi, and Singh [OSS11] (for symmetric TSP).

Asadpour et al. [Asa+17] and Oveis Gharan, Saberi, and Singh [OSS11] consider a particular distribution of spanning trees known as the *maximum entropy distribution.* We will call the algorithm that samples from the maximum entropy distribution and then finds a minimum-cost perfect matching on the odd degree vertices of the tree the *maximum entropy algorithm* for the TSP. In a breakthrough result, Karlin, Klein, and Oveis Gharan [KKO21] show that this approximation algorithm has performance ratio better than 3/2, although the amount by which the bound was improved is quite small (approximately 10^{-36}). The achievement of the paper is to show that choosing a random spanning tree from the maximum entropy distribution gives a distribution of odd degree nodes in the spanning tree such that the expected cost of the perfect matching is cheaper (if marginally so) than in the Christofides-Serdyukov analysis. Note that [KKO21] actually choose a tree plus an edge, thus working with x^* instead of $\frac{n-1}{n}x^*$. Since it is cleaner to analyze we will work with this version of the algorithm.

It has long been conjectured that there should be a 4/3-approximation algorithm for the TSP based on rounding the Subtour LP, given other conjectures about the integrality gap of the Subtour LP. The subtour LP is as follows:

$$\begin{aligned} \min \quad & \sum_{e \in E} c_e x_e \\ \text{s.t.} \quad & x(\delta(v)) = 2, & \forall v \in V, \\ & x(\delta(S)) \geq 2, & \forall S \subset V, S \neq \emptyset, \\ & 0 \leq x_e \leq 1, & \forall e \in E, \end{aligned} \tag{1}$$

where $\delta(S)$ is the set of all edges with exactly one endpoint in S and we use the shorthand that $x(F) = \sum_{e \in F} x_e$. The *integrality gap* of an LP relaxation is the worst-case ratio of an optimal integer solution to the linear program to the optimal linear programming solution. Wolsey [Wol80] (and later Shmoys and Williamson [SW90]) showed that the analysis of the Christofides-Seryukov algorithm could be used to show that the integrality gap of the Subtour LP is at most 3/2, and Karlin, Klein, and Oveis Gharan [KKO22] have shown that the integrality gap is at most $\frac{3}{2} - 10^{-36}$. It is well-known that the integrality gap of the Subtour LP is at least 4/3, and it has long been conjectured that the integrality gap is exactly 4/3. However, until the work of Karlin et al., there had been no progress on that conjecture since the work of Wolsey in 1980.

A reasonable question is whether the maximum entropy algorithm is itself a 4/3-approximation algorithm for the TSP; there is no reason to believe that the Karlin et al. [KKO21] analysis is tight. Experimental work by Genova and Williamson [GW17] has shown that the max entropy algorithm produces solutions which are very good in practice, much better than those of the Christofides-Serdyukov algorithm. It does extremely well on instances of graphic TSP, routinely producing solutions within 1% of the value of the optimal solution.

In this paper, we show that the maximum entropy algorithm can produce tours of expected cost strictly greater than 4/3 times the value of the optimal tour (and thus the subtour LP), even for instances of graphic TSP. In particular, we show:

Theorem 1. *There is an infinite family of instances of graphic TSP for which the max entropy algorithm outputs a tour of expected cost at least* $1.375 - o(1)$ *times the cost of the optimum solution.*

The instances are a variation on a family of TSP instances recently introduced in the literature by Boyd and Sebő [BS21] known as *k-donuts* (see Fig. 1). k-donuts have $n = 4k$ vertices, and are known to have an integrality gap of 4/3 under a particular metric. In contrast, we consider k-donuts under the graphic metric, in which case the optimal tour is a Hamiltonian cycle, which has cost n. The objective value of the Subtour LP for graphic k-donuts is also n; thus, these instances have an integrality gap of 1. We show that as the instance size grows, the expected length of the connected Eulerian spanning subgraph found by the max entropy algorithm (using the graphic metric) converges to $1.375n$ from below and thus the ratio of this cost to the value of the LP (and the optimal tour) converges to 1.375. We can further show that there is a bad Eulerian tour of the Eulerian subgraph such that shortcutting the Eulerian tour results in a tour that is still at least 1.375 times the cost of the optimal tour.

It thus appears that the maximum entropy algorithm is not the algorithm that will ultimately resolve the 4/3 conjecture in the affirmative, should that be possible. While this statement depends on the fact that the algorithm uses a particular Eulerian tour, all work in this area of which we are aware considers the ratio of the cost of the connected Eulerian subgraph to the LP value, rather than the ratio of the shortcut tour to the LP value. We also do not know of work which shows that there is always a way to shortcut the subgraph to a tour of significantly cheaper cost. Indeed, it is known that finding the best shortcutting is an NP-hard problem in itself [PV84].

To our knowledge, our result implies that there is currently no candidate 4/3-approximation algorithm for the TSP. Interestingly, earlier work of the authors

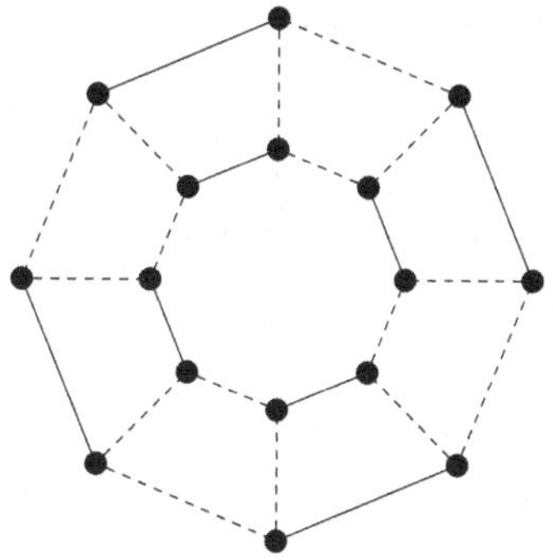

Fig. 1. Our variant on the k-donut for $k = 4$, where k indicates the number of squares of dashed edges. There are $n = 4k$ vertices. The dashed edges have $x_e = \frac{1}{2}$ and the solid edges have $x_e = 1$ in the LP solution. All edges have cost 1, as this is a graphic instance. We will refer to the outer cycle as the *outer ring*, and the inner cycle as the *inner ring*.

[JKW23] gave a 4/3-approximation algorithm for a set of instances of the TSP which includes these k-donuts.

Our work continues a thread of papers showing lower bounds on TSP approximation algorithms. Rosenkrantz, Stearns, and Lewis [RSL77] show that the nearest neighbor algorithm can give a tour of cost $\Omega(\log n)$ times the optimal, and that the nearest and cheapest insertion algorithms can give a tour of cost $2 - \frac{1}{n}$ times the optimal. Cornuéjols and Nemhauser [CN78] show that the Christofides-Serdyukov approximation factor of $\frac{3}{2}$ is essentially tight.

A full version of this paper can be found at https://arxiv.org/abs/2311.01950. Some details are omitted from this version for space reasons.

2 k-Donuts

We first formally describe the construction of a graphic k-donut instance, which will consist of $4k$ vertices. The cost function $c_{\{u,v\}}$ is given by the shortest path distance in the following graph.

Definition 1 (**k-Donut Graph**). *For $k \in \mathbb{Z}_+$, $k \geq 3$, the k-donut is a 3-regular graph consisting of $2k$ "outer" vertices $u_0, \ldots, u_{2k-1}$ and $2k$ "inner" vertices $v_0, \ldots, v_{2k-1}$. For each $0 \leq i \leq 2k-1$, the graph has edges $\{u_i, u_{i+1 \pmod{2k}}\}$, $\{v_i, v_{i+1 \pmod{2k}}\}$, and $\{u_i, v_i\}$. See Fig. 1. We call the cycle of u vertices the* outer ring *and the cycle of v vertices the* inner ring.

For clarity of notation, in the rest of the paper we will omit the "mod $2k$" when indexing the vertices of the k-donut. Thus whenever we write u_j or v_j, it should be taken to mean $u_{j \pmod{2k}}$ or $v_{j \pmod{2k}}$, respectively.

As noted by Boyd and Sebő [BS21], there is a half-integral extreme point solution x of value $4k$ as follows, which we will work with throughout this note.

Let $x_{\{u_i,v_i\}} = 1/2$ for all $0 \leq i \leq 2k-1$, $x_{\{u_i,u_{i+1}\}} = x_{\{v_i,v_{i+1}\}} = 1/2$ for all even i and $x_{\{u_i,u_{i+1}\}} = x_{\{v_i,v_{i+1}\}} = 1$ for all odd i.[1] In the rest of the paper, we will say a set $S \subseteq V$ is *tight* if $x(\delta(S)) = 2$, and S is *proper* if $2 \leq |S| \leq |V| - 2$. For a set of edges M, we'll use $c(M) = \sum_{e \in M} c_e$, and for an LP solution x, we let $c(x)$ denote the value of the LP objective function $\sum_{e \in E} c_e x_e$. For $S \subseteq V$, we use $E(S)$ to denote the subset of edges with both endpoints in S.

2.1 The Max Entropy Algorithm on the k-Donut

We now describe the max entropy algorithm, and in particular discuss what it does when specialized to the k-donut. We will work with a description of the max entropy algorithm which is very similar to the one used for half-integral TSP in [KKO20]. In [KKO20], the authors show that without loss of generality there exists an edge e^+ with $x_{e^+} = 1$. To sample a 1-tree[2] T, their algorithm iteratively chooses a minimal proper tight set S not containing e^+ which is not crossed by any other tight set, picks a tree from the max entropy distribution on the induced graph $G[S]$, adds its edges to T, and contracts S. [KKO20] shows that if no such set remains, the graph is a cycle, possibly with multiple edges between (contracted) vertices. The algorithm then randomly samples a cycle and adds its edges to T. Finally the algorithm picks a minimum-cost perfect matching M on the odd vertices of T, computes an Eulerian tour on $M \uplus T$, and shortcuts it to a Hamiltonian cycle.[3] We also remark that this algorithm from [KKO20] is equivalent to the one used in [KKO21] as one lets the error measuring the difference between the marginals of the max entropy distribution and the subtour LP solution x go to 0 (see [KKO20,KKO21] for more details).

For ease of exposition, we work with the variant in which we do not use an edge e^+ and instead contract any minimal proper tight set which is not crossed. The two distributions over trees are essentially identical, perhaps with the exception of the edges adjacent to the vertices adjacent to e^+. The performance of the two algorithms on graphic k-donuts can easily be seen to be the same as $k \to \infty$ since one can adjust the matching M with an additional cost of $O(1)$ to simulate any discrepancy between the two tree distributions. We show the essential equivalence of these two versions of the max entropy algorithm in the full version of the paper.

The reason we use this description of the algorithm is that when specialized to k-donuts, Algorithm 1 is very simple and its behavior can be easily understood without using any non-trivial properties of the max entropy algorithm. It first adds the edges with $x_e = 1$ to the 1-tree. Then, it contracts the vertices $\{u_i, u_{i+1}\}$

[1] By slightly perturbing the metric, one could ensure that x is the *only* optimal solution to the LP and thus the solution the max entropy algorithm works with. (Of course then the instance is no longer strictly graphic.).

[2] A spanning tree plus an edge.

[3] Given an Eulerian tour $(t_0, \ldots, t_\ell, t_0)$, we shortcut it to a Hamiltonian cycle by keeping only the first occurrence of every vertex except t_0 (for which we keep the first and last occurrences). Due to the triangle inequality, the resulting Hamiltonian cycle has cost no greater than that of the Eulerian tour.

Algorithm 1. Max Entropy Algorithm (Slight Variant of [KKO20])

Solve for an optimal solution x of the Subtour LP (1).
Let G be the support graph of x.
Set $T = \emptyset$. ▷ T will be a 1-tree
while there exists a proper tight set of G that is not crossed by a proper tight set **do**
  Let S be a minimal such set.
  Compute the maximum entropy distribution μ of $E(S)$ with marginals $x_{|E(S)}$.
  Sample a tree from μ and add its edges to T.
  Set $G = G/S$.
end while
▷ At this point G consists of a single cycle of length at least three, or two vertices with a set of edges F between them with $\sum_{e \in F} x_e = 2$.
if G consists of two vertices **then**
  Randomly sample two edges with replacement, choosing each edge each time with probability $x_e/2$.
else
  Independently sample one edge between each adjacent pair, choosing each edge with probability x_e.
end if
Compute the minimum-cost perfect matching M on the odd degree vertices of T. Compute an Eulerian tour of $T \uplus M$ and shortcut it to return a Hamiltonian cycle.

to a single vertex for all odd i, and does the same for $\{v_i, v_{i+1}\}$ (in other words, it contracts the 1-edges). After that, the minimal proper tight sets consist of pairs of newly contracted vertices $\{u_i, u_{i+1}\}, \{v_i, v_{i+1}\}$ for odd i. Since each of these pairs have two edges set to $1/2$ between them, the algorithm will simply choose one at random for each independently. After contracting these pairs the graph is a cycle. It follows that:

Proposition 1. *On the k-donut, the max entropy algorithm will independently put exactly one edge among every pair* $\{\{u_i, v_i\}, \{u_{i+1}, v_{i+1}\}\}$ *in* T *for every odd* i *and exactly one edge among every pair* $\{\{u_i, u_{i+1}\}, \{v_i, v_{i+1}\}\}$ *in* T *for every even* i.

We visualize these pairs in Fig. 2. The following claim is the only property we need in the remainder of the proof:

Proposition 2. *For every pair of vertices* (u_i, v_i), $0 \le i \le 2k-1$, *exactly one of* u_i *or* v_i *will have odd degree in* T, *each with probability* $\frac{1}{2}$. *Let* O_i *indicate if* u_i *and* u_{i+1} *have the same parity. Then if* $i \neq j$ *and* i, j *have the same parity, then* O_i *and* O_j *are independent.*

Proof. We prove the first part of the claim when i is odd; the case where i is even is similar. Since i is odd, the edges $\{u_i, u_{i+1}\}$ and $\{v_i, v_{i+1}\}$ are in T. Then, one of the two edges $\{u_i, v_i\}$ and $\{u_{i+1}, v_{i+1}\}$ is added to T, and regardless of the choice, u_i and v_i so far have the same parity. Finally, one edge in $\{\{u_{i-1}, u_i\}, \{v_{i-1}, v_i\}\}$ is added uniformly at random, which flips the parity of exactly one of u_i, v_i.

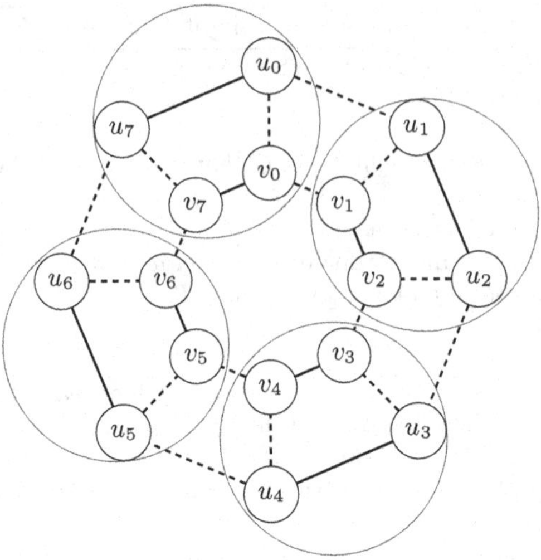

Fig. 2. One edge among the pair of dotted edges inside each red cut will be chosen independently. Then one edge among each pair of dotted edges in the cycle resulting from contracting the red sets will be chosen independently. (Color figure online)

To prove the second part of the claim, we will only do the case that both i and j are odd, as the other case is similar. To slightly simplify the notation we assume $i = 1$ perhaps after a cyclic shift of the indices. Here the event O_1 depends only on the choice of the edges among the pairs $\{\{u_0, u_1\}, \{v_0, v_1\}\}$ and $\{\{u_2, u_3\}, \{v_2, v_3\}\}$; recall the 1-tree picks one edge from each pair, independently and uniformly at random. Similarly, O_j only depends on the independent choices among $\{\{u_{j-1}, u_j\}, \{v_{j-1}, v_j\}\}$ and $\{\{u_{j+1}, u_{j+2}\}, \{v_{j+1}, v_{j+2}\}\}$. The first choice for O_1 is independent of O_j if $j \neq 2k - 1$, and the second is independent of O_j if $j \neq 3$. Since $k \geq 3$ by definition of the k-donut, at most one of the independent choices is shared among the two events O_1, O_j. The proof follows by noticing that even after fixing one of the pairs, O_1 remains equally likely to be 0 or 1.

3 Analyzing the Performance of Max Entropy

We now analyze the max entropy algorithm on graphic k-donuts. We first characterize the structure of the min-cost perfect matching on the odd vertices of T. We then use this structure to show that in the limit as $k \to \infty$, the approximation ratio of the max entropy algorithm approaches 1.375 from below.

Proposition 3. *Let T be any 1-tree with the property that for every pair of vertices (u_i, v_i) for $0 \leq i \leq 2k - 1$, exactly one of u_i or v_i has odd degree in T. (This is Proposition 2).*

Let $o_0, \ldots, o_{2k-1}$ indicate the odd vertices in T where o_i is the odd vertex in the pair (u_i, v_i). Let M be a minimum-cost perfect matching on the odd vertices of T. Define:

$$M_1 = \{(o_0, o_1), (o_2, o_3), \ldots, (o_{2k-2}, o_{2k-1})\}$$
$$M_2 = \{(o_{2k-1}, o_0), (o_1, o_2), \ldots, (o_{2k-3}, o_{2k-2})\}$$

Then,

$$c(M) = \min\{c(M_1), c(M_2)\}.$$

Proof. We will show a transformation from M to a matching in which every odd vertex o_i is either matched to $o_{i-1 \pmod{2k}}$ or $o_{i+1 \pmod{2k}}$. This completes the proof, since then after fixing (o_0, o_1) or (o_{2k-1}, o_0) the rest of the matching is uniquely determined as M_1 or M_2. During the process, we will ensure the cost of the matching never increases, and to ensure it terminates we will argue that the (non-negative) potential function $\sum_{e=(o_i,o_j)\in M} \min\{|i-j|, 2k-|i-j|\}$ decreases at every step. Note that this potential function is invariant under any reindexing corresponding to a cyclic shift of the indices.

So, suppose M is not yet equal to M_1 or M_2. Then there is some edge $(o_i, o_j) \in M$ such that $j \notin \{i-1, i+1 \pmod{2k}\}$. Without loss of generality (by switching the role of i and j if necessary), suppose $j \in \{i+2, i+3, \ldots, i+k \pmod{2k}\}$. Possibly after a cyclic shift of the indices, we can further assume $i = 0$ and $2 \leq j \leq k$. Let o_l be the vertex that o_1 is matched to. We consider two cases depending on if $l \leq k+1$ or $l > k+1$.

Case 1: $l \leq k+1$. In this case, replace the edges $\{\{o_0, o_j\}, \{o_1, o_l\}\}$ with $\{\{o_0, o_1\}, \{o_j, o_l\}\}$. This decreases the potential function, as the edges previously contributed $j+l-1$ and now contribute $1+|j-l|$, which is a smaller quantity since $j, l \geq 2$. Moreover this does not increase the cost of the matching: We have $c_{\{o_0,o_1\}} \leq 2$ and $c_{\{o_l,o_j\}} \leq |j-l|+1$, so the two new edges cost at most $|j-l|+3$. On the other hand, the two old edges cost at least $c_{\{o_0,o_j\}} + c_{\{o_1,o_l\}} \geq j+l-1$, which is at least $|j-l|+3$ since $j, l \geq 2$.

Case 2: $l > k+1$. In this case, we replace the edges $\{\{o_0, o_j\}, \{o_1, o_l\}\}$ with $\{\{o_0, o_l\}, \{o_1, o_j\}\}$. This decreases the potential function, as the edges previously contributed $j+(2k-l+1)$ and now they contribute $(2k-l)+(j-1)$. Also, the edges previously cost at least $j+(2k-l+1)$, and now cost at most $(2k-l+1)+j$. Thus the cost of the matching did not increase. □

We now analyze the approximation ratio of the max entropy algorithm without shortcutting.

Lemma 1. *If $A = T \uplus M$ is the connected Eulerian subgraph computed by the max entropy algorithm on the k-donut, then*

$$\lim_{k\to\infty} \frac{\mathbb{E}[c(A)]}{c(\mathsf{OPT})} = \lim_{k\to\infty} \frac{\mathbb{E}[c(A)]}{c(x)} = 1.375,$$

where $c(x)$ is the cost of the solution x to the subtour LP.

Proof. We know that the LP value is $4k$. Since the k-donut is Hamiltonian, we also have that the optimal tour has length $4k$. On the other hand, $c(A) = c(T)+$

$c(M)$, where T is the 1-tree and M is the matching. Note that the cost of the 1-tree is always $4k$. On the other hand, we know that $c(M) = \min\{c(M_1), c(M_2)\}$ from the previous claim. Thus, it suffices to reason about the cost of M_1 and M_2. We know that for every i, $c_{\{o_i, o_{i+1}\}} = 2$ with probability 1/2 and 1 otherwise, using Proposition 2. Thus, the expected cost of each edge in M_1 and M_2 is 1.5. Since each matching consists of k edges, by linearity of expectation, $\mathbb{E}[c(M_1)] = \mathbb{E}[c(M_2)] = 1.5k$. This implies $\mathbb{E}[c(M)] \leq 1.5k$. This immediately gives an upper bound on the approximation ratio of $\frac{4k+1.5k}{4k} = 1.375$. In the remainder we prove the lower bound.

For each i, construct a random variable X_i indicating if $c_{\{o_i, o_{i+1}\}} = 2$. By Proposition 2, the variables $\{X_0, X_2, X_4 \ldots\}$ are pairwise independent and the variables $\{X_1, X_3, X_5 \ldots\}$ are pairwise independent. Thus, for M_1, we have $\mathrm{Var}(\sum_{i=0}^{k-1} X_{2i}) = \sum_{i=0}^{k-1} \mathrm{Var}(X_{2i}) = k/4$, so the standard deviation is $\sigma(\sum_{i=0}^{k-1} X_{2i}) = \sqrt{k}/2$. We define $\mu = \mathbb{E}\left[\sum_{i=0}^{k-1} X_{2i}\right] = k/2$ to be the expected value of $\sum_{i=0}^{k-1} X_{2i}$.

Then, applying Chebyshev's inequality for M_1,

$$\begin{aligned}\mathbb{P}\left[c(M_1) \geq \left(\frac{3}{2} - \epsilon\right) k\right] &= \mathbb{P}\left[\sum_{i=0}^{k-1} X_{2i} \geq \left(\frac{1}{2} - \epsilon\right) k\right] \\ &\geq 1 - \mathbb{P}\left[\left|\sum_{i=0}^{k-1} X_{2i} - \mu\right| \geq \epsilon k\right] \geq 1 - \frac{1}{4\epsilon^2 k}.\end{aligned}$$

Choosing $\epsilon = k^{-1/4}$ and applying a union bound (the same bound applies to M_2), we obtain the chance that both matchings cost at least $\frac{3}{2}k - k^{3/4}$ occurs with probability at least $1 - \frac{1}{2\sqrt{k}}$. Even if the matching has cost 0 on the remaining instances, the expected cost of the matching is therefore at least $(1 - \frac{1}{2\sqrt{k}})(\frac{3}{2}k - k^{3/4}) \geq \frac{3}{2}k - 2k^{3/4}$. Since the cost of the 1-tree is always $4k$, we obtain an expected cost of $\frac{11}{2}k - 2k^{3/4}$ with a ratio of

$$\mathbb{E}[c(T \uplus M)] = \frac{\frac{11}{2}k - 2k^{3/4}}{\mathsf{OPT}} = \frac{\frac{11}{2}k - 2k^{3/4}}{4k} \geq \frac{11}{8} - k^{-1/4},$$

which goes to $\frac{11}{8}$ as $k \to \infty$. □

4 Shortcutting

So far, we have shown that the expected cost of the connected Eulerian subgraph returned by the max entropy algorithm is 1.375 times that of the optimal tour. However, after shortcutting the Eulerian subgraph to a Hamiltonian cycle, its cost may decrease. Ideally, we would like a lower bound on the cost of the tour after shortcutting. One challenge with this is that the same Eulerian subgraph can be shortcut to different Hamiltonian cycles with different costs, depending on which Eulerian tour is used for the shortcutting. What we will show in this section is that there is always *some* bad Eulerian tour of the connected Eulerian subgraph, whose cost does not go down after shortcutting. We highlight two important aspects of this analysis:

1. In the analysis in Sect. 3, the matching was not required to be either M_1 or M_2. Thus, we lower bounded the cost of the Eulerian subgraph for any procedure that obtains a minimum-cost matching. Here we will require that the matching algorithm always selects M_1 or M_2. From Proposition 3, we know one of these matchings is a candidate for the minimum-cost matching. However, there may be others. Therefore, we only lower bound the shortcutting for a specific choice of the minimum-cost matching.
2. Similar to above, we only lower bound the shortcutting for a specific Eulerian tour of the Eulerian subgraph. Indeed, our lower bound only holds for a small fraction of Eulerian tours.

We remark that the max entropy algorithm as described in e.g. [OSS11, KKO21] does not specify how a minimum-cost matching or an Eulerian tour is generated. Therefore, our lower bound does hold for the general description of the algorithm. Thus, despite the caveats, this section successfully demonstrates our main result: The max entropy algorithm is not a 4/3-approximation algorithm.

In the rest of this section, for space reasons, we will only sketch the argument for why shortcutting does not decrease the cost of the tour. The interested reader is invited to see https://arxiv.org/abs/2311.01950 for the full details. At a high level, we will consider Eulerian graphs resulting from adding M_1 and construct Eulerian tours whose costs do not decrease after shortcutting. We then do the same for the graphs resulting from adding M_2. These two statements together complete the proof.

4.1 Bad Tours on M_1

Recall $M_1 = \{(o_0, o_1), (o_2, o_3), \ldots, (o_{2k-2}, o_{2k-1})\}$, where o_i is the vertex in $\{u_i, v_i\}$ with odd degree in the 1-tree T. This creates graphs of the type seen in Fig. 3. By doing case analysis on which edges are chosen in the tree, it can be shown that the resulting Eulerian subgraph $T \uplus M_1$ consists of a collection of circuits of length 2, 5, or 8 arranged in a circle. Note that doubled edges (circuits of length 2) are never adjacent to one another on this circle, since they are only created between vertices (o_i, o_{i+1}) for i even.

We now describe the problematic tours on such graphs, i.e. Eulerian tours such that the resulting Hamiltonian cycle after shortcutting is no cheaper. For intuition, a reader may want to first consider the bad tour in Fig. 3. We begin at an arbitrary vertex t_0 of degree 2 and pick an edge in the clockwise direction. We now describe the procedure for picking the next vertex t_{k+1} given our current vertex t_k. If t_k has degree 2 (or has degree 4 but is being visited for the second time), there is no choice to make, as only one edge remains. So it is sufficient to describe decisions on vertices t_k of degree 4 visited for the first time. Note that since t_k has degree 4, it is at the intersection of two adjacent circuits. Let

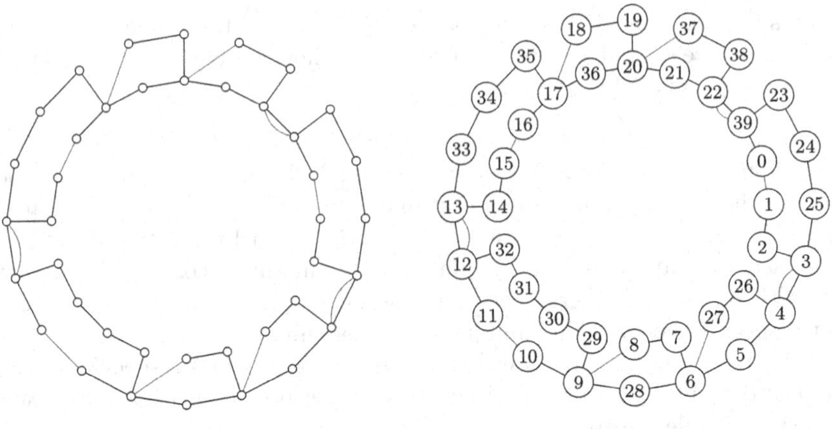

Fig. 3. On the left is an example Eulerian graph when M_1 is added. The graph consists of circuits of length 5 or 8 joined by doubled edges. On the right is the Hamiltonian cycle resulting from shortcutting the adversarial Eulerian tour we construct here, in which we always alternate the side of the circuit we traverse. One can check that every shortcutting operation does not decrease the cost.

C denote the circuit in the clockwise direction adjacent to t_k, and let C' denote the circuit in the counterclockwise direction adjacent to t_k. The next edge to pick is determined by the following two rules:

1. **Never traverse an edge in the counterclockwise direction.** Therefore, if C is a circuit of length 2, we immediately traverse one of its edges.
2. **Alternate the visited side of adjacent circuits.** Otherwise, C has length 5 or 8. For simplicity, suppose t_k is on the outer ring; the case where t_k is on the inner ring is symmetric. Let $e_{\text{outer}} = \{t_k, u\}$ and $e_{\text{inner}} = \{t_k, v\}$ be the two edges in C adjacent to t_k, where u is on the outer ring and v is on the inner ring. Let $e = \{t_j, t_{j+1}\}$ be the previous edge in the tour that was not part of a circuit of length 2. Thus, $j = k-1$, $e = \{t_{k-1}, t_k\}$ if C' is of length 5 or 8, and $j = k-2$, $e = \{t_{k-2}, t_{k-1}\}$ if C' is of length 2. Now, if t_j is on the outer ring, we take e_{inner}. Otherwise, take e_{outer}. The intuition here is that if we visited the inner ring while traversing the last circuit of length greater than 2, we now wish to visit the outer ring, and vice versa.

We call the resulting Eulerian tours B-tours because they are bad for the objective function, even after shortcutting. Indeed, we can prove the following lemma, which together with Lemma 1 demonstrates that the asymptotic cost of the tour after shortcutting is still $11/8$ times that of the optimal solution. We omit the proof of Lemma 2 for space reasons, but we hope the intuition behind the result can be seen by tracing through the example tour in Fig. 3.

Lemma 2. *Shortcutting a B-tour on a graph $T \cup M_1$ to a Hamiltonian cycle does not reduce the cost by more than 3.*

4.2 Bad Tours on M_2

Recall that $M_2 = \{(o_{2k-1}, o_0), (o_1, o_2), \ldots, (o_{2k-3}, o_{2k-2})\}$. It can be shown that in this case, the structure of $T \uplus M_2$ is one large cycle with circuits of length 2 or 3 hanging off to create some vertices of degree 4 on the large cycle; see Fig. 4.

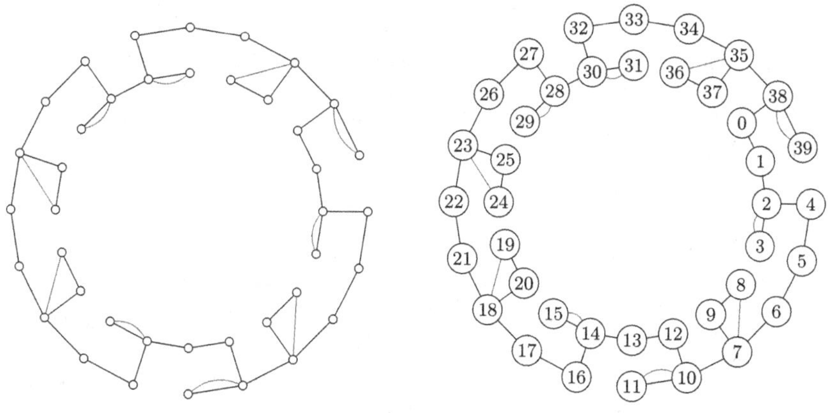

Fig. 4. An example Eulerian graph when M_2 is added. The graph consists of a single long cycle, onto which cycles of length two and three are grafted. As in the case of M_1, one can see that the length of this Hamiltonian cycle is equal to the length of the Eulerian tour that generated it.

We now describe B-tours in this instance. We will start at an arbitrary vertex t_0 on of degree 2 on the large cycle, and traverse an edge in clockwise direction. As before, it suffices to dictate the rules for degree 4 vertices visited for the first time. In this case, the only rule to produce a B-tour is: **traverse the adjacent edge in** M_2. Once again, we omit the proof of Lemma 3 for space reasons, and we hope the intuition behind the result can be seen by tracing through the example tour in Fig. 4.

Lemma 3. *The cost of the Hamiltonian cycle resulting from shortcutting a B-tour on $T \cup M_2$ is equal to the cost of $T \cup M_2$.*

5 Conclusion

We demonstrated that the max entropy algorithm as stated in e.g. [OSS11, KKO21] is not a candidate for a 4/3-approximation algorithm for the TSP. This raises the question: what might be a candidate algorithm? The algorithm in [JKW23] is a 4/3-approximation for half-integral cycle cut instances of the TSP, which include k-donuts as a special case. However, it is not clear if the algorithm can be extended to general TSP instances. One interesting direction is to find a modification of the max entropy algorithm which obtains a 4/3 or better approximation on k-donuts.

It would also be interesting to know whether one can obtain a lower bound for the max entropy algorithm which is larger than 11/8. While we have some intuition based on [KKO20, KKO21] for why the k-donuts are particularly problematic for the max entropy algorithm[4], it would be interesting to know if there are worse examples.

Acknowledgments. The first and third authors were supported in part by NSF grant CCF-2007009. The first author was also supported by NSERC fellowship PGSD3-532673-2019. The second author was supported in part by NSF grants DMS-1926686, DGE-1762114, and CCF-1813135.

References

[Asa+17] Asadpour, A., Goemans, M.X., Mądry, A., Gharan, S.O., Saberi, A.: An $O(\log n/\log\log n)$-approximation algorithm for the asymmetric traveling salesman problem. Oper. Res. **65**, 1043–1061 (2017)

[BS21] Boyd, S., Sebő, A.: The salesman's improved tours for fundamental classes. Math. Program. **186**, 289–307 (2021)

[Chr76] Christofides, N.: Worst case analysis of a new heuristic for the traveling salesman problem. Report 388. Graduate School of Industrial Administration, Carnegie-Mellon University, Pittsburgh (1976)

[CN78] Cornuejols, G., Nemhauser, G.L.: Tight bounds for Christofides' traveling salesman heuristic. Math. Program. **14**, 116–121 (1978)

[DFJ54] Dantzig, G., Fulkerson, R., Johnson, S.: Solution of a large-scale traveling-salesman problem. J. Oper. Res. Soc. Am. **2**(4), 393–410 (1954)

[GW17] Genova, K., Williamson, D.P.: An experimental evaluation of the best-of-many Christofides' algorithm for the traveling salesman problem. Algorithmica **78**, 1109–1130 (2017)

[HK71] Held, M., Karp, R.M.: The traveling-salesman problem and minimum spanning trees. Oper. Res. **18**, 1138–1162 (1971)

[JKW23] Jin, B., Klein, N., Williamson, D.P.: A 4/3-approximation algorithm for half-integral cycle cut instances of the TSP. In: Del Pia, A., Kaibel, V. (eds.) IPCO 2023. LNCS, vol. 13904, pp. 217–230. Springer, Cham (2023). https://doi.org/10.1007/978-3-031-32726-1_16

[KKO20] Karlin, A.R., Klein, N., Gharan, S.O.: An improved approximation algorithm for TSP in the half integral case. In: Proceedings of the 52nd Annual ACM Symposium on the Theory of Computing, pp. 28–39 (2020)

[KKO21] Karlin, A.R., Klein, N., Gharan, S.O.: A (slightly) improved approximation algorithm for metric TSP. In: Proceedings of the 53rd Annual ACM Symposium on the Theory of Computing, pp. 32–45 (2021)

[4] The intuition is as follows. Say an edge $e = (u, v)$ is "good" if the probability that u and v have even degree in the sampled tree is at least a (small) constant and "bad" otherwise. One can show that the ratio (in terms of x weight) of bad edges to good edges in the k-donut is as large as possible. For more details see [KKO21].

[KKO22] Karlin, A., Klein, N., Gharan, S.O.: A (slightly) improved bound on the integrality gap of the subtour LP for TSP. In: 2022 IEEE 63rd Annual Symposium on Foundations of Computer Science (FOCS), Los Alamitos, CA, USA, pp. 832–843. IEEE Computer Society (2022)

[OSS11] Gharan, S.O., Saberi, A., Singh, M.: A randomized rounding approach to the traveling salesman problem. In: Proceedings of the 52nd Annual IEEE Symposium on the Foundations of Computer Science, pp. 550–559 (2011)

[PV84] Papadimitriou, C.H., Vazirani, U.V.: On two geometric problems related to the travelling salesman problem. J. Algorithms **5**, 231–246 (1984)

[RSL77] Rosenkrantz, D.J., Stearns, R.E., Lewis, P.M., II.: An analysis of several heuristics for the traveling salesman problem. SIAM J. Comput. **6**, 563–581 (1977)

[Ser78] Serdyukov, A.: On some extremal walks in graphs. Upravlyaemye Sistemy **17**, 76–79 (1978)

[SV14] Sebő, A., Vygen, J.: Shorter tours by nicer ears: 7/5- approximation for the graph-TSP, 3/2 for the path version, and 4/3 for two-edge-connected subgraphs. Combinatorica **34**, 597–629 (2014)

[SW90] Shmoys, D.B., Williamson, D.P.: Analyzing the Held-Karp TSP bound: a monotonicity property with application. Inf. Process. Lett. **35**, 281–285 (1990)

[Wol80] Wolsey, L.A.: Heuristic analysis, linear programming and branch and bound. Math. Program. Study **13**, 121–134 (1980)

On the Number of Degenerate Simplex Pivots

Kirill Kukharenko[1(✉)] and Laura Sanità[2]

[1] Otto von Guericke University Magdeburg, Magdeburg, Germany
kirill.kukharenko@ovgu.de
[2] Bocconi University, Milan, Italy
laura.sanita@unibocconi.it

Abstract. The simplex algorithm is one of the most popular algorithms to solve linear programs (LPs). Starting at an extreme point solution of an LP, it performs a sequence of basis exchanges (called pivots) that allows one to move to a better extreme point along an improving edge-direction of the underlying polyhedron.

A key issue in the simplex algorithm's performance is *degeneracy*, which may lead to a (potentially long) sequence of basis exchanges which do not change the current extreme point solution.

In this paper, we prove that it is always possible to limit the number of consecutive degenerate pivots that the simplex algorithm performs to $n - m - 1$, where n is the number of variables and m is the number of equality constraints of a given LP in standard equality form.

Keywords: Simplex Algorithm · Linear Programming · Degeneracy

1 Introduction

The simplex algorithm first introduced in [12] has always been one of the most widely used algorithms for solving linear programming problems (LPs). Although it performs as a linear time algorithm for many real world linear programming models (see, e.g., the survey [33]) and a variant of it (the shadow simplex) has been shown to have polynomial smoothed complexity (see, e.g., [7,8,10,24,34, 39]), no polynomial version of the simplex algorithm has been found until these days, despite decades of research.

The simplex algorithm relies on the concept of *basis*, where a basis corresponds to an inclusion-wise minimal set of tight constraints which defines an extreme point solution of the underlying linear program. In each step, it performs a basis exchange by replacing one constraint currently in the basis with a different one. Such an exchange is called a *pivot*. From a geometric perspective, a basis exchange identifies a direction to move from the current extreme point. The simplex algorithm only considers pivots that yield an *improving* direction, with respect to the objective function to be optimized. As the result of pivoting, it is possible for the algorithm to either stay at the same extreme point, or to

J. Vygen and J. Byrka (Eds.): IPCO 2024, LNCS 14679, pp. 252–264, 2024.
https://doi.org/10.1007/978-3-031-59835-7_19

move to an adjacent extreme point of the underlying polyhedron. We refer to a pivot of the former type as *degenerate*, in contrast to a pivot of the latter type, which will be called *non-degenerate*. The *pivot rule* which determines how to perform these basis exchanges (i.e., deciding which inequality enters and which inequality leaves the basis) is the core of the simplex algorithm. Various pivot rules have been introduced over decades, but none of them yields a polynomial time version of the simplex algorithm so far (see, e.g., [3,13–15,17,21,35,40] and the references therein).

When analyzing the performance of the simplex algorithm, one can often assume that the underlying polyhedron $\mathcal{P}$ is non degenerate, e.g., by performing a small perturbation of the constraints right-hand-side. In this case, a necessary condition to guarantee an efficient performance of the simplex algorithm, is that there is always a way to reach an optimal solution within a polynomial number of non-degenerate pivots. Geometrically, this would imply that there is always a path between two extreme points on the 1-skeleton of $\mathcal{P}$ (the graph where vertices correspond to the extreme points of $\mathcal{P}$, and edges correspond to the 1-dimensional faces of $\mathcal{P}$), whose length is bounded by a polynomial in the input description of $\mathcal{P}$. This is a major and still open question in the field (see for example [6,11,23,31,37]).

On the other hand, even in the case when a short path exists, it is not clear how the simplex algorithm would be bound to follow it (see for instance the case of 0/1-polytopes discussed in [4]). In fact, given any LP one can apply some transformation to it to ensure the existence of a short path between an initial extreme point solution and an optimal one in the underlying polyhedron (such as, e.g., relying on an extended formulation with one extra dimension to construct a pyramid whose apex is adjacent to all extreme points of the initial polyhedron). Such standard transformations usually make the polyhedron become highly degenerate. In presence of degeneracy, an extreme point might have exponentially many distinct bases defining it, and the simplex algorithm might perform an exponential number of consecutive degenerate pivots, staying in the same vertex, (see, e.g., [9]). Such behavior is referred to as *stalling*. In some pathological cases, certain pivot rules might even provide an infinite sequence of consecutive degenerate pivots at a degenerate vertex, a phenomenon called *cycling*. Although cycling can be easily avoided by employing, for instance, Bland's rule [5] or a lexicographic rule [36], there is no known pivot rule that prevents stalling. As stated in several papers (e.g., [25,26]), it is well known that solving a general LP in strongly polynomial time can be reduced to finding a pivot rule that prevents stalling in general polyhedra. However, there are a few cases in which it is known that the issue of stalling can be handled. The most famous example is the class of transportation polytopes, for which pivot rules with polynomial bounds on the number of consecutive degenerate pivots were introduced in [9] and further developed in [1,16,29]. See also the work of [28] for a strongly polynomial version of the primal network simplex.

Our findings are inspired by [18] to a great extent. The latter work shows that, given an LP in standard equality form with a totally unimodular coefficient matrix $A \in \{-1, 0, 1\}^{m \times n}$, a vertex solution x and an improving feasible *circuit direction* for x, one can construct a pivot rule which performs at most m consecutive degenerate pivots. We generalize their proof, and show that one can employ *any* improving direction at a vertex x of a *general* polyhedron, to avoid stalling. Our result, proved in Sect. 2, is summarized in the following theorem.

Theorem 1. *Given any LP of the form* $\min\{c^T x : Ax = b, x \geq 0\}$ *with* $A \in \mathbb{R}^{m \times n}$, *there exists a pivot rule that limits the number of consecutive degenerate simplex pivots at any non-optimal extreme solution to at most* $n - m - 1$.

We stress here that the pivots considered in Theorem 1 are degenerate *simplex* pivots, meaning that each of them yields an *improving* direction at the given extreme point, though this direction is not feasible. It is important to point this out as in general, given two adjacent extreme points x, x', one can easily perform a sequence of basis exchanges that yield x' from x by identifying the common tight linearly independent constraints, and introducing each of them to the current basis in any order, until the direction $x' - x$ is seen. However, we here want a strategy that guarantees that *each* basis exchange will define an improving direction.

We then discuss some byproducts of our result in Sect. 3. In particular, we revise the analysis of the simplex algorithm by Kitahara and Mizuno [19,20] who bound the number of non-degenerate pivots it in terms of n, m and the maximum and the minimum non-zero coordinate of a basic feasible solution. Their analysis combined with our degeneracy-escaping technique show that the *total* number of simplex pivots (both degenerate and non-degenerate) can be bounded in a similar way. As a consequence, one can have a strongly-polynomial number of simplex pivots for several combinatorial LPs (see observations following Corollary 1).

Of course the drawback of the whole machinery is that it requires an improving feasible direction at a given vertex. Though it is efficiently computable, this is in general as hard as solving an LP. For some classes of polytopes though (e.g., matching or flow polytopes) finding such a direction can be easier, thus making it worthwhile to apply our pivot rule. Most importantly, we think that the main importance of our result is from a theoretical perspective: it shows that, for several polytopes, not only a short path on the 1-skeleton exists, but a short sequence of *simplex pivots* always exists (and can also be efficiently computed). That is, a short sequence of *basis exchanges* that follow *improving* directions, when performing *both* degenerate and non-degenerate pivots.

2 Antistalling Pivot Rule

The goal of this section is to prove Theorem 1. Before doing that, we state some preliminaries, and also give a geometric intuition behind the result.

2.1 Preliminaries

The simplex algorithm works with LP in standard equality form:

$$\begin{aligned} \min\ & c^T x \\ s.t.\ & Ax = b \\ & x \geq 0 \end{aligned} \tag{1}$$

For $A \in \mathbb{R}^{m \times n}$ and $B \subseteq [n]$ we use A_B to denote the submatrix of A formed by the columns indexed by B. Similarly, for $x \in \mathbb{R}^n$ and $B \in [n]$, the notation x_B is used to denote a vector consisting of the entries of x indexed by B. A *basis* of (1) is a subset $B \subseteq [n]$ with $|B| = m$ and A_B being non-singular. The point x with $x_B = A_B^{-1}b$, $x_N = \not\vdash$ where $N := [n] \setminus B$ is a *basic solution* of (1) with basis B. If additionally $x_B \geq 0$, both the basic solution x and the basis B are *feasible.* If $x_i > 0$ for all $i \in B$, then B and x are called *non-degenerate* and *degenerate* otherwise, i.e., if $x_i = 0$ for some $i \in B$. We let $\overline{A} := A_B^{-1}A_N$ and $\overline{c}_N^T := c_N^T - c_B^T\overline{A}$. In particular, $\overline{c}_N \in \mathbb{R}^N$ is the vector of so-called *reduced costs* for the basis B. The coordinates of the reduced cost vector will be addressed as $\overline{c}_{N,i}$, where the first subscript will be dropped if the basis $B = [n] \setminus N$ is clear from the context.

The simplex algorithm considers in each iteration a feasible B. If all elements of $\overline{c}_N$ are non-negative, then the basis B and the corresponding basic feasible solution x are known to be optimal. Otherwise, the algorithm *pivots* by choosing a non-basic coordinate with negative reduced cost to enter the basis, say f. It then performs a minimum ratio test to compute an index i that minimizes $\frac{x_i}{\overline{A}_{if}}$ among all indices $i \in B$ for which $\overline{A}_{if} > 0$. Such index i will be the one leaving the basis, and it corresponds to a basic coordinate which hits its bound first when moving along the direction given by the tight constraints indexed by $B \setminus \{f\}$. At each iteration there could be multiple candidate indices for entering the basis (all the ones with negative reduced cost), as well as multiple candidate indices to leave the basis (all the ones for which the minimum ratio test value is attained). The choice of the entering and the leaving coordinates is the essence of a pivot rule (see [35] for a survey). A pivot is *degenerate* if the attained minimum ratio test value is 0 (hence, the extreme point solution x does not change), and *non-degenerate* otherwise. Note that only coordinates with negative reduced cost are considered for entering the basis, since only such a choice guarantees the simplex algorithm to make progress when a non-degenerate pivot occurs. To later emphasize that we only refer to that kind of *improving* pivots, we will call them *simplex pivots.*

Finally, we let $\text{kern}(A) := \{y \in \mathbb{R}^n : Ay = 0\}$. Given (1), we call a vector $y \in \text{kern}(A)$ with $c^T y < 0$ an *improving direction.* Such y is said to be *feasible* at a basic feasible solution x if $x + \epsilon y \geq 0$ for sufficiently small positive ϵ. Note that for any $y \in \text{kern}(A)$ and any basis B of A, the following holds:

$$c^T y = c_B^T y_B + c_N^T y_N = c_B^T(-\overline{A}y_N) + c_N^T y_N = \overline{c}_N^T y_N \tag{2}$$

2.2 A Geometric Intuition

Before providing a formal proof, we give a geometric intuition on how our pivot rule works. For this, it will be easier to abandon the standard equality form. In particular, consider a degenerate vertex $x \in \mathbb{R}^d$ of a polytope $\mathcal{P}$ of dimension $d \in \mathbb{N}$ defined by inequalities $a_i^T x \leq b_i$ with $a_i \in \mathbb{R}^d, b_i \in \mathbb{R}, i \in [n]$. In this setting, degeneracy means that more than d inequalities are tight at x. Consider a subset N of the set of indices of all inequalities that are tight at x with $|N| = d$ and let $B := [n] \setminus N$. Note that here we intentionally redefine a notation used in the standard equality form, where non basic coordinates always have their corresponding constraints tight. Since x is degenerate, there is at least one inequality that is tight at x with index in B. We denote the subset of such inequalities by $W \subseteq B$. Finally, assume x is not an optimal vertex of $\mathcal{P}$ when minimizing an objective function $c \in \mathbb{R}^d$ over $\mathcal{P}$, and let $y^0 \in \mathbb{R}^d$ be an improving feasible direction at x, i.e., such that $x + \epsilon y^0 \in \mathcal{P}$ for a sufficiently small positive ϵ and $c^T y^0 < 0$. In order to find an improving edge of $\mathcal{P}$ at x, we look at the directions given by the extremal rays of the basic cone $C(N) := \{x \in \mathbb{R}^n \mid a_i^T x \leq 0, i \in N\}$. If there is an improving feasible direction among them, we are done. Otherwise, we reduce the dimension of the polytope in the following way. We pick an extremal ray $z \in C(N)$ such that $c^T z < 0$. Note that z is formed by $d-1$ inequalities from N. Let $f \in N$ be the only inequality index not used to define z. Consider a vector combination $y^0 + \alpha z$ where $\alpha \geq 0$ and note that it provides an improving direction for any positive α. Since y^0 is contained in the feasible cone $C(N \cup W) = \{x \in \mathbb{R}^n \mid a_i^T x \leq 0, i \in N \cup W\}$ but z is not, the vector $y^0 + \alpha z$ leaves $C(N \cup W)$ for sufficiently large α. Hence there has to be a number $\alpha^1 \geq 0$ such that $y^0 + \alpha^1 z$ is contained in a facet of $C(N \cup W)$. Without loss of generality assume that the latter facet is defined by the gth inequality. Note that $g \in W$ since y^0 and z are both contained in the basic cone $C(N)$ and so is $y^0 + \alpha z$ for any non-negative α. Then, we define a new improving feasible direction $y^1 := y^0 + \alpha^1 z$, $N = \big(N \setminus \{f\}\big) \cup \{g\}$, $B = \big(B \setminus \{g\}\big) \cup \{f\}$ and continue searching for an improving edge inside the facet defined by the gth inequality. Since $x + \epsilon y^1$ belongs to this facet for an $\epsilon > 0$ small enough, the letter yields dimension reduction. After at most $d-1$ such steps we are bound to encounter a one-dimensional face of $\mathcal{P}$ containing $x + \epsilon y'$, where y' is an improving feasible direction at x. For an illustration see Fig. 1.

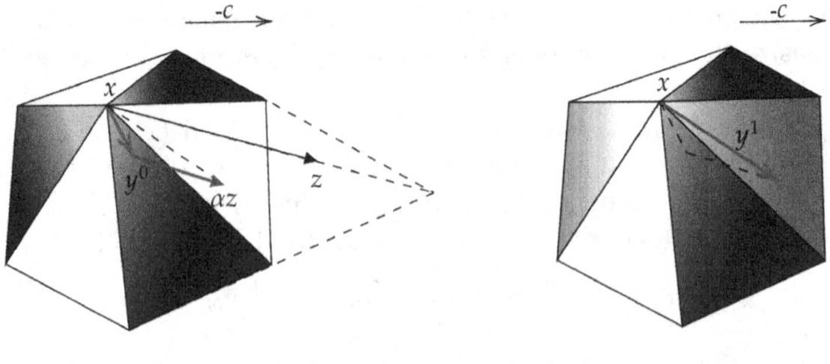

Fig. 1. To the left, facets corresponding to inequalities in N and W are colored in navy and white, respectively. To the right, the gray fading facet is defined by the fth inequality and the green one corresponds to the gth. (Color figure online)

2.3 Proof of Theorem 1

Proof. Let x be a degenerate basic feasible solution of (1) with basis B. Without loss of generality, assume $B = [m]$. Then, $x_i = 0$ for $i \in N = [n] \setminus [m]$. Assume also that $x_i = 0$ for $i \in [k]$ with $1 \le k \le m$ and $x_i > 0$ for $i = k+1, \ldots, m$. Let $S(B) := \{i \in N \mid \overline{c}_i < 0\}$. Since x is not optimal, $S(B) \neq \emptyset$.

It is well known, that a basic feasible solution x of (1) with basis B is not optimal if and only if there exists an improving feasible direction, i.e., there exists $y^0 \in \mathbb{R}^n$ such that $Ay^0 = 0$, $c^T y^0 < 0$ and $y_i^0 \ge 0$ for all $i \in [k] \cup ([n] \setminus [m])$. Consider any such y^0. Let $Q_1(y^0, B) := \{i \in [k] \mid y_i^0 > 0\}$ and $Q_2(y^0, B) := \{i \in [n] \setminus [m] \mid y_i^0 > 0\}$. Without loss of generality, let $Q_1(y^0, B) = [r]$ with $0 \le r \le k$ and $Q_2(y^0, B) = \{m+1, \ldots, m+t\}$ with $1 \le t \le n-m$. See Table 1 for an illustration.

Table 1. An illustration of entries of x, y^0, z and $y^1 := y^0 + \alpha z$. A positive, a non-negative, a negative, and a sign-arbitrary entry is denoted by $+$, 0_+, $-$ and $\star$, respectively. Without loss of generality, we here assumed that the entering index f is $m+1$, and that an index g for which the minimum in **Case III** is attained is 1.

	1		...	r		...	k		...	m	f		...	$m+t$		...	n
x	0	0	...	0	0	...	0	$+$	...	$+$	0	0	...	0	0	...	0
y^0	$+$	$+$	...	$+$	0	...	0	$\star$	...	$\star$	$+$	$+$	...	$+$	0	...	0
z	$-$	$\star$	...	$\star$	0_+	...	0_+	$\star$	...	$\star$	1	0	...	0	0	...	0
$y^0 + \alpha z$	0	0_+	...	0_+	0_+	...	0_+	$\star$	...	$\star$	$+$	$+$	...	$+$	0	...	0

Since $c^T y^0 = \overline{c}_N^T y_N^0 < 0$ due to (2), it follows that $S(B) \cap Q_2(y^0, B) \neq \emptyset$. We choose the entering index to be a non-basic one with the most negative reduced cost[1] in the support of y^0, that is $f = \operatorname{argmin}_{i \in S(B) \cap Q_2(y^0, B)} \overline{c}_i$. To detect the leaving index, we consider the following case distinction. Note that the case distinction depends on the basis B and the improving feasible direction y^0.

Case I: *There is no* $i \in [k]$ *with* $\overline{A}_{if} > 0$. In this case, the minimum ratio test for f is strictly positive, that is:

$$(*) \qquad \min_{i \in Eq^{>}(f)} \frac{x_i}{\overline{A}_{if}} > 0 \text{ where } Eq^{>}(f) := \{i \in [m] \mid \overline{A}_{if} > 0\}.$$

In particular, this means $Eq^{>}(f) \subseteq [m] \setminus [k]$. We perform a non-degenerate pivot, by selecting an index that minimizes $(*)$ as the leaving index.

Case II: $\overline{A}_{if} > 0$ *for some* $i \in [k] \setminus [r]$. In this case, we perform a degenerate pivot by selecting i as the leaving index. Let $B' := B \cup \{f\} \setminus \{i\}$. Because of

[1] We would like to point out that to prove the theorem, it suffices to choose *any* non-basic index with negative reduced cost in the support of y^0 as the entering index. The fact that f has the most negative reduced cost among such indices will only be used for the results in the next section.

degeneracy, the basic solution associated with B' is still x, and hence y^0 is still an improving feasible direction for x. Note that $|Q_2(y^0, B')| = |Q_2(y^0, B)| - 1$— since $y_f^0 > 0$ but $y_i^0 = 0$ by definition. Repeat the same for B' and y^0.

Case III: *$\overline{A}_{if} \leq 0$ for all $i \in [k] \setminus [r]$ and $\overline{A}_{if} > 0$ for some $i \in [r]$.* In this case, we are going to change our improving feasible direction. Consider the following vector $z \in \mathbb{R}^n$ which is, in fact, an improving (thought not feasible) direction for the entering variable x_f:

$$z_i = \begin{cases} -\overline{A}_{if}, & \text{for } i \in [m] \\ 1, & \text{for } i = f \\ 0, & \text{otherwise} \end{cases} \tag{3}$$

Note that $Az = 0$ and $c^T z = \overline{c}_f < 0$. Moreover $z_i \geq 0$ for each $i \in \big([r] \setminus Eq^>(f)\big) \cup \big([k] \setminus [r]\big) \cup \big([n] \setminus [m]\big)$ and $z_i < 0$ for $i \in Eq^>(f)$ by definition. Set

$$y^1 := y^0 + \alpha z \;\text{ with }\; \alpha := \min_{i \in Eq^>(f) \cap [r]} \frac{y_i^0}{|z_i|} > 0\,.$$

Observe that y^1 is a feasible direction for x, since $A(y + \alpha z) = 0$, and $y_i^1 \geq 0$ for $i \in [k] \cup \big([n] \setminus [m]\big)$, because of the choice of z and α. Furthermore, y^1 is an improving direction, because $c^T z < 0$ and hence

$$c^T y^1 = c^T y^0 + \alpha c^T z < c^T y^0 < 0\,. \tag{4}$$

Note that $Q_2(y^1, B) = Q_2(y^0, B)$. Furthermore, let $g \in Eq^>(f) \cap [r]$ be the index for which the value of α is attained. We have $y_g^1 = 0$ and $\overline{A}_{gf} > 0$. We now repeat the case distinction above for B and y^1.

The key observation is that when the third case of the above case distinction has occurred, repeating the same for B and y^1 falls into the second case (because the basis B has not changed, meanwhile $y_g^1 = x_g = 0$ with $g \in B$ and $\overline{A}_{gf} > 0$). Therefore, after each pivot, the cardinality of the support of the improving feasible direction in the non-basic indices decreases. Hence, a sequence of $|Q_2(y, B)|$ degenerate pivots would yield an improving feasible direction y' and a basis B' of x with $Q_2(y', B') = 0$, which in turn implies $0 = \overline{c}_{N'}^T y'_{N'} = c^T y'$ due to (2), yielding a contradiction. Hence the number of consecutive degenerate pivots with the suggested pivot rule cannot exceed $n - m - 1$. □

3 Exploiting the Antistalling Rule for General Bounds

Here we combine the result of the previous section with the analysis of [19,20]. The authors of the latter works give a bound on the number of non-degenerate simplex pivots that depends on n, m and the maximum and the minimum non-zero coordinate of basic solutions. The combined analysis yields a similar bound on the total number of (both degenerate and non-degenerate) simplex pivots.

We need a few additional notations. For any vector x, we let $\text{supp}(x)$ denote its support. We denote the smallest and the largest non-zero coordinate of any basic feasible solution of (1) by δ and Δ, respectively. We let $x^\star$ be an optimal basic feasible solution of (1). Finally, for a generic iteration q of the simplex algorithm with basis B^q and an improving feasible direction y^q, we let $\Delta_{\bar{c}}^q := \max_{\{i \in N^q | y_i^q > 0\}} -\bar{c}_{N^q,i}$. Note that $\Delta_{\bar{c}}^q$ is the absolute value of the reduced cost of the entering variable according to the antistalling pivot rule defined in the previous section when using y^q.

We make use of the following result from [20].

Lemma 1 (Lemma 4 of [20]). *If these exists a constant $\lambda > 0$ such that*

$$c^T(x^{q+1} - x^\star) \leq \left(1 - \frac{1}{\lambda}\right)c^T(x^q - x^\star) \tag{5}$$

holds for any consecutive distinct basic feasible solutions $x^q \neq x^{q+1}$ generated by the simplex algorithm (with any pivot rule), the total number of distinct basic feasible solutions encountered is at most

$$(n-m)\left\lceil \lambda \log_e \frac{m\Delta}{\delta} \right\rceil.$$

Given (1), apply the simplex algorithm with the antistalling pivot rule described in the previous section. In particular, at a general iteration t of the algorithm, let B^t, x^t, and y^t be respectively the basis, the basic solution, and the improving feasible direction considered by the algorithm. Let $N^t := [n] \setminus B^t$. If $x^t \neq x^{t-1}$ (i.e., we encounter x^t for the first time, as a result of a non-degenerate pivot) let $y^t := x^\star - x^t$. Observe that, once this is specified, the improving feasible direction is determined by the antistalling rule for all remaining degenerate pivots at x^t.

Lemma 2. *Let $x^t = x^{t+k}$ be the basic feasible solution associated with bases B^t and B^{t+k}, and assume $x^{t+k} \neq x^{t+k+1}$. The following holds:*

(a) $c^T y^{t+k} \leq c^T y^t$.
(b) $y_i^{t+k} = y_i^t$ for all $i \in N^{t+k} \cap \mathrm{supp}(y^{t+k})$.

Proof. The statement in (a) follows from (4). The statement in (b) follows from the construction of our antistalling rule by induction: if Case III never occurs during the k degenerate pivots, then $y^{t+k} = y^t$ and the statement holds trivially. Suppose that the situation of Case III appears for B^{t+j}, y^{t+j}. The improving feasible direction changes (as $y^{t+j+1} = y^{t+j} + \alpha z$), but among the non-basic coordinates this only affects the value of y_f^{t+j+1} (where f is the entering index). Immediately after, Case II occurs and f gets pivoted in, while a basic index i with $y_i^{t+j+1} = 0$ gets pivoted out, i.e., $B^{t+j+1} = (B^{t+j} \setminus \{i\}) \cup \{f\}$ and $N^{t+j+1} = (N^{t+j} \setminus \{f\}) \cup \{i\}$. Hence, the statement holds. □

The next result, which establishes (5) with $\lambda := \frac{(n-m)\Delta}{\delta}$ for the simplex algorithm with the antistalling pivot rule, is inspired by [20][Theorem 3].

Lemma 3. *Let $x^t = x^{t+k}$ be the basic feasible solution associated with bases B^t and B^{t+k}. Assume $x^{t-1} \neq x^t$ and $x^{t+k} \neq x^{t+k+1}$. We have*

$$c^T x^{t+k+1} - c^T x^\star \leq \left(1 - \frac{\delta}{(n-m)\Delta}\right)(c^T x^{t+k} - c^T x^\star).$$

Proof. The optimality gap for x^{t+k} can be bounded as follows:

$$\begin{aligned} c^T x^{t+k} - c^T x^\star &= c^T x^t - c^T x^\star \\ &= -c^T y^t \\ &\leq -c^T y^{t+k} \\ &= -\bar{c}^T_{N^{t+k}} y^{t+k}_{N^{t+k}} \\ &= \textstyle\sum_{i \in N^{t+k}} -\bar{c}^T_{N^{t+k},i} y_i^{t+k} \\ &= \textstyle\sum_{i \in N^{t+k} | y_i^{t+k} > 0} -\bar{c}^T_{N^{t+k},i} y_i^{t+k} \\ &\leq (n-m)\Delta_{\bar{c}}^{t+k} \Delta, \end{aligned} \tag{6}$$

where we used Lemma 2(a) for the first inequality, and (2) for the third equality. The last equality follows from $y_i^{t+k} \geq 0, i \in N^{t+k}$ which is implied by feasibility of y^{t+k} at x^{t+k}. The last inequality in turn follows from $\max_{\{i \in N^{t+k} | y_i^{t+k} > 0\}} -\bar{c}_{N^{t+k},i} = \Delta_{\bar{c}}^{t+k}$, the fact that $|N^{t+k}| = n - m$, and since $y_i^{t+k} = y_i^t = x_i^\star - x_i^t \leq \Delta$ holds for any $i \in N^{t+k} \cap \operatorname{supp}(y^{t+k})$, using Lemma 2(b).

Let x_f^{t+k} denote the entering variable at $(t+k)^{th}$ iteration. Note that $x_f^{t+k+1} \neq 0$ since $x^{t+k} \neq x^{t+k+1}$. Then,

$$\begin{aligned} c^T x^{t+k} - c^T x^{t+k+1} &= \bar{c}^T_{N^{t+k}} (x^{t+k} - x^{t+k+1})_{N^{t+k}} \\ &= -\bar{c}_{N^{t+k},f} x_f^{t+k+1} \\ &\geq \Delta_{\bar{c}}^{t+k} \delta \\ &\geq \tfrac{\delta}{(n-m)\Delta} c^T (x^{t+k} - x^\star) \end{aligned}$$

hold, where we used (2) for the first equality and (6) for the last inequality. Rearranging the terms yields the lemma statement. □

Now, combining Lemma 1 and Lemma 3 allows to conclude that the simplex algorithm with the antistalling pivot rule described in Sect. 2 encounters at most $(n-m)\lceil \frac{(n-m)\Delta}{\delta} \log_e \left(m\frac{\Delta}{\delta}\right)\rceil$ distinct basic feasible solutions. Since Theorem 1 in turn bounds the number of consecutive non-degenerate steps, combining it with the latter result yields a bound on the total number of pivots required to reach an optimal vertex. However, this bound does not take into account the number of (degenerate) pivots that might have to be performed at an optimal vertex to encounter a basis satisfying the optimality criterion. This is due to the fact that the antistalling pivot rule requires an improving feasible direction, which we do not have at an optimal vertex. Hence, we have to handle this case separately.

Let B and $B^\star$ be a non-optimal and an optimal basis, both associated with an optimal basic feasible solution $x^\star$ of (1). Since B is not optimal, there exists $f \in N$ with $\bar{c}_{N,f} = c^T z < 0$ with z as in (3). Observe that there exists $i \in B \cap N^\star$

with $\overline{A}_{if} > 0$ and $x_i = 0$: otherwise, all coordinates of $z_{N^\star}$ would be non-negative and hence $z \in C_{N^\star} := \{x \in \mathbb{R}^n \mid Ax = 0, x_{N^\star} \geq 0\}$. The latter however contradict the fact that all extreme rays of the above cone $C_{N^\star}$ have non-negative scalar products $\overline{c}_{N^\star}$ with c due to optimality of $B^\star$. Hence one could perform a simplex pivot on B with entering variable f and leaving variable i. Let $B' := \big(B \setminus \{i\}\big) \cup \{f\}$ and $N' := \big(N \setminus \{f\}\big) \cup \{i\}$. Note that $\overline{c}_{N',i} \geq 0$ and $i \in N' \cap N^\star$. Either B' is a optimal basis and we stop, or there exists $j \in N' \setminus \{i\}$ with $\overline{c}_{N',j} < 0$. In the latter case, however, we can enforce the constraint $x_i = 0$ by removing the variable x_i together with the corresponding column of A and entry c_i of c from (1). By doing so we obtain a new LP with the number of variables smaller by one that has B' and $B^\star$ as a non-optimal and an optimal basis, respectively, since $\overline{c}_{N' \setminus \{i\},j} < 0$ and $\overline{c}_{N^\star \setminus \{i\}} \geq 0$. We can set $B = B'$ and repeat the process. Since at each iteration a variable with index from $N^\star$ gets removed, after at most $n - m$ iterations $B = B^\star$. Since the number of degenerate pivots at an optimal vertex is bounded by $n - m$, we can state the following result.

Theorem 2. *Given an LP* (1) *and an initial feasible basis, there exists a pivot rule that makes the simplex algorithm reach an optimal basis in at most* $(n - m)^2 \lceil \frac{(n-m)\Delta}{\delta} \log_e \big(m\frac{\Delta}{\delta}\big)\rceil$ *simplex pivots.*

In general, the value of δ is NP-hard to compute [22]. However, observe that for integral A and b (which can be assumed without loss of generality for rational LPs), one can bound $\Delta \leq ||b||_1 \Delta_A$ and $\delta \geq \frac{1}{\Delta_A}$ due to Cramer's rule, where Δ_A is the largest absolute value of the determinant of a square $m \times m$ submatrix of A. Then the next statement a straightforward corollary of the above theorem.

Corollary 1. *For any basic feasible solution of an LP* (1) *with integral* A *and* b*, there exists a sequence of at most* $(n-m)^2 \lceil (n-m)\Delta_A^2 ||b||_1 \log_e \big(m\Delta_A^2 ||b||_1\big)\rceil$ *simplex pivots leading to an optimal basis.*

We conclude this section by observing that, using the above corollary, one can prove existence of short sequences of simplex pivots (that is, of length strongly-polynomial in n, m) between any two extreme points of several combinatorial LPs. Just to mention a few:

(a) LPs modeling matching/vertex-cover/edge-cover/stable-set problems in bipartite graphs. For matchings, the LP maximizes a linear function over a set of constraint of the form $\{A'x \leq 1, x \geq 0\}$, where the coefficient matrix A' is the node-edge incidence matrix of an undirected bipartite graph. After putting the LP in standard equality form by adding slack variables, we get constraints of the form $\{Ax = 1, x \geq 0\}$, where A is a totally unimodular matrix (and so $\Delta_A = 1$). The result then follows from Corollary 1. The same holds for vertex-cover (constraints of the form $\{{A'}^T x \geq 1, x \geq 0\}$), edge-cover (constraints of the form $\{A'x \geq 1, x \geq 0\}$), stable-set (constraints of the form $\{{A'}^T x \leq 1, x \geq 0\}$).

(b) LP relaxations for the above problems for general graphs (i.e., optimization over the *fractional* matching/vertex-cover/edge-cover/stable-set polytopes).

The same LPs as in (a), in general graphs, have a constraint matrix A' (resp. A'^T) that is not totally unimodular. However, the set of constraints defines half-integral polytopes (see [2,27]). Note that, after putting the LPs in standard equality form, the slack variables can be loosely bounded from above by the number of variables n. Hence, $\Delta \leq n$ and $\delta \geq \frac{1}{2}$, and the result follows from Theorem 2.

(c) LP for the stable marriage problem [30,38]. The (exact) LP formulation has constraints of the form $\{A'x \leq 1, B'x \geq 1, x \geq 0\}$. After putting the LP in standard equality form, the slack variables can be bounded by 1. Hence, $\Delta = \delta = 1$, and the result follows from Theorem 2.

(d) LPs modeling various flow problems (such as max flows, min cost flows, flow circulations) with unit (or bounded) capacity values. For the corresponding LPs (see [32]), the constraint matrix will be a node-arc incidence matrix of a directed graph, which is totally unimodular, and the right-hand-side vector will be bounded by the capacity values. Similarly to (a), one can rely on Corollary 1 to get the result.

Acknowledgements. The authors are grateful to Volker Kaibel for helpful discussions on the geometric interpretation of the antistalling pivot rule. We thank Alex Black and anonymous reviewers for their helpful comments. The authors express gratitude to the Deutsche Forschungsgemeinschaft (DFG, German Research Foundation) for supporting the first author within 314838170, GRK 2297 MathCoRe and providing travel funds for visiting the second author in Milan. The second author is supported by the NWO VIDI grant VI.Vidi.193.087.

References

1. Ahuja, R.K., Orlin, J.B., Sharma, P., Sokkalingam, P.: A network simplex algorithm with O(n) consecutive degenerate pivots. Oper. Res. Lett. **30**(3), 141–148 (2002)
2. Appa, G., Kotnyek, B.: A bidirected generalization of network matrices. Networks **47**(4), 185–198 (2006)
3. Avis, D., Friedmann, O.: An exponential lower bound for Cunningham's rule. Math. Program. **161**(1), 271–305 (2017)
4. Black, A., Loera, J.D., Kafer, S., Sanità, L.: On the simplex method for 0/1 polytopes (2021)
5. Bland, R.G.: New finite pivoting rules for the simplex method. Math. Oper. Res. **2**(2), 103–107 (1977)
6. Bonifas, N., Di Summa, M., Eisenbrand, F., Hähnle, N., Niemeier, M.: On subdeterminants and the diameter of polyhedra. Discrete Comput. Geom. **52**, 102–115 (2014)
7. Borgwardt, K.H.: The Simplex Method: A Probabilistic Analysis. Springer, Berlin, Heidelberg (1987)
8. Borgwardt, K.H.: A sharp upper bound for the expected number of shadow vertices in LP-polyhedra under orthogonal projection on two-dimensional planes. Math. Oper. Res. **24**(3), 544–603 (1999)

9. Cunningham, W.H.: Theoretical properties of the network simplex method. Math. Oper. Res. **4**(2), 196–208 (1979)
10. Dadush, D., Huiberts, S.: A friendly smoothed analysis of the simplex method. SIAM J. Comput. **49**(5), STOC18-449–STOC18-499 (2020)
11. Dadush, D., Hähnle, N.: On the shadow simplex method for curved polyhedra. Discrete Comput. Geom. **45**, 882–909 (2016)
12. Dantzig, G.B.: Maximization of a linear function of variables subject to linear inequalities. In: Koopmans, T.C. (ed.) Activity Analysis of Production and Allocation, pp. 339–347. Wiley (1951)
13. Disser, Y., Hopp, A.V.: On Friedmann's subexponential lower bound for Zadeh's pivot rule. In: Lodi, A., Nagarajan, V. (eds.) IPCO 2019. LNCS, vol. 11480, pp. 168–180. Springer, Cham (2019). https://doi.org/10.1007/978-3-030-17953-3_13
14. Goldfarb, D.: On the complexity of the simplex algorithm. In: Gomez, S., Hennart, J. (eds.) Advances in Optimization and Numerical Analysis, MAIA, vol. 275, pp. 25–38. Springer, Dordrecht (1994)
15. Goldfarb, D., Sit, W.Y.: Worst case behavior of the steepest edge simplex method. Discret. Appl. Math. **1**(4), 277–285 (1979)
16. Goldfarb, D., Hao, J., Kai, S.R.: Anti-stalling pivot rules for the network simplex algorithm. Networks **20**(1), 79–91 (1990)
17. Hansen, T., Zwick, U.: An improved version of the random-facet pivoting rule for the simplex algorithm. In: Proceedings of the Forty-Seventh Annual ACM Symposium on Theory of Computing, STOC 2015, pp. 209–218. ACM, Portland (2015)
18. Kabadi, S.N., Punnen, A.P.: Anti-stalling pivot rule for linear programs with totally unimodular coefficient matrix. In: Neogy, S.K., Bapat, R.B., Das, A.K., Parthasarathy, T. (eds.) Mathematical Programming and Game Theory for Decision Making, Statistical Science and Interdisciplinary Research, vol. 1, pp. 15–20. World Scientific Publishing Co., Pte. Ltd., Singapore (2008)
19. Kitahara, T., Mizuno, S.: A bound for the number of different basic solutions generated by the simplex method. Math. Program. **137**, 579–586 (2013)
20. Kitahara, T., Mizuno, S.: On the number of solutions generated by the simplex method for LP. In: Xu, H., Teo, K.L., Zhang, Y. (eds.) Optimization and Control Techniques and Applications. SPMS, vol. 86, pp. 75–90. Springer, Heidelberg (2014). https://doi.org/10.1007/978-3-662-43404-8_4
21. Klee, V., Minty, G.J.: How good is the simplex algorithm? In: Shisha, O. (ed.) Inequalities, III, pp. 159–175. Academic Press, New York (1972)
22. Kuno, T., Sano, Y., Tsuruda, T.: Computing Kitahara-Mizuno's bound on the number of basic feasible solutions generated with the simplex algorithm. Optimiz. Lett. **12**, 933–943 (2018)
23. Matschke, B., Santos, F., Weibel, C.: The width of five-dimensional prismatoids. Proc. Lond. Math. Soc. **110**(3), 647–672 (2015)
24. Megiddo, N.: Improved asymptotic analysis of the average number of steps performed by the self-dual simplex algorithm. Math. Program. **35**, 140–192 (1986)
25. Megiddo, N.: A note on degeneracy in linear programming. Math. Program. **35**(3), 365–367 (1986)
26. Murty, K.G.: Complexity of degeneracy. In: Floudas, C.A., Pardalos, P.M. (eds.) Encyclopedia of Optimization, pp. 419–425. Springer, Boston (2009)
27. Nemhauser, G.L., Trotter, L.E.: Properties of vertex packing and independence system polyhedra. Math. Program. **6**, 48–61 (1974)
28. Orlin, J.B.: A polynomial time primal network simplex algorithm for minimum cost flows. Math. Program. **78**, 109–129 (1997)

29. Rooley-Laleh, E.: Improvements to the Theoretical Efficiency of the Simplex Method. Ph.D. thesis, University of Carleton, Ottawa (1981)
30. Rothblum, U.G.: Characterization of stable matchings as extreme points of a polytope. Math. Program. **54**(1), 57–67 (1992)
31. Santos, F.: A counterexample to the Hirsch conjecture. Ann. Marth. **176**, 383–412 (2012)
32. Schrijver, A.: Combinatorial Optimization - Polyhedra and Efficiency. Springer, Berlin, Heidelberg (2003)
33. Shamir, R.: The efficiency of the simplex method: a survey. Manage. Sci. **33**, 301–334 (1987)
34. Spielman, D.A., Teng, S.H.: Smoothed analysis of algorithms: why the simplex algorithm usually takes polynomial time. J. ACM **51**(3), 385–463 (2004)
35. Terlaky, T., Zhang, S.: Pivot rules for linear programming: a survey on recent theoretical developments. Ann. Oper. Res. **46–47**(1), 203–233 (1993)
36. Terlaky, T.: Lexicographic pivoting rules. In: Floudas, C.A., Pardalos, P.M. (eds.) Encyclopedia of Optimization, pp. 1870–1873. Springer, Boston (2009)
37. Todd, M.J.: An improved Kalai-Kleitman bound for the diameter of a polyhedron. SIAM J. Discret. Math. **28**(4), 1944–1947 (2014)
38. Vate, J.H.V.: Linear programming brings marital bliss. Oper. Res. Lett. **8**(3), 147–153 (1989)
39. Vershynin, R.: Beyond Hirsch conjecture: walks on random polytopes and smoothed complexity of the simplex method. SIAM J. Comput. **39**(2), 646–678 (2009)
40. Zadeh, N.: What is the worst case behavior of the simplex algorithm? In: Avis, D., Bremner, D., Deza, A. (eds.) Polyhedral computation, CRM Proceedings and Lecture Notes, vol. 48, pp. 131–143. American Mathematical Society, Providence (2009)

On the Partial Convexification of the Low-Rank Spectral Optimization: Rank Bounds and Algorithms

Yongchun Li and Weijun Xie(✉)

H. Milton Stewart School of Industrial and Systems Engineering, Georgia Institute of Technology, Atlanta, GA 30332, USA
wxie@gatech.edu

Abstract. A Low-rank Spectral Optimization Problem (LSOP) minimizes a linear objective function subject to multiple two-sided linear inequalities intersected with a low-rank and spectral constrained domain. Although solving LSOP is, in general, NP-hard, its partial convexification (i.e., replacing the domain by its convex hull) termed "LSOP-R", is often tractable and yields a high-quality solution. This motivates us to study the strength of LSOP-R. Specifically, we derive rank bounds for any extreme point of the feasible set of LSOP-R with different matrix spaces and prove their tightness. The proposed rank bounds recover two well-known results in the literature from a fresh angle and allow us to derive sufficient conditions under which the relaxation LSOP-R is equivalent to LSOP. To effectively solve LSOP-R, we develop a column generation algorithm with a vector-based convex pricing oracle, coupled with a rank-reduction algorithm, which ensures that the output solution always satisfies the theoretical rank bound. Finally, we numerically verify the strength of LSOP-R and the efficacy of the proposed algorithms.

1 Introduction

This paper studies the Low-rank Spectral Optimization Problem (LSOP):

$$\mathbf{V}_{\text{opt}} := \min_{\boldsymbol{X} \in \mathcal{X}} \left\{ \langle \boldsymbol{A}_0, \boldsymbol{X} \rangle : b_i^l \leq \langle \boldsymbol{A}_i, \boldsymbol{X} \rangle \leq b_i^u, \forall i \in [m] \right\}, \tag{LSOP}$$

where the decision $\boldsymbol{X}$ is a matrix variable within a domain $\mathcal{X}$, $\langle \cdot, \cdot \rangle$ denotes the inner product of two equal-sized matrices, we have that $-\infty \leq b_i^l \leq b_i^u \leq +\infty$ for each $i \in [m]$, and matrices $\boldsymbol{A}_0$ and $\{\boldsymbol{A}_i\}_{i \in [m]}$ are symmetric if $\boldsymbol{X}$ is symmetric. Throughout, we let $\tilde{m}$ denote the number of linearly independent matrices in the set $\{\boldsymbol{A}_i\}_{i \in [m]}$.

Specifically, the domain $\mathcal{X}$ in LSOP is defined as below

$$\mathcal{X} := \left\{ \boldsymbol{X} \in \mathcal{S}_+^n : \text{rank}(\boldsymbol{X}) \leq k, F(\boldsymbol{X}) := f(\boldsymbol{\lambda}(\boldsymbol{X})) \leq 0 \right\}, \tag{1}$$

which consists of a low-rank and a closed convex spectral constraint. Note that (i) due to the page limit, this paper focuses on the domain of the positive semidefinite matrix space. Results for other matrix spaces, including non-symmetric,

J. Vygen and J. Byrka (Eds.): IPCO 2024, LNCS 14679, pp. 265–279, 2024.
https://doi.org/10.1007/978-3-031-59835-7_20

symmetric indefinite, and diagonal matrix spaces, can be found in our full-length version [16]; (ii) function $F(\boldsymbol{X}) : \mathcal{S}_+^n \to \mathbb{R}$ is continuous, closed, convex, and spectral which only depends on the eigenvalue or singular value vector $\boldsymbol{\lambda}(\boldsymbol{X}) \in \mathbb{R}^n$ of matrix $\boldsymbol{X}$. Thus, we can rewrite it as $F(\boldsymbol{X}) = f(\boldsymbol{\lambda}(\boldsymbol{X})) : \mathbb{R}^n \to \mathbb{R}$; and (iii) when there are multiple convex spectral constraints, i.e., $F_j(\boldsymbol{X}) \leq 0, \forall j \in J$, we can integrate them into a single convex spectral constraint by defining a function $F(\boldsymbol{X}) = \max_{j \in J} F_j(\boldsymbol{X})$. Hence, the domain $\mathcal{X}$ readily covers multiple spectral constraints. Such a set $\mathcal{X}$ naturally appears in many machine learning and optimization problems with a low-rank constraint (see Subsect. 1.1).

The low-rank constraint dramatically complicates LSOP, which often turns out to be an intractable nonconvex bilinear program. Thus, we leverage the convex hull of the domain $\mathcal{X}$, denoted by conv($\mathcal{X}$), to obtain a partial convexification for LSOP, termed LSOP-R throughout:

$$\mathrm{V}_{\mathrm{rel}} := \min_{\boldsymbol{X} \in \mathrm{conv}(\mathcal{X})} \left\{ \langle \boldsymbol{A}_0, \boldsymbol{X} \rangle : b_i^l \leq \langle \boldsymbol{A}_i, \boldsymbol{X} \rangle \leq b_i^u, \forall i \in [m] \right\} \leq \mathrm{V}_{\mathrm{opt}}, \quad \text{(LSOP-R)}$$

where the inequality is because LSOP-R serves as a convex relaxation of LSOP.

Despite being more tractable than LSOP, the optimal solution of LSOP-R may not satisfy the rank-k constraint. Therefore, an optimal solution of LSOP-R whose rank is guaranteed to be not far away from the threshold k is desirable and can fulfill various application requirements. That is, we are interested in finding the smallest integer $\widehat{k} \geq k$ such that there is an optimal solution $\boldsymbol{X}^*$ of LSOP-R satisfying rank$(\boldsymbol{X}^*) \leq \widehat{k}$. Such an integer $\widehat{k}$ is referred to as the rank bound of LSOP-R, which is shown to be useful for developing approximation algorithms for LSOP and solution algorithms for LSOP-R, as well as understanding when the relaxation LSOP-R meets LSOP (see, e.g., [6–8, 14, 20]). The analysis of rank bounds further inspires us to develop a rank-reduction algorithm. Hence, this paper aims to provide (i) theoretical rank bounds of the LSOP-R solutions and (ii) effective algorithms for solving LSOP-R while satisfying rank bounds.

1.1 Scope and Flexibility of Our LSOP Framework

In this subsection, we discuss several interesting low-rank constrained problems where the proposed LSOP framework can be applied (as a substructure). We refer readers to our full-length paper [16] for more applications.

Quadratically Constrained Quadratic Program (QCQP). QCQP has been widely studied in many application areas, including optimal power flow, sensor network problems, and signal processing, among others. QCQP of matrix form can be viewed as a special case of our LSOP:

$$\min_{\boldsymbol{X} \in \mathcal{X}} \left\{ \langle \boldsymbol{A}_0, \boldsymbol{X} \rangle : b_i^l \leq \langle \boldsymbol{A}_i, \boldsymbol{X} \rangle \leq b_i^u, \forall i \in [m] \right\}, \mathcal{X} = \{ \boldsymbol{X} \in \mathcal{S}_+^n : \mathrm{rank}(\boldsymbol{X}) \leq 1 \}. \quad \text{(QCQP)}$$

where $F(\boldsymbol{X}) = 0$.

Low-Rank Kernel Learning. Given an input kernel matrix $\boldsymbol{Y} \in \mathcal{S}_+^n$ of rank up to k, the low-rank kernel learning aims to find a rank $\leq k$ matrix $\boldsymbol{X}$ that closely approximates $\boldsymbol{Y}$, subject to additional linear constraints. The Log-Determinant divergence is a popular measure of the kernel learning quality (see, e.g., [13]), defined as $\langle \boldsymbol{Y}^{\dagger}, \boldsymbol{X}\rangle - \log\det[(\boldsymbol{X}+\alpha \boldsymbol{I}_n)(\boldsymbol{Y}+\alpha\boldsymbol{I}_n)^{-1}]$, where $\alpha > 0$ is small and $\boldsymbol{I}_n$ is the identify matrix for addressing the rank deficiency issue. Thus, the kernel learning under Log-Determinant divergence, as formulated by [13], becomes

$$\min_{z\in\mathbb{R},\boldsymbol{X}\in\mathcal{X}} \left\{\langle \boldsymbol{Y}^{\dagger}, \boldsymbol{X}\rangle - z + \log\det(\boldsymbol{Y}+\alpha\boldsymbol{I}_n) : b_i^l \leq \langle \boldsymbol{A}_i, \boldsymbol{X}\rangle \leq b_i^u, \forall i \in [m]\right\},$$
$$\mathcal{X} = \{\boldsymbol{X} \in \mathcal{S}_+^n : \operatorname{rank}(\boldsymbol{X}) \leq k, \log\det(\boldsymbol{X}+\alpha\boldsymbol{I}_n) \geq z\}. \tag{Kernel Learning}$$

Here, a closed convex spectral function is defined as $F(\boldsymbol{X}) = z - \log\det(\boldsymbol{X}+\alpha\boldsymbol{I}_n)$ for any z. The proposed LSOP naturally performs a substructure over variable $\boldsymbol{X}$ in Kernel Learning.

Fair PCA. It is recognized that the conventional Principal Component Analysis (PCA) may generate biased learning results against sensitive attributes, such as gender, race, or education level [19]. Fairness has recently been introduced to PCA. Formally, the seminal work by [20] formulates the fair PCA problem as follows, in which the substructure over $\boldsymbol{X}$ is a special case of the proposed LSOP

$$\max_{z\in\mathbb{R}_+,\boldsymbol{X}\in\mathcal{X}} \{z : z \leq \langle \boldsymbol{A}_i, \boldsymbol{X}\rangle, \forall i \in [m]\}, \mathcal{X} = \{\boldsymbol{X} \in \mathcal{S}_+^n : \operatorname{rank}(\boldsymbol{X}) \leq k, \|\boldsymbol{X}\|_2 \leq 1\}, \tag{Fair PCA}$$

where $\|\boldsymbol{X}\|_2$ denotes the spectral norm (i.e., the largest singular value) and $\boldsymbol{A}_1, \cdots \boldsymbol{A}_m \in \mathcal{S}_+^n$ denote the sample covariance matrices from m different groups.

1.2 Relevant Literature

In this subsection, we survey the relevant literature on two aspects.

Convexification of the Domain $\mathcal{X}$. There are few works on the convexification of a low-rank spectral domain $\mathcal{X}$ in (1). The work by [5] has successfully extended the perspective technique to convexify a special low-rank set $\mathcal{X}$ in which all the eigenvalues in the function $F(\boldsymbol{X}) = f(\boldsymbol{\lambda}(\boldsymbol{X}))$ are separable. Such an approach, however, may fail to cover the general set $\mathcal{X}$, e.g., the spectral norm function $F(\boldsymbol{X}) = \|\boldsymbol{X}\|_2$ in the Fair PCA which is not separable. Another seminal work [12] leverages the majorization technique on the convexification of any permutation-invariant set. We observe that our domain $\mathcal{X}$ in (1) is permutation-invariant with eigenvalues or singular values; thus, the majorization technique can be applied. This paper focuses on bounding the ranks of LSOP-R solutions, and the convexification result of [12] paves the path for the derivation of our bounds.

Rank Bounds for LSOP-R. Given a domain $\mathcal{X} = \{\boldsymbol{X} \in \mathcal{S}_+^n : \operatorname{rank}(\boldsymbol{X}) \leq 1\}$ particularly adopted in QCQP, its convex hull is the positive semidefinite matrix space and the corresponding LSOP-R feasible set is a spectrahedron. [2,9,17]

independently showed that the rank of any extreme point in the spectrahedron is in the order of $\mathcal{O}(\sqrt{2\tilde{m}})$. Recently, in the celebrated paper on Fair PCA [20], the authors also proved a rank bound for all feasible extreme points of the corresponding LSOP-R, which is $\mathcal{O}(\sqrt{2m})$, and the proof technique extended that of [17]. Another relevant work pays particular attention to the sparsity bound for the sparse optimization problem [1], which relies on a different continuous relaxation from the LSOP-R. Since the LSOP can encompass the sparse optimization when $\mathcal{S}_+^n$ in (1) denotes the diagonal matrix space, our rank bounds are, in fact, sparsity bounds under this setting. To the best of our knowledge, this is the first work to study the rank bounds for the generic partial convexification– LSOP-R. Our rank bounds recover all the ones reviewed here for QCQP and Fair PCA from a different perspective, and successfully reduce the sparsity bound of [1] when applying to the sparse ridge regression.

1.3 Contributions and Outline

We theoretically guarantee the solution quality of LSOP-R by bounding the ranks of its extreme points. Notably, our rank bounds hold for any domain $\mathcal{X}$ in the form of (1) and for any $\tilde{m}$ linearly independent inequalities, and they are attainable by the worst-case instances. Below are the major contributions of this paper, with a focus on the positive semidefinite matrix space. For a detailed review of our proposed rank bounds and application examples, please see our full-length paper [16, Table 1].

(i) In Sect. 2, we show that the rank of any extreme point in the feasible set of LSOP-R deviates at most $\lfloor\sqrt{2\tilde{m}+9/4}-3/2\rfloor$ from the original rank-k requirement in LSOP. We establish this rank bound from a novel perspective, specifically that of analyzing the rank of various faces in the convex hull of the domain (i.e., $\text{conv}(\mathcal{X})$). Besides, the rank bound gives a sufficient condition under which LSOP-R exactly solves the original LSOP.
(ii) In Sect. 3, we develop an efficient column generation algorithm with a semidefinite program (SDP)-free convex pricing oracle to solve LSOP-R. We also design a rank-reduction algorithm to ensure that the rank of the output solution satisfies the theoretical bound.
(iii) In Sect. 4, we numerically evaluate the quality of LSOP-R and the performance of our proposed algorithms.

The proofs of all the results presented in this paper can be found in the full-length paper [16], *which is available at* https://arxiv.org/pdf/2305.07638.pdf.

Notation and Definition. We use bold lower-case letters (e.g., $\boldsymbol{x}$) and bold upper-case letters (e.g., $\boldsymbol{X}$) to denote vectors and matrices, respectively, and use corresponding non-bold letters (e.g., x_i) to denote their components. We let $[n] := \{1, 2, \cdots, n\}$. We let $\boldsymbol{I}_n$ denote the $n \times n$ identity matrix. For any $\lambda \in \mathbb{R}$, we let $\lfloor\lambda\rfloor$ denote the greatest integer less than or equal to λ. For a set S, we let $|S|$ denote its cardinality. For a matrix $\boldsymbol{X} \in \mathcal{S}_+^n$, we let $\|\boldsymbol{X}\|_2$ be its largest singular value, let $\|\boldsymbol{X}\|_*$ be its nuclear norm, let $\|\boldsymbol{X}\|_F$ be its Frobenius norm, and if matrix $\boldsymbol{X}$ is square, let $\boldsymbol{X}^\dagger$ be its Moore-Penrose inverse, let $\operatorname{tr}(\boldsymbol{X})$ be the trace, let $\operatorname{diag}(\boldsymbol{X})$ be a vector including its diagonal elements, and let $\lambda_{\min}(\boldsymbol{X}), \lambda_{\max}(\boldsymbol{X})$ denote the smallest and largest eigenvalue of matrix $\boldsymbol{X}$, respectively. For a matrix $\boldsymbol{X} \in \mathcal{S}_+^n$ and an integer $i \in [n]$, we let $\|\boldsymbol{X}\|_{(i)}$ denote the sum of first i largest singular values of matrix $\boldsymbol{X}$. Note that $\|\boldsymbol{X}\|_{(n)} = \operatorname{tr}(\boldsymbol{X})$ if matrix $\boldsymbol{X}$ is positive semidefinite. Additional notation will be introduced later as needed.

2 The Rank Bound in Positive Semidefinite Matrix Space

To guarantee the solution quality of the partial convexification LSOP-R, this section derives a rank bound for all extreme points and for an optimal solution to LSOP-R, provided that $\mathcal{X}$ builds on the positive semidefinite matrix space. Notably, our results recover the existing rank bounds for two special LSOPs: QCQP and Fair PCA, and are also applicable to Kernel Learning (see the full-length paper [16]).

2.1 Convexifying the Domain

This subsection provides an explicit characterization of the convex hull of the domain $\mathcal{X}$, i.e., $\operatorname{conv}(\mathcal{X})$, which is a key component in the feasible set of LSOP-R.

Before deriving $\operatorname{conv}(\mathcal{X})$, let us define an integer $\kappa \le k$ as the strengthened rank. When necessary in the theoretical analysis of rank bounds, we replace the rank requirement k with a smaller number κ defined as below.

Definition 1 (Strengthened rank κ). *For a domain $\mathcal{X}$ in (1), we let $1 \le \kappa \le k$ be the smallest integer such that*

$$\operatorname{conv}(\mathcal{X}) = \operatorname{conv}\left(\{\boldsymbol{X} \in \mathcal{S}_+^n : \operatorname{rank}(\boldsymbol{X}) \le \kappa, F(\boldsymbol{X}) = f(\boldsymbol{\lambda}(\boldsymbol{X})) \le 0\}\right).$$

In general, computing an integer κ can be difficult, and there are unfavorable cases where $\kappa = k$. Nevertheless, for specific applications of LSOP considered in this paper, κ can be easily derived and may be strictly less than k. Below, we provide an example to illustrate that $\kappa < k$ through a simple analysis.

Example 1. Suppose a domain $\mathcal{X} = \{\boldsymbol{X} \in \mathcal{S}_+^n : \operatorname{rank}(\boldsymbol{X}) \le k, f(\|\boldsymbol{X}\|_*) \le 0\}$ with $k \ge 2$ and $f(\cdot)$ is a closed convex bounded function. Then sets $\mathcal{X}$ and $\operatorname{conv}(\mathcal{X})$ are bounded. In this example, we show below that $\kappa = 1 < k$.

For any matrix $\widehat{\boldsymbol{X}}$ in the domain $\mathcal{X}$ with rank $r \ge 2$, let its singular value decomposition be $\widehat{\boldsymbol{X}} = \boldsymbol{Q}\operatorname{Diag}(\widehat{\boldsymbol{\lambda}})\boldsymbol{P}^\top$. Next, let us construct r vectors $\{\boldsymbol{\lambda}^i\}_{i\in[r]} \subseteq \mathbb{R}_+^n$ as below, where for each $i \in [r]$,

$$\lambda_\ell^i = \|\widehat{\boldsymbol{\lambda}}\|_1, \ell = i, \quad \lambda_\ell^i = 0, \forall \ell \in [n] \setminus \{i\}.$$

Then, for each $i \in [r]$, we have that $\|\boldsymbol{\lambda}^i\|_0 = 1 \le k$ and $f(\|\boldsymbol{\lambda}^i\|_1) = f(\|\widehat{\boldsymbol{\lambda}}\|_1) \le 0$, which means that the inclusion $\boldsymbol{Q}\operatorname{Diag}(\boldsymbol{\lambda}^i)\boldsymbol{P}^\top \in \mathcal{X}$ holds for all $i \in [r]$. Also, we have $\widehat{\boldsymbol{X}} = \sum_{i\in[r]} \frac{\widehat{\lambda}_i}{\|\boldsymbol{\lambda}\|_1} \boldsymbol{Q} \operatorname{Diag}(\boldsymbol{\lambda}^i) \boldsymbol{P}^\top$ by the construction of vector $\{\boldsymbol{\lambda}^i\}_{i\in[r]} \subseteq \mathbb{R}_+^n$, implying that $\widehat{\boldsymbol{X}}$ cannot be an extreme point of the set $\operatorname{conv}(\mathcal{X})$. That is, any extreme point in $\operatorname{conv}(\mathcal{X})$ has a rank at most one. Hence, we have that $\kappa = 1$ by Definition 1. $\diamond$

We now turn to an explicit characterization of $\operatorname{conv}(\mathcal{X})$, which builds on the majorization below.

Definition 2 (Majorization). *Given two vectors $\boldsymbol{x}_1, \boldsymbol{x}_2 \in \mathbb{R}^n$, we let $\boldsymbol{x}_1 \succ \boldsymbol{x}_2$ denote that $\boldsymbol{x}_1$ weakly majorizes $\boldsymbol{x}_2$ (i.e.,* $\max_{\boldsymbol{z}\in\{0,1\}^n}\{\boldsymbol{z}^\top \boldsymbol{x}_1 : \mathbf{e}^\top \boldsymbol{z} = \ell\} \geq \max_{\boldsymbol{z}\in\{0,1\}^n}$
$\{\boldsymbol{z}^\top \boldsymbol{x}_2 : \mathbf{e}^\top \boldsymbol{z} = \ell\}$ *for all $\ell \in [n]$), and let $\boldsymbol{x}_1 \succeq \boldsymbol{x}_2$ denote that $\boldsymbol{x}_1$ majorizes $\boldsymbol{x}_2$, i.e., $\boldsymbol{x}_1$ weakly majorizes $\boldsymbol{x}_2$ and $\mathbf{e}^\top \boldsymbol{x}_1 = \mathbf{e}^\top \boldsymbol{x}_2$, where $\mathbf{e} \in \mathbb{R}^n$ denotes the all-ones vector.*

The spectral function $F(\boldsymbol{X}) = f(\boldsymbol{\lambda}(\boldsymbol{X}))$ in (1) is invariant with any permutation of eigenvalues (see, e.g., [10]), implying the permutation-invariant domain $\mathcal{X}$. This motivates us to apply the convexification result for a permutation-invariant set, as shown in the seminal work by [12].

Proposition 1. *For a domain $\mathcal{X}$ in (1), its convex hull* $\operatorname{conv}(\mathcal{X})$ *is equal to* $\{\boldsymbol{X} \in \mathcal{S}_+^n : \exists \boldsymbol{x} \in \mathbb{R}_+^n, f(\boldsymbol{x}) \leq 0, x_1 \geq \cdots \geq x_n, x_{k+1} = 0, \|\boldsymbol{X}\|_{(\ell)} \leq \sum_{i\in[\ell]} x_i, \forall \ell \in [k], \operatorname{tr}(\boldsymbol{X}) = \sum_{i\in[k]} x_i\}$, *and* $\operatorname{conv}(\mathcal{X})$ *is a closed set.*

The function $\|\boldsymbol{X}\|_{(i)}$ in Proposition 1 is known to admit a tractable semidefinite representation for each $i \in [n]$ (see, e.g., [4]).

2.2 Rank Bound

In this subsection, we derive an upper bound of the ranks of all extreme points in the feasible set of LSOP-R, which also sheds light on the rank gap between the relaxation LSOP-R and original LSOP at optimality. Let us first introduce a key lemma to facilitate the analysis of rank bound.

Lemma 1. *Given two vectors $\boldsymbol{\lambda}, \boldsymbol{x} \in \mathbb{R}^n$ with their elements sorted in descending order and $\boldsymbol{x} \succeq \boldsymbol{\lambda}$, suppose that there exists an index $j_1 \in [n-1]$ such that $\sum_{i\in[j_1]} \lambda_i < \sum_{i\in[j_1]} x_i$. Then we have $\sum_{i\in[j]} \lambda_i < \sum_{i\in[j]} x_i, \forall j \in [j_0, j_2 - 1]$, if the indices j_0, j_2 satisfy $\lambda_{j_0} = \cdots = \lambda_{j_1} \geq \lambda_{j_1+1} = \cdots = \lambda_{j_2}$ with $1 \leq j_0 \leq j_1 \leq j_2 - 1 \leq n - 1$.*

Next, we are ready to present one of our main contributions by leveraging Lemma 1. Notably, the following theorem shows that the rank bound of a face of $\operatorname{conv}(\mathcal{X})$ depends on the dimension of the face. Below is the formal definition of the face.

Definition 3 (Face & Dimension). *A nonempty convex subset F of a convex set T is called a face of T if for any line segment $[a, b] \subseteq T$ such that $F \cap (a, b) \neq \emptyset$, we have $[a, b] \subseteq F$. The dimension of a face equals the dimension of its affine hull.*

Theorem 1. *For a domain $\mathcal{X}$ in* (1)*, suppose that F is a face of* $\mathrm{conv}(\mathcal{X})$ *with dimension d. Then any point in face F has a rank at most $\kappa + \lfloor\sqrt{2d+9/4}-3/2\rfloor$, where $\kappa \le k$ follows Definition 1.*

Theorem 1 suggests that the dimension of a face of $\mathrm{conv}(\mathcal{X})$ determines the rank bound. When intersecting set $\mathrm{conv}(\mathcal{X})$ with m linear inequalities in LSOP-R, the recent work by [15, Lemma 1] established the one-to-one correspondence between the extreme points of the intersection set and no larger than m-dimensional faces of $\mathrm{conv}(\mathcal{X})$.

Lemma 2 ([15])**.** *For any closed convex set $\mathcal{T}$, if $\boldsymbol{X}$ is an extreme point in the intersection of set $\mathcal{T}$ and $\tilde{m}$ linearly independent inequalities, then $\boldsymbol{X}$ must be contained in a face of set $\mathcal{T}$ with dimensional at most $\tilde{m}$.*

Taken the above results together, we derive a rank bound for LSOP-R as follows.

Theorem 2. *Given the integer $\kappa \le k$ from Definition 1, the following results must hold:*

(i) Each feasible extreme point in LSOP-R has a rank at most $\kappa + \lfloor\sqrt{2\tilde{m}+9/4}-3/2\rfloor$; and
(ii) There is an optimal solution to LSOP-R of rank at most $\kappa + \lfloor\sqrt{2\tilde{m}+9/4}-3/2\rfloor$ if LSOP-R yields a finite optimal value, i.e., $\mathbf{V}_{\mathrm{rel}} > -\infty$.

In Theorem 2, our proposed rank bound for LSOP-R arises from bounding the rank of various faces in set $\mathrm{conv}(\mathcal{X})$, as shown in Theorem 1 through perturbing majorization constraints. The proof differs from those for two specific LSOP-Rs of QCQP [2,9,17] and Fair PCA [20]. Specifically, [17] derived the rank bound from analyzing faces of the whole feasible set of LSOP-R of QCQP, whereas our result, inspired by Theorem 1, only focuses on the faces of $\mathrm{conv}(\mathcal{X})$. In addition, the worst-case example in the full-length paper [16, Appendix B] exactly attains the proposed rank bound in Theorem 2, which confirms the tightness. We further make the following remarks about Theorem 2.

(i) Theorem 2 shows that any extreme point in the feasible set of LSOP-R violates the rank-k constraint by at most $\kappa - k + \lfloor\sqrt{2\tilde{m}+9/4}-3/2\rfloor$.
(ii) The most striking aspect is that the rank bound of Theorem 2 is independent of any domain $\mathcal{X}$, any linear objective function and any $\tilde{m}$ linearly independent inequalities in LSOP-R. We present three examples in the full-length paper [16, Subsection 2.3] to demonstrate the versatility of our rank bound;and
(iii) An extra benefit of Theorem 2 is to provide a sufficient condition about when LSOP-R is equivalent to the original LSOP, as shown below.

Proposition 2. *Given an integer $\kappa \le k$ from Definition 1, the following hold if $\tilde{m} \le (k-\kappa+2)(k-\kappa+3)/2-2$:*

(i) Each feasible extreme point in the LSOP-R has a rank at most k; and

(ii) LSOP-R achieves the same optimal value as original LSOP, i.e., $\mathbf{V}_{\mathrm{opt}} = \mathbf{V}_{\mathrm{rel}}$ if LSOP-R yields a finite optimal value, i.e., $\mathbf{V}_{\mathrm{rel}} > -\infty$.

Proposition 2 contributes to the literature of the LSOP-R objective exactness that refers to the equation $\mathbf{V}_{\mathrm{opt}} = \mathbf{V}_{\mathrm{rel}}$, as it indicates that for any LSOP, the corresponding LSOP-R problem achieves the objective exactness whenever $\tilde{m} \leq 1$ and $\mathbf{V}_{\mathrm{rel}} > -\infty$.

3 Solving LSOP-R by Column Generation and Rank-Reduction Algorithms

This section develops an efficient column generation algorithm for solving LSOP-R. Although efficient, column generation, an iterative algorithm for finding an optimal solution through a convex combination of all generated columns (variables), often yields a high-rank solution within the interior of the feasible set. To resolve this issue, we develop a rank-reduction algorithm that reduces the rank of the output from column generation while maintaining the optimality, whenever the output solution violates the theoretical rank bound of LSOP-R. Throughout this section, we suppose that LSOP-R yields a finite optimal value, i.e., $\mathbf{V}_{\mathrm{rel}} > -\infty$.

3.1 Column Generation Algorithm

The column generation algorithm does not require an explicit description of the convex hull of domain set $\mathcal{X}$, and instead uses the representation theorem 18.5 in [18] to describe the line-free convex set $\operatorname{conv}(\mathcal{X})$ via the linear programming formulation (with possibly infinite number of variables), implying an equivalent linear reformulation of LSOP-R as shown below

$$\begin{aligned}\mathbf{V}_{\mathrm{rel}} := \min_{\alpha\in\mathbb{R}_+^{|\mathcal{H}|},\gamma\in\mathbb{R}_+^{|\mathcal{J}|}} \Bigg\{ & \sum_{h\in\mathcal{H}} \alpha_h \langle \boldsymbol{A}_0, \boldsymbol{X}_h\rangle + \sum_{j\in\mathcal{J}} \gamma_j \langle \boldsymbol{A}_0, \boldsymbol{d}_j\rangle : \sum_{h\in\mathcal{H}} \alpha_h = 1, \\ & b_i^l \leq \sum_{h\in\mathcal{H}} \alpha_h \langle \boldsymbol{A}_i, \boldsymbol{X}_h\rangle + \sum_{j\in\mathcal{J}} \gamma_j \langle \boldsymbol{A}_i, \boldsymbol{d}_j\rangle \leq b_i^u, \forall i \in [m] \Bigg\},\end{aligned} \tag{2}$$

where $\{\boldsymbol{X}_h\}_{h\in\mathcal{H}}$ and $\{\boldsymbol{d}_j\}_{j\in\mathcal{J}}$ consist of all the extreme points and extreme directions in the set $\operatorname{conv}(\mathcal{X})$, respectively.

It is often difficult to enumerate all the possible extreme points $\{\boldsymbol{X}_h\}_{h\in H}$ and extreme directions $\{\boldsymbol{d}_j\}_{j\in J}$ in the set $\operatorname{conv}(\mathcal{X})$ and directly solve the (semi-infinite) linear programming problem (2) to optimality. Alternatively, the column generation algorithm starts with the *restricted master problem* (RMP) which only contains a small portion of extreme points and extreme directions and then solves the *pricing problem* (PP) to iteratively generate an improving point. The detailed implementation is presented in the full-length paper [16, Algorithm 1]. The column generation algorithm can be effective in practice since a small number of points in $\text{conv}(\mathcal{X})$ are often needed to return a (near-)optimal solution to LSOP-R (2) according to Carathéodory theorem.

As shown in [16, Algorithm 1], at each iteration, we first solve a linear program–the RMP based on a subset of extreme points $\{\boldsymbol{X}_h\}_{h\in\mathcal{H}}$ and a subset of extreme directions $\{\boldsymbol{d}_j\}_{j\in\mathcal{J}}$ collected from the set $\text{conv}(\mathcal{X})$. Then, given the corresponding optimal dual solution $(\nu^*, (\boldsymbol{\mu}^l)^*, (\boldsymbol{\mu}^u)^*)$ of RMP, the PP finds an improving point or direction for LSOP-R (2) in set $\text{conv}(\mathcal{X})$.

$$\mathbf{V}_{\mathrm{P}} := \max_{\boldsymbol{X}\in\text{conv}(\mathcal{X})} \left\langle -\boldsymbol{A}_0 + \sum_{i\in[m]} ((\mu_i^l)^* - (\mu_i^u)^*)\boldsymbol{A}_i, \boldsymbol{X} \right\rangle. \quad \text{(Pricing Problem)}$$

Besides, when the pricing problem yields a finite optimal value no larger than $\nu^* + \epsilon$, we can show that the column generation algorithm finds an ϵ−optimal solution to the LSOP-R (2), as detailed below.

Proposition 3. *The output solution of the column generation algorithm is ϵ-optimal to LSOP-R if LSOP-R yields a finite optimal value, i.e., $\mathbf{V}_{\mathrm{rel}} > -\infty$. That is, suppose that the column generation algorithm returns $\boldsymbol{X}^*$, then $\mathbf{V}_{\mathrm{rel}} \leq \langle \boldsymbol{A}_0, \boldsymbol{X}^*\rangle \leq \mathbf{V}_{\mathrm{rel}} + \epsilon$ holds.*

3.2 Efficient Pricing Oracle

In this subsection, we show that despite not describing $\text{conv}(\mathcal{X})$, the Pricing Problem is efficiently solvable and even admits the closed-form solution for some cases. Given the linear objective function, Pricing Problem can be simplified by replacing $\text{conv}(\mathcal{X})$ with the domain $\mathcal{X}$, i.e.,

$$\mathbf{V}_{\mathrm{P}} = \max_{\boldsymbol{X}\in\mathcal{S}_+^n} \left\{\left\langle \tilde{\boldsymbol{A}}, \boldsymbol{X}\right\rangle : \text{rank}(\boldsymbol{X}) \leq k, f(\boldsymbol{\lambda}(\boldsymbol{X})) \leq 0\right\}, \quad \text{(Pricing Problem-S)}$$

where we define $\tilde{\boldsymbol{A}} := -\boldsymbol{A}_0 + \sum_{i\in[m]}((\mu_i^l)^* - (\mu_i^u)^*)\boldsymbol{A}_i$ throughout this subsection.

Next, let us explore the solution structure of Pricing Problem-S. Note that matrices $\boldsymbol{A}_0$ and $\{\boldsymbol{A}_i\}_{i\in[m]}$ are symmetric, and so does matrix $\tilde{\boldsymbol{A}}$.

Theorem 3. *Suppose that $\tilde{\boldsymbol{A}} = \boldsymbol{U}\,\text{Diag}(\boldsymbol{\beta})\boldsymbol{U}^\top$ denotes the eigen-decomposition of matrix $\tilde{\boldsymbol{A}}$ with eigenvalues $\beta_1 \geq \cdots \geq \beta_n$. Then Pricing Problem-S has an optimal solution $\boldsymbol{X}^* = \boldsymbol{U}\,\text{Diag}(\boldsymbol{\lambda}^*)\boldsymbol{U}^\top$, where $\boldsymbol{\lambda}^* \in \mathbb{R}_+^n$ is defined below:*

$$\boldsymbol{\lambda}^* \in \text{argmax}_{\boldsymbol{\lambda}\in\mathbb{R}_+^n} \left\{\boldsymbol{\lambda}^\top\boldsymbol{\beta} : \lambda_i = 0, \forall i \in [k+1, n], f(\boldsymbol{\lambda}) \leq 0\right\}. \tag{3}$$

We make the following remarks about Theorem 3:

(i) For any domain $\mathcal{X}$ in (1), Pricing Problem-S involves a rank constraint and is thus nonconvex. Theorem 3 provides a striking result that Pricing Problem-S can be equivalent to solving a convex program over a vector variable of length n. Therefore, the column generation algorithm is SDP-free and can admit an efficient implementation for solving LSOP-R;

(ii) By leveraging Theorem 3, we can derive the closed-form solutions to Pricing Problem-S for a special family of the domain $\mathcal{X}$ as shown in Corollary 1;

(iii) Although the convex hull of the domain $\mathcal{X}$ may admit an explicit description based on the majorization technique (see, e.g., Proposition 1), the description involves many auxiliary variables and constraints and is semidefinite representable. Therefore, the original Pricing Problem over $\text{conv}(\mathcal{X})$ can still be challenging. Quite differently, Theorem 3 allows us to target a simple convex program for solving Pricing Problem, which significantly enhances the column generation performance in our numerical study.

Corollary 1. *For a domain* $\mathcal{X} = \{\boldsymbol{X} \in \mathcal{S}_+^n : \text{rank}(\boldsymbol{X}) \leq k, \|\boldsymbol{\lambda}(\boldsymbol{X})\|_\ell \leq c\}$, *where* $c \geq 0$, $\ell \in [1, \infty]$, *and* $1/\ell + 1/q = 1$, *the solution* $\boldsymbol{\lambda}^*$ *is optimal to problem 3 with* $\lambda_i^* = c\sqrt[\ell]{\frac{(\beta_i)_+^q}{\sum_{j\in[k]}(\beta_j)_+^q}}$ *for all* $i \in [k]$ *and* $\lambda_i^* = 0$ *for all* $i \in [k+1, n]$.

3.3 Rank-Reduction Algorithm

The column generation algorithm may not always be able to output a solution that satisfies the theoretical rank bounds. To resolve this, this subsection designs a rank-reduction algorithm to find an alternative solution of the same or smaller objective value while satisfying the desired rank bounds. The detailed implementation can be found in the full-length paper [16].

Given a (near-)optimal solution $\boldsymbol{X}^*$ of LSOP-R returned by column generation algorithm that violates the proposed rank bound, our rank-reduction algorithm runs as follows: (i) since $\boldsymbol{X}^*$ is not an extreme point of LSOP-R, then we can find a direction $\boldsymbol{Y}$ in its feasible set, along which the objective value will decrease or stay the same; (ii) we move $\boldsymbol{X}^*$ along the direction $\boldsymbol{Y}$ until a point on the boundary of the feasible set of LSOP-R; (iii) we update $\boldsymbol{X}^*$ to be the new boundary point found; and (iv) finally, we terminate the iteration when no further movement is available, i.e., $\boldsymbol{X}^*$ is a feasible extreme point in LSOP-R that must satisfy the rank bound. Hence, the rank-reduction procedure in [16, Algorithm 2] can be viewed as searching for an alternative optimal extreme point.

Following the searching procedure above, we show that the rank-reduction algorithm can output an alternative solution satisfying the rank bound in Theorem 2, as summarized below. In our numerical study, this algorithm efficiently reduces the solution rank without hurting the optimality and converges fast.

Theorem 4. *The rank-reduction algorithm always terminates, and it returns an* ϵ*-optimal solution* $\boldsymbol{X}^*$ *to LSOP-R that satisfies* $\text{rank}(\boldsymbol{X}^*) \leq \kappa + \lfloor\sqrt{2\tilde{m} + 9/4} - 3/2\rfloor$, *where integer* κ *is defined in Definition 1.*

4 Numerical Study

In this section, we numerically test the proposed rank bound, column generation algorithm, and rank-reduction algorithm on three LSOP examples: multiple-input and multiple-output (MIMO) radio communication network, optimal power flow, and matrix completion. All the testing instances are synthetic, and the generation procedure is detailed in the full-length paper [16].

4.1 MIMO Radio Network

In recent years, there has been an increasing interest in a special class of LSOP with trace and log-det spectral constraints in communication and signal processing fields. Specifically, the objective is to find a low-rank transmit covariance matrix $\boldsymbol{X} \in \mathcal{S}_+^n$, subject to the trace and log-det constraints (see, e.g., [21] and references therein). Formally, we consider LSOP built on a special domain $\mathcal{X}$ below:

$$\mathcal{X} = \{\boldsymbol{X} \in \mathcal{S}_+^n : \operatorname{rank}(\boldsymbol{X}) \leq k, \operatorname{tr}(\boldsymbol{X}) \leq U, \log\det(\boldsymbol{I}_n + \boldsymbol{X}) \geq L\}, \tag{4}$$

where $U, L \geq 0$ are pre-specified constants and here, the spectral function in (1) is defined as $F(\boldsymbol{X}) = \max\{\operatorname{tr}(\boldsymbol{X}) - U, L - \log\det(\boldsymbol{I}_n + \boldsymbol{X})\}$.

Without using Theorem 3, the original Pricing Problem can be solved as an SDP by plugging the set $\operatorname{conv}(\mathcal{X})$ in Proposition 1 when generating a new column. This column generation procedure is termed as the "*naive column generation algorithm.*"

In Table 1, we compare the computational time (in seconds) and the rank of output solution of the LSOP-R returned by three methods: MOSEK, column generation with and without the acceleration of Pricing Problem. It should be mentioned in Table 1 and all the following tables that we let "CG" denote the column generation and mark"–" in a unsolved case reaching the one-hour limit; for any output solution, we count the number of its eigenvalues or singular values greater than 10^{-10} as its rank; and we set the optimality gap of column generation algorithm to be $\epsilon = 10^{-4}$ such that the output value is no ϵ larger than $\mathbf{V}_{\text{rel}}$. It is seen that our proposed column generation algorithm performs very well in computational efficiency and solution quality, dominating the performance of the commercial solver MOSEK and the naive column generation in both respects.

Table 1. Solving LSOP-R given the domain $\mathcal{X}$ in (4)

n	m	k	MOSEK		Naive CG		Proposed Algorithms				Rank bound
							CG		Rank-reduction		
			rank	time(s)	rank	time(s)	rank	time(s)	reduced rank	time(s)	(Theorem 2)
50	5	5	48	17	3	223	3	1	2	1	7
50	10	5	29	19	5	1261	5	1	3	1	8
50	10	10	32	183	–	–	5	1	3	1	13
100	10	10	–	–	–	–	2	2	1	1	13
100	15	10	–	–	–	–	5	2	3	1	14
100	15	15	–	–	–	–	5	3	3	1	19
200	15	15	–	–	–	–	4	3	2	1	19
200	20	15	–	–	–	–	7	9	5	1	20
200	25	25	–	–	–	–	11	21	9	1	30
500	25	25	–	–	–	–	10	24	8	2	30
500	50	25	–	–	–	–	21	99	20	2	33
500	50	50	–	–	–	–	22	104	20	2	58

4.2 LSOP with Trace Constraints: Optimal Power Flow

This subsection numerically tests a special QCQP widely adopted in the optimal power flow (OPF) problem [3,11]

$$\min_{\boldsymbol{X}\in\mathcal{X}} \left\{\langle \boldsymbol{A}_0, \boldsymbol{X}\rangle : \langle \boldsymbol{A}_i, \boldsymbol{X}\rangle \leq b_i^u, \forall i \in [m]\right\}, \tag{OPF}$$

where $\mathcal{X} = \{\boldsymbol{X} \in \mathcal{S}_+^n : \text{rank}(\boldsymbol{X}) \leq 1, L \leq \text{tr}(\boldsymbol{X}) \leq U\}$ and the corresponding spectral function in (1) is $F(\boldsymbol{X}) = \max\{\text{tr}(\boldsymbol{X}) - U, L - \text{tr}(\boldsymbol{X})\}$. The computational results of solving the partial convexification of OPF are displayed in Table 2. We can see that despite the description of set $\text{conv}(\mathcal{X})$ is simple, the MOSEK and naive column generation are difficult to scale under the curse of a semidefinite program. Our column generation can efficiently return a low-rank solution that satisfies the theoretical rank bound in Theorem 2; however, the output rank is larger than one. We observe that the output solution of the column generation algorithm may not attain all m linear equations in practice, and the rank bound can be strengthened using only the binding constraints. Thus, we tailor the rank-reduction algorithm to reduce the output rank of column generation, and the reduced rank still exceeds one as presented in Table 2, implying an infeasible solution to OPF. To evaluate the relaxation value of LSOP-R, we generate a feasible solution of OPF to obtain an upper bound by replacing variable $\boldsymbol{X}$ with $\boldsymbol{x}\boldsymbol{x}^\top$ and leveraging the Burer-Monteiro method [6]. More importantly, we initialize this method by a rank-one matrix constructed by the top eigenvalue and eigenvector of the output solution from the column generation algorithm to speed up the computation. Then, we define "gap(%)" $:= 100 \times \frac{\text{upper bound}-\mathbf{V}_{\text{rel}}}{\mathbf{V}_{\text{rel}}}$. It is seen in Table 2 that the relaxation LSOP-R is close to OPF with an optimality gap no larger than 0.026%.

Table 2. Solving LSOP-R given the domain $\mathcal{X}$ in OPF

n	m	k	MOSEK		Naive CG		Proposed Algorithms					Rank bound
							CG			Rank-reduction		
			rank	time(s)	rank	time(s)	rank	gap(%)	time(s)	reduced rank	time(s)	(Theorem 2)
1000	50	1	28	160	–	–	3	0.016	42	2	3	9
1000	60	1	32	195	–	–	5	0.003	80	3	10	10
1500	60	1	27	642	–	–	3	0.007	113	2	11	10
1500	75	1	186	724	–	–	6	0.008	344	4	35	11
2000	75	1	40	1850	–	–	5	0.003	594	3	67	11
2000	90	1	12	2236	–	–	4	0.003	483	2	27	13
2500	90	1	–	–	–	–	5	0.026	1323	3	122	13
2500	100	1	–	–	–	–	4	0.012	1326	2	114	13

4.3 Matrix Completion

This subsection solves the partial convexification of matrix completion by the following bilevel optimization form that allows for the noisy observations:

$$\begin{aligned} \mathbf{V}_{\text{rel}} := \min_{z\in\mathbb{R}_+} \quad & z \\ \text{s.t.} \quad & \underline{\delta} \geq \min_{\boldsymbol{X},\delta} \{\delta : |X_{ij} - A_{ij}| \leq \delta, \forall (i,j) \in \Omega, \boldsymbol{X} \in \operatorname{conv}(\mathcal{X}(z))\}, \\ & \operatorname{conv}(\mathcal{X}(z)) := \{\boldsymbol{X} \in \mathbb{R}^{n\times p} : \|\boldsymbol{X}\|_* \leq z\} \end{aligned} \tag{5}$$

where $\underline{\delta} \geq 0$ is small and depends on the noisy level of observed entries $\{A_{ij}\}_{(i,j)\in\Omega}$ with Ω denoting the indices of the observed entries. Following the work of [22], we generate a synthetic rank-k matrix $\boldsymbol{A} \in \mathbb{R}^{n\times p}$ and the set Ω is sampled uniformly from $[n] \times [p]$ at random.

For any solution z of the problem (5), our column generation algorithm is able to efficiently solve the embedded optimization problem over $(\boldsymbol{X}, \delta)$ as it falls into our LSOP-R framework. This motivates us to use the binary search algorithm for solving problem (5). To do so, we begin with a pair of upper and lower bounds of z, where we let $z_U = \|\boldsymbol{A}_\Omega\|_*$ and $z_L = 0$ with matrix $\boldsymbol{A}_\Omega$ containing the observed entries $\{A_{ij}\}_{(i,j)\in\Omega}$ and all other zeros; at each step t, we let $z_t = (z_U + z_L)/2$ and compute $(\boldsymbol{X}_t, \delta_t)$ by the proposed column generation; and if $\delta_t \geq \underline{\delta}$, we let $z_L = z_t$, otherwise, we let $z_U = z_t$; and we terminate this searching procedure when $z_U - z_L \leq 10^{-4}$ holds.

Table 3 presents the numerical results, which show that even if being nested within the iterative binary search algorithm, our column generation algorithm is more scalable than the other two methods on the testing cases. However, our column generation fails to return a low-rank solution. The rank-reduction algorithm converges fast to reach the rank bound, as shown in Table 3. It is seen that the reduced rank is always less than the original rank k, which demonstrates the strength of the proposed partial convexification (5).

Table 3. Numerical performance of three approaches to solving problem (5)

n	p	$\|\Omega\|$	k	MOSEK		Naive CG		Proposed Algorithms				Rank bound
								CG		Rank-reduction		
				rank	time(s)	rank	time(s)	rank	time(s)	reduced rank	time(s)	[16][Theorem 4]
100	50	30	10	22	57	–	–	23	19	7	1	7
100	100	40	10	29	303	–	–	31	52	8	1	8
100	200	50	10	–	–	–	–	36	120	9	3	9
300	200	60	15	–	–	–	–	49	342	10	11	10
300	300	70	15	–	–	–	–	60	502	11	16	11
300	400	80	15	–	–	–	–	64	1434	12	29	12

Acknowledgements. This research has been supported in part by the National Science Foundation grants 2246414 and 2246417, the Office of Naval Research grant N00014-24-1-2066, and the Georgia Tech ARC-ACO fellowship. The authors would like to thank Prof. Fatma Kılınç-Karzan for her valuable suggestions on the earlier version of this paper.

References

1. Askari, A., d'Aspremont, A., Ghaoui, L.E.: Approximation bounds for sparse programs. SIAM J. Math. Data Sci. **4**(2), 514–530 (2022)
2. Barvinok, A.I.: Problems of distance geometry and convex properties of quadratic maps. Discret. Comput. Geom. **13**(2), 189–202 (1995)
3. Bedoya, J.C., Abdelhadi, A., Liu, C.C., Dubey, A.: A QCQP and SDP formulation of the optimal power flow including renewable energy resources. In: 2019 International Symposium on Systems Engineering (ISSE), pp. 1–8. IEEE (2019)
4. Ben-Tal, A., Nemirovski, A.: Lectures on modern convex optimization: analysis, algorithms, and engineering applications. SIAM (2001)
5. Bertsimas, D., Cory-Wright, R., Pauphilet, J.: A new perspective on low-rank optimization. arXiv preprint arXiv:2105.05947 (2021)
6. Burer, S., Monteiro, R.D.: A nonlinear programming algorithm for solving semidefinite programs via low-rank factorization. Math. Program. **95**(2), 329–357 (2003)
7. Burer, S., Monteiro, R.D.: Local minima and convergence in low-rank semidefinite programming. Math. Program. **103**(3), 427–444 (2005)
8. Burer, S., Ye, Y.: Exact semidefinite formulations for a class of (random and non-random) nonconvex quadratic programs. Math. Program. **181**(1), 1–17 (2020)
9. Deza, M.M., Laurent, M., Weismantel, R.: Geometry of Cuts and Metrics, vol. 2. Springer, Berlin, Heidelberg (1997). https://doi.org/10.1007/978-3-642-04295-9
10. Drusvyatskiy, D., Kempton, C.: Variational analysis of spectral functions simplified. arXiv preprint arXiv:1506.05170 (2015)
11. Eltved, A., Burer, S.: Strengthened SDP relaxation for an extended trust region subproblem with an application to optimal power flow. Math. Program. 1–26 (2022)
12. Kim, J., Tawarmalani, M., Richard, J.P.P.: Convexification of permutation-invariant sets and an application to sparse principal component analysis. Math. Oper. Res. **47**(4), 2547–2584 (2022)
13. Kulis, B., Sustik, M.A., Dhillon, I.S.: Low-rank kernel learning with bregman matrix divergences. J. Mach. Learn. Res. **10**(2) (2009)

14. Lau, L.C., Ravi, R., Singh, M.: Iterative Methods in Combinatorial Optimization, vol. 46. Cambridge University Press, Cambridge (2011)
15. Li, Y., Xie, W.: On the exactness of Dantzig-Wolfe relaxation for rank constrained optimization problems. arXiv preprint arXiv:2210.16191 (2022)
16. Li, Y., Xie, W.: On the partial convexification for low-rank spectral optimization: rank bounds and algorithms. arXiv preprint arXiv:2305.07638 (2023)
17. Pataki, G.: On the rank of extreme matrices in semidefinite programs and the multiplicity of optimal eigenvalues. Math. Oper. Res. **23**(2), 339–358 (1998)
18. Rockafellar, R.T.: Convex Analysis. Princeton University Press, Princeton (1972)
19. Samadi, S., Tantipongpipat, U., Morgenstern, J.H., Singh, M., Vempala, S.: The price of fair PCA: one extra dimension. Adv. Neural Inf. Process. Syst. **31** (2018)
20. Tantipongpipat, U., Samadi, S., Singh, M., Morgenstern, J.H., Vempala, S.: Multi-criteria dimensionality reduction with applications to fairness. Adv. Neural Inf. Process. Syst. **32** (2019)
21. Yu, H., Lau, V.K.: Rank-constrained schur-convex optimization with multiple trace/log-det constraints. IEEE Trans. Signal Process. **59**(1), 304–314 (2010)
22. Zhang, D., Hu, Y., Ye, J., Li, X., He, X.: Matrix completion by truncated nuclear norm regularization. In: 2012 IEEE Conference on Computer Vision and Pattern Recognition, pp. 2192–2199. IEEE (2012)

On the Congruency-Constrained Matroid Base

Siyue Liu[1,2] and Chao Xu[1(✉)]

[1] University of Electronic Science and Technology of China, Chengdu, China
the.chao.xu@gmail.com
[2] Carnegie Mellon University, Pittsburgh, PA 15213, USA
siyueliu@andrew.cmu.edu

Abstract. Consider a matroid where all elements are labeled with an element in $\mathbb{Z}$. We are interested in finding a base where the sum of the labels is congruent to $g \pmod{m}$. We show that this problem can be solved in $\tilde{O}(2^{4m} n r^{5/6})$ time for a matroid with n elements and rank r, when m is either the product of two primes or a prime power. The algorithm can be generalized to all moduli and, in fact, to all abelian groups if a classic additive combinatorics conjecture by Schrijver and Seymour holds true. We also discuss the optimization version of the problem.

Keywords: Matroid · Additive Combinatorics · Optimization

1 Introduction

Recently, there has been a surge of work on congruency-constrained combinatorial optimization problems, such as submodular minimization [23], constraint satisfaction problems [6], and integer programming over totally unimodular matrices [1,22]. In this paper, we consider congruency-constrained matroid base problems.

As a motivation, consider the exact matching problem, which asks for a perfect matching in a red-blue edge-colored bipartite graph with exactly k red edges [25]. A more general variant, where each edge has a weight of polynomial size and the goal is to find a perfect matching with an exact weight, has a randomized polynomial-time algorithm [20]. Exact matching is a special case of the exact matroid intersection problem, where the goal is to find a common base of a particular weight. Webb provided an algebraic formulation of the exact base problem that leads to an efficient solution when the matroid pair is Pfaffian and the weights are small [27]. For a single matroid, the exact base problem is defined as follows.

Problem 1 (Exact Matroid Base). Given a matroid $M = (E, \mathcal{I})$, a natural number t, and a label function $\ell : E \to \mathbb{N}$. Find a base B in M such that $\sum_{x \in B} \ell(x) = t$, if one exists.

J. Vygen and J. Byrka (Eds.): IPCO 2024, LNCS 14679, pp. 280–293, 2024.
https://doi.org/10.1007/978-3-031-59835-7_21

The problem is NP-hard, as the classical subset sum problem can be reduced to it. However, the subset sum and its optimization variant, the knapsack problem, can be solved in pseudopolynomial time [4]. Thus, Papadimitriou and Yannakakis questioned whether the exact matroid base could be solved in pseudopolynomial time [25]. Barahona and Pulleyblank provided an affirmative answer for graphic matroids [3]. Subsequently, Camerini and Maffioli demonstrated that for matroids representable over the reals, there exists a randomized pseudopolynomial time algorithm for the exact matroid base problem [10]. Recently, Doron-Arad, Kulik, and Shachnai showed that pseudopolynomial time algorithms for arbitrary matroids cannot exist, demonstrated through carefully constructed paving matroids [13]. For one optimization variant, where the goal is to maximize the value of a base while adhering to an upper bound on the budget, some approximation results are known [8,9].

Recent progress has involved relaxing the exact constraint to a modular constraint to achieve interesting advancements. For instance, finding a perfect matching with an even number of red edges has been shown to be solvable in deterministic polynomial time [14]. Hence, we relax the exact constraint and study the following problem:

Problem 2 (m-Congruency-Constrained Matroid Base (CCMB(m))). Let $M = (E, \mathcal{I})$ be a matroid, $g \in \{0, \ldots, m-1\}$, and $\ell : E \to \mathbb{Z}$ be a label function. Find a base B in M such that $\sum_{x \in B} \ell(x) \equiv g \pmod{m}$, if any.

As far as we know, except for results implied by the exact matroid base problem, little is known about this problem. An algorithm with a running time polynomial in m and the size of the matroid cannot exist, as it would imply the existence of a pseudopolynomial time algorithm for the exact matroid base problem. We aim for the next best thing: does there exist a fixed-parameter tractable (FPT) algorithm for CCMB(m), parameterized by m?

Similar to [21], we consider finite abelian groups instead of solely congruency constraints. For a finite abelian group G, let $\ell : E \to G$ be a label function mapping each element of the matroid to an element of G. ℓ is called a *G-labeling* of E. A base with label sum $\sum_{x \in B} \ell(x) = g$ is called a *g-base*. Our goal can be summarized as determining the complexity of the following two problems regarding finding g-bases, parameterized by some finite abelian group G.

Problem 3 (Group-Constrained Matroid Base (GCMB(m))). Given $g \in G$, a matroid $M = (E, \mathcal{I})$, and a label function $\ell : E \to G$, find a g-base, if one exists.

Problem 4 (Group-Constrained Optimum Matroid Base (GCOMB(m))). Given $g \in G$, a matroid $M = (E, \mathcal{I})$, a label function $\ell : E \to G$, and a weight function $w : E \to \mathbb{R}$, find a g-base of minimum weight, if one exists.

The closest study related to matroid bases and group constraints comes from the additive combinatorics community, where theorems considering the number of different labels attainable by the bases have been discovered [12,26]. Recently, Hörsch et al. [16] study generalized problems where the label sum of the bases is allowed to take a subset of G instead of a single element. The same problem for the common bases of two matroids is also studied.

Our Contribution. We show that GCMB(m) can be solved in $\tilde{O}(2^{4|G|}nr^{5/6})$ time for a group G, if either the size of G equals the product of two primes, or G is a cyclic group of size a power of a prime. In particular, this proves that CCMB(m) is in FPT parameterized by m if m is the product of two primes or a prime power.

These results are obtained through proximity results. For this, let $\mathcal{B}$ be the set of bases of M. Let $\mathcal{B}(g)$ be the collection of all g-bases of M. For a weight function $w : E \to \mathbb{R}$, denote by $\mathcal{B}^* := \arg\min_{B\in\mathcal{B}} w(B)$ and $\mathcal{B}^*(g) := \arg\min_{B\in\mathcal{B}(g)} w(B)$ the sets of *optimum bases* and *optimum g-bases*, respectively. We introduce the following key concepts of this paper.

Definition 1. *A G-labeling ℓ of matroid M is **k-close** if for every $g \in G$ such that a g-base exists, every base has a g-base differing from it by at most k elements. If every G-labeling is k-close for every matroid, then G is **k-close**.*

Definition 2. *A G-labeling ℓ of matroid M is **strongly k-close** if for every weight $w : E \to \mathbb{R}$ and $g \in G$ such that a g-base exists, every optimum base has an optimum g-base differing from it by at most k elements. If every G-labeling is strongly k-close for every matroid, then G is **strongly k-close**.*

Note that being strongly k-close implies being k-close by setting $w \equiv 0$. We prove that G is $(|G| - 1)$-close if either the size of G equals the product of two primes, or G is a cyclic group of size a power of a prime (Corollary 3). More generally, we show that every group is $(|G| - 1)$-close assuming a conjecture by Schrijver and Seymour (Conjecture 3). This gives rise to the desired FPT algorithm for GCMB(m) (Theorem 1). Lastly, we prove the strong k-closeness for strongly base orderable matroids (Theorem 5) and small groups, for some $k \leq |G|$ which only depends on G. These lead to FPT algorithms for GCOMB(m) for those matroids and groups.

Overview. In Sect. 2, we define the problem and provide the necessary background in matroid and group theory. In Sect. 3, we discuss established algorithms and demonstrate the existence of an FPT algorithm for GCMB(m) and GCOMB(m), provided that G is k-close and strongly k-close, respectively, for some k which only depends on G. Starting from Sect. 4, we prove the proximity results. In Sect. 4, we consider what would constitute the minimum counterexample to the statement that G is (strongly) k-close. We observe that the minimum counterexample would have to be a block matroid of rank $k + 1$. In Sect. 5, we prove the $(|G| - 1)$-closeness for certain groups. Finally, Sect. 6 discusses the optimization version and results on strong k-closeness.

Notation. We are given a matroid $M = (E, \mathcal{I})$, a group G, and a label function $\ell : E \to G$. For any $F \subseteq E$, denote by $\ell(F) := \sum_{e\in F} \ell(e)$. For a subset $H \subseteq G$, denote by $E(H) := \ell^{-1}(H) = \{e \in E \mid \ell(e) \in H\}$ all the elements with labels in H. If $H = \{g\}$ is a singleton, we shorten $E(\{g\})$ to $E(g)$. Denote by $\ell(M)$ the set $\{\ell(B) \mid B \text{ is a base of } M\}$. For $F \subseteq E$, denote the rank of F in M by $r_M(F)$, where M in the rank function $r_M(\cdot)$ can be omitted if the context is clear. Denote the rank of matroid M by $r(M) := r_M(E)$.

2 Preliminaries

Let $M = (E, \mathcal{I})$ be a matroid, where E is the ground set and $\mathcal{I}$ is the family of independent sets. For a subset $F \subseteq E$, denote by $M \setminus F$ the *deletion* minor of M defined on the ground set $E \setminus F$ whose independent sets are those of M restricted to $E \setminus F$. If $F \subseteq E$ is an independent set of M, denote by M/F the *contraction* minor of M defined on the ground set $E \setminus F$. A subset $A \subseteq E \setminus F$ is independent in M/F if and only if $A \cup F$ is independent in M. Their rank functions satisfy the relation $r_{M/F}(A) = r_M(A \cup F) - r_M(F)$. We assume the matroid is loopless throughout, which means $\{e\} \in \mathcal{I}$ for any $e \in E$.

A matroid possesses the base exchange property: For any two bases A and B, there exists a series of elements $a_1, \ldots, a_k \in A \setminus B$ and $b_1, \ldots, b_k \in B \setminus A$ such that $A - \{a_1, \ldots, a_j\} + \{b_1, \ldots, b_j\}$ is also a base for each $j = \{1, \ldots, k\}$, and $A - \{a_1, \ldots, a_k\} + \{b_1, \ldots, b_k\} = B$. The distance $d(A, B)$ between two bases A and B is the number of steps needed to exchange elements from A to B, which equals $|A \setminus B|$.

A matroid is *strongly base orderable* if for any two bases A and B, there is a bijection $f : A \to B$ such that for any $X \subseteq A$, we have $A - X + f(X)$ is a base. A function with the previous property can be taken to satisfy $f(a) = a$ for all $a \in A \cap B$. A matroid is a *block matroid* if the ground set is the union of two disjoint bases, and such disjoint bases are called *blocks*.

Given a group G and a normal subgroup $H \subseteq G$, denote by G/H the quotient group consisting of equivalence classes of the form $gH := \{g + h \mid h \in H\}$, which are called the *cosets* of H, where g_1 is equivalent to g_2 if and only if $g_1^{-1} g_2 \in H$. For a subset $F \subseteq G$, the *stabilizer* of F is defined by $\text{stab}(F) := \{g \in G \mid g + F = F\}$. It is easy to see that $\text{stab}(F)$ is a subgroup of G.

The Davenport constant [11,24] of G, denoted by $D(G)$, is the minimum value such that every sequence of elements from G of length $D(G)$ contains a non-empty subsequence that sums to 0. In other words, the longest sequence without non-empty subsequence that sums to 0 has length $D(G) - 1$. By the fundamental theorem of finite abelian groups, any finite abelian group G can be decomposed into $G = \mathbb{Z}_{m_1} \times \ldots \times \mathbb{Z}_{m_r}$ where $1 \mid m_1 \mid m_2 \mid \ldots \mid m_r$. Here $\mathbb{Z}_m$ is the group of integers modulo m, and $\times$ is the group direct product. It is shown that a lower bound of $D(G)$ is $M(G) := \sum_i^r (m_i - 1) + 1$. It was proved independently by Olson [24] and Kruswijk [2] that $D(G) = M(G)$ for *p-groups*, in which the order of every element is a power of p, and for $r = \{1, 2\}$. It also holds that $D(G) \leq |G|$, where equality holds for all cyclic groups [15].

Our paper makes progress on the following two conjectures.[1]

Conjecture 1 (Feasibility). Every finite abelian group G is $(|G| - 1)$-close.

Conjecture 2 (Optimization). Every finite abelian group G is strongly $(|G| - 1)$-close.

---

[1] In an earlier preprint version of this paper, we conjectured that every finite abelian group is (strongly) $(D(G) - 1)$-close, but this was recently disproved by [16].

Observe the conjectures are best possible when the group is cyclic. Indeed, the tight example is a block matroid of size $|G| - 1$, where the first block has all its element labeled 1 and the second has all its elements labeled 0. Then, the closest 0-base to the first block is the second block, which has distance $|G| - 1$ from it.

3 Algorithms for GCOMB

In this secton, we show how k-closeness and strong k-closeness of a group G lead to FPT algorithms for GCMB(m) and GCOMB(m), respectively. In particular, Theorem 1 shows that if there is a function f such that G is $f(G)$-close or strongly $f(G)$-close, then for a fixed group G, GCMB(m) or GCOMB(m) is in FPT, respectively.

We assume that the matroid can be accessed through an independence oracle that takes a constant time per query. Each group operation also takes $O(1)$ time.

There is a folklore algorithm which shows for any fixed group G, GCOMB(m) is solvable in polynomial time by reducing it to a polynomial number of matroid intersections.

Let $M = (E, \mathcal{I})$ be a matroid of size n and rank r. Recall $E(g) = \ell^{-1}(g)$ for $g \in G$. For a base B of matroid M, the vector $a \in \mathbb{Z}^G_{\geq 0}$ such that $a_g = |B \cap E(g)|$ is called the *signature* of B. For a target label $h \in G$, every h-base B satisfies $h = \sum_{g \in G} a_g \cdot g$ for the signature a of B. Here $a \cdot g$ is defined to be $\sum_{i=1}^{a} g$. For a fixed vector a with $\sum_g a_g = r$, $0 \leq a_g \leq |E(g)|$ and $\sum_g a_g \cdot g = h$, we can check if any base has signature a, in fact, find a minimum cost base with signature a. Indeed, one can obtain this through matroid intersection with a partition matroid, whose partition classes are $\{E(g) \mid g \in G\}$. It suffices to find an optimum base of matroid M such that its intersection with $E(g)$ has cardinality a_g for each $g \in G$. So the number of vectors we need to check is bounded above by $\binom{r+|G|-1}{|G|-1} \leq r^{|G|-1}$. Hence, we have to run the matroid intersection algorithm $r^{|G|-1}$ times in order to find the optimum g-base. Currently, the fastest algorithm for matroid intersection has running time $O(nr^{5/6} \log r)$ time assuming independence oracle takes constant time [5]. Hence the total running time is $\tilde{O}(nr^{O(|G|)})$ which is polynomial as $|G|$ is fixed. Although the algorithm is polynomial time for a fixed G, if we parameterize by G, then this algorithm is not FPT.

If G is (strongly) k-close for a constant k, then there is also a polynomial time algorithm: find an (optimum) base B, consider all possible bases B' such that $|B \setminus B'| \leq k$, and return an (optimum) base B' with the desired label. The running time is $O(r^k n^k)$ by enumerating all bases k-close to B.

We show that (strong) k-closeness gives rise to an FPT algorithm parametrized by G by combining it with the matroid intersection approach.

Theorem 1. *If G is k-close or strongly k-close, then there is a $\tilde{O}(\binom{k+|G|-1}{k}^2 nr^{5/6})$ running time algorithm for* GCMB(m) *or* GCOMB(m)*, respectively.*

Proof. The algorithm consists of two parts. First find any base B. By (strong) k-closeness, we know there is a feasible (optimum) g-base D that differs from B in at most k positions. Let a_D and a_B be the signature vectors of D and B, respectively. Let $a^+ := (a_D - a_B) \vee 0$, $a^- := (a_D - a_B) \wedge 0$ be vectors taking coordinate-wise maximum and minimum of $(a_D - a_B)$ with 0, respectively. Note that $a^+ \geq 0$ and $\sum_g a_g^+ \leq k$. Therefore, there are at most $\binom{k+|G|-1}{k}$ possible a^+'s, and similarly at most $\binom{k+|G|-1}{k}$ possible a^-'s. Hence, we only need to run the matroid intersection algorithm for at most $\binom{k+|G|-1}{k}^2$ times to search for the (optimum) g-bases which are (strongly) k-close to B. □

If we take $k = |G|-1$ and use the fact that $\binom{2k}{k} \leq 2^{2k}$, we obtain the following corollary.

Corollary 1. *If G is $(|G|-1)$-close or strongly $(|G|-1)$-close, then there is an algorithm whose running time in $\tilde{O}(2^{4|G|}nr^{5/6})$ for* GCMB(m) *or* GCOMB(m)*, respectively.*

4 Properties of Minimum Counterexamples to Conjecture 1 and Conjecture 2

In this section, we show that to prove Conjecture 1 or Conjecture 2, it suffices to prove it for a restricted type of matroids. Assume for the sake of contradiction that G is not (strongly) k-close, then there exists a counterexample of minimum size. This section shows that the minimum counterexamples are block matroids of size $(k+1)$. Hence, it suffices to prove the conjectures for those matroids. We need a lemma by Brualdi [7].

Lemma 1 ([7])**.** *Let A and B be bases in a matroid, then there exists a bijection $f : A \setminus B \to B \setminus A$, such that $A - a + f(a)$ is a base.*

We give the following lemma.

Lemma 2. *Given a matroid $M = (E, \mathcal{I})$ and any weight function $w : E \to \mathbb{R}$, let A be an optimum base such that its weight is minimized. Then, for any $a \in A$, there exists some $b \in E \setminus A$ such that $A - a + b$ is an optimum base in $M' = M \setminus a$ with weight being the restriction of w to $E \setminus a$.*

Proof. Suppose A' is an optimum base in $M \setminus a$. By Lemma 1, there exists a bijection $f : A \setminus A' \to A' \setminus A$ such that $A - e + f(e)$ is a base for any $e \in A \setminus A'$. Since A is an optimum base, we have $w(A) \leq w(A - e + f(e))$, and thus $w(e) \leq w(f(e))$ for each $e \in A \setminus A'$. For any $a \in A$, let $b = f(a)$. Then, $w(A - a + b) = w(A) - w(a) + w(b) \leq w(A) - \sum_{e \in A \setminus A'}(w(e) - w(f(e))) = w(A')$. The inequality is because $a \in A \setminus A'$ and $w(e) \leq w(f(e))$ for any $e \in A \setminus A' \setminus a$. Since A' is an optimum base in $M \setminus a$, $A - a + b$ is also an optimum base in $M \setminus a$. □

Next, we show that if G is not strongly k-close, then there has to be a G-labeling ℓ, a weight function w, an element $g \in G$, and a block matroid with blocks A, B, such that A is an optimum base, B is the closest optimum g-base to A, and $d(A, B) > k$. We call the pair of bases A, B a *witness*.

Theorem 2. *If a finite abelian group G is not strongly k-close, then the there is a block matroid M of rank $k + 1$ with blocks A and B that form a witness.*

Proof. Let matroid $M = (E, \mathcal{I})$ be the smallest matroid, in terms of the number of elements, such that G is not strongly k-close for M. Let $w : E \to \mathbb{Z}$ be a weight function and $g \in G$. Recall that $\mathcal{B}^* = \arg\min_{B \in \mathcal{B}} w(B)$ and $\mathcal{B}^*(g) = \arg\min_{B \in \mathcal{B}(g)} w(B)$ are the sets of optimum bases and optimum g-bases, respectively. There exists $A \in \mathcal{B}^*$ and $B \in \mathcal{B}^*(g)$ such that B is an optimum g-base closest to A but $|A \setminus B| > k$.

First, we argue that $A \cap B = \emptyset$. If not, let $M' = (E', \mathcal{I}')$ be the matroid obtained by contracting $A \cap B$, and let $A' = A \setminus B$, $B' = B \setminus A$, $g' = g - \ell(A \cap B)$ and $w' = w|_{E'}$. Clearly, A' is a weighted minimum base of M' and B' is a weighted minimum g'-base of M' with weight w'. Also, $d(A', B') = |A' \setminus B'| = |A \setminus B| > k$, and B' is an optimum g'-base closest to A'. However, $|E'| = |E| - |A \cap B| < |E|$, contradicting to the minimality of $|E|$. Moreover, $A \cup B = E$. Otherwise, let M' be the matroid obtained by deleting $E \setminus (A \cup B)$ and A, B stays a witness in M', contradicting to the minimality of $|E|$. Therefore, M is a block matroid which is a block matroid with blocks A and B.

Suppose $r(M) > k + 1$. Pick any $a \in A$. Let $M' = (E', \mathcal{I}') := M \setminus a$. By Lemma 2, there exists $b \in B$ such that $A' := A - a + b$ is a weighted minimum base in M'. Clearly, B is still a weighted minimum g-base in M' that is closest to A'. Moreover, $d(A', B) = |A' \setminus B| = |A| - 1 = r(M) - 1 > k$. Therefore, G is not strongly k-close for M', since A', B is a witness. Because $|E'| < |E|$, we obtain a contradiction. Therefore, $r(M) \leq k + 1$. Combining this with the fact that $r(M) = |A \setminus B| > k$, we get $r(M) = k + 1$. □

As a corollary, if G is not k-close, then there has to be a G-labeling ℓ, an element $g \in G$ and a block matroid M with blocks A, B, such that A is a base, B is the closest g-base to A, and $d(A, B) > k$. Similarly, we call the pair of bases A, B a *witness* in the unweighted setting.

Corollary 2. *If a finite abelian group G is not k-close, then the there is a block matroid M of rank $k + 1$ with blocks A and B that form a witness.*

Proof. The proof follows by setting the weight function to be 0 in the proof of Theorem 2. □

Therefore, in order to show G is k-close or strongly k-close, we just have to show it is k-close or strongly k-close for all block matroids of rank $k + 1$, respectively.

5 k-Closeness and Isolation

We discuss progress towards conjecture Conjecture 1 in this section. We will prove in Theorem 4 that whenever a certain additive combinatorics conjecture is true, any finite abelian group G is $(|G|-1)$-close.

First, we set up some notions we use later. We say a base B is *isolated* under label ℓ if it is the unique base with the label $\ell(B)$. A labeling is called *block isolating* if it isolates a base whose complement is also a base, i.e., it isolates a block. The next proposition shows that isolation and k-closeness are related concepts.

Proposition 1. *If no G-labeling of a rank $k+1$ block matroid M is block isolating, then G is k-close for M.*

Proof. Suppose not. By Corollary 2, there exists a G-labeling ℓ, $g \in G$ and witness A, B that are bases of M such that B is the closest g-base to A. Then, B is the unique base that has labeling g. Indeed, if there is any other base $B' \neq B$ with $\ell(B') = g$, then $d(A, B') < k+1 = d(A, B)$, since the block structure guarantees that B is the only farthest base from A, a contradiction. □

Equipped with Proposition 1, our goal becomes showing that block isolating labelings cannot exist.

5.1 Congruency-Constrained Base with Prime Modulus

We will start by proving that Conjecture 1 is true for any cyclic group of prime order, or equivalently congruency-constraints modulo primes. This is a special case of results for general groups which will be introduced in the next subsection. But the proof for cyclic groups of prime order is much simpler followed by a counting argument. For the sake of helping readers gain intuition, we present it as well. The main tool we are going to use is the following additive combinatorics lemma by Schrijver and Seymour.

Lemma 3 (Schrijver-Seymour [26]). *Let $M = (E, \mathcal{I})$ be a matroid with rank function r and let $\ell : E \to \mathbb{Z}_p$ for some prime number p. Let $\ell(M) := \{\ell(B) \mid B$ is a base of $M\}$. Then, $|\ell(M)| \geq \min\{p, \sum_{g \in \mathbb{Z}_p} r(E(g)) - r(M) + 1\}$.*

The following lemma states an exchange property for matroids which is most likely routine. We also give its proof here.

Lemma 4. *Given a matroid $M = (E, \mathcal{I})$, let A be a base, $A_1 \subseteq A$, and B_1 be an independent set such that $A \cap B_1 = \emptyset$. If $|A_1| + |B_1| - r(A_1 \cup B_1) \geq t$, then there exist some $A_2 \subseteq A_1$ and $B_2 \subseteq B_1$ with $|A_2| = |B_2| = t$, such that $A - A_2 + B_2$ is a base.*

Proof. By submodularity of matroid rank functions,

$$r((A \setminus A_1) \cup B_1) + r(A_1 \cup B_1) \geq r(A \cup B_1) + r(B_1) = r(M) + |B_1|.$$

By assumption, $|A_1| + |B_1| - r(A_1 \cup B_1) \geq t$. Combining these, we have

$$r((A \setminus A_1) \cup B_1) \geq r(M) - |A_1| + t = |A \setminus A_1| + t.$$

Using the fact that $A \setminus A_1$ is independent, we deduce there exists $B_2 \subseteq B_1$ with $|B_2| = t$, such that $(A \setminus A_1) \cup B_2$ is independent. Moreover, since A is a base, by adding elements from A_1, $(A \setminus A_1) \cup B_2$ can be extended to a base which has the form $A - A_2 + B_2$ for some $A_2 \subseteq A_1$ with $|A_2| = |B_2| = t$. □

Theorem 3. *For any prime p, $\mathbb{Z}_p$ is $(p-1)$-close.*

Proof. Suppose not. Then, by Proposition 1, there exists a block matroid $M = (E, \mathcal{I})$ with $E = A \cup B$, where A and B are two disjoint bases with $|A| = |B| = p$, such that A is isolated under some $\mathbb{Z}_p$-labeling ℓ.

We claim that for any $g \in G$, $r(E(g)) = |E(g) \cap A| + |E(g) \cap B| = |E(g)|$. Otherwise, letting $A_1 = E(g) \cap A$, $B_1 = E(g) \cap B$ and $t = 1$ in Lemma 4, we can find $a \in A_1$ and $b \in B_1$ such that $A - a + b$ is a base. Since $\ell(a) = \ell(b) = g$, $\ell(A - a + b) = \ell(A)$, contradicting to the fact that A is isolated.

Take any element $e \in A$ and consider the matroid $M' = M \setminus e$ with ground set $E' = E \setminus e$. Note that $r(M') = r(M)$ because B stays a base of M'. Applying Lemma 3 to M',

$$\begin{aligned}
|\ell(M')| &\geq \min\Big\{p, \sum_{g \in \mathbb{Z}_p} r(E'(g)) - r(M') + 1\Big\} \\
&= \min\Big\{p, \sum_{g \in \mathbb{Z}_p, g \neq \ell(e)} r(E(g)) + r(E(\ell(e)) \setminus \{e\}) - r(M) + 1\Big\} \\
&= \min\Big\{p, \sum_{g \in \mathbb{Z}_p, g \neq \ell(e)} |E(g)| + (|E(\ell(e))| - 1) - r(M) + 1\Big\} \\
&= \min\Big\{p, \sum_{g \in \mathbb{Z}_p} |E(g)| - p\Big\} = \min\{p, 2p - p\} = p.
\end{aligned}$$

Thus, there exists a base D of M' such that $\ell(D) = \ell(A)$. By definition, D is also a base of M. Since $e \in A$ but $e \notin D$, D is distinct from A, contradicting to the fact that A is isolated under label ℓ. □

5.2 General Group Constraints

To extend beyond prime modulus, we introduce the following conjecture of Schrijver and Seymour which is an extension of Lemma 3 to any abelian group G.

Conjecture 3 (Schrijver-Seymour [26], *see also* [12]**).** Let $M = (E, \mathcal{I})$ be a matroid with rank function r and let $\ell : E \to G$ for some finite abelian group G. Let $\ell(M) := \{\ell(B) \mid B \text{ is a base of } M\}$. Let $H = \text{stab}(\ell(M))$ be the stabilizer of $\ell(M)$. Then, $|\ell(M)| \geq |H| \cdot \min\{\sum_{Q \in G/H} r(E(Q)) - r(M) + 1, |G|/|H|\}$.

For a prime p and a cyclic group $G = \mathbb{Z}_p$, if $\ell(M) \neq \mathbb{Z}_p$, then it is easy to see $\mathrm{stab}(\ell(M)) = \{0\}$. Then, the inequality in Conjecture 3 reduces to $\ell(M) \geq \min\left\{ \sum_{g\in G} r(E(g)) - r(M) + 1,\ |G|\right\}$, which is precisely the form in Lemma 3. Otherwise, $\mathrm{stab}(\ell(M)) = \mathbb{Z}_p$ and the inequality trivially holds.

Theorem 4. *If Conjecture 3 is true for a finite abelian group G and all of its subgroups, then G is $(|G|-1)$-close.*

Proof. Let $n := |G|$. We proceed by induction on n. The theorem trivially holds when $n = 1$. Suppose the theorem does not hold for some $|G| = n$. Then by Proposition 1, there exists a block matroid $M = (E, \mathcal{I})$, $E = A \cup B$, where A and B are two disjoint bases of M with $|A| = |B| = n$ such that A is isolated under some labeling ℓ.

Take any $e \in A$, let $M' = (E', \mathcal{I}') := M \setminus e$. Let H be the stabilizer of $\ell(M')$. Since G is abelian, H is a normal subgroup of G. Denote by $gH := \{g + h \mid h \in H\}$ the coset of H with representative g. First, observe that if $g \in \ell(M')$, then $gH \subseteq \ell(M')$. This is because, by definition of H, for any $h \in H$, $g + h \in \ell(M')$. Thus $gH \subseteq \ell(M')$. Therefore, $\ell(M') = \dot{\bigcup}_{g\in R}\, gH$, where R is a collection of representatives of cosets of H. It follows that $|\ell(M')| = |R| \cdot |H|$. Let $E'(gH) = E(gH) \setminus \{e\}$ for any $g \in G$. Since A is isolated under ℓ and A is not a base of M', we know $\ell(A) \notin \ell(M')$. Thus, $|\ell(M')| < |G|$. This implies $|H| < |G|$, since $|G| > \ell(M') = |R| \cdot |H| \geq |H|$ as $R \neq \emptyset$. It also follows from $|\ell(M')| < |G|$ that $|\ell(M')| \geq |H|\,(\sum_{Q\in G/H} r(E'(Q)) - r(M') + 1)$ in Conjecture 3, and thus $|R| \geq \sum_{Q\in G/H} r(E'(Q)) - r(M') + 1 = \sum_{Q\in G/H} r(E'(Q)) - n + 1$. Then,

$$\begin{aligned}
&\sum_{Q\in G/H} \Big(|A \cap E'(Q)| + |B \cap E'(Q)| - r(E'(Q))\Big)\\
&= (2n-1) - \sum_{Q\in G/H} r(E'(Q))\\
&\geq (2n-1) - (n + |R| - 1)\\
&= n - |R|.
\end{aligned}$$

Therefore,

$$\begin{aligned}
&\max_{Q\in G/H} \Big(|A \cap E'(Q)| + |B \cap E'(Q)| - r(E'(Q))\Big)\\
&\geq \frac{n - |R|}{|G/H|} > \frac{n - |G/H|}{|G/H|} = |H| - 1,
\end{aligned}$$

where the strict inequality follows from $|\ell(M')| < |G|$. Suppose the maximum is attained at $Q_0 = g_0 H$. Then, we have

$$|A \cap E'(Q_0)| + |B \cap E'(Q_0)| - r(E'(Q_0)) \geq |H|,$$

since $|A\cap E'(Q_0)| + |B\cap E'(Q_0)| - r(E'(Q_0))$ is an integer. Let $A_1 := A \cap E'(Q_0)$, $B_1 := B \cap E'(Q_0)$. By Lemma 4, there exists a base $A - A_2 + B_2$ such that $A_2 \subseteq A_1$, $B_2 \subseteq B_1$ and $|A_2| = |B_2| = |H|$.

Let $M'' = (E'', \mathcal{I}'')$ be the matroid obtained by contracting $A \setminus A_2$ and deleting $B \setminus B_2$, i.e. $M'' = M/(A \setminus A_2) \setminus (B \setminus B_2)$. Note that M'' is a rank $|H|$ block matroid with blocks A_2 and B_2. Consider an H-labeling on $E'' = A_2 \cup B_2$, $\ell'' : E'' \to H$, such that $\ell''(e) = \ell(e) - g_0 \in H, \forall e \in E''$. Note that $|H| < |G|$. By the induction hypothesis and the assumption that Conjecture 3 is true for the subgroup H of G, we know that there is no isolating H-labeling of M''. Thus, there must be another base $D_2 \neq A_2$ of M'', such that $\ell''(D_2) = \ell''(A_2)$. Thus $\ell(D_2) = \ell''(D_2) + |H| \cdot g_0 = \ell''(A_2) + |H| \cdot g_0 = \ell(A_2)$. By the definition of contraction, $D := D_2 \cup (A \setminus A_2)$ is a base of matroid M which is distinct from A. And $\ell(D) = \ell(D_2) + \ell(A \setminus A_2) = \ell(A_2) + \ell(A \setminus A_2) = \ell(A)$, contradicting to the fact that A is isolated under label ℓ. □

The following lemma is the current status showing on which group the Schrijver-Seymour conjecture is true.

Lemma 5 ([12]). *Conjecture 3 is true for G if $|G| = pq$ for primes p, q or $G = \mathbb{Z}_{p^n}$ for a prime p and a positive integer n.*

Corollary 3. *G is $(|G| - 1)$-close if $|G| = pq$, or $G = \mathbb{Z}_{p^n}$ for any p, q primes and n a positive integer.*

6 Strong k-Closeness

This section discusses our results for strong k-closeness. Unfortunately, we have little understanding of strong k-closeness. Hence, we focus on restricted class of matroids and small groups.

For the matroids and groups concerned within this section, we obtain a tighter bounds than Conjecture 2 implies. Instead of $(|G|-1)$-closeness in the conjecture, we have $(D(G) - 1)$-closeness.

6.1 Strongly Base Orderable Matroids

We first study strongly base orderable matroids.

Theorem 5. *Every G-labeling is strongly $(D(G) - 1)$-close for strongly base orderable matroids.*

Proof. Suppose not. By noting that strongly base orderability is closed under taking minors [17], it is not hard to see from Theorem 2 that there exists a counterexample which is a rank $D(G)$ strongly base orderable block matroid M with blocks A and B as a witness. Let ℓ be a G-labeling of M, w be a weight function such that A is an optimum base and B is the optimum g-base closest from A.

Let $f : B \to A$ be the bijection obtained from strongly base orderability. Define $\ell'(X) = \ell(f(X)) - \ell(X)$ for all $X \subseteq B$. Since $|B| = D(G)$, by the definition of Davenport's constant, there is $B_1 \subseteq B$ such that $\ell'(B_1) = 0$. Let $A_1 = f(B_1)$. This implies $\ell(A_1) = \ell(B_1)$ and thus the base $B - B_1 + A_1$ has the same label as B, i.e. $B - B_1 + A_1$ is also a g-base.

Now, consider $A_2 = A - A_1$ and $B_2 = B - B_1$. One has $f(B_2) = A_2$ and therefore $B - B_2 + A_2 = A - A_1 + B_1$ is also a base. Since A is an optimum base, $w(A - A_1 + B_1) \geq w(A)$, which means $w(A_1) \leq w(B_1)$. Therefore, $w(B - B_1 + A_1) \leq w(B)$. This means $B - B_1 + A_1$ is also an optimum g-base. But $A_1 \subsetneq A$ since $\ell(A) \neq \ell(B)$, which means $B - B_1 + A_1$ is an optimum g-base closer to A, a contradiction. □

In fact, we can define a weaker property. Two bases A, B are k-replaceable if there exists a bijection $f : B \setminus A \to A \setminus B$ such that for any subset $B' \subseteq B \setminus A$ of size at most k, $B - B' + f(B')$ is a base. We say a matroid is k-replaceable if every pair of bases are k-replaceable. We have that Theorem 5 also holds for $(D(G) - 1)$-replaceable matroids.

6.2 Small Groups

Next, we consider small groups. Theorem 2 shows one only needs to check block matroids of rank $D(G)$ to know if G is strongly $(D(G) - 1)$-close.

For $G = \mathbb{Z}_2$ which has $D(G) = 2$, observe that all size 4 matroids are strongly base orderable, hence $\mathbb{Z}_2$ is 1-close. For $G \in \{\mathbb{Z}_3, \mathbb{Z}_2 \times \mathbb{Z}_2\}$ which has $D(G) = 3$, we need to check all rank 3 block matroids. Since all but $M(K_4)$, the graphic matroid on K_4, are strongly base orderable [19], we only need to test a single matroid. However, testing strongly 2-closeness of a single matroid is still computational intensive.

Instead, we introduct the notion of strong block isolation. A G-labeling ℓ *strongly isolates* a block B, if it is the unique block with label $\ell(B)$. A G-labeling is strong block isolating, if it strongly isolates a block.

Proposition 2. *If no G-labeling of a rank $k+1$ block matroid M is strong block isolating, then G is strongly k-close for M.*

Proof. Suppose not. By Theorem 2, there exists a G-labeling ℓ, $g \in G$ and witness A, B that are bases of M such that B is the closest optimum g-base to A. Fix such labeling ℓ. Since ℓ is not strong block isolating, there is an non-empty $A_1 \subseteq A$ and $B_1 \subseteq B$ such that $A - A_1 + B_1$ is a block with the same label as A. Let $A_2 = A \setminus A_1$, $B_2 = B \setminus B_1$. Then the complement of $A - A_1 + B_1$ is the base $A - A_2 + B_2$, and has the same label as B. Since $w(B) = w(A) - w(A_1) + w(B_1) - w(A_2) + w(B_2) \geq w(A) - w(A_2) + w(B_2) = w(A - A_2 + B_2)$, $A - A_2 + B_2$ is a closer optimum g-base to A, a contradiction. □

Hence, for each G in hand, we show no G-labeling of a rank $D(G)$ block matroid is strong block isolating, which implies G is strongly $(D(G) - 1)$-close. We tested $M(K_4)$ for $\mathbb{Z}_3$ and $\mathbb{Z}_2 \times \mathbb{Z}_2$ and it turned out that they are both strongly 2-close. Further, we tested for each $\mathbb{Z}_4$-labeling of a rank 4 block matroid. There are 940 non-isomorphic rank 4 matroids of size 8 [18], and for each one that is also a block matroid, we test $4^8 = 65536$ different labelings. None of them is strongly block isolating. Therefore, we get the following.

Proposition 3. *G is strongly $(D(G) - 1)$-close if $|G| \leq 4$.*

Acknowledgements. We thank the reviewers for their valuable feedback and András Imolay for noticing and correcting an error in Theorem 4.

References

1. Artmann, S., Weismantel, R., Zenklusen, R.: A strongly polynomial algorithm for bimodular integer linear programming. In: Proceedings of the 49th Annual ACM SIGACT Symposium on Theory of Computing, pp. 1206–1219. STOC 2017, Association for Computing Machinery, New York, NY, USA (2017). https://doi.org/10.1145/3055399.3055473
2. Baayen, P.: Een combinatorisch probleem voor eindige abelse groepen. In: Math. Centrum Syllabus 5, Colloquium Discrete Wiskunde Caput, vol. 3 (1968)
3. Barahona, F., Pulleyblank, W.R.: Exact arborescences, matchings and cycles, vol. 16, no. 2, pp. 91–99. https://doi.org/10.1016/0166-218X(87)90067-9
4. Bellman, R.: Notes on the theory of dynamic programming iv - maximization over discrete sets. Nav. Res. Logist. Q. **3**(1–2), 67–70 (1956). https://doi.org/10.1002/nav.3800030107
5. Blikstad, J.: Breaking $O(nr)$ for matroid intersection. In: Bansal, N., Merelli, E., Worrell, J. (eds.) 48th International Colloquium on Automata, Languages, and Programming (ICALP 2021). Leibniz International Proceedings in Informatics (LIPIcs), vol. 198, pp. 31:1–31:17. Schloss Dagstuhl – Leibniz-Zentrum für Informatik, Dagstuhl, Germany (2021). https://doi.org/10.4230/LIPIcs.ICALP.2021.31
6. Brakensiek, J., Gopi, S., Guruswami, V.: Constraint satisfaction problems with global modular constraints: algorithms and hardness via polynomial representations. SIAM J. Comput. **51**(3), 577–626 (2022). https://doi.org/10.1137/19M1291054
7. Brualdi, R.A.: Comments on bases in dependence structures. Bull. Aust. Math. Soc. **1**(2), 161–167 (1969). https://doi.org/10.1017/S000497270004140X
8. Camerini, P.M., Maffioli, F., Vercellis, C.: Multi-constrained matroidal knapsack problems. Math. Program. **45**(1), 211–231 (1989). https://doi.org/10.1007/BF01589104
9. Camerini, P.M., Vercellis, C.: The matroidal knapsack: a class of (often) well-solvable problems. Oper. Res. Lett. **3**(3), 157–162 (1984). https://doi.org/10.1016/0167-6377(84)90009-9
10. Camerini, P., Galbiati, G., Maffioli, F.: Random pseudo-polynomial algorithms for exact matroid problems **13**(2), 258–273. https://doi.org/10.1016/0196-6774(92)90018-8
11. Davenport, H.: Proceedings of the midwestern conference on group theory and number theory. Ohio State University (1966)
12. DeVos, M., Goddyn, L., Mohar, B.: A generalization of Kneser's addition theorem. Adv. Math. **220**(5), 1531–1548 (2009)
13. Doron-Arad, I., Kulik, A., Shachnai, H.: Tight lower bounds for weighted matroid problems (2023). https://arxiv.org/abs/2307.07773
14. El Maalouly, N., Steiner, R., Wulf, L.: Exact matching: correct parity and FPT parameterized by independence number. In: Iwata, S., Kakimura, N. (eds.) 34th International Symposium on Algorithms and Computation (ISAAC 2023). Leibniz International Proceedings in Informatics (LIPIcs), vol. 283, pp. 28:1–28:18. Schloss Dagstuhl – Leibniz-Zentrum für Informatik, Dagstuhl, Germany (2023). https://doi.org/10.4230/LIPIcs.ISAAC.2023.28

15. van Emde Boas, P., Kruyswijk, D.: A Combinatorial Problem on Finite Abelian Groups III. Mathematisch Centrum, Amsterdam. Afdeling Zuivers Wiskunde., Stichting Mathematisch Centrum (1969). https://books.google.co.uk/books?id=Lp38GwAACAAJ
16. Hörsch, F., Imolay, A., Mizutani, R., Oki, T., Schwarcz, T.: Problems on group-labeled matroid bases (2024). https://arxiv.org/abs/2402.16259
17. Ingleton, A.W.: Transversal matroids and related structures. In: Higher Combinatorics: Proceedings of the NATO Advanced Study Institute held in Berlin (West Germany), 1–10 September 1976, pp. 117–131 (1977)
18. Matsumoto, Y., Moriyama, S., Imai, H., Bremner, D.: Matroid enumeration for incidence geometry **47**(1), 17–43. https://doi.org/10.1007/s00454-011-9388-y
19. Mayhew, D., Royle, G.F.: Matroids with nine elements. J. Comb. Theory Ser. B **98**(2), 415–431 (2008). https://doi.org/10.1016/j.jctb.2007.07.005
20. Mulmuley, K., Vazirani, U.V., Vazirani, V.V.: Matching is as easy as matrix inversion. In: Proceedings of the Nineteenth Annual ACM Symposium on Theory of Computing, pp. 345–354. STOC '87, Association for Computing Machinery, New York, NY, USA (1987). https://doi.org/10.1145/28395.383347
21. Nägele, M., Nöbel, C., Santiago, R., Zenklusen, R.: Advances on strictly Δ-modular IPs. In: Proceedings of the 24th Conference on Integer Programming and Combinatorial Optimization (IPCO '23), pp. 393–407 (2023). https://doi.org/10.1007/978-3-031-32726-1_28
22. Nägele, M., Santiago, R., Zenklusen, R.: Congruency-Constrained TU Problems Beyond the Bimodular Case, pp. 2743–2790. https://doi.org/10.1137/1.9781611977073.108
23. Nägele, M., Sudakov, B., Zenklusen, R.: Submodular minimization under congruency constraints. Combinatorica **39**(6), 1351–1386 (2019). https://doi.org/10.1007/s00493-019-3900-1
24. Olson, J.E.: A combinatorial problem on finite abelian groups. I. J. Number Theory **1**(1), 8–10 (1969)
25. Papadimitriou, C.H., Yannakakis, M.: The complexity of restricted spanning tree problems. J. ACM **29**(2), 285–309 (1982). https://doi.org/10.1145/322307.322309
26. Schrijver, A., Seymour, P.D.: Spanning trees of different weights. In: Polyhedral Combinatorics, p. 281 (1990)
27. Webb, K.: Counting Bases. Ph.D. thesis, University of Waterloo (2004). http://hdl.handle.net/10012/1120

Online Combinatorial Assignment in Independence Systems

Javier Marinkovic[1], José A. Soto[1,2], and Victor Verdugo[3,4(✉)]

[1] Department of Mathematical Engineering, Universidad de Chile, Santiago, Chile
[2] Center for Mathematical Modeling IRL-CNRS 2807, Universidad de Chile, Santiago, Chile
[3] Institute for Mathematical and Computational Engineering, Pontificia Universidad Católica de Chile, Santiago, Chile
[4] Department of Industrial and Systems Engineering, Pontificia Universidad Católica de Chile, Santiago, Chile
victor.verdugo@uc.cl

Abstract. We consider an online multi-weighted generalization of several classic online optimization problems called the online combinatorial assignment problem. We are given an independence system over a ground set of elements and agents that arrive online one by one. Upon arrival, each agent reveals a weight function over the elements of the ground set. If the independence system is given by the matchings of a hypergraph, we recover the combinatorial auction problem, where every node represents an item to be sold, and every edge represents a bundle of items. For combinatorial auctions, Kesselheim et al. showed upper bounds of $O(\log\log(k)/\log(k))$ and $O(\log\log(n)/\log(n))$ on the competitiveness of any online algorithm, even in the random order model, where k is the maximum bundle size and n is the number of items. We provide an exponential improvement by giving upper bounds of $O(\log(k)/k)$, and $O(\log(n)/\sqrt{n})$ for the prophet IID setting. Furthermore, using linear programming, we provide new and improved guarantees for the k-bounded online combinatorial auction problem (i.e., bundles of size at most k). We show a $(1-e^{-k})/k$-competitive algorithm in the prophet IID model, a $1/(k+1)$-competitive algorithm in the prophet-secretary model using a single sample per agent, and a $k^{-k/(k-1)}$-competitive algorithm in the secretary model. Our algorithms run in polynomial time and work in more general independence systems where the offline combinatorial assignment problem admits the existence of a polynomial-time randomized algorithm that we call certificate sampler. These systems include some classes of matroids, matroid intersections, and matchoids.

Keywords: Online Algorithms · Linear Programming · Combinatorial Auctions

1 Introduction

The secretary and prophet inequality problems are fundamental in optimal stopping theory [13,21,22,25,37]. They both model online selection scenarios where

J. Vygen and J. Byrka (Eds.): IPCO 2024, LNCS 14679, pp. 294–308, 2024.
https://doi.org/10.1007/978-3-031-59835-7_22

the values (weights) of each agent are revealed online, and the decision-maker has to assign an item to only one of them in an irrevocable way. Motivated by their numerous applications in online market design and pricing [32], these problems have been extended to other combinatorial settings, such as matchings and matroids [15,16,27,36,38].

A central problem in the intersection of economics, optimization, and algorithms is the combinatorial auction problem, where the goal is to assign a bundle of items to each agent in a way that every item is assigned at most once, and the total weight of the solution is maximized, i.e., a maximum welfare allocation. On top of the intrinsic combinatorial difficulty of this setting, the question becomes even more challenging when the problem is solved online. This has been a very fruitful research area in recent years, both in the design of approximation algorithms and incentive-compatible mechanisms [2,3,5,6,8,10,11,18,34].

In this work, we consider a general setting that captures the previous optimization environments. In the *combinatorial assignment problem*, we are given a set of agents and an independence system over a ground set of elements. Each agent has a non-negative weight function over the elements. The goal of the decision-maker is to assign at most one element to each agent in a way that the total weight of the solution is maximized and such that the set of assigned elements is an independent set. When the agents are single-minded (i.e., put all the weight in a single, previously declared element), this setting captures the classic maximum weight matching problem or the problem of finding the maximum weight base in a matroid. When the agents are not necessarily single-minded, we can recover the combinatorial auction problem by taking the independence system of matchings in a hypergraph, where every node represents an item, and every edge is a bundle of items. The combinatorial assignment problem is, in general, NP-hard, as it captures k-dimensional matching [24].

In the online combinatorial assignment problem, the agents arrive sequentially, and upon arrival, they present a weight function over the elements. The decision-maker decides whether to assign an element to the agent, subject to the constraint that the set of elements assigned is an independent set. The goal is to find an assignment with total weight as close as possible to the optimal weight achievable in the *offline* setting, i.e., when the decision-maker has full information. We consider the prophet IID, prophet-secretary, and secretary models, described in detail in Sect. 2. In any of these models, an algorithm is γ-competitive if it constructs a solution with a total (expected) weight of at least a γ fraction of the optimal offline value.

At the core of our work is the k-bounded online combinatorial auction problem [9,12,26], where the agents only assign positive weight to bundles of at most k items. Translating this to our combinatorial assignment model, there is a fixed hypergraph with edges of size at most k, and each agent has a non-negative weight function over the edges.

1.1 Our Contributions and Techniques

For the k-bounded online combinatorial auction problem, Kesselheim et al. [26] showed that the competitiveness of any online algorithm, even in random-order models, is upper-bounded by $O(\log\log(k)/\log(k))$ and by $O(\log\log(n)/\log(n))$, where n is the number of items. We provide an exponential improvement on these bounds to show that the competitiveness of any online algorithm in the prophet IID setting is upper bounded by $O(\log(k)/k)$, and $O(\log(n)/\sqrt{n})$. This result is proved in Sect. 3 (Theorem 1).

From the algorithmic side, we provide new and improved guarantees for the k-bounded online combinatorial auction problem. We show algorithms that are $(1 - e^{-k})/k$-competitive in the prophet IID model, $1/(k + 1)$-competitive in the prophet-secretary model using a single sample per agent, and $k^{-k/(k-1)}$-competitive in the secretary model; these three algorithms run in polynomial time and are based on a linear programming relaxation for the maximum weight k-hypergraph matching problem (Theorem 5). Recently, Ezra et al. [17] also achieved the $k^{-k/(k-1)}$-competitiveness for secretary, but their algorithm runs in exponential time, and they leave as an open question the possibility of attaining this bound in polynomial time. Our result answers this question in the affirmative. In particular, we improve on the $1/(ek)$-competitive polynomial algorithm by Kesselheim et al. [26] for the secretary model by using a similar algorithmic approach but with a different analysis. Recently, Correa et al. [9] got a $1/(k+1)$-competitive algorithm in the prophet model with full distributional knowledge. Our $1/(k + 1)$-competitiveness in the prophet-secretary model states that we can replace the full distributional knowledge assumption with a single sample per agent at the cost of requiring the agents to arrive in random order. Furthermore, our $(1 - e^{-k})/k$-competitiveness for prophet IID states that we can surpass the $1/(k + 1)$ barrier when the agents are identically distributed.

Our new guarantees for the k-bounded combinatorial auction problem are all $1/k+o(1/k)$, and our new upper bound is $O(\log(k)/k)$, asymptotically closing the gaps (up to a logarithmic factor) in the following models: IID, prophet, prophet secretary, single-sample prophet secretary and secretary. The only models in which the gap is larger are the single-sample prophet model (and the harder order-oblivious model), in which the best lower-bound is still $\Omega(1/k^2)$ [29]. This indicates that the single-sample prophet model is much harder than its random-order counterpart and the model with full-distributional knowledge. A similar phenomenon occurs in the case of matroids [15,19,27], where $O(1)$-competitive algorithms are known only for models with full-distributional knowledge.

Our algorithms and guarantees work in more general independence systems where the offline combinatorial assignment problem admits a polynomial-time randomized algorithm that we call *certificate sampler*. In Sect. 4, we show our framework based on certificate samplers for the prophet IID (Theorem 2), prophet-secretary (Theorem 3), and secretary (Theorem 4) models and prove competitiveness guarantees parameterized on two values, γ and k. The value γ is the offline approximation guarantee of the certificate sampler, and k represents the maximum number of elements that could prevent another element from being part of an independent system. Our certificate samplers are not required to give a feasible assignment; they only need to provide guarantees on expectation. We exploit this in Sect. 5 to get polynomial-time guarantees for the three online models via linear programming and greedy algorithms in some matroids (Theorem 6), matroid intersection, and matchoids (Theorem 7). Our certificate samplers play a role similar to that of *online contention resolution schemes*, which yield state-of-the-art algorithms for some problems [1,7,16,20,33,35].

2 Preliminaries

An independence system $(E, \mathcal{I})$ is a pair where E is a finite set and $\mathcal{I}$ is a nonempty collection of subsets of E, called independent sets. The collection $\mathcal{I}$ is downward closed; namely, any subset of an independent set is independent. For an independence system $(E, \mathcal{I})$ and a set A of m agents, a function $M: A \to E \cup \{\bot\}$ is a *feasible assignment* of the agents if the following conditions hold:

(I) For every $a, a' \in A$ with $a \neq a'$, if $M(a) = M(a')$ then $M(a) = M(a') = \bot$.
(II) The set $\{M(a) : a \in A\} \setminus \{\bot\}$ is an independent set in $\mathcal{I}$.

We interpret a feasible assignment M as follows. When $M(a) = \bot$, agent a has not been assigned any element $e \in E$. Otherwise, $M(a) \in E$ is the element assigned to agent a, and condition (I) guarantees that two agents cannot be assigned to the same element in E. Condition (II) guarantees that the set of elements assigned to the agents is an independent set in $\mathcal{I}$. A *weight profile* for the agents A is a mapping $\boldsymbol{w}$ such that for every $a \in A$, $\boldsymbol{w}(a)$ is equal to a weight function $w_a : E \cup \{\bot\} \to \mathbb{R}_+$ with $w_a(\bot) = 0$. Given a weight profile $\boldsymbol{w}$ for A, we consider the problem of computing the maximum weight feasible assignment,

$$\max \Big\{ \textstyle\sum_{a \in A} w_a(M(a)) \colon M \text{ satisfies I and II} \Big\}. \tag{1}$$

We denote by $\mathsf{OPT}(\boldsymbol{w})$ the optimal value of (1). For example, when $(E, \mathcal{I})$ is the independence system of matchings in a graph with edges E, problem (1) corresponds to finding a maximum weight matching on 3-uniform hypergraphs, which is, in general, NP-hard. When $(E, \mathcal{I})$ is a matroid, we can solve (1) in polynomial time, using algorithms for matroid intersection (see, e.g., [14,28,30]). We say that the agents are single-minded if for every $a \in A$, there exists a single element $f_a \in E$ such that $w_a(f_a) > 0$, and $w_a(e) = 0$ for every $e \neq f_a$. With single-minded agents, (1) corresponds to finding the maximum weight matching, which can be solved efficiently.

When $(E, \mathcal{I})$ is the set of matchings in a hypergraph with edges E, problem (1) corresponds to the combinatorial auction problem [9,12,26]. In this case, every node in the hypergraph represents an item to be sold, and every edge represents a bundle of items. Each agent has a weight function over bundles, and (1) corresponds to finding the assignment of bundles to agents that maximize the total weight. When the maximum size of each edge (i.e., maximum bundle size) is k, the independence system is the k-hypergraph matchings, and (1) is the k-bounded combinatorial auction problem (or k-BCA for short).

In what follows, we describe the main online models considered in this work.

Prophet Model. Every agent a has a distribution $\mathcal{D}_a$ over the weight functions. The agents then arrive in an arbitrary order $a_1, \ldots, a_m$, and each a_t, independently from the rest, upon arrival reveals a random weight function $r_{a_t} \sim \mathcal{D}_{a_t}$. The decision maker decides irrevocably whether to assign agent a_t to some element $e \in E$ or not; if the agent is not assigned to an element in E, it is assigned to $\perp$. Furthermore, the assignment of agents should satisfy (I)–(II). We are in **prophet-secretary model** when the agents arrive in random order. In the **single-sample prophet secretary model**, the decision maker can access only one sample per agent, apart from the random weight function revealed upon the agent's arrival. We are in the **prophet IID model** when the distributions are all identical. An algorithm is γ-competitive if, for every instance, it constructs a feasible assignment with a total weight that is, on expectation, at least a γ fraction of the expected optimal value for the problem (1), that is, $\mathbb{E}_{\boldsymbol{r}}[\mathsf{OPT}(\boldsymbol{r})]$.

Secretary Model. Each agent a reveals, in random order, a weight function w_a, and we have to decide, irrevocably, whether to assign agent a to some element $e \in E$ or not; in case the agent is not assigned to an element in E, is assigned to $\perp$. Furthermore, the assignment of agents should satisfy (I)–(II). We say that an algorithm is γ-competitive if, for every instance, it constructs a feasible assignment with a total weight that is, on expectation, at least a γ fraction of the optimal value $\mathsf{OPT}(\boldsymbol{w})$ for the problem (1).

3 New Upper Bounds for k-BCA

In the following theorem, we give new upper bounds for the case of k-bounded combinatorial auctions in the prophet IID model.

Theorem 1. *The competitiveness of any algorithm for k-BCA in the prophet IID model is upper bounded by $O(\log(k)/k)$, $O(\log(m)/m)$ and $O(\log(n)/\sqrt{n})$, where m is the number of agents, and n is the number of items.*

Our bounds in Theorem 1 represent an exponential improvement over the previous upper bounds of $O(\log\log(k)/\log(k))$ and $O(\log\log(n)/\log(n))$ given by Kesselheim et al. [26], which is inspired by a former upper bound by Babaioff et al. for the secretary problem in independence systems [4]. Below, we describe the hypergraph $G = (V, E)$ used in our proof. Fix a number m of agents and consider an $m \times m$ table T. For each entry (i, j) with $i \neq j$, we have two different nodes corresponding to the cell $T(i, j)$; no nodes are associated with the diagonal cells of T. We call V_T this first set of nodes. Let $V_X = \{x_1, \ldots, x_m\}$ be a set of m additional nodes and set $V = V_T \cup V_X$, In total, we have $|V| = |V_T| + |V_X| = 2(m^2 - m) + m = 2m^2 - m$ nodes. The set E of edges is constructed as follows. Let E_T be the set of all subsets of nodes in V_T that can be formed by selecting an index $i \in [m]$, choosing exactly one node from each nondiagonal cell in column

i of T, and then exactly one node from each nondiagonal cell in row i of T. Let $E = \{f \cup \{x\} : f \in E_T \text{ and } x \in V_X\}$. That is, every edge in E is the union of an edge $f \in E_T$ and one node in V_X, and therefore every edge in E has size $2m-1$.

For each $i \in [m]$, we define a distribution $\mathcal{D}_i$ over weight functions as follows. Let R_i be the set of $2(m-1)$ nodes in row i, and let C be a set of size $m-1$ obtained by choosing exactly one node from every non-diagonal cell in column i, uniformly at random (i.e., with probability 1/2 each). We remark that C is a random set that depends on i. Consider the weight function $w_i : E \to \{0,1\}$ so that $w_i(e) = 1$ for every $e \in E$ such that $e = f \cup \{x_i\}$ with $f \in E_T$ and $f \subseteq R_i \cup C$, and it is equal to zero otherwise. Note that the weight function w_i is randomized, as it depends on C.

The common distribution $\mathcal{D}$ for all agents is defined as follows: every agent a picks (with replacement and independently from the rest) a label $\ell(a)$ uniformly at random in $[m]$, and reveals a weight function $w_{\ell(a)}$ distributed according to $\mathcal{D}_{\ell(a)}$. We can show that the optimal matching in G has expected weight $\Theta(m) = \Theta(k) = \Theta(\sqrt{n})$ and that no online algorithm can achieve an expected weight greater than $\log_2(m+1)$, which implies the desired bounds. The proof can be found in the full version of this article.

4 Algorithmic Framework for Independence Systems

In this section, we develop a general framework to design algorithms for the combinatorial assignment problem in independence systems, with random-order online models. To this end, we first introduce a combinatorial object in independence systems that interacts nicely with our online algorithms.

Definition 1. *A certifier for an independence system $(E, \mathcal{I})$ is a tuple $(\mathcal{S}, \mathcal{N}, \mathcal{B})$ where $\mathcal{S}$ is a finite set, and $(\mathcal{N}, \mathcal{B})$ is a digraph satisfying the following properties:*

(a) $\mathcal{N} \subseteq \mathcal{S} \times E$ and $\mathcal{B} \subseteq \mathcal{N} \times \mathcal{N}$.
(b) For all $e \in E$, and $S_1, S_2 \in \mathcal{S}$ such that $(S_1, e), (S_2, e) \in \mathcal{N}$, we have $((S_1, e), (S_2, e)) \in \mathcal{B}$.
(c) For every sequence of nodes $(S_1, e_1), (S_2, e_2), \ldots, (S_t, e_t) \in \mathcal{N}$ such that $((S_i, e_i), (S_j, e_j)) \notin \mathcal{B}$ for all $1 \le i < j \le t$, we have $\{e_1, \ldots, e_t\} \in \mathcal{I}$.

Every node in $\mathcal{N}$ is called a certificate, and if $((S, e), (S', e')) \in \mathcal{B}$ we say that (S, e) blocks the (S', e'). A sequence of certificates $(S_1, e_1), (S_2, e_2), \ldots, (S_t, e_t)$ is called a certification if (S_i, e_i) does not block (S_j, e_j) for every $1 \le i < j \le t$.

In words, from (a), we find that in the digraph $(\mathcal{N}, \mathcal{B})$ each node (certificate) consists of $S \in \mathcal{S}$ and an element $e \in E$ of the ground set. The property (b) implies that the digraph has a loop at every node (S, e), since we can take $S_1 = S_2 = S$, and that certificates for the same element $e \in E$ block each other. The property (c) also implies that only non-loops of $(E, \mathcal{I})$ can have certificates (certificates are certifications of length one).

Remark 1. Observe that for every certification $(S_1, e_1), (S_2, e_2), \ldots, (S_t, e_t)$, we have that $\{e_1, \ldots, e_t\}$ is an independent set of size t. This holds directly from part (c) in Definition 1. We use this property repeatedly in our analysis

Example 1. Consider the independence system of k-hypergraph matchings in $G = (V, E)$. Let $\mathcal{S} = E$, and let $(\mathcal{N}, \mathcal{B})$ be the following digraph: $\mathcal{N} = \{(e, e) : e \in E\}$, and $\mathcal{B} = \{((e, e), (f, f)) : e \cap f \neq \emptyset\}$. Namely, every edge defines its own certificate, and we have that (e, e) blocks (f, f) if e and f have at least one node in common. Properties (a)–(b) are satisfied by construction, and each certification is of the form $(e_1, e_1), (e_2, e_2), \ldots, (e_t, e_t)$ where for all $1 \leq i < j \leq t$, the edges e_i and e_j are disjoint, and therefore $\{e_1, \ldots, e_t\}$ is a matching.

Definition 2. *Given a certifier $(\mathcal{S}, \mathcal{N}, \mathcal{B})$ for an independence system $(E, \mathcal{I})$, a certificate sampler is a randomized algorithm $\mathcal{P}$ that for any set of agents A, receives a weight profile $\boldsymbol{w}$ for A, and outputs a certificate $\mathcal{P}_a(\boldsymbol{w}) \in \mathcal{N} \cup \{(\perp, \perp)\}$ for every $a \in A$. We denote by $\mathcal{P}(\boldsymbol{w})$ the output of $\mathcal{P}$ in input $\boldsymbol{w}$, and for every a, we denote $S_a(\mathcal{P}, \boldsymbol{w}) = S$ and $e_a(\mathcal{P}, \boldsymbol{w}) = e$ when $\mathcal{P}_a(\boldsymbol{w}) = (S, e)$.*

We remark that we allow the certificate sampler to output $(\perp, \perp) \notin \mathcal{N}$ for zero or more agents in A. We assume that $(\perp, \perp)$ does not block, nor is blocked, by any certificate in $\mathcal{N}$, i.e., $((\perp, \perp), (S, e)) \notin \mathcal{B}$ and $((S, e), (\perp, \perp)) \notin \mathcal{B}$. We do not require that $\{e_a(\mathcal{P}, \boldsymbol{w}) : a \in A\} \setminus \{\perp\}$ is an independent set, and we do not require that different $a, a' \in A$ receive different elements $e_a(\mathcal{P}, \boldsymbol{w})$ and $e_{a'}(\mathcal{P}, \boldsymbol{w})$.

Definition 3. *Consider an independence system $(E, \mathcal{I})$, and a certificate sampler $\mathcal{P}$ with certifier $(\mathcal{S}, \mathcal{N}, \mathcal{B})$. Then, $\mathcal{P}$ is a (γ, k)-certificate sampler, with $\gamma \geq 0$ and $k \in \mathbb{N}$, if for any set A, and any weight profile $\boldsymbol{w}$ for A, the following holds:*

(i) **(Approximation)** $\mathbb{E}[\sum_{a \in A} w_a(e_a(\mathcal{P}, \boldsymbol{w}))] \geq \gamma\, \mathsf{OPT}(\boldsymbol{w})$.
(ii) **(Blocking)** *For every* $(S, e) \in \mathcal{N}$, *we have* $\sum_{a \in A} \mathbb{P}[(\mathcal{P}_a(\boldsymbol{w}), (S, e)) \in \mathcal{B}] \leq k$.

In properties (i) and (ii), the probability and expectation are taken only over the internal randomness of the algorithm $\mathcal{P}$. Recall that here we use the fact that $w_a(\perp) = 0$ for all weight functions in order to evaluate the summation.

4.1 Prophet IID Model

In Algorithm 1, we provide a template that, given a certificate sampler $\mathcal{P}$ with certifier $(\mathcal{S}, \mathcal{N}, \mathcal{B})$, constructs a solution satisfying (I)–(II). The agents are presented in an arbitrary fixed order $a_1, \ldots, a_m$. Once presented, the agent a_t reveals a random weight function r_t. The algorithm then selects an index ℓ_t uniformly at random in $[m]$ and creates a new weight profile $\mathcal{R}_t$, where $\mathcal{R}_t(j) = R_{t,j} \sim \mathcal{D}$ for every $j \neq \ell_t$, and $R_{t,\ell_t} = r_t$. Then we run the certificate sampler $\mathcal{P}$ on the weight profile $\mathcal{R}_t$, and assign the element $e_{\ell_t}(\mathcal{P}, \mathcal{R}_t)$ to agent a_t as long as it is not blocked by any other previously assigned element.

Algorithm 1. Template for the prophet IID model

Input: An instance with n agents A.
Output: A feasible assignment of the agents.

1: Initialize $\mathsf{ALG}(a_t) = \perp$ for every $t \in [n]$.
2: **Selection:** Initialize an empty sequence C of certificates in $\mathcal{N}$.
3: **for** $t = 1$ to m **do**
4: Run $\mathcal{P}$ on the weight profile $\mathcal{R}_t$.
5: **if** $\mathcal{P}_{\ell_t}(\mathcal{R}_t) \neq (\perp, \perp)$ and every certificate in C does not block $\mathcal{P}_{\ell_t}(\mathcal{R}_t)$ **then**
6: Update $\mathsf{ALG}(a_t) = e_{\ell_t}(\mathcal{P}, \mathcal{R}_t)$, and append $\mathcal{P}_{\ell_t}(\mathcal{R}_t)$ to the end of C.
7: Return ALG.

Recall that we denote by $\boldsymbol{r}$ the weight profile that maps every agent a to the weight function r_a. Denote as $\mathcal{D}^A$ the distribution of the profile $\boldsymbol{r}$. By construction, the profiles $\mathcal{R}_1, \dots, \mathcal{R}_m$ are independent and identically distributed according to $\mathcal{D}^A$. The following is the main result of this section.

Theorem 2. *Let $\mathcal{P}$ be a (γ, k)-certificate sampler for an independence system $(E, \mathcal{I})$. Then, Algorithm 1 is $\gamma(1-e^{-k})/k$-competitive for the prophet IID model.*

Algorithm 1 requires $O(m^2)$ samples from the unknown distribution $\mathcal{D}$. Therefore, we recover the $\gamma(1-e^{-k})/k$ factor for the more restrictive prophet IID model with samples whose distribution $\mathcal{D}$ of the weights is unknown.

Lemma 1. *The solution* ALG *computed by Algorithm 1 satisfies (I) and (II).*

Proof. We first show by induction that the sequence C maintained by Algorithm 1 is a certification for every iteration $t \in \{1, \dots, m\}$. The base case follows since certificates are certifications of length one. We denote by (X_q, f_q) be the q-th certificate that is appended in line 6 of Algorithm 1. Suppose that for a certain agent a_t we append the $(q+1)$-th certificate to C, that is, $(X_{q+1}, f_{q+1}) = \mathcal{P}_{a_t}(\mathcal{R}_t)$. By the inductive step, (X_i, f_i) does not block (X_j, f_j) for every $1 \leq i < j \leq q$, and by the condition on line 6 we have that (X_i, f_i) does not block $(X_{q+1}, f_{q+1}) = \mathcal{P}_{a_t}(\mathcal{R}_t)$ for every $1 \leq i \leq q$, and therefore the sequence $(X_1, f_1), \dots, (X_{q+1}, f_{q+1})$ is a certification.

The image of ALG is exactly equal to the set $\{f_1, \dots, f_{|C|}\} \cup \{\perp\}$, where $|C|$ is the length of the sequence C at the end of the algorithm execution, and by Proposition 1 we have that $\{f_1, \dots, f_{|C|}\}$ is an independent set, and all its elements are different. Therefore, ALG satisfies (I) and (II). □

In the following lemma, we use the properties of $\mathcal{P}$ to lower bound the expected contribution of any agent a_t, for which $\mathcal{P}$ gives a certificate (S, e) in the algorithm. Intuitively, the lemma states that the probability that the assignment made by the algorithm on step i is compatible with (it does not block) the selection made on step t is at $(1 - k/m)$. In the analysis, we use subindices under $\mathbb{E}$ or $\mathbb{P}$ to denote the random sources over which we are taking expectations of probability: subindex t is used to represent all the randomness occurring on the

t-th iteration, that is, the choice of $\mathcal{R}_t$ (which includes $\boldsymbol{r}_{a_t}$), the choice of ℓ_t, and all the internal choices of the certificate sampler on that iteration. In this model, events depending on different subindices are mutually independent.

Lemma 2. *Suppose that in Algorithm 1, $\mathcal{P}$ is a (γ, k)-certificate sampler. Then, for every value $t \in \{2, \ldots, m\}$, and every $(S, e) \in \mathcal{N} \cup \{(\perp, \perp)\}$, we have*

$$\mathbb{E}_{1,2,\ldots,t}\left[r_t(e_{\ell_t}(\mathcal{P}, \mathcal{R}_t)) \cdot \mathbb{1}\left[\mathcal{P}_{\ell_t}(\mathcal{R}_t) = (S, e)\right] \cdot \mathbb{1}\left[\bigwedge_{i=1}^{t-1} \mathcal{A}_i(S, e)\right]\right]$$
$$\geq \mathbb{E}_t\left[r_t(e_{\ell_t}(\mathcal{P}, \mathcal{R}_t)) \cdot \mathbb{1}\left[\mathcal{P}_{\ell_t}(\mathcal{R}_t) = (S, e)\right]\right] \cdot \left(1 - \tfrac{k}{m}\right)^{t-1},$$

where $\mathcal{A}_i(S, e)$ is the event in which $\mathcal{P}_{\ell_i}(\mathcal{R}_i)$ does not block (S, e).

Proof. Since $\mathcal{P}_{\ell_i}(\mathcal{R}_i)$ depends only on the randomness of the i-th iteration, we conclude that the left-hand side of the expression to prove equals

$$\mathbb{E}_t\left[r_t(e_{\ell_t}(\mathcal{P}, \mathcal{R}_t)) \cdot \mathbb{1}\left[\mathcal{P}_{\ell_t}(\mathcal{R}_t) = (S, e)\right]\right] \cdot \prod_{i=1}^{t-1} \mathbb{E}_i\left[\mathbb{1}\left[\mathcal{A}_i(S, e)\right]\right]$$
$$= \mathbb{E}_t\left[r_t(e_{\ell_t}(\mathcal{P}, \mathcal{R}_t)) \cdot \mathbb{1}\left[\mathcal{P}_{\ell_t}(\mathcal{R}_t) = (S, e)\right]\right] \cdot \prod_{i=1}^{t-1} \mathbb{P}_i\left[\mathcal{A}_i(S, e)\right]. \quad (2)$$

Note that if ℓ is a uniformly chosen agent, then for every $i \in \{1, \ldots, t-1\}$, the random variable $\mathcal{P}_{\ell_i}(\mathcal{R}_i)$ has the same distribution as $\mathcal{P}_\ell(\boldsymbol{r})$, since both $\mathcal{R}_i$ and $\boldsymbol{r}$ have the same distribution $\mathcal{D}^A$, and ℓ_i is a uniform random agent. In particular, $\mathbb{P}_i\left[\mathcal{P}_{\ell_i}(\mathcal{R}_i) \text{ blocks } (S, e)\right] = \mathbb{E}_{\boldsymbol{r}}\left[\frac{1}{m}\sum_{\ell=1}^m \Pr[\mathcal{P}_\ell(\boldsymbol{r}) \text{ blocks}(S, e)]\right] \leq k/m$, where the inequality holds by the blocking property (ii) of the certificate sampler $\mathcal{P}$. Then, (2) can be lower bounded by $\mathbb{E}_t\left[r_t(e_{\ell_t}(\mathcal{P}, \mathcal{R}_t)) \cdot \mathbb{1}\left[\mathcal{P}_{\ell_t}(\mathcal{R}_t) = (S, e)\right]\right] \cdot \prod_{i=1}^t (1 - k/m)$, which finishes the proof of the lemma. □

Proof (Proof of Theorem 2). Observe that for every t, by fixing the weight profile R_t first, the expected contribution of agent a_t to the solution returned by the algorithm is $\mathbb{E}_{1,\ldots,t}[r_t(\mathsf{ALG}(a_t))]]$. Let $\overline{\mathcal{N}} = \mathcal{N} \cup \{(\perp, \perp)\}$. Therefore, the expected total value of the solution can be computed as

$$\sum_{t=1}^m \mathbb{E}_{1,2,\ldots,t}\left[r_t(e_{\ell_t}(\mathcal{P}, \mathcal{R}_t)) \cdot \mathbb{1}\left[\bigwedge_{i=1}^{t-1} \mathcal{P}_{\ell_i}(\mathcal{R}_i) \text{ does not block } \mathcal{P}_{\ell_t}(\mathcal{R}_t)\right]\right]$$
$$= \sum_{t=1}^m \sum_{(S,e)\in\overline{\mathcal{N}}} \mathbb{E}_{1,2,\ldots,t}\left[r_t(e_{\ell_t}(\mathcal{P}, \mathcal{R}_t)) \cdot \mathbb{1}\left[\mathcal{P}_{\ell_t}(\mathcal{R}_t) = (S, e)\right] \cdot \mathbb{1}\left[\bigwedge_{i=1}^{t-1} \mathcal{P}_{\ell_i}(\mathcal{R}_i) \text{ does not block } (S, e)\right]\right]$$
$$\geq \sum_{t=1}^m \sum_{(S,e)\in\overline{\mathcal{N}}} \mathbb{E}_t\left[r_t(e_{\ell_t}(\mathcal{P}, \mathcal{R}_t)) \cdot \mathbb{1}\left[\mathcal{P}_{\ell_t}(\mathcal{R}_t) = (S, e)\right]\right] \cdot \left(1 - \tfrac{k}{m}\right)^{t-1}$$
$$= \sum_{t=1}^m \mathbb{E}_t\left[r_t(e_{\ell_t}(\mathcal{P}, \mathcal{R}_t))\right] \cdot \left(1 - \tfrac{k}{m}\right)^{t-1}, \quad (3)$$

where the first equality holds by partitioning on the realization of $\mathcal{P}_{\ell_t}(R_t)$, the inequality holds by Lemma 2, and the last equality is obtained by recovering the expectation of $r_t(e_{\ell_t}(\mathcal{P}, \mathcal{R}_t))$. Observe that the random variable $r_t(e_{\ell_t}(\mathcal{P}, \mathcal{R}_t)) = R_{t,\ell_t}(e_{\ell_t}(\mathcal{P}, \mathcal{R}_t))$, can be obtained by first taking a random profile $\boldsymbol{r} \sim \mathcal{D}^A$ and then choosing an index $\ell \in [m]$ uniformly at random and returning $w_\ell(e_\ell(\mathcal{P}, \boldsymbol{w}))$. Then, $\mathbb{E}_t\left[r_t(e_{\ell_t}(\mathcal{P}, \mathcal{R}_t))\right] = \mathbb{E}_{\boldsymbol{w}}\left[\frac{1}{m}\sum_{\ell=1}^m w_\ell(e_\ell(\mathcal{P}, \boldsymbol{w})\right] \geq \frac{\gamma}{m}\mathbb{E}_{\boldsymbol{w}}[\mathsf{OPT}(\boldsymbol{w})]$, where the inequality is is a consequence of the approximation property (i) of the certificate sampler $\mathcal{P}$ when applied in the weight profile $\boldsymbol{w}$.

Using that $\boldsymbol{r}$ and $\boldsymbol{w}$ have the same distribution, then, from (3), the competitive ratio of our algorithm, $\mathbb{E}[\mathsf{ALG}]/\mathbb{E}_{\boldsymbol{r}}[\mathsf{OPT}(\boldsymbol{r})]$ can be lower bounded as $\frac{\gamma}{m}\sum_{t=1}^{m}(1-k/m)^{t-1} = \frac{\gamma}{k}(1-(1-k/m)^m) \geq \frac{\gamma}{k}\left(1-e^{-k}\right)$ where the equality holds by solving the geometric summation, and the inequality holds since $(1-k/m)^m \leq \exp(-k)$ for every m. This concludes the proof. □

4.2 Prophet-Secretary Model

Given a certificate sampler $\mathcal{P}$, our algorithm for the prophet-secretary model sees the agents in random order and can access a single sample s_a of the weight function for each agent a. When agent a arrives, the algorithm receives its real weight function r_a and constructs a weight profile ν by setting $\nu_b = r_b$ for every agent b that has already arrived and $\nu_b = s_b$ for those agents that haven't arrived yet. Then it runs the certificate sampler $\mathcal{P}$ on the weight profile ν and assigns the element $e_a(\mathcal{P}, \nu)$ to agent a as long as it is not blocked by any other previously assigned element. The formal description and analysis of this algorithm can be found in the full version of the article.

Theorem 3. *Given a (γ, k)-certificate sampler $\mathcal{P}$ for an independence system $(E, \mathcal{I})$, there exists a $\gamma/(k+1)$-competitive algorithm for the single-sample prophet-secretary model.*

Theorem 3 implies the same guarantee for the prophet-secretary model with full distributional access. The proof can be found in the full version and is similar to that of Theorem 2. To obtain the desired factor, we essentially replace the probability of $(1-k/m)$ that the assignment on step i does not block the selection made at a later step t, shown for the IID case in Lemma 2, by $(1-k/(m-t+i))_+$, while dealing with the fact that now choices are not fully-independent.

4.3 Secretary Model

Given a certificate sampler $\mathcal{P}$, and as it is typical for secretary problems, our algorithm performs first a learning phase where it skips a random number of elements of expected size pm, where $p \in (0,1)$. Then, for any agent a arriving later, the algorithm runs the certificate sampler $\mathcal{P}$ on the weight profile $\boldsymbol{w}$ consisting of all weight function seen so far and assigns the element $e_a(\mathcal{P}, \boldsymbol{w})$ to agent a as long as it is not blocked by any other previously assigned element. Consider $\{(p_k, \alpha_k)\}_{k\in\mathbb{N}}$ given by $(p_1, \alpha_1) = (1/e, 1/e)$, and $(p_k, \alpha_k) = (k^{-\frac{1}{k-1}}, k^{-\frac{k}{k-1}})$ when $k \geq 2$. The formal description and analysis of this algorithm can be found in the full version of the article.

Theorem 4. *Given a (γ, k)-certificate sampler for an independence system $(E, \mathcal{I})$, there exists a $\gamma\alpha_k$-competitive algorithm in the secretary model.*

Our proof is again similar to that of Theorem 2 and can be found in the full version. The main ingredient to obtain the desired factor is to replace the probability of $(1-k/m)$ that the assignment on step i does not block the selection made at a later step t, shown for the IID case in Lemma 2, by $(1-k/i)_+$.

5 Guarantees via Polynomial-Time Certificate Samplers

Recall that the k-BCA problem is captured by the optimization problem (1) in k-hypergraph matchings. In Sect. 5.1, we show how to design a polynomial-time certificate sampler, based on linear programming, for this independence system. Using this, we get the following guarantees for k-BCA.

Theorem 5. *For the online k-BCA problem, there exists algorithms that are*

(a) $(1 - e^{-k})/k$*-competitive in the prophet IID model.*
(b) $1/(k+1)$*-competitive in the single-sample prophet-secretary model.*
(c) $k^{-k/(k-1)}$*-competitive in the secretary model for every* $k \geq 2$ *and* $1/e$*-competitive for* $k = 1$.

Furthermore, all these algorithms run in polynomial time.

We prove Theorem 5 in Sect. 5.1. Then, we design certificate samplers for matroids, defining the new concept of *k-directed certifier.*

Definition 4. *A tuple* $(\mathcal{S}, \mathcal{N}, \mathcal{B})$ *for an independence system* $(E, \mathcal{I})$ *is a k-directed certifier if it satisfies conditions (a), (b), and (c) in Definition 1, and the following extra conditions also hold:*

(d) $\mathcal{S} = \mathcal{I}$ *and* $\mathcal{N} = \{(I, e) \colon I \in \mathcal{I}, e \in I\}$.
(e) For every $I, J \in \mathcal{I}$ *and every* $f \in J$*, we have* $|\{e \in I \colon ((I, e), (J, f)) \in \mathcal{B}| \leq k$.

We can construct certificate samplers for matroids admitting a k-directed certifier (the details can be found in the full version) and combined with our algorithmic framework from Sect. 4, we obtain the following result.

Theorem 6. *Let $\mathcal{M}$ be a matroid admitting a k-directed certifier. For the online combinatorial assignment problem in $\mathcal{M}$ there exist algorithms that are*

(a) $(1 - e^{-k})/k$*-competitive in the prophet IID model.*
(b) $1/(k+1)$*-competitive in the single-sample prophet-secretary model.*
(c) $k^{-k/(k-1)}$*-competitive in the secretary model for every* $k \geq 2$*, and* $1/e$*-competitive for* $k = 1$.

Furthermore, all these algorithms run in polynomial time.

We remark that Theorem 6 holds for unitary matroids ($k = 1$), graphic, transversal and matching matroids ($k = 2$), and for general k in k-sparse matroids, k-framed matroids, k-exchangeable gammoids, and k-exchangeable matroidal packings as defined in [38].

Finally, we discuss certificate samplers for *matchoid* independence systems [23,31], which generalize matchings and matroids. Given $(E_i, \mathcal{I}_i)_{i \in [t]}$ an arbitrary collection of matroids whose ground sets do not necessarily coincide, the *matchoid* associated to the collection is the independence system $(E, \mathcal{I})$ with $E = \bigcup_{i \in [t]} E_i$ and such that $X \subseteq E$ is independent if and only if for all $i \in [t]$,

$X \cap E_i \in \mathcal{I}_i$. Note that if all E_i are the same, this is the intersection of the t matroids. This construction is an ℓ-matchoid if every element $e \in E$, is active in at most ℓ of the matroids (i.e., e belongs to at most ℓ sets E_i). Just like matching on bipartite graphs can be modeled as the intersection of 2 matroids, matchings on a general k-hypergraph $G = (V, E)$ can be modeled as a k-matchoid. The following is proved in the full version of this article.

Theorem 7. *Let $(E, \mathcal{I})$ be an independence system obtained as either the intersection of ℓ matroids $(E, \mathcal{I}_1), \ldots, (E, \mathcal{I}_\ell)$ where matroid $(E, \mathcal{I}_i)$ admits a k_i-directed certifier, or as a matchoid such that for each $e \in E$, the $t(e)$ matroids involving element e in the matchoid admit a directed certifier with parameters $k_1(e), \ldots, k_{t(e)}(e)$. Let $k = \sum_{i=1}^{\ell} k_i$ in the first case, and $k = \max_{e \in E} \sum_{i=1}^{t(e)} k_i(e)$ in the second case. Then, for the online combinatorial assignment problem in $(E, \mathcal{I})$ there exists algorithms that are*

(a) $(1 - e^{-k})/k$-competitive in the prophet IID model.
(b) $1/(k+1)$-competitive in the single-sample prophet-secretary model.
(c) $k^{-k/(k-1)}$-competitive in the secretary model for every $k \geq 2$, and $1/e$-competitive for $k = 1$.

Furthermore, all these algorithms run in polynomial time.

5.1 Certificate Samplers for Hypergraph Matchings

Let $(E, \mathcal{I})$ be the system of k-hypergraph matchings in $G = (V, E)$. We consider the certifier constructed in Example 1, i.e., $\mathcal{S} = E$, $\mathcal{N} = \{(e, e) \colon e \in E\}$, and $\mathcal{B} = \{((e, e), (f, f)) \colon e \cap f \neq \emptyset\}$. Namely, every edge defines its own certificate, and we have that (e, e) blocks (f, f) if e and f have at least one node in common. Given a weight profile $\boldsymbol{w}$ for A, consider the following linear program:

$$
\begin{array}{llll}
\max & \sum_{a \in A} \sum_{e \in E} w_a(e) x_a(e) & & \mathrm{HM}(\boldsymbol{w}) \\
\text{s.t.} & \sum_{a \in A} \sum_{e \in \delta(v)} x_a(e) \leq 1 & \text{for every } v \in V, & (4) \\
& x_a(\bot) + \sum_{e \in E} x_a(e) = 1 & \text{for every } a \in A, & (5) \\
& x_a(e) \geq 0 & \text{for every } a \in A, \text{ and every } e \in E \cup \{\bot\}. &
\end{array}
$$

The variable $x_a(e)$, for each $a \in A$ and each $e \in E \cup \{\bot\}$, models whether $e \in E \cup \{\bot\}$ is assigned to $a \in A$. $\mathrm{HM}(\boldsymbol{w})$ is a linear relaxation of the problem (1) over the k-hypergraph matchings of G (we assume that the linear program has a unique optimal solution, e.g., by perturbing the weights lexicographically). Consider the following randomized algorithm.

Lemma 3. *Algorithm 2 is a $(1, k)$-certificate sampler for k-hypergraphs matchings, and it runs in polynomial time.*

Algorithm 2. Certificate sampler for hypergraph matchings

Input: A weight profile $\boldsymbol{w}$ for A
Output: A collection of certificates in $\mathcal{N}$
1: Find the unique optimal solution $x^\star$ of the linear program $\text{HM}(\boldsymbol{w})$.
2: For every a, sample $e_a \in E \cup \{\bot\}$ according to the distribution $(x^\star_a(e))_{e\in E\cup\{\bot\}}$.
3: Return (e_a, e_a) for every $a \in A$.

Proof. It is clear that the algorithm runs in polynomial time. We have that $\sum_{a\in A} \mathbb{E}[w_a(e_a)] = \sum_{a\in A} \sum_{e\in E\cup\{\bot\}} w_a(e)x^\star_a(e) \geq \mathsf{OPT}(\boldsymbol{w})$, where we have used that $w_a(\bot) = 0$ for all a and that the $\text{HM}(\boldsymbol{w})$ is a relaxation of the maximum weight feasible assignment problem (1). Thus, (i) in Definition 3 holds with $\gamma = 1$. Let (f, f) be any certificate. We have that $\sum_{a\in A} \mathbb{P}[((e_a, e_a), (f, f)) \in \mathcal{B}] = \sum_{a\in A} \sum_{e\in E: e\cap f\neq\emptyset} x_a(e) \leq \sum_{v\in V: v\in f} \sum_{a\in A} \sum_{e\in\delta(v)} x_a(e) \leq |f| \leq k$, where the first inequality holds by decomposing the inner summation over each node incident to the edge f, the second holds by exchanging the summation order, the third inequality holds by the set of constraints (4) in the linear program $\text{HM}(\boldsymbol{w})$, and the last inequality follows since every edge in the hypergraph has size at most k. Therefore, condition (ii) in Definition 3 holds. □

Proof (Proof of Theorem 5). The proof holds by calling the $(1, k)$-certificate sampler $\mathcal{P}$ of Lemma 3: The guarantee (a) in the IID model holds by Theorem 2, the guarantee (b) in the prophet-secretary model holds by Theorem 3, and the guarantee (c) in the secretary model holds by Theorem 4. □

Acknowledgments. This research was partially supported by ANID-Chile through grants Basal CMM FB210005, FONDECYT 1231669 and FONDECYT 1241846. The authors would like to thank Santiago Rebolledo for the helpful suggestions regarding Sect. 3.

Disclosure of Interests. The authors have no competing interests to declare that are relevant to the content of this article.

References

1. Adamczyk, M., Włodarczyk, M.: Random order contention resolution schemes. In: FOCS 2018, pp. 790–801 (2018)
2. Assadi, S., Kesselheim, T., Singla, S.: Improved truthful mechanisms for subadditive combinatorial auctions: breaking the logarithmic barrier. In: SODA 2021, pp. 653–661 (2021)
3. Assadi, S., Singla, S.: Improved truthful mechanisms for combinatorial auctions with submodular bidders. ACM SIGecom Exch. **18**(1), 19–27 (2020)
4. Babaioff, M., Immorlica, N., Kempe, D., Kleinberg, R.: Matroid secretary problems. J. ACM **65**(6), 1–26 (2018)
5. Babaioff, M., Lucier, B., Nisan, N., Paes Leme, R.: On the efficiency of the walrasian mechanism. In: EC 2014, pp. 783–800 (2014)

6. Baldwin, E., Klemperer, P.: Understanding preferences: demand types, and the existence of equilibrium with indivisibilities. Econometrica **87**(3), 867–932 (2019)
7. Brubach, B., Grammel, N., Ma, W., Srinivasan, A.: Improved guarantees for offline stochastic matching via new ordered contention resolution schemes. In: NeurIPS 2021, vol. 32, pp. 27184–27195 (2021)
8. Correa, J., Cristi, A.: A constant factor prophet inequality for online combinatorial auctions. In: STOC 2023, pp. 686–697 (2023)
9. Correa, J., Cristi, A., Fielbaum, A., Pollner, T., Weinberg, S.M.: Optimal item pricing in online combinatorial auctions. In: IPCO 2022, pp. 126–139 (2022)
10. Dobzinski, S.: Breaking the logarithmic barrier for truthful combinatorial auctions with submodular bidders. In: STOC 2016, pp. 940–948 (2016)
11. Dobzinski, S., Nisan, N., Schapira, M.: Approximation algorithms for combinatorial auctions with complement-free bidders. In: STOC 2005, pp. 610–618 (2005)
12. Dütting, P., Feldman, M., Kesselheim, T., Lucier, B.: Prophet inequalities made easy: stochastic optimization by pricing nonstochastic inputs. SIAM J. Comput. **49**(3), 540–582 (2020)
13. Dynkin, E.B.: The optimum choice of the instant for stopping a Markov process. Soviet Math. Dokl **4**, 627–629 (1963)
14. Edmonds, J.: Matroid intersection. Ann. Discret. Math. **4**, 39–49. Elsevier (1979)
15. Ehsani, S., Hajiaghayi, M.T., Kesselheim, T., Singla, S.: Prophet secretary for combinatorial auctions and matroids. In: SODA 2018, pp. 700–714 (2018)
16. Ezra, T., Feldman, M., Gravin, N., Tang, Z.G.: Online stochastic max-weight matching: prophet inequality for vertex and edge arrival models. In: EC 2020, pp. 769–787 (2020)
17. Ezra, T., Feldman, M., Gravin, N., Tang, Z.G.: General graphs are easier than bipartite graphs: tight bounds for secretary matching. In: EC 2022, pp. 1148–1177 (2022)
18. Feldman, M., Gravin, N., Lucier, B.: Combinatorial walrasian equilibrium. In: STOC 2013, pp. 61–70 (2013)
19. Feldman, M., Svensson, O., Zenklusen, R.: A simple O(loglog(rank))-competitive algorithm for the matroid secretary problem. Math. Oper. Res. **43**(2), 638–650 (2018)
20. Feldman, M., Svensson, O., Zenklusen, R.: Online contention resolution schemes with applications to bayesian selection problems. SIAM J. Comput. **50**(2), 255–300 (2021)
21. Gilbert, J.P., Mosteller, F.: Recognizing the maximum of a sequence. J. Am. Stat. Assoc. **61**, 35–76 (1966)
22. Hill, T.P., Kertz, R.P.: Comparisons of stop rule and supremum expectations of i.i.d. random variables. Ann. Probab. **10**, 336–345 (1982)
23. Jenkyns, T.A.: Matchoids: a generalization of matchings and matroids. PhD thesis, University of Waterloo (1975)
24. Karp, R.M.: Reducibility among combinatorial problems. In: Junger, M., et al. (eds.) 50 Years of Integer Programming 1958–2008. Springer, Berlin, Heidelberg (2010). https://doi.org/10.1007/978-3-540-68279-0_8
25. Kertz, R.P.: Stop rule and supremum expectations of i.i.d. random variables: a complete comparison by conjugate duality. J. Multivar. Anal. **19**, 88–112 (1986)
26. Kesselheim, T., Radke, K., Tönnis, A., Vöcking, B.: An optimal online algorithm for weighted bipartite matching and extensions to combinatorial auctions. In: ESA 2013, pp. 589–600 (2013)
27. Kleinberg, R., Weinberg, S.M.: Matroid prophet inequalities and applications to multi-dimensional mechanism design. Games Econ. Behav. **113**, 97–115 (2019)

28. Korte, B., Vygen, J.: Combinatorial Optimization: Theory and Algorithms. Springer, Berlin, Heidelberg (2012). https://doi.org/10.1007/978-3-642-24488-9
29. Korula, N., Pál, M.: Algorithms for secretary problems on graphs and hypergraphs. In: ICALP 2009, pp. 508–520 (2009)
30. Lawler, E.L.: Matroid intersection algorithms. Math. Program. **9**(1), 31–56 (1975)
31. Lovász, L.: Matroid matching and some applications. J. Comb. Theory Ser. B **28**(2), 208–236 (1980)
32. Lucier, B.: An economic view of prophet inequalities. ACM SIGecom Exch. **16**(1), 24–47 (2017)
33. MacRury, C., Ma, W., Grammel, N.: On (random-order) online contention resolution schemes for the matching polytope of (bipartite) graphs. In: SODA 2023, pp. 1995–2014 (2023)
34. Renato Paes Leme and Sam Chiu-wai Wong: Computing walrasian equilibria: fast algorithms and structural properties. Math. Program. **179**(1–2), 343–384 (2020)
35. Pollner, T., Roghani, M., Saberi, A., Wajc, D.: Improved online contention resolution for matchings and applications to the gig economy. In: EC 2022, pp. 321–322 (2022)
36. Rubinstein, A.: Beyond matroids: secretary problem and prophet inequality with general constraints. In: STOC 2016, pp. 324–332, July 2016
37. Samuel-Cahn, E.: Comparisons of threshold stop rule and maximum for independent nonnegative random variables. Ann. Probab. **12**(4), 1213–1216 (1983)
38. Soto, J.A., Turkieltaub, A., Verdugo, V.: Strong algorithms for the ordinal matroid secretary problem. Math. Oper. Res. **46**(2), 642–673 (2021)

Decomposing Probability Marginals Beyond Affine Requirements

Jannik Matuschke(✉)

KU Leuven, Leuven, Belgium
jannik.matuschke@kuleuven.be

Abstract. Consider the triplet $(E, \mathcal{P}, \pi)$, where E is a finite ground set, $\mathcal{P} \subseteq 2^E$ is a collection of subsets of E and $\pi : \mathcal{P} \to [0,1]$ is a *requirement function*. Given a vector of *marginals* $\rho \in [0,1]^E$, our goal is to find a distribution for a random subset $S \subseteq E$ such that $\mathbf{Pr}[e \in S] = \rho_e$ for all $e \in E$ and $\mathbf{Pr}[P \cap S \neq \emptyset] \geq \pi_P$ for all $P \in \mathcal{P}$, or to determine that no such distribution exists.

Generalizing results of Dahan, Amin, and Jaillet [6], we devise a generic decomposition algorithm that solves the above problem when provided with a suitable sequence of *admissible support candidates (ASCs)*. We show how to construct such ASCs for numerous settings, including *supermodular requirements*, Hoffman-Schwartz-type *lattice polyhedra* [14], and *abstract networks* where π fulfils a conservation law. The resulting algorithm can be carried out efficiently when $\mathcal{P}$ and π can be accessed via appropriate oracles. For any system allowing the construction of ASCs, our results imply a simple polyhedral description of the set of marginal vectors for which the decomposition problem is feasible. Finally, we characterize *balanced hypergraphs* as the systems $(E, \mathcal{P})$ that allow the *perfect decomposition* of any marginal vector $\rho \in [0,1]^E$, i.e., where we can always find a distribution reaching the highest attainable probability $\mathbf{Pr}[P \cap S \neq \emptyset] = \min\left\{\sum_{e \in P} \rho_e, 1\right\}$ for all $P \in \mathcal{P}$.

1 Introduction

Given a set system $(E, \mathcal{P})$ on a finite ground set E with $\mathcal{P} \subseteq 2^E$ and a *requirement function* $\pi : \mathcal{P} \to (-\infty, 1]$, consider the polytope

$$Z_\pi := \left\{z \in [0,1]^{2^E} \;:\; \textstyle\sum_{S \subseteq E} z_S = 1 \text{ and } \sum_{S : S \cap P \neq \emptyset} z_S \geq \pi_P \;\forall\, P \in \mathcal{P}\right\},$$

which corresponds to the set of all probability distributions over 2^E such that the corresponding random subset $S \subseteq E$ hits each $P \in \mathcal{P}$ with probability at least its requirement value π_P.[1] We are interested in describing the projection of Z_π to the corresponding marginal probabilities on E, i.e.,

$$Y_\pi := \left\{\rho \in [0,1]^E \;:\; \exists z \in Z_\pi \text{ with } \rho_e = \textstyle\sum_{S \subseteq E : e \in S} z_S \;\forall\, e \in E\right\}.$$

[1] Note that we can assume $\pi_P \in [0,1]$ without loss of generality in the definition of Z_π, but we allow negative values for notational convenience in later parts of the paper.

Proofs of results marked with (♣) can be found in the full version [24].

J. Vygen and J. Byrka (Eds.): IPCO 2024, LNCS 14679, pp. 309–322, 2024.
https://doi.org/10.1007/978-3-031-59835-7_23

For $\rho \in Y_\pi$, we call any $z \in Z_\pi$ with $\rho_e = \sum_{S \subseteq E : e \in S} z_S$ for all $e \in E$ a *feasible decomposition of* ρ *for* $(E, \mathcal{P}, \pi)$. Note that every $\rho \in Y_\pi$ fulfils

$$\textstyle\sum_{e \in P} \rho_e \geq \pi_P \qquad \forall P \in \mathcal{P} \tag{$\star$}$$

because $\sum_{S : S \cap P \neq \emptyset} z_S \leq \sum_{e \in P} \rho_e$ for any feasible decomposition z of ρ. Hence

$$Y_\pi \subseteq Y^\star := \left\{\rho \in [0,1]^E \ : \ \rho \text{ fulfils } (\star)\right\}.$$

We say that $(E, \mathcal{P}, \pi)$ is $(\star)$-*sufficient* if $Y_\pi = Y^\star$. Our goal is to identify classes of such $(\star)$-sufficient systems, along with corresponding decomposition algorithms that, given $\rho \in Y^\star$, find a feasible decomposition of ρ. Using such decomposition algorithms, we can reduce optimization problems over Z_π whose objectives and other constraints can be expressed via the marginals to optimization problems over $Y^\star$, yielding an exponential reduction in dimension.

1.1 Motivation

Optimization problems over Z_π and polytopes with a similar structure arise, e.g., in the context of *security games*. In such a game, a defender selects a random subset $S \subseteq E$ of resources to inspect while an attacker selects a strategy $P \in \mathcal{P}$, balancing their utility from the attack against the risk of detection (which occurs if $P \cap S \neq \emptyset$). Indeed, the decomposition setting described above originates from the work of Dahan, Amin, and Jaillet [6], who used it to describe the set of mixed Nash equilibria for such a security game using a compact LP formulation when the underlying system is $(\star)$-sufficient.

Two further application areas of marginal decomposition are *randomization in robust or online optimization*, which is often used to overcome pessimistic worst-case scenarios [17–19,27], and *social choice and mechanism design*, where randomization is frequently used to satisfy otherwise irreconcilable axiomatic requirements [3] and where decomposition results in various flavors are applied, e.g., to define auctions via interim allocations [2,12], to improve load-balancing in school choice [7], and to turn approximation algorithms into truthful mechanisms [21,22]. In [24, Appendix A], we discuss several applications from these three areas, including different security games, a robust randomized coverage problem, and committee election with diversity constraints. There we also show how the structures for which we establish $(\star)$-sufficiency here arise naturally in these applications and imply efficient algorithms for these settings.

1.2 Previous Results

As mentioned above, Dahan et al. [6] introduced the decomposition problem described above to characterize mixed Nash equilibria of a network security game played on $(E, \mathcal{P})$. They observed that such equilibria can be described by a compact LP formulation if $(E, \mathcal{P}, \pi)$ is $(\star)$-sufficient for all requirements π of the *affine* form

$$\pi_P = 1 - \textstyle\sum_{e \in P} \mu_e \quad \forall P \in \mathcal{P} \tag{A}$$

for some $\mu \in [0,1]^E$. They showed that this is indeed the case when E is the set of edges of a directed acyclic graph (DAG) and $\mathcal{P}$ the set of s-t-paths in this DAG and provide a polynomial-time (in $|E|$) algorithm for computing feasible decompositions in this case. Matuschke [25] extended this result by providing an efficient decomposition algorithm for *abstract networks*, a generalization of the system of s-t-paths in a (not necessarily acyclic) digraph; see Sect. 3 for a definition. He also showed that a system $(E, \mathcal{P}, \pi)$ is ($\star$)-sufficient for all affine requirement functions π if and only if the system has the *weak max-flow/min-cut property*, i.e., the polyhedron $\{y \in \mathbb{R}_+^E : \sum_{e \in P} y_e \geq 1 \ \forall P \in \mathcal{P}\}$ is integral.

While the affine setting (A) is well-understood, little is known for the case of more general requirement functions. A notable exception is the *conservation law* studied by Dahan et al. [6], again for the case of directed acyclic graphs:

$$\pi_P + \pi_Q = \pi_{P \times_e Q} + \pi_{Q \times_e P} \quad \forall P, Q \in \mathcal{P}, e \in P \cap Q, \tag{C}$$

where $P \times_e Q$ for two paths $P, Q \in \mathcal{P}$ containing a common edge $e \in P \cap Q$ denotes the path consisting of the prefix of P up to e and the suffix of Q starting with e. Dahan et al. [6] established ($\star$)-sufficiency for requirements fulfilling (C) in DAGs by providing another combinatorial decomposition algorithm. It was later observed in [25] and independently in a different context in [4] that (C) for DAGs is in fact equivalent to (A). However, this equivalence no longer holds for the natural generalization of (C) to arbitrary digraphs.

1.3 Contribution and Structure of This Paper

In this article, we present an algorithmic framework for computing feasible decompositions of marginal vectors fulfilling ($\star$) for a wide range of set systems and requirement functions, going beyond the affine setting (A). Our algorithm, described in Sect. 2, iteratively adds a so-called *admissible support candidate (ASC)* to the constructed decomposition. The definition of ASCs is based on a transitive dominance relation on $\mathcal{P}$, which has the property that a decomposition of $\rho \in Y^\star$ is feasible for $(E, \mathcal{P}, \pi)$ if and only if it is feasible for the restriction of the system to non-dominated sets.

Our algorithmic framework can be seen as a generalization of Dahan et al.'s [6] Algorithm 1 for requirements fulfilling (C) in DAGs. An important novelty which allows us to establish ($\star$)-sufficiency for significantly more general settings is the use of the dominance relation and the definition of ASCs, which are more flexible than the properties implicitly used in [6]. A detailed comparison of the two algorithms can be found in [24, Appendix B.1].

To establish correctness of our algorithm for a certain class of systems, which also implies ($\star$)-sufficiency for those systems, it suffices to show the existence of an ASC in each iteration of the algorithm. We assume that the set E is of small cardinality and given explicitly, while $\mathcal{P}$ might be large (possibly exponential in $|E|$) and is accessed by an appropriate oracle. To establish polynomial running time of our algorithm in $|E|$, it suffices to show that the following two tasks can be carried out in polynomial time in $|E|$:

(i) In each iteration, construct an ASC.
(ii) Given $\rho \in [0,1]^E$, either assert $\rho \in Y^\star$ or find a maximum violated inequality of ($\star$), i.e., $P \in \mathcal{P}$ maximizing $\pi_P - \sum_{e \in P} \rho_e > 0$.

We prove the existence and computability of admissible sets for a variety of settings, which we describe in the following.

Supermodular Requirements. A basic example for which our algorithm implies ($\star$)-sufficiency is the case where $\mathcal{P} = 2^E$ and π is a *supermodular* function, i.e., $\pi_{P \cap Q} + \pi_{P \cup Q} \geq \pi_P + \pi_Q$ for all $P, Q \in \mathcal{P}$. In Sect. 2.3, we show the existence of ASCs for this setting and observe that both (i) and (ii) can be solved when π is given by a *value oracle* that given $P \in \mathcal{P}$ returns π_P.

Abstract Networks Under Weak Conservation of Requirements. We prove ($\star$)-sufficiency for the case that $(E, \mathcal{P})$ is an abstract network and π fulfils a relaxed version of the conservation law (C) introduced by Hoffman [15]. Such systems generalize systems of s-t-paths in digraphs, capturing some of their essential properties that suffice to obtain results such as Ford and Fulkerson's [9] max-flow/min-cut theorem or Dijkstra's [8] shortest-path algorithm; see Sect. 3 for a formal definition and an in-depth discussion. In particular, our results generalize the results of Dahan et al. [6] for DAGs under (C) to arbitrary digraphs.

Lattice Polyhedra. We also study the case where $\mathcal{P} \subseteq 2^E$ is a *lattice*, i.e., a partially ordered set in which each pair of incomparable elements have a unique maximum common lower bound, called *meet* and a unique minimum common upper bound, called *join*, and where π is supermodular with respect to these meet and join operations. Hoffman and Schwartz [14] showed that under two additional assumptions on the lattice, called *submodularity* and *consecutivity*, the system defined by ($\star$) and $\rho \geq 0$ is totally dual integral (the corresponding polyhedron, which is the dominant of $Y^\star$, is called *lattice polyhedron*). These polyhedra generalize (contra-)polymatroids and describe, e.g., r-cuts in a digraph [10] or paths in s-t-planar graphs [26]. When π is monotone with respect to the partial order on $\mathcal{P}$, a two-phase (primal-dual) greedy algorithm introduced by Kornblum [20] and later generalized by Frank [10] can be used to efficiently optimize linear functions over lattice polyhedra using an oracle that returns maxima of sublattices. We show the existence and computability of admissible sets under the same assumptions by carefully exploiting the structure of extreme points implicit in the analysis of the Kornblum-Frank algorithm; see Sect. 4 for complete formal definitions and an in-depth discussion of these results.

Perfect Decompositions and Balanced Hypergraphs. We call a set system $(E, \mathcal{P})$ *decomposition-friendly* if it is ($\star$)-sufficient for all requirement functions π. Note that $(E, \mathcal{P})$ is decomposition-friendly if and only if every $\rho \in [0,1]^E$ has a feasible decomposition for $(E, \mathcal{P}, \pi^\rho)$, where $\pi^\rho_P := \min\left\{\sum_{e \in P} \rho_e, 1\right\}$ for $P \in \mathcal{P}$. We call such a decomposition *perfect*, as it simultaneously reaches the maximum intersection probability attainable under ρ for each $P \in \mathcal{P}$. In Sect. 5 we show that $(E, \mathcal{P})$ is decomposition-friendly if and only if it is a *balanced hypergraph*, a set system characterized by the absence of certain odd-length induced cycles.

1.4 Notation and Preliminaries

For $m \in \mathbb{N}$, we use the notation $[m]$ to denote the set $\{1, \ldots, m\}$. Moreover, we use the notation $\mathbb{1}_A$ to indicate whether expression A is true ($\mathbb{1}_A = 1$) or false ($\mathbb{1}_A = 0$). We will further make use of the following observation.

Lemma 1 ([25, Lemma 3]). *There is an algorithm that given $\rho \in [0,1]^E$ and $z \in Z_\pi$ with $\sum_{S:e\in S} z_S \le \rho_e$ for all $e \in E$, computes a feasible decomposition of ρ in time polynomial in $|E|$ and $|\{S \subseteq E : z_S > 0\}|$.*

2 Decomposition Algorithm

We describe a generic algorithm that is able to compute feasible decompositions of marginals for a wide range of systems. The algorithm makes use of a dominance relation defined in Sect. 2.1. We describe the algorithm in Sect. 2.2 and state the conditions under which it is guaranteed to produce a feasible decomposition. In Sect. 2.3, we provide a simple yet relevant example where these conditions are met. Finally, we prove correctness of the algorithm in Sect. 2.4.

2.1 The Relation $\sqsubseteq_{\pi,\rho}$ and Admissible Support Candidates

For $P, Q \in \mathcal{P}$ we write $P \sqsubseteq_{\pi,\rho} Q$ if either $P = Q$, or if $\pi_P \le \pi_Q - \sum_{e\in Q\setminus P} \rho_e$ and $\pi_P < \pi_Q$. We say that P is *non-dominated* with respect to π and ρ in $\mathcal{P}' \subseteq \mathcal{P}$ if $P \in \mathcal{P}'$ and there exists no $Q \in \mathcal{P}' \setminus \{P\}$ with $P \sqsubseteq_{\pi,\rho} Q$.

Lemma 2 (♣). *The relation $\sqsubseteq_{\pi,\rho}$ is a partial order. In particular, for any $\mathcal{P}' \subseteq \mathcal{P}$, there exists at least one P' that is non-dominated in $\mathcal{P}'$.*

As we will see in the analysis below, it suffices to ensure $\sum_{S:S\cap P} z_S \ge \pi_P$ for non-dominated $P \in \mathcal{P}$ to construct a feasible decomposition. This motivates the following definition. A set $S \subseteq E$ is an *admissible support candidate (ASC) for π and ρ* if the following three conditions are fulfilled:

(S1) $S \subseteq E_\rho := \{e \in E : \rho_e > 0\}$.
(S2) $|S \cap P| \le 1$ for all $P \in \mathcal{P}^{=}_{\pi,\rho} := \left\{Q \in \mathcal{P} : \sum_{e\in Q} \rho_e = \pi_Q\right\}$.
(S3) $|P \cap S| \ge 1$ for all non-dominated (w.r.t. π and ρ) P in $\{Q \in \mathcal{P} : \pi_Q > 0\}$.

We now present an algorithm, that when provided with a sequence of ASCs computes a feasible decomposition for $\rho \in Y^\star$.

2.2 The Algorithm

The algorithm constructs a decomposition by iteratively selecting an ASC S for a requirement function $\bar{\pi}$ and a marginal vector $\bar{\rho}$, which can be thought of as residuals of the original requirements and marginals, respectively, with $\bar{\pi} = \pi$ and $\bar{\rho} = \rho$ initially. It shifts a probability mass of

$$\varepsilon_{\bar{\pi},\bar{\rho}}(S) := \min\left\{\min_{e\in S} \bar{\rho}_e,\ \max_{P\in\mathcal{P}} \bar{\pi}_P,\ \delta_{\bar{\pi},\bar{\rho}}(S)\right\}$$

to S, where $\delta_{\bar{\pi},\bar{\rho}}(S) := \inf_{P \in \mathcal{P}: |P \cap S| > 1} \frac{\bar{\pi}_P - \sum_{e \in P} \bar{\rho}_e}{1 - |P \cap S|}$. Intuitively, $\varepsilon_{\bar{\pi},\bar{\rho}}(S)$ corresponds to the maximum amount of probability mass that can be shifted to the set S without losing feasibility of the remaining marginals for the remaining requirements. The residual marginals $\bar{\rho}$ are reduced by $\varepsilon_{\bar{\pi},\bar{\rho}}(S)$ for all $e \in S$, and so are the requirements of all $P \in \mathcal{P}$ (including those P with $P \cap S = \emptyset$).

Algorithm 1: Generic Decomposition Algorithm

Initialize $\bar{\pi} := \pi$, $\bar{\rho} := \rho$.
Initialize $z_\emptyset = 1$ and $z_S := 0$ for all $S \subseteq E$ with $S \neq \emptyset$.
while $\max_{P \in \mathcal{P}} \bar{\pi}_P > 0$ **do**
  Let S be an ASC for $\bar{\pi}$ and $\bar{\rho}$.
  Let $\varepsilon := \varepsilon_{\bar{\pi},\bar{\rho}}(S)$.
  Set $z_S := z_S + \varepsilon$ and $z_\emptyset := z_\emptyset - \varepsilon$.
  Set $\bar{\rho}_e := \bar{\rho}_e - \varepsilon$ for all $e \in S$.
  Set $\bar{\pi}_P := \bar{\pi}_P - \varepsilon$ for all $P \in \mathcal{P}$.
Apply Lemma 1 to z to obtain a feasible decomposition z' of ρ.
return z'

Our main result establishes that the algorithm returns a feasible decomposition after a polynomial number of iterations, if an ASC for $\bar{\pi}$ and $\bar{\rho}$ exists in every iteration. To show that a certain system is ($\star$)-sufficient, it thus suffices to establish the existence of the required ASCs.

Theorem 3. *Let* $(E, \mathcal{P})$ *be a set system and* $\pi : \mathcal{P} \to (-\infty, 1]$. *Let* $\rho \in Y^\star$. *If there exists an ASC for* $\bar{\pi}$ *and* $\bar{\rho}$ *in every iteration of Algorithm 1, then the algorithm terminates after* $\mathcal{O}(|E|^2)$ *iterations and returns a feasible decomposition of* ρ *for* $(E, \mathcal{P}, \pi)$.

Note that Theorem 3 implies that Algorithm 1 can be implemented to run in time $\mathcal{O}(\mathcal{T}|E|^2)$, when provided with an oracle that computes the required ASCs along with the corresponding values of $\varepsilon_{\bar{\pi},\bar{\rho}}(S)$ in time $\mathcal{T}$.[2] Before we prove Theorem 3, we first provide an example to illustrate its application.

2.3 Basic Example: Supermodular Requirements

Consider the case that $\mathcal{P} = 2^E$ and π is supermodular, i.e., for all $P, Q \in \mathcal{P}$ it holds that $\pi_{P \cap Q} + \pi_{P \cup Q} \geq \pi_P + \pi_Q$. Note that if π is supermodular, then $\bar{\pi}$ is supermodular throughout Algorithm 1, as subtracting a constant does not affect supermodularity. Moreover, we show in Sect. 2.4 that $\bar{\rho}$ fulfils ($\star$) for $\bar{\pi}$ throughout the algorithm. To apply Algorithm 1, it thus suffices to show existence of an ASC when $\rho \in Y^\star$ and π is supermodular. To obtain the ASC, we define $Q := \bigcup_{P \in \mathcal{P}^=_{\pi,\rho}} P$ and distinguish two cases: If $Q \cap E_\rho = \emptyset$, we let $S' := E_\rho$. Otherwise, we let $S' := (E_\rho \backslash Q) \cup \{e_Q\}$ for an arbitrary $e_Q \in Q \cap E_\rho$.

[2] In particular, note that $\varepsilon_{\bar{\pi},\bar{\rho}}(S)$ can be computed using at most $|S|$ iterations of the discrete Newton algorithm if we can solve problem (ii) from Sect. 1.3, i.e., the maximum violated inequality problem for $Y^\star$.

Lemma 4. *If $\mathcal{P} = 2^E$, π is supermodular, and $\rho \in Y^\star$, then S' is an ASC.*

Proof. Note that S' fulfils (S1) by construction and it fulfils (S2) because $P \subseteq Q$ and hence $P \cap S' \subseteq \{e_Q\}$ for all $P \in \mathcal{P}^{=}_{\pi,\rho}$. To see that S' fulfils (S3), assume by contradiction that $P \cap S' = \emptyset$ for some non-dominated $P \in \mathcal{P}$. Note that $P \cap E_\rho \subseteq Q \backslash \{e_Q\}$. Because $Q \in \mathcal{P}^{=}_{\pi,\rho}$ by standard uncrossing arguments, we obtain $\pi_P \leq \sum_{e \in P} \rho_e = \sum_{e \in Q \cap P} \rho_e = \pi_Q - \sum_{e \in Q \backslash P} \rho_e$ and thus $P \sqsubseteq_{\pi,\rho} Q$ (note that $e_Q \in Q \backslash P$ and hence $\pi_P < \pi_Q$), a contradiction. □

We remark that both the described ASC and maximum violated inequalities of $Y^\star$ can be found in polynomial time using submodular function minimization [29] when π is given by a value oracle, that given P returns π_P.

2.4 Analysis (Proof of Theorem 3)

Throughout this section we assume that $(E, \mathcal{P}, \pi)$ and ρ fulfil the conditions of the Theorem 3. In particular, $\rho \in Y^\star$ and in each iteration of the algorithm there exists an ASC. We show that under these conditions the while loop terminates after $\mathcal{O}(|E|^2)$ iterations (Lemma 6) and that after termination of the loop, $z \in Z_\pi$ (Lemma 8) and $\sum_{S : e \in S} z_S \leq \rho_e$ for all $e \in E$ (Lemma 5(a) for $k = \ell$). This implies that Lemma 1 can indeed be applied to z in the algorithm to obtain a feasible decomposition of ρ, thus proving Theorem 3.

We introduce the following notation. Let $S^{(i)}$ and $\varepsilon^{(i)}$ denote the set S and the value of ε chosen in the ith iteration of the while loop in the algorithm. Let further $\pi^{(i)}$ and $\rho^{(i)}$ denote the values of $\bar{\pi}$ and $\bar{\rho}$ at the beginning of the ith iteration (in particular, $\pi^{(1)} = \pi$ and $\rho^{(1)} = \rho$). Let $K \subseteq \mathbb{N}$ denote the set of iterations of the while loop. If the algorithm terminates, $K = \{1, \ldots, \ell\}$, where $\ell \in \mathbb{N}$ denotes the number of iterations. In that case, let $\rho^{(\ell+1)}$ and $\pi^{(\ell+1)}$ denote the state of $\bar{\rho}$ and $\bar{\pi}$ after termination.

Using this notation, we can establish the following three invariants, which follow directly from the construction of $\rho^{(i)}$ and $\varepsilon^{(i)}$ in the algorithm and the defining properties of the ASC $S^{(i)}$.

Lemma 5 (♣). *For all $k \in K$, the following statements hold true:*

(a) $\rho_e^{(k+1)} = \rho_e - \sum_{i=1}^{k} \mathbb{1}_{e \in S^{(i)}} \cdot \varepsilon^{(i)} \geq 0$ *for all* $e \in E$,
(b) $\sum_{e \in P} \rho_e^{(k+1)} \geq \pi_P^{(k+1)} = \pi_P - \sum_{i=1}^{k} \varepsilon^{(i)}$ *for all* $P \in \mathcal{P}$, *and*
(c) $S^{(k)} \neq \emptyset$ *and* $\varepsilon^{(k)} > 0$.

The next lemma shows that the while loop indeed terminates after $\mathcal{O}(|E|^2)$ iterations. Its proof follows from the fact that in every non-final iteration $k \in K$, there is an element $e \in S^{(k)}$ for which the value of $\bar{\rho}_e$ drops to 0, or there are two elements $e, e' \in S^{(k)}$ such that $e, e' \in P$ for some $P \in \mathcal{P}^{=}_{\pi^{(k+1)},\rho^{(k+1)}}$. It can be shown that the same pair e, e' cannot appear in two distinct iterations of the latter type, from which we obtain the following bound.

Lemma 6 (♣). *The while loop in Algorithm 1 takes at most $\binom{|E|}{2} + |E|$ iterations, i.e., $K = \{1, \ldots, \ell\}$ with $\ell \leq \binom{|E|}{2} + |E|$.*

The termination criterion of the while loop implies the following lemma.

Lemma 7 (♣). *It holds that* $\sum_{i=1}^{\ell} \varepsilon^{(i)} = \max_{P \in \mathcal{P}} \pi_P$.

Finally, we can use the properties of the ASCs $S^{(k)}$ to show that $z \in Z_\pi$.

Lemma 8. *After termination of the while loop, it holds that* $z \in Z_\pi$.

Proof. Note that $z_S = \sum_{i=1}^{\ell} \mathbb{1}_{S=S^{(i)}} \cdot \varepsilon^{(i)} \geq 0$ for $S \subseteq E$ with $S \neq \emptyset$ and that $z_\emptyset = 1 - \sum_{i=1}^{\ell} \varepsilon^{(i)} \geq 0$, where the nonnegativity follows from Lemma 5(c) and Lemma 7 with $\max_{P \in \mathcal{P}} \pi_P \leq 1$, respectively. This also implies $\sum_{S \subseteq E} z_S = 1$.

We will prove that $\sum_{i=k}^{\ell} \mathbb{1}_{P \cap S^{(i)} \neq \emptyset} \cdot \varepsilon^{(i)} \geq \pi_P^{(k)}$ for all $k \in [\ell + 1]$ and $P \in \mathcal{P}$, which, for $k = 1$, implies $\sum_{S : P \cap S \neq \emptyset} z_S = \sum_{i=1}^{\ell} \mathbb{1}_{P \cap S^{(i)} \neq \emptyset} \cdot \varepsilon^{(i)} \geq \pi_P^{(1)} = \pi_P$ and hence $z \in Z_\pi$. We prove the above statement by induction on k, starting from $k = \ell + 1$ and going down to $k = 1$. For the base case $k = \ell + 1$, observe that the left-hand side is 0 and $\pi_P^{\ell+1} \leq 0$ by termination criterion of the while loop.

For the induction step, let $k \in [\ell]$, assuming that the statement is already established for $k + 1$ and let $P \in \mathcal{P}$. We distinguish two cases.

- Case $P \cap S^{(k)} \neq \emptyset$: We can apply the induction hypothesis to obtain $\sum_{i=k}^{\ell} \mathbb{1}_{P \cap S^{(i)} \neq \emptyset} \cdot \varepsilon^{(i)} = \varepsilon^{(k)} + \sum_{i=k+1}^{\ell} \mathbb{1}_{P \cap S^{(i)} \neq \emptyset} \cdot \varepsilon^{(i)} \geq \varepsilon^{(k)} + \pi_P^{(k+1)} = \pi_P^{(k)}$.
- Case $P \cap S^{(k)} = \emptyset$: If $\pi_P^{(k)} \leq 0$ then the desired statement follows from $\varepsilon^{(i)} > 0$ for all $i \in [\ell]$ by Lemma 5(c). Thus, we can assume $\pi_P^{(k)} > 0$. By property (S3), there is $Q \in \mathcal{P}$ with $P \sqsubseteq_{\pi^{(k)}, \rho^{(k)}} Q$ and $Q \cap S^{(k)} \neq \emptyset$. Hence we can apply the induction step proven in the first case to Q, yielding $\sum_{i=k}^{\ell} \mathbb{1}_{Q \cap S^{(i)} \neq \emptyset} \cdot \varepsilon^{(i)} \geq \pi_Q^{(k)}$. From this, we conclude that

$$\begin{aligned}
\sum_{i=k}^{\ell} \mathbb{1}_{P \cap S^{(i)} \neq \emptyset} \cdot \varepsilon^{(i)} &\geq \sum_{i=k}^{\ell} \mathbb{1}_{P \cap Q \cap S^{(i)} \neq \emptyset} \cdot \varepsilon^{(i)} \\
&\geq \pi_Q^{(k)} - \sum_{i=k}^{\ell} \mathbb{1}_{(Q \setminus P) \cap S^{(i)} \neq \emptyset} \cdot \varepsilon^{(i)} \\
&\geq \pi_Q^{(k)} - \sum_{e \in Q \setminus P} \rho_e^{(k)} \geq \pi_P^{(k)},
\end{aligned}$$

where the first and second inequality use $\varepsilon^{(i)} > 0$ by Lemma 5(c), the third inequality uses $\rho_e^{(k)} = \sum_{i=k}^{\ell} \mathbb{1}_{e \in S^{(i)}} \cdot \varepsilon^{(i)}$ by Lemma 5(a) and the final inequality uses $P \sqsubseteq_{\pi^{(k)}, \rho^{(k)}} Q$. □

3 Abstract Networks Under Weak Conservation Law

An *abstract network* is a tuple $(E, \mathcal{P}, \preceq, \times)$, where $(E, \mathcal{P})$ is a set system, $\preceq_P$ for each $P \in \mathcal{P}$ is a linear order of the elements in P, and $\times$ is an operator that takes $P, Q \in \mathcal{P}$ and $e \in P \cap Q$ as arguments and maps them to a member of $\mathcal{P}$, such that $P \times_e Q \in \mathcal{P}$ fulfils $P \times_e Q \subseteq \{p \in P : p \preceq_P e\} \cup \{q \in Q : e \preceq_Q q\}$ for all $P, Q \in \mathcal{P}$ and $e \in P \cap Q$. Note that the definition of abstract networks does not impose any requirements on the order $\preceq_{P \times_e Q}$. In particular, it does not need to be consistent with $\preceq_P$ and $\preceq_Q$.

Abstract networks were introduced by Hoffman [15] in an effort to encapsulate the essential properties of systems of paths in classic networks that enable the proof of Ford and Fulkerson's [9] max-flow/min-cut theorem. Indeed, the set of s-t-paths in a digraph constitutes a special case of an abstract network (however, see [16] for examples of abstract networks that do not arise in this way) and the elements of $\mathcal{P}$ are therefore also referred to as *abstract paths.* The *maximum weighted abstract flow (MWAF)* problem and the *minimum weighted abstract cut (MWAC)* problem correspond to the linear programs

$$\begin{array}{rlrl} \max & \sum_{P \in \mathcal{P}} \pi_P \, x_P & \min & \sum_{e \in E} u_e \, y_e \\ \text{s.t.} & \sum_{P : e \in P} x_P \leq u_e \;\forall e \in E & \text{s.t.} & \sum_{e \in P} y_e \geq \pi_P \;\forall P \in \mathcal{P} \\ & x \geq 0 & & y \geq 0 \end{array}$$

where $u \in \mathbb{R}_+^E$ is a capacity vector and π determines the reward for each unit of flow sent along the abstract path $P \in \mathcal{P}$.

Hoffman [15] proved that MWAC is totally dual integral, if the reward function π fulfils the following weak conservation law:

$$\pi_{P \times_e Q} + \pi_{Q \times_e P} \geq \pi_P + \pi_Q \quad \forall P, Q \in \mathcal{P}, e \in P \cap Q. \tag{C'}$$

McCormick [28] complemented this result by a combinatorial algorithm for solving MWAF when $\pi \equiv 1$. This was extended by Martens and McCormick [23] to a combinatorial algorithm for solving MWAF with arbitrary π fulfilling (C') when π is given a separation oracle for the constraints of MWAC.

A combinatorial algorithm for marginal decomposition in abstract networks under affine requirements (A) based on a generalization of Dijkstra's shortest-path algorithm is presented in [25]. Here, we prove ($\star$)-sufficiency for the more general setting (C') by showing that ASCs can be constructed in this setting.

Theorem 9. *Let* $(E, \mathcal{P}, \preceq, \times)$ *be an abstract network, let* π *fulfil* (C'), *and let* $\rho \in Y^\star$. *Then* $S := \{e \in E_\rho \,:\, \text{there is no } P \in \mathcal{P}^{=}_{\pi,\rho} \text{ and } p \in P \cap E_\rho \text{ with } p \prec_P e\}$ *is an ASC for* π *and* ρ.

Proof. Note that S fulfils (S1) and (S2) by construction. It remains to show that S also fulfils (S3). For this, let $Q \in \mathcal{P}$ be non-dominated with $\pi_Q > 0$ and assume by contradiction $Q \cap S = \emptyset$. We use the notation $(P, e) := \{p \in P \,:\, p \prec_P e\}$ and $[e, P] := \{p \in P \,:\, e \preceq_P p\}$ for $P \in \mathcal{P}$ and $e \in P$.

Note that $Q \cap E_\rho \neq \emptyset$ because $\sum_{e \in Q} \rho_e \geq \pi_Q > 0$. Let $q := \min_{\preceq_Q} Q \cap E_\rho$. Observe that $q \notin S$ by our assumption, and hence, by construction of S, there must be $r \in E_\rho$ and $R \in \mathcal{P}^{=}_{\pi,\rho}$ such that $r \prec_R q$.

Let $Q' := R \times_q Q$ and $R' := Q \times_q R$. Note that $R' \cap E_\rho \subseteq [q, R]$ because $(Q, q) \cap E_\rho = \emptyset$ by choice of q as $\prec_q$-minimal element in $Q \cap E_\rho$. Using ($\star$), we obtain $\sum_{e \in [q,R]} \rho_e \geq \sum_{e \in R'} \rho_e \geq \pi_{R'}$. We conclude that

$$\pi_{Q'} + \sum_{e \in [q,R]} \rho_e \;\geq\; \pi_{Q'} + \pi_{R'} \;\geq\; \pi_Q + \pi_R \;=\; \pi_Q + \sum_{e \in R} \rho_e,$$

where the second inequality follows from (C') and the final identity is due to the fact that $\pi_R = \sum_{e \in R} \rho_e$ because $R \in \mathcal{P}_{\pi,\rho}$. Subtracting $\sum_{e \in [q,R]} \rho_e$ on both sides yields $\pi_{Q'} \geq \pi_Q + \sum_{e \in (R,q)} \rho_e$.

Using $Q' \setminus Q \subseteq (R, q)$ by construction of Q', we obtain $\pi_{Q'} \geq \pi_Q + \sum_{e \in Q' \setminus Q} \rho_e$. Note further that $\pi_{Q'} > \pi_Q$ because $r \in (R, q) \cap E_\rho$ and hence $\sum_{e \in (R,q)} \rho_e > 0$. We conclude that $Q \sqsubseteq_{\pi,\rho} Q'$, a contradiction to Q being non-dominated. □

We remark that the corresponding ASCs and hence feasible decompositions can be computed in polynomial time in $|E|$ if the abstract network is given via an oracle that solve the maximum violated inequality problem for $Y^\star$ and returns π_P and $\prec_P$ for the corresponding $P \in \mathcal{P}$.

4 Lattice Polyhedra

We now consider the case where $\mathcal{P}$ is equipped with a partial order $\preceq$ so that $(\mathcal{P}, \preceq)$ is a *lattice*, i.e., the following two properties are fulfilled for all $P, Q \in \mathcal{P}$:

- The set $\{R \in \mathcal{P} \,:\, R \preceq P,\, R \preceq Q\}$ has a unique maximum w.r.t. $\preceq$, denoted by $P \wedge Q$ and called the *meet of P and Q*.
- The set $\{R \in \mathcal{P} \,:\, R \succeq P,\, R \succeq Q\}$ has a unique minimum w.r.t. $\preceq$, denoted by $P \vee Q$ and called the *join of P and Q*.

We will further assume that $\mathcal{P}$ fulfils the following two additional properties:

$$\mathbb{1}_{e \in P \vee Q} + \mathbb{1}_{e \in P \wedge Q} \leq \mathbb{1}_{e \in P} + \mathbb{1}_{e \in Q} \quad \forall P, Q \in \mathcal{P},\ e \in E \tag{SM}$$

$$P \cap R \subseteq Q \quad \forall P, Q, R \in \mathcal{P} \text{ with } P \prec Q \prec R \tag{CS}$$

which are known as *submodularity* and *consecutivity*, respectively.

Furthermore, we assume that the requirement function π is *supermodular* w.r.t. the lattice $(\mathcal{P}, \preceq)$, i.e.,

$$\pi_{P \vee Q} + \pi_{P \wedge Q} \geq \pi_P + \pi_Q \quad \forall\, P, Q \in \mathcal{P}$$

and *monotone* w.r.t. $\preceq$, i.e., $\pi_P \leq \pi_Q$ for all $P, Q \in \mathcal{P}$ with $P \preceq Q$.

Hoffman and Schwartz [14] showed that under these assumptions (even when foregoing monotonicity of π) the system defining the polyhedron

$$Y^+ := \left\{\rho \in \mathbb{R}^E_+ \,:\, \textstyle\sum_{e \in P} \rho_e \geq \pi_P \ \forall P \in \mathcal{P}\right\},$$

which they call *lattice polyhedron*, is totally dual integral. For the case that π is monotone, Kornblum [20] devised a two-phase (primal-dual) greedy algorithm for optimizing linear functions over Y^+, which was extended by Frank [10] to the more general notions of sub- and supermodularity (still requiring monotonicity). The algorithm runs in strongly polynomial time when provided with *lattice oracle* that, given $U \subseteq E$ returns the maximum member (w.r.t. $\preceq$) of the sublattice $\mathcal{P}[U] := \{P \in \mathcal{P} \,:\, P \subseteq U\}$ along with the value of π_P. We prove the following decomposition result under the same assumptions as in [10,20].

Theorem 10 (♣). *Let $(\mathcal{P}, \preceq)$ be a submodular, consecutive lattice and let π be monotone and supermodular with respect to $\preceq$. Then $(E, \mathcal{P}, \pi)$ is $(\star)$-sufficient. Moreover, there is an algorithm that, given $\rho \in [0,1]^E$, finds in polynomial time in $|E|$ and T, a feasible decomposition of ρ or asserts that $\rho \notin Y^\star$, where T is the time for a call to a lattice oracle for $(\mathcal{P}, \preceq)$ and π.*

Our strategy for proving Theorem 10 is the following: If $\rho \in Y^\star$, we can express it as a convex combination of extreme points (and possibly rays) of Y^+. We can use the structure of these extreme points, implied by the optimality of the two-phase greedy algorithm, to construct ASCs and hence, via Algorithm 1, a feasible decomposition of each extreme point. These can then be recomposed to a feasible decomposition for ρ. The two-phase greedy algorithm allows us to carry out these steps efficiently as it implies a separation oracle for $Y^\star$.

In the remainder of this section, we show how to construct an ASC for the case that $\rho \in Y^\star$ is an extreme point of Y^+. We start by describing the properties of extreme points implied by the correctness of the two-phase greedy algorithm.

Theorem 11 ([10]). *Let $\rho \in Y^+$. Then ρ is an extreme point of Y^+ if and only if there exists $e_1, \ldots, e_m \in E$ and $P_1, \ldots, P_m \in \mathcal{P}$ with the following properties:*

(G1) $e_i \in P_i$ *for all* $i \in [m]$,
(G2) $P_i = \max_{\succeq} \mathcal{P}[E \backslash \{e_1, \ldots, e_{i-1}\}]$ *for all* $i \in [m]$,
(G3) $\pi_{P_i} > 0$ *for all* $i \in [m]$ *and* $\pi_Q \leq 0$ *for all* $Q \in \mathcal{P}[E \backslash \{e_1, \ldots, e_m\}]$,
(G4) ρ *is the unique solution to the linear system*

$$\begin{aligned} \textstyle\sum_{e \in P_i} \rho_e &= \pi_{P_i} && \forall\, i \in [m], \\ \rho_e &= 0 && \forall\, e \in E \backslash \{e_1, \ldots, e_m\}. \end{aligned}$$

We call such $e_1, \ldots, e_m \in E$ and $P_1, \ldots, P_m \in \mathcal{P}$ fulfilling these properties a *greedy support for* ρ. Indeed, note that (G4) implies $E_\rho \subseteq \{e_1, \ldots, e_m\}$. Properties (G1)–(G4) also imply that greedy supports have a special interval structure, enabling the following algorithmic and structural result.

Lemma 12 (♣). *Given a greedy support $e_1, \ldots, e_m$ and $P_1, \ldots, P_m$ of an extreme point ρ of Y^+ one can compute in time $\mathcal{O}(m)$ a set S fulfilling*

$$S \subseteq E_\rho \text{ and } |S \cap P_i| = 1 \text{ for all } i \in [m]. \tag{1}$$

The corresponding algorithm iterates through $e_1, \ldots, e_m$ in reverse order and adds element e_i to S if it does not result in $|S \cap P_i| > 1$. We now show that S as constructed above is indeed an ASC.

Theorem 13 (♣). *If S fulfils* (1) *for the greedy support of an extreme point ρ of Y^+, then S is an ASC for π and ρ.*

Proof (sketch). Note that S fulfils (S1) as $S \subseteq E_\rho$ by (1). Next, we show that S also fulfils (S2). Assume by contradiction that there is $Q \in \mathcal{P}^{=}_{\pi,\rho}$ with $|Q \cap S| > 1$. Without loss of generality, we can assume Q to be $\succeq$-maximal with this property. We distinguish three cases.

- Case 1: $Q \preceq P_m$. Note that $e_j \notin Q$ for all $j \in [m]$ with $j < m$, as otherwise $Q \prec P_m \prec P_j$ would imply $e_j \in P_m$ by (CS), contradicting (G2), which requires $P_m \subseteq E \backslash \{e_1, \ldots, e_{m-1}\}$. Therefore $Q \cap S \subseteq \{e_m\}$, from which we conclude $|Q \cap S| \leq 1$.

- Case 2: There is $i \in [m]$ with $P_i \succeq Q \succ P_{i+1}$. It can be shown that (G2) and (CS) imply $P_i \cap E_\rho \subseteq Q \cap E_\rho$ in this case. Moreover, $P_i, Q \in \mathcal{P}^{=}_{\pi,\rho}$ and monotonicity imply $\sum_{e \in P_i} \rho_e = \pi_{P_i} \geq \pi_Q = \sum_{e \in Q} \rho_e$. We conclude that in fact $P_i \cap E_\rho = Q \cap E_\rho$. Thus $P_i \cap S = Q \cap S$ and $|Q \cap S| \leq 1$ by (1).
- Case 3: There is $i \in [m]$ such that $Q \sim P_i$ (i.e., Q and P_i are incomparable w.r.t. $\preceq$). Let $i \in [m]$ be maximal with that property and define $Q_+ := Q \vee P_i$ and $Q_- := Q \wedge P_i$. Using standard uncrossing arguments we can show that $P_i, Q \in \mathcal{P}^{=}_{\pi,\rho}$ implies $Q_+, Q_- \in \mathcal{P}^{=}_{\pi,\rho}$ and $\mathbb{1}_{Q_+ \cap S} + \mathbb{1}_{Q_- \cap S} = \mathbb{1}_{P_i \cap S} + \mathbb{1}_{Q \cap S}$. Note that $|Q_+ \cap S| \leq 1$ by maximality of $Q \in \mathcal{P}^{=}_{\pi,\rho}$ with $|Q \cap S| > 1$. We will show that $|Q_- \cap S| \leq 1$, which, using the above and $|P_i \cap S| = 1$ by (1), implies $|Q \cap S| \leq |Q_+ \cap S| + |Q_- \cap S| - |P_i \cap S| \leq 1$, a contradiction.
  It remains to show $|Q_- \cap S| \leq 1$, for which we distinguish two subcases. First, if $i = m$, then $Q_- \cap S \subseteq \{e_m\}$ as shown in case 1 above. Second, if $i < m$, then maximality of i with $P_i \sim Q$ implies $Q \succ P_{i+1}$ and therefore $P_i \succ Q_- = P_i \wedge Q \succeq P_{i+1}$. Thus either $Q_- = P_{i+1}$ and hence $|Q_- \cap S| = 1$ by (1) or $P_i \succ Q_- \succ P_{i+1}$, in which case $|Q_- \cap S| \leq 1$ by case 2 above.

The proof that S fulfils (S3) follows similar lines but requires the use of some additional consequences of (G1)–(G4). □

5 Perfect Decompositions and Balanced Hypergraphs

In this section, we study decomposition-friendly systems, where every $\rho \in [0,1]^E$ has a perfect decomposition that attains requirements $\pi^\rho_P := \min\{\sum_{e \in P} \rho_e, 1\}$ for all $P \in \mathcal{P}$. We show that such systems are characterized by absence of certain substructures that hinder perfect decomposition.

A *special cycle* of $(E, \mathcal{P})$ consists of ordered subsets $C = \{e_1, \dots, e_k\} \subseteq E$ and $\mathcal{C} = \{P_1, \dots, P_k\} \subseteq \mathcal{P}$ such that $P_i \cap C = \{e_i, e_{i+1}\}$ for $i \in [k]$, where we define $e_{k+1} = e_1$. The *length* of such a special cycle $(C, \mathcal{C})$ is $|C| = k = |\mathcal{C}|$. A *balanced hypergraph* is a system $(E, \mathcal{P})$ that does not have any special cycles of odd length at least 3. Balanced hypergraphs were introduced by Berge [1] and have been studied extensively, see, e.g., the survey by Conforti et al. [5]. Our main result in this section is the following:

Theorem 14 (♣). *A set system $(E, \mathcal{P})$ is decomposition-friendly if and only if it is a balanced hypergraph. If $(E, \mathcal{P})$ is a balanced hypergraph, a perfect decomposition of $\rho \in [0,1]^E$ can be computed in polynomial time in $|E|$.*[3]

Proof (sketch). To see that every decomposition-friendly system needs to be a balanced hypergraph, consider any odd-length special cycle $(C, \mathcal{C})$ and observe that the marginals defined by $\rho_e = \frac{1}{2}$ for $e \in C$ and $\rho_e = 0$ for $e \in E \setminus C$ do not have a perfect decomposition. The existence of perfect decompositions in balanced hypergraphs can be established by a reduction to the case $\pi^\rho \equiv 1$, for which a perfect decomposition can be obtained using an integrality result of Fulkerson et al. [11]. □

[3] Note that $|\mathcal{P}|$ is bounded by $\mathcal{O}(|E|^2)$ for any balanced hypergraph [13]. Thus, the stated running time holds even when $\mathcal{P}$ is given explicitly.

References

1. Berge, C.: The rank of a family of sets and some applications to graph theory. In: Recent Progress in Combinatorics, pp. 49–57. Academic Press, New York (1969)
2. Border, K.C.: Implementation of reduced form auctions: a geometric approach. Econometrica **59**, 1175–1187 (1991)
3. Brandl, F., Brandt, F., Seedig, H.G.: Consistent probabilistic social choice. Econometrica **84**, 1839–1880 (2016)
4. Çela, E., Klinz, B., Lendl, S., Woeginger, G.J., Wulf, L.: A linear time algorithm for linearizing quadratic and higher-order shortest path problems. In: Del Pia, A., Kaibel, V. (eds.) IPCO 2023. LNCS, vol. 13904, pp. 466–479. Springer, Cham (2023). https://doi.org/10.1007/978-3-031-32726-1_33
5. Conforti, M., Cornuéjols, G., Vušković, K.: Balanced matrices. Discrete Math. **306**, 2411–2437 (2006)
6. Dahan, M., Amin, S., Jaillet, P.: Probability distributions on partially ordered sets and network interdiction games. Math. Oper. Res. **47**, 458–484 (2022)
7. Demeulemeester, T., Goossens, D., Hermans, B., Leus, R.: A pessimist's approach to one-sided matching. Eur. J. Oper. Res. **305**, 1087–1099 (2023)
8. Dijkstra, E.W.: A note on two problems in connexion with graphs. Numer. Math. **269**, 271 (1959)
9. Ford, L.R., Fulkerson, D.R.: Maximal flow through a network. Can. J. Math. **8**, 399–404 (1956)
10. Frank, A.: Increasing the rooted-connectivity of a digraph by one. Math. Program. **84**, 565–576 (1999)
11. Fulkerson, D.R., Hoffman, A.J., Oppenheim, R.: On balanced matrices. In: Pivoting and Extension: In honor of AW Tucker, pp. 120–132 (1974)
12. Gopalan, P., Nisan, N., Roughgarden, T.: Public projects, Boolean functions, and the borders of Border's theorem. ACM Trans. Econ. Comput. (TEAC) **6**, 1–21 (2018)
13. Heller, I.: On linear systems with integral valued solutions. Pac. J. Math. **7**, 1351–1364 (1957)
14. Hoffman, A., Schwartz, D.: On lattice polyhedra. In: Proceedings of the Fifth Hungarian Combinatorial Colloquium, North-Holland (1978)
15. Hoffman, A.J.: A generalization of max flow–min cut. Math. Program. **6**, 352–359 (1974)
16. Kappmeier, J.P.W., Matuschke, J., Peis, B.: Abstract flows over time: a first step towards solving dynamic packing problems. Theor. Comput. Sci. **544**, 74–83 (2014)
17. Kawase, Y., Sumita, H.: Randomized strategies for robust combinatorial optimization. In: Proceedings of the AAAI Conference on Artificial Intelligence, vol. 33, pp. 7876–7883 (2019)
18. Kawase, Y., Sumita, H., Fukunaga, T.: Submodular maximization with uncertain knapsack capacity. SIAM J. Discrete Math. **33**, 1121–1145 (2019)
19. Kobayashi, Y., Takazawa, K.: Randomized strategies for cardinality robustness in the knapsack problem. Theor. Comput. Sci. **699**, 53–62 (2017)
20. Kornblum, D.: Greedy algorithms for some optimization problems on a lattice polyhedron. Ph.D. thesis, City University of New York (1978)
21. Kraft, D., Fadaei, S., Bichler, M.: Fast convex decomposition for truthful social welfare approximation. In: Liu, T.Y., Qi, Q., Ye, Y. (eds.) WINE 2014. LNCS, vol. 8877, pp. 120–132. Springer, Cham (2014). https://doi.org/10.1007/978-3-319-13129-0_9

22. Lavi, R., Swamy, C.: Truthful and near-optimal mechanism design via linear programming. J. ACM (JACM) **58**, 1–24 (2011)
23. Martens, M., McCormick, S.T.: A polynomial algorithm for weighted abstract flow. In: Lodi, A., Panconesi, A., Rinaldi, G. (eds.) IPCO 2008. LNCS, vol. 5035, pp. 97–111. Springer, Heidelberg (2008). https://doi.org/10.1007/978-3-540-68891-4_7
24. Matuschke, J.: Decomposing probability marginals beyond affine requirements. Technical report (2023). arXiv:2311.03346
25. Matuschke, J.: Decomposition of probability marginals for security games in max-flow/min-cut systems. Technical report (2023). arXiv:2211.04922 (A prelimary version appeared under the title "Decomposition of Probability Marginals for Security Games in Abstract Networks" at IPCO 2023.)
26. Matuschke, J., Peis, B.: Lattices and maximum flow algorithms in planar graphs. In: Thilikos, D.M. (ed.) WG 2010. LNCS, vol. 6410, pp. 324–335. Springer, Heidelberg (2010). https://doi.org/10.1007/978-3-642-16926-7_30
27. Matuschke, J., Skutella, M., Soto, J.A.: Robust randomized matchings. Math. Oper. Res. **43**, 675–692 (2018)
28. McCormick, S.T.: A polynomial algorithm for abstract maximum flow. In: Proceedings of the 7th Annual ACM-SIAM Symposium on Discrete Algorithms, pp. 490–497 (1996)
29. Schrijver, A.: Combinatorial Optimization: Polyhedra and Efficiency. Springer, Heidelberg (2003)

Polynomial Algorithms to Minimize 2/3-Submodular Functions

Ryuhei Mizutani(✉) and Yuki Yoshida

The University of Tokyo, Tokyo 113-8656, Japan
ryuhei_mizutani@mist.i.u-tokyo.ac.jp, yoshida-yuki814@g.ecc.u-tokyo.ac.jp

Abstract. It is a fundamental result in combinatorial optimization that submodular functions can be minimized in polynomial-time. This paper considers the minimization problem for a more general class of set functions that contains all submodular functions. A set function is called 2/3-submodular if the submodular inequality holds for at least two pairs formed from every distinct three subsets. This paper provides two weakly polynomial-time algorithms to minimize 2/3-submodular functions, which yield an efficient algorithm to obtain color assignments of a variant of the supermodular coloring theorem.

Keywords: 2/3-submodular function · Polynomial-time algorithm · Supermodular coloring

1 Introduction

Let S be a finite set. A set function $f : 2^S \to \mathbf{R}$ is called *submodular* if the *submodular inequality* $f(T) + f(U) \geq f(T \cup U) + f(T \cap U)$ holds for every $T, U \subseteq S$. One of the most important problems related to submodular functions is submodular function minimization, which asks to find a subset $T \subseteq S$ minimizing $f(T)$. In this problem, we are given a value giving oracle of f, and the goal is to construct an algorithm to compute a minimizer of f in polynomial-time in $|S|$ and $\log \max_{T \subseteq S} |f(T)|$ if f is integer-valued. The first weakly polynomial algorithm for submodular function minimization was given by Grötschel, Lovász, and Schrijver [5], who showed the equivalence of the separation and optimization using the ellipsoid method to minimize submodular functions. The first combinatorial strongly polynomial algorithms were given much later by Iwata, Fleischer, and Fujishige [6] and by Schrijver [12]. For further improvements in running time, see e.g. [4,7,8,10].

This paper focuses on the minimization of a more general class of set functions that contains all submodular functions. A set function $f : 2^S \to \mathbf{R}$ is called *1/3-submodular* if for every three sets $T_1, T_2, T_3 \subseteq S$, there exists a pair T_i, T_j $(i \neq j)$ such that $f(T_i) + f(T_j) \geq f(T_i \cup T_j) + f(T_i \cap T_j)$. This class of

The first author was supported by Grant-in-Aid for JSPS Fellows Grant Number JP23KJ0379 and JST SPRING Grant Number JPMJSP2108.

J. Vygen and J. Byrka (Eds.): IPCO 2024, LNCS 14679, pp. 323–336, 2024.
https://doi.org/10.1007/978-3-031-59835-7_24

functions was introduced by Bérczi and Frank [2] in the context of edge covering. The minimum of two matroid rank functions [1,3] and the minimum of two submodular functions are examples of 1/3-submodular functions. While submodular functions can be minimized in polynomial-time in the value oracle model, it requires an exponential number of oracle calls to minimize a 1/3-submdoular function [2] if we are given the evaluation oracle of it.

In this paper, we consider the minimization of a stronger version of 1/3-submdoular functions. A function $f : 2^S \to \mathbf{R}$ is called *2/3-submodular* if for every distinct three sets $T_1, T_2, T_3 \subseteq S$, there exist two distinct pairs $(i,j), (k,l) \in \{(1,2), (2,3), (3,1)\}$ such that

$$f(T_i) + f(T_j) \geq f(T_i \cup T_j) + f(T_i \cap T_j), \text{ and}$$
$$f(T_k) + f(T_l) \geq f(T_k \cup T_l) + f(T_k \cap T_l).$$

The class of 2/3-submodular function is a subclass of 1/3-submodular functions, and a superclass of submodular functions. One example of 2/3-submodular functions is the rank function of a relaxation of sparse paving matroids.

Example 1. Let $\mathcal{B}$ be the base family of a uniform matroid U_n^k on $[n]$ of rank k, where $[n] = \{1, 2, \ldots, n\}$ and $n \geq k+2,\ k \geq 2$. Let $\mathcal{F} \subseteq \mathcal{B}$ be a family such that every pair $X, Y \in \mathcal{F}$ does not satisfy $|X \setminus Y| = |Y \setminus X| = 1$. Then a matroid with a base family $\mathcal{B} \setminus \mathcal{F}$ is called a *sparse paving matroid.* Let $\mathcal{F}'$ be a family such that every distinct three sets $X, Y, Z \in \mathcal{F}'$ do not satisfy $|X \setminus Y| = |Y \setminus X| = |X \setminus Z| = |Z \setminus X| = 1$. Then the rank function $r : 2^{[n]} \to \mathbf{Z}$ defined as

$$r(X) = \max\{|X \cap B| \mid B \in \mathcal{B} \setminus \mathcal{F}'\}$$

for each $X \subseteq [n]$ is not necessarily submodular but 2/3-submodular because the submodular inequality $r(X) + r(Y) \geq r(X \cup Y) + r(X \cap Y)$ does not hold if and only if $X, Y \in \mathcal{F}'$ and $|X \setminus Y| = |Y \setminus X| = 1$.

The intersecting version of 2/3-submodular functions was introduced in [9] to describe a variant of the supermodular coloring theorem. The coloring theorem in [9] showed the existence of a particular color assignment through a constructive proof. Since this proof includes the minimization of 2/3-submodular functions, whether the algorithm derived from the proof runs in polynomial-time depends on whether 2/3-submodular functions can be minimized in polynomial-time. Our main results are two weakly polynomial-time algorithms to minimize 2/3-submodular functions.

Theorem 1. *Let f be a rational-valued 2/3-submodular function on S. Then a minimizer of f can be obtained in polynomial-time in $|S|$ and $\log B$, where B is an upper bound on the numerators and denominators of the absolute values of f.*

Let $f : 2^S \to \mathbf{R}$ be a 2/3-submodular function. Throughout the paper we assume $f(\emptyset) = 0$. The *2/3-submodular polyhedron* $P(f)$ is defined as follows:

$$P(f) = \{x \in \mathbf{R}^S \mid x(T) \leq f(T) \text{ for every } T \subseteq S\}.$$

Here, we use the notation $x(T) = \sum_{s \in T} x(s)$. For $w \in \mathbf{R}_+^S$, consider the following linear program:

$$\begin{array}{lll} \text{(P)} & \max & w^\top x \\ & \text{s.t.} & x \in P(f). \end{array}$$

If f and w are rational-valued, then the polynomial solvability of (P) implies that f can be minimized in polynomial-time using the ellipsoid method through the polynomial solvability equivalence of the separation and optimization [5]. Indeed, the separation problem corresponding to (P) includes the problem to decide whether f is nonnegaive or not. Hence, by adding the some nonnegative constant to the values of f (except for $f(\emptyset)$) and using the binary search on the separation problem, we can obtain a minimizer of f.

In this paper, we provide two algorithms to compute an optimal solution of (P) in time polynomial in $|S|$. The first algorithm solves the linear program (P) with particular polynomial number of constraints using the ellipsoid method. We show that an extreme point optimal solution of (P) with such constraints is also an optimal solution of the original linear program (P). The second algorithm computes a dual optimal solution y using a dynamic programming method, and carefully constructs a primal optimal solution from y using a greedy-type algorithm. This algorithm is combinatorial and runs in $O(|S|^3)$ time. Since it requires the ellipsoid method to solve the separation problem corresponding to (P) from the optimization problem (P), even the second algorithm does not provide a combinatorial algorithm to minimize 2/3-submodular functions.

The rest of this paper is organized as follows. Section 2 presents a polynomial-time algorithm for (P) using the ellipsoid method. Section 3 provides a combinatorial cubic-time algorithm to solve (P) using a dynamic programming approach and a greedy algorithm. Section 4 describes an application to a variant of the supermodular coloring theorem.

2 Minimization via the Ellipsoid Method

This section provides a polynomial-time algorithm to compute an optimal solution of (P) via the ellipsoid method. We first show that the dual linear program of (P) has an optimal solution whose support is contained in a particular polynomial size family $\mathcal{C}'$. Then we prove that an extreme point optimal solution of $(\mathrm{P}_{\mathcal{C}'})$ is also an optimal solution of (P), where $(\mathrm{P}_{\mathcal{C}'})$ is a linear program whose constrains are restricted to $\mathcal{C}'$.

Let S be a finite set. We first show the following useful observation:

Lemma 1. *Let $f : 2^S \to \mathbf{R}$ be a 2/3-submodular function. If $X, Y \subseteq S$ satisfy $f(X) + f(Y) < f(X \cup Y) + f(X \cap Y)$, then $|X \setminus Y| = |Y \setminus X| = 1$.*

Proof. It suffices to show that any pair of $X, Y \subseteq S$ with $|X \setminus Y| \neq 1$ satisfies $f(X) + f(Y) \geq f(X \cup Y) + f(X \cap Y)$. If $|X \setminus Y| = 0$, then we have $f(X) + f(Y) = f(X \cup Y) + f(X \cap Y)$ because $X \subseteq Y$. Suppose that $|X \setminus Y| \geq 2$. Take

$x \in X \setminus Y$. Then we have $Y \cup \{x\}, (X \cap Y) \cup \{x\} \neq X, Y$. If $f(X) + f(Y) < f(X \cup Y) + f(X \cap Y)$, then since f is 2/3-submodular, we have

$$f(X) + f(Y \cup \{x\}) \geq f(X \cup Y) + f((X \cap Y) \cup \{x\}),$$
$$f((X \cap Y) \cup \{x\}) + f(Y) \geq f(Y \cup \{x\}) + f(X \cap Y).$$

By adding the above two inequalities, we obtain $f(X) + f(Y) \geq f(X \cup Y) + f(X \cap Y)$, which is a contradiction. □

Consider the following dual linear program of (P):

$$\begin{aligned} \text{(D)} \quad \min \quad & \sum_{T \subseteq S} y_T f(T) \\ \text{s.t.} \quad & y \in \mathbf{R}_+^{2^S}, \\ & \sum_{T \subseteq S} y_T \chi_T = w. \end{aligned}$$

To show that (D) has an optimal solution whose support is covered by a particular polynomial size family, we prepare some properties of optimal solutions.

Lemma 2. *There is an optimal solution y of* (D) *such that its support $\mathcal{L} = \{T \subseteq S \mid y_T > 0\}$ satisfies the following two conditions:*

- *For any pair of sets $T, U \in \mathcal{L}$, we have either $T \subseteq U$, $U \subseteq T$, or $|T \setminus U| = |U \setminus T| = 1$.*
- *For any set $T \in \mathcal{L}$, the number of $U \in \mathcal{L}$ such that $|T \setminus U| = |U \setminus T| = 1$ is at most one.*

Proof. Let $y \in \mathbf{R}_+^{2^S}$ be an optimal solution of (D) minimizing

$$\sum_{T \subseteq S} y_T |T||S \setminus T|.$$

We first show that $\mathcal{L} = \{T \subseteq S \mid y_T > 0\}$ satisfies the first condition. Suppose that $T, U \in \mathcal{L}$ satisfy $T \not\subseteq U, U \not\subseteq T$. If $f(T) + f(U) \geq f(T \cup U) + f(T \cap U)$, then set $\alpha = \min\{y_T, y_U\}$, and decrease y_T, y_U by α and increase $y_{T \cup U}, y_{T \cap U}$ by α. Let y' denote the vector obtained by this way. Then y' is an optimal solution of (D) and we have

$$\sum_{T \subseteq S} y_T |T||S \setminus T| > \sum_{T \subseteq S} y'_T |T||S \setminus T|,$$

which is a contradiction. Hence, we have $f(T) + f(U) < f(T \cup U) + f(T \cap U)$. By Lemma 1, this implies that $|T \setminus U| = |U \setminus T| = 1$. Hence $\mathcal{L}$ satisfies the first condition. To show that $\mathcal{L}$ satisfies the second condition, suppose for a contradiction that the number of $U \in \mathcal{L}$ such that $|T \setminus U| = |U \setminus T| = 1$ is at least two for $T \in \mathcal{L}$. Then by the above uncrossing argument, the number of $U \in \mathcal{L}$ such that $f(T) + f(U) < f(T \cup U) + f(T \cap U)$ is at least two, which contradicts 2/3-submodularity of f. □

Order the elements of S as $s_1, s_2, \ldots, s_n$ such that $w(s_1) \geq w(s_2) \geq \cdots \geq w(s_n)$. Define $S_i = \{s_1, \ldots, s_i\}$ for $i = 1, 2, \ldots, n$ and let $S_0 = \emptyset$. Let $\mathcal{C} = \{S_i \mid 0 \leq i \leq n\} \cup \{S_i \cup \{s_{i+2}\} \mid 0 \leq i \leq n-2\}$. To show that (D) has an optimal solution whose support is covered by $\mathcal{C}$, we first prove the following lemma:

Lemma 3. *Let y be a feasible solution of* (D) *such that $\mathcal{L} = \{T \subseteq S \mid y_T > 0\}$ satisfies the two conditions in* Lemma 2. *Then we have $\mathcal{L} \subseteq \mathcal{C}$.*

Proof. We prove the statement by induction on $|\mathcal{L}|$. The case when $\mathcal{L} = \emptyset$ is trivial. Consider the case when $|\mathcal{L}| \geq 1$. Since $\mathcal{L}$ satisfies the two conditions in Lemma 2, the number of maximal sets in $\mathcal{L}$ is at most two. We consider separately the cases when the number of maximal sets in $\mathcal{L}$ is either one or two.

(i) The case when the number of maximal sets in $\mathcal{L}$ is one.

Let m be the maximum index such that $w(s_m) > 0$. Let M be the unique maximal set in $\mathcal{L}$. Then we have $w(s) > 0$ if $s \in M$, and $w(s) = 0$ otherwise, which implies $M = S_m$. Let $w' = w - y_{S_m}\chi_{S_m}$. We define $y' \in \mathbf{R}_+^{2^S}$ as follows:

$$y'_T = \begin{cases} 0 & (T = S_m), \\ y_T & (\text{otherwise}). \end{cases}$$

Then we have $w' \geq 0$, $w'(s_1) \geq w'(s_2) \geq \cdots \geq w'(s_n)$, and

$$\sum_{T \subseteq S} y'_T \chi_T = w'.$$

Since $\mathcal{L}' = \{T \subseteq S \mid y'_T > 0\}$ satisfies $|\mathcal{L}'| < |\mathcal{L}|$, it holds by induction that $\mathcal{L}' \subseteq \mathcal{C}$, which implies $\mathcal{L} \subseteq \mathcal{C}$.

(ii) The case when the number of maximal sets in $\mathcal{L}$ is two.

Let M_1 and M_2 be the maximal sets in $\mathcal{L}$. Since $\mathcal{L}$ satisfies the two conditions in Lemma 2, $T \in \mathcal{L}$ such that $T \neq M_1, M_2$ satisfies $T \subsetneq M_1$ and $T \subsetneq M_2$, which implies $T \subseteq M_1 \cap M_2$. By the first condition in Lemma 2, it holds that $|M_1 \setminus M_2| = |M_2 \setminus M_1| = 1$. Let s_{m_1} and s_{m_2} be the unique elements in $M_1 \setminus M_2$ and $M_2 \setminus M_1$, respectively. Since every $T \in \mathcal{L}$ such that $T \neq M_1, M_2$ satisfies $T \subseteq M_1 \cap M_2$, we have $w(s_{m_1}) = y_{M_1}, w(s_{m_2}) = y_{M_2}$. It holds that $w(s) \geq y_{M_1} + y_{M_2}$ for any $s \in M_1 \cap M_2$ and $w(s) = 0$ for any $s \in S \setminus (M_1 \cup M_2)$. Hence we have $w(s) > w(s_{m_i}) > w(s')$ for every $s \in M_1 \cap M_2, s' \in S \setminus (M_1 \cup M_2)$, and $i = 1, 2$. This implies that there exists an integer m such that $M_1 \cap M_2 = S_m, m_1 = m+1, m_2 = m+2$ (we may assume without loss of generality that $m_1 < m_2$). Then we have $M_1 = S_{m+1}, M_2 = S_m \cup \{s_{m+2}\}$. Let $w' = w - y_{S_{m+1}}\chi_{S_{m+1}} - y_{S_m \cup \{s_{m+2}\}}\chi_{S_m \cup \{s_{m+2}\}}$. We define $y' \in \mathbf{R}_+^{2^S}$ as follows:

$$y'_T = \begin{cases} 0 & (T = S_{m+1}, S_m \cup \{s_{m+2}\}), \\ y_T & (\text{otherwise}). \end{cases}$$

Then we have $w' \geq 0, w'(s_1) \geq w'(s_2) \geq \cdots \geq w'(s_n)$, and

$$\sum_{T \subseteq S} y'_T \chi_T = w'.$$

Since $\mathcal{L}' = \{T \subseteq S \mid y'_T > 0\}$ satisfies $|\mathcal{L}'| < |\mathcal{L}|$, it holds by induction that $\mathcal{L}' \subseteq \mathcal{C}$, which implies $\mathcal{L} \subseteq \mathcal{C}$. □

The following corollary immediately follows from Lemma 3.

Corollary 1. *Let y be an optimal solution of* (D) *such that $\mathcal{L} = \{T \subseteq S \mid y_T > 0\}$ satisfies the two conditions in* Lemma 2. *Then we have $\mathcal{L} \subseteq \mathcal{C}$.*

Consider the following restricted dual linear program:

$$(\mathrm{D}_{\mathcal{C}}) \quad \min \sum_{T \in \mathcal{C}} y_T f(T)$$
$$\text{s.t.} \quad y \in \mathbf{R}_+^{\mathcal{C}},$$
$$\sum_{T \in \mathcal{C}} y_T \chi_T = w.$$

By Lemma 2 and Corollary 1, we obtain the following result.

Corollary 2. *Suppose that y is an optimal solution of* $(\mathrm{D}_{\mathcal{C}})$. *Define $y' \in \mathbf{R}_+^{2^S}$ as follows: $y'_T = y_T$ if $T \in \mathcal{C}$, and $y'_T = 0$ otherwise. Then y' is also an optimal solution of* (D).

Let $d_i = f(S_{i-1}) + f(S_{i-2} \cup \{s_i\}) - f(S_i) - f(S_{i-2})$ for $i = 2, \ldots, n$, and

$$\mathcal{C}' = \{S_i \mid 0 \leq i \leq n\} \cup \{S_i \cup \{s_{i+2}\} \mid 0 \leq i \leq n-2,\ d_{i+2} < 0\}.$$

We further restrict the linear program $(\mathrm{D}_{\mathcal{C}})$ to $\mathcal{C}'$:

$$(\mathrm{D}_{\mathcal{C}'}) \quad \min \sum_{T \in \mathcal{C}'} y_T f(T)$$
$$\text{s.t.} \quad y \in \mathbf{R}_+^{\mathcal{C}'},$$
$$\sum_{T \in \mathcal{C}'} y_T \chi_T = w.$$

Similar to Corollary 2, the following lemma holds for an optimal solution of $(\mathrm{D}_{\mathcal{C}'})$.

Lemma 4. *Suppose that y is an optimal solution of* $(\mathrm{D}_{\mathcal{C}'})$. *Define $y' \in \mathbf{R}_+^{\mathcal{C}}$ as follows: $y'_T = y_T$ if $T \in \mathcal{C}'$, and $y'_T = 0$ otherwise. Then y' is also an optimal solution of* $(\mathrm{D}_{\mathcal{C}})$.

Let $P_{\mathcal{C}'}(f) = \{x \in \mathbf{R}^S \mid x(T) \leq f(T)$ for every $T \in \mathcal{C}'\}$. Consider the following dual linear program of $(\mathrm{D}_{\mathcal{C}'})$:

$$(\mathrm{P}_{\mathcal{C}'}) \quad \max \quad w^\top x \qquad \text{s.t.} \quad x \in P_{\mathcal{C}'}(f).$$

By Corollary 2 and Lemma 4, we obtain the following key lemma to compute an optimal solution of (P).

Lemma 5. *Every extreme point optimal solution of* $(\mathrm{P}_{\mathcal{C}'})$ *is contained in* $P(f)$.

By Lemma 5, every extreme point optimal solution of $(\mathrm{P}_{\mathcal{C}'})$ is also an optimal solution of (P). Since an extreme point optimal solution of $(\mathrm{P}_{\mathcal{C}'})$ can be found in strongly polynomial-time by the ellipsoid method in [13], we obtain the following theorem:

Theorem 2. *Suppose that f and w are rational. Then an optimal solution of* (P) *can be computed in time polynomial in* $|S|$.

3 A Faster Algorithm

This section provides a combinatorial cubic-time algorithm to solve the linear program (P). The algorithm consists of two parts: the first part computes an optimal solution y of $(\mathrm{D}_{\mathcal{C}'})$ using a dynamic programming method, and the second part solves the linear program (P) by constructing an optimal solution of (P) from y using a greedy-type algorithm. In the dynamic programming method, we compute an optimal solution y^i of $(\mathrm{D}_{\mathcal{C}'})$ with additional constraints $y^i_{S_{j-2}\cup\{s_j\}} = 0$ for $j = i, \ldots, n$ using the previous results $y^2, \ldots, y^{i-1}$. By repeating this procedure for $i = 2, \ldots, n+1$ one by one, we finally obtain an optimal solution of $(\mathrm{D}_{\mathcal{C}'})$ because y^{n+1} is an optimal solution of $(\mathrm{D}_{\mathcal{C}'})$. By this approach, we can compute a dual optimal solution in cubic-time.

Theorem 3. *Suppose that f and w are rational. Then an optimal solution of* $(\mathrm{D}_{\mathcal{C}'})$ *can be found in time* $O(|S|^3)$.

We next provide a greedy-type algorithm to construct an optimal solution of (P) from a dual optimal solution. We first show a necessary and sufficient condition for feasible solutions of $(\mathrm{D}_{\mathcal{C}'})$ to be optimal. Then we describe the greedy algorithm that uses the necessary and sufficient condition for greedy choices.

For integers i and j, the notation $i \equiv_2 j$ means that they are either both even, or both odd, while $i \not\equiv_2 j$ means that one is even and the other is odd. For simplicity, we also use the notation $i \equiv_4 j$ to indicate that $|j-i|$ is a multiple of 4, and $i \not\equiv_4 j$ to indicate that $|j-i|$ is not a multiple of 4. Let y be a feasible solution of $(\mathrm{D}_{\mathcal{C}'})$ and let $\mathcal{L} = \{T \in \mathcal{C}' \mid y_T > 0\}$. Then y is called *uncrossing-optimal* if $\mathcal{L}$ satisfies the following conditions for every integers i, j with $2 \leq i \leq j \leq n$ and $i \equiv_2 j$:

(a) If $i \not\equiv_4 j$, $S_{i-2} \in \mathcal{L} \cup \{\emptyset\}$, and $S_{k+1}, S_k \cup \{s_{k+2}\} \in \mathcal{L}$ for every $i \leq k \leq j$ with $k \equiv_4 i$, then

$$\sum_{i \leq k \leq j, k \equiv_2 i} (-1)^{\frac{k-i}{2}} \cdot d_k \geq 0.$$

(b) If $i \not\equiv_4 j$, $S_j \in \mathcal{L}$, and $S_{k+1}, S_k \cup \{s_{k+2}\} \in \mathcal{L}$ for every $i-2 \leq k \leq j-2$ with $k \equiv_4 i-2$, then

$$\sum_{i \leq k \leq j, k \equiv_2 i} (-1)^{\frac{k-i}{2}+1} \cdot d_k \geq 0.$$

(c) If $i \equiv_4 j$ and $S_{k+1}, S_k \cup \{s_{k+2}\} \in \mathcal{L}$ for every $i-2 \leq k \leq j-2$ with $k \equiv_4 i-2$, then

$$\sum_{i \leq k \leq j, k \equiv_2 i} (-1)^{\frac{k-i}{2}+1} \cdot d_k \geq 0.$$

(d) If $i \equiv_4 j$, $S_{i-2}, S_j \in \mathcal{L} \cup \{\emptyset\}$, and $S_{k+1}, S_k \cup \{s_{k+2}\} \in \mathcal{L}$ for every $i \leq k \leq j-4$ with $k \equiv_4 i$, then

$$\sum_{i \leq k \leq j, k \equiv_2 i} (-1)^{\frac{k-i}{2}} \cdot d_k \geq 0.$$

These conditions intuitively indicate that y cannot be improved by repeating uncrossing procedures or the converse of uncrossing procedures. Hence, the conditions are necessary for y to be optimal. The following lemma shows that the converse is also true.

Lemma 6. *A feasible solution y of* ($\mathrm{D}_{\mathcal{C}'}$) *is uncrossing-optimal if and only if y is an optimal solution of* ($\mathrm{D}_{\mathcal{C}'}$).

We now describe a greedy algorithm to construct an optimal solution of (P) from that of ($\mathrm{D}_{\mathcal{C}'}$).

Greedy algorithm

Input: A 2/3-submodular function f (by a value giving oracle), a vector $w \in \mathbf{R}_+^S$, a family $\mathcal{C}'$ defined from w and f, and an optimal solution y of ($\mathrm{D}_{\mathcal{C}'}$).
Output: An optimal solution x of (P).
Step 1. Set $\mathcal{T} = \{T \in \mathcal{C}' \mid y_T > 0\} \cup \{S_{i+1} \mid 0 \leq i \leq n-2,\ d_{i+2} < 0\}$. Order the sets in $\mathcal{C}'$ as $T_1, T_2, \ldots, T_N$. Set $l = 1$.
Step 2. If $\mathcal{L} = \mathcal{T} \cup \{T_l\}$ satisfies the conditions (a),(b),(c),(d) for every integers i, j with $2 \leq i \leq j \leq n$ and $i \equiv_2 j$, then update $\mathcal{T} \leftarrow \mathcal{T} \cup \{T_l\}$.
Step 3. Update $l \leftarrow l + 1$. If $l \leq N$, then go back to Step 2. Otherwise, go to Step 4.
Step 4. Using Gaussian elimination, compute x such that $x(T) = f(T)$ for all $T \in \mathcal{T}$. Return x.

During the algorithm, let $w_{\mathcal{T}} = \sum_{T\in\mathcal{T}} \chi_T$ and define $y^{\mathcal{T}} \in \mathbf{R}_+^{\mathcal{C}'}$ as follows:

$$y_T^{\mathcal{T}} = \begin{cases} 1 & (T \in \mathcal{T}), \\ 0 & (\text{otherwise}). \end{cases}$$

To prove the correctness of the algorithm, we first show some lemmas.

Lemma 7. *$w_{\mathcal{T}}(s_i) \geq w_{\mathcal{T}}(s_{i+1})$ holds for every $i = 1, \ldots, n-1$ at any step of the algorithm.*

Proof. Suppose for a contradiction that $w_{\mathcal{T}}(s_i) < w_{\mathcal{T}}(s_{i+1})$ holds for some $1 \leq i \leq n-1$. Then $S_{i-1} \cup \{s_{i+1}\} \in \mathcal{T} \subseteq \mathcal{C}'$ and $w_{\mathcal{T}}(s_i) + 1 = w_{\mathcal{T}}(s_{i+1})$ hold because $S_{i-1} \cup \{s_{i+1}\}$ is the unique set T in $\mathcal{C}'$ such that $s_i \notin T$ and $s_{i+1} \in T$. Since $S_{i-1} \cup \{s_{i+1}\} \in \mathcal{C}'$ implies that $d_{i+1} < 0$ by the definition of $\mathcal{C}'$, we have $S_i \in \mathcal{T}$ by Step 1. This contradicts $w_{\mathcal{T}}(s_i) + 1 = w_{\mathcal{T}}(s_{i+1})$ because $s_i \in S_i$ and $s_{i+1} \notin S_i$. □

Let $(\mathrm{P}_{\mathcal{C}'}^{\mathcal{T}})$ and $(\mathrm{D}_{\mathcal{C}'}^{\mathcal{T}})$ be the linear programs obtained by replacing w with $w_{\mathcal{T}}$ in $(\mathrm{P}_{\mathcal{C}'})$ and $(\mathrm{D}_{\mathcal{C}'})$, respectively. By Lemmas 6 and 7, we obtain the following lemma:

Lemma 8. *At any step of the algorithm, $y^{\mathcal{T}}$ is an optimal solution of $(\mathrm{D}_{\mathcal{C}'}^{\mathcal{T}})$ if and only if $\mathcal{L} = \mathcal{T}$ satisfies the conditions* (a), (b), (c), (d) *for every integers i, j with $2 \leq i \leq j \leq n$ and $i \equiv_2 j$.*

To show a stronger result than Lemma 8, we prepare the following useful lemma:

Lemma 9. *Suppose that $d_i < 0$ for some $2 \leq i \leq n-1$. Then we have $d_{i+1} \geq 0$.*

Proof. Suppose for a contradiction that $d_i < 0$ and $d_{i+1} < 0$ hold for some $2 \leq i \leq n-1$. Then we have

$$\begin{aligned} &f(S_{i-1}) + f(S_{i-2} \cup \{s_i\}) + f(S_{i-1} \cup \{s_{i+1}\}) \\ &< f(S_i) + f(S_{i-2}) + f(S_{i-1} \cup \{s_{i+1}\}) \\ &< f(S_{i-2}) + f(S_{i+1}) + f(S_{i-1}), \end{aligned}$$

which implies that $f(S_{i-2} \cup \{s_i\}) + f(S_{i-1} \cup \{s_{i+1}\}) < f(S_{i-2}) + f(S_{i+1})$. This contradicts Lemma 1. □

Let $\mathcal{T}_l = \{T \in \mathcal{C}' \mid y_T > 0\} \cup \{S_{i+1} \mid 0 \leq i \leq l,\ d_{i+2} < 0\}$ for $l = -1, 0, \ldots, n-2$. Since $\sum_{T\in\mathcal{C}'} y_T \chi_T = w$, $y_{S_{i-1}\cup\{s_{i+1}\}} > 0$ implies that $y_{S_i} > 0$ for $i = 1, \ldots, n-1$ because S_i is the unique set T in $\mathcal{C}'$ such that $s_i \in T$ and $s_{i+1} \notin T$ by Lemma 9. Hence, we have the following lemma:

Lemma 10. *$S_{i-1} \cup \{s_{i+1}\} \in \mathcal{T}_{-1}$ implies $S_i \in \mathcal{T}_{-1}$ for $i = 1, \ldots, n-1$.*

Lemmas 9 and 10 yield the following result:

Lemma 11. *For each* $l = -1, 0, \ldots, n-2$, $\mathcal{L} = \mathcal{T}_l$ *satisfies the conditions* (a),(b),(c),(d) *for every integers* i, j *with* $2 \leq i \leq j \leq n$ *and* $i \equiv_2 j$.

Proof. We show the statement by induction on l. The case when $l = -1$ has already been proved because $\mathcal{T}_{-1} = \{T \in \mathcal{C}' \mid y_T > 0\}$. Consider the case when $l \geq 0$. Suppose to the contrary that $\mathcal{L} = \mathcal{T}_l$ does not satisfy one of the conditions (a),(b),(c),(d) for some integers i, j with $2 \leq i \leq j \leq n$ and $i \equiv_2 j$. Consider the case when $\mathcal{T}_l$ does not satisfy (a) for such i, j. Since $\mathcal{T}_{l-1}$ satisfies (a) for such i, j by the induction hypothesis, we have $S_{l+1} = S_{i-2}$, $S_{k+1}, S_k \cup \{s_{k+2}\} \in \mathcal{T}_{l-1}$ for every $i \leq k \leq j$ with $k \equiv_4 i$, and $d_{l+2} = d_{i-1} < 0$ by Lemma 10. Hence, the assumption of the condition (c) holds if we replace i with $i+2$ and $\mathcal{L}$ with $\mathcal{T}_{l-1}$, which implies that

$$\sum_{i+2 \leq k \leq j, k \equiv_2 i} (-1)^{\frac{k-i}{2}} \cdot d_k \geq 0$$

because $\mathcal{T}_{l-1}$ satisfies (c) for $i+2, j$ by the induction hypothesis. We also have $d_i \geq 0$ by Lemma 9. Hence, we have

$$\sum_{i \leq k \leq j, k \equiv_2 i} (-1)^{\frac{k-i}{2}} \cdot d_k = d_i + \sum_{i+2 \leq k \leq j, k \equiv_2 i} (-1)^{\frac{k-i}{2}} \cdot d_k \geq 0,$$

which is a contradiction. Next, consider the case when $\mathcal{T}_l$ does not satisfy (b) for some i, j with $2 \leq i \leq j \leq n$ and $i \equiv_2 j$. Since $\mathcal{T}_{l-1}$ satisfies (b) for such i, j by the induction hypothesis, we have $S_{l+1} = S_j$, $S_{k+1}, S_k \cup \{s_{k+2}\} \in \mathcal{T}_{l-1}$ for every $i-2 \leq k \leq j-2$ with $k \equiv_4 i-2$, and $d_{l+2} = d_{j+1} < 0$ by Lemma 10. Hence, the assumption of the condition (c) holds if we replace j with $j-2$ and $\mathcal{L}$ with $\mathcal{T}_{l-1}$, which implies that

$$\sum_{i \leq k \leq j-2, k \equiv_2 i} (-1)^{\frac{k-i}{2}+1} \cdot d_k \geq 0$$

because $\mathcal{T}_{l-1}$ satisfies (c) for $i, j-2$ by the induction hypothesis. We also have $d_j \geq 0$ by Lemma 9. Hence, we have

$$\sum_{i \leq k \leq j, k \equiv_2 i} (-1)^{\frac{k-i}{2}+1} \cdot d_k = d_j + \sum_{i \leq k \leq j-2, k \equiv_2 i} (-1)^{\frac{k-i}{2}+1} \cdot d_k \geq 0,$$

which is a contradiction. Since $\mathcal{T}_{l-1}$ satisfies the condition (c) for every i, j with $2 \leq i \leq j \leq n$ and $i \equiv_2 j$ by the induction hypothesis, $\mathcal{T}_l$ also satisfies the condition (c) for every i, j with $2 \leq i \leq j \leq n$ and $i \equiv_2 j$ by Lemma 10. Hence, it suffices to consider the case when $\mathcal{T}_l$ does not satisfy (d) for some i, j with $2 \leq i \leq j \leq n$ and $i \equiv_2 j$. Since $\mathcal{T}_{l-1}$ satisfies (d) for such i, j by the induction hypothesis, we have $S_{l+1} = S_{i-2}$ or $S_{l+1} = S_j$ by Lemma 10. Consider the case when $S_{l+1} = S_{i-2}$. In this case, we have $S_j \in \mathcal{T}_{l-1}$, $S_{k+1}, S_k \cup \{s_{k+2}\} \in \mathcal{T}_{l-1}$ for every $i \leq k \leq j-4$ with $k \equiv_4 i$, and $d_{l+2} = d_{i-1} < 0$. Hence, the assumption of the condition (b) holds if we replace i with $i+2$ and $\mathcal{L}$ with $\mathcal{T}_{l-1}$, which implies that

$$\sum_{i+2 \le k \le j, k \equiv_2 i} (-1)^{\frac{k-i}{2}} \cdot d_k \ge 0$$

because $\mathcal{T}_{l-1}$ satisfies (b) for $i+2, j$ by the induction hypothesis. We also have $d_i \ge 0$ by Lemma 9. Hence, we have

$$\sum_{i \le k \le j, k \equiv_2 i} (-1)^{\frac{k-i}{2}} \cdot d_k = d_i + \sum_{i+2 \le k \le j, k \equiv_2 i} (-1)^{\frac{k-i}{2}} \cdot d_k \ge 0,$$

which is a contradiction. Finally, consider the case when $S_{l+1} = S_j$. In this case, we have $S_{i-2} \in \mathcal{T}_{l-1} \cup \{\emptyset\}$, $S_{k+1}, S_k \cup \{s_{k+2}\} \in \mathcal{T}_{l-1}$ for every $i \le k \le j-4$ with $k \equiv_4 i$, and $d_{l+2} = d_{j+1} < 0$. Hence, the assumption of the condition (a) holds if we replace j with $j-2$ and $\mathcal{L}$ with $\mathcal{T}_{l-1}$, which implies that

$$\sum_{i \le k \le j-2, k \equiv_2 i} (-1)^{\frac{k-i}{2}} \cdot d_k \ge 0$$

because $\mathcal{T}_{l-1}$ satisfies (a) for $i, j-2$ by the induction hypothesis. We also have $d_j \ge 0$ by Lemma 9. Hence, we have

$$\sum_{i \le k \le j, k \equiv_2 i} (-1)^{\frac{k-i}{2}} \cdot d_k = d_j + \sum_{i \le k \le j-2, k \equiv_2 i} (-1)^{\frac{k-i}{2}} \cdot d_k \ge 0,$$

which is a contradiction. □

By Lemmas 8 and 11, and Step 2 of the algorithm, we obtain the following result:

Lemma 12. *$y^{\mathcal{T}}$ is an optimal solution of $(\mathrm{D}^{\mathcal{T}}_{\mathcal{C}'})$ at any step of the algorithm.*

The following lemma guarantees that the vector x computed in Step 4 of the algorithm is unique:

Lemma 13. *For the resulting family $\mathcal{T}$ in* Step 4, *$\mathcal{A} = \{\chi_T \mid T \in \mathcal{T}\}$ contains n linearly independent vectors.*

Proof. Suppose to the contrary that the maximum number of linearly independent vectors in $\mathcal{A}$ is less than n. Since $y^{\mathcal{T}}$ is an optimal solution of $(\mathrm{D}^{\mathcal{T}}_{\mathcal{C}'})$ at Step 4 by Lemma 12, $(\mathrm{P}^{\mathcal{T}}_{\mathcal{C}'})$ has an extreme point optimal solution $x^{\mathcal{T}}$ such that $x^{\mathcal{T}}(T) = f(T)$ for all $T \in \mathcal{T}$ by the complementary slackness condition. Since $\mathcal{A}$ does not contain any set of n linearly independent vectors, $x^{\mathcal{T}}(T) = f(T)$ holds for some set $T \in \mathcal{C}' \setminus \mathcal{T}$. Let $\mathcal{T}' = \mathcal{T} \cup \{T\}$. Define $w_{\mathcal{T}'}$, $y^{\mathcal{T}'}$, and $(\mathrm{P}^{\mathcal{T}'}_{\mathcal{C}'})$ and $(\mathrm{D}^{\mathcal{T}'}_{\mathcal{C}'})$ in the same way as $w_{\mathcal{T}}$, $y^{\mathcal{T}}$, and $(\mathrm{P}^{\mathcal{T}}_{\mathcal{C}'})$ and $(\mathrm{D}^{\mathcal{T}}_{\mathcal{C}'})$, respectively. By the complementary slackness condition, $x^{\mathcal{T}}$ is an optimal solution of $(\mathrm{P}^{\mathcal{T}'}_{\mathcal{C}'})$, and $y^{\mathcal{T}'}$ is an optimal solution of $(\mathrm{D}^{\mathcal{T}'}_{\mathcal{C}'})$. Hence, $y^{\mathcal{T}'}$ is uncrossing-optimal by Lemma 6. Therefore, $\mathcal{T}'$ satisfies the conditions (a),(b),(c),(d) for every i, j with $2 \le i \le j \le n$ and $i \equiv_2 j$, which contradicts Step 2 of the algorithm. □

Let $(\mathrm{P}^{\mathcal{T}})$ and $(\mathrm{D}^{\mathcal{T}})$ be the linear programs obtained by replacing w with $w_{\mathcal{T}}$ in (P) and (D), respectively. We are now ready to prove the correctness of the algorithm.

Theorem 4. *The greedy algorithm computes an optimal solution x of* (P). *The number of arithmetic steps and function evaluations in the algorithm is $O(|S|^3)$.*

Proof. For the resulting family $\mathcal{T}$ in Step 4, define $y' \in \mathbf{R}_+^{2^S}$ as follows: $y'_T = y^{\mathcal{T}}_T$ if $T \in \mathcal{C}'$, and $y'_T = 0$ otherwise. Since $y^{\mathcal{T}}$ is an optimal solution of $(\mathrm{D}^{\mathcal{T}}_{\mathcal{C}'})$ at Step 4 by Lemma 12, y' is also an optimal solution of $(\mathrm{D}^{\mathcal{T}})$ by Corollary 2 and Lemma 4. Hence, $(\mathrm{P}^{\mathcal{T}})$ has an optimal solution x' such that $x'(T) = f(T)$ for every $T \in \mathcal{T}$ by the complementary slackness condition. This implies that $x' = x$ by Lemma 13. Hence, x is a feasible solution of (P). Since $x(T) = f(T)$ holds for every $T \in \mathcal{C}'$ such that $y_T > 0$, x is an optimal solution of (P) by the complementary slackness condition.

Consider the time complexity of the algorithm. At Step 2, we can check whether $\mathcal{L} = \mathcal{T} \cup \{T_l\}$ satisfies the conditions (a),(b),(c),(d) for each i, j by using the result for the case of i and $j-4$ (if $i+4 \le j$). Hence, the number of arithmetic steps and function evaluations in Step 2 is $O(1)$ for each l, i, j. Therefore, the total number of arithmetic steps and function evaluations in Step 2 is $O(|S|^3)$. Since the time complexity of Gaussian elimination in Step 4 is $O(|S|^3)$, the total number of arithmetic steps and function evaluations in the algorithm is $O(|S|^3)$. □

By Theorems 3 and 4, we finally obtain the following result:

Theorem 5. *Suppose that f and w are rational. Then an optimal solution of* (P) *can be computed in time $O(|S|^3)$.*

Note that it still requires the ellipsoid method to minimize 2/3-submodular functions via Theorem 5 because the reduction from the separation problem to the optimization problem requires the ellipsoid method.

4 An Application to Coloring

Let S be a finite set. A function $g : 2^S \to \mathbf{R} \cup \{-\infty\}$ is called *intersecting 2/3-supermodular* if for every distinct three sets $T_1, T_2, T_3 \subseteq S$ satisfying $T_1 \cap T_2 \cap T_3 \neq \emptyset$, the supermodular inequality (the converse of the submodular inequality) holds for at least two pairs out of three pairs generating from T_1, T_2, T_3. A family $\mathcal{L} \subseteq 2^S$ is called a *g-laminar family* if at least one of the following two conditions holds for every $T_1, T_2 \in \mathcal{L}$:

- At least one of $T_1 \setminus T_2, T_2 \setminus T_1, T_1 \cap T_2$ is the empty set.
- $g(T_1) + g(T_2) \le g(T_1 \cup T_2) + g(T_1 \cap T_2)$ holds.

Define $D_g(T) = \max\{|T \cap U| \mid U = \emptyset, \text{ or } g(U) > -\infty \text{ and } T \not\subseteq U \not\subseteq T\}$ for each $T \subseteq S$. The following coloring theorem is a supermodular extension of Vizing's edge-coloring theorem [9], and an analogue to the supermodular coloring theorem [11].

Theorem 6 ([9]). *Let $g : 2^S \to \mathbf{Z} \cup \{-\infty\}$ be an intersecting 2/3-supermodular function. For a positive integer k, suppose that $\mathcal{L} = \{T \subseteq S \mid g(T) + D_g(T) > k\}$ is a g-laminar family and $\min\{|T|, k\} \geq g(T)$ holds for every $T \subseteq S$. Then there exists an assignment of colors $\pi : S \to [k]$ such that $|\pi(T)| \geq g(T)$ holds for every $T \subseteq S$.*

Here, we denote $[k] = \{1, 2, \ldots, k\}$ and $\pi(T) = \{\pi(s) \mid s \in T\}$. Theorem 6 includes Vizing's edge-coloring theorem [14,15] as a special case, which states that the chromatic index of a simple graph $G = (V, E)$ is at most $\Delta(G) + 1$, where $\Delta(G)$ is the maximum degree of G. Indeed, the following case corresponds to Vizing's theorem: $S = E$, $k = \Delta(G) + 1$, and $g(T) = |T|$ if T is a set of all the edges incident with a particular vertex, and $g(T) = -\infty$ otherwise. In this case, π will be a proper edge-coloring of G with $\Delta(G) + 1$ colors.

Although the proof of Theorem 6 in [9] is constructive, it does not, on its own, yield a polynomial-time algorithm to obtain the desired coloring. We show that in the case when g is finite, the coloring can be obtained in polynomial-time with the aid of Theorem 1. The proof of Theorem 6 in [9] provides a "sequential recoloring" algorithm to obtain the desired coloring. In this algorithm, we start with the empty coloring $\pi : \emptyset \to [k]$. For the current coloring $\pi : S' \to [k]$ of a subset $S' \subseteq S$ satisfying

$$|T \setminus S'| + |\pi(T \cap S')| \geq g(T) \tag{1}$$

for every $T \subseteq S$, take $s \in S \setminus S'$ and let $\mathcal{F}_s = \{T \subseteq S \mid s \in T\}$. Then we find maximal sets $T^* \in \mathcal{F}_s$ satisfying the inequality (1) with equality by minimizing the function $f : \mathcal{F}_s \to \mathbf{Z}$ defined as follows:

$$f(T) = 2|S| \cdot (|T \setminus S'| + |\pi(T \cap S')| - g(T)) - |T| \quad (T \in \mathcal{F}_s).$$

Then we assign a proper color to s and carefully update the colors of elements in S' so that T^* and other subsets satisfy the inequality (1) for the updated color assignment. By repeating this procedure, we finally obtain the desired coloring in Theorem 6. Given an evaluation oracle of g and a membership oracle of $\mathcal{L}$, this algorithm runs in polynomial-time if a minimizer of f can be found in polynomial-time. Since g is intersecting 2/3-supermodular, f is 2/3-submodular if we regard $\mathcal{F}_s$ as $2^{S \setminus \{s\}}$. Hence, f can be minimized in polynomial-time by Theorem 1. Thus, the algorithm runs in polynomial-time, which implies that the coloring in Theorem 6 can be obtained in polynomial-time if g is finite.

References

1. Bárász, M.: Matroid intersection for the min-rank oracle. Technical report QP-2006-03, Egerváry Research Group (2006)
2. Bérczi, K., Frank, A.: Variations for Lovász' submodular ideas. In: Grötschel, M., Katona, G.O.H., Sági, G. (eds.) Building Bridges. Bolyai Society Mathematical Studies, vol. 19, pp. 137–164. Springer, Heidelberg (2008). https://doi.org/10.1007/978-3-540-85221-6_4

3. Bérczi, K., Király, T., Yamaguchi, Y., Yokoi, Y.: Matroid intersection under restricted oracles. SIAM J. Discret. Math. **37**, 1311–1330 (2023)
4. Chakrabarty, D., Lee, Y.T., Sidford, A., Wong, S.C.: Subquadratic submodular function minimization. In: Proceedings of the 49th Annual ACM SIGACT Symposium on Theory of Computing, pp. 1220–1231 (2017)
5. Grötschel, M., Lovász, L., Schrijver, A.: The ellipsoid method and its consequences in combinatorial optimization. Combinatorica **1**, 169–197 (1981)
6. Iwata, S., Fleischer, L., Fujishige, S.: A combinatorial strongly polynomial algorithm for minimizing submodular functions. J. ACM **48**, 761–777 (2001)
7. Iwata, S., Orlin, J.B.: A simple combinatorial algorithm for submodular function minimization. In: Proceedings of the 20th Annual ACM-SIAM Symposium on Discrete Algorithms, pp. 1230–1237 (2009)
8. Lee, Y.T., Sidford, A., Wong, S.C.: A faster cutting plane method and its implications for combinatorial and convex optimization. In: Proceedings of the 2015 IEEE 56th Annual Symposium on Foundations of Computer Science, pp. 1049–1065 (2015)
9. Mizutani, R.: Supermodular extension of Vizing's edge-coloring theorem. arXiv preprint arXiv:2211.07150 (2022)
10. Orlin, J.B.: A faster strongly polynomial time algorithm for submodular function minimization. Math. Program. **118**, 237–251 (2009)
11. Schrijver, A.: Supermodular colourings. In: Lovász, L., Recski, A. (eds.) Matroid Theory, pp. 327–343. North-Holland (1985)
12. Schrijver, A.: A combinatorial algorithm minimizing submodular functions in strongly polynomial time. J. Combin. Theory Ser. B **80**, 346–355 (2000)
13. Tardos, É.: A strongly polynomial algorithm to solve combinatorial linear programs. Oper. Res. **34**, 250–256 (1986)
14. Vizing, V.G.: On an estimate of the chromatic class of a p-graph. Diskretny Analiz **3**, 25–30 (1964). In Russian
15. Vizing, V.G.: The chromatic class of a multigraph. Cybernetics **1**, 32–41 (1965)

A 4/3-Approximation for the Maximum Leaf Spanning Arborescence Problem in DAGs

Meike Neuwohner(✉)

Research Institute for Discrete Mathematics, University of Bonn, Lennéstr. 2, 53113 Bonn, Germany
work@meike-neuwohner.de

Abstract. The Maximum Leaf Spanning Arborescence problem (MLSA) is defined as follows: Given a directed graph G and a vertex $r \in V(G)$ from which every other vertex is reachable, find a spanning arborescence rooted at r maximizing the number of leaves (vertices with out-degree zero). The MLSA has applications in broadcasting, where a message needs to be transferred from a source vertex to all other vertices along the arcs of an arborescence in a given network. In doing so, it is desirable to have as many vertices as possible that only need to receive, but not pass on messages since they are inherently cheaper to build.

We study polynomial-time approximation algorithms for the MLSA. For general digraphs, the state-of-the-art is a $\min\{\sqrt{\mathrm{OPT}}, 92\}$-approximation [4,5]. In the (still APX-hard) special case where the input graph is acyclic, the best known approximation guarantee of $\frac{7}{5}$ is due to Fernandes and Lintzmayer [8]: They prove that any α-approximation for the *hereditary* 2-3-*Set Packing problem*, a special case of weighted 3-Set Packing, yields a $\max\{\frac{4}{3}, \alpha\}$-approximation for the MLSA in acyclic digraphs (dags), and provide a $\frac{7}{5}$-approximation for the hereditary 2-3-Set Packing problem.

In this paper, we obtain a $\frac{4}{3}$-approximation for the hereditary 2-3-Set Packing problem, and, thus, also for the MLSA in dags. In doing so, we manage to leverage the full potential of the reduction provided by Fernandes and Lintzmayer. The algorithm that we study is a simple local search procedure considering swaps of size up to 10. Its analysis relies on a two-stage charging argument.

Keywords: Maximum Leaf Spanning Arborescence problem · Weighted set packing · Local search

J. Vygen and J. Byrka (Eds.): IPCO 2024, LNCS 14679, pp. 337–350, 2024.
https://doi.org/10.1007/978-3-031-59835-7_25

1 Introduction

The Maximum Leaf Spanning Arborescence problem (MLSA) is defined as follows:

Definition 1 (Maximum Leaf Spanning Arborescence problem).

Input: *A directed graph* G, $r \in V(G)$ *such that every vertex of* G *is reachable from* r.
Task: *Find a spanning* r*-arborescence with the maximum number of leaves possible.*

It plays an important role in the context of broadcasting: Given a network consisting of a set of nodes containing one distinguished source and a set of available arcs, a message needs to be transferred from the source to all other nodes along a subset of the arcs, which forms (the edge set of) an arborescence rooted at the source. As internal nodes do not only need to be able to receive, but also to re-distribute messages, they are more expensive. Hence, it is desirable to have as few of them as possible, or equivalently, to maximize the number of leaves (nodes with out-degree zero).

The special case of the MLSA where every arc may be used in both directions is called the *Maximum Leaves Spanning Tree problem (MLST)*. In this setting, the complementary task of minimizing the number of non-leaves is equivalent to the *Minimum Connected Dominating Set problem (MCDS)*. Both the MLST and the MCDS are NP-hard, even if the input graph is 4-regular or planar with maximum degree at most 4 (see [10], problem ND2). The MLST has been shown to be APX-hard [9][1], even when restricted to cubic graphs [3]. The state-of-the-art for the MLST is an approximation guarantee of 2 [17].

While an optimum solution to the MLST gives rise to an optimum solution to the MCDS and vice versa, the MCDS turns out to be much harder to approximate: Ruan et al. [15] have obtained an $\ln \Delta + 2$-approximation, where Δ denotes the maximum degree in the graph. A reduction from Set Cover (with bounded set sizes) further shows that unless P = NP, the MCDS is hard to approximate within a factor of $\ln \Delta - \mathcal{O}(\ln \ln \Delta)$ [11,19]. An analogous reduction further yields the same hardness result for the problem of computing a spanning arborescence with the minimum number of non-leaves in a rooted acyclic digraph of maximum out-degree Δ.

In this paper, we study polynomial-time approximation algorithms for (special cases of) the MLSA. For general digraphs, the best that is known is a $\min\{\sqrt{\text{OPT}}, 92\}$-approximation [4,5]. Moreover, there is a line of research focusing on FPT-algorithms for the MLSA [1,2,4].

The special case where the graph G is assumed to be a dag (directed acyclic graph) has been proven to be APX-hard by Schwartges, Spoerhase and Wolff [16]. They further provided a 2-approximation, which was then improved to $\frac{3}{2}$ by Fernandes and Lintzmayer [7]. Recently, the latter authors managed to enhance

[1] Note that MaxSNP-hardness implies APX-hardness, see [13].

their approach to obtain a $\frac{7}{5}$-approximation [8], which has been unchallenged so far. In this paper, following the approach by Fernandes and Lintzmayer, we improve on these results and obtain a $\frac{4}{3}$-approximation for the MLSA in dags.

Fernandes and Lintzmayer [8] tackle the MLSA in dags by reducing it, up to an approximation guarantee of $\frac{4}{3}$, to a special case of the weighted 3-Set Packing problem, which we call the *hereditary* 2-3-*Set Packing problem.* Fernandes and Lintzmayer [8] prove it to be NP-hard via a reduction from 3-Dimensional Matching [12].

Definition 2 (weighted k-Set Packing problem).

Input: *A family $\mathcal{S}$ of sets, each of cardinality at most k, $w : \mathcal{S} \to \mathbb{Q}_{>0}$*
Task: *Compute a disjoint sub-collection $A \subseteq \mathcal{S}$ maximizing the total weight $w(A) := \sum_{s\in A} w(s)$.*

Definition 3 (hereditary 2-3-Set Packing). *An instance of the* hereditary 2-3-Set Packing problem *is an instance $(\mathcal{S}, w)$ of the weighted* 3-*Set Packing problem, where*

- $w(s) = |s| - 1$ *for all* $s \in \mathcal{S}$, *and*
- *for every* $s \in \mathcal{S}$ *with* $|s| = 3$, *all two-element subsets of* s *are contained in* $\mathcal{S}$.

Note that by the assumption that weights are strictly positive, an instance of the hereditary 2-3-Set Packing problem features only sets of cardinality 2 or 3.

Theorem 1 ([8]). *Let $\alpha \geq 1$ and assume that there is a polynomial-time α-approximation algorithm for the hereditary* 2-3-*Set Packing problem. Then there exists a polynomial-time* $\max\{\alpha, \frac{4}{3}\}$*-approximation for the MLSA in dags.*

For $k \leq 2$, the weighted k-Set Packing problem can be solved in polynomial time via a reduction to the Maximum Weight Matching problem [6]. In contrast, for $k \geq 3$, even the special case where $w \equiv 1$, the *unweighted k-Set Packing problem*, is NP-hard because it generalizes the 3-Dimensional Matching problem [12]. The technique that has proven most successful in designing approximation algorithms for both the weighted and the unweighted k-Set Packing is *local search.* Given a feasible solution A, we call a collection X of pairwise disjoint sets a *local improvement of A* if $w(X) > w(N(X, A))$, where

$$N(X, A) := \{a \in A : \exists x \in X : a \cap x \neq \emptyset\}$$

is the *neighborhood of X in A*. Note that $N(X, A)$ comprises precisely those sets that we need to remove from A in order to be able to add the sets in X.

The state-of-the-art is a $\min\{\frac{k+1-\tau_k}{2}, 0.4986\cdot(k+1)+0.0208\}$-approximation for the weighted k-Set Packing problem, where $\tau_k \geq 0.428$ for $k \geq 3$ and $\lim_{k\to\infty} \tau_k = \frac{2}{3}$ [14,18]. Note that the guarantee of 1.786 for $k = 3$ is worse than the guarantee of $\frac{7}{5}$ that Fernandes and Lintzmayer achieve for the hereditary 2-3-Set Packing problem.

In order to obtain the approximation guarantee of $\frac{7}{5}$, Fernandes and Lintzmayer perform local search with respect to a modified weight function. In addition to certain improvements of constant size, they incorporate another, more involved class of local improvements that are related to alternating paths in a certain auxiliary graph. This makes the analysis more complicated because in addition to charging arguments similar to ours, more intricate considerations regarding the structure of the auxiliary graph are required.

In this paper, we study a local search algorithm that considers local improvements consisting of up to 10 sets with respect to an objective that first maximizes the weight of the current solution, and second the number of sets of weight 2 that are contained in it. We show that this algorithm yields a polynomial-time $\frac{4}{3}$-approximation for the hereditary 2-3-Set Packing problem. In particular, this results in a polynomial-time $\frac{4}{3}$-approximation for the MLSA in dags. In doing so, we manage to tap the full potential of Theorem 1. Moreover, this work serves as a starting point in identifying, understanding, and exploiting structural properties of set packing instances that arise naturally from other combinatorial problems. Studying these instance classes may ultimately turn reductions to set packing instances into a more powerful tool in the design of approximation algorithms.

2 A 4/3-Approximation for the Hereditary 2-3-Set Packing Problem

In this section, we present a polynomial-time $\frac{4}{3}$-approximation for the hereditary 2-3-Set Packing problem. In order to define it, we formally introduce the notion of local improvement that we consider. It aims at maximizing first the weight of the solution we find, and second the number of sets of weight 2 contained in it. We first recap the notion of neighborhood from the introduction.

Definition 4 (neighborhood). *Let U and W be two set families. We define the* neighborhood *of U in W to be*

$$N(U, W) := \{w \in W : \exists u \in U : u \cap w \neq \emptyset\}.$$

Moreover, for a single set u, we write $N(u, W) := N(\{u\}, W)$.

Now, we can define the notion of local improvement we would like to consider.

Definition 5 (local improvement). *Let $(\mathcal{S}, w)$ be an instance of the hereditary 2-3-Set Packing problem and let A be a feasible solution. We call a disjoint set collection $X \subseteq \mathcal{S}$ a* local improvement *of A of size $|X|$ if*

- $w(X) > w(N(X, A))$ *or*
- $w(X) = w(N(X, A))$ *and X contains more sets of weight 2 than $N(X, A)$.*

We analyze Algorithm 1, which starts with the empty solution and iteratively searches for a local improvement of size at most 10 (and performs the respective swap) until no more exists. We first observe that it runs in polynomial time.

Algorithm 1: 4/3-approximation for hereditary 2-3-Set Packing

Input: an instance $(\mathcal{S}, w)$ of the hereditary 2-3-Set Packing problem
Output: a disjoint sub-collection of $\mathcal{S}$

```
A ← ∅
while ∃ local improvement X of A of size at most 10 do
    A ← (A \ N(X, A)) ∪ X
return A
```

Proposition 1. *Algorithm 1 can be implemented to run in polynomial time.*

Proof. A single iteration can be performed in polynomial time via brute-force enumeration. Thus, it remains to bound the number of iterations. By our definition of a local improvement, $w(A)$ can never decrease throughout the algorithm. Initially, we have $w(A) = 0$, and moreover, $w(A) \leq w(\mathcal{S}) \leq 2 \cdot |\mathcal{S}|$ holds throughout. As all weights are integral, we can infer that there are at most $2 \cdot |\mathcal{S}|$ iterations in which $w(A)$ strictly increases. In between two consecutive such iterations, there can be at most $|\mathcal{S}|$ iterations in which $w(A)$ remains constant since the number of sets of weight 2 in A strictly increases in each such iteration. All in all, we can bound the total number of iterations by $\mathcal{O}(|\mathcal{S}|^2)$.

The remainder of this section is dedicated to the proof of Theorem 2, which implies that Algorithm 1 constitutes a $\frac{4}{3}$-approximation for the hereditary 2-3-Set Packing problem.

Theorem 2. *Let $(\mathcal{S}, w)$ be an instance of the hereditary* 2-3-*Set Packing problem and let $A \subseteq \mathcal{S}$ be a feasible solution such that there is no local improvement of A of size at most* 10. *Let further $B \subseteq \mathcal{S}$ be an optimum solution. Then $w(B) \leq \frac{4}{3} \cdot w(A)$.*

Let $\mathcal{S}$, w, A and B be as in the statement of the theorem. Our goal is to distribute the weights of the sets in B among the sets in A they intersect in such a way that no set in A receives more than $\frac{4}{3}$ times its own weight. We remark that each set in B must intersect at least one set in A because otherwise, it would constitute a local improvement of size 1.

In order to present our weight distribution, we introduce the notion of the *conflict graph*, which allows us to phrase our analysis using graph terminology. A similar construction is used in [8].

Definition 6 (conflict graph). *The conflict graph G is defined as follows: Its vertex set is the disjoint union of A and B, i.e. $V(G) = A \dot{\cup} B$. Its edge set is obtained by adding, for each pair $(a, b) \in A \times B$, $|a \cap b|$ parallel edges connecting a to b.*

See Fig. 1 for an illustration. We remark that for $X \subseteq B$, $N(X, A)$ agrees with the (graph) neighborhood of X in the bipartite graph G. Analogously, for $Y \subseteq A$,

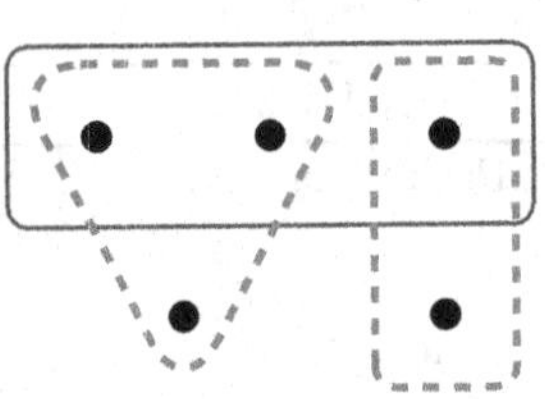

(a) The figure displays two collections A (blue, solid) and B (red, dashed) consisting of pairwise disjoint sets of cardinality 2 or 3. Black dots represent set elements.

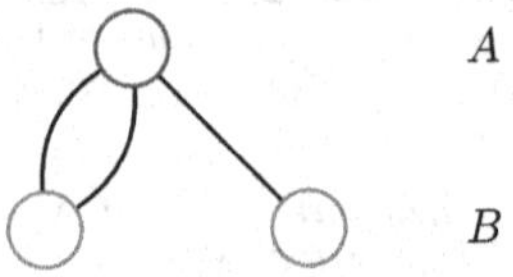

(b) The figure shows the conflict graph of $A\dot{\cup}B$. Vertices from A are drawn in blue at the top, vertices from B are drawn in red at the bottom.

Fig. 1. Construction of the conflict graph. (Color figure online)

$N(Y, B)$ equals the neighborhood of Y in G. In the following, we will simultaneously interpret sets from $A\dot{\cup}B$ as the corresponding vertices in G and talk about their degree, their incident edges and their neighbors. We make the following observation.

Proposition 2. *Let $v \in V(G)$ correspond to the set $s \in A \cup B$. Then v has at most $|s|$ incident edges in G.*

Proof. As A and B both consist of pairwise disjoint sets, each element of s can induce at most one incident edge of s.

2.1 Step 1 of the Weight Distribution

Our weight distribution proceeds in two steps. The first step works as follows:

Definition 7 (Step 1 of the weight distribution). *Let B_1 consist of all sets $v \in B$ with* exactly one neighbor *in A. Each $v \in B_1$ sends its full weight to its unique neighbor in A.*

Let further B_2 consist of those $v \in B$ with $w(v) = 2$ and exactly two incident edges, *with the additional property that they connect to* two distinct sets *from A. Each $v \in B_2$ sends half of its weight (i.e. 1) along each of its edges.*

See Fig. 2 for an illustration. Observe that in the first stage, $u \in A$ receives weight precisely from the sets in $N(u, B_1 \cup B_2)$.

We first prove Lemma 1, which tells us that we can represent the total amount of weight a collection $U \subseteq A$ receives in the first step as the weight of a disjoint set collection X with $N(X, A) \subseteq U$. The construction of X will allow us to combine X with sub-collections of $B \setminus (B_1 \cup B_2)$ to obtain local improvements.

Lemma 1. *Let $U \subseteq A$. There is $X \subseteq S$ with the following properties:*

(1.1) $N(X, A) \subseteq U$.

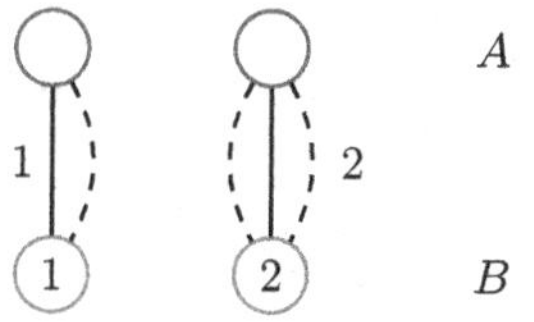

(a) Every set in B_1 sends its whole weight to its unique neighbor in A (to which it may be connected via multiple edges).

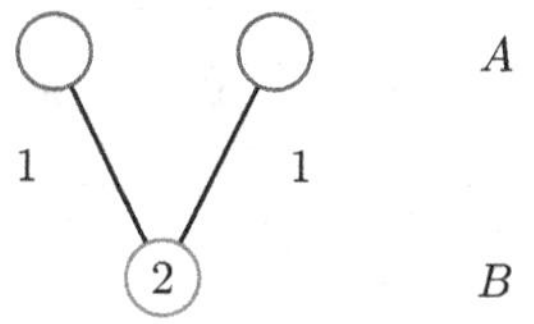

(b) Every set in B_2 sends one unit of weight to each of its neighbors in A.

Fig. 2. The first step of the weight distribution.

(1.2) $w(X)$ *equals the total amount of weight that* U *receives in the first step.*
(1.3) There is a bijection $N(U, B_1 \cup B_2) \leftrightarrow X$ *mapping* $v \in B_1 \cup B_2$ *to itself or to one of its two-element subsets.*

Proof. We obtain X as follows: We start with $X = \emptyset$ and first add those sets in $N(U, B_1 \cup B_2)$ to X that send all of their weight to U (i.e. whose neighborhood in A is contained in U). This includes all sets in $N(U, B_1)$. Second, for each set $v \in B_2$ that has one incident edge to $u \in U$ and one incident edge to $r \in A \setminus U$, we add its two-element subset $v \setminus r$ to X. By construction, (1.1)–(1.3) hold. See Fig. 3 for an illustration.

Corollary 1. *No set in* A *receives more than its own weight in the first step.*

Proof. Assume towards a contradiction that $u \in A$ receives more than $w(u)$ in the first step. Apply Lemma 1 with $U = \{u\}$ to obtain a collection $X \subseteq \mathcal{S}$ subject to (1.1)–(1.3). Then $w(X) > w(u) = w(N(X, A))$ and (1.3) and Proposition 2 imply that X is a disjoint set family with $|X| \leq 3$. Hence, X constitutes a local improvement of size at most $3 < 10$. This contradicts our assumption that there is no local improvement of A of size at most 10.

2.2 Removing "Covered" Sets

Definition 8. *Let* C *consist of those sets from* A *that receive exactly their own weights in the first step.*

The intuitive idea behind our analysis is that the sets in C are "covered" by the sets sending weight to them in the sense of Lemma 1. Hence, we can "remove" the sets in C from our current solution A and the sets in $B_1 \cup B_2$ from our optimum solution B. If we can find a local improvement in the remaining instance, we will use Lemma 1 to transform it into a local improvement in the original instance, leading to a contradiction. See Lemma 2 for an example of how to apply this reasoning. But under the assumption that no local improvement in the remaining instance exists, we can design the second step of the weight distribution in such a way that overall, no set in A receives more than $\frac{4}{3}$ times its own weight.

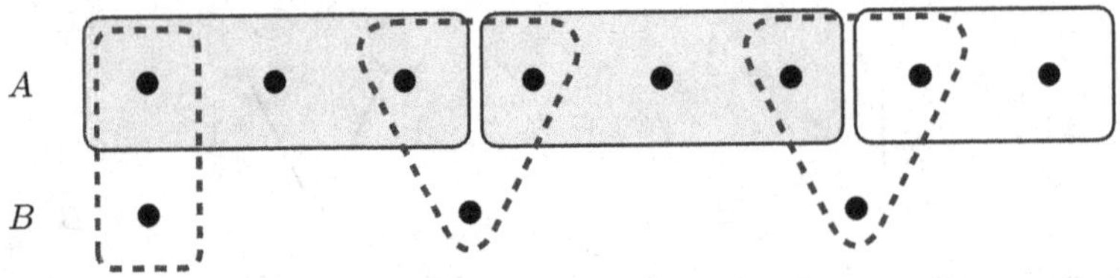

(a) The left red set is contained in B_1 and sends its whole weight to the unique set from A it intersects. The two triangular red sets are contained in B_2. The left one only intersects sets in A that are contained in U, whereas the right one also intersects a set in $A \setminus U$.

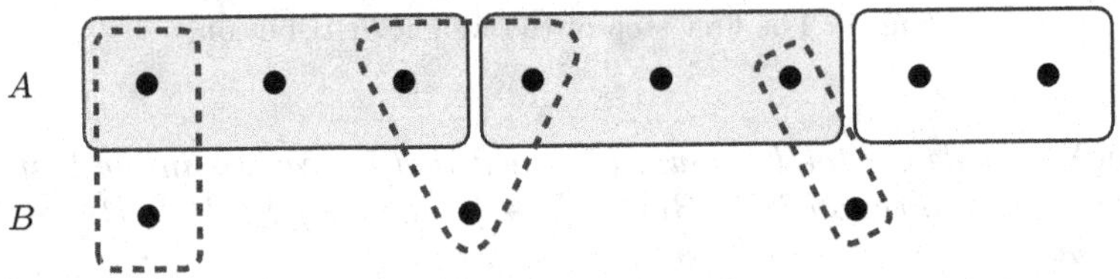

(b) The set collection X (red, dashed) we construct in the proof of Lemma 1 contains the left and the middle red set because they send all of their weight to U. For the right triangular set, we remove the element in which it intersects a set from $A \setminus U$. Then, we add the resulting set of cardinality 2 to X.

Fig. 3. Illustration of the construction in the proof of Lemma 1. Figure 3(a) shows a collection $U \subseteq A$ of sets (blue, filled,solid), the collection $N(U, B_1 \cup B_2)$ (red, dashed) of sets the sets in U receive weight from in the first step, and further sets from A (blue, not filled, solid) the sets in $N(U, B_1 \cup B_2)$ send weight to. Figure 3(b) illustrates the construction of the set collection X. (Color figure online)

2.3 Step 2 of the Weight Distribution

In order to define the second step of the weight distribution, we make the following observations:

Lemma 2. *There is no $v \in B \setminus (B_1 \cup B_2)$ with $w(N(v, A \setminus C)) < w(v)$.*

Proof. Assume towards a contradiction that there is $v \in B \setminus (B_1 \cup B_2)$ with $w(N(v, A \setminus C)) < w(v)$. Apply Lemma 1 to $U := N(v, C)$ to obtain X subject to (1.1)–(1.3). By (1.3), $X \dot{\cup} \{v\}$ consists of pairwise disjoint sets. Proposition 2 further yields $|N(v, C)| \leq |v| \leq 3$, and, thus, $|X| = |N(N(v, C), B_1 \cup B_2)| \leq 9$ by (1.3). Finally, $w(X) = w(N(v, C))$ by (1.2) and since sets from C receive exactly their own weights in the first step. Hence, (1.1) yields

$$w(X \cup \{v\}) = w(X) + w(v) > w(N(v, C)) + w(N(v, A \setminus C)) = w(N(X \cup \{v\}, A)).$$

So $X \cup \{v\}$ is a local improvement of A of size at most 10, a contradiction.

Proposition 3. *Let $v \in B \setminus (B_1 \cup B_2)$. Then*

(i) v has at least one neighbor in $A \setminus C$.

(ii) If $w(v) = 1$, then v has exactly two neighbors in A.
(iii) If $w(v) = 2$, then v has three incident edges.

Proof. (i) follows from Lemma 2. For (ii) and (iii), we remind ourselves that each $v \in B$ has at most $|v|$ neighbors/incident edges, but at least 1 neighbor in A by Proposition 2 and since $\{v\}$ would constitute a local improvement otherwise. (ii) holds since $v \in B_1$ otherwise. For (iii), we observe that in case v has at most 2 incident edges, then either v has only one neighbor in A, or two distinct neighbors to which it is connected by a single edge each. In either case, we have $v \in B_1 \cup B_2$.

Definition 9 (Step 2 of the weight distribution). *Let $v \in B \setminus (B_1 \cup B_2)$ with $w(v) = 1$.*

(a) If v has a neighbor in C, then this neighbor receives $\frac{1}{3}$ and the neighbor in $A \setminus C$ receives $\frac{2}{3}$.
(b) Otherwise, both neighbors in $A \setminus C$ receive $\frac{1}{2}$.

Now, let $v \in B \setminus (B_1 \cup B_2)$ with $w(v) = 2$.

(c) If v has degree 1 to $A \setminus C$, then v sends $\frac{1}{3}$ along each edge to C and $\frac{4}{3}$ to the neighbor in $A \setminus C$. Note that this neighbor must have a weight of 2 by Lemma 2.
(d) If v has degree 2 to $A \setminus C$, v sends 1 along each edge to a vertex in $A \setminus C$ of weight 2, $\frac{2}{3}$ along each edge to a vertex in $A \setminus C$ of weight 1, and the remaining amount to the neighbor in C.
(e) If all three incident edges of v connect to $A \setminus C$, then v sends $\frac{2}{3}$ along each of these edges.

We denote the set of vertices to which case ℓ with $\ell \in \{a, b, c, d, e\}$ applies by B_ℓ.

See Fig. 4 for an illustration.

2.4 No Set in C Receives More Than 4/3 Times Its Weight

Lemma 3. *Let $v \in B_d$ and let $u \in N(v, C)$ be the unique neighbor of v in C. If u receives more than $\frac{1}{3}$ from v, then $w(u) = 2$ and u has exactly one incident edge to $B \setminus (B_1 \cup B_2)$.*

Proof. Denote the endpoints of the two edges connecting v to $A \setminus C$ by u_1 and u_2. Assume u receives more than $\frac{1}{3}$ from v. Then $w(u_1) = w(u_2) = 1$. In particular, u_1 and u_2 are distinct by Lemma 2. Apply Lemma 1 to $U := \{u\}$ to obtain X subject to (1.1)–(1.3). Then by (1.3), $Y := X \dot\cup \{v\}$ is a disjoint collection of sets. Moreover, Proposition 2 yields

$$|X| \overset{(1.3)}{=} |N(u, B_1 \cup B_2)| \leq |u| \leq 3.$$

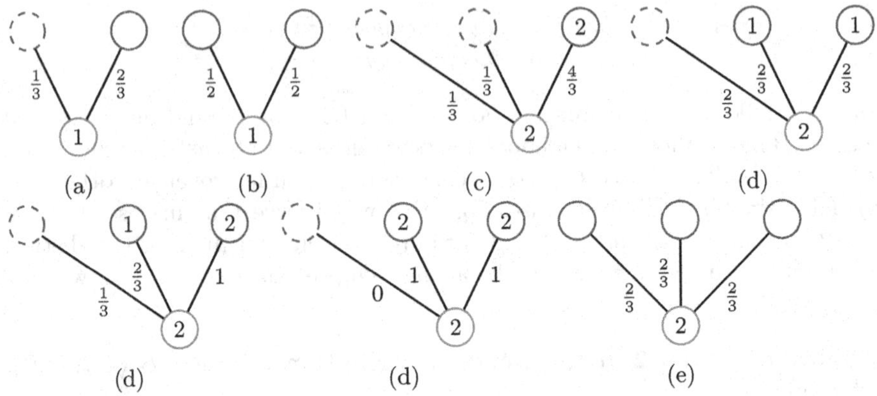

Fig. 4. Illustration of the second step of the weight distribution. Blue circles in the top row indicate sets from A, if they are dashed, the corresponding set is contained in C. Red circles in the bottom row indicate sets from $B \setminus (B_1 \cup B_2)$. The number within a circle indicates the weight of the corresponding set in case it is relevant. Even though drawn as individual circles, the endpoints in A of the incident edges of a set $v \in B \setminus (B_1 \cup B_2)$ need not be distinct. For example, in (e), two of the sets represented by the blue circles may agree, in which case the corresponding set receives $\frac{4}{3}$. (Color figure online)

Hence, $|Y| \leq 4$. By (1.2) and as $u \in C$ receives its own weight in the first step, we get $w(u) = w(X)$. Thus, $w(u_1) + w(u_2) = 1 + 1 = 2 = w(v)$ results in

$$w(N(Y, A)) \stackrel{(1.1)}{=} w(u) + w(u_1) + w(u_2) = w(X) + w(v) = w(Y).$$

As Y does not constitute a local improvement, $N(Y, A) = \{u_1, u_2, u\}$ contains at least as many vertices of weight 2 as Y. As $w(u_1) = w(u_2) = 1$, but $w(v) = 2$, this implies that $w(u) = 2$ and that all elements of X have a weight of 1. By (1.2), this implies $|X| = 2$, and by (1.3), u intersects sets from $B_1 \cup B_2$ in at least two distinct elements in total. In particular, $\{u, v\}$ is the only edge connecting u to $B \setminus (B_1 \cup B_2)$ by Proposition 2.

Lemma 4. *Each set in C receives at most $\frac{4}{3}$ times its own weight during our weight distribution.*

Proof. First, let $u \in C$ with $w(u) = 1$. Then u receives 1 in the first step and has at most one incident edge to $B \setminus (B_1 \cup B_2)$. Via this edge, u receives at most $\frac{1}{3}$, which is clear for the cases (a) and (c), and follows from Lemma 3 for case (d). Thus, u receives at most $\frac{4}{3} = \frac{4}{3} \cdot w(u)$ in total.

Next, let $u \in C$ with $w(u) = 2$. Then u receives 2 in the first step and u has at most two incident edges to $B \setminus (B_1 \cup B_2)$. If u has two incident edges to $B \setminus (B_1 \cup B_2)$, then u can receive at most $\frac{1}{3}$ via each of them: This is clear for the cases (a) and (c), and follows from Lemma 3 for case (d). Thus, u receives at most $\frac{8}{3} = \frac{4}{3} \cdot w(u)$ in total. If u has one incident edge to $B \setminus (B_1 \cup B_2)$, then the maximum amount u can receive via this edge is $\frac{2}{3}$. Again, u receives at most $\frac{8}{3}$ in total.

2.5 No Set in $A \setminus C$ Receives More than 4/3 Times Its Weight

In order to make sure that no vertex from $A \setminus C$ receives more than $\frac{4}{3}$ times its own weight, we need Lemma 5, which essentially states the following:

- If a vertex $u \in A \setminus C$ with $w(u) = 2$ receives $\frac{4}{3}$ from a vertex in B_c, then it does not receive weight from any further vertex in $B_1 \cup B_2 \cup B_c \cup B_d$.
- A vertex $u \in A \setminus C$ with $w(u) = 2$ may, in total, receive at most 2 units of weight from vertices in $B_1 \cup B_2 \cup B_d$.

Lemma 5. *Let $u \in A \setminus C$ with $w(u) = 2$. Denote the set of vertices $v \in B_d$ that are connected to u by one/two parallel edges by D_1 and D_2, respectively. Then $|N(u, B_1 \cup B_2)| + 2|N(u, B_c)| + |D_1| + 2|D_2| \leq 2$.*

Our strategy to prove Lemma 5 can be summarized as follows: We show that similar to Lemma 1, we can represent the term $2|N(u, B_c)| + |D_1| + 2|D_2|$ as the weight of a disjoint set collection Y with $N(Y, A \setminus C) \subseteq \{u\}$. Y consists of subsets of sets in $B \setminus (B_1 \cup B_2)$.

We then apply Lemma 1 to $U := N(Y, C) \cup \{u\}$ to obtain a set collection X. We argue that if $|N(u, B_1 \cup B_2)| + 2|N(u, B_c)| + |D_1| + 2|D_2| > 2 = w(u)$, then $X \cup Y$ constitutes a local improvement. In order to arrive at the desired contradiction, we need to initially restrict our attention to a minimal sub-family $\bar{Y} \subseteq N(u, B_c \cup B_d)$ with $|N(u, B_1 \cup B_2)| + 2|\bar{Y} \cap B_c| + |\bar{Y} \cap D_1| + 2|\bar{Y} \cap D_2| > 2$, which allows us to conclude that $|X \cup Y| \leq 10$.

Proof (Proof of Lemma 5). Assume towards a contradiction that

$$|N(u, B_1 \cup B_2)| + 2\, N(u, B_c)| + |D_1| + 2|D_2| \geq 3.$$

Note that $|N(u, B_1 \cup B_2)| \leq 1$ because $u \notin C$ and u receives at least one unit of weight per neighbor in $B_1 \cup B_2$. Pick an inclusion-wise minimal set $\bar{Y} \subseteq N(u, B_c \cup B_d)$ such that

$$|N(u, B_1 \cup B_2)| + 2|\bar{Y} \cap B_c| + |\bar{Y} \cap D_1| + 2|\bar{Y} \cap D_2| \geq 3. \tag{1}$$

Then

$$|N(u, B_1 \cup B_2)| + 2|\bar{Y} \cap B_c| + |\bar{Y} \cap D_1| + 2|\bar{Y} \cap D_2| = 3, \text{ or} \tag{2}$$

$$\bar{Y} \cap D_1 = \emptyset \text{ and } |N(u, B_1 \cup B_2)| + 2|\bar{Y} \cap B_c| + 2|\bar{Y} \cap D_2| = 4. \tag{3}$$

We construct a set collection Y as follows: We start with $Y = \emptyset$ and first add all sets contained in $\bar{Y} \cap (B_c \cup D_2)$ to Y. Note that for such a set v, $N(v, A \setminus C) = \{u\}$ (see Fig. 4). Second, for each $v \in \bar{Y} \cap D_1$, let v' be the set of cardinality 2 containing the element in which v intersects a set from C, and the element in which v intersects u. Add v' to Y. Then Y has the following properties:

$$N(Y, A) \subseteq C \cup \{u\} \tag{4}$$

$$|Y| = |\bar{Y} \cap B_c| + |\bar{Y} \cap D_1| + |\bar{Y} \cap D_2| \tag{5}$$

$$w(Y) = 2|\bar{Y} \cap B_c| + |\bar{Y} \cap D_1| + 2|\bar{Y} \cap D_2| \overset{(1)}{\geq} 3 - |N(u, B_1 \cup B_2)| \tag{6}$$

$$|N(Y, C)| \leq 2|\bar{Y} \cap B_c| + |\bar{Y} \cap D_1| + |\bar{Y} \cap D_2|. \tag{7}$$

The inequality (7) holds since each vertex in B_c has at most 2 neighbors in C, and each vertex in B_d has at most one neighbor in C (see Fig. 4).

Let $U := N(Y, C) \cup \{u\}$. Apply Lemma 1 to obtain X subject to (1.1)–(1.3). Then by (1.2), we get

$$w(X) \geq w(N(Y, C)) + |N(u, B_1 \cup B_2)| \tag{8}$$

because each set in $N(Y, C)$ receives its weight in the first step, and u receives at least one per neighbor in $B_1 \cup B_2$. By (1.3) and since the sets in Y constitute disjoint subsets of sets in $B \setminus (B_1 \cup B_2)$, $X \dot{\cup} Y$ is a family of pairwise disjoint sets. We would like to show that $X \cup Y$ yields a local improvement of size at most 10. By (8) and (6), we obtain

$$\begin{aligned} w(X \cup Y) = w(X) + w(Y) &\geq 3 + w(N(Y, C)) \\ &> w(u) + w(N(Y, C)) \geq w(N(X \cup Y, A)), \end{aligned}$$

where $N(X \cup Y, A) \subseteq N(Y, C) \cup \{u\}$ follows from (1.1) and (4). Thus, it remains to show that $|X \cup Y| \leq 10$. By (1.3), we have

$$\begin{aligned} |X| = |N(U, B_1 \cup B_2)| &\leq |N(u, B_1 \cup B_2)| + |N(N(Y, C), B_1 \cup B_2)| \\ &\leq |N(u, B_1 \cup B_2)| + 2|N(Y, C)|. \end{aligned} \tag{9}$$

For the last inequality, we used Proposition 2, which tells us that each set $z \in N(Y, C)$ has degree at most 3 in G. In addition, z must intersect at least one set from Y, and thus, from $\bar{Y}$. In particular, z has at least one incident edge to $B \setminus (B_1 \cup B_2) \supseteq \bar{Y}$, and, thus, at most two incident edges to $B_1 \cup B_2$. Hence, we obtain

$$\begin{aligned} |Y| + |X| &\overset{(9)}{\leq} |Y| + |N(u, B_1 \cup B_2)| + 2|N(Y, C)| \\ &\underset{(7)}{\overset{(5)}{\leq}} \underbrace{|N(u, B_1 \cup B_2)| + 5|\bar{Y} \cap B_c| + 3|\bar{Y} \cap D_1| + 3|\bar{Y} \cap D_2|}_{=:(*)}. \end{aligned}$$

If (2) holds, we can bound $(*)$ by 3 times the right-hand side of (2) and deduce an upper bound of 9. In case (3) is satisfied, we can bound $(*)$ by $\frac{5}{2}$ times the right-hand side of (3) and obtain an upper bound of 10. Thus, we have found a local improvement of size at most 10, a contradiction.

Lemma 6. *Each set $u \in A \setminus C$ receives at most $\frac{4}{3}$ times its own weight during our weight distribution.*

Proof. If $w(u) = 1$, then u cannot receive any weight in the first step because otherwise, it would receive at least 1 and be contained in C. Moreover, u has at most two incident edges and receives at most $\frac{2}{3}$ via either of them in the second step.

Next, consider the case where $w(u) = 2$. If u receives $\frac{4}{3}$ from a vertex in B_c, then by Lemma 5, there is no further vertex in $B_1 \cup B_2 \cup B_c \cup B_d$ from which

u receives weight. As u receives at most $\frac{2}{3}$ per edge in all remaining cases, u receives at most $\frac{4}{3} + 2 \cdot \frac{2}{3} = \frac{8}{3} = \frac{4}{3} \cdot w(u)$. Finally, assume that $N(u, B_c) = \emptyset$. In the first step, u can receive at most 1 in total (otherwise, $u \in C$) and this can only happen if u has a neighbor in $B_1 \cup B_2$. The maximum amount u can receive through one edge in the second step is 1, and this can only happen in situation (d). By Lemma 5, there are at most 2 edges via which u receives 1. Moreover, u can receive at most $\frac{2}{3}$ via the remaining edges. Again, we obtain an upper bound of $1 + 1 + \frac{2}{3} = \frac{8}{3}$ on the total weight received.

Combining Lemma 4 and Lemma 6 proves Theorem 2. Together with Proposition 1 and Theorem 1, we obtain Corollary 2.

Corollary 2. *There is a polynomial-time $\frac{4}{3}$-approximation algorithm for the MLSA in dags.*

Disclosure of Interests. The author has no competing interests to declare that are relevant to the content of this article.

References

1. Alon, N., Fomin, F.V., Gutin, G., Krivelevich, M., Saurabh, S.: Spanning directed trees with many leaves. SIAM J. Discret. Math. **23**(1), 466–476 (2009). https://doi.org/10.1137/070710494
2. Binkele-Raible, D., Fernau, H., Fomin, F.V., Lokshtanov, D., Saurabh, S., Villanger, Y.: Kernel(s) for problems with no kernel: on out-trees with many leaves. ACM Trans. Algorithms **8**(4) (2012). https://doi.org/10.1145/2344422.2344428
3. Bonsma, P.: Max-leaves spanning tree is APX-hard for cubic graphs. J. Discrete Algorithms **12**, 14–23 (2012). https://doi.org/10.1016/j.jda.2011.06.005
4. Daligault, J., Thomassé, S.: On finding directed trees with many leaves. In: Chen, J., Fomin, F.V. (eds.) IWPEC 2009. LNCS, vol. 5917, pp. 86–97. Springer, Heidelberg (2009). https://doi.org/10.1007/978-3-642-11269-0_7
5. Drescher, M., Vetta, A.: An approximation algorithm for the maximum leaf spanning arborescence problem. ACM Trans. Algorithms **6**(3) (2010). https://doi.org/10.1145/1798596.1798599
6. Edmonds, J.: Maximum matching and a polyhedron with 0,1-vertices. J. Res. Natl. Bureau Stand. Sect. B Math. Math. Phys. **69B**, 125–130 (1965). https://doi.org/10.6028/jres.069b.013
7. Fernandes, C.G., Lintzmayer, C.N.: Leafy spanning arborescences in DAGs. Discret. Appl. Math. **323**, 217–227 (2022). https://doi.org/10.1016/j.dam.2021.06.018
8. Fernandes, C.G., Lintzmayer, C.N.: How heavy independent sets help to find arborescences with many leaves in DAGs. J. Comput. Syst. Sci. **135**, 158–174 (2023). https://doi.org/10.1016/j.jcss.2023.02.006
9. Galbiati, G., Maffioli, F., Morzenti, A.: A short note on the approximability of the maximum leaves spanning tree problem. Inf. Process. Lett. **52**(1), 45–49 (1994). https://doi.org/10.1016/0020-0190(94)90139-2
10. Garey, M.R., Johnson, D.S.: Computers and Intractability; A Guide to the Theory of NP-Completeness. W. H. Freeman & Co., USA (1990)

11. Guha, S., Khuller, S.: Approximation algorithms for connected dominating sets. Algorithmica **20**(4), 374–387 (1998). https://doi.org/10.1007/PL00009201
12. Karp, R.M.: Reducibility among combinatorial problems. In: Miller, R.E., Thatcher, J.W., Bohlinger, J.D. (eds.) Complexity of Computer Computations: Proceedings of a symposium on the Complexity of Computer Computations. Plenum Press (1972). https://doi.org/10.1007/978-1-4684-2001-2_9
13. Khanna, S., Motwani, R., Sudan, M., Vazirani, U.: On syntactic versus computational views of approximability. SIAM J. Comput. **28**(1), 164–191 (1998). https://doi.org/10.1137/S0097539795286612
14. Neuwohner, M.: Passing the limits of pure local search for weighted k-set packing. In: Proceedings of the 2023 Annual ACM-SIAM Symposium on Discrete Algorithms (SODA), pp. 1090–1137. Society for Industrial and Applied Mathematics (2023). https://doi.org/10.1137/1.9781611977554.ch41
15. Ruan, L., Du, H., Jia, X., Wu, W., Li, Y., Ko, K.I.: A greedy approximation for minimum connected dominating sets. Theoret. Comput. Sci. **329**(1–3), 325–330 (2004). https://doi.org/10.1016/j.tcs.2004.08.013
16. Schwartges, N., Spoerhase, J., Wolff, A.: Approximation algorithms for the maximum leaf spanning tree problem on acyclic digraphs. In: Solis-Oba, R., Persiano, G. (eds.) WAOA 2011. LNCS, vol. 7164, pp. 77–88. Springer, Heidelberg (2012). https://doi.org/10.1007/978-3-642-29116-6_7
17. Solis-Oba, R., Bonsma, P.S., Lowski, S.: A 2-approximation algorithm for finding a spanning tree with maximum number of leaves. Algorithmica **77**, 374–388 (2015). https://doi.org/10.1007/s00453-015-0080-0
18. Thiery, T., Ward, J.: An improved approximation for maximum weighted k-set packing. In: Proceedings of the 2023 Annual ACM-SIAM Symposium on Discrete Algorithms (SODA), pp. 1138–1162. Society for Industrial and Applied Mathematics (2023). https://doi.org/10.1137/1.9781611977554.ch42
19. Trevisan, L.: Non-approximability results for optimization problems on bounded degree instances. In: Proceedings of the Thirty-Third Annual ACM Symposium on Theory of Computing, STOC 2001, pp. 453–461. Association for Computing Machinery, New York (2001). https://doi.org/10.1145/380752.380839

Extending the Primal-Dual 2-Approximation Algorithm Beyond Uncrossable Set Families

Zeev Nutov(✉)

The Open University of Israel, Ra'anana, Israel
nutov@openu.ac.il

Abstract. A set family $\mathcal{F}$ is **uncrossable** if $A \cap B, A \cup B \in \mathcal{F}$ or $A \setminus B, B \setminus A \in \mathcal{F}$ for any $A, B \in \mathcal{F}$. A classic result of Williamson, Goemans, Mihail, and Vazirani [STOC 1993:708-717] states that the problem of covering an uncrossable set family by a min-cost edge set admits approximation ratio 2, by a primal-dual algorithm. They asked whether this result extends to a larger class of set families and combinatorial optimization problems. We define a new class of **semi-uncrossable set families**, when for any $A, B \in \mathcal{F}$ we have that $A \cap B \in \mathcal{F}$ and one of $A \cup B, A \setminus B, B \setminus A$ is in $\mathcal{F}$, or $A \setminus B, B \setminus A \in \mathcal{F}$. We will show that the Williamson et al. algorithm extends to this new class of families and identify several "non-uncrossable" algorithmic problems that belong to this class. In particular, we will show that the union of an uncrossable family and a monotone family, or of an uncrossable family that has the disjointness property and a proper family, is a semi-uncrossable family, that in general is not uncrossable. For example, our result implies approximation ratio 2 for the problem of finding a min-cost subgraph H such that H contains a Steiner forest and every connected component of H contains zero or at least k nodes from a given set T of terminals.

1 Introduction

Let $G = (V, E)$ be graph. For $J \subseteq E$ and $S \subseteq V$ let $\delta_J(S)$ denote the set of edges in J with one end in S and the other in $V \setminus S$, and let $d_J(S) = |\delta_J(S)|$ be their number. An edge set J **covers** S if $d_J(S) \geq 1$. Consider the following problem:

<u>SET FAMILY EDGE COVER</u>
Input: A graph $G = (V, E)$ with edge costs $\{c_e : e \in E\}$ and a set family $\mathcal{F}$ on V.
Output: A min-cost edge set $J \subseteq E$ such that $d_J(S) \geq 1$ for all $S \in \mathcal{F}$.

In this problem the set family $\mathcal{F}$ may not be given explicitly, but we will require that some queries related to $\mathcal{F}$ can be answered in time polynomial in $n = |V|$. Specifically, following [8], we will require that for any edge set I, the inclusion minimal members of the **residual family** $\mathcal{F}^I = \{S \in \mathcal{F} : d_I(S) = 0\}$

J. Vygen and J. Byrka (Eds.): IPCO 2024, LNCS 14679, pp. 351–364, 2024.
https://doi.org/10.1007/978-3-031-59835-7_26

of $\mathcal{F}$ can be computed in time polynomial in $n = |V|$. We will also assume that $V, \emptyset \notin \mathcal{F}$, as otherwise the problem has no feasible solution.

Various types of set families were considered in the literature, classified according to the conditions they satisfy. In this paper we will focus on the following four conditions:

Monotonicity	$\emptyset \neq S' \subseteq S \in \mathcal{F} \Longrightarrow S' \in \mathcal{F}$
Symmetry	$S \in \mathcal{F} \Longrightarrow \bar{S} \in \mathcal{F}$, where $\bar{S} = V \setminus S$.
Disjointness	$\emptyset \neq S' \subseteq S \in \mathcal{F} \Longrightarrow S' \in \mathcal{F}$ or $S \setminus S' \in \mathcal{F}$
Uncrossability	$A, B \in \mathcal{F} \Longrightarrow A \cap B, A \cup B \in \mathcal{F}$ or $A \setminus B, B \setminus A \in \mathcal{F}$

We note that in the paper of Goemans and Williamson [5], the disjointness property was defined as follows: if A, B are disjoint and $A \cup B \in \mathcal{F}$ then $A \in \mathcal{F}$ or $B \in \mathcal{F}$. For $A = S'$ and $B = S \setminus S'$ this is equivalent to the definition given here. Let us say that a set-family $\mathcal{F}$ is:

- **Symmetric** if the Symmetry Condition holds for all $S \in \mathcal{F}$.
- **Monotone** if the Monotonicity Condition holds for all $\emptyset \neq S' \subseteq S \in \mathcal{F}$.
- **Disjointness compliable** if the Disjointness Condition holds for all $\emptyset \neq S' \subseteq S \in \mathcal{F}$.
- **Uncrossable** if the Uncrossability Condition holds for all $A, B \in \mathcal{F}$.

A set family is **proper** if it is both symmetric and disjointness compliable. It is not hard to see the following:

- If $\mathcal{F}$ is monotone then $\mathcal{F}$ is disjointness compliable and uncrossable.
- If $\mathcal{F}$ is proper then $\mathcal{F}$ is uncrossable.

Agrawal, Klein and Ravi [1] designed and analyzed a primal-dual approximation algorithm for the Steiner Forest problem, and showed that it achieves approximation ratio 2. A classic result of Goemans and Williamson [5] from the early 90's shows by an elegant proof that the same algorithm applies for Set Family Edge Cover with an arbitrary proper family $\mathcal{F}$. In fact, one of the main achievements of the Goemans and Williamson paper was defining a *generic class* of set families that models a rich collection of combinatorial optimization problems, for which the primal dual algorithm achieves approximation ratio 2. Slightly later, Williamson, Goemans, Mihail, and Vazirani [8] further extended this result to the more general class of uncrossable families by adding to the algorithm a novel reverse-delete phase. They posed an open question of extending this algorithm to a larger class of set families and combinatorial optimization problems. However, for 30 years, the class of uncrossable set families remained the most general generic class of set families for which the primal dual algorithm achieves approximation ratio 2. We present a strict generalization of this result, illustrated by several examples.

Definition 1. *We say that a set family $\mathcal{F}$ is* **semi-uncrossable** *if for any $A, B \in \mathcal{F}$ the following* **semi-uncrossability condition** *holds*

$$A \cap B \in \mathcal{F} \text{ and one of } A \cup B, A \setminus B, B \setminus A \text{ is in } \mathcal{F} \quad \text{or} \quad A \setminus B, B \setminus A \in \mathcal{F} .$$

One can see that if $\mathcal{F}$ is symmetric and if $A \cap B \in \mathcal{F}$ or $A \setminus B, B \setminus A \in \mathcal{F}$ whenever $A, B \in \mathcal{F}$, then $\mathcal{F}$ is uncrossable. Indeed, if $A, B \in \mathcal{F}$ then $\bar{A}, \bar{B} \in \mathcal{F}$ (since $\mathcal{F}$ is symmetric), and thus $\bar{A} \cap \bar{B} \in \mathcal{F}$ or $\bar{A} \setminus \bar{B}, \bar{B} \setminus \bar{A} \in \mathcal{F}$; by symmetry we get that $A \cup B \in \mathcal{F}$ or $B \setminus A, A \setminus B \in \mathcal{F}$. In particular, we get that:

If $\mathcal{F}$ is symmetric and semi-uncrossable then $\mathcal{F}$ is uncrossable.

Namely, for symmetric $\mathcal{F}$ the definitions of "uncrossable" and "semi-uncrossable" coincide. However, for non-symmetric families, these definitions are distinct. For illustration, consider the following example, see Fig. 1.

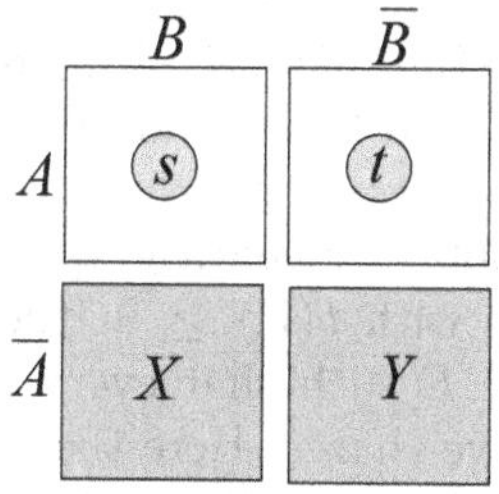

Fig. 1. An example of a semi-uncrossable family $\mathcal{F} = \{A, B, \{s\}, \{t\}\}$ that is not uncrossable.

Example 1. Let $V = X \cup Y \cup \{s, t\}$ where $X, Y \neq \emptyset$, let $A = \{s, t\}$ and $B = X \cup \{s\}$. Let

$$\mathcal{F} = \{A, B, A \cap B, A \setminus B\} = \{A, B, \{s\}, \{t\}\} .$$

This $\mathcal{F}$ is semi-uncrossable but not uncrossable. Note that an attempt to "symmetrize" $\mathcal{F}$ by adding to $\mathcal{F}$ the complements of the sets in $\mathcal{F}$ results in a family that is *not* semi-uncrossable (and hence not uncrossable), where the semi-uncrossability condition fails for $\bar{A}, \bar{B}$. Thus semi-uncrossable families strictly generalize uncrossable set families. We prove the following.

Theorem 1. Set Family Edge Cover *with semi-uncrossable $\mathcal{F}$ admits approximation ratio* 2.

The Theorem 1 algorithm is identical to the algorithm of Williamson, Goemans, Mihail, and Vazirani [8] for uncrossable families, and the proof of the approximation ratio is also similar. Hence the contribution of our paper is not technical but rather conceptual, as follows.

- We introduce a new generic class of set families that strictly contains the class of uncrossable families, and observe that the [8] algorithm achieves for this new class the same approximation ratio 2.
- We identify many combinatorial algorithmic problems that can be modeled by this new class, but not by uncrossable families.

Recently, Bansal, Cheriyan, Grout, and Ibrahimpur [2] defined a new class of set families which they called **pliable set families with property** (γ); a set family $\mathcal{F}$ is called **pliable** if for any $A, B \in \mathcal{F}$ at least two sets from $A \cap B, A \cup B, A \setminus B, B \setminus A$ belong to $\mathcal{F}$, while **property** (γ) imposes some restrictions on triples of sets from $\mathcal{F}$. They showed that the primal-dual algorithm of [5,8] achieves approximation ratio 16 for this class. Property (γ) holds for semi-uncrossable families, and the class of pliable set families with property (γ) strictly includes the class of semi-uncrossable families. But handling this more general class of families comes with the price of a worse approximation ratio – 16 instead of 2. The class of semi-uncrossable set families is sandwiched between the uncrossable families and the pliable families with property (γ), and our approximation ratio is 2 – the same as for uncrossable set families. We note that our paper is partly motivated by this novel result of [2].

Another motivation comes from the Unbalanced Generalized Point-To-Point Connection problem: given a graph $G = (V, E)$ with edge costs and "charges" $\{b_v : v \in V\}$ with $b(V) \geq 0$ (the charges may be negative), find a min-cost edge set $J \subseteq E$ such that every connected component of the graph (V, J) has a non-negative charge. Here the set family we need to cover is $\mathcal{F} = \{S \subseteq V : b(S) < 0\}$. This $\mathcal{F}$ is uncrossable when $b(V) = 0$, but in the general case only has the following weaker property: $A \cap B \in \mathcal{F}$ or $A \setminus B, B \setminus A \in \mathcal{F}$. The problem admits approximation ratio $O(\log \min\{n, 2 + b(V)\})$ [6], but has no known super-constant approximation threshold.

We illustrate application of Theorem 1 on **combinations** of several fundamental problems, among them Steiner Forest, T-Join, Point-to-Point Connection, (T, k)-Constrained Forest, k-Constrained Forest, and Steiner Network Augmentation; see formal definitions in the next section. We can combine any two of these problems by requiring that the solution will satisfy the requirements of both problems. This means that the Set Family Edge Cover formulation of both problems has the same graph with the same edge costs, and we need to cover the union of the set families of the problems.[1] Namely, if in one problem we need to cover a set family $\mathcal{A}$ and in the other a set family $\mathcal{B}$, then in the combination of the problems we need to cover the set family $\mathcal{F} = \mathcal{A} \cup \mathcal{B}$. In fact, in this way we can also combine a problem with itself. We will assume that each of $\mathcal{A}, \mathcal{B}$ is semi-uncrossable, as otherwise we should not expect that $\mathcal{A} \cup \mathcal{B}$ will be semi-uncrossable. We will prove the following.

Theorem 2. *Let $\mathcal{A}, \mathcal{B}$ be semi-uncrossable set families over a groundset V. Then $\mathcal{A} \cup \mathcal{B}$ is semi-uncrossable in the following cases:*

(i) *$\mathcal{A}$ is monotone.*
(ii) *$\mathcal{A}$ is proper and $\mathcal{B}$ is disjointness compliable.*

---

[1] This definition of "combination of problems" depends on the specific set family that models the problem. For example, the set family of the Steiner Tree problem is symmetric, but the set family of the equivalent rooted version is not. For all problems considered, we will specify the set family we choose, trying to maximize the number of "good" properties, like uncrossability and symmetry.

As was mentioned, for a symmetric family the definitions of "uncrossable" and "semi-uncrossable" coincide, hence the new applications of Theorems 1 and 2 are for non-symmetric families. Consider for example the k-CONSTRAINED FOREST problem: Given a graph $G = (V, E)$ with edge costs, find a min-cost subgraph H such that every connected component of H has at least k nodes. The set family we need to cover here is $\{S : 1 \leq |S| < k\}$. This family is not symmetric, but it is monotone. Monotone set families were introduced by Goemans and Williamson [4], where they also showed several additional graph problems that can be modeled by such families. Note that any monotone family is uncrossable, hence the problem of covering a monotone set family admits approximation ratio 2. Couëtoux, Davis, and Williamson [3] improved the approximation ratio to 1.5 by a dual fitting algorithm.

Clearly, we can achieve ratio $\alpha + \beta$ for the problem of covering the union $\mathcal{F} = \mathcal{A} \cup \mathcal{B}$ of an uncrossable family $\mathcal{A}$ and a monotone family $\mathcal{B}$, where α and β are the best known approximation ratios for covering $\mathcal{A}$ and $\mathcal{B}$, respectively. Currently, $\alpha = 2$ [8] and $\beta = 1.5$ [3], so the overall approximation ratio will be 3.5. Theorems 1 and 2 give a better approximation ratio of 2. Let us now illustrate Theorem 2 and its limitations on few examples.

Example 2. *$\mathcal{A}$ is monotone and $\mathcal{B}$ is proper, $\mathcal{A} \cup \mathcal{B}$ is semi-uncrossable but not uncrossable.*
Consider the set family $\mathcal{A} = \{S : 1 \leq |S| \leq 2\}$ of the 3-CONSTRAINED FOREST problem that is monotone, and the family $\mathcal{B} = \{S : |S \cap \{s,t\}| = 1\}$ of the st-PATH problem that is proper. Now consider the sets $A = \{s,t\}$ and $B = X \cup \{s\}$ in Example 1, where $|X|, |Y| \geq 3$. Then $A \in \mathcal{A}$, $B \in \mathcal{B}$, but among the sets $A \cap B, A \cup B, A \setminus B, B \setminus A$, only $A \cap B, A \setminus B$ belong to $\mathcal{A} \cup \mathcal{B}$, hence $\mathcal{A} \cup \mathcal{B}$ is not uncrossable. However, $\mathcal{A} \cup \mathcal{B}$ is semi-uncrossable by Theorem 2.

Example 3. *$\mathcal{A}, \mathcal{B}$ are both disjointness compliable, $\mathcal{A} \cup \mathcal{B}$ is not semi-uncrossable.*
Let $V = \{s,t,x,y\}$, $A = \{s,t\}$, $B = \{s,x\}$, and let $\mathcal{A} = \{A, A \cap B\} = \{\{s,t\},\{s\}\}$ and $\mathcal{B} = \{B, A \cap B\} = \{\{s,x\},\{s\}\}$. Then $\mathcal{A}, \mathcal{B}$ are both uncrossable and disjointness compliable, $A \in \mathcal{A}$, $B \in \mathcal{B}$, but among the sets $A \cap B, A \cup B, A \setminus B, B \setminus A$ only $A \cap B$ belongs to $\mathcal{A} \cup \mathcal{B}$.

Example 4. *$\mathcal{A}$ is proper and $\mathcal{B}$ is uncrossable, $\mathcal{A} \cup \mathcal{B}$ is not semi-uncrossable.*
Let $V = \{x,y,s,t\}$, $A = \{x,s\}$, $B = \{x,y\}$, $\mathcal{B} = \{B, \bar{B}\}$ and let $\mathcal{A} = \{S : |S \cap \{s,t\}|\} = 1$ (this is the family of the st-PATH problem). One can verify that $\mathcal{A}$ is proper, $\mathcal{B}$ is symmetric and uncrossable, $A \in \mathcal{A}$, $B \in \mathcal{B}$, but among the sets $A \cap B, A \cup B, A \setminus B, B \setminus A$ only $A \setminus B, A \cup B$ belong to $\mathcal{A} \cup \mathcal{B}$, so $\mathcal{A} \cup \mathcal{B}$ is not semi-uncrossable.

The proof of Theorem 2 is by a standard case analysis and uncrossing. Yet, Theorem 2 is sharp in the following sense.

- In the combinations considered in the theorem, the combined set family $\mathcal{F} = \mathcal{A} \cup \mathcal{B}$ may not be uncrossable even if $\mathcal{A}, \mathcal{B}$ are both uncrossable, as is shown in Example 2.

- For any combination of two semi-uncrossable set families, either the semi-uncrossability of $\mathcal{A} \cup \mathcal{B}$ can be deduced from Theorem 2, or there exists an example such that $\mathcal{A} \cup \mathcal{B}$ is not semi-uncrossable. Namely, we cannot weaken the conditions in Theorem 2. To see this, note that in the following cases $\mathcal{F} = \mathcal{A} \cup \mathcal{B}$ may not be semi-uncrossable.
  - $\mathcal{A}, \mathcal{B}$ are both disjointness compliable; see Example 3.
  - If $\mathcal{A}$ is proper and $\mathcal{B}$ is symmetric; see Example 4.

We summarize this in Table 1. Note that there is only one entry in the table that guarantees an uncrossable family – when both $\mathcal{A}, \mathcal{B}$ are proper, while there are 5 entries where we get a semi-uncrossable family that may not be uncrossable.

Table 1. Combinations that lead to semi-uncrossable families. Here, "d", "s", and "m" stand for disjointness compliable, symmetric, and monotone, respectively. Also, "+" means that the union is semi-uncrossable but is not uncrossable in general, "+*" means that the union is always uncrossable, while "–" means that the union is not semi-uncrossable in general (and there is an example for that).

	d	s	m	d,s
d	–	–	+	+
s		–	+	–
m			+	+
d,s				+*

In the next Sect. 2 we prove Theorem 2 and discuss some applications of Theorems 1 and 2 for specific problems, while Theorem 1 is proved in Sect. 3.

2 Applications

Recall that in Theorem 2 we need to prove that if $\mathcal{A}, \mathcal{B}$ are semi-uncrossable set families then $\mathcal{A} \cup \mathcal{B}$ is semi-uncrossable in the following cases:

(i) $\mathcal{A}$ is monotone.
(ii) $\mathcal{A}$ is proper and $\mathcal{B}$ is disjointness compliable.

Since each of $\mathcal{A}, \mathcal{B}$ is semi-uncrossable, we need to verify the semi-uncrossability condition only for pairs of sets A, B such that $A \in \mathcal{A}$ and $B \in \mathcal{B}$. Let $A \in A$ and $B \in B$. Assume that all sets $A \cap B, A \setminus B, B \setminus A$ are non-empty, as otherwise $\{A \cap B, A \cup B\} = \{A, B\}$ or $\{A \setminus B, B \setminus A\} = \{A, B\}$.

Part (i) is trivial – since $\mathcal{A}$ is monotone, $A \cap B, A \setminus B \in \mathcal{A}$, so the semi-uncrossability condition holds for A, B.

We prove (ii). Since $\mathcal{A}$ is symmetric, $\bar{A} \in \mathcal{A}$. Since $\mathcal{A}, \mathcal{B}$ are both disjointness compliable, the following holds:

(a) $A \cap B \in \mathcal{A}$ or $A \setminus B \in \mathcal{A}$.
(b) $\bar{A} \cap B \in \mathcal{A}$ or $\bar{A} \setminus B \in \mathcal{A}$; note that $\bar{A} \cap B = B \setminus A$ and that $\bar{A} \setminus B = V \setminus (A \cup B)$, thus this is equivalent to $B \setminus A \in \mathcal{A}$ or $A \cup B \in \mathcal{A}$ (by the symmetry of $\mathcal{A}$).
(c) $A \cap B \in \mathcal{B}$ or $B \setminus A \in \mathcal{B}$.

Let us consider the two cases in (c).

- If $A \cap B \in \mathcal{B}$, then the semi-uncrossability condition holds by (b).
- If $B \setminus A \in \mathcal{B}$ then the semi-uncrossability condition holds by (a).

In both cases the semi-uncrossability condition holds, concluding the proof of (ii) and also of Theorem 2.

We illustrate application of Theorems 1 and 2 on combinations of several fundamental problems. In all these problems we are given an (undirected) graph $G = (V, E)$ with edge costs $\{c_e : e \in E\}$, and seek a subgraph of G or an edge subset of E that satisfies a prescribed requirement. Below we indicate for each problem the additional parts of the input, the prescribed requirement, the set family we need to cover, and the (strongest) relevant properties of this family.

STEINER FOREST (SF)
Given a subpartition $\mathcal{P}$ of V, find a min-cost subgraph H of G such that every part $P \in \mathcal{P}$ is contained in the same connected component of H.
Set family: $\{S : \emptyset \neq S \cap P \neq P \text{ for some } P \in \mathcal{P}\}$; this family is proper and thus uncrossable.

T-JOIN
Given a set $T \subseteq V$ of terminals with $|T|$ even, find a min-cost subgraph H of G such that every connected component of H contains an even number of terminals.
Set family: $\{S : |S \cap T| \text{ is odd}\}$; this family is proper and thus uncrossable.

(T, k)-CONSTRAINED FOREST ((T, k)-CF)
Given a set $T \subseteq V$ of terminals and an integer $k \leq |T|$, find a min-cost spanning subgraph H of G such that every connected component of H has 0 or at least k terminals. (The case $|T| < 2k$ is the STEINER TREE problem.)
Set family: $\{S : 1 \leq |S \cap T| < k\}$; this family is disjointness compliable and uncrossable.

k-CONSTRAINED FOREST (k-CF)
This is a particular case of (T, k)-CF when $T = V$.
Set family: $\{S : 1 \leq |S| < k\}$; this family is monotone and thus uncrossable.

GENERALIZED POINT-TO-POINT CONNECTION (G-P2P)
Given integer charges $\{b(v) : v \in V\}$ with $b(V) = 0$, find a min-cost spanning subgraph H of G such that every connected component of H has zero charge.
Set family: $\{S : b(S) \neq 0\}$; this family is proper and thus uncrossable.

STEINER NETWORK AUGMENTATION (SNA)
Given a graph $G_0 = (V, E_0)$ with $E_0 \cap E = \emptyset$ and a set D of demand node pairs, find a min-cost subgraph H of G such that $\lambda_{G_0 \cup H}(u,v) \geq \lambda_{G_0}(u,v) + 1$ for all $(u,v) \in D$, where $\lambda_J(u,v)$ denotes the maximum number of edge disjoint uv-paths in a graph J.
Set family: $\{S : |S \cap \{u,v\}| = 1, d_{G_0}(S) = \lambda_{G_0}(u,v)$ for some $\{u,v\} \in D\}$; this family is symmetric and uncrossable.

Combinations of problem pairs from the list above that by Theorem 2 give a semi-uncrossable family are summarized in Table 2; we added one additional row for the st-PATH problem, that will help us to verify that the table was indeed filled correctly. Excluding the last row, we have overall 21 combinations, and 15 of them give a semi-uncrossable set family by Theorem 2. Among these, 8 combinations give a set family that is not uncrossable in general. We now will justify whether each entry should be marked by "+" (semi-uncrossable that may not be uncrossable), or by "+*" (always uncrossable), or by "–" (may not be semi-uncrossable). Interestingly, Theorem 2 "correctly filled" the table – the plus entries are exactly the ones that could be deduced from the theorem.

Table 2. Combinations that lead to semi-uncrossable families. Here, "d", "s", and "m" stand for disjointness compliable, symmetric, and monotone, respectively (uncrossability is not mentioned since all problems are uncrossable). Also, "+" means that the union is semi-uncrossable, "+*" means that the union is uncrossable, while "–" means that the union is not semi-uncrossable in general.

	SF **d,s**	T-JOIN **d,s**	G-P2P **d,s**	k-CF **m**	(T,k)-CF **d**	SNA **s**
SF **d,s**	+*	+*	+*	+	+	–
T-JOIN **d,s**		+*	+*	+	+	–
G-P2P **d,s**			+*	+	+	–
k-CF **m**				+*	+	+
(T,k)-CF **d**					–	–
SNA **s**						–
SP **d,s**	+*	+*	+*	+	+	–

We need the following observations to show that the table was indeed filled correctly.

(A) If $\mathcal{F}$ is semi-uncrossable and symmetric then $\mathcal{F}$ is uncrossable.
(B) If $\mathcal{A}, \mathcal{B}$ are both monotone then $\mathcal{A} \cup \mathcal{B}$ is also monotone and thus uncrossable.
(C) Except k-CF, the st-PATH problem is a particular case of each of the problems considered here.
(D) There is an example that the combination of st-PATH with 3-CF (and thus also with $(T,3)$-CF) gives a semi-uncrossable family that is not uncrossable.

(E) There is an example that the combination of st-PATH with SNA with $|D| = 1$ gives a set family that is not semi-uncrossable.

Observation (A) was already proved in the introduction while Observation (B) is trivial. For Observation (C), one can verify that st-PATH is a particular case of:

- SF when $\mathcal{P} = \{\{s,t\}\}$.
- T-JOIN when $T = \{s,t\}$.
- G-P2P when $b(s) = 1, b(t) = -1$, and $b(v) = 0$ otherwise.
- (T,k)-CF when $T = \{s,t\}$ and $k = 2$.
- SNA when $E_0 = \emptyset$ and $D = \{\{s,t\}\}$.

Observation (D) was proved in the Introduction, see Example 2.

We now consider Observation (E). Note that in the combined problem of st-PATH with SNA we cannot just add the pair $\{s,t\}$ to the set D of demand pairs, because we require that the augmenting graph H will contain an st-path, and not the weaker condition that $G_0 \cup H$ will contain an st-path. Consider now the families $\mathcal{A}, \mathcal{B}$ in Example 4 in the Introduction. Let $E_0 = \{xy, st\}$. Then (independently of the edge set E and costs), $\mathcal{B}$ is the family of the SNA instance with $D = \{xs\}$ and $\mathcal{A}$ is the family of the st-PATH instance. But in Example 4 the family $\mathcal{A} \cup \mathcal{B}$ is not semi-uncrossable. This shows that the combination of st-PATH with st-PATH AUGMENTATION (SNA with $|D| = 1$) gives a set family that may not be semi-uncrossable.

We use these observations to justify the last row of the table – of the st-PATH problem. The first 3 entries in the row are by Theorem 2(ii). The next two entries are by Theorem 2(i) and Observation (D). The last entry in the row is by Observation (E).

In a similar way we can fill the other entries. Note that observation (D) is used only for the plus entries, to decide whether the entry should be "+" or "+*".

The First 3 Columns: Here all combinations give a semi-uncrossable family by Theorem 2(ii). The family is also uncrossable, since the combined set family is symmetric.

The k-CF Column. The first 3 entries in the column follow from Theorem 2(i) and Observations (C) and (D). The fourth entry is by Observation (B).

The (T,k)-CF Column. All positive entries in the column of (T,k)-CF are by Theorem 2 and Observations (C) and (D); in the first 3 entries of the column, instead of Observation (C) and (D) we can use the fact that k-CF is a particular case of (T,k)-CF.

For the "–" entry, consider the following two (T,k)-CF instances with $k = 2$ and $V = \{a,a',a'',b,b',b''\}$. One has terminal set and family $T_{\mathcal{A}} = \{a,a',a''\}$ and $\mathcal{A} = \{S : |S \cap T_{\mathcal{A}}| = 1\}$ and the other $T_{\mathcal{B}} = \{b,b',b''\}$ and $\mathcal{B} = \{S : |S \cap T_{\mathcal{B}}| = 1\}$, respectively. Let $A = \{a,b,b',b''\}$ and $B = \{b,a,a',a''\}$. Then $A \in \mathcal{A}$ and $B \in \mathcal{B}$, but among the sets $A \cap B, A \cup B, A \setminus B, B \setminus A$ only $A \cap B$ belongs to $\mathcal{A} \cup \mathcal{B}$.

The SNA **Column.** All the minus entries in this column follow from Observations (C) and (E). The unique plus entry in this column is by Theorem 2(i) and Observation (D).

3 Proof of Theorem 1

In this section we prove Theorem 1. During the proof we indicate the parts that extend or fail to pliable set families; recall that a set family $\mathcal{F}$ is pliable if for any $A, B \in \mathcal{F}$ at least 2 sets from $A \cap B, A \cup B, A \setminus B, B \setminus A$ belong to $\mathcal{F}$. Also recall that the residual family of $\mathcal{F}$ w.r.t. an edge set I is $\mathcal{F}^I = \{S \in \mathcal{F} : d_I(S) = 0\}$.

One can see that if an edge e covers one of the sets $A \cap B, A \cup B, A \setminus B, B \setminus A$ then it also covers one of A, B. This implies the following.

Lemma 1. *If $\mathcal{F}$ is semi-uncrossable or if $\mathcal{F}$ is pliable then so is $\mathcal{F}^I$, for any edge set I.*

An $\mathcal{F}$**-core** is an inclusion minimal member of $\mathcal{F}$; let $\mathcal{C}(\mathcal{F})$ denote the family of $\mathcal{F}$-cores. Similarly to uncrossable families, we have the following.

Lemma 2. *If $\mathcal{F}$ is semi-uncrossable then for any $C \in \mathcal{C}(\mathcal{F})$ and $S \in \mathcal{F}$, either $C \subseteq S$ or $C \cap S = \emptyset$; in particular, the $\mathcal{F}$-cores are pairwise disjoint.*

Proof. Since $\mathcal{F}$ is semi-uncrossable $C \cap S \in \mathcal{F}$ or $C \setminus S, S \setminus C \in \mathcal{F}$. In the former case $C \subseteq S$ and in the later case $C \cap S = \emptyset$, by the minimality of C. $\square$

Note that if $\mathcal{F}$ is pliable then the cores are also pairwise disjoint. However, the first property in Lemma 2 may not hold for pliable families.

We now describe the algorithm. Consider the following LP-relaxation **(P)** for Set Family Edge Cover and its dual program **(D)**:

$$
\begin{array}{llll}
 & \min \sum_{e \in E} c_e x_e & & \\
\textbf{(P)} & \text{s.t.} \sum_{e \in \delta(S)} x_e \geq 1 & \forall S \in \mathcal{F} \\
 & \quad x_e \geq 0 & \forall e \in E
\end{array}
\qquad
\begin{array}{llll}
 & \max \sum_{S \in \mathcal{F}} y_S & & \\
\textbf{(D)} & \text{s.t.} \sum_{\delta(S) \ni e} y_S \leq c_e & \forall e \in E \\
 & \quad y_S \geq 0 & \forall S \in \mathcal{F}.
\end{array}
$$

Given a solution y to **(D)**, an edge $e \in E$ is **tight** if the inequality of e in **(D)** holds with equality. The algorithm has two phases.

Phase 1 starts with $I = \emptyset$ an applies a sequence of iterations. At the beginning of an iteration, we compute the family $\mathcal{C}(\mathcal{F}^I)$. Then we raise the dual variables corresponding to the members of $\mathcal{C}(\mathcal{F}^I)$ uniformly (possibly by zero), until some edge $e \in E \setminus I$ becomes tight, and add e to I. Phase I terminates when $\mathcal{C}(\mathcal{F}^I) = \emptyset$, namely when I covers $\mathcal{F}$.

Phase 2 applies on I "reverse delete", which means the following. Let $I = \{e_1, \ldots, e_j\}$, where e_{i+1} was added after e_i. For $i = j$ downto 1, we delete e_i from I if $I \setminus \{e_i\}$ still covers $\mathcal{F}$. At the end of the algorithm, I is output.

It is easy to see that the produced dual solution is feasible, hence $\sum_{S\in\mathcal{F}} y_S \leq$ opt, by the Weak Duality Theorem. We prove that at the end of the algorithm

$$\sum_{e\in I} c(e) \leq 2 \sum_{S\in\mathcal{F}} y_S \ .$$

As any edge in I is tight, the last inequality is equivalent to

$$\sum_{e\in I} \sum_{\delta_I(S)\ni e} y_S \leq 2 \sum_{S\in\mathcal{F}} y_S \ .$$

By changing the order of summation we get:

$$\sum_{S\in\mathcal{F}} d_I(S) y_S \leq 2 \sum_{S\in\mathcal{F}} y_S \ .$$

It is sufficient to prove that at any iteration the increase at the left hand side is at most the increase in the right hand side. Let us fix some iteration, and let $\mathcal{C}$ be the family of cores among the members of $\mathcal{F}$ not yet covered. The increase in the left hand side is $\varepsilon \cdot \sum_{C\in\mathcal{C}} d_I(C)$, where ε is the amount by which the dual variables were raised in the iteration, while the increase in the right hand side is $\varepsilon \cdot 2|\mathcal{C}|$. Consequently, it is sufficient to prove that $\sum_{C\in\mathcal{C}} d_I(C) \leq 2|\mathcal{C}|$. As the edges were deleted in reverse order, the set I' of edges in I that were added after the iteration (and "survived" the reverse delete phase), form an inclusion minimal edge-cover of the family $\mathcal{F}'$ of members in $\mathcal{F}$ that are uncovered at the beginning of the iteration. Note also that $\bigcup_{C\in\mathcal{C}} \delta_I(C) \subseteq I'$. Hence to prove approximation ratio 2, it is sufficient to prove the following purely combinatorial statement, in which due to Lemma 1 we can revise our notation to $\mathcal{F} \leftarrow \mathcal{F}'$ and $I \leftarrow I'$.

Lemma 3. *Let I be an inclusion minimal cover of a set family $\mathcal{F}$ and let $\mathcal{C} = \mathcal{C}(\mathcal{F})$ be the family of $\mathcal{F}$-cores. If $\mathcal{F}$ is semi-uncrossable then*

$$\sum_{C\in\mathcal{C}} d_I(C) \leq 2|\mathcal{C}| - 1 \ . \tag{1}$$

In the rest of this section we prove Lemma 3. Let us say that **two sets** A, B **are laminar** if they are disjoint or one of them contains the other; namely, if at least one of the sets $A \cap B, A \setminus B, B \setminus A$ is empty. A set family $\mathcal{L}$ is a **laminar set family** if its members are pairwise laminar. Let I be an inclusion minimal edge cover of a set family $\mathcal{F}$. We say that $S_e \in \mathcal{F}$ is a **witness set** for e if e is the unique edge in I that covers S_e, namely, if $\delta_I(S_e) = \{e\}$. We say that $\mathcal{L} \subseteq \mathcal{F}$ is a **witness family** for I if $|\mathcal{L}| = |I|$ and for every $e \in I$ there is a witness set $S_e \in \mathcal{L}$. By the minimality of I, there exists a witness family $\mathcal{L} \subseteq \mathcal{F}$. We will show the following.

Lemma 4. *Let I be an inclusion minimal cover of a semi-uncrossable family $\mathcal{F}$. Then there exists a witness family $\mathcal{L} \subseteq \mathcal{F}$ for I that is laminar.*

Proof. Let $A, B \in \mathcal{F}$ be witness sets of edges $e, f \in I$, respectively. Note that no edge in $I \setminus \{e, f\}$ covers a set from $A \cap B, A \cup B, A \setminus B, B \setminus A$, as such an edge covers one of A, B, contradicting that A, B are witness sets. Thus all possible locations of such e, f are as depicted in Fig. 2. We claim that one of the sets $A \cap B, A \setminus B, B \setminus A$ is a witness set for one of e, f. This follows from the following observations.

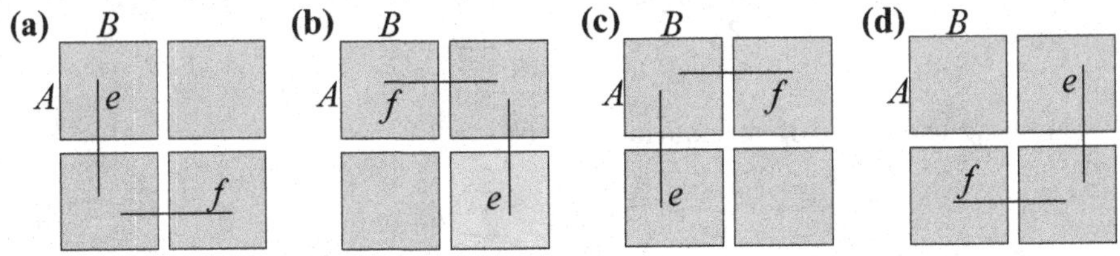

Fig. 2. Illustration to the proof of Lemma 4.

- $A \setminus B \notin \mathcal{F}$ in (a) and $B \setminus A \notin \mathcal{F}$ in (b); in both cases, $A \cap B \in \mathcal{F}$ is a witness set for one of e, f.
- $A \cup B \notin \mathcal{F}$ in (c), and $A \cap B \notin \mathcal{F}$ in (d); in both cases, $A \setminus B, B \setminus A \in \mathcal{F}$ and each of $A \setminus B, B \setminus A$ is a witness set for one of e, f.

By the minimality of I there exists a witness family for I. Let $\mathcal{L}$ be a witness family for I with $\sum_{S \in \mathcal{L}} |S|$ minimal. We claim that $\mathcal{L}$ is laminar. Suppose to the contrary that there are $A, B \in \mathcal{L}$ that are not laminar. Then there is $A' \in \{A \cap B, A \setminus B, B \setminus A\}$ and $B' \in \{A, B\}$ such that $\mathcal{L}' = (\mathcal{L} \setminus \{A, B\}) \cup \{A', B'\}$ is also a witness family for I. However, $\sum_{S \in \mathcal{L}'} |S| < \sum_{S \in \mathcal{L}} |S|$, contradicting the choice of $\mathcal{L}$. □

We note that Lemma 4 extends to pliable families by a slightly more complicated proof, see [2, Lemma 10].

Let $J = \bigcup_{C \in \mathcal{C}} \delta_I(C)$. Note that $\delta_I(C) = \delta_J(C)$ for any $C \in \mathcal{C}$. As any subfamily of a laminar family is also laminar, we conclude from Lemma 4 that if $\mathcal{F}$ is semi-uncrossable then there exists a laminar witness family $\mathcal{L} \subseteq \mathcal{F}$ for J. Note that for every $S \in \mathcal{L}$ there is $C \in \mathcal{C}$ such that $C \subseteq S$. Moreover, if $\mathcal{F}$ is semi-uncrossable then by Lemma 2 $\mathcal{L} \cup \mathcal{C}$ is laminar (this is not true if $\mathcal{F}$ is only pliable). Thus to finish the proof of Lemma 3, it is sufficient to prove the following purely combinatorial lemma, that was essentially proved in [5,8].

Lemma 5. *Let $\mathcal{L}, \mathcal{C}$ be set families such that $\mathcal{L} \cup \mathcal{C}$ is laminar, the members of $\mathcal{C}$ are pairwise disjoint, and $\mathcal{C}$ is the family of inclusion minimal members of $\mathcal{L} \cup \mathcal{C}$. Let J be an edge set such that every edge in J covers some $C \in \mathcal{C}$. If $\mathcal{L}$ is a witness family for J then*

$$\sum_{C \in \mathcal{C}} d_J(C) \leq 2|\mathcal{C}| - 1 . \tag{2}$$

Proof. Let $S = S_e$ be an inclusion minimal set in $\mathcal{L}$, let $C \in \mathcal{C}$ such that $C \subseteq S_e$ (possibly $C = S_e$), and let $\mathcal{L}_C$ be the family of sets in $\mathcal{L}$ that contain C and contain no other set in $\mathcal{C}$. We claim that then:

(i) $\delta_J(C) = \{e\}$.
(ii) $S_e \in \mathcal{L}_C$ and $|\mathcal{L}_C| \leq 2$.

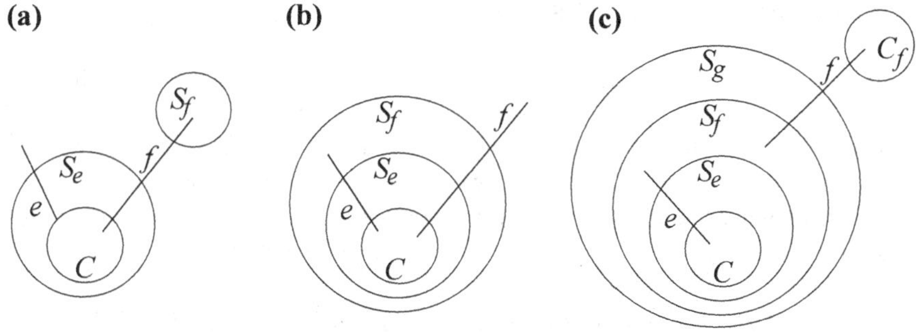

Fig. 3. Illustration to the proof of Lemma 5; possibly $S_e = C$. (a,b) The cases $S_e \cap S_f = \emptyset$ and $S_e \subset S_f$, respectively; here it is not assumed that e covers C, but in (b) it can be shown that e must cover C. (c) The case $C \subseteq S_e \subset S_f \subset S_g$, where $S_f, S_g \in \mathcal{L}_C \setminus \{S_e\}$.

We prove (i). Suppose to the contrary that there is $f \in \delta_J(C) \setminus \{e\}$. Since $\mathcal{L}$ is laminar and S_e is inclusion minimal in $\mathcal{L}$, we must have $S_e \cap S_f = \emptyset$ or $S_e \subset S_f$, see Fig. 3 (a,b). In both cases f covers S_e, contradicting that S_e is a witness set for e.

We prove (ii). It is easy to see that $S_e \in \mathcal{L}_C$. Suppose to the contrary that there are distinct $S_f, S_g \in \mathcal{L}_C \setminus \{S_e\}$. From the laminarity of $\mathcal{L}$ we may assume w.l.o.g. that $C \subseteq S_e \subset S_f \subset S_g$, see Fig. 3 (c). Since every edge in J covers some $C \in \mathcal{C}$, there is $C_f \in \mathcal{C}$ that is covered by f. Since $S_g \in \mathcal{L}_C$, we must have $C_f \cap S_g = \emptyset$. But then f must cover S_g, contradicting that S_g is a witness set for g.

The rest of the proof is by induction on $|\mathcal{C}|$. The base case $|\mathcal{C}| = 1$ is easy to verify. We prove the lemma for $|\mathcal{C}| \geq 2$, assuming that it holds for $|\mathcal{C}| - 1$.

Let C be as in (i) and let $\mathcal{C}' = \mathcal{C} \setminus \{C\}$. Define J' and $\mathcal{L}'$ as follows.

- If $\mathcal{L}_C = \{S_e\}$ and if there is $C' \in \mathcal{C}'$ such that $\delta_J(C') = \{e\}$, then $J' = J$ and $\mathcal{L}' = (L \setminus \{S_e\}) \cup \{C'\}$.
- Else, $J' = J \setminus \{e\}$ and $\mathcal{L}' = \mathcal{L} \setminus \{S_e\}$.

One can verify that the triple $\mathcal{L}', \mathcal{C}', J'$ satisfies the assumptions of the lemma and that $\sum_{C \in \mathcal{C}} d_J(C) - \sum_{C \in \mathcal{C}'} d_{J'}(C) \leq 2$. From this, using the induction hypothesis and $|\mathcal{C}'| = |\mathcal{C}| - 1$ we get

$$\sum_{C \in \mathcal{C}} d_J(C) \leq \sum_{C \in \mathcal{C}'} d_{J'}(C) + 2 \leq (2|\mathcal{C}'| - 1) + 2 = 2|\mathcal{C}| - 1 ,$$

concluding the proof. □

This concludes the proof of Theorem 1.

Remark. The essential property that enabled us to extend the 2-approximation of [8] for uncrossable families to semi-uncrossable families, by an almost identical proof, is the one in Lemma 2. If pliable families had this property, then the proof would extend to them as well.

4 Concluding Remarks

In this paper we introduced a new class of semi-uncrossable set families that extends the class of uncrossable families. The algorithm of [8] for uncrossable families extends to the more general class of semi-uncrossable families with the same approximation ratio of 2. Our work shows that often a combination of two problems with "good" uncrossing properties, such that at least one of them is non-symmetric, gives a problem with a semi-uncrossable family that is not uncrossable in general.

Jain [7] introduced the iterative rounding method, and used it to obtain approximation ratio 2 for the problem of covering a skew-supermodular (a.k.a. weakly supermodular) set function; the class of skew-supermodular set functions generalizes the class of uncrossable set families. It is an interesting question whether Jain's result can be extended to a larger class of set functions.

References

1. Agrawal, A., Klein, P., Ravi, R.: When trees collide: an approximation algorithm for the generalized Steiner problem on networks. SIAM J. Comput. **24**(3), 440–456 (1995)
2. Bansal, I., Cheriyan, J., Grout, L., Ibrahimpur, S.: Improved approximation algorithms by generalizing the primal-dual method beyond uncrossable functions. In: ICALP, pp. 15:1–15:19 (2023)
3. Couëtoux, B., Davis, J.M., Williamson, D.P.: A 3/2-approximation algorithm for some minimum-cost graph problems. Math. Program. **150**, 19–34 (2015)
4. Goemans, M.X., Williamson, D.P.: Approximating minimum-cost graph problems with spanning tree edges. Oper. Res. Lett. **16**, 183–189 (1994)
5. Goemans, M.X., Williamson, D.P.: A general approximation technique for constrained forest problems. SIAM J. Comput. **24**(2), 296–317 (1995)
6. Hajiaghayi, M.-T., Khandekar, R., Kortsarz, G., Nutov, Z.: On fixed cost k-flow problems. Theory Comput. Syst. **58**(1), 4–18 (2016)
7. Jain, K.: A factor 2 approximation algorithm for the generalized Steiner network problem. Combinatorica **21**(1), 39–60 (2001)
8. Williamson, D.P., Goemans, M.X., Mihail, M., Vazirani, V.V.: A primal-dual approximation algorithm for generalized Steiner network problems. Combinatorica **15**(3), 435–454 (1995)

Network Flow Problems with Electric Vehicles

Haripriya Pulyassary[1(✉)], Kostas Kollias[2], Aaron Schild[2], David Shmoys[1], and Manxi Wu[1]

[1] School of Operations Research and Information Engineering, Cornell University, Ithaca, USA
{hp297,david.shmoys,manxiwu}@cornell.edu
[2] Google Research, Mountain View, USA

Abstract. Electric vehicle (EV) adoption in long-distance logistics faces challenges such as range anxiety and uneven distribution of charging stations. Two pivotal questions emerge: How can EVs be efficiently routed in a charging network considering range limits, charging speeds and prices? And, can the existing charging infrastructure sustain the increasing demand for EVs in long-distance logistics? This paper addresses these questions by introducing a novel theoretical and computational framework to study the EV network flow problems. We present an EV network flow model that incorporates range constraints and nonlinear charging rates, and identify conditions under which polynomial-time solutions can be obtained for optimal single EV routing, maximum flow, and minimum-cost flow problems. Our findings provide insights for optimizing EV routing in logistics, ensuring an efficient and sustainable future.

Keywords: Electric vehicle routing · Network flow algorithms · Charge-augmented networks

1 Introduction

Electric vehicle (EV) adoption for long-distance logistics faces the challenge of range anxiety. Due to the weight of the cargo and constraints on battery sizes, charging can take a considerable proportion of the trip time and requires the use of dedicated charging facilities. Additionally, the uneven distribution and limited availability of charging stations in specific regions or along certain transportation routes introduce further complications for long-distance logistics planning. Two essential questions arise: First, how to efficiently route a single EV or a fleet of EV in a charging network given the battery constraint, charging speed and prices? Second, given the existing charging network, what is the maximum EV flow that can be supported? Answering these questions is crucial for the efficient deployment of EVs in logistics services, and assessing if the current charging infrastructure can support the growing EV demand in long-distance logistics.

To address the above-mentioned questions, we develop an EV network flow model that incorporates range constraints and nonlinear charging speeds. We

J. Vygen and J. Byrka (Eds.): IPCO 2024, LNCS 14679, pp. 365–378, 2024.
https://doi.org/10.1007/978-3-031-59835-7_27

consider a directed graph, in which a subset of nodes are equipped with electric vehicle charging stations. Chargers across different stations can vary in charging speed (e.g., DC fast chargers, level-one and level-two chargers). The charging speed, measured as the percentage of battery replenished over time, depends on the vehicle's current battery level—the higher the battery level, the slower the charging rate. We route a fleet of electric vehicles through this network, where each vehicle has a battery range limit and is directed between their origin and destination. The vehicle's routing strategy specifies the chosen path and the battery charge acquired at each visited station. A charging strategy is feasible if the battery level remains above zero, and does not exceed its range limit.

We study three fundamental problems in this setting: (i) the *single EV optimal charging strategy* problem that computes a charging strategy that minimizes its total (monetary and time) cost; (ii) the *maximum EV flow* problem that computes the maximum volume of EV fleet that can be served by the charging network given its station capacity constraints; and (iii) the *minimum-cost EV flow* problem that routes an EV fleet with a fixed volume to minimize the total (time and monetary) costs. These problems can be viewed as natural extensions of the shortest path, maximum flow, and minimum-cost flow problems in the classical network flow models, but with two new hurdles: First, given that the feasible charging strategy is a vector with continuous values, the corresponding flow vector, representing the volume of the EV fleet adopting each feasible charging strategy, is infinite-dimensional. Second, as a station's charging speed changes with the battery level of the arriving vehicle, the station's effective capacity inherently depends on the flow vector. Consequently, the EV flow problems studied in this work differ non-trivially from their standard network flow counterparts [2,7,9].

We show that the three problems described above are NP-hard in general settings. We pinpoint precise conditions under which we can provide polynomial-time algorithms to solve them. In problem (i), we employ a dynamic programming approach to solve the optimal single EV charging strategy under the assumption that the path with the minimum battery consumption between a pair of nodes is also the path with the minimum travel time cost (Assumption 1). Our algorithm generalizes the previous studies [12,24] by incorporating non-linear charging speed and general charging cost functions. In problems (ii) and (iii), we cast the flow problems as semi-infinite linear programs with an infinite number of variables and a finite set of constraints [15,21]. Our investigation reveals that there exists an optimal flow vector that utilizes only a finite number of feasible charging strategies, which are the extreme points of the feasible charging strategy set. Building on this observation, we construct a charge-augmented network, where each path represents an extreme point feasible charging strategy. When the network has capacity constraints only on the charging stations, the maximum EV flow problem and the minimum-cost EV flow problem can be solved using a polynomial-sized linear program using the charge-augmented network (the latter result also requires Assumption 1). We also provide fully polynomial time approximation schemes for solving the network flow problems in the general setting with edge capacity constraints.

Our results also contribute to the rich literature of EV routing and charging that includes path planning and charging of EV fleets in delivery problems [1,3,4,6,8,13,14,18,20,23], ride-hailing operations with EVs [5,14], infrastructure planning and battery swapping system design [16,20,22,25]. This line of literature primarily focuses on developing computational methods of path generation and charging scheduling. Our focus is different in that we aim at developing efficient network algorithms for long-distance logistics, and assessing the capacity of charging networks.

2 The Charging Network Model

We study the problem of routing electric vehicles in a directed graph $\mathcal{G} = (V, E)$, where V is the set of nodes, and E is the set of edges. The network has $|K|$ origin-destination (o-d) pairs $\{(s_k, t_k)\}_{k \in K}$. A subset of nodes $I \subseteq V$ are installed with chargers of different types. Each node $i \in I$ is equipped with $a_i \in \mathbb{N}_{\geq 0}$ number of chargers. Chargers at different nodes may have different charging speeds, which we will detail shortly.[1]

Electric vehicles are each equipped with a battery of size L. We assume that all vehicles are fully charged when departing from their origin, and origin and destination nodes $\{s_k, t_k\}_{k \in K}$ have no chargers. As a vehicle traverses an edge $e \in E$, it consumes $d_e \geq 0$ amount of battery, and takes time $\ell_e \geq 0$. A vehicle that travels between any o-d pair (s_k, t_k) selects a charging strategy (P, q), where $P : s_k = v_0\, v_1\, \ldots\, v_m\, v_{m+1} = t_k$ is an s_k-t_k path in the network, and $q = (q_v)_{v \in V}$ indicates the amount of battery charged at each node. A charging strategy (P, q) is feasible if it satisfies the following constraints:

$$b_{v_i}^{\text{in}} := \sum_{n=0}^{i-1} q_{v_n} + L - \sum_{n=0}^{i} d_{(v_{n-1}, v_n)} \geq 0, \quad \forall i = 1, \ldots, m+1, \tag{1a}$$

$$b_{v_i}^{\text{out}} := \sum_{n=0}^{i} q_{v_n} + L - \sum_{n=0}^{i} d_{(v_{n-1}, v_n)} \leq L, \quad \forall i = 1, \ldots, m, \tag{1b}$$

$$q_v = 0, \quad \forall v \in V \setminus \{P \cap I\}, \qquad q_v \geq 0, \quad \forall v \in V, \tag{1c}$$

where $b_{v_i}^{\text{in}}$ as in (1a) is the battery level of a vehicle when arriving at v_i that is non-negative, and $b_{v_i}^{\text{out}}$ as in (1b) is the battery level when leaving v_i that does not exceed L. The constraint (1c) ensures that the charge amount is non-negative, and charge can only happen at charging stations that are on the path P. For a given s_k-t_k path P, the set of feasible charging vectors q is defined as

$$Q_P = \{q \in \mathbb{R}^{|V|} : q \text{ satisfies (1a)} - \text{(1c)}\}. \tag{2}$$

[1] When a node is equipped with multiple types of chargers, we can equivalently view each type of chargers as a separate node. These nodes are connected by edges with zero distances.

Here, Q_P is a polytope. Let $\mathcal{P}_k$ be the set of all s_k-t_k paths and $\mathcal{Q}_k = \{(P,q) : q \in Q_P, P \in \mathcal{P}_k\}$. We define the set of all paths as $\mathcal{P} = \cup_{k \in K} \mathcal{P}_k$ and the set of all feasible charging strategies as $\mathcal{Q} = \{(P,q)\}_{q \in Q_P, P \in \mathcal{P}}$.

We next define the charging speed, and the cost of a feasible charging strategy. One key feature of electric vehicle charging is that the charging speed (i.e., the amount of battery charged per unit of time) decreases as the battery level of the vehicle increases. We assume that the charging speed is a piecewise constant function of a vehicle's battery level $b \in [0, L]$, and the speed of charger at node i is $r_i(b)$ given by:

$$r_i(b) = r_{ij}, \quad \forall \alpha_j \leq b < \alpha_{j+1}, \quad \forall j \in \{1, \dots, J\}, \tag{3}$$

where $\alpha_1, \dots, \alpha_{J+1}$ are fixed thresholds with $0 = \alpha_1 < \alpha_2 < \cdots < \alpha_{J+1} = L$, and $r_{ij} > 0$ is the speed at which a vehicle charges if its present battery level is between α_j and α_{j+1} (refered as the j-th interval). Here, the thresholds $\{\alpha_j\}_{j=1}^J$ are set to be the same for all chargers although their charging speeds may differ in each interval. This is without loss of generality since we can define the piecewise constant function using the combined threshold sets of all chargers.

The *cost* of a feasible charging strategy (P,q) includes the *monetary* cost for charging at each station, and the time cost of both traversing edges and charging at stations. In particular, vehicles are charged with price $\tau_{ij} > 0$ for a unit of battery obtained from charger at node i given that the vehicle's battery level is in the j-th interval, and price $\rho_i \geq 0$ for occupying a charging spot at station i for one unit of time. The cost of a charging strategy (P,q) with $P = v_0\, v_1 \cdots v_m\, v_{m+1}$ is

$$c(P,q) = \sum_{i=0}^{m} \ell_{(v_i,v_{i+1})} + \sum_{i \in I} \sum_{j=1}^{J} \left(\tau_{ij} + \frac{1+\rho_i}{r_{ij}} \right) q_{ij}, \tag{4}$$

where $\sum_{i=0}^{m} \ell_{(v_i,v_{i+1})}$ is the total latency cost of traversing the edges along path P, q_{ij} is the amount of battery at node i with battery level between α_j and α_{j+1}, and $(\tau_{ij} + \frac{1+\rho_i}{r_{ij}})q_{ij}$ is the monetary and time cost of charging q_{ij}.[2]

A *flow* vector in the charging network is defined as $x = (x_{P,q})_{(P,q)\in\mathcal{Q}}$, where $x_{P,q}$ is the non-negative flow assigned to each feasible charging strategy $(P,q) \in \mathcal{Q}$. Since the charge vector q is a real-valued vector, x has *infinite dimension.* A flow vector x is feasible if the total amount of battery charged at each station $i \in I$ is less than or equal to the amount of charge that can be dispensed per unit time. Here, we need to account for the fact that the charging speed of each charger as in (3) is a piecewise constant function of the battery level of the vehicle. Therefore, the amount of battery charge that each charger can provide

[2] Given a charging strategy (P,q), $q_{ij} = \min\left(\alpha_{j+1} - (b_i^{\text{in}} + \sum_{n=1}^{j} q_{in}), q_i - \sum_{n=1}^{j} q_{in}\right) \times \mathbb{I}\left[\alpha_j \leq b_i^{\text{in}} + \sum_{n=1}^{j} q_{in} < \alpha_{j+1}\right]$ for all $i \in I \cap P$ and all $j \in \{1, \dots J\}$, and b_i^{in} is given by (1a). Moreover, $q_{ij} = 0$ for all $i \notin P$ and all $j \in \{1, \dots, J\}$. We can check that $\sum_{j=1}^{J} q_{ij} = q_i$.

depends on x. We introduce variables $z = (z_{ij})_{i \in V, j=1,\ldots,J}$, where z_{ij} is the fraction of the total charging capacity of station i that is used to charge vehicles in the battery interval $[\alpha_j, \alpha_{j+1})$. The set of feasible flows x is characterized by the following constraints, where (5a) ensures that the amount of charge consumed by flow x at rate r_{ij} from each station $i \in I$ does not exceed the allocated amount $r_{ij} z_{ij}$, and (5b) ensures that the vector z is feasible given the station's capacity:

$$\sum_{(P,q)\in \mathcal{Q}} q_{ij} x_{P,q} \leq r_{ij} z_{ij}, \quad \forall i \in I, \quad \forall j \in \{1, \ldots, J\}, \tag{5a}$$

$$\sum_{j=1}^{J} z_{ij} = a_i, \quad \forall i \in I, \qquad x, z \geq 0. \tag{5b}$$

Our goal is to develop efficient algorithms for the following problems:
(i) *Single EV optimal charging problem*: Given a charging network, a single o-d pair (s_k, t_k), and monetary charging costs (τ, ρ), compute

$$(P^*, q^*) \in \underset{(P,q)\in \mathcal{Q}_k}{\arg\min}\ c(P, q), \tag{Single-OPT}$$

where $c(P, q)$ is given by (4).
(ii) *Maximum EV flow problem*: Given a charging network, compute

$$x^* \in \arg\max_{x} \sum_{(P,q)\in \mathcal{Q}} x_{P,q}, \quad s.t. \quad x \text{ satisfies (5a)–(5b)}. \tag{P$_{\text{max-flow}}$}$$

(iii) *Minimum-cost EV flow problem*: Given a charging network and monetary charging costs (τ, ρ), $x^\dagger$ minimizes the total cost of the flow with demand vector $(D_k)_{k\in K}$:

$$\begin{aligned} x^\dagger \in \arg\min_{x} C(x) := \sum_{(P,q)\in \mathcal{Q}} c(P, q) \cdot x_{P,q}, \\ s.t. \quad x \text{ satisfies (5a)-(5b)}, \quad \text{and} \sum_{(P,q)\in \mathcal{Q}_k} x_{P,q} \geq D_k, \quad \forall k \in K. \end{aligned} \tag{P$_{\text{min-cost}}$}$$

3 Optimal Charging Strategy for Single EV

With general battery consumption vector $d = (d_e)_{e\in E}$ and time cost vector $\ell = (\ell_e)_{e\in E}$, (i) single EV optimal charging problem is NP-hard since the *Resource Constrained Shortest Path Problem* is a special case when the network has no charging station [11]. In this section, we show that under Assumption 1, the optimal single EV optimal routing strategy can be solved in polynomial time using a dynamic programming-based approach.

Assumption 1 *For any $i, i' \in I \cup \{s_k, t_k\}_{k\in K}$, $\mathcal{P}_{ii'}$ is the set of paths that connect i and i', and* $\arg\min_{P_{ii'}\in \mathcal{P}_{ii'}} \sum_{e\in P_{ii'}} d_e = \arg\min_{P_{ii'}\in \mathcal{P}_{ii'}} \sum_{e\in P_{ii'}} \ell_e$.

Assumption 1 implies that between any pair of nodes, the path with the minimum battery consumption is also the path with the minimum time cost. In practice, this assumption holds when the changes in elevation and speed are small in the network. Given a charging network, we can verify Assumption 1 in $O(|I|^3)$ time by running the Dijkstra's algorithm to compute the shortest path between each pair of nodes using the battery consumption vector d and the time cost vector ℓ, respectively.

We construct an auxiliary network on nodes $\{(i,j)\}_{i\in I, j\in\{1,\dots,J\}} \cup \{s_k,t_k\}_{k\in K}$, where each (i,j) is the j-th copy of the original station i that has a_i number of chargers with charging rate r_{ij}. For any two stations $i,i'\in I$, we compute the path with the minimum battery consumption (which also has the minimum time cost given Assumption 1), and denote $d_{ii'}$ (resp. $\ell_{ii'}$) as the battery consumption (resp. time cost) of that path.[3] In the auxiliary network, we connect (i,j) and (i',j') for any $i,i'\in I$ such that $d_{ii'}\le L$ and any $j,j'\in\{1,\dots,J\}$. The battery consumption of such an edge $e'=((i,j),(i',j'))$ is $d_{e'}=d_{ii'}$ and the time cost is $\ell_{e'}=\ell_{ii'}$. We also connect (i,j) and (i,j') for any $j,j'\in\{1,\dots,J\}$ and any $i\in I$ with time cost $\ell_{e'}=0$ and battery consumption $d_{e'}=0$. In the auxiliary network, the total number of nodes is $|I|J$, and the number of edges is $O(|I|^2J^2)$. For each node (i,j), we denote the lower bound of battery level of charging at (i,j) as $\underline{b}_{ij}=\alpha_j$ and the upper bound as $\bar{b}_{ij}=\alpha_{j+1}$.

For any (s_k,t_k), a charging strategy (P,q) is analogously defined as $P = s_k\,(i_1,j_1)\,\dots\,(i_m,j_m)\,t_k$ being a path between the origin and destination and $q=(q_{ij})_{i\in I, j\in\{1,\dots,J\}}$. A charging strategy (P,q) is feasible if the battery level of arriving at any $(i,j)\in P$ is higher or equal to $\underline{b}_{ij}$ and less than $\bar{b}_{ij}$. Since the auxiliary network connects any pair of nodes that are within the vehicle's range constraint, we can restrict our attention to consider only charging strategies (P,q), where $q_{ij}>0$ for any $(i,j)\in P$. This is without loss of generality since any charging strategy (P',q') that violates this constraint can be improved by another strategy (P,q) such that P removes any nodes with zero charge amount, and the charge amount for the remaining nodes do not change.

To compute (P^*,q^*), we solve the following recurrence, where $A(i,j,b)$ is the minimum-cost of traveling from a node (i,j) to the destination t_k, given that the vehicle arrived at (i,j) with $b\in[0,L]$ battery level, and $N(i,j)$ is the set of neighbours of (i,j). Then,

$$A(i,j,b) = \begin{cases} \min\limits_{(i',j')\in N(i,j)} \min\limits_{x\ge 0} A(i',j',\, b+x-d_{(ij,i'j')}) + c_{ij}x + \ell_{(ij,i'j')}, & \underline{b}_{ij}\le b\le \bar{b}_{ij},\\ \infty, & \text{otherwise,}\end{cases} \tag{6}$$

where $c_{ij}=\tau_{ij}+\frac{1+\rho_i}{r_{ij}}$ for all $i\in I, j\in\{1,\dots,J\}$ is the per-unit charging cost at (i,j) as in (4).

[3] Without Assumption 1, the auxiliary network cannot be used to compute the minimum cost charging strategy because a path with the minimum battery consumption in the auxiliary network may have a high time cost in the original network.

Since the battery level b of a vehicle is a continuous quantity, solving the dynamic program as in (6) requires discretization of battery levels. Our next lemma shows that in an optimal solution with path P, a vehicle charges at a node (i_n, j_n) either to reach the maximum allowable battery level $\bar{b}_{i_n j_n}$ or to just be able to reach the next node (i_{n+1}, j_{n+1}).

Lemma 1. *There exists an optimal charging strategy (P^*, q^*) where $P^* = s_k\,(i_1, j_1) \cdots (i_m, j_m)\,t_k$ and q^* satisfies*

$$q^*_{i_n, j_n} = \begin{cases} \underline{b}_{i_{n+1} j_{n+1}} + d_{(i_n j_n, i_{n+1} j_{n+1})} - b^{\text{in}}_{i_n j_n} & \text{if } c_{i_n j_n} > c_{i_{n+1} j_{n+1}}, \\ \bar{b}_{i_n j_n} - b^{\text{in}}_{i_n j_n}, & \text{otherwise.} \end{cases} \tag{7}$$

Consequently, there always exists an optimal charging strategy satisfying the following property: the battery level b^{in}_{ij} of a vehicle arriving at any (i, j) can only take a finite number of values given by

$$B^{\text{in}}_{ij} = \{\underline{b}_{ij}\} \cup \{\bar{b}_{i'j'} - d_{(ij,i'j')} | (i', j') \in N(i, j), \underline{b}_{ij} \le \bar{b}_{i'j'} - d_{(ij,i'j')} \le \bar{b}_{ij}\}. \tag{8}$$

The proof of Lemma 1 can be found in the full version of this paper [19]. Thanks to this lemma, we only need to analyze the value of $A(i, j, b)$ for $b \in B^{\text{in}}_{ij}$ instead of the continuous battery level $b \in [0, L]$. We define an augmented network with the nodes set $\tilde{V} := \{(i, j, b)\}_{i \in I, j=1,\ldots,J, b \in B^{\text{in}}_{ij}}$. A node (i, j, b) is connected to $(i', j', \underline{b}_{i'j'})$ if $(i', j') \in N(i, j)$ and $c_{i'j'} < c_{ij}$. The cost of such edge $\tilde{e} = ((i, j, b), (i', j', \underline{b}_{i'j'}))$ is $w_{\tilde{e}} = (d_{(ij,i'j')} + \underline{b}_{i'j'} - b)c_{ij} + \ell_{(ij,i'j')}$. On the other hand, if $c_{i'j'} \ge c_{ij}$, then node (i, j, b) is connected to $(i', j', \bar{b}_{ij} - d_{(ij,i'j')})$ with cost $w_{\tilde{e}} = (\bar{b}_{ij} - b)c_{ij} + \ell_{(ij,i'j')}$. Following from (6), the value of $A(i, j, b)$ for any (i, j) and any $b \in B^{\text{in}}_{ij}$ equals to the cost of the shortest path (with respect to w) from the node (i, j, b) to node $(t_k, 0)$, and the value of $A(i, j, b) = \infty$ if such a path does not exist. Moreover, the minimum-cost charging strategy of a single EV with o-d pair (s_k, t_k) corresponds to the shortest (s_k, L)-$(t_k, 0)$ path of this augmented network, and thus can be computed by the Dijkstra's algorithm.

Theorem 1. *A single EV optimal charging strategy (P^*, q^*) can be computed in $O((|I|J)^6 \log(|I|J))$ time.*

4 Maximum and Minimum-Cost EV Flow Problems

In Sect. 4.1, we show that the maximum EV flow problem can be computed in polynomial time, and the minimum-cost EV flow problem can be computed in polynomial time under Assumption 1. In Sect. 4.2, we generalize our setting to incorporate capacity constraints on edges. We demonstrate that the flow problems become NP-hard, and we provide fully polynomial time approximation schemes.

4.1 Unconstrained Edge Capacity

The dual program of ($\text{P}_{\text{max-flow}}$) is as follows:

$$\min \sum_{i \in I} a_i y_i, \tag{$\text{D}_{\text{max-flow}}$}$$

$$\text{s.t.} \sum_{i \in I} \sum_{j=1}^{J} \pi_{ij} q_{ij} \geq 1, \qquad \forall (P, q) \in \mathcal{Q}, \tag{9a}$$

$$y_i \geq \pi_{ij} r_{ij}, \qquad \forall i \in I, \quad \forall j \in \{1, \ldots, J\}, \tag{9b}$$

$$\pi \geq 0, \tag{9c}$$

where π_{ij} and y_i are the dual variables associated with the primal constraints (5a) and (5b), respectively.

Since there are an infinite number of feasible charging strategies, and a finite number of stations, both the primal and the dual programs are semi-infinite linear programs. For each $P \in \mathcal{P}$, let $\hat{Q}_P$ denote the extreme points of the polytope Q_P, and $\hat{\mathcal{Q}} = \cup_{P \in \mathcal{P}} \{(P, \hat{q}) : \hat{q} \in \hat{Q}_P\}$. The next lemma shows that, since the set of feasible charging vectors is the union of polytopes, we can reformulate the dual program as a finite linear program, where constraints (9a) need only to be verified for all $(P, \hat{q}) \in \hat{\mathcal{Q}}$. As a result, the primal program is equivalent to a finite linear program.

Lemma 2. *The dual* ($\mathrm{D}_{\text{max-flow}}$) *is equivalent to the following linear program:*

$$\min \sum_{i \in I} a_i y_i, \tag{$\mathrm{D}'_{\text{max-flow}}$}$$

$$s.t. \sum_{i \in I} \sum_{j=1}^{J} \pi_{ij} \hat{q}_{ij} \geq 1, \qquad \forall (P, \hat{q}) \in \hat{\mathcal{Q}}, \tag{10a}$$

$$y_i \geq \pi_{ij} r_{ij}, \qquad \forall i \in I, \ \forall j \in \{1, \ldots, J\}, \tag{10b}$$

$$\pi \geq 0. \tag{10c}$$

Additionally, ($\mathrm{P}_{\text{max-flow}}$) *is equivalent to the following linear program:*

$$\max_{x,z} \sum_{(P,\hat{q}) \in \hat{\mathcal{Q}}} \hat{x}_{P,\hat{q}}, \tag{$\mathrm{P}'_{\text{max-flow}}$}$$

$$s.t. \sum_{(P,\hat{q}) \in \hat{\mathcal{Q}}} \hat{q}_{ij} \hat{x}_{P,\hat{q}} \leq r_{ij} z_{ij}, \qquad \forall i \in I, \ \forall j \in \{1, \ldots, J\}, \tag{11a}$$

$$\sum_{j=1}^{J} z_{ij} = a_i, \qquad \forall i \in I, \tag{11b}$$

$$z_{ij} \geq 0, \qquad \forall i \in I, \ \forall j \in \{1, \ldots, J\}. \tag{11c}$$

Lemma 2 indicates that the optimal flow can be achieved only by using charging strategies that are in $\hat{\mathcal{Q}}$. We know from Lemma 1 that for such charging strategies, the set of possible battery levels of vehicles arriving at each node (i, j) has to be in the (polynomial-sized) set B_{ij}^{in} as in (8). Moreover, a vehicle departing from (i, j) will either have the maximum possible battery level, $\bar{b}_{ij}$, or the minimum amount required to reach an adjacent node $(i', j') \in N(i, j)$. As a result, the set of possible battery levels of a vehicle leaving node (i, j) is given by:

$$B_{ij}^{\text{out}} = \{\bar{b}_{ij}\} \cup \{\underline{b}_{i'j'} + d_{(ij,i'j')} | (i', j') \in N(i, j), \ \underline{b}_{ij} \leq \underline{b}_{i'j'} + d_{(ij,i'j')} \leq \bar{b}_{ij}\}. \tag{12}$$

A key insight of the above discussion is that there exists an optimal solution in which every vehicle's charge level upon arrival and departure from a node belongs to a discrete, polynomial-sized set. Due to this observation, we are able to construct a polynomial-sized charge-augmented network $\hat{\mathcal{G}}$. For every $i \in I$ and $j \in \{1, \ldots, J\}$, we create $|B_{ij}^{\text{in}}| + |B_{ij}^{\text{out}}|$ copies of the node (i, j), denoted as

$$\hat{V}_{ij} = \{(i, j, b) : b \in B_{ij}^{\text{in}} \cup B_{ij}^{\text{out}}\},$$

where each node $(i, j, b) \in \hat{V}_{ij}$ represents a vehicle with battery level b at station-copy (i, j). We define the node set as $\hat{V} = \cup_{i \in I} \cup_{j=1}^{J} \hat{V}_{ij} \cup \{(s_k, J, L) : k \in K\} \cup \{(t_k, 0, 0) : k \in K\}$, where (s_k, J, L) represents that a vehicle departs from an origin s_k with full battery, and $(t_k, 0, 0)$ represents that a vehicle arrives at a destination t_k with no battery left.[4] Moreover, $\hat{\mathcal{G}}$ has three types of edges:

- *Type I (Charging) edges:* $\hat{E}_{ij} := \{((i, j, b_{ij}^{\text{in}}), (i, j, b_{ij}^{\text{out}})) | b_{ij}^{\text{in}} \in B_{ij}^{\text{in}}, b_{ij}^{\text{out}} \in B_{ij}^{\text{out}}, b_{ij}^{\text{out}} > b_{ij}^{\text{in}}\}$ represents that a vehicle gets a $b_{ij}^{\text{out}} - b_{ij}^{\text{in}}$ amount of charge from station i and battery interval j.

- *Type II (Station-copy connection) edges:* $\hat{E}_{ij,j+1} := \{((i, j, \overline{b}_{ij}), (i, j+1, \underline{b}_{i,j+1}))\}$ connects consecutive station copies.

- *Type III edges:* $\hat{E}_{ij,i'j'} := \{(i, j, b_{ij}^{\text{out}}), (i', j', b_{i'j'}^{\text{in}}) | b_{ij}^{\text{out}} \in B_{ij}^{\text{out}}, b_{i'j'}^{\text{in}} \in B_{i'j'}^{\text{in}}, b_{ij}^{\text{out}} - b_{i'j'}^{\text{in}} = d_{ij,i'j'}\}$ represents that a vehicle leaves a node (i, j) and moves to another node (i', j') with battery consumption of $b_{ij}^{\text{out}} - b_{i'j'}^{\text{in}}$ being equal to $d_{ij,i'j'}$.

Note that a vehicle does not leave the physical station by taking type I and II edges, and hence $d_{\hat{e}} = \ell_{\hat{e}} = 0$ for such edges. On the other hand, the battery consumption of a vehicle taking a type III edge $\hat{e} = ((i, j, b), (i', j', b'))$, is $d_{\hat{e}} = d_{ii'}$ (and the time cost is $\ell_{\hat{e}} = \ell_{ii'}$).

Example 1. Consider the network in Fig. 1a, with two stations, i_1 and i_2. All vehicles have battery capacity $L = 9$. The charging speed for station i_1 is 2 if the vehicle's battery level is in the range $[0, 5]$ and 1 otherwise. For station i_2, the charging speed is 3 if the vehicle's battery level is in the range $[0, 5]$ and 2 otherwise. The charge-augmented network is given in Fig. 1b. The red arcs correspond to the type I edges for each station-copy. The gray dashed arcs are type II edges connecting consecutive copies of a given station. The black arcs are type III edges. The charging speeds of the green, blue, and yellow nodes are 3, 2, and 1 respectively.

If, for some (s_k, t_k), there exists an s_k-t_k path of length at most L, an infinite amount of flow can be sent along this path. We assume that this is not the case,

[4] Setting the arriving battery level to be 0 is without loss of generality since any vehicle that arrives with a positive battery level can reduce the amount of battery charged at the last station.

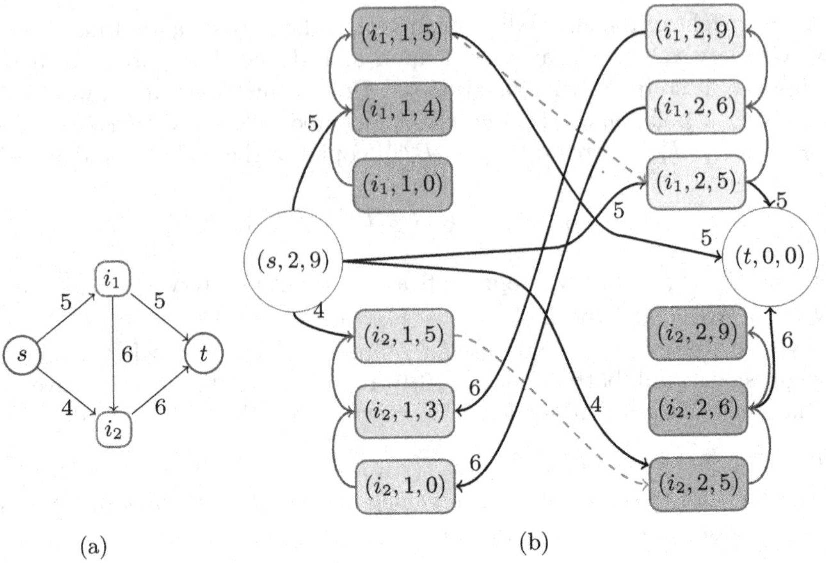

Fig. 1. (a) Original charging network $\mathcal{G}$, (b) Charge-augmented network $\hat{\mathcal{G}}$.

i.e., the length of every s_k-t_k path is strictly larger than L, for all $k \in K$. For each (s_k, t_k), if there is no (s_k, J, L)-$(t_k, 0, 0)$ path in $\hat{\mathcal{G}}$, a feasible solution does not exist. Moreover, by Lemma 2, it suffices to restrict our attention to charging strategies satisfying (7), and hence the assumption that a vehicle arrives at its destination with no battery left is without loss of generality.

Furthermore, every path connecting (s_k, J, L) and $(t_k, 0, 0)$ corresponds to a charging strategy $(P, \hat{q})$ in $\hat{\mathcal{Q}}$. Recall from Lemma 2 that we can restrict our attention to the EV flow vector that only sends positive flow along charging strategies $(P, \hat{q}) \in \hat{\mathcal{Q}}$. Therefore, the maximum EV flow vector of the charging network $\mathcal{G}$ can be equivalently computed as the maximum multicommodity flow of the charge-augmented network $\hat{\mathcal{G}}$ with station capacity imposed on type I edges. We denote the edge load vector in $\hat{\mathcal{G}}$ as $\hat{f} = (\hat{f}^k_{\hat{e}})_{\hat{e} \in \hat{E}, k \in K}$, where $\hat{f}^k_{\hat{e}}$ is the flow on edge $\hat{e}$ associated with s_k-t_k. The maximum multicommodity flow $\hat{f}^*$ of $\hat{\mathcal{G}}$ can be computed as an optimal solution of a linear program. Then, we can construct the maximum EV flow vector x^* using the flow decomposition theorem.

Theorem 2. *The vector $\hat{f}^*$ induced by the maximum EV flow vector x^* is the optimal solution of the following linear program:*

$$\max_{\hat{f}} \sum_{k \in K} \sum_{\hat{e} \in \delta^-(t_k, 0)} \hat{f}^k_{\hat{e}},$$

$$\sum_{k \in K} \sum_{\hat{e} \in \hat{E}_{ij}} \lambda_{\hat{e}} \hat{f}^k_{\hat{e}} \leq r_{ij} z_{ij}, \qquad \forall i \in I, \quad \forall j \in \{1, \ldots, J\}, \tag{13a}$$

$$\sum_{j=1}^{J} z_{ij} = a_i, \qquad \forall i \in I, \tag{13b}$$

$$\sum_{\hat{e} \in \delta^+(\hat{v})} \hat{f}_{\hat{e}}^k - \sum_{\hat{e} \in \delta^-(\hat{v})} \hat{f}_{\hat{e}}^k = 0, \quad \forall \hat{v} \in \hat{V} \setminus \{(s_k, J, L), (t_k, 0, 0)\}, \ \forall k \in K, \tag{13c}$$

$$\hat{f}_{\hat{e}}^k \geq 0, \qquad \forall \hat{e} \in \hat{E}, \quad \forall k \in K, \tag{13d}$$

$$z_{ij} \geq 0, \qquad \forall i \in I, \quad \forall j \in \{1, \ldots, J\}, \tag{13e}$$

where $\lambda_{\hat{e}}$ is the amount of battery charged when traversing the edge $\hat{e} \in \hat{E}_{ij}$.

We now turn our attention to the minimum-cost EV flow problem, where a flow of value at least D_k is routed from s_k to t_k, for each $k \in K$, so as to minimize the sum of all monetary and time-costs ($\mathrm{P}_{\mathrm{min-cost}}$). In general settings, this problem is NP-hard. This is because in the special case with unlimited station capacity, the min-cost EV flow problem reduces to routing all the flow using a single EV optimal charging strategy, which has been shown to be NP-hard (Sect. 3). Under Assumption 1, we show that a minimum-cost EV flow can be computed in polynomial-time following similar analysis as for the maximum EV flow problem building on Lemma 1. The complete proof can be found in the full version of this paper [19].

Theorem 3. *Under Assumption 1, the vector $\hat{f}^\dagger$ induced by the minimum-cost EV flow vector $x^\dagger$ is an optimal solution of the following linear program:*

$$\min_{\hat{f}} \sum_{k \in K} \sum_{\hat{e} \in \hat{E}} \gamma_{\hat{e}} \hat{f}_{\hat{e}}^k, \ \text{s.t. } \hat{f} \text{ satisfies (13a)} - -\text{(13e)}, \ \sum_{\hat{e} \in \delta^-(t_k, 0)} \hat{f}_{\hat{e}}^k \geq D_k, \ \forall k \in K, \tag{14a}$$

where $\gamma_{\hat{e}}$ is the cost of traversing one unit of flow through the edge $\hat{e} \in \hat{E}$, i.e., $\gamma_{\hat{e}} = (\tau_{ij} + (1 + \rho_i)/r_{ij})\lambda_{\hat{e}}$ for all $\hat{e} \in \hat{E}_{ij}$ and $\gamma_{\hat{e}} = \ell_{ij,i'j'}$ for all $\hat{e} \in \hat{E}_{ij,i'j'}$ and all $i, i' \in I, j, j' \in \{1, \ldots J\}$.

Thus, by solving the linear program given in Theorem 2 (resp. Theorem 3), we can compute an optimal solution to the maximum EV flow (resp. minimum-cost EV flow) problem in polynomial time.

4.2 EV Flow with Edge Capacities

In this section, we consider a generalized setting where apart from the station capacity constraints, each edge $e \in E$ also has a capacity $u_e \geq 0$. That is, an EV flow x is feasible if it satisfies (5a)–(5b), and the edge-capacity constraints:

$$\sum_{(P,q) \in \mathcal{Q}: e \in P} x_{P,q} \leq u_e, \quad \forall e \in E. \tag{15}$$

We show that the maximum EV flow and the minimum-cost EV flow problems with edge capacities are NP-hard, demonstrated by a reduction from `PARTITION`. The proof of this proposition can be found in the full version of this paper [19].

Proposition 1. *It is NP-hard to decide if there exists an EV flow x that satisfies* (5a)–(5b) *and* (15) *with* $|x| \geq D$.

Our fully polynomial-time approximation schemes build on the equivalence of the approximate separation and approximate optimization for the dual programs of ($\mathrm{P}_{\text{max-flow}}$) and ($\mathrm{P}_{\text{min-cost}}$) with edge capacity constraints (15):

$$\min \sum_{i\in I} a_i y_i + \sum_{e\in E} u_e w_e, \tag{$\mathrm{D}^{\text{cap}}_{\text{max-flow}}$}$$

$$s.t. \sum_{i\in I}\sum_{j=1}^{J} \pi_{ij} q_{ij} + \sum_{e\in P} w_e \geq 1, \qquad \forall (P,q) \in \mathcal{Q}, \tag{16a}$$

$$y_i \geq \pi_{ij} r_{ij}, \qquad \forall i \in I, \quad \forall j \in \{1,\ldots,J\}, \tag{16b}$$

$$\pi, w \geq 0, \tag{16c}$$

where w_e is the dual variable associated with constraint (15), and

$$\max \sum_{k\in K} \phi_k D_k - \sum_{i\in I} a_i y_i - \sum_{e\in E} u_e w_e, \tag{$\mathrm{D}^{\text{cap}}_{\text{min-cost}}$}$$

$$s.t. \sum_{i\in I}\sum_{j=1}^{J} (\tau_{ij} + \pi_{ij}) q_{ij} + \sum_{e\in P} (w_e + \ell_e) \geq \phi_k, \ \forall (P,q) \in \mathcal{Q}_{s_k t_k}, \forall k \in K, \tag{17a}$$

$$y_i \geq \pi_{ij} r_{ij}, \quad \forall i \in I, \quad \forall j \in \{1,\ldots,J\}, \tag{17b}$$

$$\pi, w, \phi \geq 0, \tag{17c}$$

where ϕ_k is the dual variable associated with the demand constraint in ($\mathrm{P}_{\text{min-cost}}$).

We will use the ellipsoid method to approximately solve ($\mathrm{D}^{\text{cap}}_{\text{max-flow}}$) (resp. ($\mathrm{D}^{\text{cap}}_{\text{min-cost}}$)). This requires a $(1+\varepsilon)$-approximate separation oracle that returns a separating hyperplane for any candidate solution that violates (16a) (resp. (17a)) by a factor strictly larger than $(1+\varepsilon)$. To verify if (16a) (resp. (17a)) is approximately satisfied, it suffices to compute a $(1+\varepsilon)$-approximate min-cost single EV charging strategy (P,q) where the edge cost is w_e (resp. $w_e + \ell_e$) and the charge cost is π_{ij} (resp. $\tau_{ij} + \pi_{ij}$) for every $e \in E$ and every $i \in I, j \in \{1,\ldots,J\}$. This can be implemented in $O(poly(|I|J)/\varepsilon)$ time using the FPTAS by [17]. Following similar analysis as in Lemma 2, we can equivalently reformulate the semi-infinite linear programs ($\mathrm{D}^{\text{cap}}_{\text{max-flow}}$) and ($\mathrm{D}^{\text{cap}}_{\text{min-cost}}$) as finite linear programs. This guarantees that the ellipsoid method can compute a $(1+\varepsilon)$-approximate dual solution in $O(poly(|I|J)/\varepsilon)$-time. Furthermore, by Lemma 6.5.15 in [10], we can also obtain a $(1+\varepsilon)$-approximate solution to the primal problems that compute the max EV flow and the min-cost EV flow in $O(poly(|I|J)/\varepsilon)$-time. We present the FPTAS for the min-cost EV flow problem below; the FPTAS for the maximum EV flow problem with edge capacities is similar, and can be found in the full version of this paper.

Proposition 2. *A* $(1+\varepsilon)$*-approximate solution for the maximum EV flow problem and the min-cost EV flow problem with edge capacities can be computed in* $O(poly(|I|J)/\varepsilon)$ *time.*

Algorithm 1: FPTAS for minimum-cost EV flow with edge capacities.

1 Using the ellipsoid method (with the FPTAS algorithm in [17] as the separation oracle), solve ($\mathrm{D}^{\mathrm{cap}}_{\mathrm{min-cost}}$). Let $(P^{(1)}, q^{(1)}), \ldots, (P^{(N)}, q^{(N)})$ be the charging strategies corresponding to the generated separating hyperplanes. **return** the optimal solution of the following linear program:

$$\min \sum_{m=1}^{N} \left(\sum_{i \in I} \sum_{j=1}^{J} (\tau_{ij} + (1+\rho_i)/r_{ij}) q_{ij}^{(m)} + \sum_{e \in P^{(m)}} \ell_e \right) x_{P^{(m)} q^{(m)}},$$

$$\text{s.t.} \sum_{m=1}^{N} q_{ij}^{(m)} x_{P^{(m)} q^{(m)}} \leq r_{ij} z_{ij}, \qquad \forall i \in I, \quad \forall j \in \{1, \ldots, J\},$$

$$\sum_{j=1}^{J} z_{ij} = a_i, \qquad \forall i \in I,$$

$$\sum_{m: P^{(m)} \in \mathcal{P}_k} x_{P^{(m)} q^{(m)}} \geq D_k, \qquad \forall k \in K,$$

$$\sum_{m: e \in P^{(m)}} x_{P^{(m)} q^{(m)}} \leq u_e, \qquad \forall e \in E,$$

$$x_{P^{(m)} q^{(m)}} \geq 0, \qquad \forall m \in \{1, \ldots, N\}.$$

References

1. Adler, J.D., Mirchandani, P.B.: The vehicle scheduling problem for fleets with alternative-fuel vehicles. Transp. Sci. **51**(2), 441–456 (2017)
2. Ahuja, R.K., Magnanti, T.L., Orlin, J.B.: Network flows (1988)
3. Chen, L., He, L., Zhou, Y.: An exponential cone programming approach for managing electric vehicle charging. Oper. Res. (2023)
4. Desaulniers, G., Errico, F., Irnich, S., Schneider, M.: Exact algorithms for electric vehicle-routing problems with time windows. Oper. Res. **64**(6), 1388–1405 (2016)
5. Dong, Y., De Koster, R., Roy, D., Yu, Y.: Dynamic vehicle allocation policies for shared autonomous electric fleets. Transp. Sci. **56**(5), 1238–1258 (2022)
6. Flath, C.M., Ilg, J.P., Gottwalt, S., Schmeck, H., Weinhardt, C.: Improving electric vehicle charging coordination through area pricing. Transp. Sci. **48**(4), 619–634 (2014)
7. Ford, L., Jr., Fulkerson, D.: Flows in networks (1962)
8. Froger, A., Jabali, O., Mendoza, J.E., Laporte, G.: The electric vehicle routing problem with capacitated charging stations. Transp. Sci. **56**(2), 460–482 (2022)
9. Goldberg, A.V., Tarjan, R.E.: A new approach to the maximum-flow problem. J. ACM (JACM) **35**(4), 921–940 (1988)
10. Grötschel, M., Lovász, L., Schrijver, A.: Geometric Algorithms and Combinatorial Optimization. Springer, Heidelberg (1988)
11. Johnson, D.S.: The NP-completeness column: an ongoing guide. J. Algorithms **2**(4), 393–405 (1981)

12. Khuller, S., Malekian, A., Mestre, J.: To fill or not to fill: the gas station problem. In: Arge, L., Hoffmann, M., Welzl, E. (eds.) ESA 2007. LNCS, vol. 4698, pp. 534–545. Springer, Heidelberg (2007). https://doi.org/10.1007/978-3-540-75520-3_48
13. Kınay, Ö.B., Gzara, F., Alumur, S.A.: Charging station location and sizing for electric vehicles under congestion. Transp. Sci. (2023)
14. Kullman, N.D., Goodson, J.C., Mendoza, J.E.: Electric vehicle routing with public charging stations. Transp. Sci. **55**(3), 637–659 (2021)
15. López, M., Still, G.: Semi-infinite programming. Eur. J. Oper. Res. **180**(2), 491–518 (2007)
16. Mak, H.Y., Rong, Y., Shen, Z.J.M.: Infrastructure planning for electric vehicles with battery swapping. Manag. Sci. **59**(7), 1557–1575 (2013)
17. Merting, S., Schwan, C., Strehler, M.: Routing of electric vehicles: constrained shortest path problems with resource recovering nodes. In: ATMOS 2015 (2015)
18. Parmentier, A., Martinelli, R., Vidal, T.: Electric vehicle fleets: scalable route and recharge scheduling through column generation. Transp. Sci. (2023)
19. Pulyassary, H., Kollias, K., Schild, A., Shmoys, D., Wu, M.: Network flow problems with electric vehicles (full version). arXiv preprint arXiv:2311.05040 (2023)
20. Qi, W., Zhang, Y., Zhang, N.: Scaling up electric-vehicle battery swapping services in cities: a joint location and repairable-inventory model. Manag. Sci. (2023)
21. Romeijn, H.E., Smith, R.L., Bean, J.C.: Duality in infinite dimensional linear programming. Math. Program. **53**(1–3), 79–97 (1992)
22. Schneider, M., Stenger, A., Goeke, D.: The electric vehicle-routing problem with time windows and recharging stations. Transp. Sci. **48**(4), 500–520 (2014)
23. Sweda, T.M., Dolinskaya, I.S., Klabjan, D.: Adaptive routing and recharging policies for electric vehicles. Transp. Sci. **51**(4), 1326–1348 (2017)
24. Sweda, T.M., Klabjan, D.: Finding minimum-cost paths for electric vehicles. In: 2012 IEEE International Electric Vehicle Conference, pp. 1–4. IEEE (2012)
25. Yu, J.J., Tang, C.S., Li, M.K., Shen, Z.J.M.: Coordinating installation of electric vehicle charging stations between governments and automakers. Prod. Oper. Manag. **31**(2), 681–696 (2022)

Lower Bounds on the Complexity of Mixed-Integer Programs for Stable Set and Knapsack

Jamico Schade[1(✉)], Makrand Sinha[2], and Stefan Weltge[1]

[1] Technical University of Munich, Munich, Germany
{jamico.schade,weltge}@tum.de
[2] University of Illinois at Urbana-Champaign, Champaign, USA
msinha@illinois.edu

Abstract. Standard mixed-integer programming formulations for the stable set problem on n-node graphs require n integer variables. We prove that this is almost optimal: We give a family of n-node graphs for which every polynomial-size MIP formulation requires $\Omega(n/\log^2 n)$ integer variables. By a polyhedral reduction we obtain an analogous result for n-item knapsack problems. In both cases, this improves the previously known bounds of $\Omega(\sqrt{n}/\log n)$ by Cevallos, Weltge & Zenklusen (SODA 2018).

To this end, we show that there exists a family of n-node graphs whose stable set polytopes satisfy the following: any $(1 + \varepsilon/n)$-approximate extended formulation for these polytopes, for some constant $\varepsilon > 0$, has size $2^{\Omega(n/\log n)}$. Our proof extends and simplifies the information-theoretic methods due to Göös, Jain & Watson (FOCS 2016, SIAM J. Comput. 2018) who showed the same result for the case of exact extended formulations (i.e. $\varepsilon = 0$).

Keywords: mixed-integer programming · stable set problem · knapsack problem · extended formulations

1 Introduction

Combinatorial optimization problems are often expressed by different formulations as mixed-integer programs (MIPs). A simple example is given by the *matching problem*, which is often described as

$$\max\left\{c^\intercal x : x \in \mathbb{Z}^E_{\geq 0}, \sum\nolimits_{e\in\delta(v)} x(e) \leq 1 \ \text{ for all } v \in V\right\},$$

where $G = (V, E)$ is the complete undirected graph on n nodes and $\delta(v)$ is the set of edges in G that are incident to v. This formulation is attractive in

Jamico Schade has been funded by the Deutsche Forschungsgemeinschaft (DFG, German Research Foundation) under project 451026932. Stefan Weltge has been supported by DFG under projects 451026932 and 277991500/GRK2201.

J. Vygen and J. Byrka (Eds.): IPCO 2024, LNCS 14679, pp. 379–392, 2024.
https://doi.org/10.1007/978-3-031-59835-7_28

the sense that it only consists of a small number of linear constraints, which naturally reflect the definition of a matching. However, it comes at the cost of $|E| = \Theta(n^2)$ integer variables. Since the performance of algorithms for solving integer programs is much more sensitive in the number of integer variables than in the number of constraints, the question arises whether there are MIP formulations for the matching problem that use significantly fewer integer variables, yet a reasonable (say, polynomial in n) number of constraints. Note that a formulation without any integer variables can be obtained by adding linear inequalities that completely describe the matching polytope of G, but this would require exponentially many constraints [11]. However, there is a simple (lesser-known) linear-size MIP formulation for the matching problem that only uses n integer variables: If $D = (V, A)$ is a digraph that arises from G by orienting its edges arbitrarily, a valid MIP formulation for the matching problem is

$$\max\left\{c^\intercal x : x \in \mathbb{R}^A_{\geq 0},\ y \in \mathbb{Z}^V,\ x(\delta(v)) \leq 1 \text{ and } x(\delta^{\text{in}}(v)) = y(v) \text{ for all } v \in V\right\},$$

where $\delta^{\text{in}}(v)$ denotes the set of arcs of D that enter v, and $x(F) = \sum_{e \in F} x(e)$. For the correctness of this formulation, see [8, Prop. 6.2].

As another example, consider the symmetric *traveling salesman problem* over G. Standard MIP formulations for this problem contain at least one integer variable for each edge, again resulting in $\Theta(n^2)$ integer variables. However, it is possible to come up with MIP formulations for the traveling salesman problem that use only $O(n \log n)$ integer variables and still have polynomial size, see [7, Cor. 50]. Other combinatorial optimization problems such as the *spanning tree problem* even admit polynomial-size MIP formulations without any integer variables, so-called extended formulations (see, e.g., [9,18]).

The above examples illustrate that it is usually not obvious how many integer variables are needed in small-size MIP formulations of combinatorial optimization problems. Moreover, they refer to problems for which there exist polynomial-size MIP formulations, which use considerably fewer integer variables than the standard formulations. In this work, we consider two prominent combinatorial optimization problems, for which such formulations are not known. The first is the *stable set problem* over a general undirected n-node graph $G = (V, E)$, which is usually described as

$$\max\left\{c^\intercal x : x \in \{0,1\}^V,\ x(v) + x(w) \leq 1 \text{ for all } \{v, w\} \in E\right\}, \tag{1}$$

and the second is the *knapsack problem*, typically given by

$$\max\left\{c^\intercal x : x \in \{0,1\}^n, \sum\nolimits_{i=1}^{n} a(i)x(i) \leq \beta\right\}, \tag{2}$$

where $a \in \mathbb{R}^n$ are given item sizes and $\beta \in \mathbb{R}$ is the given capacity. Both (standard) formulations have n integer variables, and no polynomial-size MIP formulations with $o(n)$ integer variables are known. Our main motivation for considering these two problems is that Cevallos, Weltge & Zenklusen [8] proved that the number of integer variables in the aforementioned MIP formulations for the matching problem and the traveling salesman problem is optimal up to logarithmic terms, while an almost quadratic gap remained for the case of the stable set problem and the knapsack problem.

To address this claim formally, let us specify what we mean by a MIP formulation for a combinatorial optimization problem. Here, we consider a combinatorial optimization problem as a pair $(\mathcal{V}, \mathcal{F})$ where $\mathcal{V}$ is a finite ground set and $\mathcal{F}$ is a family of (feasible) subsets of $\mathcal{V}$. Given weights of $w : \mathcal{V} \to \mathbb{R}$, the goal is to find a set $S \in \mathcal{F}$ maximizing $w(S) = \sum_{v \in S} w(v)$. Now, a MIP formulation for $(\mathcal{V}, \mathcal{F})$ is defined as follows. First, its feasible region Γ should only depend on $(\mathcal{V}, \mathcal{F})$ (and not on the weights to be maximized). Second, Γ should be described by linear inequalities and equations, and a subset of variables that is constrained to integer values. To this end, we represent Γ by a polyhedron $Q \subseteq \mathbb{R}^d$ and an affine map $\sigma : \mathbb{R}^d \to \mathbb{R}^k$ by setting $\Gamma = \Gamma(Q, \sigma) = \{x \in Q : \sigma(x) \in \mathbb{Z}^k\}$. Third, we want to identify each feasible subset S with a point $x_S \in \Gamma$. However, we do not require the variables to be directly associated with the elements of the ground set, and in particular allow $d \neq |\mathcal{V}|$. Finally, node weights should translate to (affine) linear objectives in a consistent way: We require that for each $w : \mathcal{V} \to \mathbb{R}$ there is an affine map $c_w : \mathbb{R}^d \to \mathbb{R}$ such that the weight of every feasible set S satisfies $w(S) = c_w(x_S)$. Note that Γ does not necessarily only contain the points x_S. However, when maximizing c_w over Γ, we require that the optimum is still attained in a point x_S, i.e., $\max\{c_w(x) : x \in \Gamma\} = \max\{c_w(x_S) : S \in \mathcal{F}\} = \max\{w(S) : S \in \mathcal{F}\}$. Notice that $\max\{c_w(x) : x \in \Gamma\}$ can be formulated as a MIP.

If (Q, σ) satisfies the above properties, we say that it is a MIP formulation for $(\mathcal{V}, \mathcal{F})$. We define the size of (Q, σ) to be the number of facets of Q (number of linear inequalities needed to describe Q) and say that it has k integer variables.

With this notion, it is shown[1] in [8] that every subexponential-size MIP formulation for the matching problem or traveling salesman problem has $\Omega(n/\log n)$ integer variables. Moreover, they proved that there exist n-node graphs and n-item instances for which any subexponential-size MIP formulation for the stable set problem or the knapsack problem, respectively, has $\Omega(\sqrt{n}/\log n)$ integer variables. In this work, we close this gap and show that, for these two problems, the standard MIP formulations (1) and (2) already use an (up to logarithmic terms) optimal number of integer variables:

Theorem 1. *There is a constant $c > 0$ and a family of graphs (knapsack instances) such that every MIP formulation for the stable set (knapsack) problem of an n-node graph (n-item instance) in this family requires $\Omega(n/\log^2 n)$ integer variables, unless its size is at least $2^{cn/\log n}$.*

We will mostly focus on proving the above result for the stable set problem. In fact, using a known polyhedral reduction, we will show that it directly implies the result for the knapsack problem. In order to obtain lower bounds on the number of integer variables in MIP formulations, the authors in [8] observed

---

[1] Actually, our definition of a MIP formulation slightly differs from the notion in [8]. However, both definitions are equivalent, see [22, Lem. A.1].

that every MIP formulation can be turned into an approximate extended formulation of the polytope $P(\mathcal{V}, \mathcal{F}) = \text{conv}\{\chi(S) : S \in \mathcal{F}\}$, where $\chi(S) \in \{0,1\}^{\mathcal{V}}$ is the characteristic vector of S. In terms of the stable set problem over a graph G, $P(\mathcal{V}, \mathcal{F})$ is the stable set polytope of G. Here, an *α-approximate extended formulation of size m* of a polytope P is (a linear description of) a polyhedron with m facets that can be linearly projected onto a polytope P' with $P \subseteq P' \subseteq \alpha P$. For instance, it is shown in [8] that, for every $\varepsilon > 0$, every MIP formulation for the stable set problem of size m with k integer variables can be turned into a $(1 + \varepsilon)$-approximate extended formulation of size $m \cdot (1 + k/\varepsilon)^{O(k)}$ of the stable set polytope of the corresponding graph (see [8, Thm. 1.3 & Lem. 2.2]). In light of this result, Theorem 1 follows from the following result about approximate extended formulations which is the main contribution of this work:

Theorem 2. *There is some constant $\varepsilon > 0$ and a family of n-node graphs H such that any $(1+\varepsilon/n)$-approximate extended formulation of the stable set polytope of H has size $2^{\Omega(n/\log n)}$.*

Note that the statement above considers approximations with very small error. Lower bounds for approximate formulations with small error are known for several prominent polytopes that arise in combinatorial optimization such as the matching polytope (see Braun & Pokutta [3], Rothvoß [21, Thm. 16], and Sinha [23]) or the cut polytope (see Braun, Fiorini, Pokutta & Steurer [2] and [7, §5.2]). For the stable set problem on the other hand, existing lower bounds apply to much larger (say, constant) errors but do not yield bounds that are close to exponential (e.g. [1]). For this reason, the aforementioned lower bound on MIP formulations in [8] did not directly rely on approximate extended formulations for stable set polytopes but instead was obtained by transferring results about the cut polytope (see Fiorini, Massar, Pokutta, Tiwary & de Wolf [12, Lem. 8]).

Unfortunately, reductions to the cut or matching polytopes can only show that the sizes of approximate extended formulations must be exponential in $\sqrt{d}$, where d is the dimension of the corresponding polytope. For the case of the matching problem and the traveling salesman problem, we have $\sqrt{d} = O(n)$ and hence these bounds are sufficient to prove optimality of the aforementioned MIP formulations. However, in the case of the stable set problem, we have $\sqrt{d} = \sqrt{n}$, which is the reason why the authors of [8] were only able to prove that subexponential-size MIP formulations for the stable set problem have $\Omega(\sqrt{n}/\log n)$ integer variables. We are able to circumvent this $\sqrt{d}$ bottleneck since we directly work with the stable set polytopes given by Theorem 2: here any extended formulation must be of size exponential in the dimension d (ignoring logarithmic factors).

We prove Theorem 2 by extending and simplifying the techniques of Göös, Jain & Watson [15] who proved Theorem 2 for the case of *exact* ($\varepsilon = 0$) extended formulations. They were inspired by connections to Karchmer-Wigderson games made in [16]. By relying on reductions from certain constraint satisfaction problems (CSPs) that arise from Tseitin tautologies, they constructed a family of stable set polytopes that they showed have $2^{\Omega(n/\log n)}$ extension complexity. Their proof uses information complexity arguments building on the work of Huynh and Nordström [17] and is significantly involved. Furthermore, the proof departs from other approximate extended formulations lower bounds for the cut and matching polytopes, which can be obtained by a fairly unified framework (see [20, Ch. 12]).

Our proof is still based on the main ideas of [15] but extending it to the approximate case is quite involved. To prove our lower bound, we follow the common information framework introduced by [6] and further developed in [4,5,23], mostly in the interest of bridging the gap to previous proofs for other approximate extended formulation lower bounds. Along the way, we simplify several parts of the information-theoretic arguments in [15] and also show that the family of stable set polytopes used in Theorem 2 are in fact simple to describe explicitly (without going through CSP reductions). Separately from this, as mentioned before, we show that this also implies a lower bound for knapsack MIP formulations via a standard polyhedral reduction.

We remark that since information complexity arguments are typically robust to approximations, we believe that with some amount of work, the approach in [15] can be made to yield Theorem 2 above. However, our proof is very much in line with the previous lower bounds for the cut and matching polytope which ultimately reduce the problem to understanding the *nonnegative rank* of a certain matrix called the *unique disjointness* matrix, possibly through a randomized reduction. Our proof in fact suggests that one may be able to prove all of these lower bounds — for matching, stable set, and knapsack polytopes — in a unified way, via a randomized reduction to unique disjointness. We leave this as an interesting open question for follow-up work.

Structure. We start by describing a family of graphs H for which Theorem 2 holds in Sect. 2. Moreover, we derive a matrix S whose nonnegative rank will give a lower bound on the size of any $(1+\frac{\varepsilon}{n})$-approximate extended formulation of the stable set polytope of H. Our proof strategy for obtaining a lower bound on the nonnegative rank of S is explained in Sect. 3. For space reasons, notions and statements on information theory as well as technical details of the main proof are deferred to the full version in [22, Sec. 4-5].

2 Instances

2.1 Graph Family

The graphs H in the statement of Theorem 2 will arise from a family of sparse graphs $G = (V, E)$ with an *odd* number of nodes and a certain connectivity property to be defined later.

To define $H = H(G)$, let us fix a set of colors $\mathcal{X} = \{0,1\}^3$. Each node $v \in V$ is lifted to several copies, so that each copy corresponds to a different coloring of the edges incident to v, i.e., the nodes of H are the pairs (v, x^v) where $v \in V$ and x^v denotes a coloring of the incident edges of v with colors in $\mathcal{X}$.

In H, all copies of a single node v form a clique C_v. Moreover, if $v, w \in V$ are connected by an edge $e \in E$, then we also draw an edge between copies of v and w in H if they label e with a different color, see Fig. 1.

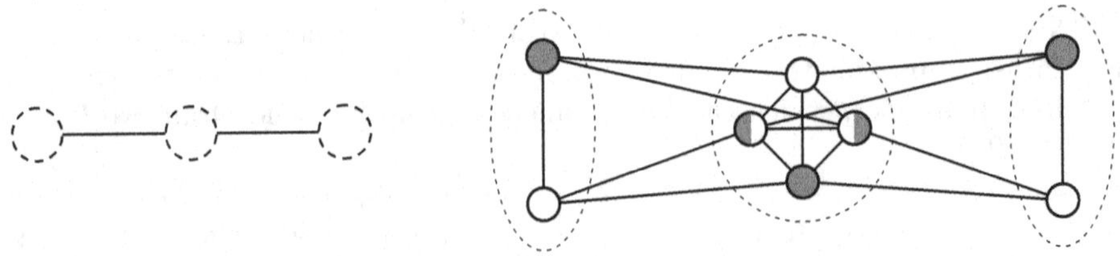

Fig. 1. Construction of the graph H (right), where G is a path on three nodes (left) and $\mathcal{X}$ consists of only two colors (white and gray).

Note that stable sets in H can be obtained in a simple way: Pick any (global) coloring $x \in \mathcal{X}^E$ of the edges of G. Now, for each node $v \in V$, select the (unique) node in C_v that colors the edges incident to v according to x. This yields a maximal stable set in H. In fact, every maximal stable set in H arises in this way.

Moreover, observe that if G has constant degree, then the number of nodes and the number of edges of H are both linear in $|V|$ (since $\mathcal{X}$ has constant size).

2.2 Nonnegative Rank and Partial Slack Matrices

Let P denote the stable set polytope of the graph H and consider any polytope P' with $P \subseteq P' \subseteq (1 + \frac{\varepsilon}{n})P$. A common approach to obtain a lower bound on the number of facets of P' is based on considering partial slack matrices of P'. A *partial slack matrix* of P' is a (nonnegative) matrix S, where each row i corresponds to some point $x^i \in P'$ and each column j corresponds to a linear inequality $a_j^\intercal x \leq b_j$ that is satisfied by all $x \in P'$, such that each entry S_{ij} is equal to the *slack* of x^i with respect to $a_j^\intercal x \leq b_j$, i.e., $S_{ij} = b_j - a_j^\intercal x^i$. From the seminal paper [24] of Yannakakis it follows that the size of any extended formulation of P is at least the *nonnegative rank* of S, which is the smallest number r_+ such that S can be written as the sum of r_+ nonnegative rank-1 matrices.

2.3 Gadget and a Particular Matrix

Setting $\mathcal{Y} = \{0,1\}^3$, we will consider a particular matrix $S = S(G, \varepsilon)$ whose rows and columns are indexed by vectors $x \in \mathcal{X}^E$ and $y \in \mathcal{Y}^E$, respectively. The entries of S are not only based on G, ε but also on a *gadget function* $g : \mathcal{X} \times \mathcal{Y} \to \{0,1\}$ defined via

$$g((x_1, x_2, x_3), (y_1, y_2, y_3)) = x_1 + y_1 + x_2 y_2 + x_3 y_3 \pmod 2. \tag{3}$$

Given a $x \in \mathcal{X}^E, y \in \mathcal{Y}^E$, one can apply this gadget to obtain a bit string in $\{0,1\}^E$ that labels each edge in E. Summing up the parity of each edge incident on v induces a *parity on a node* defined as

$$\text{parity}(v) = \sum\nolimits_{e \in \delta(v)} g(x_e, y_e) \mod 2.$$

We set

$$\text{Zeros}(x, y) = \{v \in V : \text{parity}(v) = 0\}$$

to be the set of nodes with parity zero. With this, the entries of $S \in \mathbb{R}^{\mathcal{X}^E \times \mathcal{Y}^E}$ are given by

$$S_{xy} = |\,\text{Zeros}(x, y)| - 1 + \varepsilon. \tag{4}$$

Note that $\sum_{v \in V} \text{parity}(v)$ is always even since the bit label of each edge is summed twice. This implies that $|V \setminus \text{Zeros}(x, y)|$ is even for each $x \in \mathcal{X}^E, y \in \mathcal{Y}^E$, and since the number of nodes $|V|$ is odd, it follows that $|\,\text{Zeros}(x, y)|$ is always odd and hence $|\,\text{Zeros}(x, y)| \geq 1$. In particular, we see that every entry of S is positive.

Lemma 3. *S is a partial slack matrix of every polytope P' with $P \subseteq P' \subseteq (1 + \varepsilon/|V(H)|)P$, where P is the stable set polytope of H.*

Proof. To show that S is a partial slack matrix of P', we have to define a point $z^x \in P'$ for every $x \in \mathcal{X}^E$ and associate a linear inequality to every $y \in \mathcal{Y}^E$ that is valid for P' and such that the slack of z^x with respect to the inequality associated to y is equal to S_{xy}. For $x \in \mathcal{X}^E$ we define the set

$$S_x = \{(v, x|_{\delta(v)}) : v \in V\}$$

and denote by $z^x \in \{0, 1\}^{V(H)}$ the characteristic vector of S_x. Note that S_x is a stable set in H and hence $z^x \in P \subseteq P'$. For $y \in \mathcal{Y}^E$ we define the set

$$U_y = \left\{(v, x) \in V(H) : \sum\nolimits_{e \in \delta(v)} g(x_e, y_e) \text{ is odd}\right\}$$

and consider the linear inequality

$$\sum\nolimits_{u \in U_y} z_u \leq |V| - 1 + \varepsilon. \tag{5}$$

Note that the slack of z^x with respect to this inequality is equal to

$$\begin{aligned} |V| - 1 + \varepsilon - \sum\nolimits_{u \in U_y} z_u &= |V| - 1 + \varepsilon - |S_x \cap U_y| \\ &= |V| - 1 + \varepsilon - (|V| - |\,\text{Zeros}(x, y)|) = S_{xy}. \end{aligned}$$

Thus it remains to show that (5) is satisfied by every $z \in P'$. To see this, we first claim that every stable set S in H satisfies $|S \cap U_y| \leq |V| - 1$. Indeed, if $|S| \leq |V| - 1$, then the claim is trivial. Otherwise, it is easy to see that $S = S_x$ for some $x \in \mathcal{X}^E$ and hence $|S_x \cap U_y| = |V| - |\,\text{Zeros}(x, y)| \leq |V| - 1$. This means that the linear inequality $\sum_{u \in U_y} z_u \leq |V| - 1$ is satisfied by every point $z \in P$. Since $P' \subseteq (1 + \varepsilon/|V(H)|)P$, every point $z \in P'$ must satisfy

$$\sum\nolimits_{u \in U_y} z_u \leq \left(1 + \tfrac{\varepsilon}{|V(H)|}\right)(|V| - 1) = |V| - 1 + \frac{\varepsilon}{|V(H)|} \cdot (|V| - 1) \leq |V| - 1 + \varepsilon,$$

which yields (5). □

2.4 Choice of G

We will see that, for particular choices of G, the nonnegative rank of S is large. To this end, we say that a graph G is called $(2k+3)$-*routable* if there exists a subset of $2k+3$ nodes, called *terminals*, such that for every partition of the terminals into a single terminal and $k+1$ pairs $(v_0, w_0), \dots, (v_k, w_k)$ of terminals there exist edge-disjoint paths $P_0, \dots, P_k$ such that P_i connects v_i and w_i. Note that each P_i depends on the entire partition and not only on the pair (v_i, w_i).

Our main result is the following.

Theorem 4. *There is some $\varepsilon > 0$ such that if G is $(2k+3)$-routable, then the nonnegative rank of $S = S(G, \varepsilon)$ is $2^{\Omega(k)}$.*

To obtain Theorem 2 we choose $G = (V, E)$ to be any *constant-degree* $(2k+3)$-routable graph with an odd number of nodes, and $k = \Theta(|V|/\log|V|)$. An infinite family of such graphs can be obtained by taking any sufficiently strong constant-degree expander (e.g. a Ramanujan graph) ([13,14], see also [19, §15]). Recall that H has $\Theta(|V|)$ nodes (since G has constant degree).

3 Lower Bound

3.1 Nonnegative Rank and Mutual Information

In order to prove Theorem 4, we follow an information-theoretic approach introduced by Braverman and Moitra [6] that has already been used in previous works on extended formulations (c.f. [3,5,23]). To this end, suppose that the nonnegative rank of S is r_+, in which case we can write $S = \sum_{r=1}^{r_+} R^{(r)}$ for some nonnegative rank-1 matrices $R^{(1)}, \dots, R^{(r_+)}$. Consider the discrete probability space with random variables X, Y, R and distribution $q(X = x, Y = y, R = r) = R^{(r)}_{xy}/\|S\|_1$, where x and y range over the rows and columns of S, respectively, and $r \in [r_+]$. Notice that the marginal distribution of X, Y is given by the (normalized) matrix S, i.e., $q(X = x, Y = y) = S_{xy}/\|S\|_1$. Moreover, since each $R^{(r)}$ is a rank-1 matrix, we see that X, Y are independent when conditioned on R. Thus, we obtain the following proposition.

Proposition 5. *If the nonnegative rank of S is r_+, then there is a random variable R with $|\text{supp}(R)| = r_+$, such that X and Y are independent given R and the marginal distribution of X and Y is given by the normalized matrix S.*

Note that the above also implies that R breaks the dependencies between X and Y even if we further condition on any event in the probability space that is a *rectangle*, i.e. any event where $(X, Y) \in \mathcal{S} \times \mathcal{T}$ where $\mathcal{S} \subseteq \mathcal{X}^E$ and $\mathcal{T} \subseteq \mathcal{Y}^E$ are subsets of rows and columns respectively.

To prove a lower bound on the nonnegative rank, we shall show that if the support of R was too small, then the probability of certain events in the probability space would be quite different from that given by the distribution $q(X, Y)$ above. In order to do this, instead of directly bounding the size of the support of

R, it will be more convenient for us to work with certain information-theoretic quantities that give a lower bound on the (logarithm of) the size of the support of R:

For a random variable A over a probability space with distribution p, we denote the binary *entropy* of A (with respect to p) by $\mathbf{H}_p(A)$. For any two random variables A and B, the entropy of A conditioned on B is defined as $\mathbf{H}_p(A|B) = \mathbb{E}_{p(b)}[\mathbf{H}_p(A|b)]$, where $\mathbb{E}_{p(b)}[f(b)]$ denotes the expected value of a function $f(b)$ under the distribution $p(B)$. Given an event $\mathcal{W}$, we write $\mathbf{H}_p(A|B, \mathcal{W}) = \mathbf{H}_q(A|B)$ where $q = p(\cdot \mid \mathcal{W})$ arises from p by conditioning on the event $\mathcal{W}$. The *mutual information* between A, B is defined as $\mathbf{I}_p(A : B) = \mathbf{H}_p(A) - \mathbf{H}_p(A|B)$. Further, the *conditional mutual information* is defined as $\mathbf{I}_p(A : B|C) = \mathbf{H}_p(A|C) - \mathbf{H}_p(A|BC)$. Finally, given an event $\mathcal{W}$, we define $\mathbf{I}_p(A : B|C, \mathcal{W}) = \mathbf{I}_q(A : B|C)$, where $q = p(\cdot \mid \mathcal{W})$. We will elaborate on the notation as well as the essential properties of the above quantities in the appendix.

Given q, X, Y, R as in Proposition 5, in what follows we will introduce some further specific random variables T, W, and an event $\mathcal{D}$ that satisfy the following.

Theorem 6. *There is some $\varepsilon > 0$ such that $\mathbf{I}_q(R : XY|TW, \mathcal{D}) = \Omega(k)$.*

A basic property of (conditional) mutual information [10, Thm. 2.6.4] states that if $\mathbf{I}_q(R : XY|TW, \mathcal{D}) \geq \ell$, then $|\text{supp}(R)| \geq 2^{\ell}$, and hence this directly yields Theorem 4. In the remainder of this section, we will define $T, W, \mathcal{D}$ and describe the proof outline of Theorem 6.

3.2 Random Pairings and Windows

The random variable T will denote a uniformly random partition of the terminals of G into a single terminal T_u and an ordered list of $k+1$ unordered pairs $T_0, \cdots, T_k$.

Following [15], we say that a 2×1 (horizontal) or 1×2 (vertical) rectangle w of $\mathcal{X} \times \mathcal{Y}$ is a *b-window* if the value of g on the two inputs in w is equal to b. The random variable $W = (W_e)_{e \in E}$ will denote a uniformly random window for every edge of G.

An important property of the gadget g is the following. Given a (horizontal) b-window $w = \{(x, y), (x, y')\}$, there exist unique $\tilde{x} \in \mathcal{X}$, $\tilde{y}, \tilde{y}' \in \mathcal{Y}$ such that $_{[00]}w := \{(\tilde{x}, \tilde{y}), (\tilde{x}, \tilde{y}')\}$, $_{[10]}w := \{(x, \tilde{y}), (x, \tilde{y}')\}$, and $_{[01]}w := \{(\tilde{x}, y), (\tilde{x}, y')\}$ are $\bar{b}$-windows. Thus, we may view $_{[00]}w$, $_{[10]}w$, $_{[01]}w$, $_{[11]}w := w$ as a "stretched" AND if w is a 1-window, or a "stretched" NAND if w is a 0-window, see Fig. 2 for an illustration. Since g is symmetric with respect to its inputs, we may define $_{[00]}w$, $_{[10]}w$, $_{[01]}w$, and $_{[11]}w$ analogously for vertical windows.

In what follows, we will consider the event

$$\mathcal{D} : \text{Zeros}(X, Y) = T_u \cup T_0 \text{ and } (X, Y) \in W,$$

which, in particular, enforces that the parity of the nodes is zero only on the three terminals in $T_u \cup T_0$.

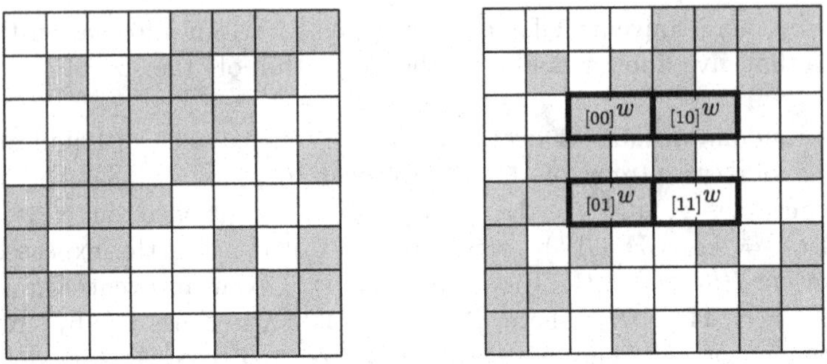

Fig. 2. Both pictures show the values of the gadget g in white (0) and gray (1), where the columns and rows corresponds to the inputs in the order $(0, 0, 0)$, $(0, 0, 1)$, $(0, 1, 0)$, etc. The right picture shows a 0-window $w = {}_{[11]}w$ and the corresponding 1-windows ${}_{[00]}w$, ${}_{[01]}w$, and ${}_{[10]}w$.

3.3 Main Argument

First, recall that for every partition t of the $(2k + 3)$ terminals of G into a single terminal t_u and pairs of terminals $t_0, \ldots, t_k$, there exist edge-disjoint paths $P_0, \ldots, P_k$ such that P_i connects the terminals t_i, which we denote by $\text{path}_i(t) = P_i$. We denote by X^i the restriction of X to the indices that correspond to edges in $\text{path}_i(T)$, i.e., $X^i = (X_e)_{e \in \text{path}_i(T)}$. Similarly, we define $X^{-i} = (X_e)_{e \notin \text{path}_i(T)}$. We use the analogous notation for other random variables as well, e.g. Y^i, W^i.

The event $\mathcal{D}$ is quite useful because it turns out that within this event, for any fixed partition $T = t$ and window $W = w$, the random variables $X^1Y^1, X^2Y^2, \ldots, X^kY^k$ are mutually independent. This allows one to use the powerful superadditivity property of mutual information (see [22, Sec. 4.4]):

$$\mathbf{I}_q \left(R : XY | TW, \mathcal{D}\right) \geq \sum_{i=1}^{k} \mathbf{I}_q \left(R : X^iY^i | TW, \mathcal{D}\right). \tag{6}$$

In view of Theorem 6, it suffices to focus on the marginal distribution of X^iY^i and show that each of the mutual information terms on the right is $\Omega(1)$. This can be generated from a much smaller rectangle (submatrix) of the slack matrix S: the event $\mathcal{D}$ fixes a window W such that for all $X, Y \in W$, the parity of all the terminals in $T_1, \cdots, T_k$ is one and $\text{Zeros}(X, Y) = T_u \cup T_0$. One could consider the event $\mathcal{D}$ as part of a larger rectangle where one is allowed to have parity zero on any of the other pair of terminals $T_1, \cdots, T_k$ as well. In contrast, in the marginal probability space for $i \in [k]$, it will suffice to consider the event $\mathcal{D}$ as part of a smaller rectangle where $\text{Zeros}(X, Y) = T_u \cup T_0 \cup T_i$ or $\text{Zeros}(X, Y) = T_u \cup T_0$. In order to do this, note that if we flip the edge labels along any path in the graph, i.e. replace the label $b_e := g(x_e, y_e)$ for each edge on the path with $\overline{b_e}$, then the parity of the end points of the path flips, while the parity of all other nodes remains unchanged. Since we are conditioning on windows as well, we will switch to a different window in order to flip the edge labels.

For this we rely on the fact that $w \mapsto {}_{[00]}w$ (as defined previously) is a bijection, which allows us to obtain various equivalent but correlated ways of generating the same event. In particular, consider the following event for each $i \in [k]$,

$$\mathcal{D}_i : \text{Zeros}(X, Y) = T_u \cup T_0 \text{ and } (X, Y) \in {}_{[00]}W^i W^{-i},$$

where ${}_{[ab]}W^i W^{-i}$ arises from W by replacing each entry w that corresponds to an edge in $\text{path}_i(T)$ by ${}_{[ab]}w$.

Note that if we have $\text{Zeros}({}_{[00]}W^i W^{-i}) = T_u \cup T_0$, then $\text{Zeros}({}_{[01]}W^i W^{-i}) = \text{Zeros}({}_{[10]}W^i W^{-i}) = T_u \cup T_0$ as well as $\text{Zeros}(W) = T_u \cup T_0 \cup T_i$. Although the event $\mathcal{D}_i$ can be considered a part of this smaller rectangle, note that $\mathcal{D}_i$ is a not a subset of $\mathcal{D}$. However, since $W \mapsto {}_{[00]}W^i W^{-i}$ is a bijection over the space of windows, the mutual information quantities remain the same. Thus, by (6) we obtain $\mathbf{I}_q\left(R : XY | TW, \mathcal{D}\right) \geq \sum_{i=1}^{k} \mathbf{I}_q\left(R : X^i Y^i | TW, \mathcal{D}_i\right)$ (see again [22, Sec. 4.4]), and hence it suffices to prove:

Theorem 7. *For $\varepsilon > 0$ small enough, there is a constant $c > 0$ such that $\mathbf{I}_q\left(R : X^i Y^i | TW, \mathcal{D}_i\right) \geq c$ holds for every $i \in [k]$.*

We elaborate on the main steps of proving the above theorem. A detailed proof is given in [22, Sec. 5]. The argument will be the same for each $i \in [k]$, so let us fix an $i \in [k]$. To obtain a bound on $\mathbf{I}_q\left(R : X^i Y^i | TW, \mathcal{D}_i\right)$, we will consider a smaller probability space p. To this end, we will treat the windows corresponding to the edges of path between the terminals in T_i, denoted W^i, differently from the windows for the rest of the edges.

Recall that any window can be seen as a "00", "01", "10" or "11" input of (a unique "stretched AND" or "stretched NAND" of) the gadget g. In particular, we may view $W^i = {}_{[11]}W^i$ as being embedded as a "11" input, which also yields three further correlated disjoint random windows that we call ${}_{[00]}W^i$, ${}_{[01]}W^i$, and ${}_{[10]}W^i$.

The probability space p arises as follows. First, we define random variables $A, B \in \{0, 1\}$ and restrict to the event that $X^i Y^i \in {}_{[AB]}W^i$ and $X_e Y_e \in W_e$ for all remaining edges e. Moreover, we fix parts of T and the set of nodes which have parity zero under the labeling given by W. We will see that for any x, y, a, b in the above probability space, we have $|\,\text{Zeros}(x, y)| = 3$ if $a = 0$ or $b = 0$ and $|\,\text{Zeros}(x, y)| = 5$ otherwise. The probability space p will be obtained by restricting to this submatrix and normalizing the entries. In view of (4), this yields

$$p(A = 0, B = 0) = \frac{2 + \varepsilon}{10 + 4\varepsilon} \approx \frac{1}{5} \text{ and } p(A = 1, B = 1) = \frac{4 + \varepsilon}{10 + 4\varepsilon} \approx \frac{2}{5}. \quad (7)$$

Moreover, it turns out that $\gamma^4 := \mathbf{I}_p\left(R : X^i Y^i | TW, A = 0, B = 0\right)$ is equal to $\mathbf{I}_q\left(R : X^i Y^i | TW, \mathcal{D}_i\right)$. We view this as an embedding of the AND function: if $\text{AND}(A, B) = 0$, then $\text{Zeros}(X, Y) = T_u \cup T_0$ while if $\text{AND}(A, B) = 1$, then because of the choice of a different window, the edge labels of the path get flipped along the path connecting the terminals in T_i and $\text{Zeros}(X, Y) = T_u \cup T_0 \cup T_i$.

The crux of the proof is to show that if $\gamma \ll 1$, then the probability of the two events in (7) must be quite different. To see this, call a triple (r, t, w) good if the distributions $p(A|R = r, T = t, W = w)$ and $p(B|R = r, T = t, W = w)$ are both close to the uniform distribution on a bit and denote by $\mathcal{G}$ the event that the triple is good. Using Pinsker's inequality and properties of the gadget, we can show that most of the contribution comes from good triples (see [22, Prop. 5.2]), i.e., $p(A = 0, B = 0) \leq p(A = 0, B = 0, \mathcal{G}) + O(\gamma)$.

We will see that, for any good (r, t, w), the conditional distribution $p(A, B|R = r, T = t, W = w)$ is very close to the uniform distribution on two bits. This by itself does not give us a contradiction to (4) as the above inequality does not imply that the probability $p(\mathcal{G})$ is large (it only shows that the probability $p(\mathcal{G} \mid A = 0, B = 0)$ is large). So, we further partition the good triples $\mathcal{G}$ into $\mathcal{G}_1$ and $\mathcal{G}_2$.

Two-Intersecting Family. The triples in $\mathcal{G}_2$ correspond to an event where all the good pairings (after fixing all but the pairing of the 5 nodes in $T_u \cup T_0 \cup T_i$) form a 2-intersecting family of $\binom{[5]}{3}$. We use bounds on the size of intersecting families to show that $p(A = 0, B = 0, \mathcal{G}_2)$ is roughly $\frac{4}{\binom{5}{3}} = \frac{2}{5}$ times $p(A = 1, B = 1)$ (see [22, Prop 5.3]).

Triples Containing Small Entries. To deal with the remaining triples in $\mathcal{G}_1$, we first note that any pairing t (along with w) chooses a rectangular submatrix $\mathcal{A}_t \times \mathcal{B}_t$ of the slack matrix S where $\mathcal{A}_t$ and $\mathcal{B}_t$ denote the set of rows and columns respectively. We show that for any good triple t (along with r, w) one can find another good t' (along with r, w) such that the two rectangles intersect and moreover, the rectangle $\mathcal{A}_{t'} \times \mathcal{B}_t$ contains entries where $|\operatorname{Zeros}(x, y)| = 1$. In light of (4), the probability of the event $|\operatorname{Zeros}(X, Y)| = 1$ in the original unconditioned probability space, denoted α, is $O(\varepsilon)$. We are able to show that the probability contribution $p(A = 0, B = 0, \mathcal{G}_1)$ can be bounded by a constant factor of α (see [22, Prop. 5.4]). We remark that if $\varepsilon = 0$, this fact is much simpler to prove as $\alpha = 0$ in this case, and one could even afford a loose bound of $2^n\alpha$ for instance. We, however, need a very precise quantitative bound here which increases the complexity of the arguments.

Overall, the above implies

$$p(A = 0, B = 0) \leq \frac{2}{5} \cdot p(A = 1, B = 1) + O(\varepsilon + \gamma) = \frac{4}{25} + O(\varepsilon + \gamma).$$

If $\gamma = O(\varepsilon)$, taking ε to be a small enough constant, the right hand side above is strictly smaller than $1/5$, which contradicts the true probability given by (7).

We note that the proof for the matching polytope proceeds along very similar lines (see [20, Ch. 12]), but is somewhat simpler compared to the present proof. The difficulty in the present proof is primarily due to the fact that the edge-disjoint paths above depend on the pairing of all the terminals.

Acknowledgements. We would like to thank Mel Zürcher for initial discussions on this topic.

References

1. Bazzi, A., Fiorini, S., Pokutta, S., Svensson, O.: No small linear program approximates vertex cover within a factor $2-\varepsilon$. Math. Oper. Res. **44**(1), 147–172 (2019)
2. Braun, G., Fiorini, S., Pokutta, S., Steurer, D.: Approximation limits of linear programs (beyond hierarchies). Math. Oper. Res. **40**(3), 756–772 (2015)
3. Braun, G., Pokutta, S.: The matching polytope does not admit fully-polynomial size relaxation schemes. In: Proceedings of the Twenty-Sixth Annual ACM-SIAM Symposium on Discrete Algorithms, pp. 837–846. SIAM (2014)
4. Braun, G., Pokutta, S.: The matching problem has no fully polynomial size linear programming relaxation schemes. IEEE Trans. Inf. Theory **61**(10), 5754–5764 (2015)
5. Braun, G., Pokutta, S.: Common information and unique disjointness. Algorithmica **76**, 597–629 (2016)
6. Braverman, M., Moitra, A.: An information complexity approach to extended formulations. In: Proceedings of the Forty-Fifth Annual ACM Symposium on Theory of Computing, pp. 161–170 (2013)
7. Cevallos, A., Weltge, S., Zenklusen, R.: Lifting linear extension complexity bounds to the mixed-integer setting. arXiv:1712.02176 (2017)
8. Cevallos, A., Weltge, S., Zenklusen, R.: Lifting linear extension complexity bounds to the mixed-integer setting. In: Proceedings of the Twenty-Ninth Annual ACM-SIAM Symposium on Discrete Algorithms (SODA), pp. 788–807. SIAM (2018)
9. Conforti, M., Cornuéjols, G., Zambelli, G.: Extended formulations in combinatorial optimization. Ann. Oper. Res. **204**(1), 97–143 (2013)
10. Cover, T.M.: Elements of Information Theory. Wiley, Hoboken (1999)
11. Edmonds, J.: Maximum matching and a polyhedron with 0, 1-vertices. J. Res. Natl. Bureau Standards B **69**(125–130), 55–56 (1965)
12. Fiorini, S., Massar, S., Pokutta, S., Tiwary, H.R., De Wolf, R.: Exponential lower bounds for polytopes in combinatorial optimization. J. ACM (JACM) **62**(2), 1–23 (2015)
13. Frieze, A.M.: Edge-disjoint paths in expander graphs. SIAM J. Comput. **30**(6), 1790–1801 (2001)
14. Frieze, A.M., Zhao, L.: Optimal construction of edge-disjoint paths in random regular graphs. Comb. Probab. Comput. **9**(3), 241–263 (2000)
15. Göös, M., Jain, R., Watson, T.: Extension complexity of independent set polytopes. SIAM J. Comput. **47**(1), 241–269 (2018)
16. Hrubes, P.: On the nonnegative rank of distance matrices. Inf. Process. Lett. **112**(11), 457–461 (2012). https://doi.org/10.1016/j.ipl.2012.02.009
17. Huynh, T., Nordström, J.: On the virtue of succinct proofs: amplifying communication complexity hardness to time-space trade-offs in proof complexity. In: Karloff, H.J., Pitassi, T. (eds.) Proceedings of the 44th Symposium on Theory of Computing Conference, STOC 2012, New York, NY, USA, 19–22 May 2012, pp. 233–248. ACM (2012). https://doi.org/10.1145/2213977.2214000
18. Kaibel, V.: Extended formulations in combinatorial optimization. Optima **85**, 2–7 (2011)
19. Nordström, J.: New wine into old wineskins: a survey of some pebbling classics with supplemental results (2009)
20. Rao, A., Yehudayoff, A.: Communication Complexity: and Applications. Cambridge University Press (2020). https://doi.org/10.1017/9781108671644

21. Rothvoß, T.: The matching polytope has exponential extension complexity. J. ACM (JACM) **64**(6), 1–19 (2017)
22. Schade, J., Sinha, M., Weltge, S.: Lower bounds on the complexity of mixed-integer programs for stable set and knapsack. arXiv:2308.16711v2 (2024)
23. Sinha, M.: Lower bounds for approximating the matching polytope. In: Proceedings of the Twenty-Ninth Annual ACM-SIAM Symposium on Discrete Algorithms, pp. 1585–1604. SIAM (2018)
24. Yannakakis, M.: Expressing combinatorial optimization problems by linear programs. In: Proceedings of the Twentieth Annual ACM Symposium on Theory of Computing, pp. 223–228 (1988)

Relaxation Strength for Multilinear Optimization: McCormick Strikes Back

Emily Schutte[1] and Matthias Walter[2(✉)]

[1] University of Luxembourg, 4365 Esch-sur-Alzette, Luxembourg
emilyschutte@live.nl

[2] Department of Applied Mathematics, University of Twente, Enschede, The Netherlands
m.walter@utwente.nl

Abstract. We consider linear relaxations for multilinear optimization problems. In a recent paper, Khajavirad proved that the extended flower relaxation is at least as strong as the relaxation of any recursive McCormick linearization (Operations Research Letters 51 (2023) 146–152). In this paper we extend the result to more general linearizations, and present a simpler proof. Moreover, we complement Khajavirad's result by showing that the intersection of the relaxations of such linearizations and the extended flower relaxation are equally strong.

Keywords: Pseudo-Boolean optimization · multilinear optimization · cutting planes · extended formulations

1 Introduction

We consider multilinear optimization problems

$$\min \sum_{I \in \mathcal{I}_0} c_I^0 \prod_{v \in I} x_v \tag{1a}$$

$$\text{s.t.} \sum_{I \in \mathcal{I}_j} c_I^j \prod_{v \in I} x_v \leq b_j \qquad \forall j \in \{1, 2, \dots, m\} \tag{1b}$$

$$x_v \in [\ell_v, u_v] \quad \forall v \in V, \tag{1c}$$

where V denotes the variables and $\mathcal{I}_0, \mathcal{I}_1, \mathcal{I}_2, \dots, \mathcal{I}_m \subseteq V$ are families of subsets thereof, where $c_I^j \in \mathbb{R}$ and $b_j \in \mathbb{R}$ are the coefficients of the *monomials*, and the right-hand side, respectively, and where $\ell, u \in \mathbb{R}^V$ are the bounds on the variables. Note that every constraint (and the objective function) is a multilinear polynomial, which means that every variable has an exponent equal to either 0 or 1 in each monomial. We refer to the excellent survey [1] by Burer and Letchford for an overview of approaches for tackling mixed-integer nonlinear optimization problems in general. One of these strategies is to introduce auxiliary variables for intermediate nonlinear terms [19]. In our case, a straight-forward linearization

J. Vygen and J. Byrka (Eds.): IPCO 2024, LNCS 14679, pp. 393–404, 2024.
https://doi.org/10.1007/978-3-031-59835-7_29

is to introduce a variable z_I for every subset I of variables that appears in any of these polynomials, which yields the equivalent problem

$$\min \sum_{I \in \mathcal{I}_0} c_I^0 z_I \tag{2a}$$

$$\text{s.t.} \sum_{I \in \mathcal{I}_j} c_I^j z_I \leq b_j \qquad \forall j \in \{1, 2, \dots, m\} \tag{2b}$$

$$z_I = \prod_{v \in I} x_v \quad \forall I \in \mathcal{E} \tag{2c}$$

$$x_v \in [\ell_v, u_v] \quad \forall v \in V, \tag{2d}$$

where $\mathcal{E} := \bigcup_{j=0}^{m} \mathcal{I}_j$ denotes the union of all involved variable subsets. It is known well [22] that there exists an optimal solution in which each x_v is at its bound, that is, $x_v \in \{\ell_v, u_v\}$ holds for all $v \in V$. Hence, by an affine transformation we can replace (2d) by $x_v \in \{0,1\}$ for all $v \in V$. We now focus on linear relaxations for constraints (2c) and the requirement that x is binary, whose set of feasible solutions is called the *multilinear set*. It is parameterized by the pair $G = (V, \mathcal{E})$, which can be interpreted as a hypergraph whose nodes index the original x-variables and whose hyperedges correspond to the (multilinear) product terms [4]. We will not rely on hypergraph concepts, but merely use these to visualize instances. However, many previous results exploit hypergraph structures such as various cycle concepts [3,5,6,8,9].

Every hypergraph $G = (V, \mathcal{E})$ gives rise to a *multilinear polytope*, defined as the convex hull

$$\text{ML}(G) := \text{conv}\{(x, z) \in \{0,1\}^V \times \{0,1\}^{\mathcal{E}} \mid z_I = \prod_{v \in I} x_v \ \forall I \in \mathcal{E}\}.$$

of the multilinear set. Its simplest polyhedral relaxation is the *standard relaxation*

$$z_I \leq x_v \qquad \forall v \in I \in \mathcal{E} \tag{3a}$$

$$z_I + \sum_{v \in I} (1 - x_v) \geq 1 \qquad \forall I \in \mathcal{E} \tag{3b}$$

$$z_I \geq 0 \qquad \forall I \in \mathcal{E} \tag{3c}$$

$$x_v \in [0,1] \quad \forall v \in V, \tag{3d}$$

which dates back to Fortet [11,12] and Glover and Wolsey [14,15]. Unfortunately, this relaxation is often very weak [18].

In this paper we relate two strengthening techniques to each other: the first is by augmenting (3) with *(extended) flower inequalities* [5,17], which are additional inequalities valid for ML(G). We will define the inequalities and the tightened relaxation in Sect. 2. The second technique works by not linearizing each variable $z_I = \prod_{v \in I} x_v$ independently, but by using auxiliary variables for two disjoint subsets $I_1, I_2 \subseteq I$ with $I = I_1 \cup I_2$ instead, and inequalities similar to (3a) and (3b) for the product $z_I = z_{I_1} \cdot z_{I_2}$. While this alone does not strengthen the relaxation, it is well known that it does so in case these variables z_{I_1}, z_{I_2} appears in several linearization steps. We will provide a concise definition of such a *recursive McCormick linearization* in Sect. 3.

In a recent paper [17], Khajavirad showed that extended flower inequalities dominate recursive McCormick linearizations in the sense that the latter can never give stronger linear programming bounds than the former. Our main contribution is presented in Sect. 4, and states that the converse statement holds in some sense as well: one may need to intersect several McCormick linearizations with each other to achieve the same strength as that of extended flower inequalities. This may be considered a big drawback, but it is worth noting that each McCormick linearization induces only a polynomial number of additional variables and constraints, while there exist exponentially many flower inequalities whose separation problem is NP-hard [7]. In fact, our result applies to linearizations that are more general than recursive McCormick linearizations, and we provide a simpler proof of Khajavirad's main result. In Sect. 5 we conclude the paper with the observation that also computationally both approaches are equivalent, and highlight an interesting open problem.

2 Flower Relaxation

To simplify our notation we will write $z_{\{v\}} := x_v$ for all $v \in V$, and denote by $\mathcal{S} := \{\{v\} \mid v \in V\}$ the corresponding set of singletons. Note that this may in principle lead to an ambiguity in case $\{v\}$ is also part of $\mathcal{E}$. However, already the standard relaxation (3) would imply $z_{\{v\}} = x_v$ in this case. Hence, we from now on assume that $|I| \geq 2$ holds for all $I \in \mathcal{E}$. Let us now define extended flower inequalities [17], which are a generalization of flower inequalities [5].

Definition 1 (extended flower inequalities, extended flower relaxation). *Let $I \in \mathcal{E}$ and let $J_1, \dots, J_k \in \mathcal{E} \cup \mathcal{S}$ be such that $J_1 \cup J_2 \cup \dots \cup J_k \supseteq I$ and $J_i \cap I \neq \emptyset$ holds for $i = 1, 2, \dots, k$. The* extended flower inequality centered at I with neighbors $J_1, J_2, \dots, J_k$ *is the inequality*

$$z_I + \sum_{i=1}^{k} (1 - z_{J_i}) \geq 1. \tag{4}$$

The extended flower relaxation $\mathrm{FR}(G) \subseteq \mathbb{R}^{\mathcal{E} \cup \mathcal{S}}$ *is defined as the intersection of $[0,1]^{\mathcal{E} \cup \mathcal{S}}$ with all extended flower inequalities.*

Note that an extended flower inequality centered at I with the neighbors $J_1, J_2, \dots, J_k$ is called a *flower inequality* if $J_i \cap J_j \cap I = \emptyset$ holds for all distinct i, j, that is, if the intersections $(J_1 \cap I), (J_2 \cap I), \dots, (J_k \cap I)$ form a partition of I. It is well known that (extended) flower inequalities are valid for the multilinear polytope [17].

A key property of the extended flower relaxation is that it is compatible with projections in the sense that the (orthogonal) projection of the extended flower relaxation is the flower relaxation of the corresponding subgraph:

Lemma 1. *Let $G = (V, \mathcal{E})$ be a hypergraph, let $\mathcal{E}' := \mathcal{E} \setminus \{I^\star\}$ be the set of hyperedges with $I^\star \in \mathcal{E}$ removed, and let $G' = (V, \mathcal{E}')$ be the corresponding hypergraph. Then the projection of* $\mathrm{FR}(G)$ *onto the z_I-variables for all $I \in \mathcal{E}' \cup \mathcal{S}$ is equal to* $\mathrm{FR}(G')$.

Proof. Let $P \subseteq \mathbb{R}^{\mathcal{E}' \cup \mathcal{S}}$ denote the projection of $\mathrm{FR}(G)$ onto all variables but $z_{I^\star}$. Clearly, every extended flower inequality (4) that does not involve $z_{I^\star}$ is present in both relaxations. This already shows $P \subseteq \mathrm{FR}(G')$.

For the reverse direction we apply Fourier-Motzkin elimination [10,13,20] to $\mathrm{FR}(G)$ and $z_{I^\star}$. To this end, we need to consider pairs of inequalities (4) in which $z_{I^\star}$ appears with opposite signs. Such a pair consists of one extended flower inequality centered at $I^\star$ with neighbors $H_1, H_2, \dots, H_k \in \mathcal{E} \cup \mathcal{S}$ and one extended flower inequality centered at another edge J with neighbors $I^\star, K_1, K_2, \dots, K_\ell \in \mathcal{E} \cup \mathcal{S}$. The sum of the two inequalities reads

$$z_{I^\star} + \sum_{i=1}^{k} (1 - z_{H_i}) + z_J + (1 - z_{I^\star}) + \sum_{i=1}^{\ell} (1 - z_{K_i}) \geq 1 + 1 \tag{5}$$

We can assume that the $H_1, H_2, \dots, H_k$ are ordered such that $H_i \cap J \neq \emptyset$ holds if and only if $i \leq k'$ holds, where $k' \in \{0, 1, \dots, k\}$ is a suitable index. We can rewrite (5) as

$$z_J + \sum_{i=1}^{k'} (1 - z_{H_i}) + \sum_{i=1}^{\ell} (1 - z_{K_i}) \qquad + \qquad \sum_{i=k'+1}^{k} (1 - z_{H_i}) \geq 1$$

which is the sum of the extended flower inequality (4) centered at J with neighbors $H_1, H_2, \dots, H_{k'}, K_1, K_2, \dots, K_\ell$ and the bound inequalities $1 - z_{H_i} \geq 0$ for all $i \in \{k'+1, k'+2, \dots, k\}$. By construction and by the choice of k' the edges (or singletons) $H_1, H_2, \dots, H_{k'}, K_1, K_2, \dots, K_\ell$ indeed cover J. Hence, the combined inequality is implied by $\mathrm{FR}(G')$, which establishes $P \supseteq \mathrm{FR}(G')$. This concludes the proof.

3 Recursive Linearizations and Their Relaxations

For a digraph and a node w we denote by $N^{\text{out}}(w)$ and $N^{\text{in}}(w)$ the *successors* and *predecessors* of w, i.e., the sets of nodes u for which there is an arc from w to u and from u to w, respectively.

Definition 2 (recursive linearizations). *Let V be a finite ground set. A* recursive linearization *is a simple digraph $\mathcal{D} = (\mathcal{V}, \mathcal{A})$ with $\mathcal{S} \subseteq \mathcal{V} \subseteq 2^V \setminus \{\emptyset\}$ that satisfies $I \supseteq J$ for each arc $(I, J) \in \mathcal{A}$ as well as $\bigcup_{J \in N^{\text{out}}(I)} J = I$ for each $I \in \mathcal{V} \setminus \mathcal{S}$. We say that $\mathcal{D}$ is a recursive linearization* of a hypergraph $G = (V, \mathcal{E})$ *if $\mathcal{E} \subseteq \mathcal{V}$ holds and if each set $I \in \mathcal{V}$ with $N^{in}(I) = \emptyset$ belongs to $\mathcal{E} \cup \mathcal{S}$. Such a linearization is called* partitioning *if for every node $I \in \mathcal{V} \setminus \mathcal{S}$ all its successors $J, J' \in N^{out}(I)$ (with $J \neq J'$) are disjoint, i.e., $J \cap J' = \emptyset$ holds. It is called* binary *if $|N^{out}(I)| = 2$ holds for each $I \in \mathcal{V} \setminus \mathcal{S}$. A recursive linearization that is both, partitioning and binary, is called* recursive McCormick linearization.

Multilinear optimization problem:

$\min x_1x_2x_3+x_2x_3x_4+x_1x_2$ s.t. $x \in \{0,1\}^4$

Hypergraph:

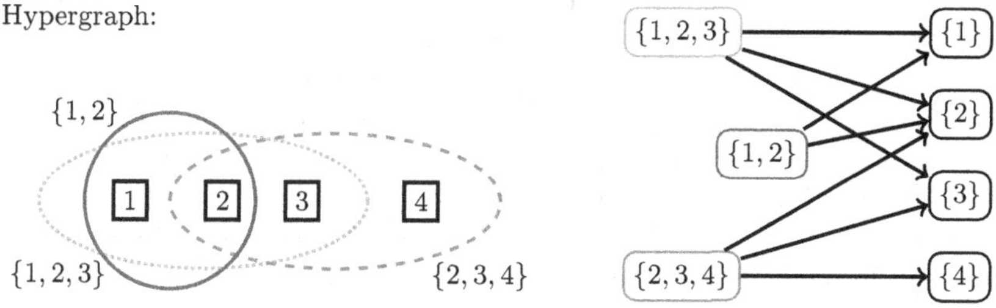

(a) A multilinear optimization problem with its hypergraph representation. Nodes are indicated as squares and hyperedges via ellipses.

(b) The standard linearization $\mathcal{D}^{(b)}$ with $z_{\{1,2,3\}} = z_{\{1\}} \cdot z_{\{2\}} \cdot z_{\{3\}}$, $z_{\{2,3,4\}} = z_{\{2\}} \cdot z_{\{3\}} \cdot z_{\{4\}}$, $z_{\{1,2\}} = z_{\{1\}} \cdot z_{\{2\}}$ and $z_{\{2,3\}} = z_{\{2\}} \cdot z_{\{3\}}$, which is non-binary but partitioning.

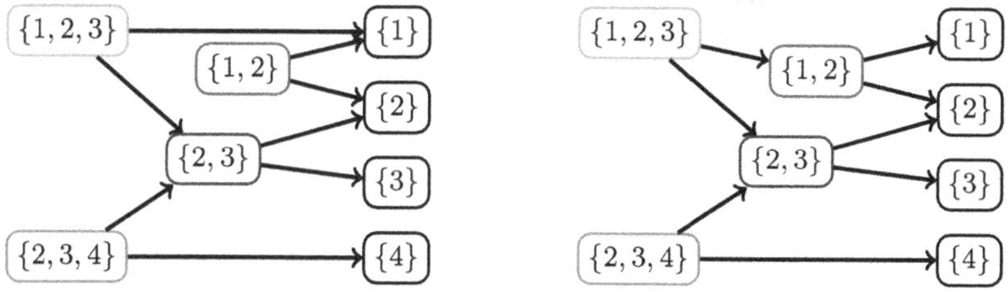

(c) A recursive McCormick linearization $\mathcal{D}^{(c)}$ with $z_{\{1,2,3\}} = z_{\{1\}} \cdot z_{\{2,3\}}$, $z_{\{2,3,4\}} = z_{\{2,3\}} \cdot z_{\{4\}}$, $z_{\{1,2\}} = z_{\{1\}} \cdot z_{\{2\}}$ and $z_{\{2,3\}} = z_{\{2\}} \cdot z_{\{3\}}$.

(d) A binary non-partitioning linearization $\mathcal{D}^{(d)}$ with $z_{\{1,2,3\}} = z_{\{1,2\}} \cdot z_{\{2,3\}}$, $z_{\{2,3,4\}} = z_{\{2,3\}} \cdot z_{\{4\}}$, $z_{\{1,2\}} = z_{\{1\}} \cdot z_{\{2\}}$ and $z_{\{2,3\}} = z_{\{2\}} \cdot z_{\{3\}}$.

Fig. 1. Three linearizations for the minimization problem in Fig. 1a with the depicted hypergraph $G = (V, \mathcal{E})$ with $V = \{1,2,3,4\}$ and $\mathcal{E} = \{\{1,2,3\},\{2,3,4\},\{1,2\},\{2,3\}\}$. The arcs that leave a node $I \subseteq V$ indicate the product that is used to represent z_I. Consider the point $z^{(1)} \in \mathbb{R}^{\mathcal{S}\cup\mathcal{E}}$ with $z^{(1)}_{\{2,3,4\}} = 0$, $z^{(1)}_{\{1,2,3\}} = z^{(1)}_{\{1,2\}} = z^{(1)}_{\{1\}} = z^{(1)}_{\{2\}} = z^{(1)}_{\{3\}} = \frac{1}{2}$ and $z^{(1)}_{\{4\}} = 1$. It is contained in $P(\mathcal{D}^{(b)})$. However, it is contained in neither $P_{\mathcal{E}}(\mathcal{D}^{(c)})$ nor in $P_{\mathcal{E}}(\mathcal{D}^{(d)})$ since in both linearizations the arc from $\{1,2,3\}$ to $\{2,3\}$ implies $z^{(1)}_{\{2,3\}} \geq \frac{1}{2}$ and since $z^{(1)}_{\{2,3,4\}} + (1 - z^{(1)}_{\{2,3\}}) + (1 - z^{(1)}_{\{4\}}) \geq 1$ implies $z^{(1)}_{\{2,3\}} \leq 0$. Consider the point $z^{(2)} \in \mathbb{R}^{\mathcal{E}\cup\mathcal{S}}$ with $z^{(2)}_{\{1,2,3\}} = 0$, $z^{(2)}_{\{2,3,4\}} = z^{(2)}_{\{1,2\}} = z^{(2)}_{\{2\}} = z^{(2)}_{\{3\}} = z^{(2)}_{\{4\}} = \frac{1}{2}$ and $z^{(2)}_{\{1\}} = 1$. It is contained in $P(\mathcal{D}^{(b)})$ and in $P_{\mathcal{E}}(\mathcal{D}^{(d)})$. However, it is not contained in $P_{\mathcal{E}}(\mathcal{D}^{(c)})$ since the arc from $\{2,3,4\}$ to $\{2,3\}$ implies $z^{(2)}_{\{2,3\}} \geq \frac{1}{2}$ and since $z^{(2)}_{\{1,2,3\}} + (1 - z^{(2)}_{\{2,3\}}) + (1 - z^{(2)}_{\{1\}}) \geq 1$ implies $z^{(2)}_{\{2,3\}} \leq 0$. Consider the point $z^{(3)} \in \mathbb{R}^{\mathcal{E}\cup\mathcal{S}}$ with $z^{(3)}_{\{1,2\}} = 0$ and $z^{(3)}_{\{1,2,3\}} = z^{(3)}_{\{2,3,4\}} = z^{(3)}_{\{1\}} = z^{(3)}_{\{2\}} = z^{(3)}_{\{3\}} = z^{(3)}_{\{4\}} = \frac{1}{2}$. It is contained in $P(\mathcal{D}^{(b)})$ and in $P_{\mathcal{E}}(\mathcal{D}^{(c)})$. However, it is not contained in $P_{\mathcal{E}}(\mathcal{D}^{(d)})$ since the arc from $\{1,2,3\}$ to $\{1,2\}$ implies $\frac{1}{2} = z^{(3)}_{\{1,2,3\}} \leq z^{(3)}_{\{1,2\}} = 0$.

A recursive linearization encodes how each variable z_I with $I \in \mathcal{E}$ is linearized by means of other variables. First, such a variable z_I is created for each node $I \in \mathcal{V}$, where $z_{\{v\}}$ is the same as the variable x_v for each $v \in V$. It is supposed to encode the product of the x_v for all $v \in V$, that is, $z_I = \prod_{v \in I} x_v$. This will actually be done by encoding (using linear inequalities) that $z_I = \prod_{J \in N^{\text{out}}(I)} z_J$ holds for each $I \in \mathcal{V}$. By induction we can assume that $z_J = \prod_{v \in J} x_v$ holds, which yields $z_I = \prod_{J \in N^{\text{out}}(I)} \prod_{v \in J} x_v$. Since I is the union of all sets $N^{\text{out}}(I)$, a variable x_v appears in this product if and only if $v \in I$ holds. Unless the linearization is partitioning, the variable may appear multiple times, but this does not harm since $x_v^2 = x_v$ holds for $x_v \in \{0, 1\}$. Notice that a linearization is partitioning if and only if, for every node $I \in \mathcal{V}$ and every $v \in I$, there is a unique path from I to $\{v\}$. Figure 1 shows three different linearizations of the same hypergraph.

A recursive linearization states the interplay between the linearization variables. The actual linear inequalities associated with a linearization are stated in the following definition.

Definition 3 (relaxation, projected relaxation). *Let $\mathcal{D} = (\mathcal{V}, \mathcal{A})$ be a recursive linearization. Its* relaxation *is the polyhedron $P(\mathcal{D}) \subseteq \mathbb{R}^{\mathcal{V}}$ defined by*

$$z_I \leq z_J \qquad \forall (I, J) \in \mathcal{A} \tag{6a}$$

$$z_I + \sum_{J \in N^{out}(I)} (1 - z_J) \geq 1 \qquad \forall I \in \mathcal{V} \setminus \mathcal{S} \tag{6b}$$

$$z \in [0, 1]^{\mathcal{V}}. \tag{6c}$$

Its projected relaxation *with respect to a given set $\mathcal{T} \subseteq \mathcal{V}$ of* target nodes *is the projection of $P(\mathcal{D})$ onto the variables z_I for all $I \in \mathcal{T} \cup \mathcal{S}$ and is denoted by $P_{\mathcal{T}}(\mathcal{D})$.*

Figure 2 shows an example hypergraph for which we can either add a quadratic number of flower inequalities or just add one additional variable, which only slightly increases the size of the LP relaxation. Moreover, as we will see in Sect. 4, both formulations will be of equal strength.

Before we turn to our main result, we show that requiring a recursive linearization to be partitioning (resp. binary) is actually a restriction.

Proposition 1. *There exists a non-partitioning recursive linearization $\mathcal{D} = (\mathcal{V}, \mathcal{A})$ for some hypergraph $G = (V, \mathcal{E})$ such that any recursive linearization $\mathcal{D}' = (\mathcal{V}', \mathcal{A}')$ with $P_{\mathcal{E}}(\mathcal{D}') \subseteq P(\mathcal{D}')$ is also non-partitioning.*

Proposition 2. *There exists a non-binary recursive linearization $\mathcal{D} = (\mathcal{V}, \mathcal{A})$ for some hypergraph $G = (V, \mathcal{E})$ such that any recursive linearization $\mathcal{D}' = (\mathcal{V}', \mathcal{A}')$ with $P_{\mathcal{E}}(\mathcal{D}') \subseteq P(\mathcal{D}')$ is also non-binary.*

The corresponding two linearizations can be found in the preprint version of the paper [21]. The propositions shows that in principle a single non-binary or

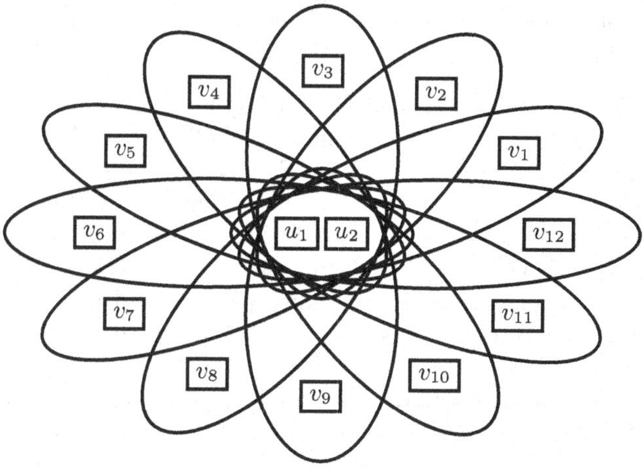

Fig. 2. An instance in which several extended flower inequalities are captured by one recursive McCormick linearization. The hypergraph $G = (V, \mathcal{E})$ has $V = \{\{v_i\} \mid i = 1, 2, \ldots, k\} \cup \{\{u_1, u_2\}\}$ and $\mathcal{E} = \{\{u_1, u_2, v_i\} \mid i = 1, 2, \ldots, k\}$ (for $k = 12$). It has $k(k-1)$ non-redundant extended flower inequalities by considering each edge as the center edge and choosing any other edge as the unique neighbor. However, it admits a recursive McCormick linearization with one additional variable $z_{\{u_1,u_2\}}$ and $2k+3$ extra inequalities ($z_{\{u_1,u_2,v_i\}} \leq z_{\{u_1,u_2\}}$ and $z_{\{u_1,u_2,v_i\}} + (1 - z_{\{u_1,u_2\}}) + (1 - z_{v_i}) \geq 1$ for $i = 1, 2, \ldots, k$ as well as those for $z_{\{u_1,u_2\}} = z_{u_1} \cdot z_{u_2}$), which turn $3k$ inequalities ($z_{\{u_1,u_2,v_i\}} \leq z_{u_1}$, $z_{\{u_1,u_2,v_i\}} \leq z_{u_2}$ and $z_{\{u_1,u_2,v_i\}} + (1 - z_{u_1} + (1 - z_{u_2}) + (1 - z_{v_i}) \geq 1$ for $i = 1, 2, \ldots, k$) from the standard relaxation redundant. By Theorem 1, both formulations have the same strength.

non-partitioning linearization may be more powerful than any single recursive McCormick linearization. However, in the next section we show that combinations of several McCormick linearizations are as powerful as the general recursive linearization.

4 Comparison of Projected Relaxations

Theorem 1. *Let $G = (V, \mathcal{E})$ be a hypergraph. Then the following polyhedra are equal:*

(i) flower relaxation $\mathrm{FR}(G)$*;*
(ii) intersection of the projected relaxations $P_{\mathcal{E}}(\mathcal{D})$ for all recursive linearizations $\mathcal{D}$ of G;
(iii) intersection of the projected relaxations $P_{\mathcal{E}}(\mathcal{D})$ for all recursive McCormick linearizations $\mathcal{D}$ of G.

The fact that (i) is contained in (iii) was established in [17]. This was proved by applying Fourier-Motzkin elimination to the relaxation of a McCormick linearization, showing that all resulting inequalities are implied by extended flower inequalities. In contrast to this, our proof works via the projection of the extended flower relaxation and is surprisingly simple.

Proof ((i) is contained in (ii)). Let $\mathcal{D} = (\mathcal{V}, \mathcal{A})$ be a recursive linearization of a hypergraph $G = (V, \mathcal{E})$. We need to show $\text{FR}(G) \subseteq P_{\mathcal{E}}(\mathcal{D})$. To this end, we consider the hypergraph $G' = (V, \mathcal{E}')$ with $\mathcal{E}' := \mathcal{V} \setminus \mathcal{S}$ and claim that $\text{FR}(G') \subseteq P(\mathcal{D})$ holds. Note that $\mathcal{E}' \supseteq \mathcal{E}$ holds, that is, G' contains all of G's hyperedges.

First, for any arc $(I, J) \in \mathcal{A}$, (6a) is equivalent to the extended flower inequality (4) centered at J with neighbor I. Second, for any node $I \in \mathcal{V} \setminus \mathcal{S}$, (6b) is equivalent to the extended flower inequality (4) centered at I with neighbors $\mathcal{N}^{\text{out}}(I)$. Third, for any $v \in \mathcal{V}$, also $0 \leq z_v \leq 1$ follows from the definition of $\text{FR}(G')$. This proves $\text{FR}(G') \subseteq P(\mathcal{D})$.

For the projection P of $\text{FR}(G')$ and for that of $P(\mathcal{D})$, both onto the variables z_I for $I \in \mathcal{E} \cup \mathcal{S}$, we thus obtain $P \subseteq P_{\mathcal{E}}(\mathcal{D})$. Successive application of Lemma 1 to $\text{FR}(G')$ for every edge from $\mathcal{E}' \setminus \mathcal{E}$ yields $P = \text{FR}(G)$. Hence, $\text{FR}(G) \subseteq P_{\mathcal{E}}(\mathcal{D})$ holds. Since $\text{FR}(G)$ is contained in each such projected relaxation, it is also contained in their intersection, which concludes the proof.

Proof ((ii) is contained in (iii)). This holds since every recursive McCormick linearization is a recursive linearization.

Proof ((iii) is contained in (i)). It suffices to show that for any hypergraph $G = (V, \mathcal{E})$ and any non-redundant extended flower inequality (4) there exists a recursive binary partitioning linearization $\mathcal{D} = (\mathcal{V}, \mathcal{A})$ whose projected relaxation with respect to $\mathcal{E}$ implies this inequality.

To this end, consider an inequality (4) centered at $I^\star$ with neighbors $J_1^\star$, $J_2^\star, \ldots, J_k^\star$. Suppose there exists a neighbor $J_{i^\star}^\star$ that is not required for all (other) neighbors to cover $I^\star$. Now consider the extended flower inequality

$$\begin{aligned} z_{I^\star} + (1 - z_{J_1^\star}) + (1 - z_{J_2^\star}) + \cdots + (1 - z_{J_{i^\star-1}^\star}) \\ + (1 - z_{J_{i^\star+1}^\star}) + \cdots + (1 - z_{J_{k-1}^\star}) + (1 - z_{J_k^\star}) \geq 1 \end{aligned}$$

in which neighbor $J_{i^\star}^\star$ was removed. Adding $1 - z_{J_{i^\star}^\star} \geq 0$ to it yields our considered inequality, which shows that the latter is redundant. Hence, from now on we assume that

$$\forall i \in \{1, 2, \ldots, k\} \quad I^\star \cap J_i^\star \setminus \bigcup_{j \neq i} J_j^\star \neq \emptyset$$

holds. We partition $I^\star$ into the sets

$$L_i := I^\star \cap \left(J_i^\star \setminus \bigcup_{j=1}^{i-1} J_j^\star \right) \qquad i = 1, 2, \ldots, k. \tag{7}$$

Note that $L_i \neq \emptyset$ holds by (4). We construct our linearization $\mathcal{D} = (\mathcal{V}, \mathcal{A})$ as follows. First, start with $\mathcal{V} := \{I^\star, L_1, L_2, \ldots, L_k\} \cup \mathcal{S}$. Second, for $i = k$, $k -$

$1, \ldots, 3, 2$, add the node $L_1 \cup L_2 \cup \cdots \cup L_{i-1}$ to $\mathcal{V}$ and the arcs from $L_1 \cup L_2 \cup \cdots \cup L_i$ to $L_1 \cup L_2 \cup \cdots \cup L_{i-1}$ and to L_i. Note that this yields a binary tree with root $I^\star$ and leaves L_i. We will later add paths from these leaves to the singletons, and for this purpose initialize the set of *unprocessed* nodes, denoted by $\mathcal{U}$, as $\mathcal{U} := \{L_1, L_2, \ldots, L_k\}$. Being unprocessed means that we still need to complete its linearization in an arbitrary manner. This is possible because all unprocessed nodes are pair-wise disjoint.

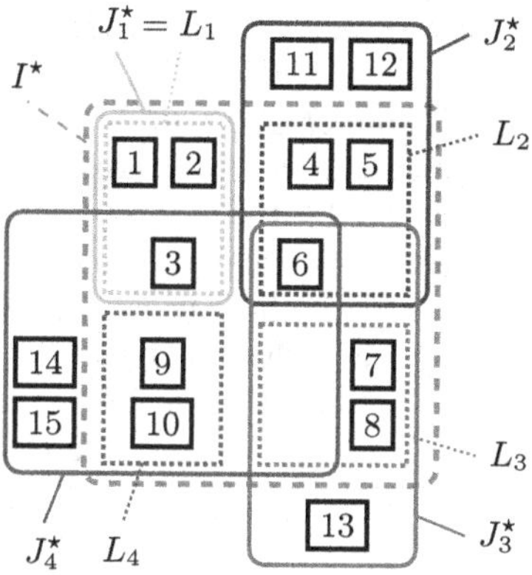

(a) Support hypergraph with a center edge $I^\star$ and four neighbors $J_1^\star$, $J_2^\star$, $J_3^\star$ and $J_4^\star$ along with the four sets L_i as defined in (7).

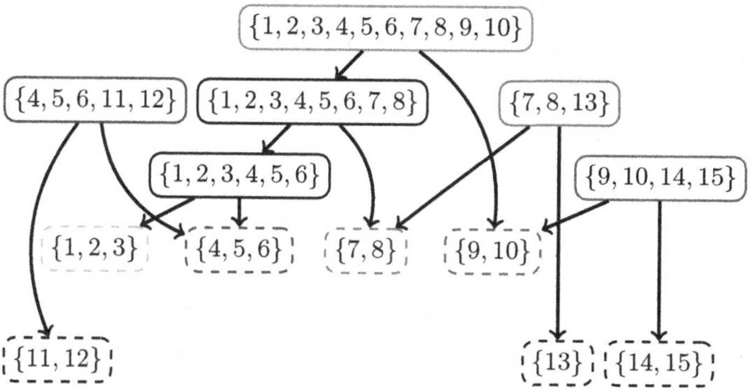

(b) Corresponding incomplete McCormick linearization from the proof. The unprocessed nodes $\mathcal{U} = \{\{1,2,3\}, \{4,5,6\}, \{7,8\}, \{9,10\}, \{11,12\}, \{13\}, \{14,15\}\}$ are depicted with a dashed border. These are pairwise disjoint and hence the completion of the linearization is arbitrary.

Fig. 3. Hyperedges for a flower inequality with four neighbors and the corresponding incomplete linearization from the proof that (iii) is contained in (i).

We now extend $\mathcal{D}$ to also include the neighbors $J_i^\star$, processing all indices $i \in \{1, 2, \dots, k\}$ one by one. Throughout this process all nodes $I \in \mathcal{V}$ will be subsets of $I^\star$ or of those $J_i^\star$ that were processed so far. If $J_i^\star = L_i$, then there is nothing to do and we continue with the next iteration $i+1$. Otherwise, if $J_i^\star \neq L_i$, then we extend $\mathcal{V}$ by the nodes $J_i^\star$ and $J_i^\star \setminus L_i$, add the arcs $(J_i^\star, L_i), (J_i^\star, J_i^\star \setminus L_i)$ to $\mathcal{A}$, and add $J_i^\star \setminus L_i$ to $\mathcal{U}$. An example of such a linearization (which is incomplete due to unprocessed nodes) is depicted in Fig. 3.

After all $J_i^\star$ have been processed, we process all unprocessed nodes: While $\mathcal{U} \neq \emptyset$, pick a node $I \in \mathcal{U}$ and remove it from $\mathcal{U}$. If $I \in \mathcal{V}$, then we continue. Otherwise, $|I| \geq 2$ must hold due to $\mathcal{S} \subseteq \mathcal{V}$. Arbitrarily partition $I = I_1 \cup I_2$ with $I_1, I_2 \neq \emptyset$. Add the nodes I_1, I_2 to $\mathcal{V}$ (unless they exist) and the arcs $(I, I_1), (I, I_2)$ to $\mathcal{A}$. Remove I from $\mathcal{U}$ and add I_1 and I_2 to it. Note that this process terminates since the cardinalities of the nodes I_1, I_2 that are added to $\mathcal{U}$ are smaller than that of the removed set I.

Now it is easily verified that $\mathcal{D}$ is a recursive McCormick linearization for G, i.e., in particular binary and partitioning. It has the following properties: $\mathcal{D}$ contains, for each $i \in \{1, 2, \dots, k\}$ a path (of length 0 or 1) from $J_i^\star$ to L_i as well as a path from $I^\star$ to L_i (of length $k-i+1$ with the inner nodes $L_1 \cup L_2 \cup \dots \cup L_j$ for $j = k, k-1, \dots, i+1$ in that order). We claim that the following inequalities are valid for $P(\mathcal{D})$:

$$(1 - z_{J_i^\star}) - (1 - z_{L_i}) \geq 0 \quad i = 1, 2, \dots, k \quad \text{(8a)}$$

$$z_{(L_1 \cup L_2 \cup \dots \cup L_j)} + (1 - z_{(L_1 \cup L_2 \cup \dots \cup L_{j-1})}) + (1 - z_{L_j}) \geq 1 \quad j = 2, 3, \dots, k \quad \text{(8b)}$$

Note that if $L_i = J_i^\star$, the two variables in (8a) are the same, and otherwise the inequality follows from (6a) due to the arc $(J_i^\star, L_i) \in \mathcal{A}$. Inequality (8b) corresponds to (6b) for the two arcs that leave $L_1 \cup L_2 \cup \dots \cup L_j$. The sum of all inequalities (8b) reads

$$z_{I^\star} + \sum_{i=1}^{k} (1 - z_{L_i}) \geq 1.$$

Adding (8a) for $i = 1, 2, \dots, k$ yields the desired flower inequality (4) centered at $I^\star$ with neighbors $J_1^\star, J_2^\star, \dots, J_k^\star$. This inequality is also implied by the projection $P_{\mathcal{E}}(\mathcal{D})$ of $P(\mathcal{D})$ onto the variables z_I for $I \in \mathcal{E} \cup \mathcal{S}$, which concludes the proof.

5 Discussion

We would like to conclude our paper with the following consequences for computational complexity. Suppose we have a family of hypergraphs $G = (V, \mathcal{E})$ for which we can solve the separation problem for extended flower inequalities (4) in polynomial time. For instance, this is the case if the degree of the linearized polynomials is bounded by a constant, since then we need to consider only flowers whose number of neighbors is bounded by a constant. The straight-forward way of using such a separation algorithm is to maintain a subset of generated extended flower inequalities and, whenever necessary, query the algorithm with a point $\hat{z} \in \mathbb{R}^{\mathcal{S} \cup \mathcal{E}}$ to check if $\hat{z} \in \text{FR}(G)$ holds or, if a violated inequality is determined, augment our subset. Note that such a point $\hat{z}$ then typically satisfies the generated inequalities, say, in the context of the Ellipsoid method [16] or when doing row generation using the Simplex method [2]. As an alternative, we could now maintain a set of recursive linearizations. For a point $\hat{z}$ we can now run the same algorithm, but apply our construction from the proof of Theorem 1 (that (iii) is contained in (i)) to obtain a new linearization that we add to our set. The disadvantage is that adding such a new recursive linearization requires us to add new variables and new inequalities. An advantage could be that a new recursive linearization can imply several flower inequalities at once, as Fig. 2 illustrates. While this discussion settles the theoretical equivalence of both dynamic approaches, we believe that the dynamic generation of recursive formulations will not be practical.

Let us consider the intersection of two projected relaxations $P_{\mathcal{E}}(\mathcal{D}^{(1)})$ and $P_{\mathcal{E}}(\mathcal{D}^{(2)})$ for linearizations $\mathcal{D}^{(j)} = (\mathcal{V}^{(j)}, \mathcal{A}^{(j)})$, $j = 1, 2$. When working with recursive linearizations one would not carry out the projection, but actually maintain variables for all nodes in $\mathcal{V}^{(1)}$ and in $\mathcal{V}^{(2)}$. The intersection is achieved by identifying the variables z_I for all $I \in \mathcal{E}$. However, if there is another variable $I \in (\mathcal{V}^{(1)} \cap \mathcal{V}^{(2)}) \setminus \mathcal{E}$, then it formally exists in both formulations and has same intended meaning in both linearizations. Hence, we can identify these variables $z_I^{(1)} = z_I^{(2)}$ with each other. First, this reduces the number of variables. Second, if some arcs (that involve I) exist in both recursive linearizations, then this may also reduce the number of inequalities. Notice that, after doing this, there might be multiple inequalities (6b) for a single $I \subseteq V$. Third, this might strengthen the formulation. In fact, *running intersection inequalities* [6] can be seen as a strengthened version of flower inequalities. We leave it as an open problem to investigate how this strengthening of intersected recursive linearizations compares to strengthening of flower inequalities to running intersection inequalities.

Acknowledgements. We thank four anonymous referees for their constructive comments that led to an improved presentation of the paper.

References

1. Burer, S., Letchford, A.N.: Non-convex mixed-integer nonlinear programming: a survey. Surv. Oper. Res. Manag. Sci. **17**, 97–106 (2012). https://doi.org/10.1016/j.sorms.2012.08.001
2. Dantzig, G.B.: Maximization of a linear function of variables subject to linear inequalities. In: Koopmans, T.C. (ed.) Activity Analysis of Production and Allocation, pp. 339–347. Cowles Commission Monograph No. 13. Wiley (1951)
3. Del Pia, A., Di Gregorio, S.: Chvátal rank in binary polynomial optimization. INFORMS J. Optim. (2021). https://doi.org/10.1287/ijoo.2019.0049
4. Del Pia, A., Khajavirad, A.: A polyhedral study of binary polynomial programs. Math. Oper. Res. **42**(2), 389–410 (2017). https://doi.org/10.1287/moor.2016.0804

5. Del Pia, A., Khajavirad, A.: The multilinear polytope for acyclic hypergraphs. SIAM J. Optim. **28**(2), 1049–1076 (2018). https://doi.org/10.1137/16M1095998
6. Del Pia, A., Khajavirad, A.: The running intersection relaxation of the multilinear polytope. Math. Oper. Res. (2021). https://doi.org/10.1287/moor.2021.1121
7. Del Pia, A., Khajavirad, A., Sahinidis, N.V.: On the impact of running intersection inequalities for globally solving polynomial optimization problems. Math. Program. Comput. **12**(2), 165–191 (2020). https://doi.org/10.1007/s12532-019-00169-z
8. Del Pia, A., Walter, M.: Simple odd β-cycle inequalities for binary polynomial optimization. In: Aardal, K., Sanità, L. (eds.) Integer Programming and Combinatorial Optimization, pp. 181–194. Springer, Cham (2022). https://doi.org/10.1007/978-3-031-06901-7_14
9. Del Pia, A., Walter, M.: Simple odd β-cycle inequalities for binary polynomial optimization. Math. Program. (2023). https://doi.org/10.1007/s10107-023-01992-y
10. Dines, L.L.: Systems of linear inequalities. Ann. Math. 191–199 (1919)
11. Fortet, R.: Applications de l'algebre de boole en recherche opérationelle. Revue Française de Recherche Opérationelle **4**(14), 17–26 (1960)
12. Fortet, R.: L'algebre de boole et ses applications en recherche operationnelle. Trabajos de Estadistica **4**, 17–26 (1960). https://doi.org/10.1007/BF03006558
13. Fourier, J.B.J.: Analyse des travaux de i'academie royale des sciences pendant i'annee 1824. Partie mathematique, Histoire de l'Academie Royale des Sciences de l'Institut de France **7**, xlvii–lv (1827)
14. Glover, F., Woolsey, E.: Further reduction of zero-one polynomial programming problems to zero-one linear programming problems. Oper. Res. **21**(1), 156–161 (1973). https://doi.org/10.1287/opre.21.1.156
15. Glover, F., Woolsey, E.: Converting the 0-1 polynomial programming problem to a 0-1 linear program. Oper. Res. **22**(1), 180–182 (1974). https://doi.org/10.1287/opre.22.1.180
16. Khachiyan, L.: A polynomial algorithm in linear programming. Dokl. Akad. Nauk SSSR **244**, 1093–1096 (1979)
17. Khajavirad, A.: On the strength of recursive McCormick relaxations for binary polynomial optimization. Oper. Res. Lett. **51**(2), 146–152 (2023). https://doi.org/10.1016/j.orl.2023.01.009
18. Luedtke, J., Namazifar, M., Linderoth, J.: Some results on the strength of relaxations of multilinear functions. Math. Program. **136**(2), 325–351 (2012). https://doi.org/10.1007/s10107-012-0606-z
19. McCormick, G.P.: Computability of global solutions to factorable nonconvex programs: Part i – convex underestimating problems. Math. Program. **10**(1), 147–175 (1976). https://doi.org/10.1007/BF01580665
20. Motzkin, T.S.: Beiträge zur Theorie der linearen Ungleichungen. Ph.D. thesis, Universität Basel (1936)
21. Schutte, E., Walter, M.: Relaxation strength for multilinear optimization: McCormick strikes back. arXiv:2311.08570 (2023)
22. Tawarmalani, M., Sahinidis, N.V.: Convex extensions and envelopes of lower semi-continuous functions. Math. Program. **93**, 247–263 (2002). https://doi.org/10.1007/s10107-002-0308-z

Online Algorithms for Spectral Hypergraph Sparsification

Tasuku Soma[1], Kam Chuen Tung[2(✉)], and Yuichi Yoshida[3]

[1] Institute of Statistical Mathematics, Tokyo, Japan
soma@ism.ac.jp
[2] University of Waterloo, Waterloo, Canada
kctung@uwaterloo.ca
[3] National Institute of Informatics, Tokyo, Japan
yyoshida@nii.ac.jp

Abstract. We provide the first online algorithm for spectral hypergraph sparsification. In the online setting, hyperedges with positive weights are arriving in a stream, and upon the arrival of each hyperedge, we must irrevocably decide whether or not to include it in the sparsifier. Our algorithm produces an (ε, δ)-spectral sparsifier with multiplicative error ε and additive error δ that has $O(\varepsilon^{-2} n \log n \log r \log(1 + \varepsilon W/\delta n))$ hyperedges with high probability, where $\varepsilon, \delta \in (0, 1)$, n is the number of nodes, r is the rank of the hypergraph, and W is the sum of edge weights. The space complexity of our algorithm is $O(n^2)$, while previous algorithms required space complexity $\Omega(m)$, where m is the number of hyperedges. This provides an exponential improvement in the space complexity since m can be exponential in n.

Keywords: Online Algorithms · Hypergraph Sparsification · Spectral Algorithms · Generic Chaining

1 Introduction

Spectral sparsification is a cornerstone of modern algorithm design. The studies of spectral sparsification date back to the seminal work of Spielman and Teng [19] for undirected graphs. Let $G = (V, E, w)$ be an undirected graph with positive edge weights $w : E \to \mathbb{R}_{>0}$. Let $\varepsilon \in (0, 1)$ be an arbitrary constant. A weighted graph $\tilde{G} = (V, E, \tilde{w})$ on the same node set V is called an *ε-spectral sparsifier* of G if

$$(1 - \varepsilon) z^\top L_G z \le z^\top L_{\tilde{G}} z \le (1 + \varepsilon) z^\top L_G z$$

for all $z \in \mathbb{R}^V$, where $\tilde{w}$ is a nonnegative weight function on the edges, and L_G and $L_{\tilde{G}}$ denote the Laplacian matrices of G and $\tilde{G}$, respectively. The number of nonzeros in $\tilde{w}$ is called the *size* of a spectral sparsifier $\tilde{G}$. Spielman Teng [19] showed that one can find an ε-spectral sparsifier with $O(\varepsilon^{-2} n \log n)$ edges in nearly linear time in the size of the input graph. Since then, there has been a

J. Vygen and J. Byrka (Eds.): IPCO 2024, LNCS 14679, pp. 405–417, 2024.
https://doi.org/10.1007/978-3-031-59835-7_30

series of works on spectral sparsification of graphs and various applications in the design of fast algorithms; see [2,21] for surveys.

Recently, the notion of spectral sparsification was extended to undirected hypergraphs and has been actively studied in the literature [1,9–11,14,17]. For a weighted hypergraph $H = (V, E, w)$ with a positive edge weight $w : E \to \mathbb{R}_{>0}$, the energy function $Q_H : \mathbb{R}^V \to \mathbb{R}$ of H is given by

$$Q_H(z) := \sum_{e \in E} w(e) \max_{u,v \in e} (z(u) - z(v))^2.$$

This is a generalization of the quadratic form of the Laplacian matrix of a graph. For $\varepsilon \in (0, 1)$, a weighted hypergraph $\tilde{H} = (V, E, \tilde{w})$ on the same node set V is called an ε-spectral sparsifier of H if

$$(1 - \varepsilon)Q_H(z) \le Q_{\tilde{H}}(z) \le (1 + \varepsilon)Q_H(z) \tag{1}$$

for all $z \in \mathbb{R}^V$. Again, the number of nonzeros in $\tilde{w}$ is called the size of a spectral sparsifier $\tilde{H}$. Since the number of hyperedges can be exponential, the existence of polynomial-size spectral sparsifiers is nontrivial. This concept was first introduced by Soma and Yoshida [17], and they showed that there exists an ε-spectral sparsifier with $O(\varepsilon^{-2}n^3)$ hyperedges and it can be found in time polynomial in the size of the input hypergraph. The current best upper bound on the size of spectral sparsifiers is $O(\varepsilon^{-2}n \log n \log r)$ by [9,14], where r is the *rank* of the hypergraph, i.e., the maximum size of a hyperedge.

However, all known algorithms for hypergraph spectral sparsification are *offline*, i.e., they first store the entire hypergraph in the working memory and then construct a spectral sparsifier. This is somewhat unreasonable because the space complexity (e.g., the size of input hypergraphs) could be exponentially larger than the size of the output sparsifier. So we are naturally led to the following question: Can we construct a spectral sparsifier of hypergraphs with smaller space complexity?

To formalize this, we study *online* spectral sparsification in this paper. In the online setting, the hyperedges $e_1, \ldots, e_m$ arrive in a stream fashion together with their weights. When edge e_i arrives, we must decide immediately whether or not to include it in the sparsifier $\tilde{H}$. Our goal is for $\tilde{H}$ to have a small number of edges and for the algorithm to use little working memory.

1.1 Our Contribution

We provide the first algorithm for online spectral hypergraph sparsification. We say that $\tilde{H}$ is an (ε, δ)-*spectral sparsifier* of H, if

$$(1 - \varepsilon)Q_H(z) - \delta\|z\|_2^2 \le Q_{\tilde{H}}(z) \le (1 + \varepsilon)Q_H(z) + \delta\|z\|_2^2$$

for all $z \in \mathbb{R}^V$. Our main contribution is the following.

Theorem 1 (Main). *There exists an online algorithm (Algorithm 1) with the following performance guarantees:*

- *The amount of working memory required is $O(n^2)$ assuming the word RAM model;*
- *With high probability (i.e., probability at least $1 - 1/n$), it finds an (ε, δ)-spectral sparsifier $\tilde{H}$ of a rank-r hypergraph H with*

$$O\left(\frac{n \log n \log r}{\varepsilon^2} \cdot \log\left(1 + \frac{\varepsilon W}{\delta n}\right)\right)$$

many hyperedges, where $W = \sum_{e \in E} w(e)$.

Remark 1 (Lower Bound). We remark that the upper bound on the number of hyperedges is tight up to logarithmic factors. In fact, in [7, Theorem 5.1] it is shown that even in the graph case it is necessary to sample $\Omega(n \log(1 + \varepsilon W/(\delta n))/\varepsilon^2)$ edges.

We can also obtain an ε-spectral sparsifier for rank-r hypergraphs if the range of edge weights is known in advance.

Corollary 1 (ε-spectral sparsifier). *Suppose that $H = (V, E, w)$ is a rank-r hypergraph and that $W_{\min} \leq w(e) \leq W_{\max}$ for every $e \in E$ for some $0 < W_{\min} \leq W_{\max}$. Then, Algorithm 1 with $\delta = O(\varepsilon W_{\min}^2 n^{-2r})$ finds an ε-spectral sparsifier with*

$$O\left(\frac{nr \log n \log r}{\varepsilon^2} \cdot \log\left(\frac{n W_{\max}}{W_{\min}}\right)\right)$$

hyperedges with high probability.

1.2 Our Techniques

We outline our algorithms below. Our starting point is the work of spectral hypergraph sparsification via *generic chaining* [14]. Lee showed that if we sample each hyperedge with probability proportional to the effective resistance of an auxiliary (ordinary) graph, then the resulting hypergraph is a spectral sparsifier with high probability. Here, the auxiliary graph is a weighted clique-graph $G = (V, F)$, where F is the multiset of undirected edges obtained by replacing every hyperedge $e \in E$ with the clique on $V(e)$. The weight of edge (u, v) coming from hyperedge e is given by $w(e)c_{e,u,v}$, where $c : F \to \mathbb{R}_{\geq 0}$ is a special *reweighting* satisfying the following conditions [11]: For each $e \in E$, (i) $\sum_{u,v \in e} c_{e,u,v} = 1$ for $e \in E$ and (ii) $c_{e,u,v} > 0$ implies $r_{u,v} = \max_{u',v' \in e} r_{u',v'}$, where $r_{u,v}$ denotes the effective resistance between u and v in G. Such a reweighting can be found by solving the following convex optimization problem

$$\begin{aligned}
\text{maximize} \quad & \log\det\left(\sum_{e \in E}\sum_{u,v \in e} w(e) c_{e,u,v} L_{uv} + J\right) \\
\text{subject to} \quad & \sum_{u,v \in e} c_{e,u,v} = 1 \quad (e \in E) \\
& c_{e,u,v} \geq 0 \quad (e \in E, u, v \in e),
\end{aligned}$$

where L_{uv} is the Laplacian of edge (u, v) and J is the all-one matrix [14]. Now we set the edge sampling probability $p_e \propto w(e) \max_{u,v \in e} r_{u,v}$. Using the generic chaining technique, [14] showed that this yields an ε-spectral sparsifier with a constant probability.

In the online setting, the entire hypergraph is not available, so we have to estimate the edge sampling probability on the fly. Inspired by the online row sampling algorithm [7], we introduce the *ridged edge sampling probability.* Let $\eta = \varepsilon/\delta$. For $i = 1, \ldots, m$, we iteratively compute a sequence of auxiliary graphs G_i. Initially, G_0 is the empty graph on V. For each $i > 0$, we construct a graph G_i from G_{i-1} by adding an edge (u, v) with weight $w_i c_{i,u,v}$ for every pair of vertices (u, v) in e_i, where w_i is the weight of e_i and $c_{i,u,v}$ is an optimal solution of the following convex optimization problem:

$$
\begin{aligned}
\text{maximize} \quad & \log\det\left(L_{G_{i-1}} + \sum_{u,v \in e_i} w_i c_{i,u,v} L_{uv} + \eta I\right) \\
\text{subject to} \quad & \sum_{u,v \in e_i} c_{i,u,v} = 1 \\
& c_{i,u,v} \geq 0 \quad (u, v \in e_i)
\end{aligned}
$$

Then, we define the η-ridged edge sampling probability by $p_i \propto \max_{u,v \in e_i} \|(L_{G_i} + \eta I)^{-1/2}(\chi_u - \chi_v)\|_2^2$, where χ_u denotes the u-th standard unit vector. Using the techniques from [8], we show that this gives an (ε, δ)-spectral sparsifier having the desired number of hyperedges with high probability. Since we only need to maintain the Laplacian of G_i, the space complexity is $O(n^2)$ as required. Note that the above convex optimization problem can be solved by projected gradient descent (up to desired accuracy), which only requires space of linear in the dimension, i.e., $O(n^2)$.

1.3 Related Work

The literature on spectral sparsification is vast. We refer the readers to [2,21] for technical details and various applications. Spectral sparsification of hypergraphs can be used to speed up semi-supervised learning with hypergraph regularizers and hypergraph network analysis; see discussion in [11,17] for further applications.

Spectral sparsification for graphs in the *semi-streaming* setting is also well-studied. This is almost identical to our online setting, but the algorithm can change the weights of edges in the output that have already been sampled. Kelner and Levin [13] initiated this line of research and provided a natural extension of the celebrated effective resistance sampling sparsification algorithm of [18].[1] Cohen et al. [7] devised an online row sampling algorithm for general tall and skinny matrices in the online setting, which includes online spectral

[1] As pointed out in [7], the original analysis has a subtle dependency issue. Later, [4] provided a complete analysis of their algorithm.

sparsification of graphs. Kapralov et al. [12] devised a fully dynamic streaming algorithm for spectral sparsification of graphs, that supports both insertion and deletion.

Beyond undirected hypergraphs, there are several works on spectral sparsification for more complex objects such as directed hypergraphs [15,17], submodular functions [16], and the sum of norms [8].

2 Preliminaries

2.1 Notations

We use $\mathbb{R}_{\geq 0}$ and $\mathbb{R}_{>0}$ to denote the set of nonnegative and positive real numbers, respectively. Given a positive integer N, we use $[N]$ to denote the set $\{1, 2, \ldots, N\}$. All logarithms are natural logarithms unless otherwise specified. Given a function $f : X \to \mathbb{R}$, its support $\mathrm{supp}(f)$ is the set $\{x \in X : f(x) \neq 0\}$. For any (finite) set X and element $u \in X$, the vector $\chi_u \in \mathbb{R}^X$ is the vector whose u-entry is 1 and all other entries are 0, and the vector $\mathbb{1}_X$ (or simply $\mathbb{1}$) is the vector whose entries are all 1. For $x, x' \in \mathbb{R}^X$, $x \perp x'$ denotes $\langle x, x' \rangle = 0$. For $0 \leq p \leq 1$, $\mathrm{Ber}(p)$ denotes the random variable that takes value 1 with probability p and 0 with probability $1 - p$.

2.2 Linear Algebra

Let $M \in \mathbb{R}^{k \times k}$ be a symmetric matrix. By the spectral theorem, M has the following decomposition $M = V \Lambda V^\top$, where V is orthogonal and $\Lambda = \mathrm{diag}(\lambda_1, \ldots, \lambda_k)$ is the diagonal matrix consisting of the eigenvalues of M. It is known that Λ is unique up to permutation. We say that M is *positive definite* (denoted $M \succ 0$) if $\lambda_i > 0$ for all i, and we say that M is *positive semidefinite (PSD)* (denoted $M \succeq 0$) if $\lambda_i \geq 0$ for all i.

Given a PSD matrix $M = V \Lambda V^\top$, its pseudoinverse $M^\dagger$ is $V \Lambda^\dagger V^\top$ where $\Lambda^\dagger_{i,i} = \lambda_i^{-1}$ if $\lambda_i > 0$ and $\Lambda^\dagger_{i,i} = 0$ if $\lambda_i \geq 0$. Accordingly, $M^{\dagger/2} = \sqrt{M^\dagger} = (\sqrt{M})^\dagger$ where $\sqrt{\cdot}$ is the usual matrix square root defined on PSD matrices.

2.3 Hypergraphs

Basic Definitions. A hypergraph $H = (V, E, w)$ is defined with a vertex set V, a (multi-)set of hyperedges $E = \{e_1, e_2, \ldots, e_{|E|}\}$, where each $e_i \subseteq V$ and $|e_i| \geq 2$, and a weight function $w : e \in E \mapsto w(e) \in \mathbb{R}_{\geq 0}$. We use w_i as a shorthand for $w(e_i)$. Unless otherwise specified, $n := |V|$ denotes the number of vertices and $m := |E|$ denotes the number of hyperedges of H. Denote by $r = \mathrm{rank}(H) := \max_{i \in [m]} |e_i|$ the *rank* of H, i.e., the size of the largest hyperedge in H. We say that H is *unweighted* if $w_i = 1$ for all $i \in [m]$.

Energy and Sparsification. Given a hyperedge $e \subseteq V$, its energy $Q_e : \mathbb{R}^V \to \mathbb{R}$ is defined as:

$$Q_e(z) := \left[\max_{u,v \in e} (z(u) - z(v))\right]^2 = \max_{u,v \in e} (z(u) - z(v))^2.$$

The energy $Q_H : \mathbb{R}^V \to \mathbb{R}$ of a hypergraph H is the weighted sum of its edge energies:

$$Q_H(z) := \sum_{i\in[m]} w_i Q_{e_i}(z) = \sum_{i\in[m]} w_i \max_{u,v\in e_i} (z(u) - z(v))^2.$$

Given a hypergraph $H = (V, E, w)$ and error parameters $\varepsilon, \delta \geq 0$ (ε for relative error, δ for absolute error), an (ε, δ)-spectral sparsifier of H is a hypergraph $\tilde{H} = (V, E, \tilde{w})$ that is supported on the hyperedge set E and its energy $Q_{\tilde{H}}$ satisfies

$$(1-\varepsilon)Q_H(z) - \delta z^\top z \leq Q_{\tilde{H}}(z) \leq (1+\varepsilon)Q_H(z) + \delta z^\top z \qquad \forall z \in \mathbb{R}^V.$$

The size of the sparsifier $\tilde{H}$ is simply $|\operatorname{supp}(\tilde{w})|$. The term "$\varepsilon$-spectral sparsifier" refers to $(\varepsilon, 0)$-spectral sparsifiers, i.e., additive error is not allowed. It corresponds to the well-established notion of spectral sparsifiers introduced in [19].

2.4 Reweighting

A reweighting of a hyperedge $e \subseteq V$ is a set of weights $\{c_{u,v}\}_{u,v\in e}$ such that $c_{u,v} \geq 0$ and $\sum_{u,v\in e} c_{u,v} = 1$. The corresponding reweighted clique-graph G is the (ordinary) graph on V with edges (u, v) having weight $c_{u,v}$ (and other edges having zero weight). A reweighting of a hypergraph H is the weighted sum of the reweightings of its hyperedges. The corresponding reweighted clique-graph G is the graph on V with edges (u, v) having weight $w(u, v) = \sum_{i\in[m]:u,v\in e_i} w_i c_{i,u,v}$, where for $i \in [m]$, $\{c_{i,u,v}\}_{u,v\in e_i}$ is a reweighting of the edge e_i.

The reason for considering reweighted clique-graphs is that the Laplacian

$$[L_G(z)]_u := \sum_{v\in V} w(u, v)(z(u) - z(v))$$

of an ordinary graph is a linear operator. More importantly, it is PSD. Its energy Q_G is then

$$\begin{aligned} Q_G(z) := \langle z, L_G(z)\rangle &= \sum_{u,v\in V} \sum_{i\in[m],u,v\in e_i} w_i c_{i,u,v}(z(u) - z(v))^2 \\ &= \sum_{i\in[m]} w_i \sum_{u,v\in e_i} c_{i,u,v}(z(u) - z(v))^2. \end{aligned}$$

The following property follows from the fact that the energy of a reweighted clique-graph of H is at most that of H.

Proposition 1 (Energy Comparison). *Let H be a hypergraph and G be a reweighted clique-graph of H. Then, $Q_H(L_G^{\dagger/2}x) \geq \|x\|_2^2$ for any $x \in \mathbb{R}^V$ such that $x \perp \mathbb{1}$.*

Proof. By the definition of a reweighting,

$$\begin{aligned} Q_H(x) &= \sum_{i=1}^{m} w_i \max_{u,v\in e} (x(u) - x(v))^2 \\ &= \sum_{i=1}^{m} \sum_{u,v\in e} w_i c_{i,u,v} \max_{u,v\in e} (x(u) - x(v))^2 \\ &\geq \sum_{i=1}^{m} \sum_{u,v\in e} w_i c_{i,u,v} (x(u) - x(v))^2 \\ &= x^\top L_G x. \end{aligned}$$

So, $Q_H(L_G^{\dagger/2} x) \geq \|x\|_2^2$.

2.5 Generic Chaining

Our sparsifier $\tilde{H}$ will be an unbiased random sample of H. Since the hypergraph energies Q_H and $Q_{\tilde{H}}$ in (1) are not quadratic forms, i.e. not of the form $z \mapsto z^\top M z$ where M is a matrix, we cannot use matrix concentration inequalities to control the deviation $|Q_{\tilde{H}}(z) - Q_H(z)|/Q_H(z)$ at all points z. Instead, we prove pointwise concentration and extend it to a uniform bound over the entire domain using Talagrand's generic chaining [20]. We summarize here certain useful facts about generic chaining. Let $(V_x)_{x\in X}$ be a real-valued stochastic process and d be a semi-metric on X. We say $(V_x)_{x\in X}$ is a *subgaussian process with respect to* d if

$$\Pr(V_x - V_y > \varepsilon) \leq \exp\left(-\frac{\varepsilon^2}{2d(x,y)^2}\right)$$

for all $x, y \in X$ and $\varepsilon > 0$. Talagrand's generic chaining relates the supremum of the process (V_x) with the following geometric quantity, called the γ-functionals:

$$\gamma_2(X, d) := \inf_{\mathcal{X}=(X_h)} \sup_{x\in X} \sum_{h\geq 0} 2^{h/2} d(x, X_h),$$

where the infimum is taken over all collections (X_h) of *admissible sequences*, meaning that $X_h \subseteq X$ and $|X_h| \leq 2^{2^h}$ for all $h \geq 0$. Intuitively, $\mathcal{X}$ is a successively finer net over which the union bound is applied.

The following lemma is key to obtaining a tail bound on the supremum of the process (V_x) and will yield high-probability success guarantees for our streaming algorithm.

Lemma 1 ([20, Theorem 2.2.27]). *Let $(V_x)_{x\in X}$ be a subgaussian process on a semi-metric space with respect to a semi-metric d and let $\Delta(X, d) := \sup_{x,y\in X} d(x, y)$ be the d-diameter of X. If $Z = \sup_{x\in X} |V_x|$, then for any $\lambda > 0$,*

$$\log \mathbb{E}[e^{\lambda Z}] \lesssim \lambda^2 \Delta(X, d)^2 + \lambda \gamma_2(X, d).$$

2.6 Concentration Inequalities

We shall use Azuma's inequality to establish the subgaussian bound required for our chaining argument.

Proposition 2 (Azuma's Inequality). *Let $\psi_0 = 0, \psi_1, \ldots, \psi_T$ be a martingale. Suppose that*

$$|\psi_i - \psi_{i-1}| \le M_i \qquad \forall i \in [T].$$

Then,

$$\Pr\left[|\psi_T| \ge \beta\right] \le 2\exp\left(\frac{-\beta^2}{2\sum_i M_i^2}\right).$$

The proof and the requisite background can be found in standard treatises such as [6].

The following special case of the Chernoff bound is also useful.

Proposition 3 (Chernoff bound). *Let $X_1, \ldots, X_T$ be independent Bernoulli random variables where $X_i = \mathrm{Ber}(p_i)$. Let $\mu := \sum_i p_i$. Then, for any $M > 0$,*

$$\Pr\left[\sum_i X_i > (\mu + M)\right] \le \inf_{\lambda > 0} \exp\left(\mu(e^\lambda - 1 - \lambda) - M\lambda\right).$$

3 Algorithm Description

In order to obtain an *unbiased* estimator $\tilde{H}$ in this setting, our options are limited. We consider the class of algorithms where the current edge e_i is sampled with probability p_i, where p_i depends on all edge arrivals and decisions so far. If sampled, the edge is added to the sparsifier $\tilde{H}$ with weight w_i/p_i.

Our proposed algorithm, Algorithm 1, has the following features:

- The sampling probability *does not* depend on previous decisions, but only on previous edge arrivals.
- It requires maintaining a reweighted graph G_i of the hypergraph H_i at all times, and uses G_i to define sampling probabilities.

Using the reweighted graph G_i, the algorithm produces *overestimates* of the "importance" of the hyperedge e_i in the entire hypergraph H, using the effective resistances of the clique edges in the reweighted graph. By virtue of it being an overestimate, it is relatively easy to analyze the success probability. The difficulty lies in choosing an appropriate reweighting, so that the number of selected hyperedges remains well-controlled. This is why it is helpful to use the log-determinant potential function to guide the search for a suitable reweighting.

Algorithm 1. Online Hypergraph Sparsification

Input: Hypergraph $H = (V, E)$ given as a stream, $\varepsilon, \delta > 0$. Let $\eta := \delta/\varepsilon$.
Initialization: let G_0 be the empty graph on V, $L_0^\eta := \eta I_n + L_{G_0} = \eta I_n$.
For $i = 1, \ldots, m$

1. Edge e_i arrives with weight w_i.
2. Compute a reweighting $c_{i,u,v}$ of edge e_i, so that
$$\log\det\left(L_{i-1}^\eta + \sum_{u,v\in e_i} w_i c_{i,u,v} L_{uv}\right)$$
is maximized.
3. Let G_i be the graph obtained from G_{i-1} by adding an edge (u, v) with weight $w_i c_{i,u,v}$ for every pair of vertices (u, v) in e_i, and let $L_i^\eta := \eta I_n + L_{G_i} = L_{i-1}^\eta + \sum_{u,v\in e_i} w_i c_{i,u,v} L_{uv}$.
4. Let $r_i := \max_{u,v\in e_i} \|(L_i^\eta)^{-1/2}(\chi_u - \chi_v)\|_2^2$ be the maximum ridged effective resistance across a pair of vertices in e_i.
5. Sample edge e_i with probability $p_i := \min(1, c r_i w_i)$, where $c = O(\varepsilon^{-2}\log n \log r)$.

4 Analyzing the Success Probability

In this section, we prove that Algorithm 1 succeeds with high probability. Given a hypergraph H, let
$$Q_H^\eta(z) := Q_H(z) + \eta z^\top z$$
be the η-ridged energy of z. We would like to control the probability that $Q_{\tilde{H}}^\eta(z)$ is within a multiplicative factor of $1 \pm \varepsilon$ from $Q_H^\eta(z)$ for all $z \in \mathbb{R}^V$. Note that the event
$$\sup_{z:Q_H^\eta(z)\le 1} |Q_{\tilde{H}}^\eta(z) - Q_H^\eta(z)| \le \varepsilon$$
is the same as
$$(1-\varepsilon)Q_H^\eta(z) \le Q_{\tilde{H}}^\eta(z) \le (1+\varepsilon)Q_H^\eta(z) \qquad \forall z \in \mathbb{R}^V,$$
which by the choice of η implies that
$$(1-\varepsilon)Q_H(z) - \delta z^\top z \le Q_{\tilde{H}}(z) \le (1+\varepsilon)Q_H(z) + \delta z^\top z \qquad \forall z \in \mathbb{R}^V,$$
i.e., that $\tilde{H}$ is an (ε, δ)-spectral sparsifier of H.

Our plan is as follows. For the desired concentration bound, we will bound the exponential moment generating function (MGF) $\mathbb{E}_{\tilde{H}}[\exp(\lambda \cdot \sup_{z:Q_H^\eta(z)\le 1} |Q_{\tilde{H}}^\eta(z) - Q_H^\eta(z)|)]$ of the energy discrepancy function. By Markov's inequality and a suitable choice of λ, we can then conclude that Q_H^η and $Q_{\tilde{H}}^\eta$ are ε-close with high probability.

The following is the main technical result of the section:

Lemma 2 (Exponential MGF Bound). *Let H be a hypergraph stream and $\tilde{H}$ be the sampled hypergraph obtained from Algorithm 1. Let*

$$Z := \sup_{z:Q^{\eta}_{H}(z)\le 1} |Q^{\eta}_{\tilde{H}}(z) - Q^{\eta}_{H}(z)|.$$

Then, for any $\lambda > 0$,

$$\mathop{\mathbb{E}}_{\tilde{H}}[\exp(\lambda Z)] \le \mathop{\mathbb{E}}_{\tilde{H}}\left[\exp\left(\frac{\lambda^2}{c}(1+Z) + \lambda\sqrt{\frac{\log n \log r}{c}}(1+Z)^{1/2}\right)\right].$$

This is an *implicit* bound on the exponential MGF of Z because Z itself appears in the bound.

The proof of Lemma 2 follows closely the chaining proofs in [8] and [14], with a few modifications. The proof in full detail can be found in the full version of the paper. In the remainder of the section, we will show how Lemma 2 implies that $Z \le \varepsilon$ with high probability.

4.1 High Probability Guarantee from Lemma 2

In order to obtain a high probability guarantee on the success probability, we must derive an *explicit* upper bound on $\mathbb{E}_{\tilde{H}} \exp(\lambda Z)$.

Suppose $c = (\kappa_1 \log n \log r)/\varepsilon^2$ and take $\lambda := (\kappa_2\sqrt{\log n \log r})/\varepsilon$, where κ_1 is an absolute constant and κ_2 depends only on ε. Let $\kappa := \kappa_2/\sqrt{\kappa_1}$. We will ensure that the parameters satisfy $\kappa^2 + \kappa\sqrt{\log n \log r} \le \lambda/2$. Then,

$$\begin{aligned}
\mathop{\mathbb{E}}_{\tilde{H}}[\exp(\lambda Z)] &\le \mathop{\mathbb{E}}_{\tilde{H}}\left[\exp\left(\frac{\lambda^2}{c}(1+Z) + \lambda\sqrt{\frac{\log n \log r}{c}}(1+Z)^{1/2}\right)\right] \qquad \text{(Lemma 2)}\\
&= \mathop{\mathbb{E}}_{\tilde{H}}\left[\exp\left(\kappa^2(1+Z) + \kappa\sqrt{\log n \log r}(1+Z)^{1/2}\right)\right]\\
&\le \mathop{\mathbb{E}}_{\tilde{H}}\left[\exp\left(\kappa^2(1+Z) + \kappa\sqrt{\log n \log r}(1+Z)\right)\right]\\
&= \exp\left(\kappa^2 + \kappa\sqrt{\log n \log r}\right) \cdot \mathop{\mathbb{E}}_{\tilde{H}}\left[\exp\left((\kappa^2 + \kappa\sqrt{\log n \log r})Z\right)\right]\\
&\le \exp\left(\kappa^2 + \kappa \log n\right) \cdot \mathop{\mathbb{E}}_{\tilde{H}}[\exp(\lambda Z)]^{\frac{\kappa^2+\kappa\sqrt{\log n \log r}}{\lambda}}\\
&\qquad \left(\text{Jensen's inequality and } \frac{\kappa^2 + \kappa\sqrt{\log n \log r}}{\lambda} \le \frac{1}{2}\right)\\
&\le \exp\left(\kappa^2 + \kappa \log n\right) \cdot \mathop{\mathbb{E}}_{\tilde{H}}[\exp(\lambda Z)]^{1/2},
\end{aligned}$$

which resolves to

$$\mathop{\mathbb{E}}_{\tilde{H}}[\exp(\lambda Z)] \le \exp(2\kappa^2 + 2\kappa \log n).$$

Markov's inequality then implies that

$$\Pr_{\tilde{H}}[Z > \varepsilon] \le \frac{\mathbb{E}_{\tilde{H}}[\exp(\lambda Z)]}{e^{\lambda\varepsilon}} \le \exp\left(2\kappa^2 + 2\kappa\log n - \lambda\varepsilon\right) = \exp\left(2\kappa^2 - (\kappa_2 - 2\kappa)\log n\right).$$

We can take $\kappa_1 \geq 16$ and $\kappa_2 \lesssim \min(\sqrt{\kappa_1}, 1/\varepsilon)$ for the above analysis to go through, and for large enough n or small enough ε (or both) satisfying $(2\kappa^2 - (\kappa_2 - 2\kappa)\log n) \leq -\log n$. We arrive at the following conclusion.

Lemma 3 (Success probability). *For Algorithm 1, let*

$$Z := \sup_{z : Q_H^\eta(z) \leq 1} |Q_{\tilde{H}}^\eta(z) - Q_H^\eta(z)|.$$

Then, we have $\Pr_{\tilde{H}}[Z > \varepsilon] \leq 1/n$. *As a corollary, with probability* $\geq 1 - 1/n$, $\tilde{H}$ *is an* (ε, δ)*-spectral sparsifier of* H.

5 Bounding the Sample Size

The expected number of edges in $\tilde{H}$ is simply $\sum_i p_i \leq \sum_i c w_i r_i$, but it is not easy to bound each r_i directly. To bound this more easily, we use a potential function

$$\Phi_i := \log\det(L_i^\eta).$$

We will show that, every time an edge gets sampled, Φ_i increases substantially. Then, we bound the value of $\Phi_m - \Phi_0$, which will in turn give a bound on the number of edges.

Since the update to L_{G_i} is no longer rank-1, but rank-$|e_i|$, we will make use of concavity of the log-determinant function, which gives us a lower bound on the potential increase.

Proposition 4. *The function* $X \mapsto \log\det(X)$ *is concave on the set of positive definite matrices.*

Proposition 5 (Potential Increase). *We have*

$$\Phi_i - \Phi_{i-1} \geq \frac{p_i \log 2}{c}.$$

Proposition 4 is standard and can be found in e.g. [3, Section 3.1.5]. The proof of Proposition 5 can be found in the full version of the paper. Now we can bound the number of sampled edges.

Lemma 4 (Number of sampled edges). *We have*

$$\mathbb{E}[|\tilde{H}|] \lesssim cn\log\left(1 + \frac{2W}{\eta n}\right),$$

where $W = \sum_{i=1}^m w_i$. *Moreover,*

$$|\tilde{H}| \lesssim cn\log\left(1 + \frac{2W}{\eta n}\right)$$

with probability at least $1 - 1/n$.

See the full version of the paper for the proof. Combining Lemma 3 and Lemma 4 yields Theorem 1.

Proof of Corollary 1. Now we prove Corollary 1. By the hypergraph Cheeger inequality [5], $Q_H(x) \gtrsim W_{\min}^2 n^{-2r}\|x\|_2^2$ for $x \in \mathbb{R}^n$ with $x \perp \mathbb{1}$. Since $\delta = O(\varepsilon W_{\min}^2 n^{-2r})$, we have $\varepsilon Q_H(x) \geq \delta\|x\|_2^2$. So an (ε, δ)-spectral sparsifier is indeed a $(2\varepsilon, 0)$-spectral sparsifier. Therefore, Algorithm 1 outputs an $(2\varepsilon, 0)$-spectral sparsifier with a constant probability. The expected number of hyperedges is immediate from Theorem 1.

6 Conclusion

To summarize, we designed and analyzed the first online algorithm for hypergraph spectral sparsification, showing that it uses significantly less space than the number of edges. We leave open the following questions concerning the performance of the algorithm:

Question 1. Can we derive a matching lower bound on the space complexity of any (online) streaming algorithm for spectral hypergraph sparsification?

Question 2. Can we improve the space complexity from $O(n^2)$ to $O(nr \operatorname{polylog} m)$, or even better? Such an algorithm would perform better when the rank of the hypergraph is small.

While this paper focused on the insertion-only setting, the fully dynamic setting is also of interest.

Question 3. Can we obtain an efficient fully dynamic algorithm (i.e., one that supports both hyperedge insertion and deletion) for spectral hypergraph sparsification?

Acknowledgements. TS is supported by JSPS KAKENHI Grant Number JP19K20212. A part of this work was done during KT's visit to National Institute of Informatics. YY is supported by JSPS KAKENHI Grant Number JP20H05965 and JP22H05001.

Disclosure of Interests. The authors have no competing interests to declare that are relevant to the content of the article.

Full version. The full version of the article can be found at arXiv:2310.02643.

References

1. Bansal, N., Svensson, O., Trevisan, L.: New notions and constructions of sparsification for graphs and hypergraphs. In: Proceedings of the IEEE 60th Annual Symposium on Foundations of Computer Science (FOCS), pp. 910–928 (2019)
2. Batson, J., Spielman, D.A., Srivastava, N., Teng, S.H.: Spectral sparsification of graphs: theory and algorithms. Commun. ACM **56**(8), 87–94 (2013)

3. Boyd, S.P., Vandenberghe, L.: Convex Optimization. Cambridge University Press, Cambridge (2004)
4. Calandriello, D., Lazaric, A., Valko, M.: Analysis of Kelner and Levin graph sparsification algorithm for a streaming setting. arXiv (2016)
5. Chan, T.H.H., Louis, A., Tang, Z.G., Zhang, C.: Spectral properties of hypergraph laplacian and approximation algorithms. J. ACM **65**(3), 1–48 (2018). https://doi.org/10.1145/3178123
6. Chung, F., Lu, L.: Concentration inequalities and martingale inequalities: a survey. Internet Math. **3**(1), 79–127 (2006)
7. Cohen, M.B., Musco, C., Pachocki, J.: Online row sampling. Theory Comput. **16**(1), 1–25 (2020). https://doi.org/10.4086/toc.2020.v016a015
8. Jambulapati, A., Lee, J.R., Liu, Y.P., Sidford, A.: Sparsifying sums of norms. arXiv (2023)
9. Jambulapati, A., Liu, Y.P., Sidford, A.: Chaining, group leverage score overestimates, and fast spectral hypergraph sparsification. In: Proceedings of the 55th Annual ACM Symposium on Theory of Computing (STOC), pp. 196–206 (2023)
10. Kapralov, M., Krauthgamer, R., Tardos, J., Yoshida, Y.: Towards tight bounds for spectral sparsification of hypergraphs. In: Proceedings of the 53rd Annual ACM SIGACT Symposium on Theory of Computing (STOC), pp. 598–611 (2021)
11. Kapralov, M., Krauthgamer, R., Tardos, J., Yoshida, Y.: Spectral hypergraph sparsifiers of nearly linear size. In: Proceedings of the 62nd IEEE Annual Symposium on Foundations of Computer Science (FOCS), pp. 1159–1170 (2022)
12. Kapralov, M., Lee, Y.T., Musco, C., Musco, C.P., Sidford, A.: Single pass spectral sparsification in dynamic streams. SIAM J. Comput. **46**(1), 456–477 (2017)
13. Kelner, J.A., Levin, A.: Spectral sparsification in the semi-streaming setting. Theory Comput. Syst. **53**(2), 243–262 (2013)
14. Lee, J.R.: Spectral hypergraph sparsification via chaining. In: Proceedings of the 55th Annual ACM Symposium on Theory of Computing, pp. 207–218 (2023)
15. Oko, K., Sakaue, S., Tanigawa, S.: Nearly tight spectral sparsification of directed hypergraphs. In: The Proceedings of the 50th International Colloquium on Automata, Languages, and Programming (ICALP) (2023)
16. Rafiey, A., Yoshida, Y.: Sparsification of decomposable submodular functions. In: Proceedings of the AAAI Conference on Artificial Intelligence, vol. 36, pp. 10336–10344 (2022)
17. Soma, T., Yoshida, Y.: Spectral sparsification of hypergraphs. In: Proceedings of the 30th Annual ACM-SIAM Symposium on Discrete Algorithms (SODA), pp. 2570–2581. SIAM (2019)
18. Spielman, D.A., Srivastava, N.: Graph sparsification by effective resistances. SIAM J. Comput. **40**(6), 1913–1926 (2011). https://doi.org/10.1137/080734029
19. Spielman, D.A., Teng, S.H.: Spectral sparsification of graphs. SIAM J. Comput. **40**(4), 981–1025 (2011)
20. Talagrand, M.: Upper and Lower Bounds for Stochastic Processes. Springer, Cham (2014)
21. Vishnoi, N.K.: $Lx = b$, laplacian solvers and their algorithmic applications. Found. Trends® Theor. Comput. Sci. **8**(1-2), 1–141 (2013). https://doi.org/10.1561/0400000054

Fast Combinatorial Algorithms for Efficient Sortation

Madison Van Dyk[1(✉)], Kim Klause[2], Jochen Koenemann[1,3], and Nicole Megow[2]

[1] Department of Combinatorics and Optimization, University of Waterloo, Waterloo, Canada
madison.vandyk@uwaterloo.ca
[2] Faculty of Mathematics and Computer Science, University of Bremen, Bremen, Germany
[3] Modeling and Optimization, Amazon, Seattle, USA

Abstract. Modern parcel logistic networks are designed to ship demand between given origin, destination pairs of nodes in an underlying directed network. Efficiency dictates that volume needs to be consolidated at intermediate nodes in a typical hub-and-spoke fashion. In practice, such consolidation requires parcel sortation. In this work, we propose a mathematical model for the physical requirements, and limitations of parcel sortation. We then show that it is NP-hard to determine whether a feasible sortation plan exists. We discuss several settings, where (near-) feasibility of a given sortation instance can be determined efficiently. The algorithms we propose are fast and build on combinatorial *witness set* type lower bounds that are reminiscent and extend those used in earlier work on degree-bounded spanning trees and arborescences.

1 Introduction

In modern parcel logistics operations, one broadly faces the problem of finding an *optimal* way to ship a given input demand between source-sink node pairs within an underlying fulfillment network, typically represented as a directed graph $D = (N, A)$. The input demand is given as a collection of *commodities* $\{(s_k, t_k)\}_{k \in \mathcal{K}}$, consisting of pairs of source and sink nodes in N. In this work, we assume each commodity k comes with a directed s_k, t_k-path, P_k, along which all packages associated with commodity k must travel. This assumption is often made in practical applications, as discussed in [15].

M. Van Dyk and J. Koenemann—This work was partially supported by the NSERC Discovery Grant Program, grant number RGPIN-03956-2017 and by an Amazon Post-Internship Fellowship. The work presented here does not relate to the authors' positions at Amazon.

J. Vygen and J. Byrka (Eds.): IPCO 2024, LNCS 14679, pp. 418–432, 2024.
https://doi.org/10.1007/978-3-031-59835-7_31

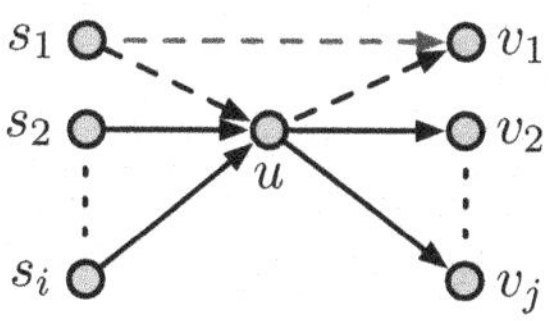

Fig. 1. Sortation introduction

This paper focuses on *parcel sortation*, an aspect of routing that has previously been left unaddressed from a theory perspective. In the simplest setting, each package that travels from s_k to t_k via some internal node u must be *sorted* at u for its subsequent downstream node. In Fig. 1, packages arrive at node u from nodes $s_1, \ldots s_i$ and travel on to downstream nodes $v_1, \ldots, v_j$. This requires sortation at node u to sub-divide the stream of incoming parcels between the j possible next stops. We refer to the physical device tasked with packages destined for one specific downstream node as a *sort point.* If all of the traffic on arc uv_1 arrives at u from s_1, then we could sort s_1, v_1 volume at s_1 to v_1 instead of to u. In practice this entails *containerizing* s_1, v_1 volume at s_1. These containers are not opened and sorted at intermediate node u, instead, they are *cross-docked*, an operation that is significantly faster and less costly than the process of sorting all individual parcels [23, 27].

Figure 1 illustrates the containerization of s_1, v_1 volume at s_1 by replacing arcs s_1u and uv_1 by s_1v_1. Sort points are required at each node, for each outgoing arc. Hence, the arc replacement operation in Fig. 1 reduces the number of required sort points at node u.

A Formal Model for Sortation. Let $\mathcal{K}$ be a set of commodities, where each commodity $k \in \mathcal{K}$ has a source s_k, sink t_k, and a designated directed path P_k in D. Let the *transitive closure* of digraph D, denoted $\mathtt{cl}(D)$, be the graph obtained by introducing short-cut arc ij whenever j can be reached from i through a directed path in D. We say that a subgraph H of $\mathtt{cl}(D)$ is *feasible* if for each commodity $k \in \mathcal{K}$, there is an s_k, t_k-dipath in $H \cap \mathtt{cl}(P_k)$. Since sort point capacity is often limited, determining the maximum number of sort points required at any warehouse is important in both short- and long-term planning. This motivates the min-degree sort point problem, defined as follows.

`min-degree-SPP`: given a directed graph D and commodity set $\mathcal{K}$, find a feasible subgraph H of $\mathtt{cl}(D)$ with minimum max out-degree.

Given a graph H, let $\Delta^+(H)$ denote the maximum out-degree of a node in H. Let Δ^* denote the minimum max out-degree in any feasible subgraph. In the corresponding decision version of the problem, we are given a positive integer *target*, T, and the problem is to determine whether or not $\Delta^* \leq T$.

We assume w.l.o.g. that D is *weakly* connected, the commodity set is non-empty ($\Delta^* \geq 1$), all commodities are non-trivial ($s_k \neq t_k$) and unique, and all nodes with no out-arcs in D serve as the sink for some commodity. Any instance can be reduced to a (set of) equivalent instances that satisfy these assumptions. Let n and m be the number of nodes and arcs in D, respectively. Let $\mathcal{S} = \{s_k : k \in \mathcal{K}\}$ be the set of sources, and $\mathcal{T} = \{t_k : k \in \mathcal{K}\}$ be the set of sinks. For each node $v \in N$, let $\mathcal{T}(v) = \{t_k : (v, t_k) \in \mathcal{K}\}$ and $\mathcal{S}(v) = \{s_k : (s_k, v) \in \mathcal{K}\}$.

1.1 Our Results

We say that an instance $\mathcal{I} = (D, \mathcal{K})$ of `min-degree-SPP` is a *tree instance* (*star instance*) if the underlying undirected graph of D is a tree (a star). We prove that `min-degree-SPP` is NP-hard, even when we restrict to the set of star instances.

Theorem 1. `min-degree-SPP` *is NP-hard, even when restricted to star instances.*

Henceforth, we focus on tree instances of `min-degree-SPP`. Note that in this class of instances, any s_k, t_k-dipath in $\texttt{cl}(D)$ can only use arcs in $\texttt{cl}(P_k)$, where P_k is the unique s_k, t_k-dipath in $D = (N, A)$.

First, we construct combinatorial lower bounds that are motivated by the witness set construction for undirected min-degree spanning trees [7]. Specifically, we define a function LB such that for all $W \subseteq N$ and $\mathcal{K}' \subseteq \mathcal{K}$, $\texttt{LB}(W, \mathcal{K}') \leq \Delta^*$. We show that in instances with a single source, this construction is the best possible. We develop an exact polynomial-time local search algorithm for single-source instances and certify its optimality by determining values of W and $\mathcal{K}'$ such that $\texttt{LB}(W, \mathcal{K}') = \Delta^+(H)$ for the graph H returned. Note that the single-source setting reduces to the problem of finding a min-degree arborescence in a directed acyclic graph. This problem can be solved via matroid intersection in $O(n^3 \log n)$ time [2,21] or by a combinatorial algorithm in $O(nm \log n)$ time [26]. Our approach uses the structure of the transitive closure, and allows us to obtain a very simple algorithm that beats the runtime of previous results by a quadratic factor. Moreover, it motivates our other algorithms in more complex settings that cannot be modeled as min-degree arborescence problems.

Theorem 2. *There is an $O(n \log^2 n)$-time exact algorithm for tree instances of* `min-degree-SPP` *with a single source.*

When there is a single source and the undirected graph of D is a tree, each node has at most one entering arc in D. We refer to tree instances in which each node in D has at most one entering arc as *out-tree* instances. We prove that $\max_{W \subseteq N, \mathcal{K}' \subseteq \mathcal{K}} \texttt{LB}(W, \mathcal{K}') \geq \Delta^* - 1$ for out-tree instances, by showing that there is a polynomial-time algorithm that returns a solution with out-degree at most one greater than the best lower bound. We also show that there are out-tree instances where $\max_{W \subseteq N, \mathcal{K}' \subseteq \mathcal{K}} \texttt{LB}(W, \mathcal{K}') = \Delta^* - 1$.

Our algorithm for multi-source out-tree instances is again combinatorial in nature, but the details are significantly more involved. In the algorithm for the single-source setting, in each iteration an arc vw is exchanged for an arc uw, where u is on the path between the root (the unique source) and v in D. However, in the multi-source setting, the same action is not sufficient to ensure a high-quality solution is found. We instead define an algorithm for `min-degree-SPP` on out-trees which takes as input a target T, and returns a feasible solution H with $\Delta^+(H) \leq T$ whenever $\Delta^* \leq T - 1$. The additional input of the target as well as a careful selection of arcs in each iteration prevent the need to backtrack. In the proof of the performance of this algorithm, we provide an explicit construction of the lower bound certificate with value at least $\Delta^* - 1$.

Theorem 3. *There is a polynomial-time additive* 1*-approximation algorithm for out-tree instances of* `min-degree-SPP`.

The analysis of the out-tree algorithm is *tight*, in that there are instances of `min-degree-SPP` where the algorithm does not produce an optimal solution. Moreover, the performance of the algorithm is bounded against a lower bound that has an inherent gap matching the proven performance. The algorithmic approach for the out-tree setting cannot easily be extended to more complex graph structures, such as stars, since an optimal solution is no longer guaranteed to be acyclic and our current algorithm heavily relies on this fact. Furthermore, the lower bound construction weakens significantly for star instances.

We also give a framework for obtaining approximation results for arbitrary tree instances. We state our findings here and defer all details to the full version [24]. As a first step, we give an efficient 2-approximation when D is a star.

Theorem 4. *There is a polynomial-time 2-approximation for star instances of* `min-degree-SPP`*, and for any* $\epsilon > 0$ *there is a star instance of* `min-degree-SPP` *where* $\max_{W \subseteq N, \mathcal{K}' \subseteq \mathcal{K}} \mathrm{LB}(W, \mathcal{K}') = \frac{2}{3}\Delta^* + \epsilon$.

A generalization of the class of star instances is the class of junction trees. A tree instance of `min-degree-SPP` is a *junction tree instance* if there is some node r in D such that r is a node in P_k for all $k \in \mathcal{K}$. Our framework for obtaining an approximation result for general tree instances builds on the approximability of junction tree instances.

Theorem 5. *Given a polytime* α*-approximation algorithm for junction tree instances, there is a polytime* $\alpha \log n$*-approximation algorithm for tree instances of* `min-degree-SPP`.

1.2 Related Work

Degree-bounded network design problems are fundamental and well-studied combinatorial optimization problems. A prominent example is the *minimum-degree spanning tree problem* which asks, given an undirected, unweighted graph, for a spanning tree with minimum maximum degree. Fürer and Raghavachari [7] introduced the problem and presented a local-search based polynomial-time algorithm for computing a spanning tree with maximum degree which is at most 1 larger than the optimal maximum degree. Their combinatorial arguments rely on *witness sets* chosen from a family of carefully-constructed lower bounds [5,25].

Various techniques have been employed in subsequent work on the weighted setting, e.g., [3,4,9,11,12,19,20], culminating in the result that one can compute a spanning tree of minimum cost that exceeds the degree bound by at most 1 [22]. Since then, generalizations have been studied such as the *degree-bounded Steiner tree* problem [13,17], *survivable network design* with higher connectivity requirements [16,17] and the *group Steiner tree* problem [6,10,14].

Directed (degree-bounded) network design problems are typically substantially harder than their undirected counterparts. Among the few nontrivial approximation results are quasipolynomial-time bicriteria approximations (with respect to cost and maximum out- or in-degree) [10] for the degree-bounded directed Steiner tree problem and approximation results for problems with intersecting or crossing supermodular connectivity requirements [1,18].

A special case in directed degree-bounded network design is the *min-degree arborescence problem* where, given a directed graph and root r, the goal is to find a spanning tree rooted at r with minimum max out-degree. This problem is NP-hard [1,7,16] in general, and polytime solvable in directed acyclic graphs [26].

2 Hardness

In this section we prove that `min-degree-SPP` is NP-hard, even in the setting where the underlying undirected graph of D forms a star. To prove this result, we will exhibit a reduction from the NP-hard problem of `Hitting-set`[8].

Theorem 1. `min-degree-SPP` *is NP-hard, even when restricted to star instances.*

Proof (sketch). Let $\mathcal{H} = (\Sigma, \mathcal{S}, b)$ be an instance of `Hitting-set`, where $b \in \mathbb{N}$, $\Sigma = \{e_1, e_2, \ldots, e_m\}$, and $\mathcal{S} = \{S_1, S_2, \ldots, S_n\}$ is a set of n non-empty subsets of Σ. The hitting set problem asks if there is a subset $R \subseteq \Sigma$ of cardinality at most b such that $R \cap S_i \neq \emptyset$ for all $i \in [n]$. We will construct an instance $\mathcal{I} = (D, \mathcal{K})$ of `min-degree-SPP` such that $\mathcal{H}$ is a `YES` instance if and only if $\Delta^* \leq c$, for some fixed integer c, where $\mathcal{I}$ and c are polynomial in the size of $\mathcal{H}$.

First, we construct a digraph $D' = (N', A')$ with node and arc sets

$$N' = \{s_i : i \in [n]\} \cup \{v\} \cup \{t_j : j \in [m]\},$$
$$A' = \{s_i v : i \in [n]\} \cup \{v t_j : j \in [m]\},$$

as shown in Fig. 2a. For each $i \in [n]$ and $e_j \in S_i$, we add a commodity with source s_i and sink t_j to form a set $\mathcal{K}'$. That is, $\mathcal{K}' = \{(s_i, t_j) : i \in [n], e_j \in S_i\}$.

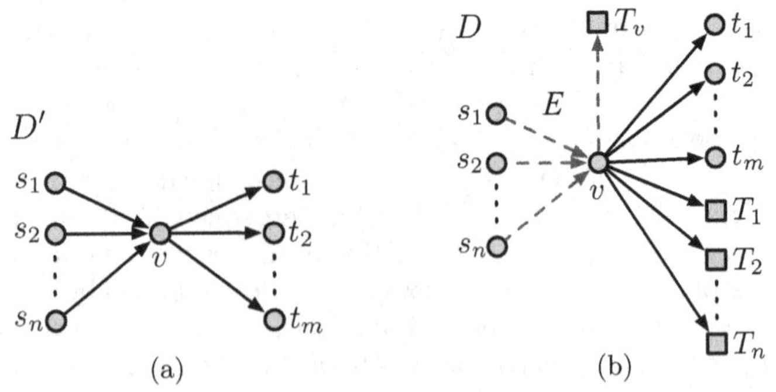

Fig. 2. Digraphs D' and D.

We now build upon the instance $\mathcal{I}' = (D', \mathcal{K}')$ to form $\mathcal{I} = (D, \mathcal{K})$. Let $c = \max\{b, \max_i |S_i|\}$. For each $i \in [n]$, we add a set of nodes, T_i, of cardinality $c - |S_i|$ to N'. Additionally, we add a set of nodes T_v with cardinality $c - b$. The arc set A is formed by adding an arc from v to all the nodes in $N \backslash N'$. That is,

$$N = N' \cup T_v \cup \{T_i : i \in [n]\},$$
$$A = A' \cup \{vt : t \in N \backslash N'\}.$$

The digraph D is given in Fig. 2b, where the square nodes indicate a set of nodes rather than a single node. Additionally, arcs entering a square node denote a set of arcs with same shared tail and differing heads – one for each node in the set represented by the square node.

We add commodities to $\mathcal{K}'$ to form the set $\mathcal{K}$. For each $i \in [n]$, we add the commodity (s_i, t) for all $t \in T_i$, and the commodity (s_i, v). Additionally, for each node $t \in T_v$ we define a commodity (v, t). Specifically,

$$\mathcal{K} = \mathcal{K}' \cup \{(s_i, v) : i \in [n]\} \cup \{(s_i, t) : t \in T_i, i \in [n]\} \cup \{(v, t) : t \in T_v\}.$$

Observe that $\mathcal{I}$ and c are polynomial in the input of $\mathcal{H}$. Furthermore, there are $c + 1$ distinct commodities with source s_i for each $i \in [n]$, and any feasible subgraph H must contain the arc set $E = \{s_i v : i \in [n]\} \cup \{vt : t \in T_v\}$ (the red dashed arcs in Fig. 2b). We now claim that the hitting set instance $\mathcal{H}$ is a YES instance if and only if $\Delta^* \leq c$. This claim is proven in the full version [24]. □

3 Combinatorial Lower Bounds

In this section we present a family of lower bounds, defined by a set of nodes W and set of commodities $\mathcal{K}'$, which we refer to as a *witness set*.

Let $D = (N, A)$ be a digraph, and consider a subset $U \subseteq N$. We define $\delta_D^+(U)$ as the set of arcs in D leaving U.

Consider a subset $W \subseteq N$ such that $s_k \in W$ and $t_k \notin W$ for some $k \in \mathcal{K}$. It follows that any feasible subgraph H must contain some arc in $\delta^+_{\mathtt{cl}(D)}(W)$. Specifically, due to the given-path structure, we know that H must contain some arc in $\delta^+_{\mathtt{cl}(P_k)}(W)$. In Fig. 3, consider the node set $W = \{s, v\}$ along with the commodity with source s and sink t_1. On the left is the base graph D, and on the right is the transitive closure. Since $s \in W$ and $t_1 \notin W$, it follows that any feasible solution H must have a non-empty intersection with the set $\{st_1, vt_1\}$.

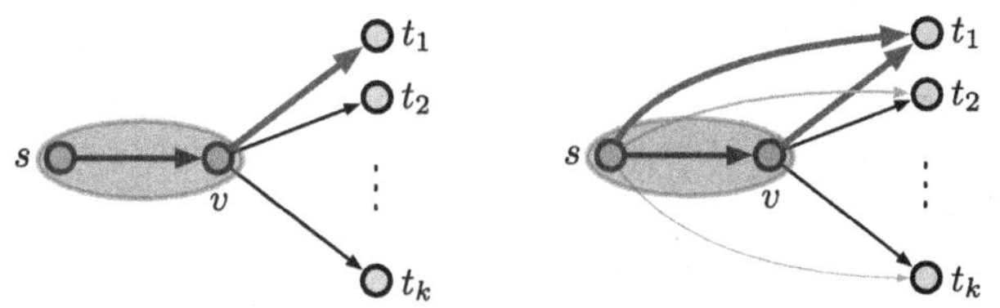

Fig. 3. A tree instance with k commodities $\{(s, t_i) : i \in [k]\}$.

We combine disjoint cuts for a fixed node set W to obtain lower bounds on the value of Δ^*. To simplify the set of cuts considered, we prove the following lemma to work with cuts in D rather than in its closure; see [24].

Lemma 1. *Let $\mathcal{I} = (D, \mathcal{K})$ be a tree instance and $W \subseteq N$. For any $k, j \in \mathcal{K}$, $\delta^+_{P_k}(W) \cap \delta^+_{P_j}(W) = \emptyset$ if and only if $\delta^+_{\mathtt{cl}(P_k)}(W) \cap \delta^+_{\mathtt{cl}(P_j)}(W) = \emptyset$.*

As is standard in proving bounds on the min-max degree [7,25], we observe the following: if ℓ distinct arcs must leave a set W of nodes in any feasible solution, then $\Delta^* \geq \lceil \ell/|W| \rceil$. The following lower bound construction shows that we can argue that such a disjoint arc set can be derived by looking at disjoint cuts of the form $\delta^+_{P_k}(W)$ for some $k \in \mathcal{K}$. Further, we can show that this lower bound can be strengthened since we must also have connectivity *within* W in order to allocate the arcs departing W to different nodes in W; see [24].

Lemma 2. *Let $\mathcal{I} = (D, \mathcal{K})$ be a tree instance of* `min-degree-SPP`*. Let $W \subseteq N$ such that $D[W]$ is weakly connected and suppose $\emptyset \neq \mathcal{K}' \subseteq \mathcal{K}$ such that $s_k \in W$ and $t_k \notin W$ for all $k \in \mathcal{K}'$, and $\delta^+_{P_k}(W) \cap \delta^+_{P_j}(W) = \emptyset$ for all distinct $k, j \in \mathcal{K}'$. Then*

$$\Delta^* \geq \left\lceil \frac{|\mathcal{K}'| + |W| - |\mathcal{S}(\mathcal{K}')|}{|W|} \right\rceil,$$

where $\mathcal{S}(\mathcal{K}')$ denotes the set of sources for commodities in $\mathcal{K}'$.

We define the functions LB and $\mathtt{LB}^w$, where for each $W \subseteq N$ and $\mathcal{K}' \subseteq \mathcal{K}$,

$$\mathtt{LB}_{\mathcal{I}}(W, \mathcal{K}') := \begin{cases} \left\lceil \frac{|\mathcal{K}'|+|W|-|\mathcal{S}(\mathcal{K}')|}{|W|} \right\rceil & \text{if } W, \mathcal{K}' \text{satisfy conditions of Lemma 2 for } \mathcal{I} \\ 0 & \text{otherwise} \end{cases}$$

$$\mathtt{LB}^w_{\mathcal{I}}(W, \mathcal{K}') := \begin{cases} \left\lceil \frac{|\mathcal{K}'|}{|W|} \right\rceil & \text{if } W, \mathcal{K}' \text{ satisfy conditions of Lemma 2 for } \mathcal{I} \\ 0 & \text{otherwise} \end{cases}$$

We drop the subscript $\mathcal{I}$ when the instance is clear from context. Since we have $|W| - |\mathcal{S}(\mathcal{K}')| \geq 0$ for all inputs W and $\mathcal{K}'$, we obtain the following corollary.

Corollary 1. *Let $\mathcal{I} = (D, \mathcal{K})$ be a tree instance. For all $W \subseteq N$ and $\mathcal{K}' \subseteq \mathcal{K}$, $\Delta^* \geq \mathtt{LB}^w(W, \mathcal{K}')$.*

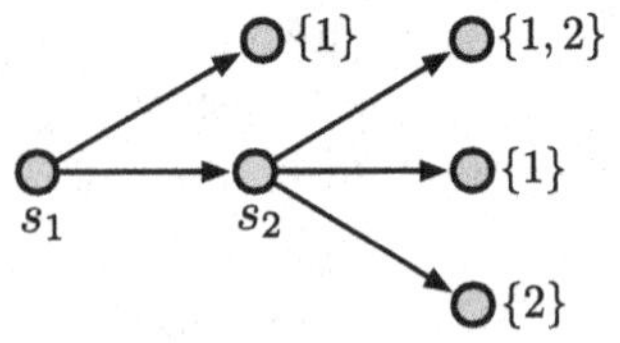

In the single-source setting, we prove that for each instance $\mathcal{I}$, Δ^* is equal to $\max_{W \subseteq N, \mathcal{K}' \subseteq \mathcal{K}} \mathtt{LB}(W, \mathcal{K}')$. A natural next step is to ask if the lower bound construction is also exact for multi-source out-tree instances. However, this is not the case, since there are instances of `min-degree-SPP` on out-trees where $\max_{W \subseteq N, \mathcal{K}' \subseteq \mathcal{K}} \mathtt{LB}(W, \mathcal{K}') = \Delta^* - 1$. Consider the out-tree instance on the right, where there are two sources, s_1 and s_2. The remaining nodes are sinks labelled with each of the corresponding indices of sources that are matched to it. That is, a node v labelled with $\{1, 2\}$ indicates that there are commodities (s_1, v) and (s_2, v) in $\mathcal{K}$. In this instance, $\Delta^* = 3$, whereas $\max_{W \subseteq N, \mathcal{K}' \subseteq \mathcal{K}} \mathtt{LB}(W, \mathcal{K}') = 2$.

4 (Near-)Optimal Algorithms for Out-Trees

We first present a simple combinatorial algorithm that solves single-source tree instances. (With a single source, D necessarily forms an out-tree with root s.) We then present a polynomial-time algorithm for out-trees with multiple sources that returns a feasible solution with max out-degree at most one more than optimal. The algorithms are purely combinatorial and the analysis relies on bounds via the witness sets introduced in Sect. 3.

4.1 Simple Algorithm for Single-Source Setting

We show that for each single-source instance $\mathcal{I}$, $\Delta^* = \max_{W \subseteq N, \mathcal{K}' \subseteq \mathcal{K}} \texttt{LB}(W, \mathcal{K}')$. We prove this result by presenting a simple and efficient local search algorithm that returns an optimal solution, H, as well as a witness set $W, \mathcal{K}'$ such that $\Delta^+(H) = \texttt{LB}(W, \mathcal{K}')$. The running time is $O(n \log^2 n)$, beating previous results. We state here the results and defer the proofs to the full version [24].

Algorithm 1: Local search algorithm for the single-source setting.

1 Start with the feasible solution D.
2 Select a max out-degree node v^*, and move an arbitrary arc departing v^* to the nearest predecessor u of v^* (in D) with $\deg^+(u) \leq \deg^+(v^*) - 2$.
3 Repeat Step 2 until no such predecessor exists.

Theorem 2. *There is an $O(n \log^2 n)$-time exact algorithm for tree instances of* `min-degree-SPP` *with a single source.*

4.2 Additive 1-Approximation Algorithm for Out-Trees

First, we describe how the multi-source setting differs from the single-source setting, making an extension of Algorithm 1 non-trivial. Consider the instance given in Fig. 4a. The sinks are the set of leaves, and each sink t is labelled with the set of indices of sources for which there is a commodity with that source and sink t. Observe that we can no longer select departing arcs to shift arbitrarily: the arc vw cannot be shifted since there is a commodity with source v and sink w, while the other arcs departing v can be shifted. Furthermore, it is no longer the case that in order to decrease the degree of the highest node, we only need to shift a single arc. In Fig. 4b we see that in the potential second iteration of Algorithm 1, no arc can be moved from v since s_2 has out-degree 3, and all arcs departing v serve commodities with source s_2.

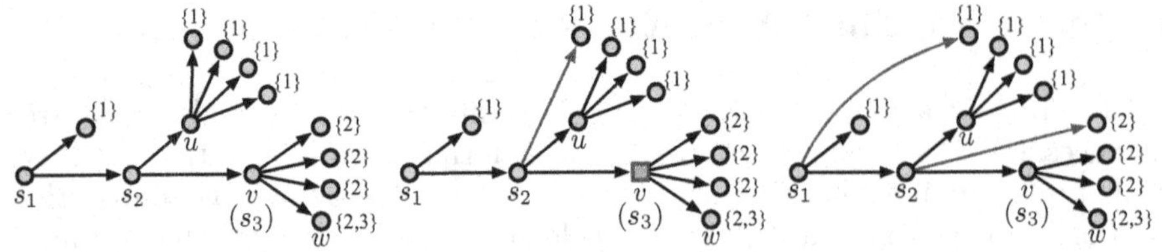

(a) Out-tree instance (b) Terminal stage for alg. (c) Solution ($\Delta^* = 3$).

Fig. 4. Challenges in extending the single-source algorithm to multiple sources.

Definition 1 ($\mathcal{S}(a)$)**.** *Let $D = (N, A)$ be an out-tree, and let $a \in A$. The* set of sources that require a, *denoted $\mathcal{S}(a)$, is the set of sources s_k for which a is on the unique s_k, t_k-dipath in D. That is, $\mathcal{S}(a) := \{s_k : a \in A(P_k), k \in \mathcal{K}\}$.*

Definition 2 (Blocking source). *Let $D = (N, A)$ be an out-tree and let $a = vw$ be an arc in A. The* blocking source *of a, denoted $b(a)$, is the unique source s in $S(a)$ such that the s, v-dipath in D has the fewest arcs.*

Consider the instance in Fig. 5, where nodes w_i for $i \in [5]$ are sinks, and each is labelled with the corresponding set of indices of sources. For each arc a in D, the chart on the right indicates the set $\mathcal{S}(a)$ and value of $b(a)$.

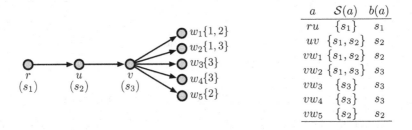

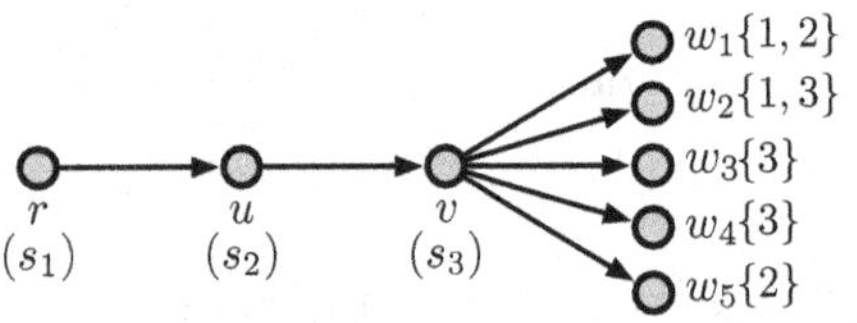

a	$\mathcal{S}(a)$	$b(a)$
ru	$\{s_1\}$	s_1
uv	$\{s_1, s_2\}$	s_2
vw_1	$\{s_1, s_2\}$	s_2
vw_2	$\{s_1, s_3\}$	s_3
vw_3	$\{s_3\}$	s_3
vw_4	$\{s_3\}$	s_3
vw_5	$\{s_2\}$	s_2

Fig. 5. Example of blocking sources

We will present an algorithm that takes as input a target, T, and returns a feasible solution H with $\Delta^+(H) \leq T$ whenever $T \geq \Delta^* + 1$. We now define the contraction subroutine used to generate subinstances of `min-degree-SPP`.

Contraction of an Instance for Target T. Let $\mathcal{I} = (D, \mathcal{K})$ be a feasible out-tree instance of `min-degree-SPP` with root r, and let $T \in \mathbb{Z}_{>0}$. For any node $v \in N$, let $N_D^+(v)$ be the set of nodes reached by arcs departing v in D.

Suppose there is more than one non-leaf node. Let v be a non-leaf node where all nodes in $N_D^+(v)$ are leaves. That is, the subgraph of D rooted at v is a claw, denoted C_v. Such a node can be found efficiently by starting at r and moving to a descendant that is not a leaf. For any pair of nodes $v, w \in N$, let $d(v, w)$ denote the number of edges in the unique path between v and w in D. Breaking ties arbitrarily, let A_v^T denote the $\min\{T, |\delta_D^+(v)|\}$ arcs $a \in \delta_D^+(v)$ with the smallest values of $d(b(a), v)$. Let $B_v^T = \delta_D^+(v) \setminus A_v^T$. We write A_v and B_v when T is clear from context. In the example in Fig. 5, we see that C_v is a claw, and when T is 3, $A_v = \{vw_2, vw_3, vw_4\}$ and $B_v = \{vw_1, vw_5\}$ since $b(vw_2), b(vw_3), b(vw_4) = s_3 = v$ and $b(vw_1), b(vw_5) = s_2$ which is further from v.

By definition, for any pair of arcs $a \in A_v$ and $a' \in B_v$, $b(a)$ is on the unique path in D between $b(a')$ and v. Let $V(A_v)$ denote the heads of the arcs in A_v and let $V(B_v)$ denote the heads of the arcs in B_v. For the same example and target, $V(A_v) = \{w_2, w_3, w_4\}$ and $V(B_v) = \{w_1, w_5\}$.

We define the instance obtained from $\mathcal{I}$ by *contracting v for target T*, denoted $\mathcal{I}_v^T = (D_v^T, \mathcal{K}_v^T)$ as follows. An example is given in Fig. 6. For all $v \neq r$, let $p(v)$ denote the parent node of v in D.

Definition 3 **($\mathcal{I}_v^T$).** *The instance obtained from $\mathcal{I}$ by contracting v for target T is $\mathcal{I}_v^T = (D_v^T, \mathcal{K}_v^T)$, where $E := \{p(v)w : vw \in B_v\}$, $D_v^T := \left(D \setminus \delta_D^+(v)\right) \cup E$, and the commodity set is $\mathcal{K}_v^T = \{(s_k, t'_k) : k \in \mathcal{K}\}$ where*

$$t'_k = \begin{cases} v, & if t_k \in V(A_v) \\ t_k, & otherwise. \end{cases}$$

We defer the proofs of Lemmas 3–5 to the full version [24].

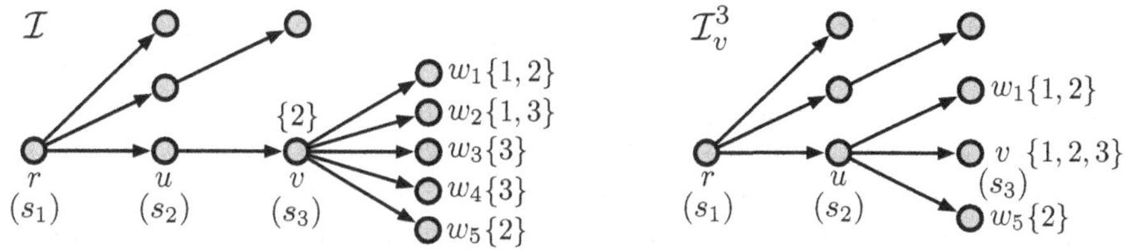

Fig. 6. Example for contraction process.

Lemma 3. *If $T \geq |\mathcal{T}(v)|$, then $\mathcal{I}_v^T$ is a feasible instance of* `min-degree-SPP`.

Algorithm. For a given target T, our algorithm returns a feasible solution H with $\Delta^+(H) \leq T$ whenever $T \geq \Delta^* + 1$. To guarantee the performance of the algorithm if no such solution is produced (the algorithm outputs the empty set), we show that in this case there is a witness set $W, \mathcal{K}'$ such that $\mathtt{LB}^w(W, \mathcal{K}') \geq T$.

When there is a single non-leaf node, r, either the graph itself is the desired solution with max out-degree at most T, or the set $W = \{r\}, \mathcal{K}' = \mathcal{K}$ is a witness set certifying that $\Delta^* \geq T$. Otherwise, we find a node, v, with only leaves as descendants. Then either the subtree rooted at v already provides a witness set, or we apply the target algorithm to the instance $\mathcal{I}_v^T$. If the algorithm produces a feasible solution H_v to $\mathcal{I}_v^T$ with $\Delta^+(H_v) \le T$, we then extend this subgraph to a feasible solution H for $\mathcal{I}$ with $\Delta^+(H) \le T$ by simply adding back the arcs in the set A_v.

Algorithm 2: `out-tree`$(\mathcal{I}, T)$

Input: A target $T \in \mathbb{Z}_{\ge 1}$, and an out-tree instance $\mathcal{I} = (D, \mathcal{K})$

$L \leftarrow \{v \in N : \delta_D^+(v) = \emptyset\}$
$I \leftarrow N \backslash L$
if $|I| = 1$ **then**
  if $|L| > T$ **then**
    return $\emptyset$
  return D
Let v be a non-leaf where $N_D^+(v) \subseteq L$
if $|\mathcal{T}(v)| > T$ **then**
  return $\emptyset$
$H_v \leftarrow$ `out-tree`$(\mathcal{I}_v^T, T)$
if $H_v = \emptyset$ **then**
  return $\emptyset$
else
  return $H_v \cup A_v$

If Algorithm 2 returns the empty set, we argue by induction that there is a witness set $W_v, \mathcal{K}'_v$ for the instance $\mathcal{I}_v^T$ such that $\texttt{LB}^w_{\mathcal{I}_v^T}(W_v, \mathcal{K}'_v) \ge T$. We then extend this pair to a witness set $W, \mathcal{K}'$ for $\mathcal{I}$ such that $\texttt{LB}^w_{\mathcal{I}}(W, \mathcal{K}') \ge T$. Note that we cannot necessarily set $W = W_v$ and $\mathcal{K}' = \mathcal{K}'_v$, since the commodity paths may differ in $\mathcal{I}$ and $\mathcal{I}_v^T$. For each commodity k, the corresponding source-sink pair is (s_k, t_k) in $\mathcal{I}$ and (s_k, t'_k) in $\mathcal{I}_v^T$. Let Q_k denote the unique s_k, t'_k-dipath in D_v^T, and recall that P_k denotes the unique s_k, t_k-dipath in D. The following lemmas relate the cut sets in $\mathcal{I}$ and $\mathcal{I}_v^T$ for a fixed commodity and node set.

Lemma 4. *Let $X \subseteq N \backslash \{v\}$ such that $D[X]$ is weakly connected, and let $k \in \mathcal{K}$. If $s_k \in X$ and $t'_k \notin V(B_v)$, then $\delta^+_{P_k}(X) = \delta^+_{Q_k}(X)$.*

Lemma 5. *Suppose $|\mathcal{T}(v)| \le T$. Let $X \subseteq N$ such that $D[X]$ is weakly connected, and let $k \in \mathcal{K}$. If $s_k \in X$, $t'_k \in V(B_v)$, and $t'_k \notin X$, then either $\delta^+_{P_k}(X) = \delta^+_{Q_k}(X)$ or $\delta^+_{Q_k}(X) = p(v)t'_k$.*

Proposition 6. *For each $T \in \mathbb{Z}_{>0}$, Algorithm 2 either returns a solution, H, to instance $\mathcal{I}$ with $\Delta^+(H) \le T$, or there is a witness set $W, \mathcal{K}'$ such that $\texttt{LB}^w(W, \mathcal{K}') \ge T$, certifying that $\Delta^* \ge T$.*

Proof. Let $\mathcal{I} = (D, \mathcal{K})$ denote an instance of `min-degree-SPP` where D is an out-tree with root r. Let T be a positive integer. We prove the result by induction on the number of non-leaf nodes in D. Recall that we may assume that we are working with minimal, non-trivial instances, in the sense that all $k \in \mathcal{K}$ have $s_k \neq t_k$, $\mathcal{K} \neq \emptyset$, and any leaf node $l \neq r$ must be a sink for some commodity.

Suppose there is a single non-leaf node, r. Then each node in $\delta_D^+(r)$ is a leaf. If $|\delta_D^+(r)| \leq T$, then the graph D is a solution with $\Delta^+(D) \leq T$. Otherwise, $|\mathcal{T}(r)| = |N_D^+(r)| > T$, and the witness set $W = \{r\}$, $\mathcal{K}' = \mathcal{K}$ gives the lower bound $\texttt{LB}^w(W, \mathcal{K}') = \lceil |\mathcal{T}(r)|/1 \rceil \geq T$.

Suppose the result holds for all out-tree instances with at most d non-leaf nodes, and consider an instance with $d+1$ non-leaf nodes. Let v be a non-leaf node where all nodes in $N_D^+(v)$ are leaves.

If $|\mathcal{T}(v)| > T$, then setting $W = \{v\}$ and $\mathcal{K}' = \{k \in \mathcal{K} : s_k = v\}$ gives $\texttt{LB}^w(W, \mathcal{K}') > T$, so suppose $|\mathcal{T}(v)| \leq T$. Let A_v, B_v, E, and $\mathcal{I}_v^T = (D_v^T, \mathcal{K}_v^T)$ be defined as above, where $A_v \cup B_v = \delta_D^+(v)$ and for any pair $a \in A_v$ and $a' \in B_v$, $b(a)$ is on the unique path in D between $b(a')$ and v. By Lemma 3, it follows that $\mathcal{I}_v^T$ is a feasible out-tree instance with at most d non-leaf nodes. Let Δ_v^* denote the min-max degree of a feasible subgraph of $\texttt{cl}(D_v)$ for instance $\mathcal{I}_v^T$. By induction, the algorithm either returns a feasible subgraph $H_v \subseteq \texttt{cl}(D_v)$ for $\mathcal{I}_v^T$ with $\Delta^+(H_v) \leq T$, or there is a witness set $W_v, \mathcal{K}_v'$ certifying that $\Delta_v^* \geq T$.

Suppose the algorithm returns a feasible subgraph $H_v \subseteq \texttt{cl}(D_v)$ for $\mathcal{I}_v^T$ with $\Delta^+(H_v) \leq T$. Let $H = H_v \cup A_v$, as in the example in Fig. 7. We claim that H is a feasible solution for $\mathcal{I}$ with $\Delta^+(H) \leq T$. Observe that $\deg_{H_v}^+(v) = 0$ and $|A_v| \leq T$, so $\Delta^+(H) \leq T$. It remains to argue that H is feasible. Let $k \in \mathcal{K}$. If $t_k \notin V(A_v)$, then (s_k, t_k) is a commodity in $\mathcal{I}_v^T$, and so an s_k, t_k-dipath is present in $H_v \subseteq H$. Otherwise $t_k \in V(A_v)$, and so there is a commodity (s_k, v) in $\mathcal{K}_v^T$. By feasibility of H_v, H contains an s_k, v-dipath. With the addition of arc vt_k in A_v we obtain an s_k, t_k-dipath in H. Therefore, H is feasible and $\Delta^+(H) \leq T$.

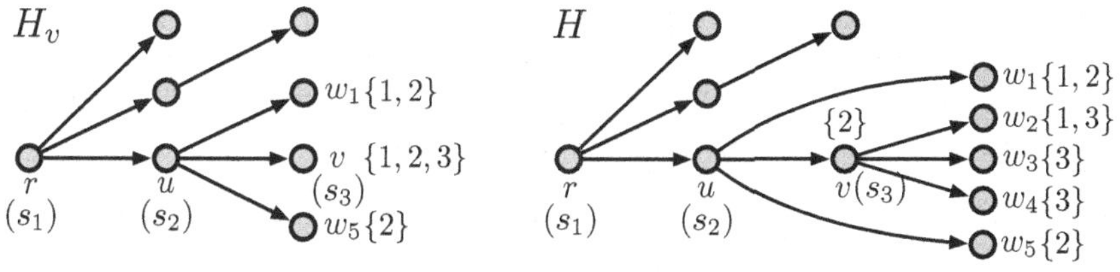

Fig. 7. Extending feasible H_v for $\mathcal{I}_v^3$ to a feasible solution H for $\mathcal{I}$.

Suppose instead, there is a witness set $W_v, \mathcal{K}_v'$ with $\texttt{LB}^w_{\mathcal{I}_v^T}(W_v, \mathcal{K}_v') \geq T$. That is, $D_v^T[W_v]$ is weakly connected, $\emptyset \neq \mathcal{K}_v' \subseteq \mathcal{K}_v^T$ such that $s_k \in W_v$ and $t_k' \notin W_v$ for all $k \in \mathcal{K}_v'$. Moreover, for all distinct $k, j \in \mathcal{K}_v'$, $\delta_{Q_k}^+(W) \cap \delta_{Q_j}^+(W) = \emptyset$. Note that we may assume W_v does not contain any leaves in D_v^T, since removing a leaf from W_v can only increase the lower bound. So, $W_v \subseteq N \backslash \{v\}$.

We now construct a witness set $W, \mathcal{K}'$ for $\mathcal{I}$ such that $\Delta^* \geq \texttt{LB}^w_{\mathcal{I}}(W, \mathcal{K}') \geq T$. Consider the same set of nodes W_v and commodities $\mathcal{K}_v'$ in D instead of D_v^T.

Since no leaf node in D_v^T is in W_v, $D[W_v]$ is weakly connected. Moreover, for each $k \in \mathcal{K}'_v$, it holds $s_k \in W_v$ and $t_k \notin W_v$.

If $\delta^+_{P_k}(W) = \delta^+_{Q_k}(W)$ for all $k \in \mathcal{K}'_v$, then it follows that $W = W_v$ and $\mathcal{K}' = \mathcal{K}'_v$ gives $\mathtt{LB}^w_{\mathcal{I}}(W, \mathcal{K}') = \mathtt{LB}^w_{\mathcal{I}^T_v}(W, \mathcal{K}'_v) \geq T$ as required. So, suppose there is some commodity $k \in \mathcal{K}'_v$ such that $\delta^+_{P_k}(W) \neq \delta^+_{Q_k}(W)$. By Lemmas 4 and 5, it follows that there is a commodity $\ell \in \mathcal{K}'_v$ with $t'_\ell \in V(B_v)$ and $p(v) \in W_v$.

Let $W = W_v \cup \{v\}$. Since $p(v) \in W_v$ and $D[W_v]$ is weakly connected, it follows that $D[W]$ is weakly connected. Let $\bar{\mathcal{K}} = \{(b(vt_k), t_k) : t_k \in V(A_v)\}$. By definition of $\bar{\mathcal{K}}$ and B_v, it follows that s_k is on the unique dipath in D between s_ℓ and v for all $k \in \bar{\mathcal{K}}$. Since $D[W]$ is weakly connected and contains both s_ℓ and v, it follows that $s_k \in W$ for all $k \in \bar{K}$.

If there is a commodity $k \in \mathcal{K}'_v$ with $t_k = v$, let k_v denote this commodity, and otherwise let k_v denote the empty set. Let $\mathcal{K}' = (\mathcal{K}'_v \cup \bar{\mathcal{K}}) \setminus k_v$. Then $s_k \in W$ for all $k \in \mathcal{K}'$, and $t_k \notin W$ for each $k \in \mathcal{K}'$ since k_v was removed. We also claim that $\delta^+_{P_k}(W) \cap \delta^+_{P_j}(W) = \emptyset$ for any $k, j \in \mathcal{K}'$. If $k \in \mathcal{K}'$ with $t_k \in N^+_D(v)$, then $\delta^+_{P_k}(W) = vt_k$. Otherwise, $t_k \notin V(C_v)$ and $\delta^+_{P_k}(W) = \delta^+_{Q_k}(W) \notin \delta^+_D(v)$. Therefore, the disjointness of the cuts follows.

To prove that $\mathtt{LB}^w_{\mathcal{I}}(W, \mathcal{K}') \geq T$, it remains to argue that the cardinality of $|\mathcal{K}'|$ is at least $(T-1)|W| + 1$. Observe that by definition, if $B_v \neq \emptyset$, then $|A_v| = T$. By induction, we have $|\mathcal{K}'_v| \geq (T-1)|W_v| + 1$. Therefore, we have $|\mathcal{K}'| \geq (T-1)|W_v| + T - 1 + 1 = (T-1)|W| + 1$, and so $\mathtt{LB}_{\mathcal{I}}(W, \mathcal{K}') \geq T$ as required. □

Theorem 3. *There is a polynomial-time additive* 1*-approximation algorithm for out-tree instances of* `min-degree-SPP`.

Proof. By executing Algorithm 2 for all possible target values, we determine the smallest value of T for which 2 gives a feasible subgraph H with $\Delta^+(H) \leq T$. Since the algorithm did not return a feasible subgraph for the target $T-1$, there is a witness set $W, \mathcal{K}'$ with $\mathtt{LB}^w(W, \mathcal{K}') \geq T - 1$ (constructible in polytime), certifying that $\Delta^* \geq T - 1$. Hence, $\Delta^+(H) \leq \Delta^* + 1$. □

There are out-tree instances where $\max_{W,\mathcal{K}'} \mathtt{LB}(W, \mathcal{K}')$ is equal to $\Delta^* - 1$. Furthermore, the analysis of Algorithm 2 is tight since there are out-tree instances for which the algorithm does not return a solution with out-degree at most T, even if $\Delta^* = T$; see [24]. It remains open whether or not out-tree instances are NP-hard. In the full version [24], we relate the hardness of these instances to that of a natural packing problem.

5 Conclusion

In this paper, we propose a mathematical model to determine an allocation of sort points that provides a feasible sortation plan. Even in the case where the underlying undirected physical network forms as tree, it is NP-hard to determine whether a feasible sort point allocation exists. We focus on the natural objective of minimizing the maximum number of sort points required at a warehouse, and define the directed min-degree problem of `min-degree-SPP`. As a main result, we give a fast combinatorial additive 1-approximation algorithm for out-trees.

Finding good approximation algorithms for arbitrary graphs remains a challenging open problem. Further, alternative objectives of interest include varying degree constraints on nodes and determining the minimum number of sort points required in any feasible solution. Both are highly relevant in applications.

Acknowledgements. We thank Sharat Ibrahimpur and Kostya Pashkovich for valuable discussions on lower bounds and the link to matroid intersection.

References

1. Bansal, N., Khandekar, R., Nagarajan, V.: Additive guarantees for degree-bounded directed network design. SIAM J. Comput. **39**(4), 1413–1431 (2009)
2. Chakrabarty, D., Lee, Y. T., Sidford, A., Singla, S., Wong, S.C.: Faster matroid intersection. In: FOCS, pp. 1146–1168. IEEE Computer Society (2019)
3. Chaudhuri, K., Rao, S., Riesenfeld, S.J., Talwar, K.: A push-relabel algorithm for approximating degree bounded MSTs. In: Bugliesi, M., Preneel, B., Sassone, V., Wegener, I. (eds.) ICALP 2006. LNCS, vol. 4051, pp. 191–201. Springer, Heidelberg (2006). https://doi.org/10.1007/11786986_18
4. Chaudhuri, K., Rao, S., Riesenfeld, S.J., Talwar, K.: What would Edmonds do? Augmenting paths and witnesses for degree-bounded MSTs. Algorithmica **55**(1), 157–189 (2009)
5. Chvátal, V.: Tough graphs and Hamiltonian circuits. Discrete Math. **5**, 215–228 (1973)
6. Dehghani, S., Ehsani, S., Hajiaghayi, M., Liaghat, V.: Online degree-bounded steiner network design. In: SODA, pp. 164–175. SIAM (2016)
7. Fürer, M., Raghavachari, B.: Approximating the minimum-degree steiner tree to within one of optimal. J. Algorithms **17**, 409–423 (1994)
8. Garey, M.R., Johnson, D.S.: Computers and Intractability. A Guide to the Theory of NP-Completeness. W.H. Freeman & Co, New York (1990)
9. Goemans, M.X.: Minimum bounded degree spanning trees. In: FOCS, pp. 273–282. IEEE Computer Society (2006)
10. Guo, X., Kortsarz, G., Laekhanukit, B., Li, S., Vaz, D., Xian, J.: On approximating degree-bounded network design problems. Algorithmica **84**(5), 1252–1278 (2022)
11. Koenemann, J., Ravi, R.: A matter of degree: improved approximation algorithms for degree-bounded minimum spanning trees. SIAM J. Comput. **31**(6), 1783–1793 (2002)
12. Koenemann, J., Ravi, R.: Primal-dual meets local search: approximating MSTs with nonuniform degree bounds. SIAM J. Comput. **34**, 763–773 (2005)
13. Könemann, J., Ravi, R.: Quasi-polynomial time approximation algorithm for low-degree minimum-cost steiner trees. In: Pandya, P.K., Radhakrishnan, J. (eds.) FSTTCS 2003. LNCS, vol. 2914, pp. 289–301. Springer, Heidelberg (2003). https://doi.org/10.1007/978-3-540-24597-1_25
14. Kortsarz, G., Nutov, Z.: Bounded degree group steiner tree problems. In: Gąsieniec, L., Klasing, R., Radzik, T. (eds.) IWOCA 2020. LNCS, vol. 12126, pp. 343–354. Springer, Cham (2020). https://doi.org/10.1007/978-3-030-48966-3_26
15. Lara, C.L., Koenemann, J., Nie, Y., de Souza, C.C.: Scalable timing-aware network design via Lagrangian decomposition. Eur. J. Oper. Res. **309**(1), 152–169 (2023)

16. Lau, L.C., Naor, J., Salavatipour, M.R., Singh, M.: Survivable network design with degree or order constraints. SIAM J. Comput. **39**(3), 1062–1087 (2009)
17. Lau, L.C., Singh, M.: Additive approximation for bounded degree survivable network design. SIAM J. Comput. **42**(6), 2217–2242 (2013)
18. Nutov, Z.: Approximating directed weighted-degree constrained networks. Theoret. Comput. Sci. **412**(8), 901–912 (2011)
19. Ravi, R., Marathe, M.V., Ravi, S.S., Rosenkrantz, D.J., and Hunt III, H.B.: Many birds with one stone: multi-objective approximation algorithms. In: STOC, pp. 438–447. ACM (1993)
20. Ravi, R., Singh, M.: Delegate and conquer: An LP-based approximation algorithm for minimum degree MSTs. In: Bugliesi, M., Preneel, B., Sassone, V., Wegener, I. (eds.) ICALP 2006. LNCS, vol. 4051, pp. 169–180. Springer, Heidelberg (2006). https://doi.org/10.1007/11786986_16
21. Schrijver, A.: Combinatorial Optimization-Polyhedra and Efficiency. Springer, Heidelberg (2003)
22. Singh, M., Lau, L.C.: Approximating minimum bounded degree spanning trees to within one of optimal. In: STOC, p]. 661–670. ACM (2007)
23. Van Belle, J., Valckenaers, P., Cattrysse, D.: Cross-docking: State of the art. Omega, **40**(6), 827–846 (2012). Special Issue on Forecasting in Management Science
24. Van Dyk, M., Klause, K., Könemann, J., Megow, N.: Fast combinatorial algorithms for efficient sortation. CoRR, abs/2311.05094 (2023)
25. Win, S.: On a connection between the existence of k-trees and the toughness of a graph. Graphs Comb. **5**, 201–205 (1989)
26. Yao, G., Zhu, D., Li, H., Ma, S.: A polynomial algorithm to compute the minimum degree spanning trees of directed acyclic graphs with applications to the broadcast problem. Discret. Math. **308**(17), 3951–3959 (2008)
27. Zheng, K.: Sort optimization and allocation planning in amazon middle mile network. Informs (2021)

A New Branching Rule for Range Minimization Problems

Bart van Rossum[1(✉)], Rui Chen[2], and Andrea Lodi[2]

[1] Econometric Institute, Erasmus University Rotterdam, Rotterdam, The Netherlands
vanrossum@ese.eur.nl
[2] Cornell Tech, New York, USA
{rui.chen,andrea.lodi}@cornell.edu

Abstract. We consider range minimization problems featuring exponentially many variables, as frequently arising in fairness-oriented or bi-objective optimization. While branch-and-price is successful at solving cost-oriented problems with many variables, the performance of classical branch-and-price algorithms for range minimization is drastically impaired by weak linear programming relaxations. We propose range branching, a generic branching rule that directly tackles this issue and is compatible with any problem-specific branching scheme. We show several desirable properties of range branching and show its effectiveness on a series of fair capacitated vehicle routing instances. Range branching significantly improves multiple classical branching schemes in terms of computing time, optimality gap, and size of the branch-and-bound tree, allowing us to solve many more large instances than classical methods.

Keywords: Range minimization · Branch-and-price · Vehicle routing · Fairness

1 Introduction

A growing stream of literature focuses on fairness-oriented optimization [6–8,10, 11]. A popular fairness measure is the range, defined as the maximum minus the minimum of the quantities of interest [7,8,10,11]. In many practical applications, the range and hence the fairness of the solution can be significantly improved with a minor loss in efficiency [8,10]. Indeed, Fig. 1 presents an example where the range between the distances of vehicle routes can be reduced by 26.2% by tolerating an efficiency loss of 0.05% compared to the most cost-efficient solution. In addition to fairness applications, range minimization is often encountered as one of the objectives in bi-objective optimization [6] or even as a stand-alone problem [9].

We explicitly focus on range minimization problems featuring a number of columns that is prohibitively large for mixed-integer programming solvers (e.g., Gurobi [5]). This is commonly observed in, among other things, railway crew scheduling, vehicle routing, and airline crew pairing problems [4]. Here, formulations with a huge number of variables are preferred as they typically display

J. Vygen and J. Byrka (Eds.): IPCO 2024, LNCS 14679, pp. 433–445, 2024.
https://doi.org/10.1007/978-3-031-59835-7_32

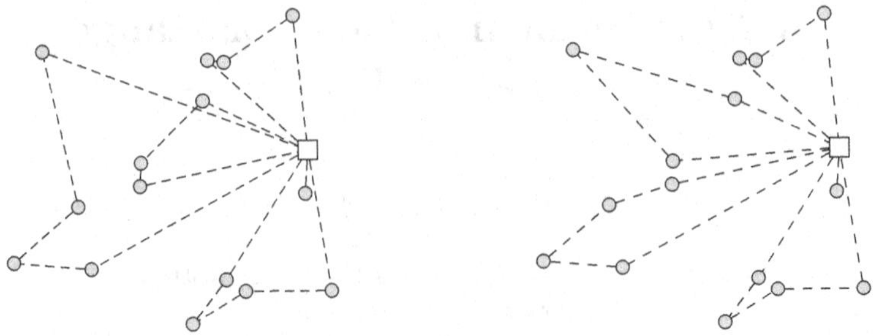

(a) Total distance: 6,178. Range: 2,170. (b) Total distance: 6,181. Range: 1,601.

Fig. 1. Efficient (left) and fair (right) vehicle routing solutions.

a tighter linear programming (LP) relaxation and a less symmetric structure than their compact counterparts [1]. When a classical cost-based objective is used, branch-and-price is the state-of-the-art solution approach for this type of problem. At each node of the branch-and-bound tree, the LP relaxation is solved through column generation. Here, only a small subset of all possible columns is initially considered, and a pricing algorithm is used to solve the pricing problem of identifying columns that potentially improve the objective value. A branching scheme is used to obtain integer solutions, and all branching decisions must be compatible with the pricing algorithm. We refer to [1] for a detailed introduction to branch-and-price.

Unfortunately, formulations for range minimization problems often suffer from poor LP relaxations that drastically impair the performance of classical branch-and-price algorithms [10]. As a result, most researchers resort to heuristic methods instead. Consider, for example, the capacitated vehicle routing problem (CVRP) with route balancing, which can be phrased as a bi-objective problem where range minimization is one of the objectives. While branch-and-price (-and-cut) is highly effective for the classical cost-oriented CVRP [2], few efficient branch-and-price-based algorithms exist for the range-oriented counterpart. Instead, many heuristics have been proposed [7].

Our main contribution is to propose an efficient branch-and-price framework for large-scale range minimization problems. In particular, we propose a dedicated branching rule, which we refer to as *range branching*, that directly tackles the poor LP relaxation of range minimization formulations. It can be used on top of any problem-specific branching scheme, and allows one to effectively leverage branch-and-price techniques commonly applied to cost-oriented optimization problems. We show some promising properties of range branching, and evaluate its effectiveness on a series of fair CVRP instances. We find that range branching leads to a more than 12-fold reduction in computing time on instances with 20 customers, and allows us to solve 70% more instances with 25 customers.

The remainder of this paper is structured as follows. We formally define the range minimization problem in Sect. 2 and present range branching in Sect. 3. In

Sect. 4, we present the results of computational experiments on the fair capacitated vehicle routing problem, and we conclude in Sect. 5.

2 Range Minimization Problem

We formally define the range minimization problem in Sect. 2.1, and discuss mathematical formulations for this problem in Sect. 2.2. In Sect. 2.3, we present some examples related to fair capacitated vehicle routing.

2.1 Problem Description

Throughout this work, we consider problems with the following structure. Given a set of N columns (implicitly), the problem requires to select a subset of them, with decision variables $\boldsymbol{x} = (x_1, \ldots, x_N) \in \{0,1\}^N$ being the indicator variables of the selected columns.[1] Moreover, we assume that some (linear) constraints $A\boldsymbol{x} \leq \boldsymbol{b}$ must be respected. These constraints can include efficiency lower bounds, constraints on the number of selected columns, and other problem-specific constraints (e.g., in the case of the CVRP, regular customer partition constraints). To summarize, the feasible region of the decision variables is given by $\mathcal{X} = \{\boldsymbol{x} \in \{0,1\}^N : A\boldsymbol{x} \leq \boldsymbol{b}\}$.

The objective is to minimize the range of the payoffs of selected columns. In particular, we assume that each column i has an associated payoff value p_i. Without loss of generality, we assume non-negative and bounded payoffs, i.e., for all i we have $p_i \in [0, M]$ for some (large) constant M. For notational convenience, given $\boldsymbol{x} \in [0,1]^N$ and $\boldsymbol{p} \in [0,M]^N$, define $p_{\min}(\boldsymbol{x}) = \min_{i:x_i>0} p_i$ and $p_{\max}(\boldsymbol{x}) = \max_{i:x_i>0} p_i$ to be the minimum and maximum payoff of any (possibly fractional) solution $\boldsymbol{x}$, respectively. We use a definition that covers fractional solutions because fractional solutions are frequently encountered during the course of the branch-and-price algorithm. The goal of the range minimization problem is to determine an integer solution $\boldsymbol{x} \in \mathcal{X}$ minimizing the difference between the largest and smallest payoffs, i.e., minimizing $p_{\max}(\boldsymbol{x}) - p_{\min}(\boldsymbol{x})$.

Finally, we assume that the large number of columns N prohibits the use of standard mixed-integer programming solvers, and that an exact branch-and-price algorithm is used to solve the problem. This algorithm consists of a problem-specific branching scheme, used to obtain integer solutions, and a pricing algorithm, used to identify columns that potentially improve the objective value. The range branching rule we propose will operate on top of the problem-specific branching scheme. Since the range branching decisions induce constraints involving the payoff value p_i as well as constraints limiting the domain of p_i, we assume that these constraints are amendable to the pricing algorithm. We will show that this is the case for the CVRP application under consideration.

[1] For simplicity, we do not assume additional auxiliary variables, but auxiliary variables can be easily incorporated into our approach.

2.2 Mathematical Formulation

We now discuss a general class of formulations for the range minimization problem. To this end, we first define the concept of *range-respecting solutions.*

Definition 1 (Range-Respecting Solution). *Let a vector* $(\bar{\boldsymbol{x}}, \bar{\eta}, \bar{\gamma}) \in [0,1]^N \times \mathbb{R}^2$. *We say this vector is* range-respecting *(with respect to* $\mathbf{p}$*) if* $\bar{\eta} \geq p_{\max}(\bar{\boldsymbol{x}})$ *and* $\bar{\gamma} \leq p_{\min}(\bar{\boldsymbol{x}})$, *and* range-violating *(with respect to* $\mathbf{p}$*) otherwise.*

Note that the range of a solution $\bar{\boldsymbol{x}} \in \mathcal{X}$ is equal to $\bar{\eta} - \bar{\gamma}$ with the smallest $\bar{\eta}$ and largest $\bar{\gamma}$ that make $(\bar{\boldsymbol{x}}, \bar{\eta}, \bar{\gamma})$ range-respecting. We are now ready to define a class of formulations for the range minimization problem.

Definition 2 ((Weak) Valid Formulation). *Let* $A, \boldsymbol{b}$ *be given. Let* $\boldsymbol{w} = (C, \boldsymbol{d}, \boldsymbol{e}, F, \boldsymbol{g}, \boldsymbol{h})$ *be a tuple of constraint matrices and coefficient vectors, and consider the following mixed binary linear program:*

$$\min_{\boldsymbol{x},\eta,\gamma} \quad \eta - \gamma \tag{1a}$$

$$s.t. \quad A\boldsymbol{x} \leq \boldsymbol{b} \tag{1b}$$

$$C\boldsymbol{x} + \eta \boldsymbol{d} \leq \boldsymbol{e} \tag{1c}$$

$$F\boldsymbol{x} + \gamma \boldsymbol{g} \leq \boldsymbol{h} \tag{1d}$$

$$\boldsymbol{x} \in \{0,1\}^N. \tag{1e}$$

Auxiliary constraints (1c)–(1d), as defined by $\boldsymbol{w}$, *model the values of* η *and* γ *representing the maximum and minimum payoff, respectively. We say program (1) is a* valid formulation *for the range minimization problem when* $\boldsymbol{z} = (\boldsymbol{x}, \eta, \gamma)$ *is feasible in (1) if and only if* $\boldsymbol{x}$ *satisfies* $\boldsymbol{x} \in \{0,1\}^N$, $A\boldsymbol{x} \leq \boldsymbol{b}$, *and* $\boldsymbol{z}$ *is range-respecting.*

Let $\mathcal{X}_{LP}(\boldsymbol{w}) \subseteq \mathbb{R}^{N+2}$ *be the feasible region of the LP relaxation of (1), obtained by relaxing constraints (1e). If, in addition, for each range-respecting vector* $\boldsymbol{z} = (\boldsymbol{x}, \eta, \gamma)$ *for which* $\boldsymbol{x} \in [0,1]^N$ *and* $A\boldsymbol{x} \leq \boldsymbol{b}$, *it holds that* $\boldsymbol{z} \in \mathcal{X}_{LP}(\boldsymbol{w})$, *we say program (1) is a* weak *valid formulation. Similarly, we say* $\boldsymbol{w}$ *is* (weak) valid *when it defines a (weak) valid formulation.*

Informally, a valid formulation is weak when it admits all fractional range-respecting solutions. In this case, the auxiliary constraints modeling the range do not tighten the description of the feasible region of $\boldsymbol{x}$. Our experience suggests that many such formulations are also weak in the sense that their LP relaxations admit many fractional range-violating solutions. As a result, the LP bounds of such formulations are poor and the performance of branch-and-price algorithms is hampered.

In particular, several issues arise when attempting to come up with a "strong" formulation for range minimization. First, it is impractical to use strong constraints of the type $p_i x_i \leq \eta$ for all $i = 1, \ldots, N$, as this would require simultaneous column-and-row generation and yield a prohibitive number of constraints.

Instead, one typically resorts to aggregated constraints of the type $\sum_{i \in I} p_i x_i \leq \eta$ where I is a subset of columns of which at most one column will be selected in any feasible solution. Second, it is not clear how to correctly model the minimum payoff γ without a big-M constraint or without resorting to problem-specific information. Third, problem-specific valid inequalities are unlikely to be of use, as they are typically aimed at tightening the formulation of the problem-specific feasible region $\mathcal{X}$ and do not necessarily cut off fractional range-violating solutions.

2.3 Fair Capacitated Vehicle Routing

In this section, we introduce an application of the range minimization problem which we call the *fair capacitated vehicle routing problem (F-CVRP)*:

Definition 3 (Fair Capacitated Vehicle Routing Problem). *Let C be a set of customers, where customer $i \in C$ has demand $d_i > 0$. Let $G = (\{0\} \cup C, A)$ be the complete directed graph on the set of customers and a central depot, and let c_a and p_a denote the cost and payoff of traversing arc $a \in A$, respectively. Moreover, let K homogeneous vehicles with capacity Q be given, and let B be an upper bound on the total cost of all routes. The goal of the fair capacitated vehicle routing problem is to select a set of K routes, starting and ending at the depot, such that each customer is visited exactly once, the sum of customer demands along each route does not exceed Q, the total cost of traversed edges does not exceed B, and the range of the route payoffs is minimized.*

We present two formulations for the F-CVRP. The vehicle-index formulation explicitly assigns routes to vehicles. The last-customer formulation relies on the insights that (i) each route has a unique last customer and (ii) no customer can appear as last customer on more than one route.[2] We observe that the direct branch-and-price implementation of both formulations suffers from poor LP relaxations.

Vehicle-Index Formulation. Let R denote the index set of feasible routes. For each $r \in R$, let c_r and p_r denote the cost and the payoff of route r, respectively, and let binary parameter a_{ir} indicate whether customer i is visited by route r. We introduce the binary decision variable x_{rk}, indicating whether or not route r is assigned to vehicle k. Then the following mixed binary linear program is a valid formulation of the F-CVRP:

$$\min \quad \eta - \gamma \tag{2a}$$

$$\text{s.t.} \quad \sum_{k \in K} \sum_{r \in R} a_{ir} x_{rk} = 1 \qquad \forall i \in C \tag{2b}$$

[2] Our last-customer formulation builds on the last-customer formulation for the min-max multiple traveling salesman, as presented by N. Bianchessi, C. Tilk, and S. Irnich at the 2023 International Workshop on Column Generation in Montréal.

$$\sum_{k\in K}\sum_{r\in R} c_r x_{rk} \leq B \tag{2c}$$

$$\sum_{r\in R} x_{rk} = 1 \qquad \forall k \in \{1,\ldots,K\} \tag{2d}$$

$$\sum_{r\in R} p_r x_{rk} \leq \eta \qquad \forall k \in \{1,\ldots,K\} \tag{2e}$$

$$\sum_{r\in R} p_r x_{rk} \geq \gamma \qquad \forall k \in \{1,\ldots,K\} \tag{2f}$$

$$x_{rk} \in \mathbb{B} \qquad \forall k \in \{1,\ldots,K\},\ r \in R. \tag{2g}$$

Due to the symmetry of the formulation in index k, this formulation always admits a fractional solution with objective value 0 (by setting $x_{r1} = x_{r2} = \ldots = x_{rK}$ for all $r \in R$). There exist techniques that partially break the symmetry among vehicles, but completely eliminating this symmetry in branch-and-price remains challenging [3].

Last-Customer Formulation. We use the same parameters as in the vehicle-index formulation. In addition, for each route $r \in R$, we let the binary parameter b_{ir} indicate whether customer i is the last customer on route r. We introduce the binary decision variable x_r, indicating whether or not route r is selected. Then the following mixed binary linear program is a valid formulation of the F-CVRP:

$$\min \quad \eta - \gamma \tag{3a}$$

$$\text{s.t.} \quad \sum_{r\in R} a_{ir} x_r = 1 \qquad \forall i \in C \tag{3b}$$

$$\sum_{r\in R} c_r x_r \leq B \tag{3c}$$

$$\sum_{r\in R} x_r = K \tag{3d}$$

$$\sum_{r\in R} p_r b_{ir} x_r \leq \eta \qquad \forall i \in C \tag{3e}$$

$$M\left(1 - \sum_{r\in R} b_{ir} x_r\right) + \sum_{r\in R} p_r b_{ir} x_r \geq \gamma \qquad \forall i \in C \tag{3f}$$

$$x_r \in \mathbb{B} \qquad \forall r \in R. \tag{3g}$$

The big-M constraints (3f) contribute to poor LP bounds. In addition, one can obtain fractional solutions with low objective values by reversing the orientation of a route. This last effect, however, can be largely mitigated by enforcing that the index of the last customer exceeds that of the first customer within the pricing problem.

It is readily seen that both formulations are weak.

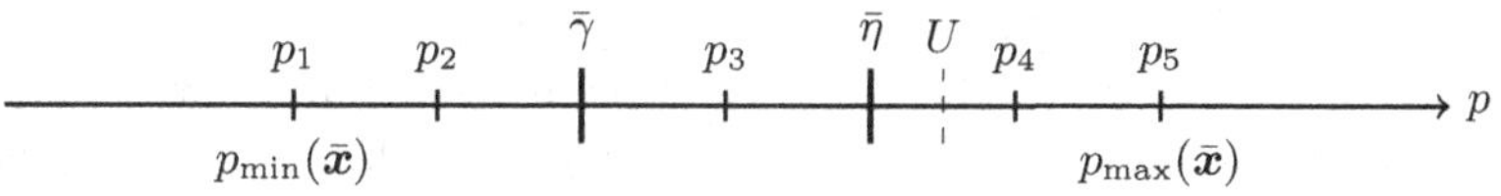

Fig. 2. Example of a range-violating solution (with $\bar{x}_1, \bar{x}_2, \bar{x}_3, \bar{x}_4, \bar{x}_5 \in (0,1)$) that is cut off by applying range branching with cutoff value U.

Proposition 1 (Weak F-CVRP Formulations). *The vehicle-index and last-customer formulation are weak valid formulations for the F-CVRP.*

3 Range Branching

We propose *range branching*, a generic branching strategy for range minimization problems. We define the range branching rule, to be used on top of problem-specific branching schemes, in Sect. 3.1, and show some desirable properties of range branching in Sect. 3.2. In Sect. 3.3, we discuss some practical considerations regarding the implementation of range branching.

3.1 Range Branching Rule

The goal of the branching rule is to eliminate all range-violating fractional solutions from consideration and thereby improve the LP bound in branch-and-price algorithms for range minimization problems. To this end, we propose to (i) branch on the values of η and γ and (ii) make use of a combination of variable bounding and variable fixing.

We now formally define the range branching rule. Consider a fractional solution $(\bar{\boldsymbol{x}}, \bar{\eta}, \bar{\gamma})$ encountered at any node in the branch-and-price tree. Recall that $p_{\min}(\bar{\boldsymbol{x}}) = \min_{i:\bar{x}_i>0} p_i$ and $p_{\max}(\bar{\boldsymbol{x}}) = \max_{i:\bar{x}_i>0} p_i$. When this solution is range-respecting, i.e., $\bar{\gamma} \leq p_{\min}(\bar{\boldsymbol{x}}) \leq p_{\max}(\bar{\boldsymbol{x}}) \leq \bar{\eta}$, our branching rule does not create any branches. Instead, the problem-specific branching scheme is invoked to construct branches that cut-off the incumbent fractional solution. Suppose that the incumbent fractional solution $(\bar{\boldsymbol{x}}, \bar{\eta}, \bar{\gamma})$ is range-violating, for example, $\bar{\eta} < p_{\max}(\bar{\boldsymbol{x}})$. We wish to cut off this solution by branching on η. In particular, we select a cutoff value $U \in (\bar{\eta}, p_{\max}(\bar{\boldsymbol{x}}))$ and create two child nodes. In the left child node, we impose $\eta \leq U$. In addition, in the left child node and its descendants, we apply variable fixing by setting to zero all x_i for which $p_i > U$ and prevent the generation of columns with payoffs $p_i > U$ in the pricing problem. In other words, we apply the locally valid inequality $\sum_{i:p_i>U} x_i = 0$ in the left node. In the right child node, we only impose $\eta \geq U$. The case $\bar{\gamma} > p_{\min}(\bar{\boldsymbol{x}})$ is treated similarly. This branching rule preserves all range-respecting solutions, but cuts off range-violating incumbent solutions. Figure 2 illustrates a range-violating solution upon which we apply range branching. Note that range branching is to be applied at every node in the branch-and-price tree, even those where problem-specific branching decisions are already being enforced.

With proper choices of the cutoff values (see Sect. 3.3), the use of this branching rule has several effects on the branch-and-price algorithm. First, we typically observe that the branch in which variable fixing is applied becomes infeasible and can be pruned, especially in early branching iterations. The branch where only variable bounding on η and γ is used encounters a higher objective value and thereby drives up the overall LP bound. Second, the variable fixing has an ambiguous effect on the efficiency of the pricing problems. On the one hand, incorporating the constraints $p_i \leq U$ or $p_i \geq L$ restricts the feasible region of the pricing problem, which can yield a speed-up. On the other hand, actually enforcing this constraint can complicate the pricing problems, e.g., by prohibiting the use of efficient dominance rules in labeling algorithms. Which effect dominates the other is highly specific to the problem and payoff structure under consideration.

3.2 Properties of Range Branching

We define a *range-branching-free (RBF) node solution* to be any solution vector $\boldsymbol{z} = (\bar{\boldsymbol{x}}, \bar{\eta}, \bar{\gamma})$ encountered at any node in the branch-and-bound/price tree for which the range branching rule does not detect any branching opportunities. We refer to the corresponding node as *RBF node*. We say $\boldsymbol{z}$ is a RBF node solution with respect to $\boldsymbol{w}$ when it is obtained by solving the range formulation defined by $\boldsymbol{w}$. By construction, all RBF node solutions are range-respecting.

Proposition 2 (RBF Node Solutions are Range-Respecting). *Let $\boldsymbol{z} = (\bar{\boldsymbol{x}}, \bar{\eta}, \bar{\gamma})$ be a RBF node solution with respect to some valid $\boldsymbol{w}$. Then, $\boldsymbol{z}$ is range-respecting.*

Together with Definition 2, it follows that any RBF node solution is feasible in the LP relaxation of any weak formulation for this range minimization problem.

Corollary 1 (Formulation-Independence). *Let $\boldsymbol{z} = (\bar{\boldsymbol{x}}, \bar{\eta}, \bar{\gamma})$ be a RBF node solution with respect to some valid $\boldsymbol{w}$. Then, $\boldsymbol{z} \in \mathcal{X}_{LP}(\boldsymbol{w}')$ for all weak valid $\boldsymbol{w}'$.*

As a result, the objective value at any RBF node is at least as high as the best possible root node bound of any weak valid formulation for range minimization.

Corollary 2 (Strong LP Bound). *Let $\boldsymbol{z} = (\bar{\boldsymbol{x}}, \bar{\eta}, \bar{\gamma})$ be a RBF node solution with respect to some valid $\boldsymbol{w}$, and denote its objective value by $z_{RBF} = \bar{\eta} - \bar{\gamma}$. Moreover, let $z^*_{LP}(\boldsymbol{w}') = \min_{(\boldsymbol{x},\eta,\gamma)\in\mathcal{X}_{LP}(\boldsymbol{w}')}(\eta - \gamma)$ be the LP bound of the valid formulation defined by $\boldsymbol{w}'$. Then, $z_{RBF} \geq z^*_{LP}(\boldsymbol{w}')$ for all weak valid $\boldsymbol{w}'$.*

Recall that the LP lower bound in branch-and-price is defined as the minimum objective over all leaf nodes in the branch-and-bound tree. At some point in the course of the branch-and-price algorithm, all leaf nodes in the branch-and-bound tree will be RBF nodes or descendants from RBF nodes. From this point onward, the LP lower bound will be at least as high as the best possible root node bound of any weak valid formulation.

3.3 Practical Considerations

Cutoff Values. As discussed in Sect. 3.1, the success of range branching can be partially explained by its differential effect on the child branches it creates. In one branch, we apply both variable bounding and fixing, after which we can typically prune by infeasibility. In the other branch, we apply variable bounding, thereby driving up the LP bound of this branch. These effects are likely to be small, however, when cutoff values close to $\bar{\gamma}$ and $\bar{\eta}$ are used.

In practice, we find that the following relaxation of range branching works well. Let $\alpha \in [0, 1)$. At a node with incumbent solution $(\bar{\boldsymbol{x}}, \bar{\eta}, \bar{\gamma})$, set $U_\alpha = (1+\alpha)\bar{\eta}$ and $L_\alpha = (1-\alpha)\bar{\gamma}$. We check whether routes with payoff above U_α or below L_α are used, i.e., when range violations of more than a factor α occur. In this case, we perform range branching with U_α and L_α as cutoff values. A value of $\alpha = 0.025$ seems to perform well in our experiments. The results in Sect. 3.2 hold only for the case of $\alpha = 0$. From a computational point of view, however, it would be interesting to consider layered branching schemes with decreasing values of α.

Further Improvements. Several interesting refinements of range branching remain. First, it is clear that the branching scheme can be tightened when all p_i are integer-valued (e.g., branch by creating nodes $\eta \leq U$ and $\eta \geq U+1$ with integer-valued U). Second, one could incorporate a restricted master heuristic throughout the branch-and-price algorithm and use valid upper bounds on the range to locally restrict the domain of η and γ. Third, if one uses big-M constraints to model the minimum, this value of M can be refined throughout the course of the algorithm based on the variable domain of γ. Fourth, our theoretical analysis suggests that it is possible to solve range minimization problems through range branching without the use of auxiliary constraints of the type (1c)–(1d), i.e., without an explicit valid formulation.

4 Computational Experiments

We evaluate the effectiveness of range branching by performing extensive computational experiments on instances of the fair capacitated vehicle routing problem. We describe the set-up of these experiments in Sect. 4.1 and discuss the results in Sect. 4.2.

4.1 Set-Up

Similar to [8], we construct instances based on CVRP instance `X64` [12]. For each number $|C| \in \{15, 20, 25\}$, we construct 20 instances by selecting $|C|+1$ customer locations from the original instance, where the first customer location serves as the depot. Each instance has $K = 5$ vehicles with capacity Q chosen such that at least K vehicles are needed to serve all demand. We set the budget B equal to 110% of the cost of the cost-efficient solution. The payoff of a route is equal to the route's distance, i.e., we aim to balance the length of the routes.

We consider the vehicle-index and last-customer formulations for the F-CVRP presented in Sect. 2.3. For the first formulation, we solve a pricing problem per vehicle. The classical branching scheme consists of branching on customer-vehicle combinations first and arcs second. For the second formulation, we solve a pricing problem per potential last customer. The classical branching scheme consists of branching on the last customer first and arcs second. For both formulations we solve the pricing problems using bi-directional labelling and ng-route relaxation (see [2] for an introduction) and solve up to 16 pricing problems in parallel. We provide the range of the efficient solution as an upper bound to the branch-and-price algorithm, and impose a time limit of one hour. Based on preliminary computational experiments, we implement range branching with cutoff values defined by $\alpha = 0.025$. Recall that range branching is applied at any branch-and-price node, even those where classical branching decisions are being enforced. When the incumbent fractional solution is already range-respecting, the problem-specific classical branching scheme is invoked.

4.2 Results

Table 1 presents the results for the three sets of instances. For each combination of number of customers, formulation, and branching scheme, we present the number of instances solved to optimality within the time limit, the computing time, the optimality gap, the number of branch-and-bound nodes, and the percentage-wise reduction in range of the best solution found compared to the most cost-efficient solution. The results are averaged over all 20 instances. In case not all instances are solved, the computing times and number of branch-and-bound nodes provide lower bounds on the true values.

Table 1. Computational performance of vehicle-index and last-customer formulations with classical and range branching schemes.

$\|C\|$	Formulation	Branching	# Solved	Time (s)	Gap (%)	# Nodes	Δ Range (%)
15	Vehicle	Classical	20/20	207.1	0	6,619.4	49.4
		Range	20/20	175.4	0	5,687.8	49.4
	Customer	Classical	17/20	916.9	4.0	15,726.8	48.7
		Range	20/20	3.6	0	385.8	49.4
20	Vehicle	Classical	9/20	2,769.5	33.3	14,446.9	34.2
		Range	16/20	1,099.0	12.3	6,386.6	44.8
	Customer	Classical	1/20	3,494.3	67.4	21,279.1	25.4
		Range	19/20	290.6	0.0	3,430.9	53.5
25	Vehicle	Classical	0/20	3,600.0	94.0	11,349.4	8.6
		Range	6/20	3,053.2	25.2	983.8	38.6
	Customer	Classical	0/20	3,600.0	94.1	13,220.6	5.1
		Range	14/20	1,741.1	5.8	1,728.1	54.0

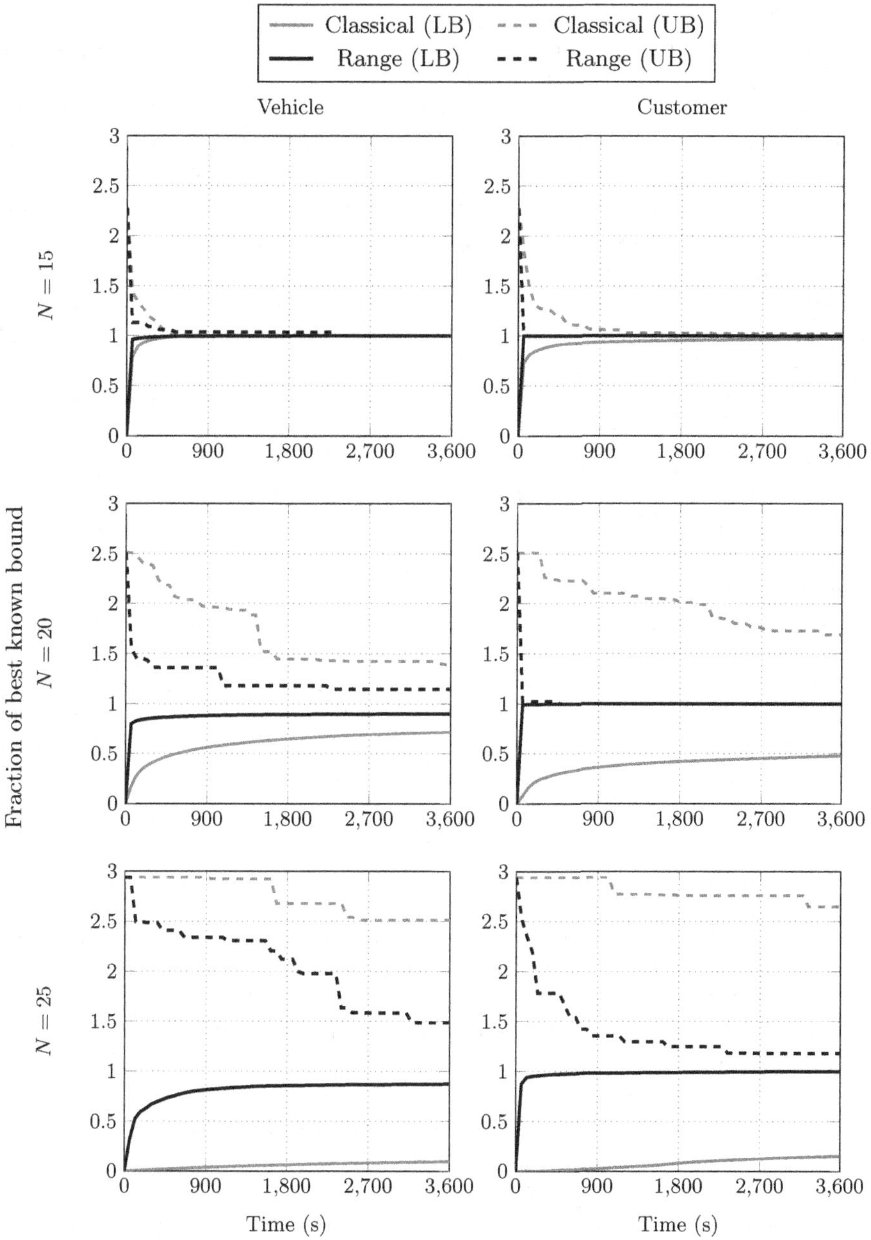

Fig. 3. Progression of lower and upper bound of vehicle index formulation (left) and last customer formulation (right) for the classical and range branching scheme.

The results in Table 1 show that range branching outperforms classical branching in all cases. For all sizes and formulations, range branching leads to a strong reduction in computing time, optimality gap, and number of branch-and-bound nodes. In addition, it allows us to solve 70% more instances with $|C| = 25$ customers, the largest instances we consider. The last-customer formulation seems to particularly benefit from range branching, an effect that appears to be largely driven by a huge reduction in the size of the branch-and-bound tree. While it is not competitive with the vehicle-index formulation under the classical branching scheme, it dominates this formulation on all instance sets when range branching is used. Finally, we note that fairness gains up to 54% can be attained compared to the most cost-efficient solution.

Figure 3 displays the progression of the lower and upper bounds, for each combination of number of customers, formulation, and branching scheme, throughout the course of the branch-and-price algorithm. These relative values are expressed in terms of the best known lower bound, and subsequently averaged over all instances. We find that the performance of range branching is largely driven by its ability to quickly drive up the lower bound to near its final value. This is well in line with the theoretical properties of range branching derived in Sect. 3.2. On the larger instances, the lower bounds of the classical branching scheme are still slowly increasing after an hour of computation, whereas the bounds with range branching reach near-optimal values early in the algorithm. This is especially true for the last-customer formulation that attains near-optimal lower bounds in the first few iterations of the algorithm. This might explain why it outperforms the vehicle-index formulation. In addition to improved lower bounds, range branching also has a positive effect on the upper bound evolution throughout the course of branch-and-price. Hence, range branching appears to be the method of choice even on large instances that are unlikely to be solved to optimality within the time limit.

5 Conclusion

In this work, we propose a generic branch-and-price approach for range minimization problems. Our approach relies on range branching, a branching rule that effectively cuts off fractional nodes underestimating the true range of the solution. We show several desirable properties of the branching rule and its excellent computational performance in extensive experiments on instances of the fair capacitated vehicle routing problem. Using our new branching rule, we are able to solve many more large instances to optimality, compared to using classical branching strategies.

Several promising directions for future research remain. First, it would be interesting to apply our approach to other problems, such as railway crew scheduling or airline crew pairing. Second, Sect. 3.3 contains many suggestions for different ways of implementing range branching, including, for example, layered branching schemes with increasingly tight cutoff values. Finally, we point out that our approach can be readily applied to problems where the objective consists of minimizing the maximum payoff, or maximizing the minimum payoff.

References

1. Barnhart, C., Johnson, E.L., Nemhauser, G.L., Savelsbergh, M.W., Vance, P.H.: Branch-and-price: column generation for solving huge integer programs. Oper. Res. **46**(3), 316–329 (1998)
2. Costa, L., Contardo, C., Desaulniers, G.: Exact branch-price-and-cut algorithms for vehicle routing. Transp. Sci. **53**(4), 946–985 (2019)
3. Darvish, M., Coelho, L.C., Jans, R.: Comparison of symmetry breaking and input ordering techniques for routing problems. CIRRELT (2020)
4. Desaulniers, G., Desrosiers, J., Solomon, M.M.: Column Generation, vol. 5. Springer, New York (2006)
5. Gurobi Optimization, LLC: Gurobi Optimizer Reference Manual (2023). https://www.gurobi.com
6. Jozefowiez, N., Semet, F., Talbi, E.G.: An evolutionary algorithm for the vehicle routing problem with route balancing. Eur. J. Oper. Res. **195**(3), 761–769 (2009)
7. Matl, P., Hartl, R.F., Vidal, T.: Workload equity in vehicle routing problems: a survey and analysis. Transp. Sci. **52**(2), 239–260 (2018)
8. Matl, P., Hartl, R.F., Vidal, T.: Workload equity in vehicle routing: the impact of alternative workload resources. Comput. Oper. Res. **110**, 116–129 (2019)
9. Puerto, J., Ricca, F., Scozzari, A.: Range minimization problems in path-facility location on trees. Discret. Appl. Math. **160**(15), 2294–2305 (2012)
10. van Rossum, B., Chen, R., Lodi, A.: Optimizing fairness over time with homogeneous workers. In: 23rd Symposium on Algorithmic Approaches for Transportation Modelling, Optimization, and Systems (ATMOS 2023). Schloss Dagstuhl-Leibniz-Zentrum für Informatik (2023)
11. Tsang, M.Y., Shehadeh, K.S.: Convex fairness measures: theory and optimization. arXiv preprint arXiv:2211.13427 (2022)
12. Uchoa, E., Pecin, D., Pessoa, A., Poggi, M., Vidal, T., Subramanian, A.: New benchmark instances for the capacitated vehicle routing problem. Eur. J. Oper. Res. **257**(3), 845–858 (2017)

Sensitivity Analysis for Mixed Binary Quadratic Programming

Diego Cifuentes, Santanu S. Dey(✉), and Jingye Xu

H. Milton Stewart School of Industrial and Systems Engineering, Georgia Institute of Technology, Atlanta, USA
{dfc3,sdey30,jxu673}@gatech.edu

Abstract. We consider sensitivity analysis for Mixed Binary Quadratic Programs (MBQPs) with respect to changing right-hand-sides (rhs). We show that even if the optimal solution of a given MBQP is known, it is NP-hard to approximate the change in objective function value with respect to changes in rhs. Next, we study algorithmic approaches to obtaining dual bounds for MBQP with changing rhs. We leverage Burer's completely-positive (CPP) reformulation of MBQPs. Its dual is an instance of co-positive programming (COP), and can be used to obtain sensitivity bounds. We prove that strong duality between the CPP and COP problems holds if the feasible region is bounded or if the objective function is convex, while the duality gap can be strictly positive if neither condition is met. We also show that the COP dual has multiple optimal solutions, and the choice of the dual solution affects the quality of the bounds with rhs changes. We finally provide a method for finding good nearly optimal dual solutions, and we present preliminary computational results on sensitivity analysis for MBQPs.

Keywords: Sensitivity Analysis · Mixed Binary Quadratic Programming · Copositive programming · Duality Theory

1 Introduction

A mixed binary quadratic program (MBQP) has the form:

$$
\begin{aligned}
z(\mathbf{b}) := \min_{\mathbf{x}\geq 0} \; & \mathbf{x}^\top Q\mathbf{x} + 2\mathbf{c}^\top \mathbf{x} \\
\text{s.t.}\; & \mathbf{a}_i^\top \mathbf{x} = b_i, \;\; \forall i \in \{1,\ldots,m\} \\
& x_j \in \{0,1\}, \forall j \in \mathcal{B},
\end{aligned} \tag{1}
$$

where Q is a symmetric matrix with rational entries of size $n \times n$, $\mathbf{b} \in \mathbb{Q}^m$, $\mathbf{c} \in \mathbb{Q}^n$, $\mathbf{a}_i \in \mathbb{Q}^n$ for all $i \in \{1,\ldots,m\}$, and $\mathcal{B} \subseteq \{1,\ldots,n\}$ is the set of variables restricted to be binary. This is a very general optimization model that captures mixed binary linear programming [8,22], quadratic programming [3], and several instances of mixed integer nonlinear programming models appearing in important application areas such as power systems [23].

We would like to acknowledge the support from ONR grant #N000142212632.

J. Vygen and J. Byrka (Eds.): IPCO 2024, LNCS 14679, pp. 446–459, 2024.
https://doi.org/10.1007/978-3-031-59835-7_33

Many practical optimization problems related to operational decision-making involve solving similar MBQP instances repeatedly. Moreover, they typically need to be solved within a short time window. This is because, unlike long-term planning problems, for such problems the exact problem data becomes available only a short time before a good solution is required to be implemented in practice. See, for examples, problems considered and discussed in [15,17,25]. In many of these applications, the constraint matrix remains the same, as these represent constraints related to some invariant physical resources, while the right-hand-side changes from instance to instance.

The practical consideration discussed above motivates us to study sensitivity analysis of MBQPs with respect to changing right-hand-sides. We are particularly interested in the change of the optimal objective value for (MBQP) given the fact that most modern solvers are able to find a good solution of (MBQP) in a reasonable time while finding it more challenging to provide matching dual bounds. Therefore we aim to provide methods to infer high-quality dual bounds from sensitivity analysis of already-solved instances. The pioneering results on sensitivity for integer programs (IPs) with changing right-hand-sides were obtained by Cook et al. [9]. See [7,10,11,13,19] for many advances in this line of research. However, these results yield trivial bounds in the case of binary variables since they rely on the infinity norm of integer constrained variables, which is a constant for all non-zero binary vectors.

We consider an alternative approach in this paper. Specifically, we leverage Burer's completely-positive (CPP) reformulation of MBQP [5]. The advantage of the CPP reformulation is that, although still challenging and NP-hard in general to solve, is a convex problem. Thus, one can examine its dual, which is an instance of copositive programming (COP). The optimal dual variables can provide bounds on $z(\mathbf{b})$, i.e., they allow to bound the optimal objective function of the MBQP as the right-hand-side changes. Details of the CPP reformulation and the COP dual problem are presented in the next section. This approach of using Burer's CPP reformulation of MBQPs [5] to obtain shadow price information was first considered in [16] for the electricity market clearing problem.

Section 2 presents all our results and Sect. 3 provides future avenues of research.

2 Main Results

Notation. Given a positive integer n, we let $[n]$ denote the set $\{1, \dots, n\}$. For $u \in \mathbb{R}$, we denote its absolute value by $|u|$. For a discrete set $\mathcal{B}$, we use $|\mathcal{B}|$ to denote its cardinality. We let $\mathbb{S}^n$ to be the set of symmetric $n \times n$ matrices, and $\mathbb{S}^n_+$ to be the cone of $n \times n$ positive-semidefinite (PSD) matrices. We denote a matrix M being PSD by $M \succeq 0$ and denote M not being a PSD matrix by $M \not\succeq 0$. We let $\mathbb{S}^n_p$ be the set of symmetric $n \times n$ matrices with non-negative entries. We let $\mathcal{CP}$ to be the cone of completely positive matrices, i.e., $\mathcal{CP} = \{M \in \mathbb{S}^n \mid M = BB^\top \text{ and } B \text{ is a } m \times n \text{ entry-wise nonnegative matrix for some integer } m\}$. We let $\mathcal{COP}$ to be the cone of copositive matrices, i.e., $\mathcal{COP} = \{M \in \mathbb{S}^n \mid \mathbf{x}^\top M \mathbf{x} \geq 0, \forall \mathbf{x} \geq 0\}$. We use $\mathbf{e}_i$ to denote the i-th standard basis vector.

2.1 Complexity

We begin our study by establishing formally the difficulty of approximating $z(\mathbf{b} + \Delta\mathbf{b})$ for varying $\Delta\mathbf{b}$, assuming that we know the exact value of $z(\mathbf{b})$.

Definition 1. *An algorithm is called (α, β)-approximation for some $\beta \geq 1 \geq \alpha > 0$ if it takes $(A, \mathbf{b}, \mathbf{c}, Q, \mathcal{B}, z(\mathbf{b}), \Delta\mathbf{b})$ as input, where A, $\mathbf{b}$, $\mathbf{c}$, Q, $\mathcal{B}$ represents an instance of (1), $z(\mathbf{b})$ is its optimal objective function value, $\Delta\mathbf{b}$ is the change in right-hand-side, and it outputs a scalar p satisfying*

$$\alpha|\Delta z| \leq p \leq \beta|\Delta z|,$$

where $\Delta z = z(\mathbf{b}) - z(\mathbf{b} + \Delta\mathbf{b})$.

We note that unlike the traditional definition of approximation for optimization, the two-sided bound is necessary. Otherwise, if the lower bound α is not specified, an algorithm can "cheat" by returning $p = 0$ to achieve $p \leq \beta|\Delta z|$. Similarly, if the upper bound β is not specified, an algorithm can "cheat" by returning p to be a very large number to achieve $p \geq \alpha|\Delta z|$.

Our main result of this section is the following.

Theorem 1. *It is NP-hard to achieve (α, β)-approximation for any $\beta \geq 1 \geq \alpha > 0$ for general MBQPs.*

Our proof of Theorem 1 is based on a reduction from the edge chromatic number problem, using the fact that deciding whether the edge chromatic equals the max degree of a graph or one more than the max degree of a graph is NP-complete.

2.2 Strong Duality

We first present the results from [5], which are the starting point for our analysis. Burer's reformulation makes the following assumption:

$$\mathbf{x} \geq 0,\ \mathbf{a}_i^\top \mathbf{x} = b_i,\ \forall i \in [m] \quad \Longrightarrow \quad 0 \leq x_j \leq 1 \text{ for all } j \in \mathcal{B}. \tag{A}$$

As mentioned in [5], if $0 \leq x_j \leq 1$ for some $j \in \mathcal{B}$ is not implied, then we can explicitly add a constraint of the form $x_j + w_j = 1$ where $w_j \geq 0$ is a slack variable. Thus, this assumption is without any loss of generality.

Consider the following CPP problem:

$$\begin{aligned} z_{\mathcal{CP}}(\mathbf{b}) := \min\ & \langle C, Y \rangle \\ \text{s.t. } & \langle T, Y \rangle = 1, \\ & \langle A_i, Y \rangle = 2b_i, \forall i \in [m] \\ & \langle AA_i, Y \rangle = b_i^2, \forall i \in [m] \\ & \langle N_j, Y \rangle = 0, \forall j \in \mathcal{B} \\ & Y \in \mathcal{CP}, \end{aligned} \tag{2}$$

where $A_i = \begin{bmatrix} 0 & \mathbf{a}_i^\top \\ \mathbf{a}_i & 0 \end{bmatrix}$, $AA_i = \begin{bmatrix} 0 & 0 \\ 0 & \mathbf{a}_i\mathbf{a}_i^\top \end{bmatrix}$, $T = \begin{bmatrix} 1 & 0 \\ 0 & 0 \end{bmatrix}$, $N_j = \begin{bmatrix} 0 & -\mathbf{e}_j^\top \\ -\mathbf{e}_j & 2\mathbf{e}_j\mathbf{e}_j^\top \end{bmatrix}$, $C = \begin{bmatrix} 0 & \mathbf{c}^\top \\ \mathbf{c} & Q \end{bmatrix}$.

Burer [5] proves the following result.

Theorem 2 (Burer's reformulation [5]). *Given a feasible MBQP in the form (1) satisfying assumption* (A)*, then we have that* $z(\mathbf{b}) = z_{\mathcal{CP}}(\mathbf{b})$.

Let us now consider the dual program to (2)[1]:

$$\begin{aligned} z_{\mathcal{COP}}(\mathbf{b}) := \sup \ & -\Big(\sum_{i=1}^{m} 2b_i\alpha_i + b_i^2\beta_i\Big) - \theta \\ \text{s.t. } & C + \sum_{i=1}^{m}\Big(\alpha_i A_i + \beta_i AA_i\Big) + \Big(\sum_{j\in\mathcal{B}} \gamma_j N_j\Big) + \theta T = M \\ & M \in \mathcal{COP}. \end{aligned} \tag{3}$$

Given an optimal solution to the dual (3), say $(\alpha^*, \beta^*, \gamma^*, \theta^*, M^*)$, and a perturbation to the right-hand-side of (1) by $\Delta b \in \mathbb{R}^m$, we can obtain a lower bound to $z(\mathbf{b} + \Delta b)$ as:

$$z(\mathbf{b} + \Delta\mathbf{b}) \geq -\Big(\sum_{i=1}^{m} 2(b_i + \Delta b_i)\alpha_i^* + (b_i + \Delta b_i)^2\beta_i^*\Big) - \theta^*, \tag{4}$$

since this follows from weak duality.

If there is a positive duality gap between (2) and (3), then we do not expect the bound (4) to be strong. Understanding when strong duality holds is the topic of this section. Our results of this section are aggregated in the next theorem:

Theorem 3 (Strong duality). *Consider a MBQP in the form (1) satisfying the assumption* (A)*. Let* $\mathbb{P} = \{\mathbf{x} : \mathbf{a}_i^\top \mathbf{x} = b_i, \forall i \in [m], \mathbf{x} \geq 0\} \neq \emptyset$ *denote the feasible region of (1) that is assumed to be non-empty. Suppose* l *is a finite lower bound on* $z(\mathbf{b})$*. Then:*

(a) When $\mathbb{P}$ *is bounded, there is a strictly copositive feasible solution of (3) which implies strong duality holds between (2) and (3) by the Slater condition.*
(b) When Q *is PSD or* $\mathbb{P}$ *is bounded, there is a closed-form formula to construct a feasible solution to (3) whose objective function value is* $l - \epsilon$ *where* ϵ *can be any arbitrarily small positive number. In particular, when* l *is the optimal value of (MBQP), this is a constructive proof that strong duality holds between (2) and (3).*
(c) There exists examples where Q *is not PSD and* $\mathbb{P}$ *is unbounded, such that there is a positive duality gap between (2) and (3).*
(d) The optimal solution of (3) is not attainable in general even if $\mathbb{P}$ *is bounded and there is no duality gap.*

Note that part (a) of Theorem 3 was shown in [4], where the authors prove that strong duality holds between (2) and (3) in a non-constructive way when $\mathbb{P}$ is bounded. Their argument utilizes a recent result from [18]. Also [20] explores similar questions in a recent presentation. To the best of our understanding, parts (b), (c), and (d) of Theorem 3 were not known before.

[1] For convenience, we have written the dual variables with 'negative sign'.

We note that in part (b) of Theorem 3 the promised closed form solutions can achieve additive ϵ-optimal solutions for any $\epsilon > 0$. Is this an artifact of our proof technique? Clearly when $Q = 0$ and $\mathcal{B} = \emptyset$, the optimal solution of (3) can be achieved, as the dual optimal solution can be achieved by linear programming duality (set $\beta = 0$). Part (d) shows that even if strong duality holds, the optimal solution is not attainable in (3) in general. Therefore, part (b) of Theorem 3 is the best we can hope for, as an optimal solution is not always achievable. Moreover, Part (d) indicates that there is no Slater point in (2) in general when $\mathbb{P}$ is bounded. One can further show via simple examples that it is possible there is no Slater point in (3) when $\mathbb{P}$ is unbounded and Q is PSD.

Proof Sketch for Part (a) **of Theorem 3:** For the proof of this part, we show that $\hat{M} := C + \hat{\lambda} \cdot H$ is a feasible solution of (3) and a strictly copositive matrix when $\mathbb{P}$ is bounded, where $\hat{\lambda}$ is sufficiently large positive quantity and

$$H = T + \sum_{i=1}^{m} AA_i, \tag{5}$$

that is, $\hat{\alpha_i} = 0 \; \forall i \in [m]$, $\hat{\beta_i} = \lambda \; \forall i \in [m]$, $\hat{\gamma_j} = 0 \; \forall j \in \mathcal{B}$, $\hat{\theta} = 0$. Thus $\hat{M}$ is a Slater point leading to the required result.

Proof Sketch for Part (b) **of Theorem** 3: The closed form solution promised in part (b) of Theorem 3 is built using *specific building blocks* or combinations of values for the variables $\alpha, \beta, \gamma, \theta$. In particular we consider the following two building blocks:

(i) Building block 1. For all $i \in [m]$, consider the following combination: ($\hat{\alpha_i} = -b_i, \hat{\beta_i} = 1, \hat{\theta} = b_i^2$) and all other variables are zero; let the resulting matrix be:

$$KK_i = \sum_{i=1}^{m} \left(\hat{\alpha_i} A_i + \hat{\beta_i} AA_i\right) + \left(\sum_{j \in \mathcal{B}} \hat{\gamma_j} N_j\right) + \hat{\theta} T = -b_i A_i + AA_i + b_i^2 T. \tag{6}$$

Note that KK_i is the the matrix associated to the quadratic form obtained by homogenizing $(b_i - \mathbf{a}_i^\top \mathbf{x})^2$.

(ii) Building block 2. For all $j \in \mathcal{B}$, consider the following combination: $\tilde{\alpha_i} = -fb_i \; \forall i \in [m], \tilde{\beta_i} = f \; \forall i \in [m], \tilde{\theta} = (f \sum_{i=1}^m b_i^2) + r$, and $\tilde{\gamma_j} = -g$; let the resulting matrix be

$$\begin{aligned} G_j(f, g, r) &= \sum_{i=1}^{m} \left(\tilde{\alpha_i} A_i + \tilde{\beta_i} AA_i\right) + \left(\sum_{j \in \mathcal{B}} \tilde{\gamma_j} N_j\right) + \tilde{\theta} T \\ &= f\left(\sum_{i=1}^{m} KK_i\right) - gN_j + rT, \end{aligned} \tag{7}$$

where f, g, r are parameters.

The closed form solution that we construct for (3) is of the form:

$$U(f_1, f_2, g, r, \tau) = C + f_1\Big(\sum_{i=1}^{m} KK_i\Big) + \sum_{j \in \mathcal{B}} G_j(f_2, g, r) + \tau H - lT, \tag{8}$$

where H is defined in (5). We specify values for the parameters f_1, f_2, g, r, τ such that the above matrix is copositive and has objective value of $l - \epsilon$.

Lets us first compute the objective function value of $U(f_1, f_2, g, r, \tau)$. Observe that for fixed values of (f_1, f_2, g, r, τ) we have that $\alpha_i = -b_i \cdot (f_1 + f_2|\mathcal{B}|)$, $\beta_i = (f_1 + f_2|\mathcal{B}|) + \tau$ for all $i \in [m]$ and $\theta = \sum_{i=1}^{m} b_i^2 \cdot (f_1 + f_2|\mathcal{B}|) + r|\mathcal{B}| + \tau - l$. Thus, the objective value of $U(f_1, f_2, g, r, \tau)$ is

$$-\Big(\sum_{i=1}^{m} 2b_i\alpha_i + b_i^2\beta_i\Big) - \theta = l - r \cdot |\mathcal{B}| - \tau \cdot \Big(1 + \sum_{i=1}^{m} b_i^2\Big).$$

We show that we may choose r and τ to be arbitrarily small positive numbers, thus obtaining an objective value of $l - \epsilon$.

Next consider the question of showing that $U(f_1, f_2, g, r, \tau)$ is copositive. It is easy to verify that $KK_i \succeq 0$ and therefore KK_i is copositive. We also show that for sufficiently large f, g, r we have that $G_j(f, g, r) \in \mathcal{COP}$ and H is also copositive. However, due to presence of the terms C and $-lT$ in $U(f_1, f_2, g, r, \tau)$, one has to additionally verify its copositivity. Consider a non-negative vector $\mathbf{y} := [t; \mathbf{x}] \geq 0$ and we need to verify that $\mathbf{y}^\top U(f_1, f_2, g, r, \tau)\mathbf{y} \geq 0$. In the case when $t = 0$, this follows from the fact that $Q \succeq 0$ or H is strictly copositive when $\mathbb{P}$ is bounded. In the case when $t > 0$, the building blocks KK_i's and $G_j(f, g, r)$'s behave like *augmented Lagrangian penalties* (see [12, 14] for strong duality results for general mixed integer convex quadratic programs) of the original constraints of the MBQP, implying the required inequality. In particular, given $\epsilon > 0$, we partition $\mathbf{y}$ in the following cases and assert that $\mathbf{y}^\top U(f_1, f_2, g, r, \tau)\mathbf{y} \geq 0$:

- If $|\mathbf{a}_i^T \mathbf{x}\frac{1}{t} - b_i| > \epsilon$, then it is easy to see that $y^\top KK_i y = |\mathbf{a}_i^T x - t\mathbf{b}_i|^2$ and this "penalty" yields that $\mathbf{y}^\top U(f_1, f_2, g, r, \tau)\mathbf{y} \geq 0$ for the selected values of parameters (f_1, f_2, g, r, τ).
- If $\frac{x_j}{t} \in (\epsilon, 1 - \epsilon) \cup (1 + \epsilon, \infty)$ (i.e., $\frac{x_j}{t}$ is far from being binary), then $y^\top G_j(f, g, r)y \gg 0$ and this "penalty" yields that $\mathbf{y}^\top U(f_1, f_2, g, r, \tau)\mathbf{y} \geq 0$ for the selected values of parameters (f_1, f_2, g, r, τ).
- The remaining case is when $\frac{1}{t}\mathbf{x}$ is "almost" feasible for MBQP, where we refer such region as MBQP(ϵ). This case is handled by Theorem 4 stated below that may be of independent interest.

For the remaining case, consider the following perturbation of the original MBQP:

$$\begin{aligned} \zeta(\mathbf{b}, \epsilon) := \min_{\mathbf{x}, \varepsilon^{(i)}} \ & \mathbf{x}^\top Q\mathbf{x} + 2\mathbf{c}^\top \mathbf{x} \\ & \mathbf{a}_i^\top \mathbf{x} = b_i + \varepsilon_i^{(1)}, \forall i \in [m] \\ & x_j + \varepsilon_j^{(2)} \in \{0, 1\}, \forall j \in \mathcal{B} \\ & \left\|\varepsilon^{(r)}\right\|_\infty \leq \epsilon, \forall r \in \{1, 2\} \\ & \mathbf{x} \geq 0. \end{aligned} \tag{MBQP(ϵ)}$$

We prove the following result:

Theorem 4 (Local stability). *Let l be a lower bound on $z(\mathbf{b})$, i.e., a lower bound on $\zeta(\mathbf{b}, 0)$. When Q is PSD or $\mathbb{P}$ is bounded, there exists $t_1 > 0, t_2 \geq 0$ that depends on $A, \mathbf{b}, \mathbf{c}, Q, \mathcal{B}$ such that if $0 \leq \epsilon < t_1$, then $\zeta(\mathbf{b}, \epsilon) \geq l - \epsilon t_2$.*

We note that if we were only considering the case where Q is PSD, then the above result could possibly be obtained using disjunctive arguments. Since we also allow for non-PSD Q matrices (when $\mathbb{P}$ is bounded), our proof of Theorem 4 requires the use of result from [24] characterizing the optimal solution of quadratic programs.

Finally, note that by the construction of (2), the inequality $z(\mathbf{b}) \geq z_{\mathcal{CP}}(\mathbf{b})$ trivially holds. Part (b) of Theorem 3 shows that $z_{\mathcal{COP}}(\mathbf{b}) \geq z(\mathbf{b}) - \epsilon$ for any positive ϵ when Q is PSD or $\mathbb{P}$ is bounded. Since $z_{\mathcal{CP}}(\mathbf{b}) \geq z_{\mathcal{COP}}(\mathbf{b})$ as a consequence of weak duality, we arrive at the following observation.

Remark 1 (Alternative proof of Burer's Theorem). The proof of part (b) of Theorem 3 provides an alternative proof of Theorem 2 in the case when Q is PSD or $\mathbb{P}$ is bounded.

Proof for Part (c) of Theorem 3: We will provide an example where $Q \not\succeq 0$ and $\mathbb{P}$ is unbounded, such that (2) is feasible and has finite value while (3) is infeasible. Consider the following instance:

$$\min\{x_1^2 - x_2^2 | x_1 - x_2 = 0, x_1 \geq 0, x_2 \geq 0\}.$$

This problem is feasible and its optimal value is zero. Hence, (2) is also feasible and has value zero by Theorem 2. The COP dual is:

$$\begin{aligned} &\max -\theta \\ &\text{s.t} \begin{bmatrix} 0 & 0 & 0 \\ 0 & 1 & 0 \\ 0 & 0 & -1 \end{bmatrix} + \theta \begin{bmatrix} 1 & 0 & 0 \\ 0 & 0 & 0 \\ 0 & 0 & 0 \end{bmatrix} + \alpha \begin{bmatrix} 0 & 1 & -1 \\ 1 & 0 & 0 \\ -1 & 0 & 0 \end{bmatrix} + \beta \begin{bmatrix} 0 & 0 & 0 \\ 0 & 1 & -1 \\ 0 & -1 & 1 \end{bmatrix} =: M \in \mathcal{COP} \end{aligned}$$

We claim that the dual is infeasible. Let $\mathbf{y} = \begin{bmatrix} 0 \\ 1 \\ 1+\epsilon \end{bmatrix}$, where $\epsilon > 0$. Then

$$\mathbf{y}^\top M \mathbf{y} = 1 - (1+\epsilon)^2 + \beta(1 + (1+\epsilon)^2 - 2(1+\epsilon)) = -2\epsilon + (\beta - 1)\epsilon^2$$

When ϵ is small enough, $\mathbf{y}^\top M \mathbf{y} < 0$. This completes the proof.

Remark 2 (Local stability not satisfied when $Q \not\succeq 0$ and $\mathbb{P}$ is unbounded). It is instructive to see that the above example does not satisfy the local stability property. Indeed, for any positive value of ϵ, it is straightforward to verify that $\zeta(0, \epsilon) = -\infty$, even though $\zeta(0, 0) = 0$. Hence, the sufficient conditions for local stability Theorem 4 cannot be further relaxed.

Part (d) **of Theorem** 3: We prove this result for the copositive dual corresponding to the standard integer programming formulation for finding the stable set number of a clique with 6 nodes, i.e., we consider the following MBQP:

$$\begin{aligned} &\min -2 \sum_{j\in[6]} x_i \\ &\text{s.t } x_u + x_v + s_{uv} = 1, \forall u < v, \; u, v \in [6] \\ &\quad \mathbf{x} \in \{0,1\}^6, \mathbf{s} \geq 0 \end{aligned}$$

2.3 How Good Is the Closed Form Solution of Theorem 3 for Sensitivity Analysis?

Previous works consider solving (3) using cutting-plane techniques [1,2,16,20] as a way to solve the original MBQP. However, in this paper we take a different perspective. We believe that with the success of modern state-of-the-art integer programming solvers, the original MBQP may be (in most cases) best solved directly using an integer programming solver. The key attraction therefore of Theorem 3 is to be able to build a closed-form solution (8) of the dual (3) using the optimal solution (or best known lower bound) of MBQP. One can therefore directly start conducting sensitivity analysis after solving the original MBQP and building the closed-form dual solutions.

However, conducting sensitivity analysis using dual solutions is challenging due to the presence of multiple ϵ-optimal solution. First note that, given an ϵ-optimal dual solution $(\alpha^*, \beta^*, \gamma^*, \theta^*)$ guaranteed by strong duality verified in Theorem 3, we have that $z(\mathbf{b}) = -\left(\sum_{i=1}^{m} 2b_i\alpha_i^* + b_i^2\beta_i^*\right) - \theta^* + \epsilon$. Subtracting the right-hand-side of the above from the right hand-side of (4), we obtain that the predicted change in the objective function value using the dual solution $(\alpha^*, \beta^*, \gamma^*, \theta^*)$ when the right-hand-side changes form $\mathbf{b}$ to $\mathbf{b} + \Delta\mathbf{b}$ is:

$$\text{Predict}(\alpha^*, \beta^*, \gamma^*, \theta^*) := -\sum_{i=1}^{m} 2\Delta b_i\alpha_i^* - \sum_{i=1}^{m}((\Delta b_i)^2 + 2b_i\Delta b_i)\beta_i^* - \epsilon. \quad (9)$$

Next consider the building block KK_i in (6), used to construct the closed for solution in part (b) of Theorem 3, corresponding to $(\hat{\alpha_i} = -b_i, \hat{\beta_i} = 1, \hat{\theta} = b_i^2)$. The objective function value of this block is $-(2b_i\hat{\alpha_i} + b_i^2\hat{\beta_i}) - \hat{\theta} = 0$. Moreover, $KK_i \succeq 0$. Thus, we arrive at the following observation:

Proposition 1. *Let* $\mathbf{b}^i \in \mathbb{R}^m$ *be the vector with* i^{th} *component equal to* b_i *and zeros everywhere else. If* $(\alpha^*, \beta^*, \gamma^*, \theta^*, M^*)$ *is an* ϵ*-optimal solution of (3), then* $(\alpha^* - \mathbf{b}^i, \beta^* + \mathbf{e}_i, \gamma^*, \theta^* + b_i^2, M^* + KK_i)$ *is also an* ϵ*-optimal solution of (3).*

On the other hand, substituting $(\alpha^* - \mathbf{b}^i, \beta^* + \mathbf{e}_i, \gamma^*, \theta^* + b_i^2, M^* + KK_i)$ in place of $(\alpha^*, \beta^*, \gamma^*, \theta^*, M^*)$ in (9) we obtain:

$$\text{Predict}(\alpha^* - \mathbf{b}^i, \beta^* + \mathbf{e}_i, \gamma^*, \theta^* + b_i^2) = \text{Predict}(\alpha^*, \beta^*, \gamma^*, \theta^*) - (\Delta b_i)^2.$$

Thus, we arrive at the following conclusion:

Remark 3. Since Predict(·) is a valid lower bound on $z(\mathbf{b}+\Delta\mathbf{b})-z(\mathbf{b})$, we wish that this lower bound is as high as possible. Therefore, if $(\alpha^*,\beta^*,\gamma^*,\theta^*,M^*)$ is an ϵ-optimal solutions of (3), then the lower bound obtained using the dual optimal solution $(\alpha^*-\mathbf{b}^i,\beta^*+\mathbf{e}_i,\gamma^*,\theta^*+b_i^2,M^*+KK_i)$ for the right-hand-side vector $\mathbf{b}+\Delta\mathbf{b}^i$ is worse than that obtained by $(\alpha^*,\beta^*,\gamma^*,\theta^*,M^*)$.

Therefore, in order to obtain the best possible sensitivity results, we would like the contribution of KK_i's in the dual optimal matrix to be as small as possible. The main role of KK_i's is to ensure that the constructed solution is in $\mathcal{COP}$. However, as an artifact of our proof of part (b) of Theorem 3, the contribution of the KK_i's in the closed-form solution is much higher than what is really needed to ensure copositivity. This fact was empirically verified by preliminary computations.

By examining the structure of optimal solution $U(f_1,f_2,g,r,\tau)$ in (8) and noting that the second building block $G_j(f,g,r)$ is a linear combination of KK_i's, N_j's and T, we may try to find good dual solutions, with small contribution of KK_i's and fixed values of τ and r, as follows:

$$\begin{aligned}&\min_{p,\gamma}\sum_{i=1}^m w_i p_i\\&\text{s.t. } C+\sum_{i=1}^m p_i KK_i+\sum_{j\in\mathcal{B}}\gamma_j N_j+\tau H-(l+r)T\in\mathcal{COP},\end{aligned}\tag{10}$$

where w_i's are some non-negative weights. Note that part (b) of Theorem 3 guarantees that the above problem finds an ϵ (whose value depending on τ and r) optimal dual solution. In our computations, we solved a variant of the above optimization problem.

2.4 Preliminary Computations

Modifications to (10). The following changes are made to improve the quality of the bound and the computational cost.

Linear Penalty. Consider a new building block corresponding to $\alpha_i=-1,\theta=2b_i$ and all other variables zero. This block is associated to the homogenization of the linear function $(b_i-\mathbf{a}_i^\top\mathbf{x})$, and it does not contribute to the objective function, just like KK_i. Setting, $K_i=2b_iT-A_i$, we solve the following problem:

$$\begin{aligned}&\min_{p,\gamma,\delta}\sum_{i=1}^m w_i^{(1)}p_i+w_i^{(2)}\delta_i\\&\text{s.t. } C+\sum_{i=1}^m p_i KK_i+\sum_{i=1}^m \delta_i K_i+\sum_{j\in\mathcal{B}}\gamma_j N_j+\tau H-(l+r)T\in\mathcal{COP},\end{aligned}\tag{11}$$

Although including K_i into the problem is not necessary, we have empirically observed that it leads to tighter sensitivity bounds due to the added degree of freedom.

Solving a Restriction of (11). Solving a copositive program is challenging. Therefore, we replaced the restriction of being in the copositive cone in (11) with a restriction of being in the $\mathbb{S}_+ + \mathbb{S}_P$. This leads to a semidefinite program, which can be solved in polynomial time. However, the resulting problem can become infeasible. To mitigate this problem, we consider two more changes:

1. Allowing non-optimal dual solutions: Instead of fixing l, we let l become a variable. We also penalize finding a poor quality dual solution by changing the objective of (11) to: $\min_{p,\gamma,\delta,l} \; -l + \sum_{i=1}^m w_i^{(1)} p_i + w_i^{(2)} \delta_i$. In this way, we may increase the chances of finding a feasible solution, However the dual solution we find may be have lesser objective than the known optimal value of original MBQP.
2. McCormick inequalities: The Y variable in (2) satisfies the following well-known McCormick inequalities:
$$Y_{ij} \le Y_{1,i}, \quad Y_{ij} \le Y_{1,j}, \quad Y_{ij} \ge Y_{1,i} + Y_{1,j} - 1, \quad Y_{ij} \ge 0. \tag{12}$$
We add new columns to (11) corresponding to these inequalities.

Preliminary Experimental Results

Instances. In our preliminary experiments, we generate three classes of instances, which we refer to as (COMB), (SSLP), and (SSQP).

The first class of instances are a weighted stable set problem with a cardinality constraint:

$$\min\{-\mathbf{c}^\top x \mid \sum_{j=1}^n x_j \le p, \; x_i + x_j \le 1 \; \forall (i,j) \in E\} \tag{COMB}$$

We generate random instances in the following way. The underlying graph is a randomly generated bipartite graph $(V_1 \cup V_2, E)$ with $|V_1| = |V_2| = 10$ and each edge $(i,j) \in E$ is present with probability d where $d \in \{0.3, 0.5, 0.7\}$. Each entry of c is uniformly sampled from $\{0, \dots, 10\}$. The right-hand-side of the cardinality constraint is $p = 3$. Twenty instances were generated for each choice of d. For this class of instances, we performed sensitivity analysis with respect to the right-hand-side of the cardinality constraint, where we increased the value of p by $\Delta p \in \{1, \dots, 10\}$.

The next class of instances contains continuous variables that are "turned on or off" using binary variables. The instances have the following form:

$$\min\left\{-2\mathbf{c}_x^\top \mathbf{x} + 2\mathbf{c}_y^\top \mathbf{y} \mid \mathbf{a}_i^\top \mathbf{x} \le b_i \; i \in [m], \; x_i \le y_i \; i \in [n], \; x \ge 0, \; y \in \{0,1\}^n\right\} \tag{SSLP}$$

We generate instances in the following way. We set $n = 20$ and $m = 5$. Each entry of $\mathbf{c}_x$ is uniformly sampled from $\{0, \dots, 10\}$ and $\mathbf{c}_y = (3, \dots, 3)$ is a constant vector. Each entry of $\mathbf{a}_i$ is uniformly sampled from $\{0, \dots, 10\}$ and then each entry of $\mathbf{a}_i$ is zeroed out with probability $d \in \{0.3, 0.5, 0.7\}$. Finally, $b_i = \lfloor \frac{1}{2} \mathbf{a}_i^\top \mathbf{e} \rfloor$ for all $i \in [m]$. Twenty instances were generated for each probability.

For this class of instances, we focus on sensitivity with-respect to right-hand-side of $\mathbf{a}_i^\top x \leq b_i, \forall i \in [m]$ when $\Delta\mathbf{b} \in \{0,1,2,3\}^m$.

The last class of instances is similar, except that the objective is quadratic:

$$\min\left\{-2\mathbf{c}_x^\top\mathbf{x} + 2\mathbf{c}_y^\top\mathbf{y} + \mathbf{x}^\top Q\mathbf{x} \mid \mathbf{a}_i^\top\mathbf{x} \leq b_i\ i\in[m],\ x_i \leq y_i\ i\in[n],\ x\geq 0,\ y\in\{0,1\}^n\right\} \tag{SSQP}$$

We generate $\mathbf{c}_x$, $\mathbf{c}_y$, $\mathbf{a}_i$, $\mathbf{b}$ as before. The matrix Q is randomly generated such that $Q = \sum_{i\in\{1,2\}} \mathbf{u}_i\mathbf{u}_i^\top$ where each entry of $\mathbf{u}_i$ is uniformly sampled from $\{-1,0,1\}$.

Experiments Conducted. Our experiments are implemented in Julia 1.9, relying on Gurobi version 9.0.2 and Mosek 10.1 as the solvers. We solve on a Windows PC with 12th Gen Intel(R) Core(TM) i7 processors and 16 RAM. We compare our method with other known methods. Those methods consider certain convex relaxation of (MBQP) and obtain dual variables of constraints to conduct sensitivity analysis via weak duality. In this case, we consider three convex relaxations, which we call Shor1, Shor2, Cont.

First, we consider the Shor relaxation of (MBQP):

$$\min_{Y\in\mathbb{S}_+\cap\mathbb{S}_P} \left\{\langle C,Y\rangle \;\middle|\; \begin{array}{l}\langle T,Y\rangle = 1, \langle N_j,Y\rangle = 0, \forall j\in\mathcal{B}\\ \langle A_i,Y\rangle = 2b_i, \forall i\in[m]\end{array}\right\}. \tag{Shor1}$$

Next, we consider the relaxation of the problem obtained by augmenting Shor1 with redundant quadratic constraints and McCormick inequalities:

$$\min_{Y\in\mathbb{S}_+\cap\mathbb{S}_P} \left\{\langle C,Y\rangle \;\middle|\; \begin{array}{l}\langle T,Y\rangle = 1, \langle N_j,Y\rangle = 0, \forall j\in\mathcal{B}\\ \langle A_i,Y\rangle = 2b_i, \langle AA_i,Y\rangle = b_i^2, \forall i\in[m]\\ \text{McCormick inequalities (12)}\end{array}\right\}. \tag{Shor2}$$

Finally, and assuming that Q is PSD, we obtain a convex relaxation simply by relaxing the binary variables to be continuous variables in $[0,1]$:

$$\min_{\mathbf{x}\geq 0}\left\{\mathbf{x}^\top Q\mathbf{x} + 2\mathbf{c}^\top\mathbf{x} \;\middle|\; \mathbf{a}_i^\top\mathbf{x} = b_i,\ \forall i\in[m], x_j\in[0,1] \text{ and } \forall j\in\mathcal{B}\right\}. \tag{Cont}$$

The relaxations Shor1 and Shor2 are solved using Mosek, while the relaxation Cont is solved using Gurobi. Notice that these problems may have multiple optimal solutions, so different solvers might lead to different solutions.

We choose relaxation Shor1 as a baseline and measure the goodness of those predictions by *relative gap*. Given the rhs change $\Delta\mathbf{b}$, the ground-truth $z(\mathbf{b}+\Delta\mathbf{b})$, the prediction p_1 by Shor1 and the prediction p_2 by some method, then

$$\text{relative gap} = \frac{z(\mathbf{b}+\Delta\mathbf{b}) - p_2}{z(\mathbf{b}+\Delta\mathbf{b}) - p_1}.$$

This is always a non-negative number and a smaller relative gap indicates a better performance of the given method.

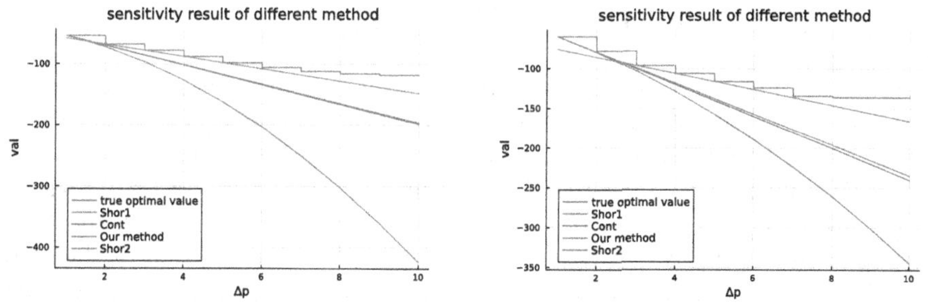

(a) a (COMB) instance with $d = 0.3$ (b) a (COMB) instance with $d = 0.5$

Fig. 1. Two randomly generated (COMB) instances with $d = 0.3$ and $d = 0.5$. The x-axis corresponds to Δp and y-axis corresponds to optimal value of the new program or different predicted value of different methods.

Table 1. Average relative gap (COMB) – all densities

Δk	1	2	3	4	5	6	7	8	9	10	time (s)
Shor1	1	1	1	1	1	1	1	1	1	1	7.30
Shor2	1.33	2.04	2.7	2.71	2.76	2.79	2.87	2.88	2.94	3.03	10.17
our method	0.83	0.02	0.00	0.11	0.19	0.26	0.32	0.38	0.41	0.44	8.35
Cont	0.97	1.07	1.09	1.08	1.07	1.06	1.05	1.05	1.04	1.04	0.00

Table 2. Average relative gap for (SSLP) – all densities

$\|\|\Delta \mathbf{b}\|\|_\infty$	≤ 1	≤ 2	≤ 3	time (s)
Shor1	1	1	1	3.63
Shor2	1.20	1.48	1.64	7.21
our method	0.59	0.55	0.60	5.82
Cont	1.00	1.00	1.00	0.00

Table 3. Average relative gap for (SSQP) – all densities

$\|\|\Delta \mathbf{b}\|\|_\infty$	≤ 1	≤ 2	≤ 3	time (s)
Shor1	1	1	1	3.58
Shor2	1.24	1.39	1.54	7.08
our method	0.52	0.48	0.53	5.90
Cont	1.00	1.00	1.00	0.00

Results and Discussion. Figure 1 shows an example of bounds obtained using different methods. Tables 1, 2, and 3 summarize the relative gaps obtained in the experiments mentioned above.

We observe that our method provides the tightest sensitivity bounds in all cases. Also note that method Shor2 provides the worst bounds. This is easier to observe in Fig. 1. This is interesting, because the SDP from Shor2 is quite similar to Burer's formulation (with additional McCormick inequalities). This discrepancy is most likely due to the fact that these problems have multiple optimal solutions - similar to the discussion in Sect. 2.3. The naive Shor2 approach finds an optimal dual which does not give good bounds after $\mathbf{b}$ is perturbed. On the other hand, our method attempts to find a good dual solution (with respect to producing good bounds for changing rhs) inside the ϵ-optimal face of the dual.

3 Conclusion and Future Directions

We proved sufficient conditions for strong duality to hold between Burer's reformulation of MBQPs and its dual. One direction of research is to extend such strong duality results for reformulations of more general QCQPs [6].

We have proposed a SDP-based algorithm to conduct sensitivity analysis of general (MBQP) which provides much better bounds than existing methods. This algorithm is motivated by the structure of ϵ-optimal solution of the COP dual. However, the sizes of instances we can currently perform sensitivity analysis are limited by the SDP solver. One possible future direction is to develop a more scalable solver for the SDP in (11), for instance, using the techniques from [21].

References

1. Anstreicher, K.M.: Testing copositivity via mixed-integer linear programming. Linear Algebra Appl. **609**, 218–230 (2021). https://doi.org/10.1016/j.laa.2020.09.002. https://www.sciencedirect.com/science/article/pii/S0024379520304171
2. Badenbroek, R., de Klerk, E.: An analytic center cutting plane method to determine complete positivity of a matrix. INFORMS J. Comput. **34**(2), 1115–1125 (2022)
3. Bomze, I.M., De Klerk, E.: Solving standard quadratic optimization problems via linear, semidefinite and copositive programming. J. Glob. Optim. **24**, 163–185 (2002)
4. Brown, R., Neira, D.E.B., Venturelli, D., Pavone, M.: Copositive programming for mixed-binary quadratic optimization via Ising solvers. arXiv preprint arXiv:2207.13630 (2022)
5. Burer, S.: On the copositive representation of binary and continuous nonconvex quadratic programs. Math. Program. **120**(2), 479–495 (2009). https://doi.org/10.1007/s10107-008-0223-z
6. Burer, S., Dong, H.: Representing quadratically constrained quadratic programs as generalized copositive programs. Oper. Res. Lett. **40**(3), 203–206 (2012)
7. Celaya, M., Kuhlmann, S., Paat, J., Weismantel, R.: Improving the Cook et al. proximity bound given integral valued constraints. In: Aardal, K., Sanitá, L. (eds.) IPCO 2022. LNCS, vol. 13265, pp. 84–97. Springer, Cham (2022). https://doi.org/10.1007/978-3-031-06901-7_7
8. Conforti, M., Cornuéjols, G., Zambelli, G.: Integer programming models. In: Conforti, M., Cornuéjols, G., Zambelli, G. (eds.) Integer Programming. GTM, vol. 271, pp. 45–84. Springer, Cham Springer (2014). https://doi.org/10.1007/978-3-319-11008-0_2
9. Cook, W., Gerards, A.M.H., Schrijver, A., Tardos, É.: Sensitivity theorems in integer linear programming. Math. Program. **34**(3), 251–264 (1986). https://doi.org/10.1007/BF01582230
10. Del Pia, A., Ma, M.: Proximity in concave integer quadratic programming. Math. Program. **194**(1–2), 871–900 (2022)
11. Eisenbrand, F., Weismantel, R.: Proximity results and faster algorithms for integer programming using the Steinitz lemma. ACM Trans. Algorithms (TALG) **16**(1), 1–14 (2019)
12. Feizollahi, M.J., Ahmed, S., Sun, A.: Exact augmented Lagrangian duality for mixed integer linear programming. Math. Program. **161**, 365–387 (2017)

13. Granot, F., Skorin-Kapov, J.: Some proximity and sensitivity results in quadratic integer programming. Math. Program. **47**(1–3), 259–268 (1990)
14. Gu, X., Ahmed, S., Dey, S.S.: Exact augmented Lagrangian duality for mixed integer quadratic programming. SIAM J. Optim. **30**(1), 781–797 (2020)
15. Gu, X., Dey, S.S., Xavier, Á.S., Qiu, F.: Exploiting instance and variable similarity to improve learning-enhanced branching. arXiv preprint arXiv:2208.10028 (2022)
16. Guo, C., Bodur, M., Taylor, J.A.: Copositive duality for discrete energy markets (2021)
17. Johnson, E.S., Ahmed, S., Dey, S.S., Watson, J.P.: A K-nearest neighbor heuristic for real-time DC optimal transmission switching. arXiv preprint arXiv:2003.10565 (2020)
18. Kim, S., Kojima, M.: Strong duality of a conic optimization problem with a single hyperplane and two cone constraints. arXiv preprint arXiv:2111.03251 (2021)
19. Lee, J., Paat, J., Stallknecht, I., Xu, L.: Improving proximity bounds using sparsity. In: Baïou, M., Gendron, B., Günlük, O., Mahjoub, A.R. (eds.) ISCO 2020. LNCS, vol. 12176, pp. 115–127. Springer, Cham (2020). https://doi.org/10.1007/978-3-030-53262-8_10
20. Linderoth, J., Raghunathan, A.: Completely positive reformulations and cutting plane algorithms for mixed integer quadratic programs. INFORMS Annual Meeting (2022)
21. Majumdar, A., Hall, G., Ahmadi, A.A.: Recent scalability improvements for semidefinite programming with applications in machine learning, control, and robotics. Ann. Rev. Control Robot. Auton. Syst. **3**, 331–360 (2020)
22. Nemhauser, G.L., Wolsey, L.A.: Integer and Combinatorial Optimization, vol. 55. Wiley, Hoboken (1999)
23. Shahidehpour, M., Yamin, H., Li, Z.: Market Operations in Electric Power Systems: Forecasting, Scheduling, and Risk Management. Wiley, Hoboken (2003)
24. Vavasis, S.A.: Quadratic programming is in NP. Inf. Process. Lett. **36**(2), 73–77 (1990). https://doi.org/10.1016/0020-0190(90)90100-C. https://www.sciencedirect.com/science/article/pii/002001909090100C
25. Xavier, Á.S., Qiu, F., Ahmed, S.: Learning to solve large-scale security-constrained unit commitment problems. INFORMS J. Comput. **33**(2), 739–756 (2021)

Author Index

J. Vygen and J. Byrka (Eds.): IPCO 2024, LNCS 14679, pp. 461–462, 2024.
https://doi.org/10.1007/978-3-031-59835-7

Printed in the United States
by Baker & Taylor Publisher Services